1994

W9-CZM-968

EVERGLADES

The Ecosystem and Its Restoration

Editors

Steven M. Davis
John C. Ogden

Associate Editor

Winifred A. Park

S^t_L

St. Lucie Press

Printed and bound in the U.S.A. Printed on acid-free paper.

Library of Congress Cataloging-in-Publication Data

Everglades : the ecosystem and its restoration / edited by Steven Davis
and John Ogden.
    p.   cm.
   Includes bibliographical references and index.
   ISBN 0-9634030-2-8
    1. Swamp ecology--Florida--Everglades. 2. Restoration ecology--
Florida--Everglades. I. Davis, Steven M. II. Ogden, John C.,
1938– .
QH105.F6E94   1994
574.5'26325'0975939--dc20   93-44643
                                          CIP

Direct all inquiries to St. Lucie Press, Inc., 100 E. Linton Blvd., Suite 403B, Delray Beach, Florida 33483.
   Phone: (407) 274-9906
   Fax: (407) 274-9927

$S\overset{t}{L}$

Published by
St. Lucie Press
100 E. Linton Blvd., Suite 403B
Delray Beach, FL 33483

Cover photo: White ibis and anhinga settle down for the night in a willow head at the Arthur R. Marshall Loxahatchee National Wildlife Refuge. (Photo by S. M. Davis.)

DEDICATION

This book is more than a scientific treatise; it is also an expression of the high respect for an ecosystem that has developed among the many researchers, including the authors of the chapters, who have seen the Everglades in all its moods. We especially recognize three individuals who, as a result of long hours afield, were among the first within the scientific community to have developed a personal commitment to the preservation of the Everglades: Daniel B. Beard, Arthur R. Marshall, and J. Walter Dineen.

We also dedicate this book to the next generation who will see in their lifetimes either the restoration or the demise of the Everglades. To Jennifer, Laura, Nicholas, and the children and families of all the authors: May yours be an Everglades that once again is wide reaching, abounding with water, and teeming with life.

JCO
SMD

CONTENTS

V TOWARD ECOSYSTEM RESTORATION

FOREWORD

In a quiet place in our home, a sanctuary from the vicissitudes of life, I have a comfortable chair, a good reading light, and three books: *The Complete Plays of William Shakespeare,* the King James Version of the Bible, and Marjory Stoneman Douglas's *The Everglades: River of Grass.* Let me quote some of Marjory's poetry:

> "There are no other Everglades in the world. They have always been one of the unique regions of the earth, remote, never wholly known. Nothing anywhere else is like them: their vast glittering openness, wider than the enormous visible round of the horizon, the racing free saltness and sweetness of their massive winds, under the dazzling blue heights of space. They are unique also in simplicity, the diversity, the related harmony of the forms of life they enclose.

> "The miracle of the light pours over the green and brown expanse of sawgrass and of water—shining and slow moving below, the grass and water that is the meaning and central fact of the Everglades of Florida.

> "It is a river of grass.

> "Where do you begin? Because when you think of it, history, the recorded time of the earth and man, is in itself something like a river. To try to present it whole is to find oneself lost in the sense of continuing change. The source can be only the beginning in time and space and the end is the future and the unknown.

> "So it is with the Everglades, which have the quality of long existence in their own nature.

> "They were changeless. They were changed."

No one else writes like Marjory Stoneman Douglas. Marjory recognized the complexity of the system, and yet, in hindsight, it was far less complex—even simplistic—in 1947 compared to the Everglades system as it exists today. It took only 40 years and $500 million to get us where we are today.

Four pillars must now be raised, as a framework for the future, for recovering the simpler and viable Everglades known to Marjory. First, we need to turn to this book for the technical knowledge that is so essential for designing the grand restoration programs that we have been impatiently waiting to see implemented. The chapters in this book present a picture of the Everglades ecosystem that is both more complex and better understood than many of us have recently recognized. And, above all else, this book tells us that we know enough to boldly move forward with an agenda for restoration.

Second, we need a goal that everyone recognizes and most accept. I say this with some frustration. When, with Secretary Morton and the stout support of the Florida congressional delegation and the congressional appropriations committees, I established Everglades National Park's research center, I, frequently frustrated by the scene of an ecosystem literally dying, would shout at the Everglades staff, "Tell me what to ask for. What must you have?" And that, my friends, is still the key question 20 years later. Where is it that we are going?

Do you remember the scene in Lewis Carroll's masterpiece, *Alice in Wonderland*? Alice, perpetually lost, stopped at a crossroad, spotted the Cheshire Cat stretched out on the branch of a nearby tree. "Cheshire Cat," asked Alice, "would you please tell me which way I ought to go from here?" The wise cat responded, "That depends a good deal on where you want to get to." Alice: "But I don't know where I am going." The cat responded, falling off to sleep, "Then it doesn't matter which way you go."

Third, we need to reinforce the commitment of the scientists, managers, and even famously noncommittal politicians to continue and even increase the expensive, time-consuming research, censusing, and management efforts needed as we improve our designs and evaluate our results.

Fourth, we must be imaginative and even bold; we need to take risks in achieving our goals before time literally runs out. Boldness means moving forward in large experimental steps, in the absence of complete information from field studies. It would easily take our lifetimes and hundreds of millions of dollars for us to truly comprehend all the ramifications of man's past actions and all the alternatives now available. But even if these funds and qualified scientists were available, time is not!

The continuing Everglades paradox for the administrators, scientists, and amateur Everglades "buffs" is how to enhance and protect the whole when so much of our effort is focused on the pieces. It is a sad truth, but one worth acknowledging, that even winning all the "brush fires" and tackling all the individual problems will not necessarily protect the Everglades: the whole is indeed more than just the sum of the parts.

This book focuses on water—water as a catalyst, not a commodity, and the driving force of this system. The technical recognition of the importance of water begets an earthshaking political recognition, namely that the Everglades, along with the traditional agricultural and urban components, is a legitimate user of water. That concept leads to a near heretical statement of fact: future urban water needs will have to be met from alternatives such as reverse osmosis, and agriculture must implement clean water practices and learn to live with far less water.

We must tell this story, the Everglades story, in terms the American people can understand. They will save the Everglades. There are tough choices ahead. Money must be spent—but above all, we must not give up.

Late in Sir Winston Churchill's life, he was invited back to his boyhood school to accept honors and accolades. Churchill had hated that school. Sent so very young, bored with books, born with a lisp, awkward at sports, frequently teased and punished, he had managed to survive. His long life was marked by tenacity, but he bore real childhood scars.

He rose to the Headmaster's eloquent citation. He stumbled to the podium and stared down at the rows of young men. His voice, barely audible, rose as he said, "Never give up," and he returned to his seat. He paused, turned, and spoke, "never, never give up."

I count on you to never give up!

Nathaniel P. Reed
Hobe Sound, Florida
November 1993

CONTRIBUTORS

G. Thomas Bancroft
Everglades System Research Campaign
National Audubon Society
Miami, Florida

Oron L. Bass, Jr.
Everglades National Park
Homestead, Florida

Robert E. Bennetts
Florida Cooperative Fish and Wildlife
  Research Unit
University of Florida
Gainesville, Florida

Robin D. Bjork
Tavernier, Florida

Michael J. Bodle
South Florida Water Management District
West Palm Beach, Florida

Laura A. Brandt
Department of Wildlife and Range Sciences
University of Florida
Davie, Florida

Joan A. Browder
Southeast Fisheries Center
National Marine Fisheries Service
Miami, Florida

Michael W. Collopy
Cooperative Research Center
Bureau of Land Management
Forest Sciences Laboratory
Corvallis, Oregon

James Davidson
Office of the Vice President for
  Agriculture and Natural Resources
University of Florida
Gainesville, Florida

Steven M. Davis
South Florida Water Management District
West Palm Beach, Florida

Donald L. DeAngelis
Environmental Sciences Division
Oak Ridge National Laboratory
Oak Ridge, Tennessee

J. Walter Dineen (deceased)
South Florida Water Management District
West Palm Beach, Florida

Michael J. Duever
Ecosystem Research Unit
National Audubon Society
Naples, Florida

Patrick J. Dugan
IUCN World Conservation Union
Gland, Switzerland

Anne-Marie Eklund
Southeast Fisheries Center
National Marine Fisheries Service
Miami, Florida

Robert J. Fennema
Everglades National Park
Homestead, Florida

Amy P. Ferriter
South Florida Water Management District
West Palm Beach, Florida

Peter C. Frederick
Department of Wildlife and Range
  Sciences
University of Florida
Gainesville, Florida

Patrick J. Gleason
James M. Montgomery Consulting
  Engineers, Inc.
Lake Worth, Florida

Lance H. Gunderson
Department of Zoology
University of Florida
Gainesville, Florida

Wayne Hoffman
Tavernier Science Center
National Audubon Society
Tavernier, Florida

Crawford S. Holling
Arthur R. Marshall Chair in Ecological
  Sciences
Department of Zoology
University of Florida
Gainesville, Florida

Susan D. Jewell
Arthur R. Marshall Loxahatchee National
  Wildlife Refuge
Boynton Beach, Florida

Robert A. Johnson
Everglades National Park
Homestead, Florida

Janet A. Ley
South Florida Water Management District
West Palm Beach, Florida

Stephen S. Light
South Florida Water Management District
West Palm Beach, Florida

William F. Loftus
Everglades National Park
Homestead, Florida

Thomas K. MacVicar
South Florida Water Management District
West Palm Beach, Florida

Edward Maltby
Department of Geography
University of Exeter
Exeter, Devon
Great Britain

Jennifer E. Mattson
Florida Cooperative Fish and Wildlife
  Unit
University of Florida
Gainesville, Florida

Frank J. Mazzotti
Department of Wildlife and Range
  Sciences
University of Florida
Davie, Florida

Jean M. McCollom
Ecosystem Research Unit
National Audubon Society
Naples, Florida

Carole C. McIvor
Arizona Cooperative Fish and Wildlife
  Unit
University of Arizona
Tucson, Arizona

John F. Meeder
Southeast Environmental Research
  Program
Florida International University
Miami, Florida

Linda C. Meeder
Princeton, Florida

Calvin J. Neidrauer
South Florida Water Management District
West Palm Beach, Florida

John C. Ogden
Everglades National Park
Homestead, Florida

Winifred A. Park
South Florida Water Management District
West Palm Beach, Florida

Randall W. Parkinson
Department of Oceanography
Florida Institute of Technology
Melbourne, Florida

William A. Perkins
Battelle Pacific Northwest Laboratories
Richland, Washington

George V. N. Powell
RARE Center for Tropical Bird
  Conservation
Philadelphia, Pennsylvania

John R. Richardson
Florida Cooperative Fish and Wildlife
  Unit
University of Florida
Gainesville, Florida

William B. Robertson, Jr.
Everglades National Park
Homestead, Florida

James A. Rodgers, Jr.
Florida Game and Freshwater Fish
  Commission
Wildlife Research Laboratory
Gainesville, Florida

Richard J. Sawicki
Tavernier Science Center
National Audubon Society
Tavernier, Florida

Tommy R. Smith
National Wildlife Research Center
Tais, Saudi Arabia

George H. Snyder
Everglades Research and Education
  Center, IFAS
University of Florida
Belle Glade, Florida

James R. Snyder
Big Cypress National Preserve
Ochopee, Florida

Marilyn G. Spalding
Department of Infectious Diseases
University of Florida
Gainesville, Florida

Peter A. Stone
Ground Water Protection Division
South Carolina Department of Health
  and Environmental Control
Columbia, South Carolina

Allan M. Strong
Department of EEO Biology
Tulane University
New Orleans, Louisiana

David R. Swift
South Florida Water Management District
West Palm Beach, Florida

Lenore P. Tedesco
Department of Geology
Indiana University/Purdue University of
  Indianapolis
Indianapolis, Indiana

Daniel D. Thayer
South Florida Water Management District
West Palm Beach, Florida

Carl J. Walters
Resource Ecology
Resource Management Science
University of British Columbia
Vancouver, British Columbia
Canada

Harold R. Wanless
Department of Geological Sciences
University of Miami
Coral Gables, Florida

Peter S. White
Department of Biology and North
  Carolina Botanical Garden
University of North Carolina at
  Chapel Hill
Chapel Hill, North Carolina

I

The EVERGLADES ISSUES in a BROADER PERSPECTIVE

1

INTRODUCTION

Steven M. Davis

John C. Ogden

Home to a diverse mixture of Caribbean and temperate flora, the keystone American alligator, the Florida panther, and hundreds of thousands of nesting and overwintering wading birds, the Everglades has long been recognized as one of the world's special places. Marjory Stoneman Douglas (1947) successfully focused national attention on the Everglades one-half century ago when she described its natural and cultural history in her book, *The Everglades: River of Grass*. In that same year, Everglades National Park opened to the public. The park's role as protector of a portion of the Everglades has been internationally recognized through its designation as an International Biosphere Reserve, a World Heritage Site, and a Ramsar Convention Wetland of International Importance. However, Everglades National Park includes only one-fifth of the original river of grass that once encompassed more than one million hectares. Nearly half of the Everglades has been drained for agriculture and development. The marsh that remains upstream of Everglades National Park is mostly dissected into shallow, diked impoundments known as Water Conservation Areas, where the South Florida Water Management District controls levels and flows in efforts to balance environmental concerns with the needs of a burgeoning human population (Plate 1).* From a broad perspective, changes in the Everglades ecosystem during the 20th century have been threefold. The geographic extent of the system has been reduced. The spatial and temporal patterns of the major physical driving forces such as hydrology, fire, and nutrient supply have been altered in the remaining system. Abundance of wildlife has declined. Most conspicuous and alarming among the biological changes have been the plummeting of Everglades wading bird populations to less than one-fifth of their abundance during the 1930s and the near extinction of the Florida panther.

Although this book describes details of the past and present Everglades, its

---

* Plate 1 appears following page 432.

justification will only be realized in the future Everglades. In fact, we cannot describe the future: our preferred view is one of a restored and sustainable Everglades, one that mimics as closely as possible the appearance and behavior of the system as if drainage and development had not occurred. What this book does provide is a careful examination of what the Everglades is now, how it has changed during a century of involvement with modern humans, and some guidelines for restoring the system's lost attributes. The picture presented is not drawn from one set of hands; indeed, its fundamental strength is the diversity of authorship: 57 authors currently involved in Everglades research, from 27 institutional affiliations. The book is an outgrowth of the 1989 Everglades Symposium in Key Largo, six adaptive environmental assessment workshops (Holling, 1978; Walters, 1986), the intensive communication and interaction among contributors during a period of nearly five years, referee review of all chapters, and integration, revision, and updating of topics through 1992. It is a result of the joint sponsorship of Everglades National Park and the South Florida Water Management District. Its purpose is to promote discussion and share information on topics of importance to Everglades ecosystem restoration.

A few words are needed concerning the scope of the book—what it is and what it is not—and its relevance to Everglades science and restoration. Certainly this book is neither the first attempt nor should be the last to review what is known about the Everglades. While the present book demonstrates a broad expansion in our understanding of the Everglades, earlier Everglades reviews (Gleason, 1974, 1984; Gunderson and Loftus, 1993) leave strong contributions to the Everglades topic. One important way this book is different is in its focus on the interrelated roles of ecosystem size, disturbance patterns, and hydrology as determinants of large-scale ecosystem restoration. The emphasis on hydrologic aspects rather than water quality is due to the comprehensive treatment of water quality in the draft Everglades Surface Water Improvement and Management (SWIM) Plan (South Florida Water Management District, 1992b). Similarly, the problem of exotic plant species in the Everglades has been examined by Doren et al. (1991) and Geiger (1981). Consequently, water quality and an exotic of major concern, *Melaleuca,* are addressed here in summary chapters in the context of hydrologic restoration determinants. Emphasis on the importance of all three restoration aspects is found throughout the book, and indeed, any successful Everglades ecosystem restoration plan must address issues of water quality and exotics as well as hydrological determinants. A note on the rapidly emerging field of restoration ecology (Jordan et al., 1987) as it is applied here: We propose that large-scale ecosystem restoration requires an approach different from those more frequently practiced on a smaller scale, and thus the emphasis on restoring ecosystem driving forces rather on wetland creation, supplemental planting, or revegetation.

The boundaries of the ecosystem to be treated in this volume were the subject of much discussion and debate in early meetings. In the largest scale, south Florida's greater Everglades wetland system encompasses the Kissimmee River and floodplain, Lake Okeechobee, and the freshwater marshes south of Lake Okeechobee. The marshes interconnect with the Big Cypress Swamp to the west and ultimately flow through the mangrove and salt marsh estuaries into Florida Bay at the downstream end of the system, as delineated in the Davis (1943) vegetation map included with this volume. At the level of complexity that we deemed necessary to examine ecological relationships as bases for restoration guidelines, treatment of the greater Everglades system is beyond the scope of any single volume. This book focuses on that part of the greater system that is characterized as the River of Grass, the extensive mosaic of freshwater marshes that stretch from the shores of Lake Okeechobee to the mangrove interface with Florida Bay. For works on the ecology and restoration of other parts of the greater system, the reader is referred to the U.S. Army Corps of Engineers (1991) and Toth (1993, in press) for the Kissimmee River, the South Florida Water Management District (1992a) for Lake Okeechobee, Duever et al. (1986) for Big Cypress Swamp, and Tilmant (1989) for Florida Bay.

An integrating theme throughout the book focuses on the stability, dynamics, and persistence of Everglades biota, as affected by the size of the ecosystem, its habitat diversity, and by the spatial and temporal patterns of major physical driving forces. Following the examination of Everglades issues in a broader perspective, the chapters are organized to develop this theme as a technical foundation for an emerging ecosystem restoration program for the Everglades. The theme is established in an introductory synthesis of the ecological concepts of stability, dynamics, and persistence as manifested in the spatial and temporal patterns of ecosystems, based upon examples from other systems. This sets the stage for the examination of Everglades ecosystem function in increasing levels of complexity. The spatial and temporal characteristics of the physical forces that have created, and continue to drive, the Everglades ecosystem—geology, sea level, climate, hydrology, and fire—are described in six chapters and integrated in a synthesis chapter. This leads to the examination of vegetation response to ecosystem driving forces in chapters that analyze community components and determinants of composition, invasion by the exotic species *Melaleuca quinquenervia,* vegetation response to eutrophication, the ecology of periphyton communities, and ecological implications of landscape-level change. A synthesis chapter relates vegetation pattern and process to physical driving forces and biological processes. The level of complexity continues to increase as faunal responses to both physical driving forces and vegetation patterns and processes are examined in chapters on the Everglades small-fish assemblage, the American alligator, the snail kite, interactions between the Florida panther and white-tailed deer, and five chapters on wading bird dynamics. The chapters on wading birds examine nesting colony patterns, nutrient transport, foraging habitat, relationships of foraging patterns and colony locations to hydrology, and reproductive success. The section ends with a synthesis chapter on faunal patterns and processes.

The final section provides overviews of the Everglades, past and present, as foundations for ecosystem restoration. Topics include a cross-scale characterization of the structure and dynamics of the Everglades ecosystem, the adaptive environmental assessment process utilized in the series of workshops which led to broad restoration goals, an initial screening of water policy alternatives for ecosystem restoration, and essential elements of composite water policies. In reality, the greatest success in the development of syntheses of topics is found in the individual chapters, due to the close collaboration among the authors. In the final chapter, the ecological concepts and restoration implications provided by the individual chapters are summarized and integrated. In conclusion, a series of restoration hypotheses are proposed that represent a composite vision from the chapters. We hope these will provide guidelines that will contribute to the restoration and perpetuation of a functional Everglades ecosystem in the 21st century.

It should be noted that the restoration guidelines developed here do not necessarily reflect official policy of the National Park Service or the South Florida Water Management District. The "hands-off" approach that these agencies have taken in their support of this book has assured the unencumbered development of ecological concepts and restoration guidelines as they apply to the Everglades. Consequently, the conclusions expressed by the authors in this volume should be considered their own and not by implication to reflect agency policy.

ACKNOWLEDGMENTS

From its conception in 1988 by a group of Everglades scientists who spawned the idea of bringing their work together in a way relevant to decision makers, through its growth during nearly five years of increasingly organized collaboration, and into its maturity with the production of this volume, the effort represented here has proceeded in the midst of a virtual war of politics and litigation over the future of the Everglades. We cannot give enough credit to participants and supporters for their unwavering dedication to a collective vision of Everglades science and restoration that transcended the more immediate issues associated with Everglades management and protection. Much of the focus and organization of this book were provided by an Editorial Board consisting of G. T. Bancroft, D. L. DeAngelis, M. J. Duever, D. M. Fleming, L. H. Gunderson, C. S. Holling, W. F. Loftus, T. K. MacVicar, M. Maffei, W. B. Robertson, C. J. Walters, and P. S. White. Special acknowledgment must go to the South Florida Water Management District and U.S. National Park Service for their generous staff and financial support.

The commitment of the agency leaders who supported the unimpeded publication of this volume is a testament to their long-term vision of the Everglades.

LITERATURE CITED

Davis, J. H. 1943. The natural features of southern Florida. *Fla. Geol. Soc. Geol. Bull.,* No. 25, 311 pp.

Doren, R. F., T. D. Center, R. L. Hofstetter, R.L . Myers, and L. D. Whiteaker. 1991. Proc. Symp. on Exotic Pest Plants, Tech. Rep. NPS/NREVER/NRTR-91/06, National Park Service, U.S. Department of the Interior, Washington, D.C., 387 pp.

Douglas, M. S. 1947. *The Everglades: River of Grass* (revised ed., 1988), Pineapple Press, Sarasota Fla., 448 pp.

Duever, M. J., J. E. Carlson, J. F. Meeder, L. C. Duever, L. H. Gunderson, L. A. Riopelle, T. R. Alexander, R. L. Myers, and D. P. Spangler. 1986. *The Big Cypress National Preserve,* Res. Rep. No. 8, National Audubon Society, New York, 444 pp.

Geiger, R. K. (Ed.). 1980. Proc. of Melaleuca Symp., Publ. F81T1, Division of Forestry, Florida Department of Agriculture and Consumer Services, Tallahassee, 140 pp.

Gleason, P. J. (Ed.). 1974. *Environments of South Florida: Present and Past,* Memoir No. 2, Miami Geological Society, Miami, 452 pp.

Gleason, P. J. (Ed.). 1984. *Environments of South Florida: Present and Past II,* Memoir No. 2, 2nd ed., Miami Geological Society, Miami, 551 pp.

Gunderson, L. H. and W. F. Loftus. 1993. The Everglades. in *Biodiversity of the Southeastern United States/Lowland Terrestrial Communities,* W. H. Martin, S. G. Boyce, and A. C. Echternacht (Eds.), John Wiley & Sons, New York, pp. 191–255.

Holling, C. S. (Ed.). 1978. *Adaptive Environmental Assessment and Management,* John Wiley & Sons, New York, 377 pp.

Jordan, W. R. III, M. E. Gilpin, and J. D. Aber (Eds.). 1987. *Restoration Ecology. A Synthetic Approach to Ecological Restoration,* Cambridge University Press, Cambridge, 342 pp.

South Florida Water Management District. 1992a. Surface Water Improvement and Management Plan Update for Lake Okeechobee, Vol. 1: Planning Document, and Vol. 2: Appendices, South Florida Water Management District, West Palm Beach, 97 + 274 pp.

South Florida Water Management District. 1992b. Draft Surface Water Improvement and Management Plan for the Everglades. Supporting Information Document, South Florida Water Management District, West Palm Beach, 472 pp.

Tilmant, J. T. (Ed.). 1989. Symp. on Florida Bay, a subtropical lagoon. *Bull. Mar. Sci.,* 44:1–524.

Toth, L. A. 1993. The ecological basis of the Kissimmee River restoration plan. *Fla. Sci.,* 56: 25–51.

Toth, L. A. In press. Principles and guidelines for restoration of river/floodplain ecosystems— Kissimmee River, Florida. in *Rehabilitating Damaged Ecosystems,* 2nd ed., J. Cairnes (Ed.), Lewis Publishers, Boca Raton, Fla.

U.S. Army Corps of Engineers. 1991. Environmental Restoration of the Kissimmee River, Florida. Final Integrated Feasibility Report and Environmental Impact Statement, U.S. Army Corps of Engineers, Jacksonville, Fla., 264 pp. + appendix.

Walters, C. 1986. *Adaptive Management of Renewable Resources,* Macmillan, New York, 374 pp.

2

ECOSYSTEMS as PRODUCTS of SPATIALLY and TEMPORALLY VARYING DRIVING FORCES, ECOLOGICAL PROCESSES, and LANDSCAPES: A THEORETICAL PERSPECTIVE

Donald L. DeAngelis

Peter S. White

ABSTRACT

The general features of an ecosystem can be predicted from the underlying geology and exogenous abiotic driving forces, such as mean annual temperature, insolation, precipitation, and nutrient input. A more precise prediction of the ecosystem depends on knowing the natural variability of the driving forces, both temporally and spatially. These driving forces control processes of material accumulation and loss that result in the structural aspects of the ecosystem, for example, soil, vegetation, and landscape pattern. There are three general types of variability: gradual change, such as changes in climate and sea level; disturbances, such as hurricanes, freezes, and fires; and natural periodicities, such as the cycles of dry and wet seasons.

Species are adapted in a number of ways, such as phenotypic plasticity and mobility on a heterogeneous landscape, in order to maintain relative stability in spite of this natural variability in conditions. However, rapid changes outside the normal range of variability and reduction of the size and heterogeneity of the landscape on which the biotic community exists can threaten the integrity of the system.

© St. Lucie Press CCC 0-9634030-2-8 1/94/$100/$.50

INTRODUCTION

The task of ecological theory is to find ways of organizing information that lead to principles or generalizations about ecological systems. These generalizations help us understand how the system under consideration works and how it compares with other systems.

Ecological theory has changed dramatically over the past few decades. The earlier concept that an ecosystem tends to stay near a steady state (homeostasis) has given way to an appreciation of the highly dynamic character of a system in which changes may occur on many time scales. Spatial extent and heterogeneity are now recognized as such crucial aspects of ecological systems that the traditional ecosystem concept has been expanded to the perspective of a larger landscape. The ecosystem is conceived as a dynamic system in which temporally and spatially varying abiotic forces, interacting with biological processes, create a continually changing landscape (or seascape) and a diverse biological community.

While it is true that the Everglades is a unique ecosystem, behind its many singular characteristics are processes that are common to all ecosystems or have analogs in all other ecosystems. This overview will review the general processes.

At a simplistic level, the biological community of an ecosystem derives from the fundamental characteristics of the species and the abiotic characteristics of the environment (Figure 2.1). Fundamental species characteristics include life cycle parameters such as maximum possible life span and age at fecundity, mobility, and tolerances to temperature, drought, salinity, and other conditions. Abiotic environmental characteristics include spatial extent and heterogeneity, the hydrologic regime, nutrient availability, and a variety of disturbances. The combination of the fundamental characteristics of a given species, the abiotic environmental characteristics, and the other species present determine the success of a particular species.

Actually, this view is overly simplistic. Four major factors that complicate this view will be discussed (Figure 2.1):

1. *The importance of history:* Different ecological communities can occur in the same environment, even when the species available for potential invasion are the same, depending on the history of disturbances. For example, Olson (1958) showed that plant succession on sand dunes could follow alternate paths, in part depending on the biotic history (e.g., which species colonizes first) and fire history of a particular site.

2. *Fundamental species characteristics are changeable:* The species characteristics that occur in a given environment at a particular time do not include all that may occur in the future. Invasions of new species can happen continually. Moyle (1986), for example, documented the introductions of exotic fish into North America. In some states in the United States more than 40% of the species were introduced during the past century or so. In addition, evolutionary changes within species present in an environment will occur over sufficiently long time scales. The diverse honeycreeper community of the Hawaiian Islands is an example of the adaptive radiation of birds over millions of years to fill many ecological niches (Carlquist, 1974).

3. *Feedback between biota and the environment:* Not only does the environment influence the species community that becomes established, but this

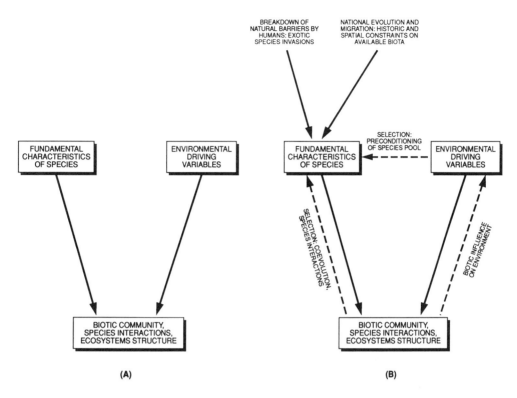

Figure 2.1 (A) The fundamental characteristics of the species available for colonization p.us the properties of a given environment influence the type of biotic community that develops. (B) Additional factors determine the actual community that develops. These include the constraints determining species invasions, environmental preconditioning of the available species pool, the influence of biotic feedback on the environment, coevolution, and other species interactions.

community, in turn, affects the environment by modifying such abiotic characteristics as temperature, humidity, incident light, and nutrient availability. In the North American taiga, for example, colonization by black spruce (*Picea mariana*) of a site initially dominated by white spruce (*P. glauca*) and hardwoods can lower summer soil temperature and slow nutrient cycling. Because black spruce litter is poor in nutrients and slow to decompose, litter accumulates and acts as an insulator, preventing rapid warming of the soil during the summer (Van Cleve et al., 1983).

4. *Species interactions:* Species will interact with each other in ways that are difficult to anticipate. The fundamental properties of the species provide some indication of potential interactions, but what actually occurs is usually impossible to predict. For example, a reduction in lobsters in St. Margaret's Bay, Nova Scotia, was hypothesized as the cause a few decades ago of the destruction of more than 90% of the subtidal seaweed beds. The causal sequence suggested was the following: the decrease in lobsters reduced predation on sea urchins, which changed their feeding habits from detritus to live seaweed. The reduction of seaweed, which provides a protective habitat for the larval lobsters, accelerated the reduction in the lobster population, causing a further increase in sea urchins (Mann, 1977). Such complex causal chains among species are common.

Detailed consideration of all of these aspects of interactions in an ecosystem is impossible here. Instead, the focus will be on the more direct effects of abiotic driving forces on species populations. This emphasis may be appropriate for the Everglades ecosystem because many of the chief concerns in management of the Everglades involve the persistence of populations. The main threats to these populations are changes in abiotic driving forces, such as water conditions, which the health of key species populations tracks very closely. This does not mean that ecosystem-level processes (such as net primary and secondary productivity, nutrient cycling rates, and so forth) are not important issues. Restoration of populations may also require preserving natural ecosystem processes.

The ABIOTIC ENVIRONMENT

Classification of Driving Forces

An ecosystem is ultimately the product of exogenous forces of climate and geology, including insolation, temperature, precipitation, overland and ground water flow, sea level rise and fall, storm intensity and frequency, and nutrient input through precipitation and the weathering of rock substrate. These forces can be termed *driving forces*. They are all at least partly physical in character, although most can be strongly influenced by biological structures and processes (e.g., biogenic limestone deposition, modification of land surface albedo by soil and plants, modification of water flow and evaporation). The driving forces govern the rates of a variety of *processes,* both biological and physical, that directly build, destroy, or change the physical and biological *structures* of an ecosystem. For example, in terrestrial ecosystems, soil organic matter is a structure that is formed by the deposition of litter and its processing by the detrital community. This process of soil organic matter build-up is counteracted by gradual processes of oxidation as well as by severe fires and erosion.

The simple average values of the main driving variables sometimes determine the basic ecosystem type. The plot developed by Holdridge (1947), based on temperature and precipitation (Figure 2.2), provides some rough categorization of broad ecosystem types. It is noteworthy that these two factors alone are often insufficient to predict ecosystem type without additional characteristics (such as the seasonality of temperature and precipitation, the types and patterns of disturbances, and topography) which influence microclimate and the movement of water. If the driving forces were unchanging, an ecosystem would change very little through time, except for processes of species invasion and biological evolution. However, the driving forces undergo various types of temporal variation, and the spatial patterns of these variations create the unique characteristics of an ecosystem and determine the types of organism life cycles that can exist there. The seasonality of precipitation in the Everglades, for example, is a key characteristic, as it limits biota to types that have adaptations to either avoid or cope with changes in water availability. The significance of disturbances is that they may prevent the system from undergoing succession to reach the state predicted by the Holdridge scheme. As an example, pastoralism in the Mediterranean region has denuded land that

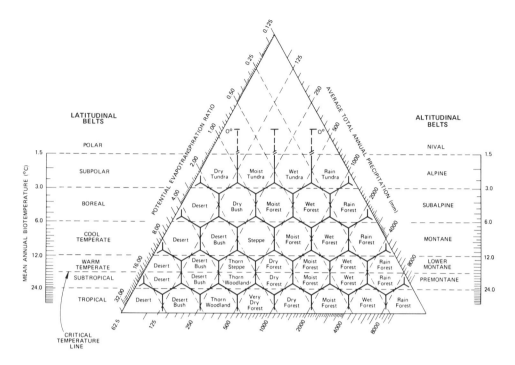

Figure 2.2   Diagram showing the determination of ecosystem type through a combination of mean annual temperature and average total annual precipitation. (Modified from original diagram by Holdridge, 1947.)

would be forested under undisturbed conditions (Hillel, 1991). One or two other abiotic factors, such as nutrient input, may also modify the development of specific ecosystems. The driving forces that shape and change ecosystems through time can be divided into three general types:

1. Gradual, continuous changes in external conditions (e.g., geological and climatic changes)
2. Temporally discrete events or "disturbances" (e.g., fires, storms, droughts, floods)
3. Natural periodicities (e.g., annual temperature cycles, annual precipitation cycles)

This classification is somewhat arbitrary, and some abiotic forces could fall equally well under one category or another. For example, (a) a flood is in some sense a disturbance, but it may also be simply an exaggeration in the annual hydroperiod of ecosystems that are normally submerged for parts of a year; (b) sea level is a gradually changing condition, but it also undergoes significant tidal and annual rises and falls. The division of driving forces into three types is a useful organizational tool as long as it is recognized that the distinctions are not rigid and may change with changing context.

The temporal variations in the driving forces affect the balances of the ecological processes, causing changes in ecological structure. Such changes are natural. However, in many ecosystems human activities are now pushing the driving forces beyond their normal ranges.

Temporal and Spatial Scales of Driving Forces

All ecosystems consist of processes that occur over wide ranges of temporal and spatial scales. A hierarchical approach that orders the processes in the ecosystem according to the temporal and spatial scales on which they act is essential in organizing such information (O'Neill et al., 1986). This ordering makes it possible to decide which processes are important and which processes can safely be approximated as constants when trying to understand ecosystem behavior on particular spatial and temporal scales (Powell, 1989).

This ordering greatly simplifies evaluation of the driving forces. For example, some geological processes, such as bedrock formation, generally act on time scales of millions of years. For this reason, bedrock geology can be assumed to be relatively constant on the time scales on which a particular ecosystem might be expected to persist. Other geological conditions, such as climate and sea level, can undergo substantial changes over periods of thousands of years. On much shorter time scales, say a few decades, climate and sea level may be regarded as effectively constant so that ecosystem dynamics can be studied without considering them as changing driving forces.

Spatial scales also can be fit into a hierarchical scheme. Variations in bedrock type and elevation (along with large-scale hydrologic patterns) over large spatial scales influence plant community type on scales of kilometers or greater, for example. This will be referred to here as the *macroscale*. Other sources of macroscale variability, such as increase in the amplitude of the annual temperature cycle as one moves inland away from the coast, may also contribute to variations in plant communities on this scale.

However, irregularities in microbedrock topography at scales of a few square meters (termed the *microscale* here) or several hectares (termed the *mesoscale* here) create conditions for microscale and mesoscale patterns in vegetation within the macroscale regions dominated by a particular plant community. Whittaker (1977) classified biotic diversity at these three scales as *alpha* (within a given habitat), *beta* (within a large enough spatial area to include an environmental gradient), and *gamma* (over an entire region).

Using this concept of scale, the effects of different types of variation can be discussed in detail.

Gradually Changing Environmental Conditions

Primary examples of abiotic conditions that change over a long time scale are climate change and sea level rise. These two driving forces may be correlated.

The processes of climate change and sea level rise, as well as other gradually changing conditions (e.g., nutrient loading, acid deposition), can affect some of the basic physical structures or features of an ecosystem (the physical size of a favorable habitat, the shoreline of a coastal ecosystem, the soil and subsoil,

Table 2.1 Main Structural Aspects of the Everglades Ecosystem, Processes that Shape These Structures, and Driving Forces that Affect Process Rates

Structural aspects	Processes	Driving forces
Shoreline	Build-up of marl levees and mud banks (+) Erosion and seawater transgression (–)	Sea level rise Freshwater flow
Vegetation	Primary production (+) Mortality (–)	Insolation Fire, storms
Soil and peat	Mortality of vegetation and deposition of peat and marl (+) Oxidation, including fire (–)	Climate change (hydroperiod)
Spatial pattern of water flow	Precipitation (+) Evaporation (–)	Climate change
Tree islands	Accretion of matter through eddy currents; vegetation growth (+) Erosion (–)	Climate change (hydroperiod)
Nutrient status	Precipitation, nutrient transport from upstream (+) Outflow, peat deposition (–)	Climate change Upstream nutrient level

the spatial pattern of water flow, and nutrient trophic status) by altering the rates of the main processes that control these structures (Table 2.1). (Here we will consider the pattern of water flow to be a physical structure in some sense, determined by the differences between precipitation and evaporation, as mediated by topography.)

Nutrient or trophic status refers to the available nutrients in the system, and systems are commonly classed as oligotrophic, mesotrophic, or eutrophic for low, medium, and high levels of nutrients, respectively. This physical feature of the ecosystem may not seem as conspicuous as the other physical structures discussed previously. However, it is reasonable to include nutrient status in the list of structural aspects in Table 2.1 because it partly determines the material structure of the system through its influence on the plant community (Grime, 1979; Tilman, 1988). The level of nutrients in the system is determined, similar to other physical features, by forces building up this level (various inputs) and forces depleting this level (various losses).

In considering examples of continuous driving forces, one can see that the structural features of an ecosystem arise from competing processes of accretion and deterioration. These structures may appear to be in equilibrium on short enough time scales, although on sufficiently long time scales they are certainly changing. Changing environmental conditions that affect the rates of one or more of the processes involved could lead in many cases to rapid structural changes.

Disturbances

Disturbances are defined as relatively discrete events in time, although events classified as disturbances range from lightning strikes to prolonged droughts. Disturbances of a given type can be assumed to be characterized by a temporal

Table 2.2 Some Characteristics of Disturbances

Descriptor	Definition
Frequency	Mean number of events per time period
Return interval	The inverse of frequency; mean time period disturbances
Disturbance area	Mean area disturbed per event
Rotation period	Mean time needed to disturb an area equivalent to the study area
Intensity	Physical force of disturbance event
Synergism	Effects of occurrence of other disturbances (e.g., drought increases fire frequency and intensity)

From Pickett and White, 1985.

scale that is the mean time interval between disturbances. This mean time interval is the inverse of the frequency of disturbance events. Disturbances can be classified by several other criteria in addition to frequency, including spatial extent of the disturbance, rotation period, and intensity (Table 2.2). Intensity can be a complex function of the duration of a disturbance and the state of the system affected. For example, the intensity of a freeze disturbance depends on absolute temperature, duration, preconditioning of plants, and possible repetition. Fire intensity in the Everglades commonly has binary levels: fires that affect only vegetation and fires that also burn peat. *Severity* is a term that may be roughly synonymous with intensity, although it usually refers to the result of a disturbance to the ecosystem and may also include the aspect of spatial area. In addition to the differentiation between intensity or severity of disturbances within a given category, different types of disturbances have different intrinsic severities in relation to the system they affect. For example, a hurricane, even a severe one that affected an entire ecosystem, would not have the devastating effect of an intense fire of equal spatial extent. Thus, it can be misleading to lump all disturbances together. All disturbances are not equivalent.

In general, there is a negative relationship between the spatial area of a disturbance and its frequency; that is, major disturbances affecting large areas tend to occur less frequently than small disturbances affecting small spatial areas. For example, fires in a forest stand occur much less frequently than individual tree-fall events (Shugart, 1984). Disturbances are generally somewhat random; however, because storms, droughts, fires, and other disturbances tend to be seasonal phenomena, there are biases regarding the probabilities of when they occur within a year. In addition, many weather phenomena, such as El Niño and the weather phenomena related to it, recur on fairly regular cycles of several years. The periodicity in precipitation results in periodicities in other disturbances, such as fire frequency and flood frequency.

Disturbances play an organizing role as much as they do a disruptive role in an ecosystem. A disturbance regime may be instrumental in keeping certain plant types (especially late successional species) from becoming dominant. For example, fire slows or prevents the establishment of late successional tree species that are not fire-adapted relative to other, faster growing species (Spurr and Barnes, 1980). The net effect of such disturbances can then be "good" in the sense of preserving diversity by preventing dominance by a few species. However, this favorable effect on diversity works well only on landscapes sufficiently large that individual disturbance events affect only relatively small portions (patches) of the entire landscape and in which the rotation period (Table 2.2) is sufficiently long that a full range of successional patches can be present at once (Whittaker and Levin, 1977). The communities of flora and fauna will be determined in part by how much time the community has to undergo secondary succession before it is disturbed again. A short rotation period will favor early successional species, while a long rotation time favors late successional species. Diversity is generally highest in a medium disturbance regime (Connell, 1978; Huston, 1979; Bazzaz, 1983; Denslow, 1985).

A *hierarchical pattern* of disturbances relates to the above ideas. Disturbances that affect large spatial areas tend to occur less frequently than those that affect small areas (Figure 2.3). Woods and Whittaker (1981) found that the spatial area needed to preserve some important ecosystem type, or a species population, is much larger than the spatial scale of any destructive disturbance that might affect the ecosystem. These authors commented on spatial scale with respect to stability of forest ecosystems: "No forest community would be stable if areas containing only one or a few trees were considered as units. At a large spatial scale, almost any forest might satisfy our criteria for stability." Under *destructive disturbances* (which

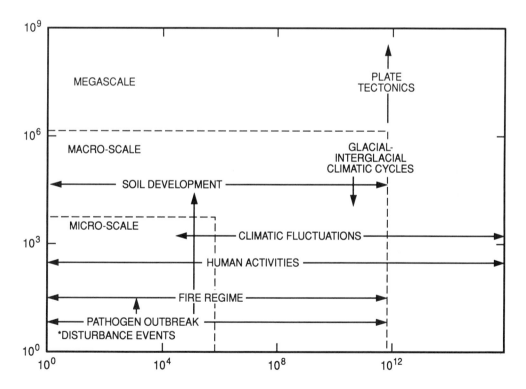

Figure 2.3 A hierarchical scaling of biotic disturbance regimes. The horizontal axis represents spatial area in square meters and the vertical axis represents time in years. Asterisk indicates disturbance events such as wildfire, wind damage, clear cut, flood, and earthquake. (From Delcourt et al., 1982.)

is, of course, a relative notation), disturbances that disrupt normal annual periodicities are included, in addition to fires, severe droughts, and severe storms. The result of the hierarchy of scales is that disturbances create a spatial patchiness in the ecosystem, with small patches generally exceeding large patches in number. This is generally true in forest ecosystems, where small patches of open area (gaps), caused by the fall of individual canopy trees, greatly outnumber the larger openings created by fires and severe storms (Shugart, 1984).

Because ecosystems are subjected to an array of natural disturbances, one might ask whether anthropogenic disturbances constitute much of a problem for ecosystems. Reiners (1983) noted that while some human disturbances closely mimic natural disturbance, in general "...human-induced disturbances contrast sharply in kind, scale, intensity, and frequency; for example, plowing grasslands or distributing exotic biocides are essentially unique disturbances in kind compared with natural disturbances. Human disturbances may or may not be more intensive than natural disturbances. For example, flood control has reduced the frequency as well as the amplitude of floods in some riverine systems. In contrast, fire suppression policy in the U.S.A. has led to a higher frequency of crown fires in forests, totally different in their impact from ground surface fires."

Natural Periodicities

The third type of driving force is natural periodicity, which usually means an annual cycle. Because most organisms focus on the annual cycles of temperature and precipitation for reproduction, these can have a magnified importance. The wet-dry cycle is of particular importance in many ecosystems, as it results in a spatially varying annual period of growth that partially determines the floristic and faunal communities of the ecosystem.

It is difficult to characterize annual precipitation pattern by a few indices, because such a wide variety of patterns is possible. Several aspects of the yearly precipitation cycle are variable, including (1) the length of the wet period, (2) the intensity of the wet season, (3) the amount of precipitation during the dry season, and (4) the amount of precipitation falling as snow versus rain. A great variety of different patterns may exist within a spatial region, depending on the topography.

The characteristics of the annual precipitation cycle may greatly affect breeding success of some species. The dependence of reproductive success in populations on the details of annual abiotic periods is quite common in ecosystems. Many freshwater temperate fish species, for example, are dependent on the details of the scenario of water temperature change through the spawning season (Ridgway and Friesen, 1992). Reversals in the steady increase in water temperature can lead to nest abandonment.

In general, annual periodicities lead to annual occurrences of dynamic spatial features in ecosystems. These include such phenomena as moving rain fronts in arid zones and ocean fronts separating two types of water conditions. It has been said that "...if a concentration of 'energy' [variance] were to be found at a certain spatial scale in a physical phenomenon—say, a front—then one might expect to find a concentration of variance at the same scale in biologically interesting quantities" (Powell, 1989). Ocean fronts are described by Harris (1986): "Sustained primary production may also occur at oceanic fronts and ring margins where physical processes serve not only to concentrate biological activity by virtue of reduced horizontal and vertical diffusivities, but also to produce a sustained flow of nutrients from the mixing of water masses." Ward (1971) described the spatiotemporal pattern of food availability to the African weaverbird (*Quelea quelea*) in the Sahel region. Predictable rains travel in moving fronts several hundred kilometers wide. The weaverbird colonies move in advance of the front, first feeding on dry seeds and then settling in the vicinity to lay eggs and raise young on a diet of insects made available by the rains.

Variability in seasonal patterns, such as rainfall pattern and especially the occurrence of large rainfall events (which can be termed "disturbances") during the normal dry season, adds a risk to populations, especially with respect to breeding, as exemplified by wading birds in the Everglades. However, spatial nonuniformities, such as microscale and mesoscale elevational irregularities, tend to moderate interannual rainfall variability. If there is a sufficient amount of spatial heterogeneity, then even if the rainfall pattern varies from year to year, each year there will be a variety of hydrologic patterns in the landscape. Individual wading birds will almost certainly find areas of adequate food availability in some part of their foraging ranges.

Interactions between Driving Forces

It is often said that every process or component of an ecosystem affects every other process or component. The problem that confronts ecologists is to determine the interactions that are crucial and must be studied in detail. Figure 2.4 illustrates, in an abstract way, how each of the general types of driving forces can affect the others. Examples of each of the interactions noted in Figure 2.4 are as follows.

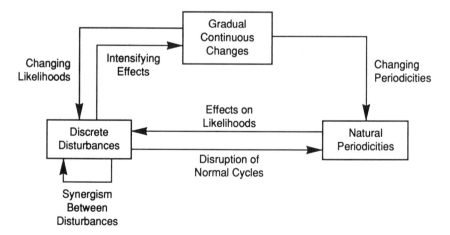

Figure 2.4   Three general types of alterations in the environmental driving forces can occur: gradual continuous change, discrete disturbances, and natural periodicities. These types of alterations are not independent and can mutually influence each other.

Natural periodicities, such as the alternation of wet and dry seasons or seasonal sea level fluctuations, condition the intra-annual probabilities of certain disturbances, such as droughts, fires, storms, and floods. Discrete disturbances, such as unseasonable storms, can upset natural periodicities. Disturbances also affect the probability of other disturbances. Both droughts and electrical storms, for example, contribute to the initiation of fires. Disturbances can also intensify the effects of gradual continuous changes. Storms may be the specific events in which sea level transgressions occur during a time of rising sea level.

Gradual changes in climate and sea level will change disturbance frequencies and the details of annual periodicities. A particular instance of possible importance is the change toward a drier climate for the southeastern United States, which has been predicted by some climate models. This drying trend could increase the frequency of fires in some ecosystems.

RESPONSES of SPECIES POPULATIONS to DRIVING FORCES

A key element in population ecology theory is the concept of stability. Population stability is a complex concept, but it generally refers to the ability of a population to remain relatively constant through time. A weaker form of stability is *persistence,* which simply means the ability of a population to maintain itself in a region, although it may suffer severe fluctuations.

There are two general ways in which interactions among individuals and between individuals and their environment create conditions of population instability in the sense of high variability in numbers and vulnerability to crashes. The first involves changes in the abiotic environment (the driving forces) to which the populations are not well adapted, while the second involves the destabilizing effects of biotic feedbacks (Figure 2.5).

Figure 2.5 Schematic showing two main types of factors that cause instabilities in biotic communities and populations: strong biotic feedbacks and strong changes in the environment.

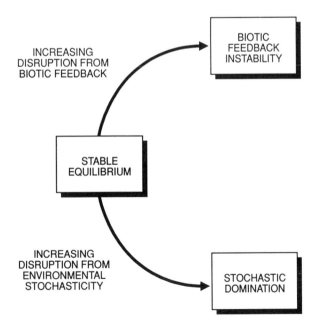

The three main types of abiotic driving forces have already been described. First, gradual environmental change may put populations under increasing stress, which can lead to gradual or sometimes sudden population collapses. Noy-Meir (1975) showed that a gradual increase in grazing pressure in a grassland could, beyond some critical threshold, lead to a catastrophic decline in primary production. Second, an increased disturbance rate may inflict heavy mortalities from which a population cannot recover. Heavy hunting pressure on the passenger pigeon (*Ectopistes migratorius*), for example, eventually forced the population down to low levels, from which it did not recover (Halliday, 1980). Finally, large variations in the annual climatic cycle may disrupt sensitive parts of the life cycle, such as reproduction and early life stages, for example in many fish populations (Cushing, 1988). It must also be added that the changes in each of these three driving forces can favor as well as harm species populations. Gradual environmental change may cause the ranges of some species to expand or allow new invaders to be established. For example, Davis (1981) used the pollen record to trace the expansion of many tree species after the retreat of the glaciers during the early Holocene in North America. Disturbances may have a net positive effect on a species by having a stronger negative effect in decreasing the number of competitors or predators of some species other than the species in question. Wave disturbances on rocky intertidal substrates can act in this way to decrease predation and competition and are believed to be important in structuring the communities of such substrates (Dayton, 1971). A major challenge for theory is to predict the changes within the community that might occur through a gradual change in conditions or a change in the pattern of disturbances.

The other major source of population instability—biotic feedbacks—can affect populations and communities when the feedback becomes unbalanced and leads to violent population oscillations or drastic population decreases. For example, intraspecific competition and cannibalism, which often plays a homeostatic role in populations, may become too strong under some circumstances, resulting in older age classes or larger size classes, suppressing younger members and causing the age structure of the population to oscillate through time. This has been described in fish populations (Carpenter, 1988). Interactions with other species (prey, predators, competitors), if strong, can cause similar imbalances, leading to oscillations from prey overexploitation or to competitive displacements. The well-known fluctuations in snowshoe rabbits (*Lepus bairdi*) and lynxes (*Lynx lynx*) in the boreal region may be an example of such predator–prey interactions (May, 1973). The instabilities induced by strong biotic feedbacks may be influenced by abiotic conditions. As discussed earlier, natural disturbance regimes may in some cases limit predation pressure by exerting a stronger negative effect on predators than on their prey. In this way, abiotic disturbances can sometimes reduce biotic feedback instabilities.

Whether, and to what extent, changes in abiotic environmental conditions or biotic feedbacks will destabilize a population or community depends on the fundamental characteristics of the species within the community. Low population variability in the face of environmental disturbances depends on aspects of stability termed *resistance* and *resilience*. Resistance is the tendency of a population to stay within some range of abundance values, despite external perturbations and stresses. Resilience is the rate at which a population returns following a perturbation away from equilibrium. For example, long lifetimes of individuals in a species contribute to the ability of a population to resist disturbances that primarily cause mortalities to younger members, because a large fraction of the population is likely to be old and skilled enough in foraging to be unaffected by the disturbances. Goodman (1974) developed a mathematical model of colonial offshore-feeding bird species in which the adults were able to survive through years with both good and poor food availability, but whose offspring generally did not survive in the poor years. Adaptation to a wide spectrum of habitats and food types, especially when coupled with high motility, is another useful mechanism of resistance, because disturbances are unlikely to affect all habitats and all food resources in a region simultaneously. For example, a grizzly bear utilizes a diverse array of foods and may have a home range of more than 260 km^2 (Van Gelder, 1982). Mutualistic social behavior, parental care, and sophisticated behavioral modes all contribute to population resistance to some kinds of disturbances. For example, in areas with a high probability of drought conditions, wildebeests (*Connochaetes taurinus*) are organized into migratory herds that are well adapted to moving to more favorable areas if the current habitat becomes too unproductive (Estes, 1969). On the other hand, when social characteristics are obligate, they may actually lead to vulnerability to certain disturbances, because the population may not be able to recover if it is reduced to such a low numerical level that the mutualistic mechanisms cannot be used. This may have occurred in the passenger pigeon (*Ecopistes migratorius*) after populations were reduced to low levels by hunting (Halliday, 1980). Population resilience following mortality-inducing disturbances will be positively correlated with high dispersal ability and high fecundity. These latter characteristics enable individuals to recolonize sites and quickly reestablish populations. Sessile animals such as corals and mussels, therefore, disperse in the motile larval stage (Lissner et al., 1991).

Characteristics that may provide resistance to gradual environmental changes, such as climatic change and habitat shrinkage or alteration, include genetic diversity and phenotypic plasticity. Genetic diversity makes it possible for some individuals in a population to be predisposed to surviving the change, while phenotypic plasticity of individuals means that it is possible that they may adapt to the changing conditions.

High dispersal ability of a population favors population survival because at least some individuals may reach sanctuaries before environmental change is sufficient to exclude them from their former territories. Many North American temperate tree species were able to survive in the lower Mississippi Valley and northern Florida during the Ice Ages (Davis, 1981). Both the motility of individuals and the dispersal ability of the population as a whole, whether through individual organism movements or the movements of propagules, are important species characteristics relative to survival and stability of species populations under changing environmental driving forces. Both of these qualities may not suffice, however, without adequate spatial extent of suitable habitat. Willis (1974) and Karr (1982), for example, have documented the loss of bird species after Barro Colorado Island, Panama, was created by the rise of waters in Gatun Lake.

Wiens (1989) noted two general conceptual models concerning a population on a landscape or region and how the stability of the population is related to the spatial scale of the available habitat of the region. To help discuss these two models, think of the region as consisting of many smaller subregions, for simplicity, like squares on a checkerboard. The first conceptual model might be called the *model of semi-independent subpopulations*. This pictures the region as consisting of subpopulations on many of the squares that are favorable habitats. The subpopulations are small and they may fluctuate greatly through time because of demographic stochasticity (that is, in small subpopulations, births and deaths do not always equal each other over short time periods) and environmental stochasticity (a local storm, flood, or other event may severely affect a subpopulation on a small square). However, if the region consists of a very large number of habitat squares and the population fluctuations on the individual squares are more or less independent of each other, the total regional population does not fluctuate much at all, because the subpopulations will normally do well enough on a sufficient number of squares to compensate for the losses on other squares. Subsequently, subpopulations from the favored squares can recolonize the squares where losses occurred. This reflects the Central Limit Theorem of probability theory. Thus, the population integrates over a mosaic of subregions, some of which may be favorable and some unfavorable within a given year.

The other conceptual model that Wiens (1989) noted for population viability on a landscape is that individuals themselves may integrate over wide spatial areas containing a mosaic of habitats. The favorability of a given subregion or square in the region may be changing from day to day or year to year, but the individual can move to areas that are favorable. This hypothesis differs from the first in that in this case, each individual in the population requires that a sufficiently large region of land be available such that some of the checkerboard squares will always have favorable conditions.

These ideas can be applied to wading birds and other mobile organisms. The second conceptual model relating landscape to wading bird population success probably applies because the birds shift colony location within the region from year to year in response to changing patterns of food availability, and the birds within a given colony integrate over a large part of the total region. As in the examples cited earlier, the availability of food for wading birds in the Everglades is a dynamic factor, changing spatially and temporally across the squares of the checkerboard. An individual bird and an individual colony must make use of many of these squares to maintain its own energetic and nutrient demands over time. The second conceptual model predicts that any reduction in area, if it consists of area that is critical to many of the individual birds for even a short period of time, can have a greater than proportional effect on the total population. In the worst possible case, a few squares of land that are crucial to all birds at some phase of their annual cycle might be removed. This would devastate the population.

Ecologists have demonstrated many cases in which individual animals integrate over a large area of landscape. Hansson (1979) noted that small mammals in Scandinavia require different parts of the landscape. Forested habitats may provide abundant food in summer, but earlier successional stages provide more seeds needed for mammal survival in the winter. Birds such as gallahs (*Cacatua roseicapilla*) and little corellas (*C. sanguinea*) in Australia forage over shrub-desert habitats, grasslands, or grain fields for several kilometers around nesting sites. Use of a particular habitat type at a given time depends on seed availability, which depends on rainfall (Wiens, 1989).

Grey teals (*Anas gibberifrons*) breeding in the area of the Murray River and its tributaries in Australia during times of flooding are faced with great irregularity of inundations. They persist by being nomadic and taking advantage of floodwaters wherever flooding may occur over a wide area of the river floodplains (Andrewartha and Birch, 1984). Terborgh (1989) noted that seed abundance for granivorous birds in the southwestern North American deserts is highest after the end of the winter rains and then declines, as flocks of sparrows and finches search out remaining pockets of seeds. The seed crop varies greatly from year to year with rainfall, and the birds must be able to effectively locate the areas of desert that received the highest rainfall in a particular year.

It has been estimated that since it became an island, Barro Colorado Island in Panama may have lost 50 to 60 bird species (Karr, 1982). Many of these birds seem to have had plenty of natural habitat on the island, so their extinctions are not explained by island biogeography theory. However, Karr suggested that the reason for the extinctions lay in the fact that the small size of the island resulted in the loss of some parts of the mosaic of habitats that were essential for certain periods of time (Morton, 1987).

Environmental heterogeneity in the form of mesoscale patchiness may also be essential in the persistence of communities of species competing for the same resources. Smith and Vrieze (1979) studied habitat use by three rodent species in the Everglades. The species shared tropical hardwood hammocks during the wet breeding season, but segregated into three different habitats (hammock, prairie interstices, and a third unknown site) during the dry season. Spatial extent and heterogeneity thus smooth out or "spread the risks" of fluctuations in annual periodicities that affect organisms and populations.

CONCLUSIONS

What can be said about the stability of a biological community? Stability is a relative concept and depends on spatial and temporal scale. Stability is the tendency for species to persist and not fluctuate wildly. Communities occupying small, spatially homogeneous areas are subject to two basic types of disruption: (1) environmental stresses or disturbances causing extinctions because of high mortalities or lack of successful reproduction in a population and (2) overly strong biotic couplings, resulting in drastic population oscillations and competitive exclusions. Spatial extent and heterogeneity can decrease the total effect of these disruptions. If the spatial extent of a system is large compared to the spatial scale on which disruptive forces act, then the ecological community will be subjected to such destabilizing forces only in some areas of the overall region at a given time. Animals can move to areas not disrupted and return after the disruption. The conditions under which the spatial area of a habitat preserve is sufficient to ensure community persistence have been explored in an abstract way with mathematical models, but much more detailed models of specific systems are needed to produce reliable predictions.

ACKNOWLEDGMENTS

The authors appreciate the comments of R. V. O'Neill, W. M. Post, R. H. Gardner, and D. S. Shriner on earlier versions of this chapter. Research was sponsored by the National Science Foundation's Ecosystem Studies Program under Interagency Agreement No. 40-689-78 with the U.S. Department of Energy under Contract DE-AC05-840R21400 with Martin Marietta Energy Systems, Inc. Environmental Sciences Division.

LITERATURE CITED

Andrewartha, H. G. and L. C. Birch. 1984. *The Ecological Web,* University of Chicago Press, Chicago.

Barndorff-Nielsen, O. E., P. Blaesild, J. L. Jensen, and M. Sorensen. 1985. The fascination of sand. in *A Celebration of Statistics,* A. C. Atkinson and S. E. Feinberg (Eds.), Springer-Verlag, New York, pp. 57–87.

Bazzaz, F. 1983. in *Disturbance and Ecosystems: Components of Response,* H. A. Mooney and M. Godron (Eds.), Springer-Verlag, Berlin.

Carlquist, S. 1974. *Island Biogeography,* Columbia University Press, New York.

Carpenter, S. R. 1988. Transmission of variance through lake food webs. in *Complex Interactions in Lake Communities,* S. R. Carpenter (Ed.), Springer-Verlag, New York, pp. 119–140.

Connell, J. H. 1978. Diversity in tropical rain forests and coral reefs. *Science,* 199:1302–1310.

Cushing, D. H. 1988. The study of stock and recruitment. in *Fish Population Dynamics,* 2nd ed., J. A. Gulland (Ed.), John Wiley & Sons, New York, pp. 105–128.

Davis, M. B. 1981. Quaternary history and the stability of forest communities. in *Forest Succession,* D. C. West, H. H. Shugart, and D. B. Botkin (Eds.), Springer-Verlag, New York, pp. 132–153.

Dayton, P. K. 1971. Competition, disturbance, and community organization: The provision and subsequent utilization of space in a rocky intertidal community. *Ecol. Monogr.,* 41:351–389.

Delcourt, H. R., P. A. Delcourt, and T. Webb III. 1982. Dynamic plant ecology: The spectrum of vegetational change in space and time. *Q. Sci. Rev.,* 1:153–176.

Denslow, J. S. 1985. Disturbance-mediated coexistence of species. in *The Ecology of Natural Disturbance and Patch Dynamics,* S. T. A. Pickett and P. S. White (Eds.), Academic Press, New York, pp. 307–323.

Estes, R. D. 1969. Territorial behavior of the wildebeest (*Connochaetes taurinus* Burchell, 1823). *Zeit. Tierpsychol.,* 26:284–370.

Goodman, D. 1974. Natural selection and a cost-ceiling on reproductive effort. *Am. Nat.,* 108: 247–268.

Grime, J. P. 1979. *Plant Strategies and Vegetation Processes,* Wiley-Interscience, New York.

Halliday, T. R. 1980. The extinction of the passenger pigeon *Ectopistes miqratorius* and its relevance to contemporary conservation. *Biol. Conserv.,* 17:157–162.

Hansson, L. 1979. On the importance of landscape heterogeneity in northern regions for the breeding populations of homeotherms: A general hypothesis. *Oikos,* 33:182–189.

Harris, G. P. 1986. *Phytoplankton Ecology: Structure, Function, and Fluctuations,* Chapman and Hall, London.

Hillel, D. J. 1991. *Out of the Earth: Civilization and the Life of the Soil,* The Free Press, New York.

Holdridge, L. R. 1947. Determination of world plant formations from simple climatic data. *Science,* 105:367–368.

Huston, M. A. 1979. A general hypothesis of species diversity. *Am. Nat.,* 113:81–101.

Karr, J. R. 1982. Avian extinction in Barro Colorado Island, Panama: A reassessment. *Am. Nat.,* 119:220–239.

1⁵, 1⁸3

Karr, J. R. and K. E. Freemark. 1985. Disturbance and vertebrates: An integrative perspective. in *The Ecology of Natural Disturbance and Patch Dynamics,* S. T. A. Pickett and P. S. White (Eds.), Academic Press, New York, pp. 153–168.

Kushlan, J. A. 1979. Design and management of continental wildlife reserves: Lessons from the Everglades. *Biol. Conserv.,* 15:281–290.

Lissner, A. L., G. L. Taghon, D. R. Diener, S. C. Schroeter, and J. D. Dixon. 1991. Recolonization of deep-water hard-substrate communities: Potential impacts from oil and gas development. *Ecol. Appl.,* 1:258–267.

Mann, K. H. 1977. Destruction of kelp beds by sea-urchins: Cyclic phenomenon or irreversible degradation? *Helgo. Wiss. Meerunders.,* 30:455–467.

May, R. M. 1973. *Stability and Complexity in Model Ecosystems,* Princeton University Press, Princeton, N.J.

Morton, E. S. 1987. Reintroducing recently extirpated birds into a tropical forest preserve. in *Endangered Birds, Management Techniques for Preserving Threatened Species,* S. A. Temple (Ed.), The University of Wisconsin Press, Madison, pp. 379–386.

Moyle, P. B. 1986. Fish introductions into North American. in *Ecology of Biological Invasions of North America and Hawaii,* H. A. Mooney and J. A. Drake (Eds.), Springer-Verlag, New York, pp. 27–43.

Noy-Meir, I. 1975. Stability of grazing systems: An application of predator-prey graphs. *J. Ecol.,* 63:459–481.

Olson, J. 1958. Rates of succession and soil changes on southern Lake Michigan sand dunes. *Bot. Gaz.,* 119:125–170.

O'Neill, R. V., D. L. DeAngelis, J. B. Waide, and T. F. H. Allen. 1986. *A Hierarchical Concept of the Ecosystem,* Princeton University Press, Princeton, N.J., 253 pp.

Pickett, S. T. A. and P. S. White. 1985. Natural disturbance and patch dynamics: An introduction. in *The Ecology of Natural Disturbance and Patch Dynamics,* S. T. A. Pickett and P. S. White (Eds.), Academic Press, New York, pp. 3–13.

Powell, T. M. 1989. Physical and biological scales of variability in lakes, estuaries, and the coastal ocean. in *Perspectives in Ecological Theory,* J. Rouqhgarden, R. M. May, and S. A. Levin (Eds.), Princeton University Press, Princeton. N.J., pp. 157–176.

Reiners, W. A. 1983. Disturbance and basic properties of ecosystem energetics. in *Disturbance and Ecosystems: Components of Response,* H. A. Mooney and M. Godron (Eds.), Springer-Verlag, Berlin, pp. 83–98.

Ridgway, M. S. and T. G. Friesen. 1992. Annual variation in parental care in smallmouth bass, *Micropterus dolomieui. Environ. Biol. Fishes,* 35:243–255.

Shugart, H. H. 1984. *A Theory of Forest Dynamics,* Springer-Verlag, New York.

Smith, A. T. and J. M. Vrieze. 1979. Population structure of Everglades rodents: Responses to a patchy environment. *J. Mammol.,* 60:778–794.

Spurr, S. H. and B. V. Barnes. 1980. *Forest Ecology,* 3rd ed., John Wiley & Sons, New York, 687 pp.

Terborgh, J. 1989. *Where Have All the Birds Gone?* Princeton University Press, Princeton, N.J., 207 pp.

Tilman, D. 1988. *Plant Strategies and the Dynamics of Plant Communities,* Princeton University Press, Princeton, N.J.

Van Cleve, K., C. T. Dyrness, L. A. Viereck, J. Fox, F. S. Chapin III, and W. Oechel. 1983. Taiga ecosystems in interior Alaska. *BioScience,* 33:39–44.

Van Gelder, R. G. 1982. *Mammals of the National Parks,* The Johns Hopkins University Press, Baltimore.

Ward, P. 1971. The migration patterns of *Quelea quelea* in Africa. *Ibis,* 113:275–297.

Whittaker, R. H. 1977. Evolution of species diversity in land communities. *Evol. Biol.,* 10:1–67.

Whittaker, R. H. and S. A. Levin. 1977. The role of mosaic phenomena in natural communities. *Theor. Popul. Biol.,* 12:117–139.

Wiens, J. A. 1989. *The Ecology of Bird Communities,* Vol. 2: Processes and Variations, Cambridge University Press, Cambridge, 316 pp.

Willis, E. O. 1974. Populations and local extinctions of birds on Barro Colorado Island, Panama. *Ecol. Monogr.,* 44:153–169.

Woods, K. D. and R. H. Whittaker. 1981. Canopy-understory interaction and the internal dynamics of mature hardwood and hemlock-hardwood forests. in *Forest Succession: Concepts and Applications,* D. C. West, H. H. Shugart, and D. B. Botkin (Eds.), Springer-Verlag, New York, pp. 305–323.

Yentsch, C. S. 1980. Phytoplankton growth in the sea. A coalescence of disciplines. in *Primary Productivity of the Sea,* P. Falkowski (Ed.), Plenum Press, New York, pp. 17–32.

3

WETLAND ECOSYSTEM PROTECTION, MANAGEMENT, and RESTORATION: An INTERNATIONAL PERSPECTIVE

E. Maltby

P. J. Dugan

ABSTRACT

The critical attention currently directed toward wetland ecosystems is not restricted to scientists. Rather, wetlands pose some of today's most contentious, difficult, and politically sensitive environmental questions and rival tropical rain forests for priority on the world conservation agenda. Examples are examined to illustrate the two main themes which underpin the concern for wetlands worldwide: a rapidly diminishing resource base and a growing awareness of wetland values.

The Everglades is important not only at a national level to the United States, but also at an international level to the world community. Everglades National Park is one of only three sites to be recognized on three international lists—International Biosphere Reserve, World Heritage, and Ramsar. Everglades National Park is a key standard bearer for the efficacy of international conservation designation and the viability of supporting conventions to which the United States government is a signatory member. No other subtropical or tropical wetland enjoys such overt recognition of biological importance.

Conservation and sound management of the Everglades are of paramount importance, particularly in view of the practical problems experienced elsewhere in restoration of degraded wetland ecosystems. The United States has taken a world lead in wetland regulation, yet wetland loss continues. The option of regulation and a goal of "no net loss" is of limited applicability in the developing world. However, the development of such strategies and the growing awareness of the role that wetlands play in our daily lives is in itself a powerful and important message to the global community, especially to those countries where a limited national budget precludes public investment in restoration of damaged ecosystems.

Details of three recent initiatives in wetland management are given, from the World Conservation Union, World Wide Fund for Nature, and the Commission of the European Communities. Together with the major investment being made in an attempt to maintain the unique mosaic of habitats in the Everglades, these serve as powerful examples of the degree of interest worldwide in the conservation and wise use of wetlands.

BACKGROUND

Natural processes and natural systems are important not only to maintain plant and animal life, but also to sustain the living standards of the human population. Damage to or degradation of natural systems is normally either expensive or irreparable.

The philosophy of conservation for sustainable utilization of natural resources worldwide is discussed in the "World Conservation Strategy," prepared by the World Conservation Union (IUCN) with the assistance of the United Nations Environment Program (UNEP), World Wide Fund for Nature (WWF), Food and Agriculture Organization of the United Nations (FAO), and United Nations Educational, Scientific and Cultural Organization (UNESCO). The three main concepts are

1. Maintenance of essential ecological processes
2. Preservation of genetic diversity
3. Sustainable utilization of species and ecosystems

It is increasingly apparent that the extinction of species and degradation or loss of particular ecosystems by human intervention is undesirable. "People have come to recognize that, in the long term, the natural heritage is as important as the cultural heritage. A protected area is a storehouse of nature and natural processes in the same manner as a museum preserves culturally important items and concepts. No civilized country would ever contemplate destroying its museums and other components of its cultural heritage nor its natural heritage" (UNESCO, 1988). In fact, protected areas can go well beyond the museum analogy by maintaining an entire ecosystem complex for the world community, complete with wide-ranging environmental and ecological benefits.

The critical attention currently directed toward wetland ecosystems is not restricted to scientists. National and international conservation agencies, government and nongovernmental organizations, politicians and lawyers, a vast array of commercial interests, and diverse sectors of the general public are also involved. Wetlands generate some of today's most contentious, difficult, and politically sensitive environmental questions and rival tropical rain forests for priority on the world conservation agenda. Only the most sanguine of observers would maintain that their future depends more on processes and cycles in the natural world than on trends in economic, social, and political development and the outcome of legislative and administrative debate. Two main themes underpin the concern for wetlands worldwide:

1.   A rapidly diminishing resource base
2.   A growing awareness of wetland values

The DIMINISHING RESOURCE BASE

Over a long historical time scale, the once expansive areas of European wetlands have been reduced to tiny residual and often isolated fragments due to agricultural, urban, and industrial expansion; eradication of disease; and regulation of river flow (Maltby, 1986; Hollis and Jones, 1991). The dramatic loss of over half the wetland area present in the United States in colonial times, and its causes and regional variations, have been well documented (Tiner, 1984). Throughout the developed world, loss of wetland habitat and impairment of ecosystem functioning continues, despite increasing recognition of adverse ecological, environmental, and even economic impacts; unprecedented public support for conservation; and legal instruments available for their protection.

Conversion of the English fenland to an intensive agricultural landscape was applauded by the majority of contemporary writers on the grounds of increased agricultural productivity and the decline in malaria (Darby, 1983). The full cost of the regional drainage program, however, has yet to be met. Progressive peat subsidence has required larger investment in pumping equipment and drainage works. In places, the peat has been lost entirely, exposing relatively infertile clays and marine sediments which require applications of expensive and potentially environmentally damaging fertilizers to maintain crop yields. The area now has some of the most serious nitrate pollution problems anywhere in Europe; this factor is of uppermost concern to the present government, in view of the possibilities of legal action over water quality by the European Commission.

The wetland complex of Lake Karla, Greece, once an important fishery and habitat for migratory birds, has been completely converted to agriculture over the last 30 years. In Volos, a coastal town which receives the drainage waters from the former wetland, there has been rioting in the streets over the "stench and filth" of the water; this reflects the extreme consequences of water quality decline due to agricultural development (P. Gerakis, personal communication).

Many of the most important remaining wetlands are in the tropics and Third World nations. Large-scale wetland loss has occurred more recently here than in the developed world, but this is where the greatest future losses are likely to occur and, thus, where much of the international conservation effort has been targeted.

River regulation and economic development schemes already have significantly reduced the area and altered the functioning of floodplain wetlands in Africa, Asia, and South America. In the Senegal River delta, 2400 ha of floodplain has been lost as a result of dike construction and an associated irrigation scheme, which not only has failed but also has impaired other ecological and environmental functions (Braakhekke and Marchand, 1987). Water management schemes continue to threaten Africa's important floodplain wetlands. Studies of the Senegal Valley, Niger Inner Delta, Sudd (Nile marshes), Tana Delta, Kafue Flats, Logone Floodplain, and Benone Valley demonstrate the scale of environmental change that has resulted or is likely to result from dams, flood control, irrigation, and canal structures (Drijver and Marchand, 1985). Recent analyses conclude that the existing traditional, multifunctional use of these wetland complexes actually has a higher financial margin of profit than agricultural conversion projects, largely because of the scale of capital investment necessary in engineering. Projects generally have ignored the "shadow-price of the work done by nature" (Marchand, 1987). Traditional uses of wetlands have been reduced, which may result in economic losses and adverse social impacts, in addition to loss of nature conservation values both beyond as well as within the wetland boundaries. Reduction of coastal fisheries of the eastern Mediterranean (Rzoska, 1976) and increased erosion of the Nile Delta, resulting from closure of the Aswan Dam and regulation of the Nile River (Stanley, 1988), are poignant reminders of unforeseen consequences which have yet to serve as a "learning experience."

Almost 20% of the internationally important wetlands in Central and South America are threatened by drainage for farming or ranching (Scott and Carbonell, 1986). Mangrove swamps throughout the tropics and subtropics are under increasing pressure to be converted to aquaculture and agriculture. Pressures for new land for population resettlement and a growing interest in peat mining threaten both coastal and inland wetlands in Southeast Asia, South America, and the Caribbean. At no time has the need for conservation action within the tropical and subtropical wetlands of the world been more urgent. The scale of loss and degradation of Everglades ecosystems to date is significant by comparison with any of the above examples and indeed dwarfs most, despite occurrence in the nation with the most sophisticated program for wetland protection.

The Everglades marsh complex originally encompassed over 10,000 km^2. Much of this area has been altered by conversion to urban or agricultural land, by hydrological modification, and by a range of ecological changes. Current efforts at large-scale "restoration" of the Everglades system, such as in the Kissimmee River and Water Conservation Area 2, add another layer of modification to the landscape. Only a small residual area of the original Everglades marsh is included within the 5633 km^2 of Everglades National Park. No other subtropical or tropical marsh ecosystem has suffered this scale or complexity of modification. Nowhere else in the world is there so much interest in and debate surrounding the prevention of further degradation and its reversal through appropriate restoration methods. Nevertheless, the experience of competing demands for land and water in Europe, and increasingly within the floodplains and peat swamp forests of the tropics, points to several key factors relevant to the Everglades:

1. Wetland ecosystems are particularly vulnerable to degradation when they occur in the lower parts of large catchments where incompatible land use activities and abstractions can occur.
2. On-site restoration efforts may be futile, unless an integrated land management approach is adopted in the larger area affecting the character of wetland ecosystems.

3. Successful maintenance and restoration of wetland ecosystems, at scales similar to that of the Everglades, may require major political, social, and economic intervention, as well as scientific input.

Examples from Central and South America underpin these key points. The Usumacinta River rises in the highlands of Guatemala and Mexico and flows to the Bay of Campeche in the Gulf of Mexico, where its delta supports an area of some 17,000 km^2 of wetland. In March 1992, the Mexican government announced plans to construct the first of a series of hydropower dams at Boca del Cero. Major concerns surround the fate of traditional Indian cultures (some of which are descendants of the Mayans), the rich biodiversity of the river basin (which includes jaguars, ocelots, howler and spider monkeys, harpy and solitary eagles, and crocodiles), together with the survival of the delta itself, which would be denied sediment, nutrients, and the seasonal fluctuations in river flow essential for maintaining the highly productive wetland ecosystems (International Rivers Network, 1992).

Results from an extensive survey on the mercury concentration in different environmental compartments (including river water, sediment, fish, and human hair) have revealed wide-ranging and heavy contamination levels in the Madeira River, southwest Amazon (Malm et al., 1990). The contamination is the result of gold mining, and the high levels occurring in both fish and human hair show the direct environmental and health threat for local ecosystems and people in this part of the Amazon. It is still not known how far the pollution threat may extend through the food web and biophysical pathways in the river basin, eventually impacting ecosystems and populations well beyond the source.

INCREASING AWARENESS of WETLAND VALUES

Copious examples of the diverse values of wetland ecosystems to society appear in the scientific literature. Wetland processes result in ecosystem functioning which is significant for:

1. Direct and indirect human use
2. Welfare of wildlife and maintenance of biological diversity
3. Environmental maintenance and quality

Recognizing the widest benefits to society of ecosystem functioning is pivotal to the argument for wetland protection and management (Maltby, 1991).

Experience summarized by Adamus et al. (1987) shows that wetland ecosystems may be valuable specifically for:

*   Groundwater recharge/discharge
*   Flood flow alteration
*   Sediment stabilization/shoreline anchoring
*   Sediment/toxicant retention
*   Nutrient processing/water quality

- Production and food chain support
- Wildlife habitat
- Fisheries
- Heritage/recreation/science

Unfortunately, we still have limited knowledge of:

1. Precisely how specific wetland ecosystems function, especially those in the tropics (much of what is known comes from United States- or European-based research)
2. Compatibility among and between different functions
3. The consequences of particular human actions on ecosystem functioning
4. The dependence of human communities on wetland functioning

While such knowledge may be perceived as largely of academic or strategic interest in supporting conservation measures in the developed world, it is, in contrast, of crucial importance in the largely tropical and subtropical wetlands of the developing world, where both human and wildlife populations are commonly dependent on the ecosystem for the support of life (Maltby, 1986).

POLITICAL PERCEPTIONS and REALITIES

While governmental awareness of the importance of maintaining natural processes and ecosystems has increased dramatically in recent years, effective changes in policy have generally been tempered by regional and global social, economic, and political considerations. The immediacy and scale of human needs is often cited as an argument for relegating environmental issues. On July 17, 1989, *The Times* (London) reported on the July 16 Paris summit of the seven most prosperous industrial democracies, which promised "a global crusade against pollution." It went on to state: "President Bush, who has described 1989 as the year the planet cried enough...called the Paris summit a 'watershed for the environment' and said 'urgent action is required to preserve the Earth.'" The linked themes of environment and development were raised to the top of the world agenda by the United Nations Conference on Environment and Development, held in Rio de Janeiro in June 1992. Over 100 governments made a commitment to *Agenda 21,* a text which "addresses the pressing problems of today and also aims at preparing the world for the challenges of the next century." Over 150 states also signed a Global Convention on Biological Diversity, which commits parties to comprehensive action to safeguard national and global living resources and to use them sustainably. Despite earlier statements, the United States was noticeable in its absence from the list of signatories to the Convention on Biological Diversity, with the Bush Administration explaining its position in terms of conflicting economic concerns.

Despite the pioneering role of United States agencies and scientists in promoting wetland protection (which has been based on regulation of "defined" wetlands), together with the statement by President Bush in 1988 advocating "no net loss" of wetlands, recent trends have threatened to undermine these earlier initiatives. Reacting to a need to "better balance economic and environmental concerns" (*Miami Herald,* August 10, 1991), the Bush administration successfully prevented U.S. federal agencies from adopting the most recent science-based revision (1989) of the Federal Manual for Wetland Delineation. However, they did not succeed in replacing it with a nonscience-based manual drafted by former Vice President Quayle's now defunct Council on Competitiveness. This new manual would have considerably reduced the area of wetland previously falling under federal protection. In the case of south Florida, areas around the edges of the Everglades in Dade, Broward, Palm Beach, Monroe, Collier, Hendry, and Lee counties would undoubtedly have lost protection. The U.S. federal agencies all have now agreed to use the earlier (1987) version of the manual. The Bush administration was reacting to several circumstances, including, in particular, growing pressure from the regulated community (which was becoming increasingly well organized by construction, farm, and forestry lobbying groups), as well as a White House that was actively reducing the effect of federal regulations in general (J. Larson, personal communication).

Concern that protection of wetlands precludes essential economic and social development lies behind the formation of a "wise use" group in the United States, unfortunately reinterpreting the wise use meaning of the Ramsar Convention on Wetlands of International Importance and advocating conversion to nonwetland usage. However, the world community is today increasingly aware that good environmental management is invariably good economics. One of the strongest proponents of this argument is U.S. Vice President Gore. It remains to be seen whether changes in the political structure of the government will be sufficient to carry public opinion and maintain or even enhance policies favoring wetland protection.

ECOSYSTEM CONSERVATION

If wetland ecosystems are to contribute fully to meeting the needs of society on a sustainable basis, the resource base must be conserved and enhanced. Throughout the developed world where only relatively small areas remain intact, conservation has usually involved establishing protected areas, often excluding people, with largely cosmetic management and restoration applied to a greater or lesser extent. The strategy is comparatively expensive and is generally inappropriate to large tropical and subtropical wetland areas, located predominantly in the poorest countries. In the Inner Niger Delta, Northern Mali, 10,000 families rely on an annual fish catch of 100,000 tonnes, worth $5 million. Many African rural communities survive by linking cattle movement and crops to flood cycles and hunting of herbivores, which also migrate according to wetland cycles. An estimated 1.5 million people depend on the wetland cycle in the middle Senegal River in Senegal and Mauritania, the Inner Niger Delta in Mali, and the Jonglei region of the Sudd (Maltby, 1986).

Such are the complex couplings between human population, wetlands, their "upland" catchments, adjacent permanent bodies of water, and an atmosphere in which management confined by the boundaries of the wetland area exclusively is an entirely inadequate strategy for conservation. Reduced freshwater flows into Lake Ichkeul, Tunisia, North Africa, which was caused by damming of headwaters a considerable distance from the wetland, has resulted in large salinity increases and loss of biological diversity (Hollis, 1986). In Spain, the Tablas de Damiel Wetland National Park has been seriously impacted as a result of lowering the water table by groundwater abstraction for irrigation on agricultural land outside the park. The famous Donana National Park is similarly threatened by competing demands for water elsewhere in the catchment (Llamas, 1988). All of these examples are wetlands of international importance—all are Ramsar sites, and in the case of Ichkeul the wetland is also a world Biosphere Reserve and World Heritage site.

The Everglades is no exception to other wetlands of international significance in terms of the linkage of the wetland ecosystem to the welfare of adjacent, as well as wider, human populations. Water supply, aquifer recharge, prevention of saltwater intrusion, and maintenance of recreational and aesthetic values illustrate just some of the utilitarian importance of maintaining the natural functioning and ecological integrity of the Everglades ecosystem. However, the conflicting jurisdictional and management goals of the wide range of institutions and interest groups within the drainage basin have yet to achieve a condition in which current and projected human activities can be regarded in any sense as sustainable.

The problem is particularly acute when the rivers which support wetlands rise not only hundreds or thousands of kilometers from the wetland itself, but also cross different administrative, political, or international boundaries. Thus, Angola controls the water supply of Botswana's Okavango swamps. Maintenance of the Sundarbans mangrove forest in Bangladesh depends on water regulation in India. Wetlands of the Lower Mekong Basin depend on decisions made in Laos, Kampuchea, Thailand, and Vietnam.

The lesson learned increasingly from such worldwide experiences is that wetland conservation and management efforts must be carried out within an appropriate hydrological, ecological, and socioeconomic framework. The entire catchment of the wetland and its management must be examined, including the impact of particular land use and other human activities on the characteristics of the wetland ecosystem, even though they may be remote from the source of intervention. It is also necessary to take full account of the communities using the resource or dependent on it. Such communities may be local or distant and may vary from individual settlements to an international or even global scale in regard to such issues as migratory waterfowl or endangered species and atmospheric carbon balance. Consideration also must be given to the practical realities of the capacity of a national government to implement and sustain prescribed management measures for wetland maintenance or restoration.

SIGNIFICANCE of the EVERGLADES

The importance of the Everglades to the people of the United States was confirmed when National Park status for a part of the remaining ecosystem was authorized by Congress in 1934. This established protection of some half million hectares of Everglades wetlands which once covered over 10,000 km² from Lake Okeechobee to the southwest tip of Florida.

The importance of Everglades National Park to the world community has been recognized through its nomination and acceptance on three international lists:

1. Biosphere Reserve (1976)
2. World Heritage (1979)
3. Wetland of International Importance, under the terms of the Ramsar Convention (1987; Convention on Wetlands of International Importance especially as Waterfowl Habitat)

There are only two other sites in the world which appear on all three lists: Lake Ichkeul in Tunisia and Srebarna Lake in Bulgaria. Lake Ichkeul already has suffered major degradation because of the reduction of water inflow by upstream dams, while Srebarna is a relatively small and not easily accessed area (600 ha).

Everglades National Park, therefore, is a key standard bearer for the efficacy of international conservation designation and the viability of supporting conventions to which the United States government is a signatory member. No other subtropical or tropical wetland enjoys such overt recognition of its biological importance.

The task force on the criteria and guidelines for the choice and establishment of biosphere reserves recognized two elements of great importance:

1. The integrity of the essential characteristics of many ecosystems cannot be safeguarded unless the protected areas are large and varied.
2. Successful stewardship will also depend on adequate control of the use of land and water in surrounding areas.

It is further recognized that they are places where government decision makers, scientists, managers, and local people cooperate to develop a model program for managing land and water to meet human needs, while conserving natural processes and biological resources. Additionally "Biosphere reserves must have adequate long term legislative, regulatory or institutional protection" (UNESCO, 1984).

The World Heritage listing introduces the specific notion of a "world heritage whose importance transcends all political or geographical boundaries" (UNESCO, 1988). Inclusion of the Everglades on the World Heritage list was made on the basis of:

1. An outstanding example representing major stages of the evolutionary history of the earth
2. An outstanding example of biological evolution (a subtropical biome where temperate North America meets tropical America)
3. Significant natural habitats exist, where populations of rare or endangered plants and animals still survive

"Everglades National Park is a superlative example of viable biological processes and whose examples of rare and endangered species are of universal interest and significance" (United States nomination document).

By adopting the convention, each nation recognizes that it holds in trust for the rest of mankind those parts of the world heritage that are found within its boundaries. A state party is obliged to "integrate the protection of that heritage into comprehensive planning programs, and to take appropriate legal, administrative, scientific, technical and financial measures to implement this policy."

Placement on the Ramsar list is recognition of the global importance of a wetland site. Seven criteria for inclusion have been specified:

1. Supports ten endangered species and three threatened species
2. Outstanding example of a wetland community characteristic of its biogeographic region
3. Plays a major role in its region as a habitat of plants and of aquatic and other animals of scientific and economic importance
4. Has special value for maintaining genetic and ecological diversity because of the quality and peculiarities of its flora and fauna
5. Outstandingly important, well situated, and well equipped for scientific research and education
6. Offers special opportunities for promoting public understanding and appreciation of wetlands and is open to people from all countries
7. Physically and administratively capable of being effectively conserved and managed

Among the management objectives listed in the same nomination document is to "promote and coordinate cooperative regional resources planning, protection and management with priority given to quantity, quality, distribution and periodicity of a reasonable water supply" (Whisenant, 1987).

It is explicitly stated under the convention that the contracting parties have a general obligation to include wetland conservation considerations within their national land use planning. There is a requirement to formulate and implement this planning so as to promote, as far as possible, the wise use of wetlands in their territories. The contracting parties have interpreted this wise use requirement to mean the maintenance of the ecological character of wetlands.

There is, therefore, not only great ecological importance associated with the Everglades but also great significance in the politics of conservation. At a time when the Ramsar Convention is seeking to attract more contracting parties from the poorer tropical nations of the world, it is imperative that one of the wetland "showpieces" of the world, with arguably the most sophisticated regulatory mechanisms and certainly the greatest investment in wetland research and management, be a true model of conservation success.

The level of designation of Everglades National Park is a manifestation of the dramatic change in attitudes and perceptions concerning wetlands over the last few decades. There is increasing realization of the importance of the goods and services provided by wetlands, often at no direct cost, and of the serious implications of the resources lost when these ecosystems are degraded or lost. Their drainage is no longer viewed as a "progressive, public-spirited endeavour" (Baldock, 1984), except by narrow sectoral interest groups. Instead, the growing European and other worldwide tourist interest in Everglades National Park is a manifestation of the rapid growth of a "conservation" ethic and "wildlife" recreation. This trend is likely to continue as the relative costs of air travel decrease and the level of education and information about the Everglades increases. It is a trend further supported by the ease and proximity of accessibility to the various biotopes possible in Everglades National Park and the political security of the location as compared with many other tropical or subtropical wetlands.

Many scientists and natural resource managers come to Everglades National Park to see "how things should be done," as did the Petroleum Corporation of Jamaica, in regard to their proposed development of the Negril Morass (Maltby, 1986), and various Brazilian institutions, in regard to management and protection of the Pantanal (J. Ogden, personal communication). Thus scientific, management, and recreational interests all contribute to the pressure for sound ecological conservation and, where necessary, restoration of the Everglades ecosystem.

PRACTICE and PROBLEMS of
RESTORATION EXPERIENCE and TECHNOLOGIES

Restoration can be defined as "bringing back into a former, normal or unimpaired state or condition" (*Webster's Dictionary*). The implication is that an ecosystem has been altered or degraded in a way which conflicts with defined conservation or management objectives. It is essential that restoration goals be identified as a first step. Normally, these are linked to particular functions or combinations of functions. Goals define the project and help in recognizing those elements of the ecosystem which must be included to provide the management and planning team with a clear framework for operation and implementation. Examples abound from the United States and are well covered by Zelazny and Feierabend (1988). In contrast, the experience of wetland restoration in tropical and subtropical ecosystems is virtually nonexistent and largely limited to examples where water supply has been restored.

Thus, the Ndaiel basin, a Ramsar site in the Senegal delta, is the focus of a restoration effort designed to restore flooding by the Senegal River. The natural flooding regime of the Ndaiel was disrupted in the 1970s as a consequence of hydroagricultural development and road construction in the delta. As the potential contribution of water and pasture resources of the Ndaiel to rural development in the delta has become more fully appreciated, funds are now being sought for construction of the necessary canals.

In a similar vein, in a number of other African countries there is today a growing debate over the potential for rehabilitating riverine floodplains by means of modifications in the water-release regimes of hydrodams. While few such proposals have been implemented, this is likely to be a major area of debate over the next decade.

The most ambitious wetland restoration project outside the United States is almost certainly Lake Hornborga in Sweden. At the beginning of the 19th century, Lake Hornborga covered more than 30 km², with a probable maximum depth of 3 m. From 1803 to 1933, it was lowered progressively to provide more arable land around the perimeter. The residual lake bottom area became colonized by emergent macrophytes dominated by *Phragmites* reed, and the southern part of the lake was invaded by willow and birch. Continued sedimentation and lack of the natural rejuvenating processes of wave action and ice movement led to marked degradation of the ecosystem as an important habitat for breeding and migratory bird species and as a fishery.

The main aims of practical restoration efforts in Lake Hornborga are to:

1. Re-establish a functioning, self-maintained wetland ecosystem
2. Transform most reed belts into open water or a mosaic of open water and emergent vegetation
3. Promote the establishment of a rich submergent vegetation
4. Bring into practice such a water regime, which—within certain limits—will allow natural variations induced by weather within each year, as well as between different years
5. Eliminate the conducting effects of the channels and increase water circulation in the lake
6. Promote management practices in the surrounding wet meadows which are favorable to birds and the scenic beauty of the countryside
7. Purchase land around the lake, where options are available
8. Declare the lake and its surroundings a Nature Reserve

A key part of the restoration program involves large-scale removal of *Phragmites*. The rationale of reed removal is to:

1. Reduce the detritus-forming biomass, thereby reducing sedimentation
2. Create a more diverse wetland habitat with more open water
3. Facilitate establishment of other species enabling improved water flow

Fire, mowing, and rotary cultivation techniques have been used in field experiments to remove large areas of reed. As a result, the wetland character of the northern part of the lake has been restored to the extent that several species of birds have increased in numbers. It is anticipated that some 13 km² of presently reed-dominated land will be transformed into open water or a mosaic of aquatic vegetation.

The current works of the Swedish Natural Environmental Protection Board are aimed at raising water levels by 0.85 m, which would reduce the need for new retaining structures and maximize the area of surrounding marshy meadow and wet forest. However, previous plans and current vociferous criticism from limnologists of the present actions have called for a greater increase in water level of 1.0 m and then a further 0.5 m (Bjork, 1988). The debate centers on differences of opinion between restoring a lake ecosystem with long-term viability and restoring a complex wetland ecosystem which is more important as a wildlife habitat (Larsson, 1989). This inevitably raises the question of which stage restoration in a dynamic ecological succession should address and emphasizes the importance of "purpose" when devising management strategies. Restoration of the Hornborga ecosystem is expensive. It is estimated to cost at least $5 million to raise the water level by 0.85 m, reduce the reed beds and other invaders, and compensate landowners. Increasing the water depth by 1.5 m would at least triple this cost (T. Larsson, personal communication).

EXTENDING to the TROPICS

Over the breadth of the tropics, there are many other wetland systems similar in characteristics to the Everglades, some of which are many times larger (notably the Llanos of Venezuela and the Pantanal of Brazil). These systems are all characterized by a wide range of threats and pressures—within the wetland from overuse and poor management of resources and in the catchment from disruption of the hydrological system. Some of these problems are largely concerned with changes in the quantity of water arriving at the wetland due to dams and diversion upstream, but others are more strongly impacted by changes in water quality (e.g., the Pantanal, where mercury pollution is a serious problem). To solve these problems and restore degraded areas, awareness of the problem and a commitment to address it are, of course, necessary. This commitment will, in many cases, only be forthcoming if there is a clear local or national incentive to restore the wetland system, as well as the money to accomplish this. In most countries of the tropics, pressure on the largest wetland systems has not yet reached a critical level, although in the drier parts of Africa, there is now very substantial pressure on many of the floodplain systems as water is abstracted and diverted upstream for urban, industrial, and agricultural use. As these pressures grow, it is clear that in the future the need for enhancing protection, and where this is absent rehabilitation, of wetlands will grow. However, the lessons of the Everglades suggest that the time to act is now, so that the problems that are beginning to appear do not become critical. In the Everglades, it is clear that the action required is expensive and that money has yet to be found. If the United States cannot act clearly and effectively, it is unrealistic to expect poorer nations to do so.

PROBLEMS and REALITIES

The need for restoration is generally the result of sectoral competition for land use within the wetland catchment which conflicts with conservation objectives. Thus, the agriculturalist sees a wetland as an area of potentially fertile soil, given suitable drainage and fertilizer application; the hydrologist may view it as part of the system affording flood protection and water supply to meet industrial, urban, agricultural, and nature conservation needs; the ecosystem manager wants to retain the elements and functioning in an undisturbed, natural state. The reality is that governments at both national and regional levels are organized sectorally, and integrated wetland management, which might resolve conflicts, is not currently attainable from a practical organizational standpoint. Where wetland resources are the responsibility of different national or state agencies with contrasting objectives, it is doubtful whether conflicts of interest can ever be avoided.

A great deal of attention has been drawn to the scientific problems of restoration. The main issues are discussed by Larson and Neill (1987), Odum (1988), and Zedler (1988). The major difficulties encountered concern:

1. Definition of the predisturbance condition of the ecosystem
2. Agreement on thresholds for impacts/intervention which are tolerable in maintaining particular ecosystem characteristics
3. Dealing with highly complex ecosystems in which the interactions of species and elements are at best imperfectly understood
4. Handling ecosystems which are spatially and temporally dynamic
5. Coping with progressive environmental changes, e.g., atmospheric pollution, climate change, sea level movement

If we wait for the perfect science base, then it is clear that wetland degradation will continue and rehabilitation will become progressively more problematic. Management strategies should proceed on the basis of the best available knowledge, with "fine tuning" or even more significant alteration in light of increased knowledge.

Restoration is extremely expensive. Costs are not necessarily restricted to scientific and technical aspects; they may involve litigation and land purchase or easements, which may be especially high in areas of multiple ownership and large catchments. It is possible for the rich nations of the world to afford the costs of earlier environmental mismanagement. However, the reality is that the countries of the tropics and subtropics simply cannot afford such restoration programs.

The important message is that it is far better to avoid such costs in the first place by taking the appropriate steps to manage wetland habitats in a more effective manner. This also will require enormous financial assistance from the international community, but in this respect the technical experience of the United States can play an invaluable role.

VALUE of UNITED STATES INITIATIVES

Recent initiatives in the United States are based on a growing recognition of the social and economic costs of wetland loss. For example, Section 404 of the Clean Water Act and the "Swampbuster" provision of the Food Security Act recognize that the activities of individuals, while to their personal benefit, have ecological and hydrological impacts with public costs. Today in the United States, where for decades the public has paid higher prices for clean water and other services once provided free by wetlands, the public costs of wetland loss are judged unacceptable by the public and the government.

There are obvious limitations to applying the precise measures adopted in the United States to combat wetland loss elsewhere. Most of the legal instruments for regulating use and much of the scientific methodology for assessing wetland functions are tied closely to the special administrative and legal structures in the United States. Even the goal of "no net loss" is of limited applicability in most other countries. However, the development of these instruments and the growing awareness of the role that wetlands play in our daily lives send a powerful message to the global community and, in particular, to those countries where the small size of national budgets precludes a large public investment on the level made by the United States in restoring degraded wetlands and in replacing services, such as water purification and flood control.

WORLD CONSERVATION UNION WETLANDS PROGRAM

The World Conservation Union (IUCN) Wetlands Program was initiated in 1984/ 1985 in response to growing concern about the rate of wetlands loss worldwide and the need for more concerted action to address the diverse array of management problems faced by wetlands. However, because of the economic cost of wetland restoration, this program has focused attention on improving management of remaining wetlands and, in doing so, emphasizing the benefits yielded and stressing the true social and economic cost of wetland loss.

The core of the program is a series of field projects which develop the methodologies for wetland management, particularly in countries of the developing world, where wetlands are used intensively by local communities which depend upon them for their well being. Thus, in West Africa, IUCN is working with local communities and national institutions to design and implement management regimes for wetland systems in the Senegal Valley, the inner delta of the Niger River in Mali, the floodplain of the Logone River in northern Cameroon, and the coastal wetland complex of Guinea Bissau. In Central America, the program supports local institutions in managing mangrove resources in Panama, Costa Rica, Nicaragua, and Honduras and freshwater wetland systems in Costa Rica and Nicaragua. Similar activities are underway in eastern and southern Africa and Southeast Asia, with special focus on tidal wetland systems and freshwater floodplains, the two broad categories of wetlands that are most intensively used by human communities. Related strategies and policy initiatives draw upon the results of these projects, and conclusions are presented in a form useful for government decision makers and planners.

The field projects of the program are based on biological, hydrological, and socioeconomic analyses. The last of these is of special importance in the developing world, and field projects in Latin America, Africa, and Asia are designed based on the premise that effective management of wetland ecosystems requires understanding of the socioeconomic issues that underly resource utilization. On the basis of this understanding, several IUCN field projects include management activities which will diversify the options available for wetland communities to use natural resources, thus relieving pressure on them. Similarly, by opening a dialogue with the people in these communities, IUCN activities are designed to give the users of the resources a greater say in their management.

WORLD WIDE FUND for NATURE

The World Wide Fund for Nature (WWF) has taken a recent (1989) major initiative in one country—Greece—to develop an action plan for wetland conservation and management. Underlying the strategy recommended is the premise that: "Conservation of wetlands must be based on the careful planning and effective control and regulation of all human activities in and around them. Integrated management is necessary of the wetlands themselves, their catchment and even of a wider area outside the watershed" (Maltby, 1991; Gerakis, 1992). Efforts to rehabilitate damaged wetlands are a high priority in order to recover lost values. Among the recommendations made by the WWF working group is the establishment of Integrated Management Agencies in each wetland group, which would include the participation of central and local governments, as well as scientists.

COMMISSION of the EUROPEAN COMMUNITIES

Directorate General XI (Environment) of the Commission of the European Communities (CEC) has identified coastal wetlands of Mediterranean type as a major cause for conservation concern. It is assumed that each wetland should be a self-regulating system, characterized by primary biotic conditions, and that no wetland area ranging in this region can be sacrificed for uses and functions other than those provided by wetlands. These wetlands are under great pressure and in some cases have already been degraded to a large extent. A strategy of integrated management has been proposed as a means by which to achieve conservation—including the retention and positive management of a wetland system for the maintenance of its present interests and diversity, as well as the enhancement of degraded wetlands.

Collaboration must involve authorities and bodies concerned primarily with conservation and wetland protection and also concerned with other forms of use, within both the wetland and the surrounding area. It is proposed that the wetlands be managed by executive bodies that are able and willing to plan and enforce balanced and sustainable use, with participation of local people and authorities. The most appropriate instrument on which to base integrated management is an integrated management plan. The approach has yet to be implemented, but there is no doubting the importance of the preparatory action on the part of the CEC in highlighting the needs and possible institutional mechanisms for achieving sound wetland management and restoration in some of the most important yet threatened sites in Europe.

CONCLUSIONS

The attention given to wetlands conservation today is a direct consequence of the loss of wetlands which has occurred during the recent past. As these resources have been lost, their true value to human society has come to be more fully appreciated, and in the United States today, government and private organizations are prepared to pay large sums to maintain wetlands.

In this movement for wetlands conservation, the Everglades, and the major investment being made to maintain its unique mosaic of wetland habitats, serves as a powerful example. However, care must be taken in interpreting this trend. Not only is the Everglades a unique ecosystem, but the efforts being made to repair the damage of past investments, which previously did not consider the complexity of the hydrological and biological system, are also unique. No other country has yet spent the sums of money on wetland restoration that are now being devoted to the Everglades. Indeed, while in some countries in Europe some wetland sites are the subject of intensive, but smaller scale, restoration measures, most countries in the developing world can scarcely contemplate investments of this scale in any sector.

The Everglades today, therefore, serves as both an example and a warning. It demonstrates clearly the value which the United States now places on wetlands and the extent to which it is prepared to invest to maintain this resource of national and international importance. It also, however, highlights to most other countries that once their wetlands are degraded, their economies will be unable to cover the costs of rehabilitation. The challenge to these countries, and indeed increasingly also the United States, is to design and implement programs which protect remaining wetland systems and manage them as an integral component of overall development investment.

LITERATURE CITED

Adamus, P. R. and L. T. Stockwell. 1983. A Method for Functional Assessment, Vol. 1: Critical Review and Evaluation Concepts, FHWA-IP 82 83, U.S. Department of Transportation Highway Administration, Washington, D.C.

Adamus, P. R., E. J. Clairain, Jr., D. R. Smith, and R. E. Yang. 1987. Wetland Evaluation Technique, Environmental Laboratory, Waterways Experiment Station, U.S. Army, Vicksburg, Miss.

Baldock, D. 1984. *Wetland Drainage in Europe,* IIED/IEEP, London.

Bjork, S. 1988. Redevelopment of lake ecosystems: A case study approach. *Ambio,* 17:90–98.

Braakhekke, W. G. and M. Marchand. 1987. *Wetlands: The Community's Wealth,* The Environmental Bureau, Brussels.

Darby, H. C. 1983. *The Changing Fenland,* Cambridge University Press, Cambridge, 267 pp.

Drijver, C. A. and M. Marchand. 1985. *Taming the Floods: Environmental Aspects of Floodplain Development in Africa,* Centre for Environmental Studies, State University of Leiden, Netherlands.

Gerakis, A. (Ed.). 1992. *Conservation and Management of Greek Wetlands,* Thessaloniki Workshop Proc., IUCN, Gland, Switzerland, 493 pp.

Hollis, G. E. (Ed.). 1986. *Modelling and Management of the Internationally Important Wetlands at Geraet el Ichkeul,* Tunisia IWRB, Slimbridge, U.K.

Hollis, G. E. and T. A. Jones. 1991. Europe and the Mediterranean Basin. in *Wetlands,* M. Finlayson and M. Moser (Eds.), Facts on File, Oxford, pp. 27–56.

International Rivers Network. 1992. *Special Briefing Flooding Mayan Heritage,* International Rivers Network, Berkeley, Calif.

IUCN, UNEP, WWF. 1985. *World Conservation Strategy,* Gland, Switzerland.

Larson, J. S. and C. Neill (Eds.). 1987. *Mitigating Freshwater Wetland Alterations in the Glaciated North-Eastern United States: An Assessment of the Science Base,* Publ. 87-1, Environmental Institute, University of Massachusetts, Amherst, 143 pp.

Larsson, T. 1989. Restoration of Hornborga Lake, Sweden. *Ambio,* 18:253.

Llamas, M. R. 1988. Conflict between wetland conservation and groundwater exploitation: Two case histories in Spain. *Environ. Geol. Water Sci.,* 11:241–251.

Malm, O., W. C. Pfeiffer, C. M. M. Souza, and R. Reuther. 1990. Mercury pollution due to gold mining in the Madeira River Basin, Brazil. *Ambio,* 19:11–15.

Maltby, E. 1986. *Waterlogged Wealth,* Earthscan, London, 200 pp.

Maltby, E. 1991. The world's wetlands under threat—Developing wise use and international stewardship. in *Environmental Concerns,* J. A. Hansen (Ed.), Danish Academy of Technical Sciences, Elsevier Applied Science, London, pp. 109–136.

Marchand, M. 1987. The productivity of African floodplains. *Int. J. Environ. Stud.,* 29:201–211.

Odum, W. E. 1988. Predicting ecosystem development following creation and restoration of wetlands. in *Increasing Our Wetland Resources,* J. Zelazny and J. S. Feierabend (Eds.), National Wildlife Federation, Washington, D.C.

Scott, D. A. and M. Carbonnell. 1986. *A Directory of Neotropical Wetlands,* IUCN, Gland, Switzerland.

Stanley, D. J. 1988. Subsidence in the northeastern Nile Delta: Rapid rates, possible causes, and consequences. *Science,* 240:497–500.

Tiner, R. W. 1984. Wetlands of the United States: Current Status and Trends, U.S. Fish and Wildlife Service, Washington, D.C.

UNESCO. 1988. *Operational Guidelines for the Implementation of the World Heritage Convention,* MSS WHC/2/revised, Intergovernmental Committee for the Protection of the World Cultural and Natural Heritage, UNESCO, Paris, 33 pp.

Whisenant, K. A. 1987. Why is it so difficult to replace lost wetland functions. in *Increasing Our Wetland Resources,* J. Zelazny and J. S. Feierabend (Eds.), National Wildlife Federation, Washington, D.C.

4

WATER CONTROL in the EVERGLADES: A HISTORICAL PERSPECTIVE

Stephen S. Light

J. Walter Dineen*

ABSTRACT

This chapter traces the evolution of human efforts to alter the vast, pulsing water and landscape known as the Everglades. It describes the pattern of construction and the regulation schedules that govern control of water in the Water Conservation Areas and the eastern boundary of Everglades National Park. The Water Conservation Areas are spatially and functionally central to water management in the Everglades. They are also the most viable candidates for ecological revitalization, with the potential to provide substantial downstream benefits to Everglades National Park, including Florida Bay.

The evolution of water structures in the Everglades began about the turn of the century and was part of a populist movement designed to settle the southern portion of the Florida peninsula. Early efforts at water control included the Everglades Drainage District works, consisting of 440 mi (70.8 km) of canals and levees, and the Okeechobee Flood Control District, which constructed a federally subsidized dike around the southern rim of Lake Okeechobee. However, these efforts were a prelude to a massive federal project (Central and Southern Florida Project for Flood Control and Other Purposes) which was authorized after the massive flooding during 1948.

---

* J. Walter Dineen passed away before this manuscript could be completed. Walt devoted his life and 26 years of his professional career to understanding the Everglades. He was a man who was admired and respected by all who love the Everglades. Walt was a dear friend and he is missed.

This project employed four principal technologies: levees, water storage areas, channel improvements, and large-scale pumps to supplement gravity drainage. The project effectively circumscribed all the remnant Everglades north of the park and has been continually expanded and enlarged upon throughout much of its 40-year history. Through a series of time steps of 3–5 years (1952–90), the chronology of structural installations and modifications is traced, reflecting an evolution of objectives from flood control to water supply and, more recently, to redress adverse environmental impacts. Water flows and levels in the remaining system are controlled by regulation schedules for Lake Okeechobee and the Water Conservation Areas, which have evolved parallel to structural modifications. As a result, Shark River Slough has gained a series of alternative flow regimes in recent decades.

The environmental consequences of the construction of appurtenant works in the Everglades suggests that nothing short of a major reconfiguration of the structural works of the Central and Southern Florida Project system is needed to repair the Everglades while providing for the human needs of the region in the future.

INTRODUCTION

The first reclamation efforts in the Everglades began in 1881. Just over 100 years later, Governor Graham formally embraced a system-wide restoration initiative entitled "Save Our Everglades." More has happened to the Everglades between these two benchmarks than in the preceding 50 centuries. Some life-long natives of the region say that land use history in the Everglades falls into two epochs— "BC" and "AD"—meaning before canals and after drainage (Brown, 1948).

Prior to the U.S. and Seminole Wars of 1835–42 the nature, extent, and potential of the region were virtually unknown. Images of the region started a slow and gradual transformation once Florida entered the Union (1845) and Congress passed the Swamplands Act of 1850. By the turn of this century, the grand unfeigned wilderness was poised for rapid settlement. The portrait of the region as an "impenetrable swamp" began to fade. New visions for the extensive watery wilderness trimmed in gleaming white beaches began to emerge. The homesteaders and wealthy northern developers discovered and exploited south Florida's natural capital (e.g., rich muck soils and subtropical climate) as an agricultural breadbasket and winter playground for the nation. Turning these flights of fancy into reality required massive and relentless alteration of this "prolific paradise" through drainage. Proponents of reclamation looked past the natural devastation that occurred and likened the canal digging to the laying of water mains for a new city, to be tapped for the benefit of its citizenry (Dovell, 1947). For much of the 20th century, water management generated unrestrained growth. The region's bounty seemed inexhaustible, which in turn triggered more landscape alteration at the expense of what the Native Americans in the region called "Pa-hay-okee" or grassy lake.

The Everglades is now considered to be one of the most threatened ecosystems in the nation. Populations of wading birds (an important indicator of ecosystem health) have declined to levels that verge on complete collapse of nesting activities in the Everglades for some species. Based on Everglades National Park research efforts, in 1987 Superintendent M. V. Finley reported that the mean annual breeding population of wood storks between 1980 and 1987 was only 374, down from 2370 (86% decrease) in 1960. Historically, large mixed colonies of wading birds in Shark River Slough were estimated to contain a minimum of 250,000 pairs. During the 1960s these populations were estimated at 150,000 nesting pairs. In 1987, no large nesting colonies of white ibis formed. Fewer that 1000 pairs of great egrets and 500 pairs of snowy egrets have nested in mainland colonies of Everglades National Park in most years since 1980 (Finley, 1987). Water management practices over the past 50 years have been found at fault:

> The alteration of natural water flow volume and hydroperiods in the Everglades caused by diversions for agriculture and urban use, and the construction of an extensive system of canals and levees for flood control, resulted in serious declines in abundance of most species of wading birds (National Audubon Society, 1992).

The purpose of this chapter is to trace the evolution of human efforts to alter the vast, pulsing water and landscape known as the Everglades. More specifically, the pattern of construction and the regulation schedules that govern control of water in the Water Conservation Areas and the eastern boundary of Everglades National Park will be described. The Water Conservation Areas lie north of the park, southeast of Lake Okeechobee, and west of the urban corridor abutting the Atlantic Ocean. The eastern boundary of Everglades National Park includes the East Everglades and C-111 basin (Figure 4.1).

The rationale for focusing primarily on the Water Conservation Areas in an Everglades restoration volume may be obvious to many but still deserves recitation. The Water Conservation Areas are not only spatially central, but are also functionally central to water management in the Everglades (Gunderson, 1992). The Water Conservation Areas contain most of the last remnants of the original sawgrass marshes, wet prairies, and hardwood swamp forests outside Everglades National Park. This unique ecological landscape, while exceptionally resilient, has become increasingly vulnerable to excessive human intervention and invasion by exotic species. In addition, the area represents one of the most viable candidates for ecological revitalization, with the potential for providing substantial downstream benefits to Everglades National Park including Florida Bay. Residential and agricultural developments dominate the eastern boundary of the park, particularly in the north along L-31N. The C-111 basin, on the other hand, is largely undeveloped and in mostly public ownership.

The northern Everglades (qua Water Conservation Areas) was increasingly furrowed and balkanized during this century in attempts to achieve multiple water control objectives for economic development (e.g., water supply, flood protection, resource recreation, and wildlife management). However, since Governor Graham launched the program in August 1983, there has been a growing political commitment to "Save Our Everglades." The reconfiguration of the Water Conservation Areas appurtenant works appears essential to the effective recoupling of the remnant Everglades with Everglades National Park and Florida Bay to the south and Lake Okeechobee to the north. The goal of such an effort is ultimately rooted in

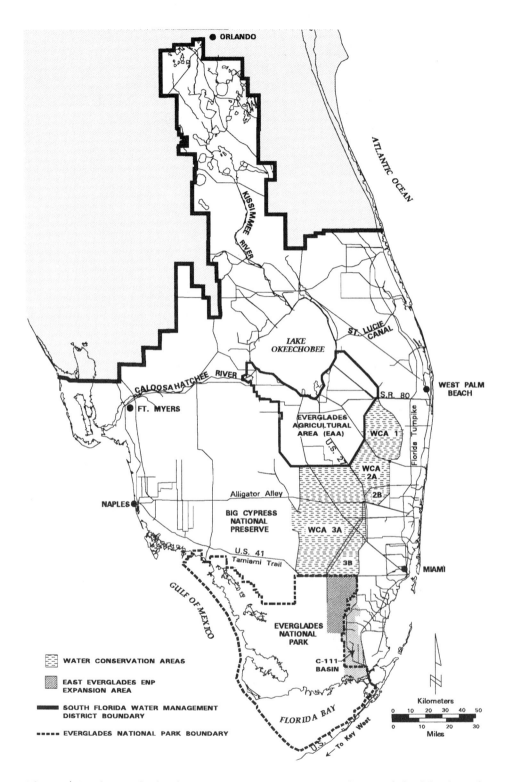

Figure 4.1 The South Florida Water Management District and major federal lands in the region.

the belief that systems restoration* will not be complete until the remnants of the entire Kissimmee–Okeechobee–Everglades watershed are recovered. Finally, the Water Conservation Areas have been the site of many significant environmental battles (e.g., the Jetport in Dade County, management of the deer population, minimum deliveries to Everglades National Park, water quality impacts from the Everglades Agricultural Area on sawgrass) over the future destiny of the region and attempts to control growth. Although current attention is riveted on the remediation of water quality problems, the more expansive issue of defining water requirements and sources for the Everglades *vis à vis* urban and agricultural uses looms even larger on the horizon.

The KISSIMMEE–OKEECHOBEE–EVERGLADES DRAINAGE BASIN

Although this chapter is focused on the Everglades, the Everglades is part of a much larger watershed—the Kissimmee–Okeechobee–Everglades system (Figure 4.2). The size of the main watershed is 10,890 mi² (28,205 km²) or approximately 310 mi (449 km) north to south and 62 mi (100 km) east to west. Prior to drainage and the installation of water structures, the natural system was connected hydrologically. The Kissimmee River discharges into Lake Okeechobee. Historically, during wet cycles the lake would overflow its south bank, providing additional flow to the Everglades.

The Water Conservation Areas consist of five surface water management impoundments comprising 1372.1 mi² (3554 km²) in Palm Beach, Broward, and Dade counties. Their design was an attempt to accomplish seven objectives:

- Receive and store agricultural runoff from the Everglades Agricultural Area
- Prevent waters accumulated in the Everglades from overflowing into urban and agricultural lands
- Recharge regional ground water and prevent saltwater intrusion
- Store and convey water supply for agricultural irrigation, municipal and industrial use, and natural system requirements in Everglades National Park
- Enhance fish and wildlife and recreation
- Receive regulatory releases from Lake Okeechobee
- Dampen the effect of hurricane-induced wind tides by maintaining marsh vegetation in the system

Water Conservation Areas 1, 2, and 3 are managed as surface water reservoirs, with a combined storage capacity of 1,882,000 acre-ft (2.321·10⁹ m³). Water Conservation Areas 2B and 3B primarily recharge and maintain groundwater levels in coastal areas to the east. They overlay parts of the Biscayne Aquifer, which underlies approximately 3000 mi² (7770 km²) of Dade, Broward, and southern Palm Beach counties. This aquifer, which ranges in depth from 200 ft (61 m) at the coast

---

* Restoration in this chapter connotes the process of rehabilitating or revitalizing ecosystem functions or values in at least the natural remnants of the historic Kissimmee–Okeechobee–Everglades system.

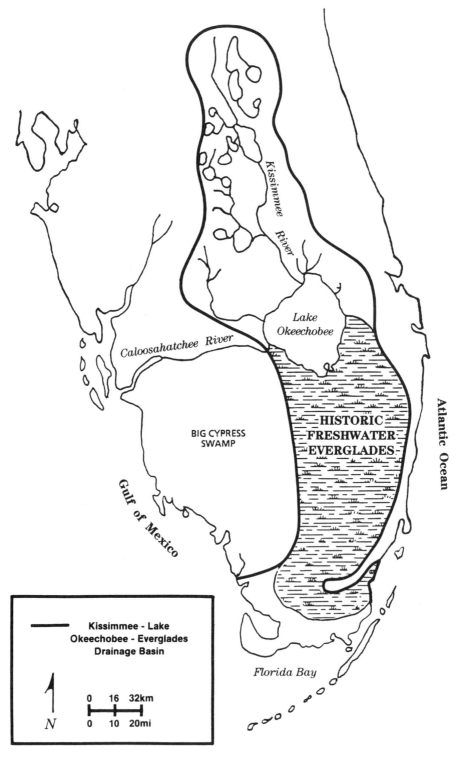

Figure 4.2   Location of Kissimmee–Okeechobee–Everglades watershed in south Florida.

to a few feet thick along its inland border, is surficial and highly permeable (Fernald and Patton, 1984). To the immediate north of the Water Conservation Areas is the Everglades Agricultural Area, which consists of 1181 mi^2 (3059 km^2) in western Palm Beach and Martin, Hendry, and Glades counties (Figure 4.1). The Everglades Agricultural Area is served by 15 project canals and 25 water control structures (Cooper and Roy, 1991). South of the Water Conservation Areas is Everglades National Park (1684.5 mi^2 [4363 km^2]), located in western Dade, northwestern Monroe, and southwestern Collier counties. As a natural preserve, project works rarely penetrate the area; however, surface water inflows to the basin are made by way of project water control structures and canals.

Everglades National Park has received national and international classification as an outstanding subtropical preserve (Maltby and Dugan, 1994). At the turn of the 20th century, efforts began to preserve this unique area. First, plume hunting was banned. During subsequent decades, state and federal governments were urged to establish the southern Everglades as a tropical preserve. Congress acted in 1935, and in 1947 the Florida legislature appropriated money to purchase and dedicate Everglades National Park (Blake, 1980). The Everglades National Park Act requires the National Park Service to preserve intact the unique flora and fauna and to protect the essential primitive natural conditions that prevailed when the park was established (Finley, 1988).

The EARLY YEARS

It took a decade after Florida achieved statehood in 1855 for the state legislature to pass an act creating an entity designated as the Internal Improvement Fund and Trustees to manage the lands ceded to the state from the federal government (Bottcher and Izuno, 1992). The Swamplands Act of 1850 authorized the transfer of 20 million acres (8,094,000 ha) to Florida for the purpose of drainage and reclamation (Blake, 1980). The Internal Improvement Fund was plagued by financial difficulties, the War Between the States, and economic depression during much of its history. In 1881, Governor Bloxham temporarily revived the fund from bankruptcy when he negotiated the sale of 4 million acres (1,618,800 ha) to Philadelphia millionaire Hamilton Disston (Blake, 1980).

Disston's efforts were impressive but short lived. In little over a decade he succeeded in draining over 50,000 acres (20,235 ha). His agricultural experiments proved that the land in the northernmost reaches of the Kissimmee River was quite productive (e.g., sugarcane and rice and construction of processing mills) (Blake, 1980). In attempts to lower the surface water elevations throughout the Kissimmee–Okeechobee system, he connected headwater lakes, opened the Kissimmee River for navigation, and linked the Caloosahatchee River to Lake Okeechobee (Figure 4.3). Disston is also credited with the first attempt to drain the Everglades by excavating 11 mi (17.7 km) of canal south of Lake Okeechobee in the direction of Miami. This grand assay came to an abrupt end with the panic of 1893, which devastated Disston and his enterprises. He took his own life. Despite eventual failure, Disston's efforts substantiated in good measure the Buckingham Smith Report of 1848, which predicted that with proper drainage, the region could

Figure 4.3 Disston water works 1881–94.

become a source of tropical products "of no trifling advantage to the whole country" (Smith, 1848).

CUT 'n TRY ERA

By 1891 considerable knowledge had been assembled regarding the reclamation of mucklands in the Everglades. A consensus was emerging that a large-scale drainage program would work (Wright, 1912). The railroad interests still controlled the Internal Improvement Fund. Reclamation would have to wait for new political leadership. The depression of the 1890s created a political environment that was ripe for change. The populist notions of curbing the excesses of big business (i.e., the landed aristocracy and the railroad interests) through government control swept liberal reform candidate William S. Jennings into office as governor in 1901. He was decidedly anti-railroad and pro-land reclamation.

The flood of 1903 devastated the crops in south Florida and caused considerable damage to farms in the region. Governor Jennings beseeched President Theodore Roosevelt to provide disaster relief. The president sent no money but transferred more land in south Florida from federal to state ownership (Blake, 1980). At the time, Florida's constitution allowed governors only one term in office, so the fate of reclamation would rest on the shoulders of Jennings's successor—Napoleon Bonaparte Broward.

The election of 1905 became a referendum on the fate of the Everglades and the Internal Improvement Fund. Broward won and immediately took personal responsibility for the Everglades drainage project. He became planning engineer, construction superintendent, and promoter of the Everglades reclamation project (Johnson, 1974). He had a simple policy and plan—cut 'n try. To fund and manage the construction, the legislature (with Broward's prompting) created the Everglades

Drainage District in 1907, consisting of 7150 mi^2 (18,518 km^2). By 1917, four major muck scalped canals dissected the Everglades from Lake Okeechobee to the Atlantic Ocean (Allison et al., 1948). After considerable enlargement by the U.S. Army Corps of Engineers, these canals would later become the backbone of the system known as the Central and Southern Florida Project for Flood Control and Other Purposes (Figure 4.4):

- West Palm Beach Canal   42 mi (67.6 km)
- Hillsboro Canal   51 mi (82.1 km)
- North New River Canal   58 mi (93.4 km)
- Miami Canal   85 mi (136.8 km)

By 1931, the outlet from Lake Okeechobee to the Caloosahatchee River was improved, and the completion of the St. Lucie Canal east to the Atlantic Ocean provided another way of controlling lake levels. The Bolles and Cross canals became connectors to the four major canals south of Lake Okeechobee, bringing the total miles of canal excavated to 440 (708 km). The Everglades Drainage District also constructed 47 mi (76 km) of sand and muck levee around the southern rim of Lake Okeechobee (Allison et al., 1948). The drainage program in the Everglades probably helped stimulate urban growth. The population in the lower east coast soared from 22,961 in 1900 to 228,454 by 1930 (Dietrich, 1978). Within a similar time frame (1915–28) another major alteration in the Everglades was taking place—construction of the Tamiami Trail, which linked Miami with Naples on the west coast (Figure 4.5). Although there are no data to measure the magnitude of the hydrologic change, debate exists over the impact that the Tamiami Trail has had on water flow to the southern Everglades. The presence of the road clearly is a departure from pre-existing conditions. On the other hand, designers realized that flow could not be stopped, so numerous culverts and bridges were (and still are) installed.

Financial woes continually plagued the Everglades Drainage District. The state capital was tied up in the potential of its land and water. Financial support was difficult to obtain. State revenues were slim, and financial lenders were shaky and dubious of the success of the drainage venture. The Great Depression brought more financial hardship (Blake, 1980). The Everglades Drainage District faltered. During the 1930s and well into the 1940s, construction was abandoned and maintenance ceased on Everglades Drainage District works.

Killer hurricanes struck the region in 1926 and 1928, claiming over 2500 lives. Material losses were estimated to be over $75 million (Figure 4.6) (Blake, 1980). Attention shifted from Everglades drainage to controlling flooding around Lake Okeechobee. The Mississippi River and Tributaries Act of 1928 changed the position of the federal government on flood control. In 1930, the Corps of Engineers became a major participant with the state (i.e., Okeechobee Flood Control District) in controlling flooding around Lake Okeechobee. Florida agreed to share a portion of the costs to increase discharge from the lake, improve canal works, and reconstruct and enlarge the levees around it. Historic levels for Lake Okeechobee were estimated to be 20 ft (6.1 m) mean sea level (MSL) at capacity. Following the floods, control levels on Lake Okeechobee were dropped from a

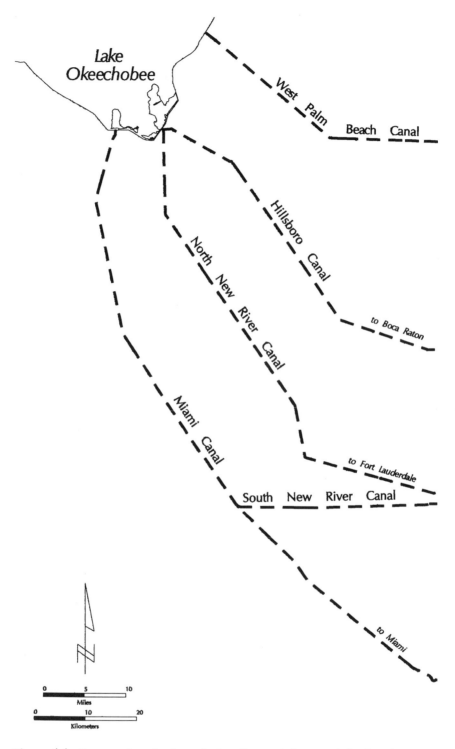

Figure 4.4 Four muck-scalped canals that dissected the Everglades by 1917.

Figure 4.5 Location of Tamiami Trail.

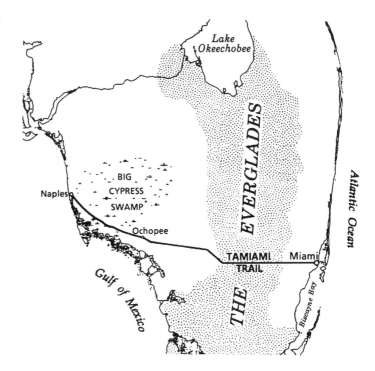

Figure 4.6 Areal extent of flood damage around Lake Okeechobee from 1926 and 1928 hurricanes.

maximum of 19 ft to 17 ft (5.8 m to 5.2 m) to provide additional freeboard. The U.S. Army Corps of Engineers raised lake levees to over 30 ft (9.15 m), and a series of hurricane gates were installed to better regulate water levels (Allison et al., 1948).

The effect of levees on the fledgling and anemic agricultural area south of Lake Okeechobee was dramatic. Coupled with price supports from the Sugar Act of 1934, the new flood control works gave farmers the perception of security that they needed to double sugarcane production—410,000 tons ($3.72 \cdot 10^8$ kg) to 873,000 tons ($7.92 \cdot 10^8$ kg) in the 10 years between 1931 and 1941 (Clarke, 1977).

The ERA of TURNING GREEN LINES into RED

The depression of the 1930s brought more than a decade of misery to the Everglades Drainage District. Debt, interest, and nonpayment of taxes were crippling and forced it into bankruptcy. The war years that followed saw Everglades Drainage District works fall into further disrepair. Water hyacinth (*Eichhornia crassipes*) choked large stretches of the Hillsboro and Miami canals. Landowners abandoned lock and control structures. Instead of discharging to the Atlantic, up to 20% per year of the flow from the Glades backed up into Lake Okeechobee (Allison et al., 1948).

Major hurricanes pounded the Everglades again in 1947–48. In one year 108 in. (274 cm) of rain fell and millions of acres of land were under water for up to 6 months (Figure 4.7). To dramatize the severity of conditions to the nation, the state published what became known as the "weeping cow" book, with a heifer standing belly-deep in water that completely blanketed the landscape depicted on the cover.

In response to the state's plight and plea, the Jacksonville district engineer came to the conclusion that one comprehensive plan to cover all phases of the water problem was the only solution. The U.S. Army Corps of Engineers published the Comprehensive Plan in 1948, and Congress responded (PL 80-858) in the same year with authorization for a massive endeavor called the Central and Southern Florida Project for Flood Control and Other Purposes (i.e., House Document 643). The project would ultimately lower water tables by 4–5 ft (1.2–1.5 m) east of the protective levee. The security against flooding inherent in this project, coupled with technological advances in refrigeration, concrete block construction, and mosquito control, became a powerful engine for growth. From 1950 to 1990 the population in Dade, Broward, Palm Beach, and Monroe counties skyrocketed from approximately 750,000 to close to 4 million residents (Marth and Marth, 1993; Dietrich, 1978).

At the behest of the federal government and based on a 21-member gubernatorial-appointed task force, the state legislature consolidated its water management and coordinating functions into one entity—the Central and Southern Florida Flood Control District (Bellamy, 1992). The responsibilities of the Everglades Drainage District and the Okeechobee Flood Control District were folded into this new institution which had primary responsibility for implementing the federal plan, referred to in the vernacular as "turning green lines into red lines." The U.S. Army Corps of Engineers designed and constructed the project. The Central and Southern Florida Flood Control District (and its successor, the South Florida Water Manage-

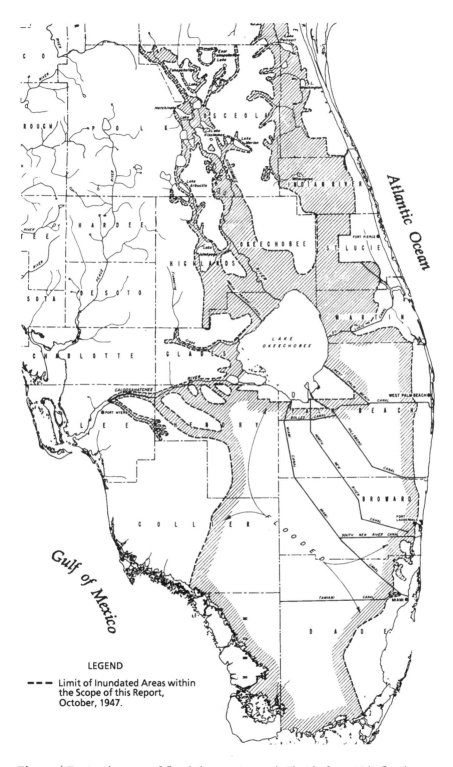

Figure 4.7 Areal extent of flood damage in south Florida from 1947 flood.

ment District) was the local sponsor of the project, with responsibility for operation and maintenance of the facilities built for the project.

The Central and Southern Florida Project for Flood Control and Other Purposes put to use four principal technologies to carry out the U.S. Army Corps of Engineers design: levees, water storage areas, channel improvements, and large-scale pumps to supplement gravity drainage. In effect, the federal project superseded existing control works and significantly expanded them. The Central and Southern Florida Project, incorporating all the Everglades north of Everglades National Park, has been continually expanded and enlarged upon throughout its over 40-year history.

Before one can fully appreciate the operation of the system, the ecological impacts of the project, and the potentials for restoration, it is necessary to understand the expansion of structural changes in a time-step fashion. Then one can focus on the basic operational rules—regulation schedules—that have governed the operation of the system.

CHRONOLOGY of STRUCTURAL CHANGES

1952–54: The Eastern Perimeter Levee

The first order of business for the Central and Southern Florida Project was the construction of a series of 9- to 18-ft (2.7- to 5.5-m) high perimeter levees and borrow canals (L-40, 36, 35A, 35, 37, 33, 30, and 31) that stretched from Palm Beach to Dade County (Figure 4.8) The levee is roughly 100 mi (161 km) long and runs parallel to the coastal ridge. This series of levees became the eastern boundaries of the Water Conservation Areas, which stopped the sheet flow from the Everglades toward the urbanizing coastal area. The levees were designed to standard project flood specifications with the assumption that rooted emergent vegetation would continue in the Water Conservation Areas to help control wind tides and wave run-up (U.S. Army Corps of Engineers, 1992).

In retrospect, the perimeter levees and flood control structures became the only effective barrier to agricultural expansion and urban development. The notable exception was the 8.5-mi² (22-km²) area in Dade County just south of Tamiami Trail that protrudes into the Northeast Shark River Slough. Increased flood protection from the Everglades and lowering of the water table east of the perimeter levees had considerable influence (among several other key factors) on the demographic shifts that followed in the region. As Ehrlich and Ehrlich (1992) stated quite poignantly, "...if the region was an independent nation it would be one of the fastest growing in the world."

1954–59: Everglades Agricultural Area

The next step was to secure the western and northern boundaries of the Water Conservation Areas and install pump stations (Figure 4.9). Flood protection for eight of the nine basins was designed to remove three-quarters of an inch (1.9 cm) of runoff per day from agricultural lands. The 4.5-mi² (11.7-km²) portion of the S-4 basin that included the city of Clewiston was designed to handle 4 in. (10.2 cm)

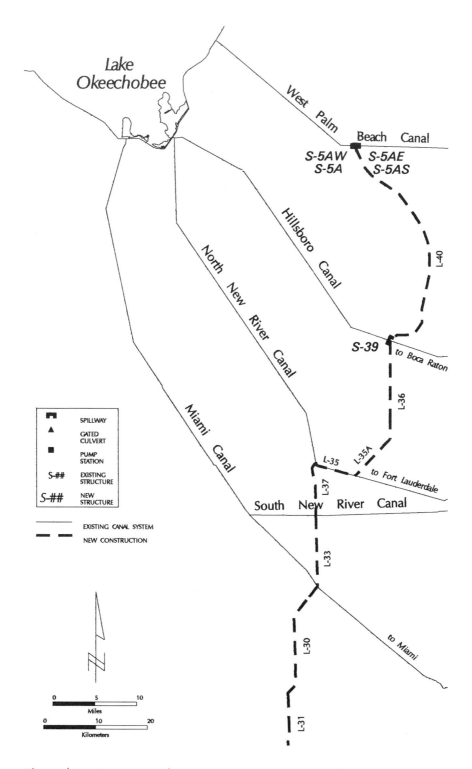

Figure 4.8   Construction of eastern perimeter levee (1952–54).

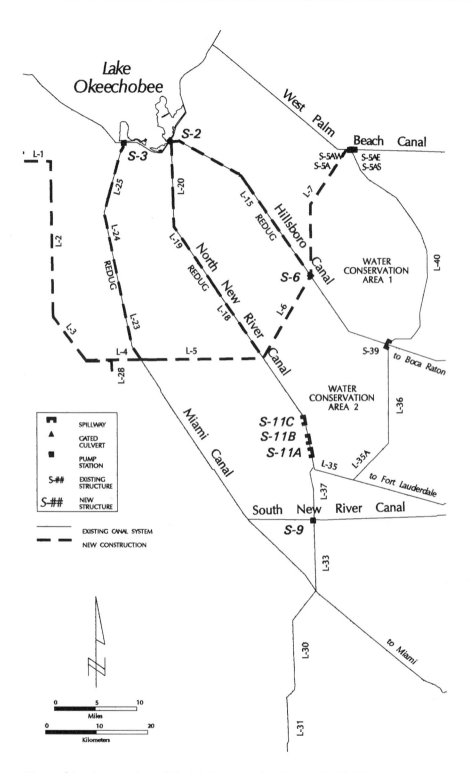

Figure 4.9   Construction of Everglades Agricultural Area (1954–59).

of runoff per day (Cooper, 1989). Levees 7, 6, 5, and 4 and a short run of L-28 partitioned the 700,000 acres (283,290 ha) of rich muck lands (named the Everglades Agricultural Area) to the north from the remainder of the Everglades enveloped in the Water Conservation Areas. In addition, the old Everglades Drainage District canals (Hillsboro, North New River, and Miami) were deepened within the Everglades Agricultural Area to provide better flood conveyance from the agricultural area into the Water Conservation Areas. Agricultural production, which had hovered under 50,000 acres (20,235 ha) in previous decades, began to soar. Sugarcane production increased over sixfold—to 300,000 acres (121,410 ha) by 1975 (Clarke, 1977).

Three very large water control structures (S-11A, S-11B, and S-11C) were constructed during this time to allow discharge from what was to become Water Conservation Area 2 into Water Conservation Area 3. These structures were consistent with the original intent of the project—to hold water away from the coastal area and divert it to the southwest, toward Everglades National Park. The U.S. Army Corps of Engineers constructed the S-9 pump station in western Broward County during this period. It pumped water west from the South New River Canal into Water Conservation Area 3 from the C-11 basin to the east—one of the few examples where backpumping from suburbs was designed and installed in the project (Cooper and Roy, 1991). Other examples of backpumping into the Water Conservation Areas are the privately operated drainage districts such as the Acme and the North Springs Improvement districts in Palm Beach and Broward counties. Acme and North Springs have the capacity to pump water (200–400 cfs [61–122 $m^3 \cdot sec^{-1}$]) into Water Conservation Areas 1 and 2A, respectively (Cooper and Roy, 1991).

1960–63: The Water Conservation Areas

This period saw substantial portions of the project completed and vast areas of sawgrass prairie delimited by levees and borrow canals (Figure 4.10). The Water Conservation Areas hold water on the surface away from the Biscayne Aquifer to augment water supply along the east coast and in Dade County and Everglades National Park. To help address water supply concerns in the park, discharge points from the Water Conservation Areas were aligned in the direction of the park. Management of fish and wildlife in the Water Conservation Areas was turned over to appropriate federal and state agencies. The U.S. Fish and Wildlife Service secured a lease from the Central and Southern Florida Flood Control District in 1961 and created a refuge in Water Conservation Area 1, named the Loxahatchee and later called the Arthur R. Marshall Loxahatchee National Wildlife Refuge. The Florida Freshwater Game and Fish Commission took charge of Water Conservation Areas 2 and 3.

Water Conservation Area 1

The Hillsboro Canal became the dividing line between Water Conservation Areas 1 and 2. The levees around Water Conservation Areas 1, 2, and 3 were completed in successive years, 1961–63, respectively. Water Conservation Area 1

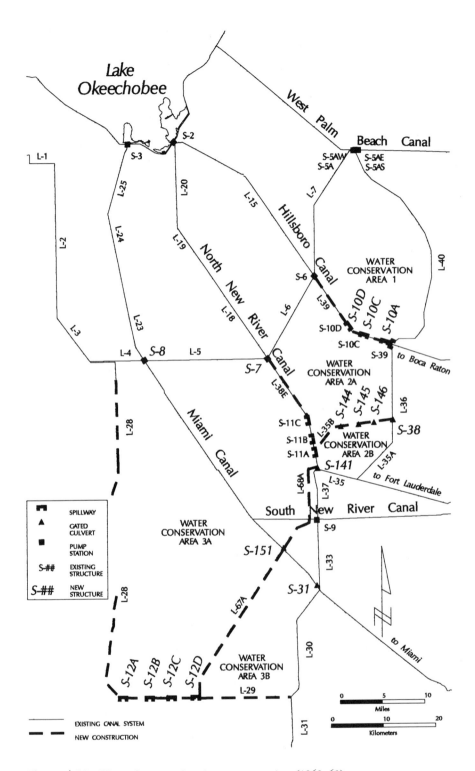

Figure 4.10   Water Conservation Area construction (1960–63).

became totally enclosed with the completion of the L-39 dike along its southern boundary (Cooper and Roy, 1991). Control structures S-10A, 10C, and 10D were installed and designed to regulate discharges from Water Conservation Area 1 into Water Conservation Area 2.

Water Conservation Area 2

The North New River flowing into Fort Lauderdale became the southern boundary of Water Conservation Area 2. This design was modified with the installation of a partitioning levee to create two units, Water Conservation Areas 2A and 2B. The latter was located over the Biscayne Aquifer and was considered too porous to hold water as needed for the project (U.S. Army Corps of Engineers, 1992). Structures S-144, 145, and 146 were constructed to allow discharge from Water Conservation Area 2A into Water Conservation Area 2B. Collectively, these three structures allow only about 630 cfs (192 $m^3 \cdot sec^{-1}$) to enter Water Conservation Area 2B (Cooper and Roy, 1991). Outlet structures S-141 and S-38 convey water from Water Conservation Area 2B into Water Conservation Area 3 and to canals in the eastern coastal area, respectively. The latter structure and in some instances S-141 are used to direct water during drought to recharge coastal groundwater levels and retard saltwater intrusion into the coastal wellfields. Regulatory releases through S-38 occur if two conditions are met: water is not needed in Water Conservation Area 3A and the coastal canals can accept additional water without flooding. Structure S-10E was subsequently added to the project to provide increased hydroperiod to the northern part of Water Conservation Area 2A.

Water Conservation Area 3

Just as Water Conservation Area 2 was partitioned into A and B in 1962, the design of Water Conservation Area 3 was modified to address the problem of excessive seepage to the east through the aquifer. L-68A was installed to direct water away from Water Conservation Area 2B and for additional seepage protection of eastern lands. L-68A intersected with L-67A in the vicinity of pump station S-9. The levee subdivided Water Conservation Area 3 into 3A and 3B. In addition to holding water away from the porous Water Conservation Area 3B, the L-67A borrow canal later became an important conveyance channel for transporting water through Water Conservation Area 3A down to Everglades National Park. It indirectly connects the Miami Canal to S-333 and the S-12 structures, and its flow is affected by their operation (Cooper and Roy, 1991).

Tamiami Trail

Levee 29, across the top of Everglades National Park, was constructed along the north side of the Tamiami Trail. Four control structures (S-12A, 12B, 12C, and 12D) regulate discharge from Water Conservation Area 3A into the park. The four sets of structures consist of six 25-ft (7.6-m) wide vertical-lift gates theoretically capable of 8000 cfs (2440 $m^3 \cdot sec^{-1}$) flow from each set (U.S. Army Corps of Engineers, 1992). These structural measures effectively limited water release to only the western part of the Shark River Slough. The fate of the northeast portion of the

Shark River Slough (10 mi [16.1 km] wide at Tamiami Trail) had been part of a political concession to the development interests, which left the door open for land speculation and a potential Southwest Dade Flood Control Project in that area. This component of the project was deauthorized in the 1970s.

Two other important water control structures (S-151 and S-31) were built. S-151 straddles the Miami Canal at the L-67A dike between Water Conservation Areas 3A and 3B, while S-31 is farther downstream where the Miami Canal leaves Water Conservation Area 3B. These two structures improved discharge capacity down the Miami Canal to the coastal communities (Cooper and Roy, 1991).

The last major structural component of this period was the completion of the two legs of L-28, which forms the western border separating Water Conservation Area 3A from the Big Cypress watershed. The original design called for a continuous levee, until it was discovered that water was flowing in from the west. With much of the property west of Water Conservation Area 3 in private ownership, engineers were faced with the alternative of having to pump water over the levee from the west to avoid flooding or leaving a 7.5-mi (12.1-km) gap in the dike. The latter concept was chosen (U.S. Army Corps of Engineers, 1992).

1965–73: Water Supply and Minimum Deliveries for Everglades National Park

Between 1965 and 1973, portions of the Central and Southern Florida Project were revamped and extended in an attempt to satisfy water demands for Everglades National Park (Figure 4.11).

Increasing Downstream Conveyance to the Park

The portion of the Everglades Agricultural Area under sugarcane production was expanding rapidly. By 1973, the number of farms growing sugarcane had increased to over 120, with more than 200,000 acres (80,940 ha) producing 800,000 tons ($7.26 \cdot 10^8$ kg) of raw sugar per year (Clarke, 1977). These changes did not put new demands on the system. S-7 and S-8 had been properly sized to remove 3/4 in. of rain per day within respective basins. However, downstream conveyances were improved to move water more rapidly to the park. L-38 was resized and a parallel L-38A was dug. C-123 was also dug and L-68A was redug (J. Lane, personal communication).

Minimu Deliveries for Everglades National Park

Wrangling during the early 1960s over who was responsible for and how much water should be delivered to Everglades National Park was resolved (at least temporarily) in 1970 by congressional action (PL 91-282). Minimum flow to the park was set at 315,000 acre-ft ($3.88 \cdot 10^8$ m^3) per year, with 260,000 acre-ft ($3.20 \cdot 10^8$ m^3) for Shark River Slough and 37,000 acre-ft ($45,621 \cdot 10^3$ m^3) guaranteed supplemental water delivery to Taylor Slough. The remaining 18,000 acre-ft ($22.194 \cdot 10^3$ m^3) was allocated to the park's eastern panhandle (C-111 basin). This annual allocation was divided into minimum monthly deliveries said to reflect discharge characteristics during the 1940s and 1950s (Table 4.1).

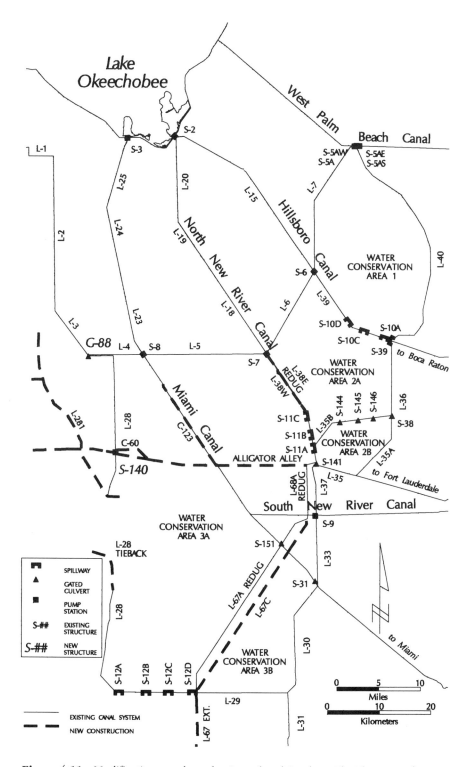

Figure 4.11 Modifications made to the Central and Southern Florida Project for water supply purposes (1965–73).

Table 4.1 Minimum Delivery Schedule
for Shark River Slough and Taylor Slough

	Minimum delivery schedule (acre-ft) ($m^3 \cdot 10^3$ in parentheses)			
	Shark River Slough		Taylor Slough	
Jan.	22,000	(27,126)	740	(912)
Feb.	9,000	(11,097)	370	(456)
Mar.	4,000	(4,932)	185	(228)
Apr.	1,700	(2,096)	185	(228)
May	1,700	(2,096)	370	(456)
Jun.	5,000	(6,165)	6,660	(8,212)
Jul.	7,400	(9,124)	7,400	(9,124)
Aug.	12,200	(15,043)	2,960	(3,650)
Sep.	39,000	(48,087)	5,920	(7,299)
Oct.	67,000	(82,611)	7,770	(9,580)
Nov.	59,000	(72,747)	3,700	(4,562)
Dec.	32,000	(39,456)	740	(912)
Totals	260,000	(320,580)	37,000	(45,619)

Water Supply for South Dade

The initial concept for the south Dade County area contained in the Corps Plan of 1948 called for levee 31 to protect lands to the east. To carry runoff away from the south Dade and Homestead area, a grid of canals on an east-west and north-south alignment was envisioned (House Document 643). These canals would empty into Biscayne Bay, Florida Bay, and Barnes and Long sounds.

During the planning of the Flood Control Project's south Dade works, the National Park Service requested that water being drained from Taylor Slough headwaters along the eastern boundary of Everglades National Park be directed to the slough, rather than routed to Barnes Sound via C-111. The Everglades National Park–South Dade Conveyance System was authorized in 1968 (PL 90-483) "...for the purpose of improving the supply and distribution of water supplies to Everglades National Park, and for expanding agricultural and urban needs in Dade County" (U.S. Army Corps of Engineers, 1992). L-31 was realigned and L-31W was added around a 5000-acre (2024-ha) agricultural area known as the Frog Pond on the eastern boundary of Everglades National Park. The purpose of L-31W was to provide 10-year flood protection for agricultural and industrial areas to the east (U.S. Army Corps of Engineers, undated). Pump station 332 ensured that minimum monthly deliveries made to Taylor Slough complied with PL 91-282. The project was intended to convey 160 cfs (4.5 $m^3 \cdot sec^{-1}$) to Taylor Slough delivered at S-332 and 75 cfs (2.1 $m^3 \cdot sec^{-1}$) to the Everglades National Park panhandle delivered at S-18C (U.S. Army Corps of Engineers, 1988). Structure 174 (S-174) controls releases from L-31N borrow canal. Structure 175 (S-175) maintains water levels in L-31W borrow canal (Figure 4.12).

In the C-111 basin, the S-18C spillway was installed for three purposes: to pass flood flows from the C-111 basin, to maintain a freshwater head against saltwater

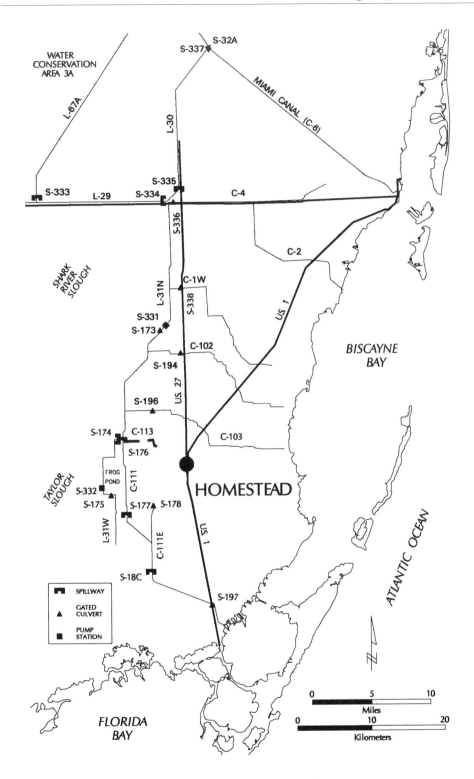

Figure 4.12 The South Dade Conveyance System.

intrusion, and to supply water to the panhandle of Everglades National Park through gaps cut in the south berm adjacent to the canal downstream of the structure (U.S. Army Corps of Engineers, 1992).

Since completion of the South Dade Conveyance System in 1983, water levels in L-31N and L-31W have been a source of controversy between Everglades National Park and development interests along its eastern frontier. These canals traverse residential and agricultural lands (e.g., crops such as tomato, lime, avocado, and exotic fruits and vegetables) in an area known as the East Everglades.

C-111 Removable Plug Controversy

No discussion of the C-111 basin during this period would be complete without mentioning the controversy surrounding discharges to Barnes Sound through S-197. The initial design for the Central and Southern Florida Project in the panhandle of Everglades National Park called for several canals (including C-111) to discharge directly into Florida Bay. What ensued was born of concerns for the environmental impacts of the project on the park. The project was reconfigured to eliminate the canals that would penetrate the boundaries of the park. Instead, gaps in C-111 were provided to permit sheet flow to move south through the panhandle of the park toward Florida Bay. During the process of construction, an earthen plug was placed across the mouth of C-111 where it intersected U.S. Highway 1. As completion of the project neared, the environmental community raised issues about the potential of saltwater intrusion into the canal and the park. S-197 was proposed as a temporary solution—a gated culvert in C-111. This structure remains a problem today. Removal of the plug to allow flood flows into Barnes Sound is unacceptable to the environmental community. They are equally concerned about the poor distribution of flows from C-111 gaps into Florida Bay. The U.S. Army Corps of Engineers is currently attempting to address these deficiencies through a general design memorandum, which will address the environmental concerns while maintaining the authorized flood protection for that area (U.S. Army Corps of Engineers, 1988).

Modifications in Water Conservation Area 3

The L-67A canal was enlarged to increase water conveyance capacity in order to facilitate water delivery from Water Conservation Area 3A to Everglades National Park. Rather than place the spoil on the east side (as was done during the initial construction), the dredged material, due to construction limitations, was placed on the western bank in the marsh. To permit flow from the marsh into the canal, 100-ft (30 m) breaks were created among the 500-ft (152-m) spoil islands. The levee breaks ensured that fish in the marsh could reach the canal during periods of declining water levels.

Other significant changes during this period included the construction of L-67C and the extension of L-67 for several miles south of the Tamiami Trail in the heart of the Shark River Slough. During the years immediately after the 1963 completion of the Water Conservation Areas and while they were filling with water, flow to Shark River Slough was interrupted. Drought conditions exacerbated the deficiency of water supply to Everglades National Park during the period. To help alleviate problems during low flow conditions, L-67C was dug parallel to L-67A, 2500 ft (762 m) to the east. L-67C intercepted water that would otherwise disappear into the aquifer under Water Conservation Area 3B and redirected it to the southwest toward the park. When engineers discovered that the S-12s actually discharged at only 40% of design capacity, a 9.5-mi (15.3-km) L-67 canal extension was excavated in 1967 to enable low-flow discharges further south into the center of the slough (U.S. Army Corps of Engineers, 1992).

Finally, at the northern end of the system in the Everglades, G-88 and G-89 structures were installed to convey water from Hendry County to the S-8 and the S-140 pump stations, respectively. Discharge was exceeding capacity in the northern reaches of L-1, 2, and 3 and flooding occurred frequently in developing private lands in that area. Unfortunately, this alteration reduced water flowing into the northwest corner of Water Conservation Area 3A—an area already starved for water due to the project.

1974–79: Improving Water Conveyance to the South

By 1973 the Central and Southern Florida Project in the Everglades was essentially complete. However, the next 6 years witnessed further modifications for water supply purposes (Figure 4.13) The S-151 culverts and the L-67A canal were enlarged for greater water supply conveyance. The capacity of S-151 toward S-31 was enlarged from 600 to 1100 cfs (16.8 to 30.8 $m^3 \cdot sec^{-1}$). A new structure, the S-337, delivered water to southern Dade County via the enlarged L-31 and L-30.

Other significant alterations during this period included S-333, a gated spillway constructed where L-67 intersects L-29 and S-334 at the eastern end of L-29. Numerous culverts along this section of the Tamiami Trail allowed dispersed discharge into Northeast Shark River Slough. The S-333 and S-334 structures as installed were originally intended to supply water to the basins in southern Dade County (Cooper and Roy, 1991).

1980–85: Modifications to Protect the Environment

During the first half of the 1980s, additional structural modifications were undertaken for environmental purposes to (1) reduce nonpoint-source pollution to Lake Okeechobee from the Everglades Agricultural Area and (2) further improve deliveries to Everglades National Park (Figure 4.14).

Interim Action Plan

The original drainage design for the Everglades Agricultural Area called for removing excess water from the area to both the Water Conservation Areas and Lake Okeechobee. At the insistence of the Florida Department of Environmental Regulation, the district applied for a permit to operate pump stations S-2 and S-3 to backpump agricultural runoff to Lake Okeechobee. The agreement was referred to as the interim action plan. Storm water from the Everglades Agricultural Area was determined to have an adverse impact on the lake.

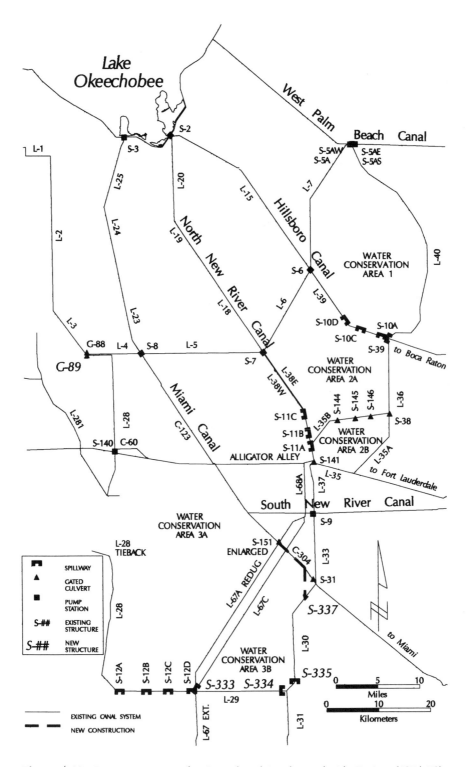

Figure 4.13 Improvements to the Central and Southern Florida Project (1974–79).

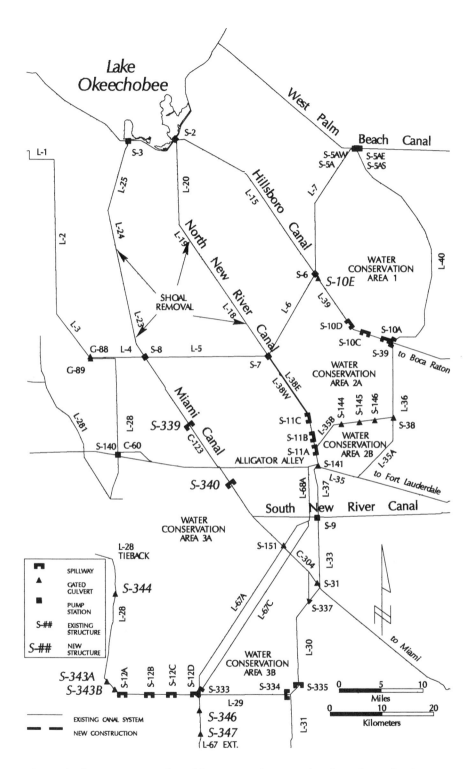

Figure 4.14 Environmental modifications to the Central and Southern Florida Project (1980–85).

Increased Water Supply Conveyance

To increase conveyance (up to 1000 cfs [28 m³·sec⁻¹]) south from the Everglades Agricultural Area, shoal spots in the Miami and North New River canals were removed and the C-123 was excavated through the middle of Water Conservation Area 3A in 1972 (Cooper and Roy, 1991). When engineers found that C-123 significantly overdrained Water Conservation Area 3A, two gated-sheetpile barrier dams (S-339 and S-340) were added to hold water in the marsh during low flow periods (U.S. Army Corps of Engineers, 1992). These were the first two structures in the entire system installed solely for environmental purposes.

Crisis in the Park

Unusually heavy rains associated with an El Niño weather system persisted through January and February 1983, requiring heavy, undesirable, and off-season regulatory releases to Everglades National Park. By March, an environmental emergency was declared in the park and a seven-point plan for immediate action was developed:

- Fill in L-28 canal and remove substantial segments of the levee
- Fill in L-67 extension and completely remove its levee
- Divert as much flood water as ecologically acceptable to Water Conservation Area 3B
- Distribute water deliveries from Water Conservation Area 3A along the full length of Tamiami Trail from L-28 to L-30
- Establish rigorous water quality monitoring
- Defer any implementation of new drainage districts
- Field test a new delivery schedule for the park based on current rainfall and normal runoff

The South Florida Water Management District and the U.S. Army Corps of Engineers acted swiftly to the measures proposed. Gated culverts S343A and S-343B (installed through L-29 between L-28 and S-12A) facilitated discharges into the Big Cypress. S-344, another gated culvert, was cut through L-28 to increase regulatory discharges from Water Conservation Area 3A and to extend the hydroperiod in the Big Cypress Preserve during dry periods (U.S. Army Corps of Engineers, 1992; Cooper and Roy, 1991).

The operation of S-333 and S-334 helped to direct water from L-67 through the northeast region of the Shark River Slough via a series of small culverts under U.S. 41, rather than discharge only though the S-12s. This effort was intended to more closely mimic the historic distribution of flow across Tamiami Trail. In conjunction with S-333 and S-151, the S-12s remain the prime outlets for discharges from Water Conservation Area 3A.

The L-67 extension and borrow canal were originally constructed to provide a means of supplying water directly to the Shark River Slough inside Everglades National Park and to help prevent water from spreading eastward onto private property during large S-12 releases (U.S. Army Corps of Engineers, 1992). S-346 and S-347 are culverts located in the L-67 extended borrow canal. They are equipped with risers and stoplogs (U.S. Army Corps of Engineers, 1992). These structures, which are normally closed, were intended to force water out of the borrow canal to increase sheet flow into the park. The levee has not been degraded.

In 1984 Congress temporarily set aside the allocation system imposed in 1970 for water deliveries. This gave the district, the park, and the Corps of Engineers the opportunity to explore ways to restore historic flow patterns to the park. This topic is discussed further in the subsequent section.

1985–90: The Proposed Northeast Shark River Slough General Design Memorandum

During this period, two additional steps were taken to help restore natural hydrologic conditions within Everglades National Park. In 1989, Congress authorized the Everglades Protection and Expansion Act to purchase 107,600 acres (43,546 ha) and the construction of modifications to enhance restoration (Figure 4.15).

The U.S. Army Corps of Engineers published a draft General Design Memorandum in 1990 which, if approved, would change the way water enters Everglades National Park. Changes were proposed in the physical system of the canals and structures and the methods of operation for the structures:

There are five major modifications being considered: (1) construction of structures in L-67A to allow water to be passed from WCA-3A to 3B; (2) construction of structures in L-29 to allow water to pass from WCA-3B to the Northeast Shark Slough; (3) removal

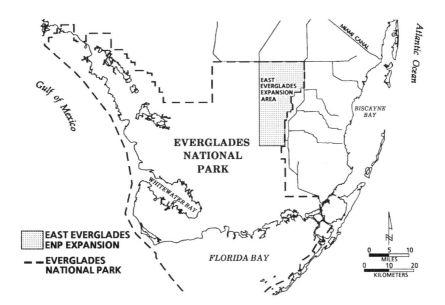

Figure 4.15 Location of land authorized for purchase by the National Park Service in 1989.

of L-67 extension and filling of the borrow canal; (4) construction of a levee around the 8.5 square mile residential area in the East Everglades; and (5) construction of two pumping stations to control seepage and discharge from the protected area and L-31N (Cooper and Roy, 1991; see also U.S. Army Corps of Engineers, 1990).

The Corps of Engineers plan and impact statement, which was deadlocked for over 2 years, was recently approved by the Assistant Secretary of the Army for Civil Work. The confab was over endangered species. Although the Corps, the South Florida Water Management District, and Everglades National Park agreed to the plan, the U.S. Fish and Wildlife Service objected on the grounds that this approach would jeopardize the continued existence of the snail kite (*Rostrhamus sociabilis*) and adversely modify its critical habitat. When experimental deliveries began in the mid-1980s, concerns about their impact on the snail kite spawned concentrated studies. In 1988, these efforts resulted in initiation of a consultation under Section 7 of the Endangered Species Act. The 1990 report issued by the Fish and Wildlife Service concluded that drydowns in unusually dry years (the preferred plan) would jeopardize the future of the snail kite (National Audubon Society, 1992). The Fish and Wildlife Service would give priority to single-species management alternatives, while Everglades National Park and a recent Audubon report favor "multispecies" or ecosystem approaches to managing endangered species.

REGULATION for the WATER CONSERVATION AREAS and EVERGLADES NATIONAL PARK

The operational criteria for water management in the Everglades center around regulation schedules for the Water Conservation Areas and minimum flow requirements for Everglades National Park. Regulation of water levels is based on rule curves initially developed by the U.S. Army Corps of Engineers. These schedules have been modified over the years. They basically dictate when discharge structures will be opened and closed.

Although the Water Conservation Areas serve many purposes, the regulation schedules are driven by two objectives which are generally at odds with natural system requirements: (1) minimizing flood risk during hurricane season (June–October) and (2) maximizing storage during the dry season (November–May). The minimum flow requirements established for Everglades National Park were an attempt by Congress to ensure that federal water reserve rights were adequately addressed in any state water-allocation scheme. These issues will be addressed separately later.

Discussion of the operating criteria for the Water Conservation Areas, Everglades National Park, and the C-111 basin touches on four major points: sources of inflows, sources of outflows, the operating schedule and any unique features, and significant ecological issues related to the operation schedule.

Water Conservation Areas and Shark River Slough

The Water Conservation Areas make up the largest remnants of the original Everglades. The topographic gradient from Lake Okeechobee to Florida Bay was approximately 1 ft (0.3 m) every 10 mi (16.1 km). The Water Conservation Areas function as surge tanks which attempt to modulate the wide swings in hydrologic patterns that are common to the region, while attempting to maintain the indigenous sawgrass marshes and hardwood hammocks which have thrived in the ephemeral character of the hydrologic system of the region.

Although the Water Conservation Areas are a receiving body for agricultural and some urban runoff, in addition to occasionally receiving discharges from Lake Okeechobee, it must be noted that rainfall is their dominant input (roughly 70% in an average year). They serve to recharge the region's principal drinking water aquifer and protect against saltwater intrusion along the coast. Increasingly, the Water Conservation Areas are being recognized for their ecological value—not just for the wild and endangered species that inhabit the Glades, but for the environmental benefits they provide for Everglades National Park basin and Florida Bay.

The regulatory schedule and the more recent operational policies attempt to reconcile all these various and competing objectives and uses. In the Water Conservation Areas, the relationship of water for various projects varies by season and multiyear hydrologic cadences. When water levels fall below minimum levels (e.g., 7.5 ft [2.3 m] MSL in Water Conservation Area 3A), releases from Lake Okeechobee or other Water Conservation Areas are made to meet water supply demands.

Water Conservation 1

Water Conservation Area 1 is enclosed by levees (Figure 4.14). The ground contours range from 13 to 15 ft (4.0 to 4.6 m) MSL. The major inputs are S-5A and S-6. Inside the levee system is a continuous canal that allows water to flow through the area without reaching the interior of the marsh, which is fed by rainfall. From an ecological perspective, the interior borrow canal has resulted in significant water quality and vegetative changes along the periphery, while the internal portions of the marsh have been spared (Rhoads, 1993).

The major discharges from Water Conservation Area 1 are S-10A, C, D, and E along the southwestern boundary. S-39 provides discharges to the east for water supply and regulatory purposes. Water Conservation Area 1 (also known as the A. R. Marshall Loxahatchee National Wildlife Refuge) has functioned under four regimes since it became operational in 1960 (Cooper and Roy, 1991). The initial range was from 17 ft (5.2 m) during November and December to 14 ft (4.3 m) at the end of the dry season. The intent was to provide storage for rainfall and hurricanes. In 1969 the bottom of the schedule was raised to 15 ft (4.6 m), which led to adverse environmental impacts. In 1972 the schedule was reset at the original range. In 1975, zones were added to accommodate different water conditions as rainfall patterns oscillated and to accommodate specific bird species (Figure 4.16). The current regulation schedule allows for a wet and dry *year* regime. Minimum pools are designed to allow the marsh an opportunity to periodically dry out every 2 or 3 years under historical rainfall patterns (Cooper and Roy, 1991). The U.S. Fish and Wildlife Service is still looking for ways to improve the schedule for ecological reasons. Discussions are currently underway between the Fish and Wildlife Service and the U.S. Army Corps of Engineers.

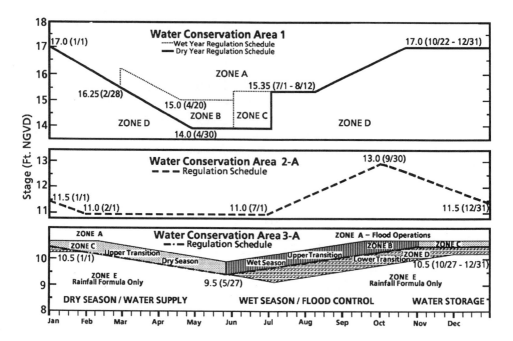

Figure 4.16 Regulation schedules for Water Conservation Areas 1, 2A, and 3A.

Water Conservation Area 2A

Water Conservation Area 2A is also totally enveloped by levees, but, in contrast to Water Conservation Area 1, it has gravity in addition to pumped inflow (Figure 4.14). Discharges into this area are made through the S-10 structures and the S-7 pump station. S-10E is the discharge structure of first priority (Cooper and Roy, 1991). Although it does not have near the capacity of the other S-10s, it does provide a means of directing water into the driest parts of Water Conservation Area 2A. The major outputs are the S-11s which direct water into Water Conservation Area 3A and three culverts (S-144, S-145, and S-146) which are used to provide water into Water Conservation Area 2B. In addition, S-38 provides discharge for water supply purposes into coastal Broward County during drought conditions and for regulatory purposes in some instances (Cooper and Roy, 1991).

The design for Water Conservation Area 2A, as with the others, was intended to allow for the continued growth of emergent vegetation. The ground elevation contours range from 12 to 11 ft (3.7 to 3.4 m) MSL, north to south. However, after its completion in 1961, the regulation scheme for this area was set too high (13–14.5 ft [4.0–4.4 m] MSL) and the vegetation communities began to convert to those found in deep water marshes. Tree islands began to disappear. As a result, a series of drawdowns was conducted in 1974 and 1980 under Walt Dineen's leadership. By 1980, the periodic drawdowns were not working and the district proposed steps that would return the area to a more natural state. A permanent drawdown schedule (11–13 ft [3.4–4.0 m]) was proposed and has been in effect ever since (Figure 4.16). Under this schedule, the S-11 structures would open at a much lower elevation and Water Conservation Area 2A would dry out more frequently, resulting in significant ecological benefit but lowered water storage capacity (Rhoads, 1993).

Water Conservation Area 3A

Water Conservation Area 3A is leveed on three sides (Figure 4.14). On the western border it is only partially leveed. There is a 7.5-mi (12.1-km) gap in L-28, which allows overland flow from the Big Cypress. The major inputs to Water Conservation Area 3A are through the S-11 structures, S-8 and S-9 pump stations in the Miami Canal, and S-140 pump station, which drains water from Hendry County. At times, there is also some gravity drainage into the area by way of the S-150 spillway. Ground elevation ranges from 10 to 7 ft (3.0 to 2.1 m) MSL.

The initial operating schedule and configuration of structures overdrained the northern portion of the area. This enabled a large deer herd to flourish until wet cycles returned and large deer mortalities resulted. Substantial changes were made in deer management in the 1980s, when Governor Graham resolved that the deer herd would be managed in a manner consistent with whatever water levels were in place. Since then, with a reduced herd and no major floods, there has not been a problem (Rhoads, 1993).

Shark River Slough

Water provisions to Everglades National Park have been a source of constant controversy and uncertainty since the park was established. Since the turn of the century, water entering the park through Shark River Slough has been under eight different hydrologic regimes (Gunderson, 1992; MacVicar, 1993):

1. Unregulated flow through an undisturbed system (1900–20)
2. Unregulated flow through culverts under Tamiami Trail (1925–60)
3. Overland flow at least partially interrupted during construction of Water Conservation Areas (1960–62)
4. Water delivered to the western reach of the park at Shark River Slough, only after the regulation schedule was exceeded through culverts in the Tamiami Trail (December 1963–March 1965)
5. A three-zone schedule based on stage, to determine monthly allocation (1965–66)
6. Static portion of water allocated to the park each year during the 1970s
7. S-12 structures left open so that deliveries became a function of hydraulic gradients between Water Conservation Area 3A and the park (Flow-Through Plan) (June 1983–85)
8. The Rainfall Plan (1985–present), which linked timing and quantities of deliveries to upstream weather conditions and flow that was spatially redistributed closer to the way it was prior to construction of Water Conservation Area 3

Under the leadership of the South Florida Water Management District, the Rainfall Plan, an experimental water delivery plan, was proposed in 1985. Based on this plan, water allocations to the Shark River Slough are determined as a function of rainfall, evaporation, and water level in Water Conservation Area 3A and the previous week's discharge. Discharge rates are calculated on a weekly basis (Cooper and Roy, 1991). As far as is possible, 55% of the calculated discharge is released to the Northeast Shark River Slough by way of S-333 and the L-29 borrow canal culverts. The remaining 45% of the target is discharged to the west through the S-12 structures (U.S. Army Corps of Engineers, 1992). The current rule curve for the S-12 structures includes three transition zones which allow for phased operation of structures, rather than the abrupt shifts in water delivery that proved harmful for wildlife nesting habits in Everglades National Park (Figure 4.16). The rainfall plan is part of an experimental program authorized by Congress (PL 99-190 and subsequent legislation) which permits "a series of iterative field tests for the purpose of collecting hydrologic and biologic data with the ultimate goal being the development of an optimum delivery plan for the ENP...." (U.S. Army Corps of Engineers, 1992).

A 2-year field test of the Rainfall Plan began in July 1985. Strict limits were placed on the flow rate through S-333 to make sure that the reintroduction of flow into the Northeast Shark River Slough would not cause problems in the residential and agricultural lands east of L-31. The Rainfall Plan was deemed successful in re-establishing significant overland flows to a vital part of the slough system that feeds Everglades National Park (Neidrauer and Cooper, 1988). The plan remains in effect today while additional improvements are being studied.

Taylor Slough and the C-111 Basin

Taylor Slough and the C-111 basin form the eastern hydrologic boundaries for water delivered to Everglades National Park. The South Dade Conveyance System, which links this basin with the works to the north, also illustrates the abrupt transition from metropolitan Dade County to the natural resources of the park.

Minimum water deliveries to the park for the eastern component of the project are controlled by two structures: S-332 and S-18C (U.S. Army Corps of Engineers, 1992). Based on a 1970 formula, these structures accept a minimum of 37,000 and 18,000 acre-ft ($45,621$ and $22,194 \cdot 10^3$ m^3) for delivery to Taylor Slough and the park panhandle, respectively. To divert water to Taylor Slough, L-31W was constructed with two gated culverts: S-174 and S-175. S-174 directs water into L-31W. S-175 serves two purposes: (1) to hold water levels up for water supply to the park and (2) to provide an outlet to release flood flows. S-18C is used to control upstream stage levels in C-111 and was designed to pass flood flows up to 40% of a standard project flood. The South Florida Water Management District is developing a water management plan for Taylor Slough similar to the Rainfall Plan for Shark River Slough. The goal of the plan would be to provide water supply and environmental benefit to Everglades National Park by restoring the rainfall–runoff response of the slough that existed prior to alteration of the slough hydrology by the Central and Southern Florida Project (South Florida Water Management District, 1990). This and other approaches remain under study.

The Frog Pond, an agricultural area, lies at the headwaters of the Taylor Slough. Water levels in L-31W have a significant effect on groundwater levels in the slough.

During the 1980s, controversy raged over this issue. It was resolved in favor of the park in 1988, when the district rescinded its resolution allowing seasonal draw-downs to facilitate agricultural planting (U.S. Army Corps of Engineers, undated).

CONCLUSION

Water management and the construction of appurtenant works in the Everglades has gone through a dramatic evolution over the past 90 years. It began with the works of the Everglades Drainage District, which became the foundation for the federal flood control project that followed. As the needs and perceptions of the citizenry and leadership of the region have changed over the past three decades, structures were installed or modified and rule curves were reshaped to address water supply and environmental mitigation and restoration objectives.

The flood protection which the Central and Southern Florida Project secured helped spur phenomenal urban and agricultural growth. However, it has come at the expense of the unique natural endowments of the region. As will be documented more thoroughly in subsequent chapters, past water management practices and the construction of related works have resulted in:

- The loss of the transitional glades, which provided an early-season feeding habitat for wading birds
- Modification of flow pattern (attenuated to pulsed), which reduced hydroperiods
- Unnatural pooling and overdrainage as a result of canals and levees
- Accelerated reversal from muck building to rapid oxidation
- Abandonment of wading bird nesting areas in Everglades National Park due to change in hydroperiod
- Unnatural and reduced flow of fresh water to Florida Bay

Futhermore, as the Central and Southern Florida Project was designed and constructed, appropriate consideration was not given to the concluding statement by Mason J. Young, Corps of Engineers Division Director, in House Document 643, which authorized the federal flood control project. He warned that if the region was to continue to prosper "…the conservation of water resources is as important and urgent as the provision of additional drainage and the elimination of flood damage….Until the need for freshwater has been satisfied, only the irreducible minimum that cannot be conserved should be discharged to coastal waters…."

Unfortunately, as recent records of freshwater discharge reveal (Figure 4.17), an annual average of 3.3 million acre-ft ($4069 \cdot 10^6$ m³) may have been discharged to the Atlantic Ocean between 1980 and 1989, while the combined discharges to the Shark River and Taylor sloughs and C-111 averaged 813,100 acre-ft ($1002 \cdot 10^6$ m³) (South Florida Water Management District, 1993). It should be recognized that there is some uncertainty in these estimated water budgets, possibly leading to over-estimation of coastal outflows. Nevertheless, budget results indicate that capture of flows presently lost to tide water represents an immense potential additional water supply for Everglades ecological restoration.

The extent of land and water alteration and its adverse impacts on the ecology

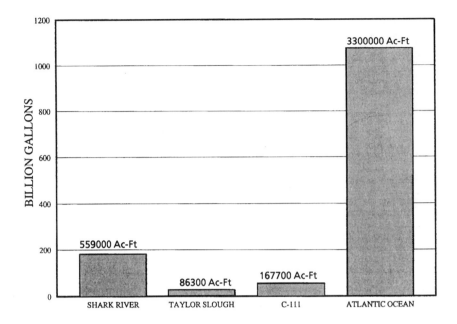

Figure 4.17   1980–89 records showing total discharge to the Atlantic Ocean, Shark River Slough, and Taylor Slough.

of the region are just a few key indicators of how far out of balance the natural system has been driven by man's colonization of the peninsula. These alterations and the attendant ecological degradation began long before the Central and Southern Florida Project was constructed. Daniel Beard's reconnaissance of the Everglades (1938) for the U.S. Department of the Interior recognized that the "hand of man," through exploitation of the biotic resources of the region, had already had a significant impact. In 1938, Beard hoped that with proper attention Everglades National Park would fully recover in 50 years (Beard, 1938). Unfortunately, just the reverse has been the case. The Central and Southern Florida Project and the park were intended by statute to be complementary features of a unified federal interest in the region. Unfortunately, that intent has yet to be realized.

For close to a century, the Everglades has been furrowed, drained, balkanized, and invaded by exotic species. The challenge now faced is nothing short of reconfiguring the structural works of the Central and Southern Florida Project to repair the Everglades, while providing for the human needs of the region.

ACKNOWLEDGMENTS

The authors would like to express their indebtedness and substantial reliance on the various historians and scientific researchers of the Everglades. Without the diligent efforts of Blake (1980), Bottcher and Izuno (1992), Carter (1974), Cooper (1989, 1991), Dovell (1947), and the U.S. Army Corps of Engineers (Jacksonville District, 1988, 1990, 1992), this synthesis would not have been possible.

The authors would also like to acknowledge the contributions of those who helped review various drafts, including Tom MacVicar, Pete Rhoads, Jorge Marban, Steve Davis, Joe Schweigart, and Lance Gunderson. Special thanks to Winnie Park, who worked tirelessly to prepare the graphics.

LITERATURE CITED

Allison, R. V. et al. 1948. Soils, Geology and Water Control in the Everglades Region, Agriculture Experiment Station, University of Florida, Gainesville, 168 pp.

Beard, D. B. 1938. Wildlife Reconnaisance: Everglades National Park Project, National Park Service, U.S. Department of the Interior.

Bellamy, J. 1992. Personal communications regarding history of Flood Control District, June 1992.

Blake, N. M. 1980. *Land into Water—Water into Land. A History of Water Management in Florida,* University Presses of Florida, Tallahassee, 344 pp.

Bottcher, A. B. and F. T. Izuno. 1992. *Everglades Agricultural Area: Water, Soil, Crop and Environmental Management,* University Presses of Florida, Gainesville, 479 pp.

Brown, A. H. 1948. Haunting heart of the Everglades. *National Geographic,* February:145–173.

Carter, L. J. 1974. *The Florida Experience: Land and Water Policy in a Growth State,* Johns Hopkins University Press, Baltimore.

Clarke, M. J. 1977. *An Economic and Environmental Assessment of the Florida Everglades Sugarcane Industry,* Johns Hopkins University, Baltimore, 140 pp.

Cooper, R. M. 1989. An Atlas of the Everglades Agricultural Area Surface Water Management Basins, South Florida Water Management District, September 1989, 85 pp.

Cooper, R. M. and J. Roy. 1991. An Atlas of Surface Water Management Basins in the Everglades: The Water Conservation Areas and Everglades National Park, South Florida Water Management District, West Palm Beach, 88 pp.

Dietrich, T. S. 1978. The Urbanization of Florida's Population: An Historical Perspective of County Growth 1830–1970, Bureau of Economic and Business Research, University of Florida, Gainesville.

Dovell, J. E. 1947. A History of the Everglades of Florida, Ph.D. dissertation, University of North Carolina, Chapel Hill.

Ehrlich, P. R. and A. H. Ehrlich. 1990. *The Population Explosion,* Simon and Schuster, New York, 320 pp.

Fernald, E. A. and D. J. Patton. 1984. Water Resources: Atlas of Florida, Florida State University, Tallahassee, 291 pp.

Finley, M. V. 1987. Superintendent of Everglades National Park, Statement before the South Florida Water Management District Board, September 10, 1987, 12 pp.

Finley, M. V. 1988. Superintendent of Everglades National Park, Statement before the South Florida Water Management District Board, June 16, 1988, 4 pp.

Gunderson, L. 1992. Spatial and Temporal Dynamics in the Everglades Ecosystem with Implications for Water Deliveries to Everglades National Park, Ph.D. dissertation, University of Florida, Gainesville, 241 pp.

Johnson, L. 1974. *Beyond the Fourth Generation,* University of Florida Press, Gainesville, 78 pp.

MacVicar, T. 1993. Personal communication based on presentation at Technical Information Series, South Florida Water Management District, May 1993.

Maltby, E. and P. J. Dugan. 1994. Wetland ecosystem protection, management, and restoration: An international perspective. in *Everglades: The Ecosystem and Its Restoration,* S. M. Davis and J. C. Ogden (Eds.), St. Lucie Press, Delray Beach, Fla., chap. 3.

Marth D. and M. J. Marth (Eds.), 1993. *Florida Almanac. 1992–1993,* Pelican Publishing Company, Gretna, Fla.

National Audubon Society. 1992. Report of the Advisory Panel on the Everglades and Endangered Species, Audubon Conservation Report No. 8, National Audubon Society, New York, 44 pp.

Neidrauer, C. J. and R. M. Cooper. 1988. A Two-Year Field Test of the Rainfall Plan, Tech. Publ. 88-2, South Florida Water Management District, West Palm Beach.

Rhoads, P. 1993. Personal communications based on presentation at Technical Information Series, South Florida Water Management District, May 1993.

Secretary of the Army. 1949. Comprehensive Report on Central and Southern Florida for Flood Control and Other Purposes, House Document 643, U.S. Government Printing Office, 60 pp.

Smith, B. 1848. Reconnaisance of the Everglades. Report to Secretary of the United States Treasury, Senate Report 242, 30th Congress, First session, August 12, 1848.

South Florida Water Management District. 1990. The Taylor Slough Rainfall Plan, South Florida Water Management District, West Palm Beach, January 1990, 17 pp.

South Florida Water Management District. 1993. Working Document in Support of the Lower East Coast Regional Water Supply Plan, South Florida Water Management District, West Palm Beach, March 1993.

U.S. Army Corps of Engineers (Jacksonville District). 1988. Fact Sheet: Canal 111 General Design Memorandum, Jacksonville, Fla., July 1988, 2 pp.

U.S. Army Corps of Engineers (Jacksonville District). 1990. General Design Memorandum: Modified Water Deliveries to Everglades National Park, Florida. Parts I and II, Jacksonville, Fla., July 1990.

U.S. Army Corps of Engineers (Jacksonville District). 1992. Water Control Plan for Water Conservation Areas—Everglades National Park and ENP–South Dade Conveyance System Central and Southern Florida Project, October 1992, 15 pp.

U.S. Army Corps of Engineers (Jacksonville District). Undated. Background Paper: Frog Pond Agricultural Area, Jacksonville, Fla., 8 pp.

U.S. Congress. 1911. The Everglades of Florida: Acts, Reports, and Other Papers Relating to the Everglades of the State of Florida and Their Reclamation, Senate Document No. 89, U.S. Government Printing Office, Washington, D.C.

Wilkinson, A. 1989. *Big Sugar,* Alfred A. Knopf, New York, 263 pp.

Wright, J. O. 1912. *The Florida Everglades,* Tallahassee, Fla., 83 pp.

5

EVERGLADES AGRICULTURE: PAST, PRESENT, and FUTURE

G. H. Snyder

J. M. Davidson

ABSTRACT

Agriculture began in the Everglades, south of Lake Okeechobee, after the drainage projects of the 1906–27 era and intensified after the water control projects of the early 1950s, which created the Everglades Agricultural Area. Today in south Florida, more than $750 million is earned annually from production of sugarcane, vegetables, sod, and rice and over 20,000 full-time equivalent jobs are provided. The future of agriculture in the Everglades is, however, uncertain due to possible changes in federal farm programs, the loss of organic soils as a result of drainage, and concerns about nutrients in drainage water from the Everglades Agricultural Area. These latter concerns may be significantly alleviated by the development of an agriculture more compatible with a periodic wet season, high water tables, and flooding.

The PAST

For most of Florida's history, the region has possessed vast land resources, but has been sparsely populated and economically impoverished (Johnson, 1974). When Florida became a state in 1845, it only contained about 50,000 persons, nearly half of whom were slaves who, of course, were unable to participate fully in the economic development of the state. Ninety percent of the population lived north of Gainesville and almost no one lived in the interior of the peninsula. The state had little commerce and almost entirely lacked any kind of industry, being largely dependent on a scattered agricultural economy. The 1920 census still counted less than 1 million Floridians, at a time when the population of the United States as a whole exceeded 100 million (Stockbridge and Perry, 1926).

Encouraging Settlement

The state's only ready resource was land, and land could be used to entice new settlers in the hopes of developing a viable economic base for the state. However, because much of the land in the nearly uninhabited and unknown interior of the peninsula was seasonally flooded, it was only natural that the drainage of these areas became a priority for the state. Within 10 months after statehood, the legislature petitioned the state's congressional delegation to have the federal government study and survey south Florida and prepare a report on its value and the feasibility of reclamation (reclamation being synonymous with drainage). Senator James D. Westcott, Jr. became the champion of Everglades drainage, and it was upon his urging that T. Buckingham Smith was appointed by the federal Secretary of the Treasury, Robert J. Walker, to "procure authentic information in relation to what are generally called the 'Ever Glades' on the peninsula of Florida" (Dovell, 1947).

A year later, Smith submitted his report (Dovell, 1947). To reclaim the Everglades, he said it would be necessary to lower the level of Lake Okeechobee by draining it through the Caloosahatchee River to the west and the Loxahatchee or St. Lucie rivers to the east. Additionally, he suggested that several drains through the Everglades would be needed to carry surplus waters to tidewater. As it turned out, this was essentially the means used to drain the Everglades.

Following reclamation, Smith predicted that the land would be profitable for the production of coffee, sugar, cotton, rice, tobacco, sisal hemp, citrus, bananas, figs, olives, pineapples, coconuts, and other tropical crops and fruits. He concluded his report by stating:

> The Ever Glades are now suitable only for the haunt of noxious vermin, or the resort of pestilent reptiles. The statesman whose exertions shall cause the millions of acres they contain, now worse than worthless, to teem with the products of agricultural industry; that man who thus adds to the resources of his country...will merit a high place in public favor, not only with his own generation, but with posterity. He will have created a state!

T. Buckingham Smith did not live to meet many of the present generation who, having never seen the Everglades that Smith knew and having never been required to earn a living off the land, now decry Smith's vision of the future.

Following Smith's report, the federal government passed the Swamp Lands Act of 1850, which granted swamp and overflowed lands to the state, after surveying (Johnson, 1974). The act specifically required that proceeds of sales of these lands be applied "as far as necessary" to the reclamation of the lands by levees and drains (Douglas, 1947; Elvove, 1943; Johnson, 1974). The state of Florida gained title to over 20 million acres (8,100,000 ha) of land as a result of the act. Early on, most of the acquired lands were in the northern part of the state, because Indians inhabiting the southern part made surveying difficult until 1858. During the Civil War and the desolate days of the Reconstruction period that followed, the Everglades lay largely forgotten (Blassingame, 1974). The state had no money for drainage, and there was little settlement or agriculture in the Everglades.

Drainage

In about 1890, a wealthy railroad tycoon named Henry B. Plant considered building a railroad from Tampa to Miami, which would require crossing the Everglades. He sent J. E. Ingraham with a surveying party to investigate the route. On March 20, 1892, the party reached the edge of the sawgrass (*Cladium jamaicense* Crantz) glades east of Fort Myers. There, an Indian woman told them that an Indian could reach Miami in 4 more days, but that a white man might take 10. The surveying party reached Miami after an arduous journey of 18 days. Following this adventure, one of the surveyors, Alonzo Church, wrote:

> My advice is to let every discontented man make a trip through the Everglades; if it don't kill, it will certainly cure him...A day's journey in slimy, decaying vegetable matter which coats and permeates everything it touches, and no water with which to wash it off will be good for him, but his chief medicine will be his morning toilet. He must rise with the sun when the grass and leaves are wet with dew and put on his shrinking body clothes heavy and wet with slime, and scrape out of each shoe a cupful of black and odorous mud; it's enough to make a man swear to be contented forever afterwards with a board for a bed and a clean shirt once a week (Blassingame, 1974).

Today, in a few hours we travel in air-conditioned cars on a ribbon of asphalt across the Everglades between Tampa and Miami, longing for the Everglades that Church knew, but probably with no intention of enduring the suffering he knew.

In spite of Church's firsthand opinion of the Everglades, the Ingraham expedition, and its 1883 predecessor sponsored by the *New Orleans Times-Democrat* (which crossed the Everglades from Lake Okeechobee to the Shark River in 26 days), once more got people talking about the region. The inauguration of William D. Bloxham as governor for his second term in 1887 marked the beginning of positive state interest in the reclamation of swamp and overflowed lands for the purpose of securing settlement of the lands and to bring immigration to the state (Dovell, 1947). Because the state lacked funds for reclamation of the Everglades, it always tried to use land grants to entice the private sector to do the reclamation. In 1898 the state, through the Internal Improvement Fund, entered into an agreement with the Florida East Coast Drainage and Sugar Company for lands south and east of Lake Okeechobee on the basis of 20,000 acres (8094 ha) of land for each 200,000 yd^3 (152,920 m^3) of excavation for drainage purposes, plus 25 cents per acre. The terms of the contract were never met (Hanna and Hanna, 1948), although enough was accomplished by 1904 to impress Charles C. Elliott, who was Engineer in Charge of Irrigation and Drainage Investigations of the Office of Experiment Stations of the U.S. Department of Agriculture (Dovell, 1947). He concluded that with sufficient drainage, the Everglades could be profitably utilized for the growth of subtropical fruits and that in view of the interest taken in growing fruits and vegetables for the northern winter markets, Everglades reclamation warranted further attention. William S. Jennings followed William Bloxham as governor. In his 1903 speech to the state legislature, he dwelled at length on the Everglades, emphasizing that the state-owned lands were not salable because they had not been drained and that they could not be drained because they could not be sold. He prepared a comprehensive plan for draining the Everglades, but his second term ran out before the plan was enacted. Nevertheless, he believed that the purpose of the 1850 Land Grant Act of Congress was to have wetlands drained and reclaimed, and he had acted upon that policy.

In 1904, Napoleon Bonaparte Broward ran for governor on the pledge to do everything in his power to reclaim the Everglades (Dovell, 1947). He won. To finance his drainage plan, he got the legislature in 1905 to create a Board of Drainage Commissioners that could levy yearly drainage taxes in drainage districts of their own creation. The board created the Everglades Drainage District, 60 mi (97 km) wide and 150 mi (241 km) long, which encompassed all of the Everglades, adjoining prairie, and adjacent timber lands. In time, seven subdrainage districts embracing an area of 95,400 acres (38,608 ha) were created along the southern lakeshore between Moore Haven and Canal Point as integral units of the Everglades Drainage District (Bestor, 1943). With funding seemingly secure because of the legal right to tax property owners in the affected region, drainage of the Everglades began in earnest. In 1906, dredging began on what would become the North New River Canal. However, money ran short, in part because of suits by affected landowners who contested being taxed (Johnson, 1974) and because the only additional source of funds for drainage was the sale of large blocks of state-owned swamp and overflow lands on or near Lake Okeechobee (Hanna and Hanna, 1948).

Governor Broward and former governor Jennings encouraged Richard J. Bolles, who had made a fortune selling farm tracts in Oregon, to buy into the Everglades. In 1908, Bolles bought 500,000 acres (202,350 ha) of overflowed state land for two dollars an acre, to be paid over an 8-year period; half of the payment was to be devoted by the state "solely and exclusively for drainage and reclamation purposes" (Hanna and Hanna, 1948). He planned to profit from his investment by selling farm tracts, as he had done in Oregon. His salesmen traveled the rural communities of the midwest touting the Everglades as the "tropical paradise," the "promised land," the "land of destiny," and the "Magnet whose climate and agriculture would bring the human flood." He also employed various sales companies to sell his land. By 1911, there were 50 such agencies in Chicago alone. Some land companies in Washington, D.C. maintained a storefront vegetable patch utilizing muck brought from the Everglades. Bolles himself built a hotel in present-day Lake Harbor, where the Miami Canal enters Lake Okeechobee, to accommodate prospective land purchasers, and transported them to the hotel by boat from Fort Myers. Next to the hotel was a demonstration farm. In most places, however, the land was not drained by the time the settlers began arriving, and a flood of lawsuits eventually were filed against the land salesmen by people who felt they had been sold "land by the gallon."

Through land sales, improved tax collection, and the sale of bonds, the Everglades Drainage District acquired sufficient funds to continue the drainage projects. No general state revenues were used for Everglades drainage until after the creation of the Central and Southern Florida Flood Control District in 1949 (Johnson, 1974). By mid-1927, 433 mi (697 km) of canal, 54 mi (87 km) of levee, and 14 canal locks had been completed. The St. Lucie Canal was placed in operation at 70% of capacity. The outlet to the west, through the Caloosahatchee River, had been completed in 1883 by Hamilton Disston, a Philadelphia entrepreneur to whom the Internal Improvement Fund in 1881 had eagerly sold 4 million acres (1,618,800 ha) of state-owned land adjacent to Lake Okeechobee in order to bail the fund out of bankruptcy (Hanna and Hanna, 1948). He also dug 13 mi (21 km) of canal south from present-day Lake Harbor in an aborted effort to connect Lake Okeechobee with the Shark River (Will, 1977). Eventually, 8.5 mi (14 km) of this canal was incorporated into the Miami Canal. Disston died in 1896 and his heirs declined to continue developing his Florida holdings, which eventually were either forfeited for unpaid taxes or were liquidated at a fraction of the value they once commanded.

Early Everglades Agriculture

Over the years, the term *Everglades* has been used to designate various regions of the Florida peninsula, ranging from the Kissimmee River valley southward, including areas now considered to be east coast communities. As used henceforth in this text, *Everglades,* and in particular *Everglades agriculture,* refer only to the region of organic soils south of Lake Okeechobee in the present-day Everglades Agricultural Area (Figure 5.1), unless otherwise indicated.

As drainage of the Everglades began in earnest, agriculture began in the region. Of course, the Seminole Indians had carried on an agriculture of their own after being driven deep into the Everglades during the Seminole Wars. There are numerous reports about crop production in or near villages on hammocks in the Everglades (Dovell, 1947). Sugarcane was a common crop found in most gardens. The Indians made a flour from "coontie" (*Zamia* sp.) and were known to cultivate corn, bananas, tobacco, pumpkins, squash, and melons. However, in terms of the quantity of agricultural production, the excitement that was generated over farming in the Everglades, and the extent to which Everglades products reached the outside world, nothing in the way of agricultural production prior to the beginning of the 20th century drainage projects could compare with that which was to come.

Soils

In the early days, three types of organic soils were recognized in the Everglades (Baldwin and Hawker, 1915). The soils closest to Lake Okeechobee were known as "custard apple muck," because they were covered with a dense growth of custard apple (*Annona glabra* L.). Proceeding away from the lake, the next belt of soils was known as the "willow and elder" soils, because, again, willow (*Salix* spp.) and elderberry (*Sambucus canadensis* L.) were the dominant vegetation. The majority of the Everglades organic soils, however, were known as "sawgrass" peats or mucks, because sawgrass was the dominant vegetation.

The custard apple muck soils, now known as the Torry mucks (McCollum et al., 1976), constitute only about 3% of the Everglades north of the Broward County line. They are distinctively different from the willow and elder and sawgrass mucks in that the Torry soils contain 50% or more mineral matter by weight. The remainder of the Everglades organic soils contain 15% or less mineral matter. The Torry mucks have considerably more native fertility than the sawgrass mucks, which formed under oligotrophic (low nutrient status) conditions. Most of the pioneers in the early part of this century settled on custard apple mucks along the shore of Lake Okeechobee.

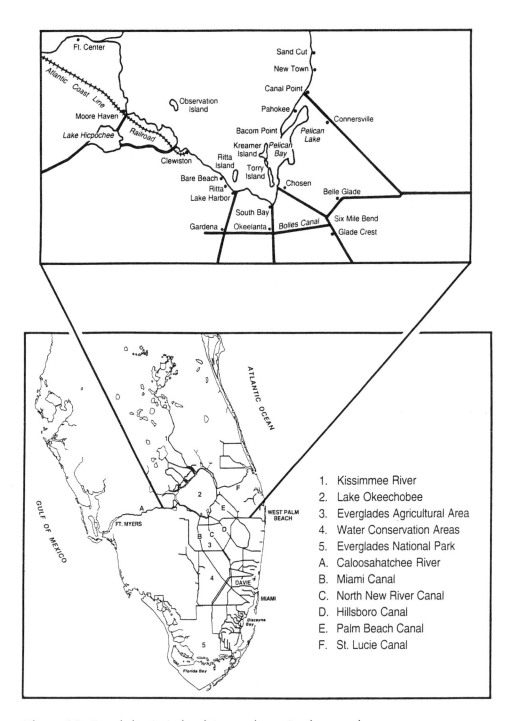

Figure 5.1   Everglades Agricultural Area and associated geography.

Early Settlements in the Sawgrass

In spite of the fact that the earliest reports on farming in the Everglades came from experience gained on the fertile Torry mucks, the first two major settlements in the upper Everglades were on the sawgrass mucks. The first (by a few months) was Okeelanta, which was located at the intersection of the Bolles and North New River canals. A larger settlement was located even deeper into the sawgrass Everglades, approximately 1.5 mi (2.4 km) south of "six-mile bend" (Figure 5.1). The first two settlers of this community, known as Glade Crest, were Canadians who arrived in the last months of 1913 (Will, 1984). By January 1914, Glade Crest had 20 inhabitants, and by the spring of 1915 there were 72, which probably represented the peak population. These settlers purchased 5- and 10-acre (2- and 4-ha) tracts sight unseen from Holland and Butterworth, two former land salesmen for R. J. Bolles who, after leaving Bolles, bought all of Section 20, Township 44, and Range 38 from the Southern States Land and Timber Company. They enticed people to the area by telling them that a family of four could make a living off of one acre, because four crops could be grown in a year, frost had never been known to damage the most tender vegetation, and there were even no mosquitoes. In the first winter of the settlement there were 15 frosts and freezes between December and April. The sawgrass had to be cleared by hand, and on a long, hard day, one man could clear a patch about 50 ft^2 (4.6 m^2). Worst of all, it was learned that nothing would grow on sawgrass land but sawgrass (Will, 1984). Most other plants turned yellow and died, a condition termed "reclaiming disease." Only after production of Irish potatoes did the land support the growth of some other crops. Eventually it was discovered that the sawgrass soils were too highly deficient in copper to support most crop production (Allison et al., 1927) and that the copper sulfate used to prevent potato blight provided sufficient copper to overcome the deficiency in crops that followed the potatoes. Over the years, it was found that many other fertilizer elements were required for crop production on the sawgrass muck, despite claims by the land companies that no fertilizer would ever be needed. Some vegetables were produced in Glade Crest, but by the fall of 1915 families began leaving. Following a flood in 1920, most remaining settlers left and everyone was gone by 1921. There was a little poultry farm near Glade Crest, but it too was abandoned after the hurricane and flood of 1926 (Will, 1984).

The town of Okeelanta, located on sawgrass land 4 mi (6.4 km) south of Lake Okeechobee, was formed by 4860 stockholders of the Okeelanta Corporation in the spring of 1913 (Hanna and Hanna, 1948). The first settlers, a party of four, were accompanied by Dr. Thomas E. Will, who became a prominent citizen of the Everglades and whose son, Lawrence, chronicled much of the early history of the area. Dr. Will had been ousted as president of the Kansas State Agricultural College because of his Populist political association (Douglas, 1947). He was not interested in land speculation, but rather in building a good place to live. The party arrived in October 1913 and began clearing the sawgrass to farm the land. By 1920, the community had a population of over 200. A town hall and school were built. Irish potatoes, beans, corn, tomatoes, eggplants, and other vegetables were grown and shipped down the North New River Canal to Fort Lauderdale. However, the farmers of Okeelanta had experiences similar to those of Glade Crest: frosts, infertile sawgrass muck, arduous land clearing by hand, floods, and muck fires. In the early

1920s, most of the settlers left Okeelanta for their former homes in the northeast and midwest, for Miami, or for lake region towns that faired better than those in the interior of the Everglades. Dr. Will continued his struggle to make Okeelanta and other settlements in the sawgrass regions successful, until his death in 1936. Just 6 months earlier he wrote:

> This country has a tremendous future once the cloud lifts…As to my holding on…with everything down here shot, I've never felt that I would be justified in running off and leaving everybody in the lurch. This has cost me a professional career, and every cent of such money I had; and has meant 27 years of hard work and fierce fighting; but if only we can get out, and I can say with a clear conscience, "The Glades is at least ready to occupy and use," I'll feel amply repaid (Dovell, 1947).

In fact, there still are no settlements of consequence in the sawgrass muck region.

Early Settlements on Lake Okeechobee

The farms and towns situated on the Torry muck bordering Lake Okeechobee and on islands in the southern part of the lake were more successful than those on the sawgrass lands. This probably was due to the more favorable nature of the soil, frost protection provided by the warm waters of the lake, the opportunity to augment incomes by fishing and by engaging in the commerce created by the activities associated with the arrival of new settlers, and perhaps because these settlements were not part of any highly promoted land sale schemes. Towns such as Belle Glade, Pahokee, Canal Point, Clewiston, Moore Haven, and South Bay survived and prospered. Surnames of the earliest settlers and farmers of the lake area (i.e., Beardsley, Boe, Boynton, Burke, Challancin, Erickson, Friend, Harrington, Hooker, Kirchman, Schlecter, Stein, Thomas, Wedgworth, Wilson, and Zywicki) still identify prominent citizens engaged in agriculture in the upper Everglades.

Although there were no real settlements along the south rim of the lake during the first decade of this century, by 1921 there were 16 settlements on or near Lake Okeechobee, with a total estimated population of 2000 (Dovell, 1947). Two railroads, one from the east and one from the west, reached Lake Okeechobee by 1918, and the two lines met at Lake Harbor in 1928.

Corporate Farming

Land promoters in the early part of the century sold farms in small parcels, often 5 and 10 acres (2 and 4 ha) in size. By 1910, the Bolles Company had sold 10,000 10-acre (4-ha) farms, the Everglades Land Company had sold 2000 small farms, and the Everglades Plantation Company and the Everglades Land Sales Company each sold about 1000 farms (Dovell, 1947). Most were sold to speculators who hoped to resell them at a profit "when they come out from under the water." However, a private report, commissioned by the Everglades Land Sales Company in 1912, held that agricultural development in 10-acre tracts was impractical and that the bulk of the Everglades should be developed in large tracts for orchards, sugarcane, and other produce for which the lands might be suitable. The Everglades Land Sales Company chose not to make this report public.

Sugarcane was one of the first crops to be grown in large acreages in the Everglades. It had been grown by the Seminole Indians, and many settlers had small plots of sugarcane for their own use. Interest in larger scale production intensified in the last years of World War I (Dovell, 1947). In 1917, the Southern States Land and Timber Company had a 10-acre (4-ha) experimental planting of sugarcane varieties near Canal Point (Will, 1977). In 1919, 60 acres (24 ha) of sugarcane was grown in Pelican Bay, along the lakeshore, and the crop was sent to a syrup mill in Jacksonville. Also in 1919, the Pennsylvania Sugar Company began a 1000-acre (405-ha) plantation 16 mi (26 km) northwest of Miami on the Miami Canal and later erected a sugar mill. Due to soil deficiencies and poor water control, however, the operation was ended. In 1920, the U.S. Department of Agriculture established a cane breeding station near Canal Point because the area is relatively free from frosts. The station still produces sugarcane varieties for Florida, Texas, and Louisiana. In 1920, 125 acres (51 ha) of sugarcane was planted on the southeast lakeshore under the supervision of Miami Judge John C. Gramling (Dovell, 1947) for the Moore Haven Sugar Corporation, which established the first sugar mill in the upper Everglades (Will, 1977). In 1922, the mill produced some brown (raw) sugar, but high expenses and poor water control in the cane fields bankrupted the company. Frank G. Bryant, who was instrumental in founding the town of Lake Worth, organized the Florida Sugar and Food Products Company in about 1920, along with G. T. Anderson. At a point 2 mi (3 km) down the Palm Beach Canal from Canal Point, they erected a 400-ton (363-tonne) sugar mill and produced their first brown sugar in 1923. At that time, there was about 900 acres (364 ha) of sugarcane in the Everglades, 800 acres (324 ha) of which was near Canal Point (Will, 1977). Yields were very good, but between start-up costs and floods in 1922 and 1924, the company ran into debt and the operation was virtually abandoned by 1925. B. G. Dahlberg, head of the Celotex Company of Chicago, was interested in using sugarcane bagasse for manufacturing wallboard (Dovell, 1947). He purchased outright or on option approximately 100,000 acres (40,470 ha) along the lakeshore, combining that holding with the Bryant operation, and the sugar mill built by the Pennsylvania Sugar Company, to form the Southern Sugar Company, headquartered in Clewiston. The Southern Sugar Company helped create a residential and resort city in Clewiston and moved the Pennsylvania Sugar Company sugar mill from its location in Pennsuco (north of Hialeah) to Clewiston. Cane grinding began at the Clewiston mill in 1929, and the mill was expanded in 1930. However, in the same year the company was thrown into receivership because of the collapse of the stock market (the company had planned to raise capital by selling stock), a decline in sugar prices, and crop losses due to high water. In 1931, Charles Stewart Mott, a former vice-president of General Motors and the largest stockholder in the Southern Sugar Company, along with Clarence Bitting, bought the property at a court insolvency sale and reorganized it into the United States Sugar Corporation, which is still in operation today.

Another early large-scale farming activity in the Everglades was initiated in 1924 by the New England-based Brown Paper Company (Will, 1984). Their Brown Farm, located on 72,000 acres (29,138 ha) 14 mi (22.5 km) down the Hillsboro Canal from "six-mile bend," was started primarily for the production of peanuts, because the company had learned that hydrogen released during the manufacture of paper produced a superior cooking fat when injected into peanut oil. This company used

crawler-tracked tractors and wagons, built its own planting and cultivating machines especially for use on organic soils, and employed a chemist, Dr. H. P.
Vannah, who helped solve some of the plant nutrition problems encountered when
farming on sawgrass muck. The company developed Shawano Village, complete
with a waterworks, an electric power plant, and recreation for its employees. In
addition to peanuts, miscellaneous other vegetables, potatoes, and sugarcane were
produced. However, the Brown Paper Company suffered greatly during the
depression, and the Brown Farm was disbanded in 1931. It is reported (Will, 1984),
nevertheless, that the experiments and activities of the Brown Farm had a very
positive impact on Everglades agriculture and on economic development in the
region.

Livestock

Many people interested in the development of the Everglades were of the
opinion that the raising and grazing of livestock had the most favorable possibility for the area (Baldwin and Hawker, 1915). The first experiences with cattle
grazing on the sawgrass muck lands, however, were discouraging. A fine Jersey
heifer that was ferried to Okeelanta in 1917 and grazed on native grasses eventually
weakened and died (Dovell, 1947). Other settlers who brought family cows into
the Everglades found that they produced milk for only a short time and failed to
reproduce (Kidder, 1979). In a 1924 study, malnutrition was observed in many
cattle whose food consisted solely of grazing crops, and sterility was found in 60%
of the cows around the south and southeast shores of Lake Okeechobee (Tedder,
1925). If a calf was born, it generally failed to survive for a year (Figure 5.2). Dr.
A. L. Shealy came to the Everglades from the University of Florida in Gainesville
in 1925 to study the livestock situation, but did not find an answer to the problems
(Kidder, 1979). Thus, there was little successful livestock production in the early
days of Everglades agriculture.

Suitcase Farmers

In the 1920s, the remainder of the agriculture in the Everglades was largely in
the hands of "suitcase farmers," who arrived in the fall, leased land, and lived out
of their suitcases until the crop was harvested and sold in 60 days or so (Dovell,
1947). Their goal was to market vegetable crops in the height of winter when other
regions in the continental United States were not producing. There were, of course,
some resident farmers living in lakeshore communities, and some suitcase farmers
did well and settled in the area. Even in the 1930s, it was estimated that 75% of
the farms in the upper Everglades were operated on a tenant basis (Stephan, 1944).

The Everglades Experiment Station

In Governor Cary A. Hardee's first address to the legislature, he (as several of
his predecessors had done) made a strong plea for the establishment of an
agricultural experiment station in the Everglades (Dovell, 1947). He pointed out
that the entire justification for draining the Everglades rested on the assumption that
when drained, the soil would be available for agriculture and that the problem of

Figure 5.2 Copper deficiency resulted when cattle grazed on grasses grown on sawgrass muck. (Photo credit: R. W. Kidder.)

reclamation would not be solved until agriculture was placed on a sound basis. In 1921, the Florida legislature created an agricultural experiment station under the direction of the State Boards of Control and Education (Newell, 1921). By 1923, one section of land along the Hillsboro Canal east of Belle Glade was drained, two buildings were constructed, and the facility was turned over to the State Board of Education. Dr. R. V. Allison was appointed as the first resident director and soils specialist in 1926. The experiment station, now known as the Everglades Research and Education Center, is a part of the University of Florida Institute of Food and Agricultural Sciences (IFAS).

The Hoover Dike

The hurricane of 1926 and especially the hurricane of 1928, in which over 2000 persons died in the upper Everglades (Figure 5.3), led to the creation of the Hoover Dike around the south shore of Lake Okeechobee. Because the federal government did not want to set a precedent by approving a project that was primarily for flood control, the dike project eventually was approved and signed by President Hoover in 1930 as a navigation project (Johnson, 1974). In the process of digging a navigable channel around the southern perimeter of Lake Okeechobee, the spoil from the channel became the Hoover Dike. With the completion of the Hoover Dike, lakeshore residents had assurance that the lake could be kept within bounds. After nearly 75 years of sporadic efforts and failures, intensive settlement of the Everglades finally got underway (Elvove, 1943).

Everglades Agriculture following the 1928 Hurricane

Through a combination of somewhat improved water control, individual and corporate enterprise, and government-sponsored agricultural research and extension by the U.S. Department of Agriculture/Agricultural Research Service Canal Point Field Station and the Everglades Experiment Station in Belle Glade, agricultural production advanced on several fronts during the 1930s and 1940s.

Figure 5.3 Bodies discovered several weeks after the 1928 hurricane, badly decomposed and unidentifiable, were burned in heaps containing several dozen or more persons. (Photo credit: R. V. Allison.)

Sugarcane

Sugarcane production in Florida was limited by the Jones–Costigan Act of 1934 and other federal legislation that followed, because marketing quotas were imposed based on prior production and Florida production was small prior to 1934 (Sitterson, 1953). In the 1929–30 harvest season, about 7000 acres (2833 ha) of sugarcane was harvested in Florida, part of which was in the Fellsmere area 80 mi (129 km) north of Belle Glade. By the 1933–34 season, harvested acreage more than doubled to 14,800 acres (5990 ha). Under the various sugar acts, however, acreage increased to only 21,000 (8499 ha) during the 1939–40 season and to 37,800 (15,298 ha) during the 1949–50 season. Mr. Clarence R. Bitting, president of U.S. Sugar, spent many years trying unsuccessfully to get a larger quota for Florida. Finally, because of an impending shortage of sugar during the Second World War, sugar quotas were suspended in 1942, but during the war years Florida producers were not able to increase production because of shortages of labor and capital facilities (Salley, 1986). Florida had only two sugar mills prior to the war, the U.S. Sugar mill in Clewiston and a small sugar mill in Fellsmere, built in the early 1930s by men associated with the Punta Alegre Sugar Company in Cuba and operated by the Fellsmere Sugar Producers Association. During the Second World War, a sugar mill which had originally been located at Vieques Island just east of Puerto Rico was dismantled and reconstructed at Okeelanta and operated as the Okeelanta Growers and Processors Cooperative (Hanna and Hanna, 1948). The men behind the operation were Puerto Rican and had considerable experience in the sugar business. The first crop of sugar was processed in the Okeelanta mill in 1947.

Vegetables

The most active segment of agriculture in the Everglades during the 1930s and 1940s in terms of total acreage, number of farmers, and dollar value was the vegetable industry. In 1929, there were approximately 17,000 acres (6880 ha) of vegetables in the upper Everglades (Dovell, 1947), whereas the reported harvested acreage of sugarcane in the 1929–30 season was only 7000 (2833 ha) (Salley, 1986). In 1943, it was estimated that of 110,000 acres (44,517 ha) in production in the upper Everglades, 75,000 acres (30,353 ha) was in vegetables, 30,000 (12,141 ha) acres was in sugarcane, and the remaining 5000 acres (2024 ha) was in pasture and miscellaneous minor crops (Elvove, 1943). Vegetable crops were grown primarily to supply northern markets during the winter. Some of the principal crops were snap beans, potatoes, celery, cabbage, English peas, and tomatoes (Stephan, 1944). Lesser amounts of lima beans, escarole, peppers, eggplant, and onions also were grown. Unfortunately, vegetable farming was always risky and markets unpredictable. Mature crops sometimes had to be left to rot in the field because the market value did not cover harvesting costs. Vegetable crops are easily damaged by excess water, by frosts, and by insect and disease pests that thrive in subtropical Florida. Never knowing the price they would receive at harvest, or whether there would be any harvest at all, Everglades vegetable growers truly were gamblers. Some made a fortune one year, only to lose it, and all they could borrow, the next. For this reason, it was difficult for small farmers to survive. Nevertheless, some survived and prospered in almost Horatio Alger fashion (Hanna and Hanna, 1948). Farms became larger and more diversified in order to stabilize financial returns (Stephan, 1944). Inclusion of such commodities as sugarcane and cattle, which were less subject to production failures and to sudden price depressions, helped offset the hazards of vegetable production.

Livestock

Through studies conducted at the University of Florida Everglades Experiment Station during the years following the 1928 hurricane, it was found that cattle grazing on grasses in the sawgrass muck suffered from copper deficiency, in part because there was an abnormally high level of molybdenum in the forage. When copper was added to the animals' diet, it acted to alleviate molybdenum toxicity (Kidder, 1979). However, even after the copper/molybdenum problem was corrected, many problems still plagued cattle production in the Everglades: frozen winter pastures, foot rot, anaplasmosis, horn flies, horseflies, deerflies, and mosquitoes were so plentiful that they sometimes killed cattle by suffocation. The situation was so discouraging that around 1935, Dr. H. H. Hume, Associate Director of the Florida Agricultural Experiment Stations, suggested that research on cattle production should be discontinued and the area be declared unfit for livestock (Kidder, 1979).

Figure 5.4 The Everglades Experiment Station was flooded for months following the 1928 hurricane. (Photo credit: R. V. Allison.)

The PRESENT

The foundation for present-day agricultural production in the Everglades was laid in the late 1940s and early 1950s, when the State of Florida and the federal government joined forces in a comprehensive flood control project that addressed water control problems encountered by both farmers in the Everglades and coastal urban populations. Up to this time, only drainage had been stressed in the Everglades. However, drainage often had been inadequate during very wet periods (Figure 5.4) and excessive during times of drought (Figure 5.5). Furthermore, there were conflicts of interest between farmers in the Everglades and farmers and residents of coastal areas. For example, during excessively wet periods when water was drained from the Everglades via the various major canals, downstream farmers and residents could be flooded by the outpouring of water from the Everglades. In one instance in the early 1930s, Broward County coastal interests built a dam on the North New River Canal near Andy Town to reduce the flow of water down the canal. This provided dramatic flood relief for Davie and Fort Lauderdale, but upon discovery it was blasted away by Everglades farmers (Forman, 1979). The coastal residents built another dam a little farther down the canal and guarded it with rifles and shotguns. A similar incident involving dynamite and armed guards occurred more than a decade later (Johnson, 1974).

The Everglades Agricultural Area

Although many people were aware of the inadequacies of the pre-1950s flood control system, it was the 1947 hurricanes that rallied forces to rectify the situation. The year 1947 was unusually rainy even before the hurricanes. The upper Everglades recorded over 50 in. (127 cm) of rainfall between June 1 and October 20, 1947, and the Fort Lauderdale area had over 70 in. (178 cm), including rain from hurricanes on September 17 and October 11 (Johnson, 1974). Property damage in the Everglades and east coast areas as a result of flooding was estimated to be $50 million. This flooding came only a few years after the 1944–45 drought, which had been the worst ever recorded in the Everglades (Johnson, 1974). Everglades

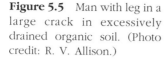

Figure 5.5 Man with leg in a large crack in excessively drained organic soil. (Photo credit: R. V. Allison.)

vegetation and soil burned for months (Figure 5.6), spreading acrid smoke over coastal communities. Saltwater intruded landward into freshwater aquifers. These incidents, following shortly on the heels of one another, drove home the point that a water control system was needed to buffer the cycles of drought and flood that occur in south Florida and to accommodate the sometimes conflicting interests of Everglades and coastal residents and farmers.

Based on an extensive survey of the Everglades under the supervision of Mr. Lewis Jones of the U.S. Department of Agriculture Soil Conservation Service (Jones, 1948), it was decided that only in the upper Everglades were the organic soils sufficiently deep and of the proper physical composition to warrant development for farming. This area was designated as the Everglades Agricultural Area. Organic-soil-dominated areas to the east and south of the Everglades Agricultural Area were surrounded by dikes and designated as Water Conservation Areas. A dike was constructed along the eastern edge of the Everglades to protect coastal communities from floodwaters. Spillways were constructed where major canals entered the Intracoastal Waterway. A system of canals, dikes, and pumping stations was installed such that in periods of dry weather, water could be drawn from Lake Okeechobee and distributed throughout the Everglades Agricultural Area, transported to the east coast to refill aquifers and retard saltwater intrusion, and transported to Everglades National Park at the tip of the peninsula. Growers in the Everglades Agricultural Area developed their own on-farm canal systems which were tied into the major canals, so they could maintain desired water tables

Figure 5.6 Muck-soil fires were common during dry seasons, when water control emphasized drainage but made no provision for irrigation. (Photo credit: R. V. Allison.)

within their farms. During wet periods, Everglades growers could drain water from their farms into the major canals, and that water could be moved into Lake Okeechobee or the Water Conservation Areas for storage. The Central and Southern Flood Control District (now the South Florida Water Management District) was created by the Florida legislature in June 1949 to manage and operate the system. They have done a good job of managing water on a quantity basis, minimizing damage due to flooding and reducing the effects of droughts. Certainly, the reduction in water-related risks associated with farming in the Everglades Agricultural Area has contributed to the rapid increase in farmed acreage in the latter decades of the 20th century.

Livestock

Cattle production increased considerably during the 1950s, in response to improved drainage, problem solving by the Everglades Experiment Station, and the establishment of the Glades Livestock Market (Kidder, 1979). In spite of some derision, the Experiment Station clearly demonstrated that St. Augustine grass (*Stenotaphrum secundatum* [Walt.] Kuntze), hitherto known only for its use in lawns, was the best forage for the Everglades. Live-weight cattle gains exceeded 1800 lb·acre^{-1}·yr^{-1} (1606 kg·ha^{-1}·yr^{-1}) in controlled tests on St. Augustine grass (Kidder, 1952), and 900–1000 lb of gain per acre annually (803–892 kg·ha^{-1}·yr^{-1})

have been routinely recorded (Chapman et al., 1963). In time, most cattlemen turned to a cow-calf operation and found that St. Augustine grass pasture in the Everglades would sustain one cow-calf unit per acre (0.4 ha). Dairies and confined feeding of beef cattle, on the other hand, never prospered in the Everglades because the soft organic soil could not stand up under the trampling of confined herds and the traffic generated around milking barns (Kidder, 1979).

Sugarcane

Following the gains of the 1950s, cattle production declined steadily in the 1960s as pastures were plowed to make way for sugarcane following Castro's revolution in Cuba. When Castro rose to power in 1959, only 47,000 acres (19,020 ha) of sugarcane was harvested in Florida, and part of that was in the Fellsmere area. By the 1962–63 milling season, almost 138,000 acres (55,849 ha) was harvested, and over 228,000 acres (92,272 ha) was harvested during the 1964–65 season, the last year of production in Fellsmere. Harvested cane acreage exceeded 300,000 acres (121,410 ha) in the 1978–79 season and 400,000 acres (161,880 ha) in 1986–87 (Salley, 1986).

Florida had only three sugar mills at the time of the Cuban revolution: U.S. Sugar's mill in Clewiston and mills in Okeelanta and Fellsmere which were owned by the Okeelanta Sugar Refinery, Inc., a corporation organized by the operator of five sugar mills in Cuba (Salley, 1986). In 1964, the Okeelanta corporation sold out to the South Puerto Rico Sugar Company, which shortly thereafter was bought by Gulf and Western Industries, Inc. In 1984, the "Okeelanta" mill, along with 90,000 acres (36,423 ha) in the Everglades and a mill and 240,000 acres (97,128 ha) in the Dominican Republic, was sold by Gulf and Western to the Fanjul family in Palm Beach, a family with generations of experience in the sugar industry (Berman and Khalaf, 1990). The Fanjuls had previously brought a mill from Louisiana to a site east of Pahokee (Everglades News, 1960) and began producing sugar in 1961 as the Osceola Farms Company.

Four sugar mills began production in the Everglades in a single year—1962. U.S. Sugar opened up a second mill in Bryant, near Pahokee. The Glades County Sugar Growers Cooperative Association began sugar production at a mill in Moore Haven. This mill was purchased by Gulf and Western and closed in 1977. A mill began production near Belle Glade under the ownership of the Sugarcane Growers Cooperative of Florida. This mill has set several world records for daily sugar production. Talisman Sugar Corporation began production in 1962 in a mill along highway U.S. 27, in the southern portion of the Everglades Agricultural Area. Talisman bought, and then shut down in 1971, the Florida Sugar Corporation mill located approximately 10 mi (16 km) east of Belle Glade. That mill had been brought from Louisiana and began production in 1961.

The Atlantic Sugar Association, a farmers' cooperative mill located approximately 20 mi (32 km) east of Belle Glade, began milling operations in 1964. Thus, there have been as many as nine sugarcane mills in operation in the Everglades since the Cuban revolution, and cane is still being processed in seven of them: U.S. Sugar's mills in Clewiston and Bryant, the Okeelanta mill, Talisman, Osceola, the Belle Glade Cooperative, and the Atlantic Cooperative. Together, these mills ground over 13.5 million tons (12.3 million tonnes) of sugarcane in the 1988–89 season, producing over 1.5 million tons (1.36 million tonnes) of sugar and 92 million gallons (348 million L) of blackstrap molasses, thereby making Florida the highest cane-sugar-producing state in the Union (The Florida Sugar Cane League, Clewiston).

Most sugar produced in the Everglades is refined elsewhere. However, there are two refineries in south Florida (the Okeelanta Sugar Factory and Refinery and the Everglades Sugar Refinery, Inc.) which market a considerable portion of their output to local Florida markets (Polopolus and Alvarez, 1990).

In Florida, sugarcane generally is planted in the fall and winter (Coale, 1987). Furrows are created on a 5-ft (1.5-m) spacing, and fertilizer is placed in the furrows as indicated by soil tests (Sanchez, 1990). Cane stalks are manually laid in the furrows and cut into pieces containing three to six nodes. If needed, insecticide is placed in the furrows before they are closed. The crop is subirrigated by maintaining a water table at a depth of approximately 20 in. (51 cm) through a system of field canals, water control structures, and pumps. Weeds are controlled by mechanical cultivation and/or with herbicides during the early growth of the cane. Some growth occurs during the winter, but growth is especially rapid from May through September (Kidder and Rice, 1979). In the late fall or winter, the cane is burned to facilitate harvesting and then cut either by hand or by machine. Hand harvesting is still generally preferred because low-sucrose-yielding tops and other trash can be discarded in the field and not transported to the mill. However, growers can harvest by machine (Figure 5.7), and several large companies mechanically harvest all of their cane. The harvested cane is transported to the mill

Figure 5.7 Mechanical harvesting of sugarcane in the Everglades Agricultural Area. (Photo credit: F. J. Coale.)

either by truck or by rail. New growth arises from the cane stubble, and this "ratoon" crop is harvested the following year. Yield generally declines with successive ratoon crops; thus, after two to four ratoons, the stubble is disked out, and the field is replanted the following fall. In some cases, "successive planting" is practiced in which the field is replanted immediately following the final harvest. Most sugarcane in Florida is produced by large companies, using mechanization wherever feasible. Sugarcane production in Florida is considered by many to be the most cost efficient in the world.

Vegetables

Vegetable production in the Everglades has increased steadily in terms of dollar value during the past two decades, although in terms of carloads, production is almost unchanged (Hutcheson, 1990). The value of vegetables produced in the Everglades was estimated to be $41 million in the 1961–62 season, $72 million in the 1971–72 season, and $151 million in the 1987–88 season, but the number of carloads has varied from 20,000 to 30,000 throughout the period. Estimates of acreage devoted to vegetables are somewhat difficult to express, because several successive crops of vegetables may be harvested off the same acre in a single season. It is estimated that 75,000 acres (30,353 ha) of vegetables is harvested annually in the Everglades (Schueneman, 1990).

In spite of the relative stability in the total quantity of vegetables produced in the Everglades during the past 20 years, the composition of the production has changed considerably. In the 1961–62 season, over 3700 carloads of beans was produced, but less than 100 carloads of beans was produced annually in the 1980s. Lettuce production, on the other hand, was less than 1000 carloads annually in the early 1960s, whereas it exceeded 5000 carloads in the 1987–88 season. In the 1980s, producers of sweet corn switched almost entirely to the shrunken-2 "super sweet" types developed by Professor Emil Wolf at the Everglades Research and Education Center. This has permitted them to more easily market their fresh corn during times of plentiful supply. In general, vegetable production in the Everglades today consists primarily of "leaf" crops such as lettuce, cabbage, endive, escarole, parsley, and celery, along with major quantities of sweet corn and radishes. Vegetable crops such as tomatoes, peppers, and beans, which were very important in the 1930s and 1940s, are considered minor crops in the Everglades today.

Because vegetables are grown for the winter markets, essentially no vegetables are produced in the Everglades during the summer months. Because several vegetable commodities are produced in the Everglades, a variety of production methods are employed. Fields generally are thoroughly prepared before planting to provide a level, weed-free seedbed. Most vegetable crops are direct seeded, except celery, which is grown in seedbeds and then transplanted into the field. Vegetables are subirrigated, although celery transplants are overhead irrigated once or twice after being placed in the field, and overhead irrigation is used to moisten dry soil before lettuce is seeded. Radishes require 30 days or less from planting to maturity, whereas celery requires approximately 90 days in the seedbed and another 90 days or more in the field. Crops such as radishes, carrots, and, in some cases, celery, are mechanically harvested, but hand harvesting is used for lettuce, sweet corn, and some celery. Because vegetable buyers generally require a cosmetically perfect product, fertilizers and pesticides are used to produce a product that the market will accept. Harvested vegetables are either packed in the field or are transported to packing houses where they are cleaned, graded, packed, and chilled. From there, they are shipped by truck throughout the eastern United States. One processing plant in South Bay produces shredded lettuce, sliced carrots, and other vegetables throughout the year for several fast-food chains and "black tie" restaurants.

Sod

The production of lawngrass sod has increased considerably in the upper Everglades since the hurricanes of 1947 (White and Busey, 1987). Prior to those hurricanes, sod production in south Florida was centered in the Davie area. However, the extensive and lengthy flooding that followed the hurricanes wiped out the Davie sod fields, forcing growers to take new planting stock from wherever they could find it. One source was St. Augustine grass pastures in the Everglades Agricultural Area, even though most of these pastures were in cv. 'Roselawn,' a pasture-type grass. After the Davie sod fields were replanted, ranchers in the Everglades Agricultural Area continued selling sod, although at a reduced price. Some ranchers began planting lawngrass-type St. Augustine grass, but the Everglades Agricultural Area had a reputation for poor-quality sod well into the 1960s. Today, however, most high-quality St. Augustine grass sod in south Florida is grown in the Everglades area, and the total value of this production exceeded $37 million in the 1987–88 season (Hutcheson, 1990).

Following land preparation, St. Augustine grass sod is established from sprigs (McCarty and Cisar, 1989). Sprinkler irrigation may be used for a short time after sprigging, but subirrigation is employed for most of the growth period. The grass is mowed regularly and receives periodic rolling and fertilization. Pests are chemically controlled as needed, but generally infestations are minor. Weed control often is the major concern during the grow-in period. Approximately one year after sprigging, the sod is ready for lifting. The sod is cut with oscillating knives that slice off an approximately 1-in. (2.5-cm) combined layer of crown, root, and soil, and the sod pieces are stacked on wooden pallets (Figure 5.8). The key factor for the producer is that the sod pieces hold together during the lifting and stacking operation, and that criterion generally governs the time when the sod is harvested. Sod pieces generally are 16 × 24 in. (41 × 61 cm) in size. The cutting machines leave parallel ribbons of uncut sod approximately 2 in. (5 cm) in width, which provide the starting material for the next crop.

Rice

Interest in rice production in the Everglades dates back to T. Buckingham Smith's 1848 report on the potential for agricultural production in the region (Dovell, 1947). Serious interest in rice production did not occur until more than a century later, in the 1950s. However, these attempts were plagued by diseases and other cultural problems and were totally abandoned when the rice virus disease

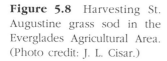

Figure 5.8 Harvesting St. Augustine grass sod in the Everglades Agricultural Area. (Photo credit: J. L. Cisar.)

Hoja blanca was discovered in the Everglades, because the disease was thought to be a threat to the major rice-producing areas of the United States (Green and Panzer, 1958). In fact, the disease eventually was detected in other states, resistant varieties of rice were quickly developed, and the disease has never caused major production losses in the United States. Interest in rice production was renewed in the late 1970s (Alvarez et al., 1989), and acreage has increased since then to about 15,000 acres (6070 ha) in 1989. Two mills for processing rice have been built in the Everglades Agricultural Area (Figure 5.9).

Most rice in the Everglades is grown in rotation with sugarcane or vegetables during the summer, when the land would otherwise be idle; therefore, rice complements, rather than competes with, traditional crops. Fields generally are carefully leveled before planting and then disked and rolled. The rice is seeded in rows 7 in. (18 cm) apart with a grain drill. When grown in rotation with adequately fertilized vegetables or sugarcane, rice needs no conventional fertilization (Snyder and Jones, 1989). However, iron is applied with the seed in certain iron-deficient soils (Snyder and Jones, 1988), and silicon from calcium silicate slag has been shown to substantially increase rice yields in sawgrass mucks distant from Lake Okeechobee (Snyder et al., 1986). When the seedlings are tall enough to survive flooding (generally 3–5 weeks after seeding), temporary pumps are used to move water into the field to provide a shallow flood, which is maintained until a few weeks before harvest (approximately 3 months later). Because rice is grown in the summer when rainfall generally is plentiful, water for the flood often comes from drainage of adjacent upland crops. Temporary dikes are formed around the field perimeters as needed, and rock roads often can serve to contain the flood. In some instances, chemical weed control is used prior to flooding, but flooding generally provides sufficient weed control. Stinkbugs are controlled with insecticides. How-

Figure 5.9   Rice (foreground) and a rice mill/drier (background) in the Everglades Agricultural Area. (Photo credit: G. H. Snyder.)

ever, small acreages of rice have been grown in the Everglades without fertilizers or pesticides, to be sold as organically grown. Rice is harvested with a combine, following removal of the flood. The fields may be reflooded to grow a ratoon crop which can be harvested approximately 2 months later.

Economic Impact

At present, the principal crops in the Everglades, in terms of acreage, are (in descending order) sugarcane, vegetables, lawngrass sod, and rice. Together, these crops had an aggregate farm and mill-gate value of over $754 million in the 1988–89 season (Schueneman, 1990). Certainly, most of the hopes and promises of the early promoters of Everglades drainage for agricultural production have been fulfilled. The region leads the nation in sugarcane production, accounting for 22% of the sugar produced nationally (Buzzanell, 1990). Approximately 4000 full-time employees, 3000 part-time employees, and 10,000 seasonal offshore cane cutters were employed by the sugar industry in Florida during the 1988–89 season, and the aggregate payroll exceeded $230 million (Florida Sugar Cane League, Clewiston). Approximately one-third of the Florida employees live outside of the Everglades (survey conducted by the authors). Thousands of south Florida vendors supply the sugar industry. The majority are located outside of the Everglades, and perhaps a third do most of their business in the Everglades. Thus, the economic impact of the sugar industry is felt throughout south Florida. Mulkey and Clouser (1988) estimated that the sugar industry generates more than 18,000 full-time-equivalent jobs in Florida (15,000 in south Florida). Receipts from vegetable sales exceeded

$141 million in the 1988–89 season (compared to $549 million for sugar products), sod sales exceeded $42,000,000, and rice sales exceeded $6 million (Schueneman, 1990). In addition, $6 million was generated in 1988–89 by growing field corn for hybrid seed, an opportunity that arose because of the 1988 droughts in the Midwest. Although the economic impact on south Florida of the vegetable, sod, rice, and corn seed production enterprises has not been determined, if it is proportional to that of the sugar industry, then approximately 5500 full-time-equivalent south Florida jobs depend on these Everglades enterprises.

Rich (1971) pointed out that for a regional economy to prosper, there must be a net flow of money into the region. Passing money from hand to hand does not increase the wealth of the community. He emphasized the economic importance to a community of producing a product that can be sold to outsiders and suggested that agriculture, along with tourism, make good major industries for south Florida. Because the major agricultural products produced in the Everglades are sold out of state (sugar, vegetables), it is clear that Everglades agriculture meets the conditions set by Rich for creating a prosperous local community.

The FUTURE

The demise of Everglades agriculture has long been predicted. Farmers in the Everglades have learned to cope with problems of drought, floods, soil infertility, frosts and freezes, a variety of pest problems, and uncertain market prices. These problems have been sufficiently overcome such that over $750 million worth of agricultural products are shipped out of the Everglades annually. Other problems, however, appear on the horizon, and it remains to be seen whether Everglades farmers will be able to survive and prosper in the years to come.

Subsidence

The organic soils of the Everglades were formed under anaerobic conditions when, due to insufficient oxygen because of flooding, microorganisms were unable to completely decompose plant remains to carbon dioxide, water, and mineral constituents. Therefore, partially decomposed organic matter accumulated, forming peat soils which in some places were 12 ft (3.7 m) thick. When the soils were drained, the land surface began falling (subsiding) (Figure 5.10) for a number of reasons: loss of buoyancy, peat shrinkage, fires, wind erosion, and, most importantly, aerobic microbiological decomposition (oxidation). This is not a condition unique to the Everglades; all organic soils subside when drained. In the Everglades, the rate of subsidence averages about 1 in. (2.5 cm) per year, the exact rate being directly related to the depth of the water table; the deeper the water table, the more rapid the rate of subsidence.

In 1951, Stephens and Johnson (1951) thoroughly discussed the factors of subsidence in the Everglades and made predictions about the effect of subsidence on Everglades agriculture. Their predictions were unpopular at the time they were made (Johnson, 1974) and even today are not well received by growers. However, the rate of subsidence predicted in 1951 appeared to be on target in 1969 (Johnson,

Figure 5.10 Subsidence post at the Everglades Research and Education Center, showing the approximate location of the soil surface in 1924, 1954, and 1984. The post, which rests on bedrock, is 9 ft tall. The soil elevation in 1984 was approximately 4 ft above bedrock. Thus approximately 5 ft (60 in.) of subsidence occurred during the 60-year period from 1924 to 1984. (Photo credit: G. H. Snyder.)

1974), and a recent study by the Soil and Water Conservation Service showed that subsidence has continued as expected through 1988 (Smith, 1990). Most of the soils in the Everglades Agricultural Area are underlain by hard, dense, limestone rock (Figure 5.11) which cannot be farmed by conventional techniques. Therefore, when the organic soils are gone, it appears that the agriculture supported by the soils will also cease to exist. Stephens and Johnson predicted that whereas in 1912 over 95% of the soils in the Everglades Agricultural Area were >5 ft (1.5 m) in thickness, by the year 2000, 45% of the soils will be <1 ft (0.3 m) in thickness over limestone bedrock and 87% will be <3 ft (0.9 m) in thickness. Proceeding on the assumption that a soil 1 ft in thickness over bedrock no longer can be economically used (Johnson, 1951), they predicted that a large part of the Everglades Agricultural Area will have to be abandoned by the year 2000 (Stephens and Johnson, 1951). Today, some farmers are in fact growing sugarcane on soil somewhat less than 1 ft in thickness over bedrock, but already areas have been abandoned, and only a few years of conventional production can be expected from such soil.

Subsidence will not cause a sudden cessation of agriculture in the Everglades, because soil depths vary throughout the region. Additionally, growers can be expected to make changes in production methods in response to decreasing soil depths. It has been suggested that wetland crops might be used in shallow soils or even be used in deeper soils as a means of decreasing subsidence (Laurent et al., 1983). A cattle industry may again surface in the Everglades, because pasture grasses can be grown on more shallow, wetter soils that are not suitable for currently grown crops. Nevertheless, because of subsidence, acreage in production in the Everglades Agricultural Area is likely to decrease during the forthcoming years. At this writing, the state of Florida is exploring various plans for dealing with nutrient loads in drainage water from the Everglades Agricultural Area. Will millions of dollars be spent creating a method for dealing with today's problem, without

Figure 5.11   A hard, dense sheet of limestone rock underlies most of the organic soils in the Everglades Agricultural Area. (Photo credit: G. H. Snyder.)

considering that agricultural activity in the Everglades Agricultural Area will likely diminish, or change dramatically, in the future?

The Sugar Industry

Because over two-thirds of the dollar value of Everglades agriculture is generated by sugar production, and over 80% of the cropland is in sugarcane, it is clear that the general well being of Everglades agriculture depends on the well being of sugarcane production.

Government Programs

For many years, sugar production (cane sugar and beet sugar) has been regulated by the federal government. From 1934 through 1974, sugar production was regulated by the Sugar Act (Sugar y Azucar, 1990). Florida production was severely limited by that act because of allotments imposed until the Second World War (Salley, 1986). In 1977, following an interim sugar program, sugar was included in the federal farm bill, and it remains there today. Everglades agriculture in no small way depends on the treatment sugar receives in farm bills enacted by Congress. The present farm bill extends through the 1996 crop.

Most countries protect their domestic sugar producers in one manner or another. Direct subsidies are often paid to producers, so that the domestic price of sugar can remain relatively low. Most international trade in sugar results from bilateral trade agreements. Sugar produced in excess of domestic consumption or bilateral agreements is sold on the international market at depressed prices, because open market sales are mostly composed of excess production. In the

United States, the domestic sugar (cane and beet) industry is "protected" by the imposition of quotas on sugar imports. When the domestic price of sugar is deemed to be too high, quotas are increased to increase supply and thereby decrease the domestic price. Conversely, when the domestic price is below the market stabilization price (about $0.22 per pound raw sugar), imports are restricted. By using this quota system, the federal government avoids any drain on the United States treasury. The government, however, is ideologically opposed to trade restrictions and is in favor of free trade. Sugar imports also are used as foreign aid for sugar-producing countries (Smith, 1988). The government currently is involved in trade negotiations as part of the General Agreement on Trade and Tariffs (GATT). The goal of the United States and other countries negotiating for a reduction in agricultural trade barriers is "tariffication," i.e., the conversion of all nontariff trade barriers, such as quotas, to tariffs that eventually can be phased out through gradual annual reductions (Sugar y Azucar, 1990). Thus, it is possible that in time the sugar industry will lose its protection. Leaders of the Florida sugar industry and several market studies contend that United States sugar producers can compete in an open market if foreign producers are not subsidized by their governments (Dorschner, 1990; Miedema, 1987; Smith, 1988). Of course, sugar producers want protection until then.

The Sugar Market

Although the consumption of sweeteners in the United States has increased in recent years, the consumption of cane and beet sugar has not fully participated in this trend (Polopolus and Alvarez, 1990). The decade beginning in the mid-1970s saw a dramatic increase in the use of high-fructose corn syrup and glucose, at the expense of cane and beet sugar. Since 1986, the contribution from these sources to total caloric sweetener consumption has stabilized somewhat, but still remains slightly in favor of corn syrup and glucose (Polopolus and Alvarez, 1991). In addition, there has been a phenomenal increase in the consumption of noncaloric sweeteners, especially since the introduction of aspartame in 1981, and a number of other substitutes are in various stages of development and marketing. On the supply side, the suspension of the Cuban sugar quota, which led to the great expansion of the Florida sugar industry since the Cuban revolution, is just that: a temporary suspension of a quota. If normal trade relations were again resumed with Cuba, the quota could be reinstated, or in some other way Cuba could once again become a factor in the United States sugar picture. In addition, the effect that economic restructuring in former "eastern-block" European nations will have on agricultural productivity is not known, but many of these countries have the potential to expand sugar beet production. Thus, the outlook for a strong sugar market is not bright.

Sugar Industry Vilification

By any measure, the domestic sugar industry and the Florida sugarcane industry in particular does not enjoy good "public relations." Sugar is blamed for a variety of health problems and people are urged to avoid it (Hunter, 1982), even though the Food and Drug Administration does not support this position (Moore, 1989). Many federal legislators vehemently oppose any government subsidy of the domestic sugar industry (Smith, 1988; Sugar y Azucar, 1989). Despite data and efforts to the contrary (Germinsky, 1990; Orsenigo and Barber, 1986), the Florida sugar industry continues to be blamed for a variety of south Florida environmental problems. The *Miami Herald* calls sugar "Florida's second most destructive industry," sapping our resources, enriching a few at everyone's expense, enslaving poor people, playing havoc with our foreign policy, mistreating its workers, and polluting the Everglades (Dorschner, 1990). The belief that a few already wealthy people profit from government protection of the domestic sugar industry (Berman and Khalif, 1990) provides a simplistic rallying cry for elimination of all government programs, even though doing so might destroy the entire United States domestic sugar production industry.

In Florida, many thousands of people directly or indirectly owe their livelihood to the sugarcane industry. If the industry were to suddenly collapse, the majority of these people would suffer severe, and in many cases permanent, economic hardship, not to mention the emotional distress that comes from loosing one's job in an area devoid of employment opportunities for those who have spent most of their lives working in a single industry. Simple observation reveals that a sizable percentage of those employed in the sugar industry are minorities (blacks and Hispanics), who may face additional employment hurdles. Small landowners, including families that have farmed the area for years and even generations, also would likely be bankrupted by elimination of the sugar industry. The few very wealthy "sugar barons," on the other hand, generally have insulated themselves from economic hardship by owning farms out of state and in foreign countries and by diversifying into nonagricultural enterprises. In addition, other Everglades agricultural enterprises count on sugarcane production to provide stability for their operation. These enterprises could be weakened by a loss of the sugar industry.

Environmental Problems

Water control facilities installed after the 1947 hurricane have resulted in improved handling of water in south Florida on a quantity basis. Floods and droughts have been mitigated, and Everglades farmers have been able to maintain water tables that both maximize crop production and minimize soil subsidence (Snyder et al., 1978). However, the water control system was not designed with water quality issues in mind. These issues now are at the forefront of public thinking in Florida.

Various state agencies have declared phosphorus to be the principal cause of algal blooms in Lake Okeechobee and the reason that cattails (*Typha* spp.) have replaced sawgrass in portions of certain Water Conservation Areas. Drainage from the Everglades Agricultural Area is seen as a major source of phosphorus. Pumping of drainage water from the Everglades Agricultural Area into Lake Okeechobee has been dramatically reduced, and proposals are on the table for filtering Everglades Agricultural Area drainage through man-made marshes within the Everglades Agricultural Area before the water enters the Water Conservation Areas. Both of these plans adversely affect Everglades agriculture. Perhaps of even greater consequence, by virtue of being blamed for environmental problems in Lake Okeechobee and Water Conservation Areas, Everglades agriculture is considered expendable by vocal members of the public. The Everglades Coalition has declared that flooding the Everglades Agricultural Area would bring more economic benefit to south Florida than a continuation of agriculture (McLachlin, 1990). Marjory Stoneman Douglas is quoted as saying of Everglades farmers: "I don't care where they go as long as they get the hell out of Florida" (Kleinberg, 1990). The chairman of the South Florida Water Management District has stated that saving the Everglades is more important than preventing an economic collapse of the surrounding farm communities (Newman, 1991). Clearly, Everglades farmers have powerful foes and receive little sympathy from the general public.

Possible Future Scenarios

It is possible that agriculture in the Everglades reached its zenith in the 1980s. Because virtually all of the Everglades Agricultural Area has been planted, there is no room for future expansion. Land for conventional agriculture has been lost because of subsidence and will continue to be lost at an accelerating rate in the future. Farmers are facing mounting costs for mandated pollution controls, at a time when the future profitability of their principal commodity, sugar, is threatened. In response to these problems, there may be a shift to other commodities such as rice or cattle, and it is possible that methods will be developed for growing sugarcane on more shallow, wetter soils. These measures, and the development of an agriculture based on water-tolerant crops (Morton and Snyder, 1976; O'Hair et al., 1982; Laurent et al., 1983; Morton et al., 1988) could reduce both subsidence and the need for drainage, making Everglades agriculture more environmentally acceptable. It is unlikely, however, that these alternatives will be as profitable as present-day sugarcane production. Vegetable production is limited primarily by the market, and it is probable that there will be sufficient soil of suitable depth to sustain vegetable production at its current level for many years. However, it remains to be seen whether vegetable growers can survive fluctuating market conditions without the stabilizing influence of sugarcane production, and in an unprotected market, even sugarcane would be subject to cyclical world price fluctuations. In time, some lands may be abandoned, as has been long predicted, or growers may find a way to profit from the potential for using their lands for hunting preserves. Writing in 1956, Dr. R. V. Allison foresaw the day when:

> As the use of Everglades lands for agricultural purposes approaches the sunset of their experience in this field of production there is little doubt that transition into a wildlife area of ultimate world fame will follow, perhaps in an easy and natural manner...There can be no doubt that by the end of the present century the population of the coastal areas to the east and west will be getting very heavy indeed and that the "Playground" outlook for these sections could be immeasurably benefited by having such a huntsman and fisherman's paradise spreading out in vast proportions within the main body of the Florida peninsula at this latitude...It is also quite certain that demands for water will increase several times over during the next few decades in south Florida...Naturally, as the Everglades returns to an essentially wild life area in the years to come, its capacity for receiving and holding water...would be tremendous. Such a visualization for the area in the future then, becomes simply a gradual change of duty for this great peat

and muck land section of the State that will enable it to deliver a diversity of services and values which, while substantially different from the agricultural types of the present and of the immediate future, may prove to be of equal or even greater importance to the economy of the State at that not too distant time (Allison, 1956).

Because soils are more shallow over bedrock in the southernmost Everglades Agricultural Area, it is there that natural changes in land use are most likely to occur first. The creation of water filtration marshes in this region could serve as a gradual bridge between present-day agriculture and the return of the Everglades Agricultural Area to a wildlife area, as envisioned by Allison.

Thus, the outlook for Everglades agriculture is not bright. However, it is certain that growers, many of whom have roots in the Everglades going back for generations, will use every means at their disposal to remain in business in the future, as they have done in the past.

ACKNOWLEDGMENTS

The authors wish to express their appreciation to the following persons who assisted in the preparation of this manuscript: Dr. J. Alvarez, Dr. R. H. Cherry, Mr. S. M. Davis, Dr. C. W. Deren, Dr. J. M. Duxbury, Mrs. M. H. Mosely, Dr. J. R. Orsenigo, Dr. W. H. Patrick, Jr., Mr. G. H. Salley, Dr. C. A. Sanchez, Dr. T. J. Schueneman, Mr. J. C. Stephens, Mr. M. Ulloa, Dr. V. H. Waddill, and Mr. F. Williamson, Jr. A portion of Figure 5.1 is adapted from Will (1977) by permission of The Glades Historical Society.

LITERATURE CITED

Allison, R. V. 1956. The influence of drainage and cultivation on subsidence of organic soils under conditions of Everglades reclamation. *Soil Crop Sci. Soc. Fla. Proc.,* 16:21–31.

Allison, R. V., O. C. Byran, and J. H. Hunter. 1927. The stimulation of plant response on the raw peat soils of the Florida Everglades through the use of copper sulfate and other chemicals. *Fla. Agric. Exp. Stn. Bull.,* No. 190, University of Florida, Gainesville.

Alvarez, J., G. H. Snyder, and D. B. Jones. 1989. The integrated program approach in the development of the Florida rice industry. *J. Agron. Educ.,* 18:6–11.

Baldwin, M. and H. W. Hawker. 1915. Cumulose soils. in USDA Bureau of Soils Soil Survey of the Ft. Lauderdale Area, Florida, pp. 31–52.

Berman, P. and R. Khalaf. 1990. The Fanjuls of Palm Beach: The family with a sweet tooth. *Forbes,* 145(10):54–69.

Bestor, H. A. 1943. Reclamation problems of sub-drainage districts adjacent to Lake Okeechobee. *Soil Sci. Soc. Fla. Proc.,* V-A:157–165.

Blassingame, W. 1974. *The Everglades from Yesterday to Tomorrow,* G. P. Putnam's Sons, New York, 126 pp.

Buzzanell, P. 1990. Sugar and Sweetener Situation and Outlook, No. SSRV14N5, Economic Research Service, U.S. Department of Agriculture, March 1990, 65 pp.

Chapman, H. L., Jr., R. W. Kidder, C. E. Haines, R. J. Allen, Jr., V. E. Green, Jr., and W. T. Forsee, Jr. 1963. Beef Cattle Production on Organic Soils of South Florida, Bull. 663, IFAS, University of Florida Agricultural Experiment Station, Gainesville.

Coale, F. J. 1987. A Traveler's Guide to Florida Sugarcane, Agronomy Facts No. 216, Florida Cooperative Extension Service, University of Florida (IFAS), Gainesville.

Dorschner, J. 1990. Big sugar. *The Miami Herald, Tropic Section,* Jan. 28, 1990, pp. 8–18.

Douglas, M. S. 1947. *The Everglades: River of Grass,* Rinehart, New York.

Dovell, J. E. 1947. A History of the Everglades of Florida, Ph.D. thesis, University of North Carolina, Chapel Hill, 598 pp.; condensed in *Soil Sci. Soc. Fla. Proc.,* IV-A:132–161, 1942.

Elvove, J. T. 1943. The Florida Everglades—A region of new settlement. *J. Land Public Util. Econ.,* 19:464–469.

Everglades News. 1960. New Osceola Farms Company to plant 4,000 acres of cane 5 miles east of Canal Point. *The Everglades News,* Oct. 28, 1960, p. 1.

Forman, C. 1979. The Forman family: Everglades pioneers. *Broward Legacy,* 3(3 and 4):2–9.

Germinisky, R. A. 1990. Florida's sugar industry preserves and protects the environment. *1990 Sugar Azucar Yb.,* 57:6–8.

Green, V. E., Jr. and J. D. Panzer. 1958. Hoja blanca: How the disease was discovered in Florida. *Rice J.,* 61:16–26.

Hanna, A. J. and K. A. Hanna. 1948. *Lake Okeechobee, Wellspring of the Everglades,* Bobbs-Merrill, New York, 380 pp.

Hunter, B. T. 1982. *The Sugar Trap and How to Avoid It,* Houghton Mifflin, New York, 239 pp.

Hutcheson, C. E. 1990. Personal communication involving examination of file records on Palm Beach County agriculture compiled by the Palm Beach County Cooperative Extension Service, West Palm Beach, Fla.

Johnson, L. 1951. Panel discussion: Soil and water conservation operation and planning in the Everglades area as influencing soil subsidence and ultimate land use changes. *Soil Sci. Soc. Fla. Proc.,* 11:99–101.

Johnson, L. 1974. *Beyond the Fourth Generation,* University Presses of Florida, Gainesville, 230 pp.

Jones, L. A. 1948. Soils, Geology, and Water Control in the Everglades Region, Bull. 442, University of Florida Agricultural Experiment Station, Gainesville.

Kidder, G. and E. R. Rice. 1979. Growth rates of Florida sugarcane during seven growing seasons. *Proc. Am. Soc. Sugarcane Technol.,* 9:8–12.

Kidder, R. W. 1952. Ton of beef per acre of grass in 12 months. *Breeder's Gaz.,* 117(11):8.

Kidder, R. W. 1979. *From Cattle to Cane,* available from the Belle Glade Historical Society, Belle Glade, Fla.

Kleinberg, E. 1990. "River of grass" author turns 100. *The Palm Beach Post,* April 7, 1990, Section A, pp. 1, 16.

Laurent, M. F., S. F. Shih, and G. H. Snyder. 1983. Stopping subsidence with aquatic crops. *IFAS Coop. Ext. Serv. Spec. Publ.,* No. 23, University of Florida.

McCarty, L. B. and J. L. Cisar. 1989. Basic guidelines for sod production in Florida. *Fla. Coop. Ext. Serv. Bull.,* No. 260, University of Florida (IFAS), Gainesville.

McCollum, S. H., V. W. Carlisle, and B. G. Volk. 1976. Historical and current classification of organic soils in the Florida Everglades. *Soil Crop Sci. Soc. Fla. Proc.,* 35:173–177.

McLachlin, M. 1990. Coalition wants to wipe out sugar in the Everglades. *The Palm Beach Post,* Jan. 21, 1990, Section A, pp. 1, 10.

Miedema, B. J. 1987. USSPG calls for USTR to act against EEC sugar subsidies. *Sugar News,* 24(3):1–3.

Moore, D. 1989. The big scoop on sugar. *New Choices,* June:71–73.

Morton, J. F. and G. H. Snyder. 1976. Aquatic crops vs. organic soil subsidence. *Proc. Fla. State Hortic. Soc.,* 89:125–129.

Morton, J. F., C. A. Sanchez, and G. H. Snyder. 1988. Chinese waterchestnuts in Florida—Past, present, and future. *Proc. Fla. State Hortic. Soc.,* 101:139–144.

Mulkey, D. and R. Clouser. 1988. The Economic Impact of the Florida Sugar Industry, Staff Paper No. 341, Food and Resource Economics Department, University of Florida (IFAS), Gainesville.

Newell, W. 1921. *Univ. Fla. Agric. Exp. Stn. Annu. Rep.,* University of Florida, Gainesville, pp. 15R–17R.

Newman, J. 1991. Water district: Sacrifice farms for Everglades. *The Palm Beach Post,* Dec. 12, 1991, Section B, p. 1.

O'Hair, S. K., G. H. Snyder, and J. F. Morton. 1982. Wetland taro: A neglected crop for food, feed, and fuel. *Proc. Fla. State Hortic. Soc.,* 95:367–374.

Orsenigo, J. R. and C. E. Barber. 1986. Environmental aspects of the Florida sugarcane industry. in Proc. Inter-American Sugar Cane Seminars, Vanguard, Miami, pp. 67–72.

Polopolus, L. C. and J. Alvarez. 1990. Florida's Sugar Industry and the Emerging Sweetener Consumption Patterns, University of Florida IFAS Cooperative Extension Service Food and Resource Economics Rep. No. 92, Jan.–Feb.

Polopolus, L. C. and J. Alvarez. 1991. *Marketing Sugar and Other Sweeteners,* Elsevier Science, New York, pp. 138–139.

Rich, E. R. 1971. The population explosion in Dade county. in *The Environmental Destruction of South Florida,* W. R. McCluney (Ed.), University of Miami Press, Coral Gables, Fla., pp. 41–43.

Salley, G. H. 1986. *A History of the Florida Sugar Industry,* G. H. Salley, Miami; available from the Florida Sugar Cane League, Clewiston.

Sanchez, C. A. 1990. Soil Testing and Fertilizer Recommendations for Crop Production on Organic Soils in Florida, Tech. Bull. 876, University of Florida (IFAS), Gainesville.

Schueneman, T. J. 1990. Personal communication involving examination of file records on Everglades agriculture compiled by the Palm Beach County Cooperative Extension Service, Belle Glade, Fla.

Sitterson, J. C. 1953. *Sugar Country,* University of Kentucky Press, Lexington.

Smith, A. 1988. Of cane, corn, congressmen, and communists: The strange story of sugar. *Adam Smith's Money World,* transcript of show no. 425, Journal Graphics, New York.

Smith, G. 1990. The Everglades agricultural area revisited. *Citrus Vegetable Mag.,* 53(9):40–42.

Snyder, G. H. and D. B. Jones. 1988. Prediction and prevention of Fe-related rice seedling chlorosis. *Soil Sci. Soc. Am. J.,* 52:1043–1046.

Snyder, G. H. and D. B. Jones. 1989. Phosphorus and potassium fertilization of rice on Everglades Histosols. *Soil Crop Sci. Soc. Fla. Proc.,* 48:22–25.

Snyder, G. H., H. W. Burdine, J. R. Crockett, G. J. Gascho, D. S. Harrison, G. Kidder, J. W. Mishoe, D. L. Myhre, F. M. Pate, and S. F. Shih. 1978. Water Table Management for Organic Soil Conservation and Crop Production in the Florida Everglades, Bull. 801, University of Florida Agricultural Experiment Station, Gainesville.

Snyder, G. H., D. B. Jones, and G. J. Gascho. 1986. Silicon fertilization of rice on Everglades Histosols. *Soil Sci. Soc. Am. J.,* 50:1259–1263.Stephan, L. L. 1944. Vegetable production in the northern Everglades. *Econ. Geogr.,* 20:78–101.

Stephens, J. C. and L. Johnson. 1951. Subsidence of organic soils in the upper Everglades region of Florida. *Soil Sci. Soc. Fla. Proc.,* 11:191–237.

Stockbridge, F. P. and J. H. Perry. 1926. *Florida in the Making,* de Bower Publishing, New York, 351 pp.

Sugar y Azucar. 1989. Facts about sugar. *Sugar y Azucar,* April:4–12.

Sugar y Azucar. 1990. Facts about sugar. *Sugar y Azucar,* May:7–10.

Tedder, G. E. 1925. Report of the Everglades Experiment Station. in *Univ. Fla. Agric. Exp. Stn. Annu. Rep.,* W. Newell (Ed.), University of Florida, Gainesville, pp. 95R–97R.

Wanless, H. R. 1989. The inundation of our coastlines: Past, present, and future with a focus on south Florida. *Sea Frontiers,* 35:264–271.

White, R. W. and P. Busey. 1987. History of turfgrass production in Florida. *Proc. Fla. State Hortic. Soc.,* 100:167–174.

Will, L. E. 1977. *A Cracker History of Okeechobee,* The Glades Historical Society, Belle Glade, Fla.

Will, L. E. 1984. *Swamp to Sugar Bowl: Pioneer Days in Belle Glade,* The Glades Historical Society, Belle Glade, Fla.

6

CHANGES in FRESHWATER
INFLOW from the EVERGLADES
to FLORIDA BAY INCLUDING
EFFECTS on BIOTA and BIOTIC
PROCESSES: A REVIEW

Carole C. McIvor

Janet A. Ley

Robin D. Bjork

ABSTRACT

The freshwater Everglades and estuarine Florida Bay ecosystem are closely linked by marine and freshwater hydrologic cycles and by organisms that depend on both systems during different times of the year or periods of their life cycles. Impounding of water in the Water Conservation Areas and diversion of water away from Shark River Slough and Taylor Slough for purposes of urban use and flood control have significantly reduced the volume of fresh water to Florida Bay. As a result, bay waters are now more saline in more locations and for longer periods of time than under premanaged conditions. The filling of passes and shallow banks between several of the Keys for construction of the Flagler Railroad in the early 1900s reduced circulation in Florida Bay, thereby exacerbating anthropogenically generated salinity anomalies. Delivery of fresh water to Florida Bay differs from premanaged conditions in both volume and timing. Numerous effects on biota and biotic processes in the bay and southern Everglades ecotone have been documented or implicated, including reduced recruitment of pink shrimp, snook, and redfish; lowered reproductive success of ospreys and great white herons; and shifts in distribution of West Indian manatees, American crocodiles, and many of the wading birds that historically nested in the estuarine ecotonal area. One species, however, the gray snapper (*Lutjanus griseus*), exhibits enhanced recruitment in years of higher salinity in Florida Bay. Reduced freshwater inflow is also implicated as one of a complex series of factors in the mass mortality of seagrasses in the bay that has occurred since 1987. Similarly, hypersalinity is likely a factor in dieback of mangroves in some Florida Bay localities. Excessive amounts and unnatural timing of freshwater delivery can also adversely affect biota. A sudden release of greatly elevated volumes of fresh water from the C-111 canal resulted in the mortality of many estuarine organisms in Manatee Bay when salinities dropped from near marine to zero in a few hours and remained low for an 8-day period. These observations provide powerful evidence that productivity of Florida Bay is declining under current management practices.

INTRODUCTION

Florida Bay

Florida Bay is a lagoonal estuary bordered on the north by the Florida mainland and on the southeast by the Florida Keys (Figure 6.1). It is about 2200 km² in total area, with 1800 km² included within Everglades National Park (Tilmant, 1989;

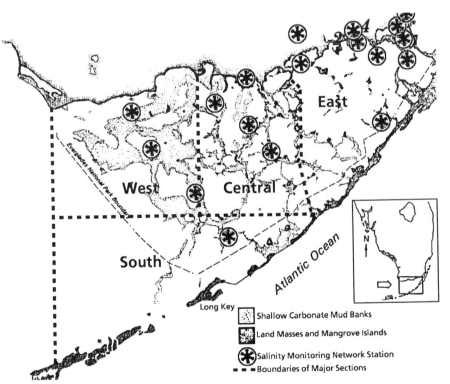

Figure 6.1   Base map of Florida Bay showing descriptive subregions. (From Costello et al., 1986.)

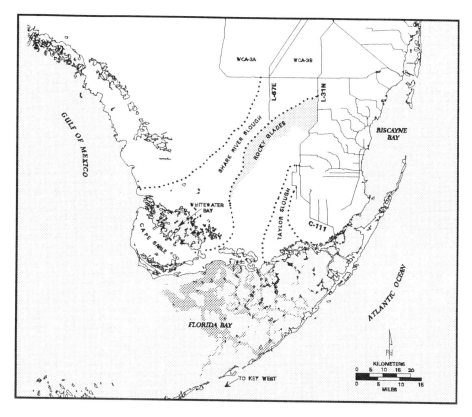

Figure 6.2 Regional features and sources of freshwater inflow to Florida Bay. (From Schomer and Drew, 1982.)

Robblee et al., 1991). The bay is shallow, with an average depth of less than one meter (Tilmant, 1989). Direct inflow of fresh surface water to Florida Bay occurs via sheet flow across marl prairies of the southern Everglades and from 20 creek systems fed by Taylor Slough and the C-111 canal. Water from Shark River Slough, the largest drainage in the Everglades, flows into Whitewater Bay (Figure 6.2). Subsequent exchange with Florida Bay occurs as this lower salinity water mass flows around Cape Sable and into the western bay (Fourqurean et al., 1993).

Florida Bay is effectively divided into a series of basins by a complex network of anastomosing carbonate mud banks that restrict circulation (Sogard et al., 1989). Seagrass (primarily *Thalassia testudinum, Halodule wrightii,* and *Syringodium filiforme*) meadows covered >80% of the bottom of Florida Bay prior to the seagrass die-off that began in 1987 (Zieman et al., 1989; Robblee et al., 1991). Mangrove (*Rhizophora mangle, Avicennia germinans,* and *Laguncularia racemosa*) isles, many of which are overwashed at high tide, cover <2% of the area of the bay (Enos, 1989). Hard-bottom areas of calcium carbonate rock, overlain by a thin layer of carbonate sediment, are most common in the southern portion of the bay. These areas are characterized by the occurrence of sponges, octocorals, and macroalgal patches (Butler et al., 1992).

At the northern edge of Florida Bay lies the interface or ecotone between the bay and the freshwater Everglades wetland ecosystems. In northeastern Florida Bay, the ecotone is typified by shallow marl soils dominated by scrub forests of red mangrove and by open flats of broadhead spikerush (*Eleocharis cellulosa*) (Tabb et al., 1967; Odum et al., 1982). Annual (Schomer and Drew, 1982) and short-term, wind-driven (Holmquist et al., 1989) fluctuations in water level in Florida Bay strongly influence the hydrology of about a 5-km-wide band of the southern end of the Florida peninsula (Bjork and Powell, 1993).

Estuaries are nursery grounds for many species of juvenile fishes and invertebrates whose adult forms occur, spawn, and are harvested offshore (Gunter, 1961; Costello and Allen, 1966; McHugh, 1967; Weinstein, 1979; Browder and Moore, 1981; Davis and Dodrill, 1989). In fact, the nursery value of estuaries is often advanced as one of the key arguments for their preservation and restoration worldwide. Florida Bay serves as a nursery ground for at least 22 species of commercially and recreationally harvested species (Table 6.1).

Although all commercial fishing was banned after 1985 from the portion of Florida Bay within Everglades National Park (Tilmant, 1989), the area is heavily used by recreational anglers, many of whom hire local fishing guides. Thus, the bay is an important source of revenue in the south Florida economy.

This review was prompted by the position of Florida Bay as the estuary most influenced by hydrological and other anthropogenic modifications made in the southern Everglades watershed, plus recent concerns over the health of the bay (Butler et al., 1992; Carlson et al., 1990; Robblee et al., 1991; Durako et al., in press; South Florida Water Management District, 1992). Furthermore, Florida Bay, the Everglades, and the ecotonal areas between them are tightly linked ecologically through the interdependencies of many organisms. An obvious example of these

Table 6.1   Aquatic Fauna of Commercial and Recreational Importance Using Florida Bay as a Nursery Ground

Fishes
  Drums: Sciaenidae
    Cynoscion nebulosus
    Cynoscion arenarius
    Bairdiella chrysura
    Leiostomus xanthurus
    Menticirrhus americanus
    Menticirrhus saxatilis
    Pogonias cromis
    Sciaenops ocellatus
  Porgies: Sparidae
    Lagodon rhomboides
    Archosargus probatocephalus
  Grunts: Haemulidae
    Haemulon sciurus
  Snappers: Lutjanidae
    Lutjanus griseus
    Lutjanus synagris

Tarpons: Elopidae
    Megalops atlanticus
    Elops saurus
Snooks: Centropomidae
    Centropomus undecimalis
Barracuda: Sphyraenidae
    Sphyraena barracuda
Requiem sharks: Carcharhinidae
    Carcharhinus leucas
Mullets: Mugilidae
    Mugil cephalus
    Mugil curema
Crustaceans
  Pink shrimp
    Penaeus duorarum
  Florida lobster
    Panulirus argus
  Stone crab
    Menippe mercenaria

linkages is exhibited by the wading birds (Ciconiiformes) that shift between freshwater and estuarine and marine foraging habitats, depending upon water levels (Bancroft et al., 1994). Many of the concerns over conditions in Florida Bay, as in the Everglades, revolve around the quantity, quality, and timing of freshwater delivery. Restoration of the Everglades cannot be achieved without concurrent restoration of the estuaries, including Florida Bay, and the fresh/saltwater interface between these two major habitats. The issue of freshwater inflow to estuaries is briefly reviewed from a global perspective in the following section, before exploring the issue for Florida Bay.

Role of Freshwater Inflow to Estuaries

Many alterations to inland watersheds are manifested and often amplified at the margin between the land and estuary (Simenstad et al., 1992; Bjork and Powell, 1993). Undisturbed estuarine communities are highly productive because organisms are adapted to, and dependent upon, gradients generated when freshwater inflow meets the sea. Altering patterns and amounts of freshwater inflow influence the balance among plant and animal communities and often lead to catastrophic decreases in valuable fisheries and other biotic processes (e.g., Nichols et al., 1986).

Conflicts between water resource management and the quality of estuarine habitats are not confined to Florida. In California, flow from two major tributaries to San Francisco Bay (San Joaquin and Sacramento rivers) has been reduced to less than 40% of historic levels (Nichols et al., 1986). Projects upstream in the watersheds now hold most of the historic flows in reservoirs for use by agricultural and urban consumers during dry California summers. Three major ecological impacts have resulted. Collapse of the salmonid fishery is directly attributable to the construction of Shasta Dam. Reduced freshwater inflow along with overfishing are responsible for the decline in abundance of sturgeon, sardines, flatfish, crabs, and shrimp. In addition, greatly diminished freshwater inflow may also reduce the capacity of this estuary to dilute, transform, or flush contaminants that are discharged into San Francisco Bay (Nichols et al., 1986).

Even if changes in volume of discharge are small, alteration of timing of freshwater flow can lead to negative impacts on the estuarine ecosystem. Twenty-eight major dams have been constructed along the Columbia River in the Pacific Northwest since 1900. Management policies have reduced springtime freshwater flows to 50% of former levels, while fall discharges have been artificially increased by 10–50% (Simenstad et al., 1992). These changes in the timing of inflow have been related to declines in salmon fisheries in the region.

Other examples of how changes in freshwater flow affect the estuarine ecosystems located downstream can be found throughout the world. Since 1941, flow from the Colorado River into the Gulf of California has virtually been eliminated (Flanagan and Hendrickson, 1976). Cessation of freshwater inflow and development of hypersaline conditions in the upper Gulf of California are implicated in the near extinction of the totoaba (*Cynoscion macdonaldi*), the largest species of drum, although overfishing may also have played a role in the decline.

Completion of the Aswan High Dam completely eliminated flows from the Nile River to the Mediterranean Sea during the dry season. As a result, the sardine fishery in the area was reduced by 95% due to elimination of the nutrient plume from the river delta (Alleem, 1972). As a result of greatly reduced sediment supply from operation of the Aswan High Dam, coastal erosion now exceeds deltaic sedimentation. This situation, coupled with delivery of nutrient-laden agricultural runoff and urban wastewater to the lower delta, threatens ecologically critical coastal lagoons and the wetland species they support (Stanley and White, 1993).

As these examples indicate, problems associated with altered freshwater inflows to estuaries are worldwide, not local, in nature. The Florida Bay ecosystem, however, is unique in many respects, as described in the following overview of conditions related to freshwater inflow to the bay system.

CURRENT CONDITIONS in FLORIDA BAY: SALINITY, FRESHWATER FLOW, and CIRCULATION

Salinity

A salinity monitoring network was established in Everglades National Park and Florida Bay during the mid to late 1980s. Currently there are 18 permanently affixed monitoring stations (Figure 6.1). Results of these efforts are critical to the task of analyzing salinity conditions and long-term salinity changes in the bay. These Everglades National Park databases are not yet available; thus, salinity data have been compiled from observations of several investigators who conducted studies from one to several years in duration in Florida Bay.

In all but years of highest rainfall, portions of Florida Bay become hypersaline (greater than 35 parts per thousand [ppt]) because evaporation exceeds additions of rainfall and freshwater inflow (Schomer and Drew, 1982; Fourqurean et al., 1993). According to published and unpublished reports, hypersalinity has characterized Florida Bay periodically since the 1950s (Table 6.2). In fact, in 12 of the 17 years for which data are available between 1956 and 1986, some areas of Florida Bay remained hypersaline all year (Robblee et al., 1989). The highest recorded salinity values occurred in 1956, when levels of 70 ppt were measured in the central bay (Finucane and Dragovitch, 1959, as reported in Schmidt and Davis, 1978).

An isohaline map for Florida Bay (Figure 6.3) was prepared from monthly sampling at a network of stations during 1989–90 (Fourqurean et al., 1993). This was a very dry period and mean salinities exceeded 36 ppt for all bay locations. The highest salinities occurred in the north-central bay near Rankin Lake. Salinity levels gradually moderated toward the northeast, south, and west. This contour pattern, with peak salinities in mid-bay, may be typical for most years as an annual average, although actual salinity values may vary.

Freshwater Inflow

Taylor Slough was historically the major contributor of fresh water to Florida Bay (Van Lent et al., 1993) (Figure 6.2). Overall water management alterations in

Table 6.2 Compilation of Florida Bay Salinity Studies and Range of Measurements

Authors	Period of study	Location of investigation[a]	Salinity maximum (ppt)	Salinity minimum (ppt)	Subject of study
Finucane and Dragovitch (1959)	Mar 55–May 57	C	70.0	34.4	Red tide
McCallum and Stockman (1959)	Dec 56–Feb 58	E, C, W	57.8	5.9	Geology
Lloyd (1964)	Jan 58–Nov 58	Baywide	39.4	11.4	Mollusks
Tabb (1967)	Dec 64–Nov 65	E, C, W	65.3	19.8	Hydrology
Tabb (unpublished) in Schmidt and Davis (1978)	Apr 65–Jun 66	Baywide	67.7	0.0	Salinity
Costello et al. (unpublished) in Schmidt and Davis (1978)	Jan 67–Jan 68	W, C, S	44.8	26.0	Pink shrimp
Schmidt (unpublished) in Schmidt and Davis (1978)	Apr 73–Sep 76	Baywide	66.6	1.7	Fish
Sogard et al. (1987)	Jan 84–Nov 85	W, C, S	44.0	17.0	Fish
A. Powell et al. (1987)	Mar 84–Sep 85	W, S	39.0	21.0	Plankton
Thayer et al. (1987)	May 84–May 85	W, C, S	42.0	27.0	Fish
Holmquist et al. (1989)	Dec 84–Nov 86	W, C, S	50.0	17.0	Invertebrates
Montague et al. (1989)	Mar 86–Sep 87	E	46.6	0.1	Benthos
Ley (1992)	May 89–May 90	E	57.5	0.2	Fish
Fourqurean et al. (1993)	Jun 89–May 90	Baywide	59.0	18.0	Nutrients

[a] See Figure 6.1 for locations. E = east, W = west, S = south, C = central.

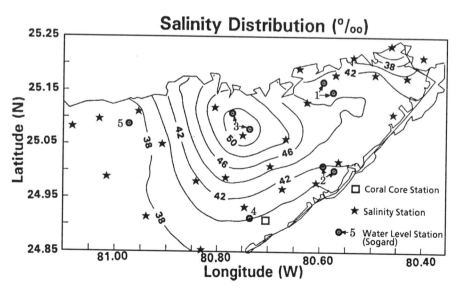

Figure 6.3 Average salinity for Florida Bay based on data collected from 25 stations (marked by asterisks) from June 1989 to August 1990. (From Fourqurean and Zieman, 1992.)

the Taylor Slough and C-111 drainage basins have changed the natural patterns of freshwater inflow to Florida Bay (Johnson and Fennema, 1989). These changes have resulted in shortened periods of flow through the Taylor Slough basin and may be contributing factors to hypersaline conditions and abrupt salinity changes in northeastern Florida Bay (Van Lent et al., 1993). Currently, freshwater inflow to Florida Bay increases in June from direct rainfall, surface flow, and groundwater discharge when the rainy season begins (Mazzotti and Brandt, 1989). Salinity patterns as related in part to sources of fresh water will be discussed in greater detail by region of the bay.

Circulation and Flushing

The causes of salinity patterns in Florida Bay include not only freshwater inflow but also circulation and flushing patterns. The anastomosing shallow mud banks divide Florida Bay into discrete basins. These banks act as barriers, effectively limiting water exchange between basins, particularly in interior sections of the bay. Tidal flushing (as indicated by tidal amplitude) is greatest in western Florida Bay adjacent to the Gulf of Mexico, second highest in the portions of the eastern bay influenced by the Atlantic Ocean (i.e., Plantation Key southward), and minimal in the remainder of the bay (Figures 6.3 and 6.4). Reduced flushing, when coupled with low freshwater inflow and high evaporation, leads to conditions of hypersa-

Figure 6.4 Monthly variation in mean water level at the six sampling sites in Florida Bay (see Figure 6.3 for locations). Ranges associated with each value are mean low tide and mean high tide level for that month. (From Sogard et al., 1989.)

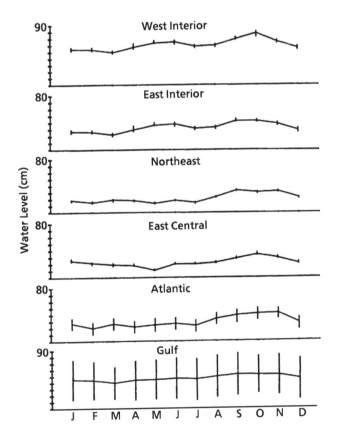

linity which are particularly pronounced in the central bay (Figure 6.3). An additional factor confounding historical comparisons of salinity patterns with those of the present is the absence of hurricanes for the last 32 years. With no major hurricanes, both basins and banks have undoubtedly become shallower, further restricting circulation.

Proxy data relating to changes in patterns of circulation and water exchange come from preliminary analysis of a coral core from Peterson Key Basin in southern Florida Bay (Figure 6.3). Based on oxygen and carbon isotopic variation, Swart et al. (1992) conclude that Peterson Key Basin experienced decreased tidal exchange with waters of Hawk Channel following 1908–10, the years when passages between Upper and Lower Matecumbe Keys were filled for construction of the Flagler Railroad. Other annual variations in the isotopic values in the 107-year record seem to correspond to the passage of hurricanes. In summary, the carbon and oxygen isotopic data seem to relate more to water circulation and exchange in this portion of Florida Bay than to freshwater inflow to the bay per se, though further analyses are proposed.

Regions of Florida Bay

For discussion purposes, Florida Bay can be divided into four regions based on general arrangement of mud banks and islands and on patterns of water movement from and into adjacent systems (Figure 6.1). Definition of regions of the bay follows that of Costello et al. (1986).

Eastern Florida Bay

On an annual basis, the greatest range of salinity occurs in eastern Florida Bay. In Little Madeira Bay at the mouth of the Taylor River, a range of 52 ppt was recorded in a single year (Robblee et al., 1989). Because this is where surface fresh water enters the bay most directly, a range of 35 ppt (from fresh to salt water) would be expected in the course of a normal water year. Added to the inherent natural variability in salinity at this location, the salinity regime under managed conditions may be particularly sensitive to operational policies of the South Florida Water Management District, which determines timing and distribution of flows through the L-31W/C-111 canal systems (Van Lent et al., 1993). Finally, the lack of tidal passes in this area prevents mixing with water from the Atlantic Ocean, which results in extended periods of lowered or raised salinity.

Central Florida Bay

This area has a long history of hypersaline conditions; 70 ppt was recorded in the 1950s (Finucane and Dragovitch, 1959). Direct inflow of surface fresh water to this area is limited by the absence of creeks (only one—McCormick Creek) and the occurrence of a low ridge along the northern shoreline (Craighead, 1964; Schomer and Drew, 1982; Wanless et al., 1994). In order for surface water from the Everglades to enter central Florida Bay, water stages in the marshes must exceed ridge elevation. Hypothetically, groundwater seepage may occur through the highly porous Miami Limestone Formation into this portion of Florida Bay (Schomer and Drew, 1982), although the frequency and magnitude are unknown. Lack of circulation due to mud banks, coupled with high evaporation rates, tends to concentrate not only salt but also dissolved organic material in the central Florida Bay area (Fourqurean et al., 1993).

Western Florida Bay

Western Florida Bay is strongly influenced by tidal exchange with the Gulf of Mexico (Sogard et al., 1989). This tidal influence maintains salinity levels near that of sea water in all but the driest years (e.g., 1989–90) (Powell et al., 1989b; Fourqurean et al., 1993). Near the northern coastline, the water is turbid and appears to be influenced by runoff from rivers along the southwest coast of Florida (Fourqurean et al., 1993). Apparently, this runoff flows south and east around Cape Sable, mixes with Gulf waters, and enters western Florida Bay.

Southern Florida Bay

The water in the southern portion of Florida Bay is historically much clearer and is influenced by the Gulf Stream and Atlantic Ocean (Fourqurean et al., 1993). Tidal exchange is relatively low in comparison with the northwestern region (Figures 6.3 and 6.4). Recent studies have indicated that net nontidal flows are primarily from Florida Bay into the Atlantic through passes in the Florida Keys (N. Smith, 1992). Salinity is very near that of normal sea water, i.e., 35–41 ppt (Schomer and Drew, 1982).

Salinities Based on Predictive Models

All physical factors—rainfall, freshwater flow, evaporation, circulation, and flushing—must be accounted for in any attempt to predict estuarine salinities in Florida Bay based on upstream freshwater databases. Tabb (1967) investigated the relation between groundwater level in northern Taylor Slough (Homestead Well, station S-196A) and salinity data collected between 1963 and 1966 at several sites in Florida Bay. He found significant negative correlations between 16-day antecedent groundwater level in northern Taylor Slough and salinity at the mouth of the Taylor River in Little Madeira Bay ($r = -0.85$, $p < 0.01$) and between 14-day antecedent groundwater level at Homestead Well and salinity in eastern Joe Bay ($r = -0.91$, $p < 0.01$). Based on these data, he developed regression models to predict salinity at a given site (Taylor River: $Y = -6.889 \cdot X + 49.4964$ and Joe Bay: $Y = -8.033 \cdot X + 48.8991$, where X = groundwater level at Homestead Well lagged 14 or 16 days).

Sculley (1986) retested the relation using salinity data collected between 1979 and 1986. He derived weaker relations (Taylor River: $r = -0.58$, Joe Bay: $r = -0.66$), but concluded that there still existed a strong linear association between the variables. He also found the same lag time for Joe Bay (14-day antecedent groundwater level). Both researchers agreed that the relation tended to break down when groundwater levels became lower than 1.8 ft (0.55 m) at the Homestead Well

(Tabb, 1967) and that the linear relation between ground water and salinity diminished with distance from shore. White (1983) used Tabb's model to hindcast salinities in Madeira Bay over the period 1933–81 and compared those values to salinity data reported in the literature. She found a high level of accuracy in the predictions.

EVIDENCE for LONG-TERM CHANGES in FRESHWATER INFLOW

Several lines of evidence suggest that freshwater inflow from the Everglades to Florida Bay has decreased. A survey documenting anecdotal observations by long-time residents of the Keys indicates that freshwater springs formerly occurred in Florida Bay (K. K. DeMaria, Nature Conservancy and Center for Marine Conservation, Key West, personal communication). Physical evidence indicating that changes have occurred in Florida Bay includes the natural record stored in the bands of corals, as described in the next section. In addition, mathematical models simulate conditions which may have historically existed.

Biological Evidence: Coral Banding

Frequency and intensity of fluorescent banding in hard corals are highly correlated with terrestrial runoff into nearshore waters (Isdale, 1984). Fulvic acids present in runoff become part of the limestone skeleton of coral and are responsible for this fluorescence. Fluorescent banding is almost certainly a record of periodic freshwater discharges from coastal rivers into nearshore waters. In 1986, Smith et al. (1989) obtained a 105-year-old core from a brain coral (*Solenastrea bournoni*), from Peterson Key Basin, 6 km west-northwest of Lignumvitae Key in southern Florida Bay (Figure 6.3). These investigators correlated coral fluorescence data with measurements of freshwater flow at a station 64 km upstream of the Shark River mouth for 1940–85. Significant relations between fluorescence and freshwater flow were clearly apparent at the annual level, and the regression model explained 57% of the variation in annual flow of the Shark River. Fluorescence intensity was greater in earlier years of the record, with an obvious and persistent decrease first seen in 1932. A 4- to 6-year periodicity seen in the pre-1932 fluorescence record was hypothesized to be a result of rainfall periodicity and was lost when the flows were diverted by canal construction in the latter period (Figure 6.5). Using this correlation, Smith et al. (1989) then hindcast a chronology of annual flow for the Shark River back to 1881, the age of the oldest coral band. Based on this data set, the authors calculated that flows from the Shark River Slough to Florida Bay have been diminished by approximately 59% compared to premanaged conditions.

Evidence from Simulation Models

There are currently at least two simulation models that predict water depth and length of hydroperiod in grid cells in the Everglades region based on rainfall, evapotranspiration, net inflow from surrounding cells, and additions or removals by canals. Both models are driven by historical rainfall data. The Natural System

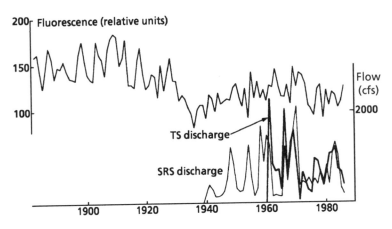

Figure 6.5   Average annual mean fluorescence of coral bands and total annual discharges from Shark River Slough (SRS) and Taylor Slough (TS). (From Smith et al., 1989.)

Model (Fennema et al., 1994) was developed to plan water policy and operations. The Adaptive Environmental Assessment Model (Walters et al., 1992; Walters and Gunderson, 1994) was developed to permit quick comparisons of the effects of alternative management policies. Groundwater flows are ignored in the Adaptive Environment Assessment Model; the Natural System Model accounts for flows into and extraction from major well fields along the eastern edge of the area being studied. Both also hindcast water depths and length of hydroperiod in a more

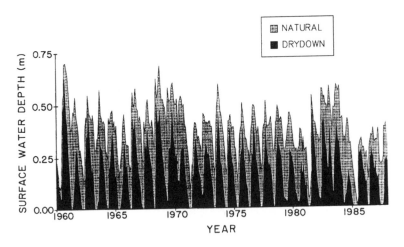

Figure 6.6   Simulated water depths in the south end of the Shark River Slough using the Adaptive Environmental Assessment Model, for two extreme scenarios: natural = all canals and diversions removed south of Lake Okeechobee; drydown = all flow to Everglades National Park diverted out of the system at the Tamiami Trail, lower end of Water Conservation Area 3A. (From Walters et al., 1992.)

natural system, unmodified by water management structures. However, they do not account for historic changes in other types of land use. Both models are able to closely reproduce seasonal and interannual variations in water depths at key index stations in the southern Everglades. Importantly, both models indicate that in most years before major water management controls, there must have been much greater flow to the southern part of the Shark River Slough, particularly late in the dry season (Walters et al., 1992) (Figure 6.6). Similarly, flows to Taylor Slough were much greater under the "natural" system, as predicted by the Natural System Model (Van Lent et al., 1993). Unfortunately, neither model as presently configured predicts either freshwater flow or salinity at the estuarine or Florida Bay interface, a variable necessary in order to reconstruct "natural" system estuarine conditions.

Recently, a first step was made to generate salinity predictions under a "natural" system scenario. Bjork and Powell (1993) compared salinity predictions generated by White (1983) with salinity predictions (using Tabb's model for the same site) based on water levels generated by the Natural System Model for the Homestead Well (provided by R. Fennema). The results (Figure 6.7) provide a first glimpse at

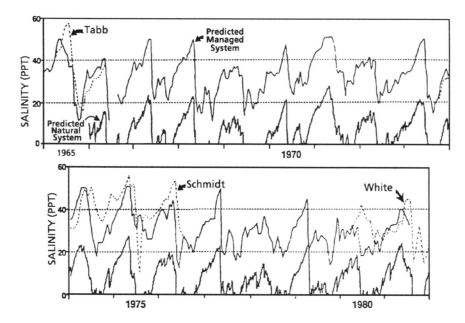

Figure 6.7 Salinities at Little Madeira Bay (north-central Florida Bay): actual vs. modeled. Lines on the graph represent the following values. (1) Actual values (dashed lines) illustrate published salinity data collected at Little Madeira Bay, from Tabb (1967) for 1965–66, Schmidt and Davis (1978) for 1973–76, and White (1983) for 1980–81. (2) Predicted managed system values (top line on graph) are calculated salinity values for Little Madeira Bay determined by Tabb's (1967) regression equation (see text) and actual groundwater levels recorded at the Homestead Well for 1965–81. (3) Predicted natural system values (lower line on the graph) are hindcasted historical salinities for Little Madeira Bay, again determined using Tabb's equation but using water levels generated by the Natural System Model (see text) for the Homestead Well.

what north-central Florida Bay might have looked like before water diversion began. The results suggest that (1) in the natural system Little Madeira Bay never experienced hypersaline conditions, even in drought years, and (2) salinities in the managed system exceeded those of the natural system at this location by 20–30 ppt.

CHANGES in FRESHWATER INFLOW and RELATION to the BIOTIC COMMUNITY

Freshwater inflow affects biota and biotic processes (e.g., productivity) in two primary ways: (1) by delivering nutrients needed by primary producers from upland sources (Funicelli, 1984) and (2) by moderating (or diluting) saline conditions to produce a gradient of salinities in estuaries (Lee and Rooth, 1973). The classic estuary (a semi-enclosed body of water in which fresh water and salt water mix, sensu Pritchard, 1967) can be divided into several zones based on a salinity characterization (Figure 6.8). Flora and fauna are often distributed in a predictable manner along complex gradients generated by the mixing of fresh and salt water (Weinstein, 1979; T. Smith, 1992; Robertson and Blaber, 1992).

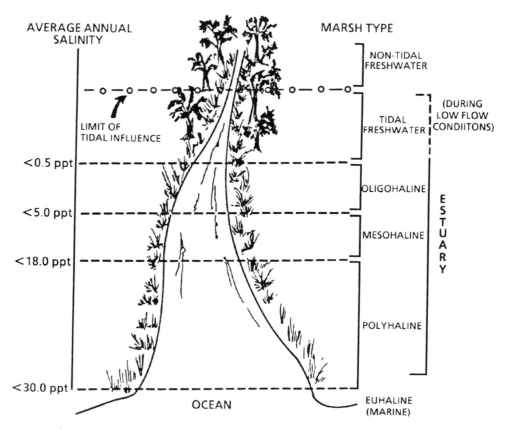

Figure 6.8 Classic estuarine salinity zones based on salinity characterization. (From Odum et al., 1984.)

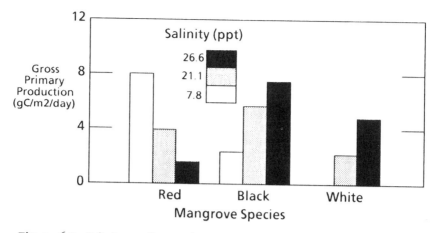

Figure 6.9 Salinity gradient and mangrove productivities for the 3 species found in south Florida. (Modified from Hicks and Burns, 1975, as cited in Odum et al., 1982.)

Mangroves

The three mangrove species found in south Florida—red mangrove (*Rhizophora mangle*), black mangrove (*Avicennia nitida*), and white mangrove (*Laguncularia racemosa*)—have different optimal zones along this salinity gradient, as indicated from data of Hicks and Burns (1975) (Figure 6.9). Figure 6.10 summarizes the hypothetical relation between position in the estuary (and correlated parameters) and net primary productivity of Florida mangrove forests. In support of the idea that red mangroves grow best at moderate salinities, Cintron et al. (1978) found

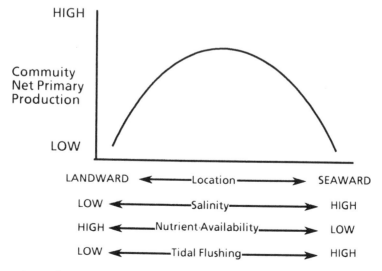

Figure 6.10 Summary of the hypothetical relation between position in the estuary (and correlated factors) and net primary production of Florida mangrove forests. (Modified from Odum et al., 1982.)

more dead than living red mangrove trees where interstitial soil salinities exceeded 65 ppt. The two other mangrove species can tolerate, although not necessarily thrive at, soil salinities up to 80–90 ppt (Odum and McIvor, 1990). One hypothesis currently being investigated to explain mangrove dieback on overwash isles in Florida Bay is that soil salinity on these isles is outside the range of tolerance of the trees (T. Armentano, Everglades National Park, personal communication).

Plankton

Freshwater inflow does not seem to directly influence the standing stock of phytoplankton as measured by concentrations of chlorophyll *a* in Florida Bay. This lack of a relation with freshwater inflow is because (1) phytoplankton are phosphorus limited in the bay and (2) very little phosphorus occurs in freshwater runoff from the Everglades to the bay (Fourqurean et al., 1993). Rather, the main source of phosphorus for phytoplankton is the Gulf of Mexico. Thus, under "normal" conditions, there is a decreasing gradient in chlorophyll *a* concentrations from southwest to northeast in Florida Bay (Fourqurean et al., 1993).

Seagrasses

Salinity tolerance of seagrasses in south Florida is similarly known. Shoal grass (*Halodule wrightii*) is broadly euryhaline. Turtle grass (*Thalassia testudinum*) thrives only in intermediate salinities of 20–40 ppt, with a maximum productivity at approximately 30 ppt. Manatee grass (*Syringodium filiforme*) tolerates only a narrow range of salinity of 32–34 ppt, near that of sea water (Zieman, 1982). While it occurs in small patches throughout Florida Bay (Powell et al., 1991), widgeongrass (*Ruppia maritima*) is normally found in small tidal streams and shallow embayments (e.g., Whitewater Bay), where salinity is seasonally low and salinity variation is pronounced. The standing crop of this species is highly variable at a given locale on relatively short time frames, presumably as a result of highly variable salinity regimes (Montague and Ley, in press). However, regional differences in nutrient availability may also be an important factor controlling its distribution (Powell et al., 1991). In general, a gradient of decreasing seagrass standing crop runs from northwest to east Florida Bay and correlates well with decreasing sediment depth (Zieman et al., 1989) and phosphorus limitation (Fourqurean et al., 1993). Standing stock and productivity of all seagrasses in Florida Bay and tributary streams are thought to be phosphorus limited (Fourqurean and Zieman, 1992; Powell et al., 1989b; Montague and Ley, in press).

It has been argued that reduced freshwater inflow into Florida Bay is implicated in the recent and widespread mortality of turtle grass in the bay (Robblee et al., 1991) (Figure 6.11). Although the die-off may be attributable to many causes, long-term salinity stress (based on tolerance of turtle grass) may have exacerbated susceptibility to other factors. To date, die-off of approximately 40,000 ha of seagrass has occurred; many areas have recolonized with other seagrass species, but die-off and water turbidity problems continue (M. Robblee, personal communication).

The scenario of mass mortality of Florida Bay seagrasses is complex, and multiple factors are undoubtedly involved. Briefly, abnormally high water tempera-

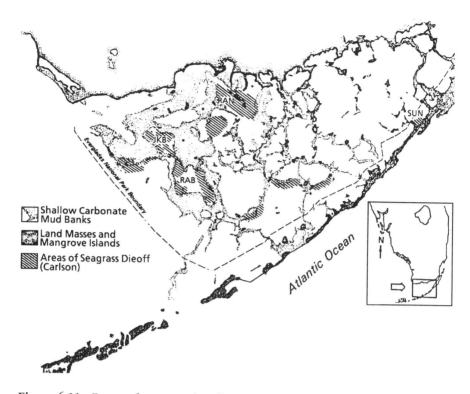

Figure 6.11 Extent of seagrass die-off in Florida Bay. RAN = Rankin Lake; JKB = Johnson Key Basin; RAB = Rabbit Key Basin; SUN = Sunset Cove. (From Robblee et al., 1991.)

ture, recent reduced frequency of hurricanes, hypersalinity, a pathogenic slime mold (*Labyrinthula* sp.) related to the causative agent of eelgrass wasting disease, and chronic hypoxia of turtle grass roots and rhizomes are potential causes of the phenomenon currently under investigation (Robblee et al., 1991). A model summarizing these hypothesized relationships is reproduced in Figure 6.12. A different explanation which emphasizes excess nutrients rather than hypersalinity has been proposed by Tomasko and LaPointe (in review).

Fish and Invertebrates

Whereas many euryhaline fishes and crustaceans tolerate hypersaline conditions, little evidence has been found in the literature to indicate that animals characteristic of Florida Bay reach their peak numbers or optimal development under such conditions. One exception may be rainwater killifish (*Lucania parva*), which Sogard et al. (1989) found to be most abundant in interior sites in Florida Bay, where salinities are often hypersaline.

There is abundant literature on fish and macroinvertebrate communities in different subregions of Florida Bay with different salinity characteristics. However, these studies cannot be used to infer the effect of salinity (or freshwater inflow) on faunal abundances because any effect of salinity is confounded by differences

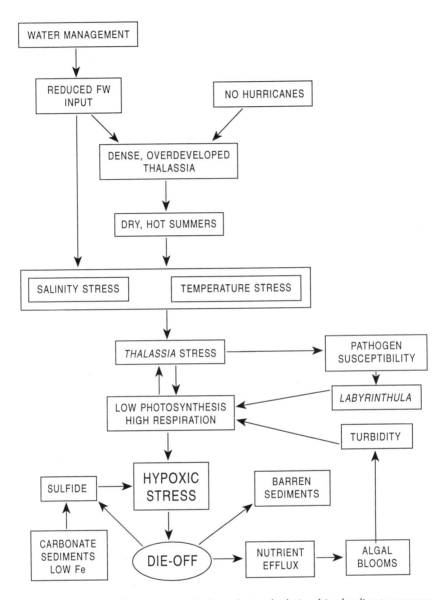

Figure 6.12 Model summarizing the hypothesized relationships leading to seagrass die-off. (From Durako et al., in press.)

in recruitment potential of fauna to basins with different circulation and flushing patterns (e.g., Sogard et al., 1989; Thayer and Chester, 1989). The best evidence for a connection between freshwater inflow and faunal abundance or recruitment thus comes from studies where catch rate or recruitment success of a given species in Florida Bay is correlated with some rainfall or runoff variable in the watershed. Both Higman (1967) and Rutherford et al. (1989a) found a positive correlation between catch rates of spotted sea trout (*Cynoscion nebulosus*) in Florida Bay and rainfall either 2 or 3 years earlier. Laboratory studies cited by Rutherford et al.

(1989a) show that larval survival and growth as well as juvenile metabolic scope for growth in spotted sea trout are greatest at 28 ppt.

Tilmant et al. (1989a) determined that recreational fishermen caught significantly more red drum (*Sciaenops ocellatus*) in years following high rainfall ($r = 0.814$, $n = 10$). The nursery habitat of juvenile common snook (*Centropomus undecimalis*) is most often in low-salinity tidal marshes and streams (Gilmore et al., 1983). Highest recruitment of this species to the fishery also follows years of higher than average rainfall (Tilmant et al., 1989b). Investigators in Texas have also observed that freshwater inflow (as measured by river discharge) is correlated with regional variations in fish harvest. Chapman (1966), cited in Day et al. (1988), showed that Texas estuaries with higher than average river discharge had higher average fish yields per unit area and that fisheries harvest was higher in wet years (Figure 6.13).

A final example of the requirement of important bay species for freshwater inflow is provided by pink shrimp (*Penaeus duorarum*). In a preliminary study, Browder (1985) reported that three variables—standardized fishing effort, water level at well station P-35 in the Shark River Slough, and shrimp catch per unit effort—explained 88% of the variation in the 14 years of combined quarterly pink shrimp data from the Tortugas fishing grounds. Water level, lagged one quarter, explained 40% of the variation in landings not explained by fishing effort. Models containing longer term databases currently include surface water and well water levels at several Everglades National Park sites from June to October each year to predict landings from the Tortugas grounds for the following November to October (P. Sheridan, National Marine Fisheries Service, unpublished data). These annual forecasts have been within 10% of actual landings for the past 3 years. The pink shrimp fishery in the Tortugas has declined by 45% in recent years (J. Hunt, Florida Department of Natural Resources, personal communication). The models of Browder (1985) and Sheridan (P. Sheridan, unpublished) would suggest that reduced freshwater inflow to Florida Bay is at least partly responsible.

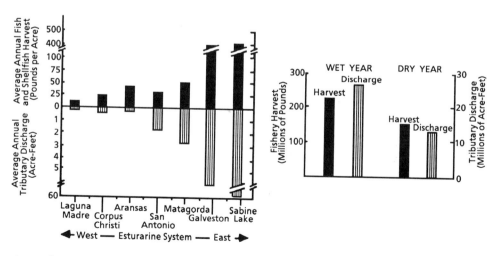

Figure 6.13 Comparison of the relationship between freshwater discharge and fisheries harvest at seven Texas estuaries. (From Chapman, 1966, as cited in Day et al., 1988.)

An exception to this pattern of enhanced recruitment following years of above-average rainfall is provided by gray snapper (*Lutjanus griseus*). For this species, there is a significant inverse relation between catch rates and rainfall in upland marshes 3 years prior (Rutherford et al., 1989b). Gray snapper does not require a low-salinity nursery habitat to complete its life cycle.

Crocodiles

About one-half of the U.S. population of American crocodiles (*Crocodylus acutus*), a federally listed endangered species, nest in Florida Bay within Everglades National Park (Mazzotti and Brandt, 1994). Over half of the nesting activity within the park occurs around Little Madeira and Joe bays in northeastern Florida Bay, with the remainder in the greater Flamingo area in western Florida Bay. Laboratory studies have suggested significantly reduced survival of hatchling crocodiles in salinities greater than 18–27 ppt (Dunson, 1970; Evans and Ellis, 1977; Ellis, 1981; Dunson, 1982). Optimum growth occurs at about 9 ppt (Dunson 1982). Young crocodiles (up to about 200 g or 4 months of age) in the wild can live and grow even in hypersaline conditions, but they are dependent on periodic access to fresh water (Mazzotti, 1983), either low-salinity water in their nursery area or a surface lens of water from rainfall. This initial growth period for the hatchlings (August–December) is a time when low salinities would naturally occur in Florida Bay. However, salinities exceeding 30 ppt have been regularly recorded in northeastern Florida Bay during that time of year (Ley, 1992). Reduced growth rates and, hence, decreased survival of young crocodiles have occurred during years of low rainfall (Moler, 1991). First-year survival estimates are lowest for Everglades National Park (<10%) compared to nearby nesting areas at Turkey Point (25–30%) (F. Mazzotti, unpublished data) and Key Largo (20%) (Moler, 1991). Mazzotti (personal communication) suggests that lower juvenile survival estimates in Everglades National Park are related to greater distance from hatching sites to low-salinity nursery habitats.

Although tolerant of high salinities, adult crocodiles prefer a low-salinity habitat, spending most of their time in waters <10 ppt (Mazzotti, 1983). Although crocodile nesting in Everglades National Park was reported to be stable from 1978 through 1986 (Mazzotti, 1989), there has been a shift in nesting from eastern to western Florida Bay (F. Mazzotti, personal communication). Given the salinity preference of adults, nursery requirements of juveniles, and elevated salinities in northeastern Florida Bay from the 1988–90 drought, this distributional shift in nesting could be interpreted as a move toward less saline, more stable conditions.

Crocodile nests located in low-elevation marl sites along creek banks are susceptible to flooding from rapid increases in ground water. Such inundation results in embryonic mortality (Mazzotti, 1989; Mazzotti and Brandt, 1989). If increased freshwater flow to Florida Bay elevated groundwater levels in the marl prairies and creeks of the C-111 and Taylor River drainages, nest flooding would be more likely. Therefore, proposed freshwater increases to Florida Bay could negatively influence crocodile reproduction. However, the impact of water discharges on ground water in the nesting area is still not clearly understood. There has been a marked decline in creek bank nests in recent years (F. Mazzotti, personal communication) and a shift to the more successful sites located on shorelines and islands (Mazzotti, 1989). With the return to a more "natural" system of freshwater input, the benefits of maintaining a low-salinity nursery habitat and even possibly decreasing dispersal distances by extending the nursery habitat closer to hatching sites could outweigh the potential loss of creek nests, should greater releases cause increased nest flooding (F. Mazzotti, personal communication).

Manatees

Florida Bay provides a year-round habitat for the Florida manatee (*Trichechus manatus latirostris*), a federally listed endangered species. During recent surveys, Everglades National Park waters accounted for up to 20% of the west coast population and 9% of the state-wide population (Snow, 1991). Submergent aquatic vegetation, such as seagrasses, is a major component of the diet of manatees. Although manatees tolerate marine and hypersaline conditions, they are most frequently found in fresh or brackish waters. Therefore, effects of changes in freshwater flow on salinity patterns, seagrasses, and overall quality of foraging habitat in Florida Bay (along with water temperature) are important influences on distribution and abundance of manatees in the area (R. Snow, Everglades National Park, personal communication).

Preliminary data show interannual variability in distribution of manatees within northeastern Florida Bay, which may in part be related to freshwater flow and the resultant distribution of food resources (R. Snow, personal communication). Relative abundance of manatees in the northeast bay was higher in 1992 compared to 1990 following several years of drought (R. Snow, personal communication). Accounts of manatee distribution and abundance from the 1930s suggest a decline in the use of northern Florida Bay; in particular, the northeastern bay, "including the streams and estuaries bordering it," was the most important area for manatees in southern Florida (Beard, 1938).

Observations of manatees in northern Florida Bay are now extremely rare (Snow, 1991, 1992). During 1979–81 (a period of higher rainfall compared to 1991), significant numbers of manatees were observed in the southern and southeastern Whitewater Bay area (S. Bass, Everglades National Park, personal communication). In recent years, few manatees have been observed there, and in some areas submerged vegetation is scarce (R. Snow, personal communication).

Birds

Ospreys (*Pandion haliaetus*), which are resident in Florida Bay, have shown a dramatic decline in population: nesting numbers have decreased from over 200 pairs in the 1970s (Kushlan and Bass, 1983) to about 70 pairs in the 1990s (Bowman et al., 1989; J. Curnutt and J. Mercador, University of Tennessee, unpublished data). Subpopulations of ospreys that breed and forage exclusively in Florida Bay had lower reproductive success than subpopulations that forage in the Atlantic Ocean, suggesting lowered food availability in the bay (Bowman et al., 1989).

Interestingly, the Florida Bay population of bald eagles (*Haliaeetus leucocephalus*), which also depends on food resources from the bay, has not shown such a decline. In contrast to ospreys, a synthesis of data from 1959 through 1990 indicates that bald eagle nesting abundance and reproduction in Florida Bay have been stable throughout that period; reproductive success compares favorably to other U.S. populations (Curnutt, 1991). These data suggest that bald eagles inhabiting Florida Bay have not been significantly affected by changes in habitat conditions throughout the past three decades.

Differences in population stability and reproduction between bald eagles and ospreys in Florida Bay, both top-level carnivores dependent on Florida Bay food resources, may be related to differential plasticity in their diets. Bald eagles are opportunistic feeders and eat, alive or as carrion, numerous foods including fish, waterbirds, and terrapins (Peterson, 1978). In contrast, ospreys are obligate piscivores. The flexibility in diet of bald eagles would allow them greater stability when some of their food sources are depressed.

Wading birds are good indicator organisms of ecosystem health because they respond quickly to changes in both local and regional hydrological conditions and associated foraging conditions, as reviewed by Ogden (1994). Nesting of wading birds has historically (i.e., before 1946) been concentrated in the mangrove–freshwater ecotone along headwaters of major rivers feeding into Whitewater Bay and the lower Gulf coast, i.e., East River to Lostman's River (Ogden, 1994). Between the 1930s and the 1960s, many of these colonies collapsed, and most birds now nest in the interior Everglades north of the Shark River Slough, largely in the Water Conservation Areas (Frederick and Collopy, 1988; Ogden, 1994). One likely explanation for such a change in abundance and location of nesting birds is decreased quantity and availability of food in the estuarine ecotone due to lowered freshwater inflow (Powell et al., 1989a; Walters et al., 1992; Ogden, 1994; Bjork and Powell, 1993). This zone is highly dynamic, and processes that regulate secondary productivity of small fish and invertebrates in the estuary are not well understood (Lorenz and Powell, 1992). Potential factors leading to this decline in productivity include a compression of the low-salinity zone (Bjork and Powell, 1993); wide ranges in salinity within the ecotone exceeding physiological tolerances of some of the prey organisms (Tabb and Roessler, 1989); fewer total nutrients as a result of lowered inflow, thereby limiting both primary and secondary productivity (Roessler, 1967, as cited in Tabb and Roessler, 1989; Powell et al., 1989b); and loss of permanent flooding in the deeper and southern portions of the Shark River and Taylor sloughs (Ogden, 1994). In contrast, wading birds nesting in northwest Florida Bay, which depend on mainland estuaries around Cape Sable peninsula, have remained relatively stable since the 1930s (Ogden, 1994). Ogden (1994) suggests that this foraging habitat may be the least changed because it is not directly in the path of major freshwater discharges from the Shark River Slough.

Clutch size in great white herons (*Ardea herodias occidentalis*) nesting in eastern Florida Bay from 1981 through 1984 was compared to similar data from 1923, which predates major human alterations to Everglades hydrology. Herons that fed naturally in Florida Bay during the recent period had significantly smaller clutches and produced significantly fewer fledglings than those of 1923, suggesting that habitat quality in eastern Florida Bay was reduced from the 1923 levels (Powell and Powell, 1986).

Seagrass flats are the primary foraging areas for herons and, therefore, serious effects to the Florida Bay population may result from seagrass die-off. It has been suggested that the instability of this population, as well as the roseate spoonbill (*Ajaia ajaja*) population which nests in Florida Bay and depends on the south Florida ecotone for foraging, is likely related to manipulation of south Florida hydrology (Powell et al., 1989a).

ENTIRE ECOSYSTEMS and FRESHWATER FLOW

In addition to the community and population effects described above, the influence of freshwater flow occurs at the scale of the entire Florida Bay ecosystem. Particular attention should be drawn to one such hypothesis offered by Browder and Moore (1981) regarding the influence of freshwater flow on quality of habitats that serve as nursery grounds. Fauna using the estuary are thought to recruit to their preferred salinity (or habitat) zones. Motile forms, with salinity optima defined by physiological tolerance limits, move with the salinity zone as it changes position as a result of the mixing of constantly changing volumes of salt and fresh water. Zones in the estuary (as defined by habitat type, e.g., mangrove shoreline, creek channel) are stable over longer time scales than are those defined by salinity of water masses. Distribution of some organisms is probably defined more by specific habitat requirements (e.g., seagrass canopy for protection from predators, primary foraging habitat) than by narrow salinity requirements. As previously described, habitat features such as seagrasses are in turn influenced by the salinity regime to which they are subjected. Nevertheless, all organisms residing in the estuary must locate areas with acceptable combinations of both salinity and habitat. The dynamic nature of the overlap of the fixed habitat zone and constantly moving salinity zone, and how this overlap relates to the production of commercial or recreational fishery species, is illustrated conceptually in Figure 6.14.

To test the hypothesis of Browder and Moore, a 4-year study was recently conducted in northeastern Florida Bay (Ley, 1992; Montague and Ley, in press). They found that the standard deviation of mean salinity was the best environmental

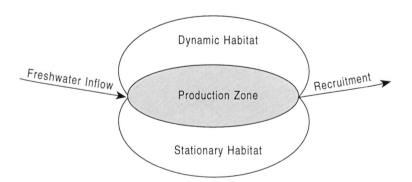

Figure 6.14   Conceptual model of the influence of freshwater inflow on recruitment of fisheries organisms. (From Browder and Moore, 1981.)

correlate with mean submerged plant biomass and benthic animal density; for every 3-ppt increase in standard deviation, total benthic plant biomass decreased by an order of magnitude. Abundance of larger species of fishes was also significantly lower in areas of rapid salinity variation. To explain these findings, Ley and Montague (in press) hypothesized that sudden, wide fluctuations in salinity may inhibit development of abundant seagrasses (and associated invertebrate fauna) due to physiological stress that reduces seagrass growth and survival. Sparse seagrass development results in reduced habitat quality in upstream areas of northeastern Florida Bay; these areas are potentially prime nursery grounds for several species of fish (Ley, 1992). The validity of this hypothesis was illustrated in late summer of 1988, when a large volume of fresh water was suddenly introduced into the saline habitats of Manatee Bay and Barnes Sound from a point source, the C-111 canal (Figure 6.2). For over a year prior to the release, Manatee Bay had experienced salinities ranging from 20 to 35 ppt. On August 15, 1988, an earthen dam was removed from the C-111 canal outfall to relieve upland flooding. Over an 8-day period, 2065 ft$^3 \cdot$sec^{-1} (58 m^3/sec) of fresh water was introduced to the estuary, and salinity dropped to zero in Manatee Bay (D. Haunert, personal communication). Ecologists and fishermen observed extensive seagrass, echinoderm, sponge, crustacean, and gastropod mortality associated with the event.

Finally, brief comments are in order regarding potential cascading effects of documented declines in the biota on the environmental health of Florida Bay. Relationships between these phenomena with freshwater inflow are not fully documented. As discussed earlier, however, long-term salinity stress may have increased the vulnerability of seagrasses to other agents of mortality. Connections are most likely indirect, occurring via complex geochemical, hydrological, and ecological relationships. The origin and predicted fate of present blooms of planktonic algae in Florida Bay (continuing since November 1991) remain under investigation. Factors implicated in these blooms include nutrients released from seagrass die-off, dissolved organic matter washed from the southwestern mangrove zone into Florida Bay following the passage of Hurricane Andrew in August 1992 (J. Zieman, personal communication), and nutrients transported from phosphate-rich rivers to the north (e.g., Peace River) mixing with nitrogen-rich flows from Shark River Slough (B. LaPointe, personal communication). The algal bloom has been directly implicated in the mass mortality of most species of sponges in affected portions of Florida Bay (Butler et al., 1992). Because loggerhead sponges (*Spheciospongia vesparium*), one of the affected species, provide habitat for juvenile spiny lobsters (*Panulirus argus*) (Butler et al., 1992), a decline in the lobster fishery in south Florida is highly likely.

SUMMARY and CONCLUSIONS

Evidence that freshwater flow to Florida Bay has been reduced compared to historical flows was reviewed in this chapter. Because rainfall patterns have not changed significantly from the late 1800s (Duever et al., 1994), changes in freshwater inflow to the estuary must of necessity be due to water management practices in the watershed. Impounding of water in the Water Conservation Areas and diversion of water away from the Shark River and Taylor sloughs for urban use and flood control have modified the salinity regime of Florida Bay, making it on average more salty or even hypersaline. Significant portions of Florida Bay are now hypersaline during drought years. Smaller portions of the bay remain hypersaline at the end of the dry season even in years of normal rainfall. Hypersalinity is exacerbated by continued reduced tidal flushing caused by filling of passages between some of the Florida Keys for railroad construction in 1910–11. Further, the lack of significant hurricanes passing over Florida Bay in the past 32 years has allowed mud banks to accrete, thereby reducing internal circulation. Extreme hypersaline conditions (>60 ppt) are detrimental to even euryhaline organisms normally found in estuaries where salinity levels are variable (Tabb and Roessler, 1989). Evidence from Florida Bay, Texas estuaries, and elsewhere strongly supports the idea that secondary productivity of many fishes and invertebrates is enhanced by adequate amounts and "natural" timing of freshwater inflow. There is strong evidence to indicate that food availability for osprey and several species of wading birds in Florida Bay is inadequate to maintain normal reproductive outputs in these species. All these observations suggest that estuarine productivity in Florida Bay has deteriorated compared to historical levels. Reference to Figure 6.10 is instructive in integrating these observations and conclusions: the authors suggest that in many times and places, the Florida Bay ecosystem is being pushed past the "seaward" end of the gradient to hypersalinity, where productivities are relatively low due to salinity-induced stress.

In a recent review of existing databases from Florida Bay, a panel of experts concluded that whereas the "preponderance of evidence indicates that the ecosystem of northeastern Florida Bay would benefit by restoring the amount and timing of freshwater flows through Taylor Slough and the...C-111 basin, the benefits to the entire Bay...are uncertain (Boesch et al., 1993)." We agree, but submit that if restoration of flows to Shark River Slough proceeds in concert with those to Taylor Slough, this dual approach would have a higher probability of ameliorating hypersaline conditions in more extensive areas of the bay.

There is obviously an ecological need for a water management system that delivers sufficient fresh water to the marshes to mimic historic hydropatterns and that avoids both sudden pulses of fresh water and excessively long periods of below-normal flows to Florida Bay. Balancing freshwater needs of the urban and agricultural users with those of the greater Everglades ecosystem presents significant challenges for managers of water, environmental quality, endangered species, biodiversity, and ecosystem health.

ACKNOWLEDGMENTS

The authors thank T. Armantano, S. Bass, K. K. DeMaria, J. Hunt, B. LaPointe, F. Mazzotti, M. Robblee, P. Sheridan, R. Snow, and J. Zieman for providing unpublished data or findings for inclusion in this chapter. J. Fourqurean, G. V. N. Powell, P. Sheridan, P. Vohs, and J. van Armen provided insightful and very helpful comments on an earlier draft. We thank J. van Armen for assistance with graphics.

LITERATURE CITED

Alleem, A. A. 1972. Effect of river outflow management on marine life. *Mar. Biol.,* 15:200–208.

Bancroft, G. T., A. M. Strong, R. J. Sawicki, W. Hoffman, and S. D. Jewell. 1994. Relationships among wading bird foraging patterns, colony locations, and hydrology in the Everglades. in *Everglades: The Ecosystem and Its Restoration,* S. M. Davis and J. C. Ogden (Eds.), St. Lucie Press, Delray Beach, Fla., chap. 25.

Beard, D. B. 1938. Wildlife Reconnaissance: Everglades National Park Project, National Park Service Report, Everglades National Park, Homestead, Fla.

Bjork, R. D. and G. V. N. Powell. 1993. Relationships between Hydrological Conditions and Quality and Quantity of Foraging Habitat for Roseate Spoonbills and Other Wading Birds in the C-111 Basin, Draft final report to the South Florida Research Center, Everglades National Park, Homestead, Fla., April 1993.

Bowman, R., G. V. N. Powell, J. A. Hovis, N. C. Kline, and T. Wilmers. 1989. Variations in reproductive success between subpopulations of the osprey (*Pandion haliaetus*) in south Florida. *Bull. Mar. Sci.,* 44(1):245–250.

Browder, J. A. 1985. Relationship between pink shrimp production on the Tortugas grounds and water flow patterns in the Florida Everglades. *Bull. Mar. Sci.,* 37:839–856.

Browder, J. A. and D. Moore. 1981. A new approach to determining the quantitative relationship between fishery production and the flow of fresh water to estuaries. in Proc. Natl. Symp. on Freshwater Inflow to Estuaries, Vol. I, R. Cross and D. Williams (Eds.), FWS/OBS-81/04, Office of Biological Services, U.S. Fish and Wildlife Service, Washington, D.C., pp. 403–430.

Butler, M. J., IV, W. F. Herrnkind, and J. H. Hunt. 1992. Sponge mass mortality and Hurricane Andrew: Catastrophe for juvenile spiny lobsters in south Florida? in Symp. on Florida Keys Regional Ecosystem (abstract), University of Miami, Coral Gables, Fla.

Carlson, P. R., M. J. Durako, T. R. Barber, L. A. Yarbro, Y. deLama, and B. Hardin. 1990. Catastrophic Mortality of the Seagrass *Thalassia testudinum* in Florida Bay, Annual Completion Report, Grant CM-257, Office of Coastal Zone Management, Florida Department of Environmental Regulation, Tallahassee, 52 pp.

Chapman, C. 1966. The Texas basins project. in Symp. on Estuarine Fisheries, R. Smith, A. Swartz, and W. Massman (Eds.), *Am. Fish. Soc. Spec. Publ.,* 3:83–92.

Cintron, G., A. E. Lugo, D. J. Pool, and G. Morris. 1978. Mangroves of arid environments in Puerto Rico and adjacent islands. *Biotropica,* 10:110–121.

Costello, T. J. and D. M. Allen. 1966. Migrations and geographic distribution of pink shrimp, *Penaeus duorarum,* of the Tortugas and Sanibel grounds, Florida. *Fish. Bull.,* 65(2):449–459.

Costello, T. J., D. M. Allen, and J. H. Hudson. 1986. Distribution, seasonal abundance and ecology of juvenile northern pink shrimp, *Penaeus duorarum,* in the Florida Bay area. *NOAA Tech. Mem.,* NMFS-SEFC-161:1–84.

Craighead, F. C. 1964. Land, mangroves and hurricanes. *Fairchild Trop. Gard. Bull.,* 19(4):1–28.

Curnutt, J. L. 1991. Population Ecology of the Bald Eagle (*Haliaeetus leucocephalus*) in Florida Bay, Everglades National Park, Florida 1959–1990, 1991 Annual Report, South Florida Research Center, Everglades National Park, Homestead, Fla., Vol. 1, Section 5.

Davis, G. E. and J. W. Dodrill. 1989. Recreational fishery and population dynamics of spiny lobsters, *Panulirus argus,* in Florida Bay, Everglades National Park, 1977–1980. *Bull. Mar. Sci.,* 44:78–88.

Day, J. W., Jr., C. A. S. Hall, W. M. Kemp, and A. Yanez-Arancibia. 1988. *Estuarine Ecology,* Wiley-Interscience, New York, 558 pp.

Duever, M. J., J. F. Meeder, L. C. Meeder, and J. M. McCollom. 1994. The climate of south Florida and its role in shaping the Everglades ecosystem. in *Everglades: The Ecosystem and Its Restoration,* S. M. Davis and J. C. Ogden (Eds.), St. Lucie Press, Delray Beach, Fla., chap. 9.

Dunson, W. A. 1970. Some aspects of electrolyte and water balance in three estuarine reptiles, the diamondback terrapin, American and "salt water" crocodiles. *Comp. Biochem. Physiol.,* 32:161–174.

Dunson, W. A. 1982. Osmoregulation of crocodiles: Salinity as a possible limiting factor to *Crocodylus acutus* in Florida Bay. *Copeia,* pp. 374–385.

Durako, M. J., T. R. Barber, J. B. C. Bugden, P. R. Carlson, J. W. Fourqurean, R. D. Jones, D. Porter, M. B. Robblee, L. A. Yarbro, R. T. Zieman, and J. C. Zieman. In press. Seagrass die-off in Florida Bay. in Proc. of the Gulf of Mexico Symposium.

Ellis, T. M. 1981. Tolerance of hypersalinity by the Florida crocodile, *Crocodylus acutus, J. Herpetol.,* 15(2):187–192.

Enos, P. 1989. Islands in the bay—a key habitat of Florida Bay. *Bull. Mar. Sci.,* 44(1):365–386.

Evans, D. H. and T. M. Ellis. 1977. Sodium balance in the hatchling American crocodile, *Crocodylus acutus. Comp. Biochem. Physiol.,* 58:159–162.

Fennema, R. J. C. J. Neidrauer, R. A. Johnson, T. K. MacVicar, and W. A. Persins. 1994. A computer model to simulate natural Everglades hydrology. in *Everglades: The Ecosystem and Its Restoration,* S. M. Davis and J. C. Ogden (Eds.), St. Lucie Press, Delray Beach, Fla., chap. 10.

Finucane, J. H. and A. Dragovitch. 1959. Counts of red tide organisms, *Gymnodinium breve,* and associated oceanographic data from Florida west coast 1954–1957. *U.S. Fish Wildl. Serv. Spec. Sci. Rep. Fish.,* 289:209–295.

Flanagan, C. A. and J. R. Hendrickson. 1976. Observations on the commercial fishery and reproductive biology of the totoaba, *Cynoscion macdonaldi,* in the northern Gulf of California. *Fish. Bull.,* 74(3):531–544.

Fourqurean, J. W. and J. C. Zieman. 1992. Phosphorus limitation of primary production in Florida Bay: Evidence from C:N:P ratios of the dominant seagrass *Thalassia testudinum. Limnol. Oceanogr.,* 37(1):162–171.

Fourqurean, J. W., R. D. Jones, and J. C. Zieman. 1993. Processes influencing water column nutrient characteristics and phosphorus limitation of phytoplankton biomass in Florida Bay, FL, USA: Inferences from spatial distributions. *Estuarine Coastal Shelf Sci.,* 36:295–314.

Frederick, P. C. and M. W. Collopy. 1988. Reproductive Ecology of Wading Birds in Relation to Water Conditions in the Florida Everglades, Tech. Rep. No. 30, Florida Cooperative Fish and Wildlife Research Unit, School of Forestry Research and Conservation, University of Florida, Gainesville, 259 pp.

Funicelli, N. A. 1984. Assessing and managing effects of reduced freshwater inflow to two Texas estuaries. in *The Estuary as a Filter,* V. S. Kennedy (Ed.), Academic Press, Orlando, pp. 435–446.

Gilmore, R. G., C. J. Donohoe, and D. W. Cooke. 1983. Observations on the distribution and biology of east-central Florida populations of the common snook, *Centropomus undecimalis. Fla. Sci. Spec. Suppl.,* 45(3/4):313–336.

Gunter, G. 1961. Some relations of estuarine organisms to salinity. *Limnol. Oceanogr.,* 6(2): 182–190.

Hicks, D. B. and L. A. Burns. 1975. Mangrove metabolic response to alterations of natural freshwater drainage to southwestern Florida estuaries. in *Proc. Int. Symp. on the Biology and Management of Mangroves,* G. Walsh, S. Snedaker, and H. Teas (Eds.), University of Florida, Gainesville, pp. 238–255.

Higman, J. B. 1967. Relationships between catch rates of sport fish and environmental conditions in Everglades National Park, Florida. *Proc. Gulf Carib. Fish. Inst.,* 19:129–140.

Holmquist, J. G., G. V. N. Powell, and S. M. Sogard. 1989. Sediment, water level, and water temperature characteristics of Florida Bay's grass-covered mudbanks. *Bull. Mar. Sci.,* 44(1): 348–364.

Isdale, P. J. 1984. Fluorescent bands in massive corals record centuries of coastal rainfall. *Nature,* 310:578–579.

Johnson, R. A. and R. J. Fennema. 1989. Conflicts over Flood Control and Wetland Preservation in the Taylor Slough and Eastern Panhandle Basins of Everglades National Park, South Florida Research Center, Homestead, Fla.

Kushlan, J. A. and O. L. Bass, Jr. 1983. Decreases in the southern Florida osprey population, a possible result of food stress. in *Biology of Bald Eagles and Ospreys,* D. M. Bird (Ed.), Harpel Press, Ste-Anne-de-Bellevue, Quebec, pp. 187–200.

Lee, T. N. and C. Rooth. 1973. Water Movements in Shallow Coastal Bays and Estuaries, Sea Grant Coastal Zone Management Bull. #3, University of Miami, Coral Gables, Fla., 19 pp.

Ley, J. A. 1992. Influence of Freshwater Flow on the Use of Mangrove Prop Root Habitat by Fish, Ph.D. dissertation, University of Florida, Gainesville, 171 pp.

Lloyd, M. 1964. Variation in the oxygen and carbon isotope ratios of Florida Bay mollusks and their environmental significance. *J. Sedimentol. Petrol.,* 36(1):84–111.

Lorenz, J. J. and G. V. N. Powell. 1992. Influence of Hydrology on Fish Populations in the Mangrove Dominated Zones of the C-111 and Taylor Slough Basins and Biscayne Bay: Second Annual Progress Report, National Audubon Society, Tavernier, Fla., 40 pp.

Mazzotti, F. J. 1983. The Ecology of *Crocodylus acutus* in Florida, Ph.D. dissertation, The Pennsylvania State University, University Park, 161 pp.

Mazzotti, F. J. 1989. Factors affecting the nesting success of the American crocodile, *Crocodylus acutus,* in Florida Bay. *Bull. Mar. Sci.,* 44:220–228.

Mazzotti, F. J. and L. A. Brandt. 1989. Evaluating the Effects of Groundwater Levels on the Reproductive Success of the American Crocodile in Everglades National Park, Final Report to the U.S. Army Corps of Engineers and South Florida Water Management District, 120 pp.

Mazzotti, F. J. and L. A. Brandt. 1994. Ecology of the American alligator in a seasonally fluctuating environment. in *Everglades: The Ecosystem and Its Restoration,* S. M. Davis and J. C. Ogden (Eds.), St. Lucie Press, Delray Beach, Fla., chap. 20.

McCallum, J. and K. Stockman. 1959. Salinity in Florida Bay, Geological Miscellaneous No. 21, Shell Oil Company, Houston, 14 pp.

McHugh, J. L. 1967. Estuarine nekton. in *Estuaries,* G. H. Lauff (Ed.), American Association for the Advancement of Science, Washington, D.C., pp. 581–620.

Moler, P. E. 1991. American Crocodile Population Dynamics, Final Report, Study No. 7532, Florida Game and Fresh Water Fish Commission, Bureau of Wildlife Research, Tallahassee, 24 pp.

Montague, C. L. and J. A. Ley. In press. The abundance of benthic vegetation and associated fauna along salinity gradients in northeastern Florida Bay: A possible result of salinity fluctuations. *Estuaries.*

Montague, C. L., R. D. Bartleson, and J. A. Ley. 1989. Assessment of Benthic Communities along Salinity Gradients in Northeastern Florida Bay, Final Report to the South Florida Research Center and South Florida Water Management District, West Palm Beach, 156 pp.

Nichols, F. H., J. E. Cloern, S. N. Louma, and D. H. Peterson. 1986. The modification of an estuary. *Science,* 231:567–573.

Odum, W. E. and C. C. McIvor. 1990. Mangroves. in *Ecosystems of Florida,* R. L. Myers and J. J. Ewel (Eds.), University of Central Florida Press, Orlando, pp. 517–548.

Odum, W. E., C. C. McIvor, and T. J. Smith III. 1982. The Ecology of Mangroves of South Florida: A Community Profile, FWS/OBS-81/24, Office of Biological Services, U.S. Fish and Wildlife Service, Washington, D.C., 144 pp.

Odum, W. E., T. J. Smith III, J. K. Hoover, and C. C. McIvor. 1984. The Ecology of Tidal Freshwater Marshes of the United States East Coast: A Community Profile, FWS/OBS-83/17, Office of Biological Services, U.S. Fish and Wildlife Service, Washington, D.C., 177 pp.

Ogden, J. C. 1994. A comparison of wading bird nesting colony dynamics (1931–1946 and 1974–1989) as an indication of ecosystem conditions in the southern Everglades. in *Everglades: The Ecosystem and Its Restoration,* S. M. Davis and J. C. Ogden (Eds.), St. Lucie Press, Delray Beach, Fla., chap. 22.

Peterson, D. W. 1978. Southern bald eagle. in *Rare and Endangered Biota of Florida,* Vol. II: Birds, H. W. Kale II (Ed.), University Presses of Florida, Gainesville, pp. 27–30.

Powell, G. V. N. and A. H. Powell. 1986. Reproduction by great white herons *Ardea herodias* in Florida Bay as an indicator of habitat quality. *Biol. Conserv.,* 36:101–113.

Powell, A. B., D. E. Hoss, W. F. Hettler, D. S. Peters, L. Simoneaux, and S. Wagner. 1987. Abundance and Distribution of Ichthyoplankton in Florida Bay and Adjacent Waters, Report SFRC-87/01, South Florida Research Center, Everglades National Park, Homestead, Fla., 45 pp.

Powell, G. V. N., R. D. Bjork, J. C. Ogden, R. T. Paul, A. H. Powell, and W. B. Robertson. 1989a. Population trends in some Florida Bay wading birds. *Wilson Bull.,* 101:436–457.

Powell, G. V. N., W. J. Kenworthy, and J. W. Fourqurean. 1989b. Experimental evidence for nutrient limitation of seagrass growth in a tropical estuary with restricted circulation. *Bull. Mar. Sci.,* 44(1):324–340.

Powell, G. V. N., J. W. Fourqurean, W. J. Kenworthy, and J. C. Zieman. 1991. Bird colonies cause seagrass enrichment in a subtropical estuary: Observational and experimental evidence. *Estuarine Coastal Shelf Sci.,* 32:567–579.

Pritchard, D. W. 1967. What is an estuary: Physical viewpoint. in *Estuaries,* G. H. Lauff (Ed.), American Association for the Advancement of Science, Washington, D.C., pp. 3–5.

Robblee, M. B., J. L. Tilmant, and J. Emerson. 1989. Quantitative observations on salinity. *Bull. Mar. Sci.,* 44(1):523 (abstract).

Robblee, M. B., T. B. Barber, P. R. Carlson, Jr., M. J. Durako, J. W. Fourqurean, L. M. Muehlstein, D. Porter, L. A. Yarbro, R. T. Zieman, and J. C. Zieman. 1991. Mass mortality of the tropical seagrass *Thalassia testudinum* in Florida Bay (USA). *Mar. Ecol. Prog. Ser.,* 71:297–299.

Robertson, A. I. and S. J. M. Blaber. 1992. Plankton, epibenthos and fish communities. in *Coastal and Estuarine Studies: Tropical Mangrove Ecosystems,* A. I. Robertson and D. M. Alongi (Eds.), American Geophysical Union, Washington, D.C., pp. 173–224.

Roessler, M. 1967. Observations on Seasonal Occurrence and Life Histories of Fishes in Buttonwood Canal, Everglades National Park, Ph.D. dissertation, University of Miami, Coral Gables, Fla., 154 pp.

Rutherford, E. S., J. T. Tilmant, E. B. Thue, and T. W. Schmidt. 1989a. Fishery harvest and population dynamics of spotted seatrout, *Cynoscion nebulosus,* in Florida Bay and adjacent waters. *Bull. Mar. Sci.,* 44(1):108–125.

Rutherford, E. S., J. T. Tilmant, E. B. Thue, and T. W. Schmidt. 1989b. Fishery harvest and population dynamics of gray snapper, *Lutjanus griseus,* in Florida Bay and adjacent waters. *Bull. Mar. Sci.,* 44(1):139–154.

Schmidt, T. W. and G. E. Davis. 1978. A Summary of Estuarine and Marine Water Quality Information Collected in Everglades National Park, Biscayne National Monument, and Adjacent Estuaries from 1879 to 1977, Report T–519, South Florida Research Center, Everglades National Park, Homestead, Fla., 59 pp.

Schomer, N. S. and R. D. Drew. 1982. An Ecological Characterization of the Lower Everglades, Florida Bay, and the Florida Keys, FWS/OBS-82/58.1, Office of Biological Services, U.S. Fish and Wildlife Service, Washington, D.C., 246 pp.

Sculley, S. P. 1986. Florida Bay Salinity Concentration and Ground Water Stage Correlation and Regression, South Florida Water Management District, West Palm Beach.

Simenstad, C. A., D. A. Jay, and C. R. Sherwood. 1992. Impacts of watershed management on land-margin ecosystems: The Columbia River estuary. in *Watershed Management: Balancing Sustainability and Environmental Change,* R. J. Naiman (Ed.), Springer-Verlag, New York, pp. 266–306.

Smith, N. P. 1992. Long-term net transport through tidal channels in the Florida Keys. in 1992 Symp. on Florida Keys Regional Ecosystem (abstract), University of Miami, Coral Gables, Fla.

Smith, T. J., III. 1992. Forest structure. in *Coastal and Estuarine Studies: Tropical Mangrove Ecosystems,* A. I. Robertson and D. M. Alongi (Eds.), American Geophysical Union, Washington, D.C., pp. 101–136.

Smith, T. J., III, J. H. Hudson, M. B. Robblee, G. V. N. Powell, and P. J. Isdale. 1989. Freshwater flow from the Everglades to Florida Bay: A historical reconstruction based on fluorescent banding in the coral *Solenastrea bournoni. Bull. Mar. Sci.,* 44(1):274–282.

Snow, R. W. 1991. The Distribution and Relative Abundance of the Florida Manatee in Everglades National Park, Annu. Rep. South Florida Research Center, Everglades National Park, Homestead, Fla., 26 pp.

Snow, R. W. 1992. The Distribution and Relative Abundance of the Florida Manatee in Everglades National Park: An Interim Report of Aerial Survey Data from March 1991 through February 1992, South Florida Research Center, Everglades National Park, Homestead, Fla., 30 pp.

Sogard, S. M., G. V. N. Powell, and J. G. Holmquist. 1987. Epibenthic fish communities on Florida Bay banks: Relations with physical parameters and seagrass cover. *Mar. Ecol. Prog. Ser.,* 40:25–39.

Sogard, S. M., G. V. N. Powell, and J. G. Holmquist. 1989. Spatial distribution and trends in abundance of fishes residing in seagrass meadows on Florida Bay mudbanks. *Bull. Mar. Sci.,* 44(1):179–199.

South Florida Water Management District. 1992. Surface Water Improvement and Management Plan for the Everglades, West Palm Beach, 202 pp.

Stanley, D. J. and A. G. White. 1993. Nile Delta: Recent geological evolution and human impact. *Science,* 260:628–634.

Swart, P. K., P. Kramer, J. J. Leder, and J. H. Hudson. 1992. A 120 year record of natural and anthropogenic variations in Florida Bay based on oxygen and carbon isotopic variations in a coral *Solenastrea bournoni.* in Symp. on Florida Keys Regional Ecosystem, RSMAS, University of Miami, Miami, Nov. 16–20, 1992, abstract.

Tabb, D. C. 1967. Predictions of Estuarine Salinities in Everglades National Park, Florida by the Use of Ground Water Records, Ph.D. dissertation, University of Miami, Coral Gables, Fla., 107 pp.

Tabb, D. C. and M. Roessler. 1989. History of studies on juvenile fishes of coastal waters of Everglades National Park. *Bull. Mar. Sci.,* 44(1):23–34.

Tabb, D. C., T. R. Alexander, T. M. Thomas, and N. Maynard. 1967. The Physical, Biological, and Geological Character of the Area South of C-111 Canal in Extreme Southeastern Everglades National Park, Florida, University of Miami, Coral Gables, Fla.

Thayer, G. W. and A. J. Chester. 1989. Distribution and abundance of fishes among basin and channel habitats in Florida Bay. *Bull. Mar. Sci.,* 44:200–219.

Thayer, G. W., D. R. Colby, and W. F. Hettler, Jr. 1987. Utilization of the red mangrove prop root habitat by fishes in south Florida. *Mar. Ecol. Prog. Ser.,* 35:25–38.

Tilmant, J. T. 1989. A history and an overview of recent trends in the fisheries of Florida Bay. *Bull. Mar. Sci.,* 44:3–22.

Tilmant, J. T., E. S. Rutherford, and E. B. Thue. 1989a. Fishery harvest and population dynamics of red drum (*Sciaenops ocellatus*) from Florida Bay and adjacent waters. *Bull. Mar. Sci.,* 44(1):126–138.

Tilmant, J. T., E. S. Rutherford, and E. B. Thue. 1989b. Fishery harvest and population dynamics of the common snook (*Centropomus undecimalis*) from Florida Bay and adjacent waters. *Bull. Mar. Sci.,* 44(1):523–524 (abstract).

Tomasko, D. A. and B. E. LaPointe. In review. The potential role of nutrient enrichment of the eastern Gulf of Mexico on a massive die-off of seagrass in Florida Bay (U.S.A.). *Bull. Mar. Sci.,* in review; Symp. on Florida Keys Regional Ecosystem, RSMAS, University of Miami, Miami, Nov. 16–20, 1992.

Van Lent, T. J., R. Johnson, and R. Fennema. 1993. Water Management in Taylor Slough and Effects on Florida Bay, Draft Report, South Florida Research Center, Everglades National Park, National Park Service, Homestead, Fla., 88 pp.

Walters, C. J. and L. H. Gunderson. 1994. A screening of water policy alternatives for ecological restoration in the Everglades. in *Everglades: The Ecosystem and Its Restoration,* S. M. Davis and J. C. Ogden (Eds.), St. Lucie Press, Delray Beach, Fla., chap. 30.

Walters, C., L. Gunderson, and C. S. Holling. 1992. Experimental policies for water management in the Everglades. *Ecol. Appl.,* 2(2):189–202.

Wanless, H. R., R. W. Parkinson, and L. P. Tedesco. 1994. Sea level control on stability of Everglades wetlands. in *Everglades: The Ecosystem and Its Restoration,* S. M. Davis and J. C. Ogden (Eds.), St. Lucie Press, Delray Beach, Fla., chap. 8.

Weinstein, M. P. 1979. Shallow marsh habitats as primary nurseries for fishes and shellfish, Cape Fear River, North Carolina. *Fish. Bull.,* 77(2):339–357.

White, D. 1983. Oceanographic Monitoring Study 1980–1983, South Florida Research Center, Everglades National Park, Homestead, Fla., 144 pp.

Zieman, J. C. 1982. The Ecology of the Seagrasses of South Florida: A Community Profile, FWS/OBS-82/25, Office of Biological Services, U.S. Fish and Wildlife Service, Washington, D.C., 158 pp.

Zieman, J. C., J. W. Fourqurean, and R. L. Iverson. 1989. Distribution, abundance and productivity of seagrasses and macroalgae in Florida Bay. *Bull. Mar. Sci.,* 44(1):292–311.

II

SPATIAL and TEMPORAL CHARACTERISTICS of ECOSYSTEM DRIVING FORCES

7

AGE, ORIGIN, and LANDSCAPE EVOLUTION of the EVERGLADES PEATLAND

Patrick J. Gleason

Peter Stone

ABSTRACT

The Everglades—a huge freshwater marshland and, in large part, peatland—developed in recent geologic time during a globally controlled convergence of both climatic change and sea level rise within a shallow bedrock trough located in south Florida. The recession of glaciers in northern North America at the end of the Pleistocene period and the change to a subtropical climate in south Florida provided both the abundant precipitation and the seasonal rainfall climate necessary for the generation of the Everglades wetland ecosystem. The rising sea level has undoubtedly retarded runoff and downward leakage out of the trough and helped to retain water within the Everglades basin. This, in turn, has allowed thick accumulations of peat (3–3.7 m) to develop within the deeper parts of the basin. The eastern coastal ridge, which was necessary to retain water within and in part defines the Everglades basin, owes its origin to marine geologic deposition which last occurred during the Sangamon interglacial age (about 125,000 years before present [YBP]), when sea level was up to 8 m above the present level. Repeated alterations between freshwater and marine conditions are revealed for interglacial times by the limestone rock record, with freshwater limestone layers occurring within the generally marine limestone sequence. Over the past 5000–6000 years, the southern end of the Everglades trough, open to Florida Bay and the Gulf of Mexico, has had the continually rising sea again move across it, into the Everglades and coastal salt marshes, transgressing over previously freshwater habitats.

Formation of the Everglades peatland, which is supported by a freshwater wetland system of seasonal flooding and dominated by marshes with scattered tree islands, began on a large scale by about 5000 YBP. Portions of the present Everglades area had become short-flooded, seasonal, calcitic mud marshes around 6500 YBP or even earlier. Extensive areas of marsh peats, with basal dates in the 5000–4500 YBP range, indicate the widespread development or rapid expansion of wetland environments with long peat-forming hydroperiods early in the history of the Everglades.

The Everglades began to develop after Paleo-Indians had already entered present-day south Florida. The young geologic age of the Everglades peatland is indicated by its development contemporaneously with the rise elsewhere of the Mayan, Egyptian, and Sumerian cultures.

Transgressive sequences of marine-brackish sediments overlying freshwater sediments are very evident within Whitewater Bay, along the Gulf coast from Cape Sable to the Harney River, and throughout Florida Bay, indicating that freshwater environments of about the same age and related to the early Everglades were once considerably more extensive to the south and southwest. Peat deposits underlying open water on the south end of Lake Okeechobee indicate that the Everglades marsh also once extended northward, into present Lake Okeechobee.

The common occurrence of charcoal fragments in peats, especially sawgrass peats, is evidence for the pervasiveness of fire in the prehistoric Everglades and its associated long-term ecological influence. However, no widely correlated charcoal or ash layer has been noted.

A somewhat drier (or more strongly seasonal) hydrologic regime prevailed in the central Everglades in the period about 3000 to roughly 2000 YBP (and later in some places), as indicated by a calcitic mud layer intercalated in marsh peats. Interbedding of peats and calcitic muds at various locations and depths within the Everglades, especially Taylor Slough, suggests that other fairly long-term hydrologic fluctuations may have occurred in the past. Calcitic muds (often called marl) are very important to the paleohydrologic record because they record hydrologic conditions below the "wetness" threshold for marsh peat formation and accretion. Calcitic muds are formed by precipitation of calcite by blue-green algae in submerged algal mats ("periphyton"). Calcitic mud deposits occur in areas which received runoff high in calcium and bicarbonate concentration from nearby or distant limestone terrains.

Limited age dating of forest peats underlying the most common type of tree island in the northeastern Everglades suggests that these small rounded bayhead tree islands formed in the past 1200 years or so. Patches of water lily marsh peats that dislodge and float in deeper water areas of the Everglades appear to have initiated these small bayhead tree islands. Floating peat islands also give rise to "alligator holes" or similar depressions when they are displaced laterally. Tree islands and both floating islands and their emergence holes are convergence points for Everglades wildlife. Development of floating peat islands may cause much of the patchy mosaic pattern of the marsh vegetation in the northeastern Everglades, as well as most of the forest patches.

A continuation in the rise of sea level that has occurred since the end of the last glacial maximum could result in a significant reduction in the area of the

Everglades in the future. If, as a consequence of the "greenhouse effect," sea level experiences a 70-cm rise within the next century, as predicted by some, then significant reductions in the area of the Everglades will be experienced much more quickly. In the past, however, interglacial periods such as the present Holocene were followed by a resurgence of glaciers. Human-caused overdrainage coupled with drought would then be the likely scenario. Local climatic impacts will be superimposed on global climatic forces. Clear prediction of the long-term future of climate or sea level is not possible, although the present upward rise in sea level is undeniable. For the foreseeable future, in any case, the fate of the Everglades ecosystem is in human hands, whether for good or for bad.

INTRODUCTION

An overview of the factors which gave rise to the Everglades (Figure 7.1) and then continued to influence the development of the peatland during the past 5000 years is presented in this chapter. The Everglades, as a wetland environment, spread out upon a long dry and emerged landscape. The favorable conditions which allowed the Everglades to form were due to a unique convergence of globally controlled and interrelated forces: climate and sea level. These forces acted upon the broad shallow Everglades bedrock basin to create the seasonal conditions of rainfall, temperature, and hydrology favorable to the proliferation of marshland plant communities. The limestone floor of the Everglades basin was largely shaped by deposition during Pliocene and Pleistocene times within a shallow pseudo-atoll basin (Petuch, 1986, 1987) and in response to changes in sea level during the Ice Age (mainly during the Pleistocene). Extensive sawgrass marshes, water lily sloughs, and numerous tree islands, bordered downflow by coastal mangrove swamps and upflow by an enlarging lake, all developed within this basin.

Evidence for the picture of Everglades development presented here comes from many sources. A global perspective assists in elucidating climatological and sea level shifts contributing to the actual formation and development of the Everglades basin. Analogy with other peatlands such as the Okefenokee Swamp and marllands such as Big Cypress Swamp was used in interpretation of causality in wetland formation. The Everglades sediment record and radiocarbon dating of peats and some calcitic muds have revealed past vegetational stages and changes, including the development of tree islands. Most discussion here will relate to the Everglades peatland, because it is from this deposit that most information has been garnered. Modern observations in the Everglades have been applied to older sediments in interpreting their original environments of formation and the sequence of past conditions.

The term Everglades is used here to denote the freshwater wetland occupying the contiguous basin aside but mostly south of Lake Okeechobee. The term is also used to refer to the geographic area encompassed (e.g., Everglades basin). Included in this definition of the Everglades are nearby former extensions of freshwater marshes to the north in areas now inundated and within Lake Okeechobee and tentatively also those forming the freshwater peats and calcitic muds buried under saltwater sediments in adjacent mangrove swamps,

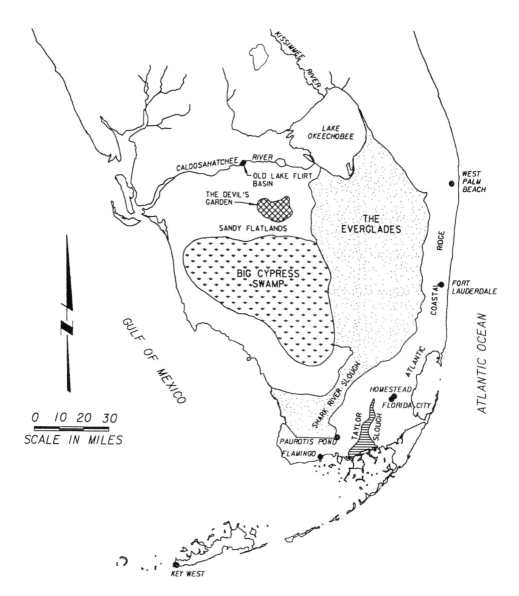

Figure 7.1 Natural surface features of south Florida. (Adapted from Schroeder and Klein, 1954.)

Whitewater Bay, Florida Bay, and nearshore Gulf of Mexico (further research there may show additional similarities to Big Cypress Swamp). Also included are the associated and contiguous wet prairie area in Taylor Slough and adjacent areas of the southernmost tip and southeastern edge of the peninsula, which are underlain by freshwater marls and some peats. The prehistory of this wetland—peatland, marlland, and some flooded rockland—can best be understood if put into the context of geologic and human events. From a geologic perspective, the Everglades, although it appears timeless and ancient, is actually very young.

The EVERGLADES in GEOLOGIC and HISTORICAL TIME

The Holocene Epoch

The Everglades peatland formed during the latter half of the Holocene epoch (Table 7.1), that is, during the last 5000 years (radiocarbon years, or YBP). The Holocene epoch followed the Pleistocene epoch. The Pleistocene epoch was

Table 7.1   Geologic Chronology Important to the Everglades Bedrock Basin

Period/epoch	Age[a]	Years before present[b]	Sediment- and rock-forming events	Terraces and associated higher sea level stands (above mean sea level)
Holocene	(Probably interglacial)	0–10,000	Formation/development of Everglades peatlands (beginning ca. 5000 YBP)	Rising sea level; coastal marshes advancing inland
Quaternary Period				
Pleistocene "Ice Age"	Wisconsin (glacial)	10,000–67,000[b]	Erosion	Low sea level
	Sangamon (interglacial)	67,000[b]–128,000	Miami Limestone, Anastasia Formation, Key Largo Limestone, Coffee Mill Hammock Formation (Brooks Unit 6)[d]	Princess Ann (6 m/18 ft) Pamlico (8 m/25 ft)
	Illinoisan (glacial)	128,000–180,000	Erosion	Low sea level
	Yarmouth (interglacial)	180,000–230,000	Fort Thompson Formation (Brooks Unit 3, 4, and 5)[d]	Talbot (12 m/40 ft) Penholoway (21 m/70 ft)
	Kansan (glacial)	230,000–300,000	Erosion	Low sea level
	Aftonian (interglacial)	300,000–330,000	Caloosahatchee Formation (Brooks Unit 2)[d]	Wicomico (27 m/90 ft)
	Nebraskan unnamed in U.S.	330,000–470,000 470,000–2,000,000	Caloosahatchee Formation (Brooks Unit 1)	Okefenokee (37 m/120 ft–43 m/140 ft)
Tertiary Period				
Pliocene		2,000,000–5,000,000	Caloosahatchee Formation, Tamiami Formation	

[a] Major glacial advances in North America separated by major periods of glacial ice recession.

[b] Age dates for Pleistocene ages are from Fairbridge (1968). All are approximate, and stage boundaries are interpreted in part; for example, note that many investigators consider the Sangamon to extend no younger than around 100,000 YBP.

[c] Adapted from Brooks (1968, 1974) and Duever et al. (1979). The correlation of geologic formations to sea level stands is subject to debate.

[d] Mitterer (1975) determined the ages of zones corresponding to Q zones of Perkins (1977), which in turn correlate with Brooks (1974). Units as follows: Unit 6 (Q5) = 134,000 YBP; Unit 5 (Q4) = 180,000 YBP; Unit 4 (Q3) = 236,000 YBP; Unit 3 (Q2) = 324,000 YBP.

largely the time of the great Ice Age (although that began in previous Pliocene times), during the last part of which the huge Laurentide ice mass formed over Canada and the northern United States. The Holocene epoch began about 10,000 YBP and marks a time of glacial retreat, significant change in climate and vegetation, sea level rise (sea level was 35–40 m below current sea level at the beginning of the Holocene epoch) (Bloom, 1983), and many important extinctions of land mammals such as the mastodon (Lundelius, 1988; Webb, 1974).

Even after the enormous environmental change that gave rise to the Holocene postglacial (or interglacial) conditions, there continued an oscillating sequence of lesser climatic changes which have been recognized primarily in cooler European latitudes from pollen stratigraphy. These changes reflect significant fluctuations in temperature there, although in subtropical regions, such as south Florida, climatic variations were more likely marked by precipitation changes. Table 7.2 shows a chronology of Holocene stages based primarily on European paleobotanical indicators and their approximate correlation with archeological cultures in both Europe and south Florida (Fairbridge, 1968; Milanich and Fairbanks, 1980). Climatic changes shown in Table 7.2 have not been correlated with or shown to exist in south Florida, but they suggest that climatic changes may have influenced south Florida with a more subtle effect. Profound changes in nonflooded "upland" vegetation occurred in south Florida in postglacial times, as recently as 5000–7000 YBP, and these probably relate in part to climatic changes (Watts, 1975, 1980; Watts and Hansen, 1988). As will be shown later, evidence does exist for some prolonged oscillations in hydrology and therefore probably climate, as evidenced by alternating strata of peat and marl in the southern Everglades and likely by the existence of organic muds (mucks) interbedded with peats around Lake Okeechobee. The origin of the Everglades in mid-Holocene time presumably also had a climatic component with an increase in precipitation, in addition to the effect of rising sea level.

The youth of the Everglades is accentuated by the advances that major civilizations were making at the time that the Everglades peatland first began to form (about 5500–5000 YBP). This time marks the height of the Sumerian civilization between the Tigris and Euphrates Rivers, where writing and wheeled vehicles were in use, the smelting of gold and silver was known, and cities, agriculture, and animal husbandry existed, as did bread and even beer. In Egypt, the use of numerals additionally had developed and the Great Sphinx of Giza dates from about this time. In nearby Central America, the first date in the Mayan chronology (3372 B.C.) dates back to the beginning of the Everglades peatland.

Indian Cultures in South Florida

Paleo-Indians in south Florida predate the Everglades peatland by at least 5000 years. They predate the earlier Everglades marl wetland by thousands of years (with some possible limited marl areas of exception). Archeological evidence from Jefferson County, Florida provided an excellently preserved bison skull with a chert fragment protruding from the right fronto-parietal area. Dating of the skull at about 11,000 YBP documents an association between the bison and Paleo-Indians (Webb et al., 1984). Investigations at Warm Mineral Spring and Little Salt Spring in Sarasota

Table 7.2 Holocene Chronology of Mid-Latitude Climatic Variation and Comparison of European and South Florida Indian Cultures

Years before present[a] (YBP)	Blytt–Sernander[a] climate classification key dates (YBP)	Mid-latitude[a] mean (century) temperature departures (°C)	Northern[a] Europe archeological cultures	South Florida indian cultures[b]	
				Okeechobee Basin	Circum-Everglades
0–1000	Late Subatlantic				
	600	−1	Historic	Seminole	Seminole
	1000	+0.5	Viking	Belle Glade plain pottery	Glades III
1000–2300	Early Subatlantic				
	1600	+1	Roman	Belle Glade	
	2000	−0.5		plain pottery	
	2300	+1	Iron Age	Corn cultivation (Fort Center)	
2300-3700	Late Subboreal				
	3000	−1	Bronze Age	Sand- and fiber-tempered pottery (3000 YBP)	Sand- and fiber-tempered pottery (3000 YBP)
3700–5300	Early Subboreal				
	4100	+2	Neolithic		
	4300	−0.5			
5300–6600	Main Atlantic				
	5500	+2.5	Mesolithic	Onset of Everglades peatland development (ca. 5000 YBP)	
	6500	+1		Calcitic mud deposition near Lake Okeechobee	
6600–7500	Early Atlantic				
	7000	+0.5			
7500–8700	Late Boreal				
	7500	+0.5			
	7800	+1			
8700–9800	Early Boreal				
	8800	−0.5			
	9000				
9800–10,300	Preboreal	+1		(Paleo-Indians in south Florida)	

[a] Adapted from Fairbridge (1968). Refers to European climatic variations. These cultures and temperature departures have not been detected in south Florida, but are presented to show that finding evidence of climatic variation during the Holocene, principally as a consequence of rainfall variations in south Florida, is not unlikely.

[b] Adapted from Milanich and Fairbanks (1980).

County indicate the existence of people in south Florida as far back as 10,000–12,000 YBP (Clausen et al., 1975a, 1975b, 1979; Cockrell and Murphy, 1978). Sparse findings associated with Paleo-Indians suggest that they were small groups of hunters, perhaps moving from water hole to water hole over the much drier landscape which comprised south Florida at the end of the last ice age (Clausen et al., 1979). Most Paleo-Indian sites may be submerged below the Gulf of Mexico and the Atlantic Ocean due to the much lower sea level and therefore much wider peninsula at that time. Evidence of Paleo-Indians has not been found in the

Everglades proper, but their nearby locations (including Cutler, near Miami on the Atlantic Coastal Ridge; Carr, 1986) suggest that they may have wandered across fairly barren pre-Everglades terrain, where limestone crusts formed the ground surface. Archeological sites have been useful in documenting the kinds of plants and animals that lived at the time of these earliest people in south Florida.

As late as 7000–5000 YBP, humans at least seasonally or occasionally occupied peninsular interior sites and focused upon shallow depressions, possibly for water or subsistence (slough site) (Clausen et al., 1979; Beriault et al., 1981). By the end of this time, people were occupying small topographic elevations in the Everglades basin, very possibly for protection from flooding (Carr and Sandler, 1992; Masson et al., 1988, the latter with a date of 4840 YBP for an occupation level from a present tree island). Within the Everglades, these prehistoric encampments occurred mainly on what are now tree islands (Carr, 1979). Larger sites are situated along the Everglades periphery, near the Miami River, the New River in Fort Lauderdale (Carr and Beriault, 1984), and at Belle Glade (Willey, 1949). The most complex early Indian culture in interior south Florida developed not far from the Everglades in the Lake Okeechobee basin at the Fort Center site located on Fisheating Creek (Sears, 1982; Milanich and Fairbanks, 1980). Initial occupation of this site was about 3000 YBP, while the Everglades peatland was still rapidly expanding. Indians of the Fort Center site practiced corn (maize) horticulture at least as early as 2400 YBP (Sears and Sears, 1976).

ENVIRONMENTAL CHANGES LEADING to the ESTABLISHMENT of the EVERGLADES

Geologic Processes Create the Everglades Basin

The extended seasonal period of flooding (hydroperiod) required for development of the Everglades marshland owes its existence in large part to a limestone bedrock depression, the development of which may have been reef-controlled (Petuch, 1986). This limestone depression conforms to a pseudo-atoll lagoon surrounded by fossil reefs (Petuch, 1986, 1987). The youngest reef tract is Pliocene in age and underlies the Atlantic Coastal Ridge and other topographic highs that surround the Everglades. Petuch (1987) speculates that the pseudo-atoll may have formed around a meteorite impact feature which occurred in late Eocene (Ocala) time in the position of the present southern Everglades. The Pliocene reef tract may have produced the original topographic high along the southeastern coast, which was later covered by oolitic rock and sand in subsequent Pleistocene marine stages to form the coastal ridge.

Geologic processes that created the Everglades basin occurred during the past 5 million years. The oldest rocks are from the Tamiami Formation, comprising the western side and underlying the southern side of the Everglades trough. Rocks and sediments immediately flooring the southern and eastern portions of the Everglades were deposited during the most recent period of marine submergence, the much younger Sangamon interglacial period. The Fort Thompson Formation comprising the floor of the northern Everglades was, in part, contemporaneous, but also was deposited during one or more earlier interglacial periods. Deposition of the sandy-shelly Anastasia Formation and the oolitic Miami Limestone occurred at a higher elevation nearer to the present coast; this ridge isolated the large shallow depression behind it, although, as indicated above, the Pliocene coral reef tracts along the coast occurred previously. Much later, after nearly 100,000 years of exposure as land, these rocks and sediments along the eastern side of the Everglades basin in effect blocked the drainage of fresh water to the Atlantic Ocean. Shallow fresh water accumulating seasonally in the long trough extending south from Lake Okeechobee to the Gulf of Mexico gave rise then to a vast marshland—the Everglades. Similar sequences have been recognized in south Florida (as described below) and in the southeastern United States, where freshwater wetlands evolved on a long drained and exposed depression of interglacial age in Georgia and elsewhere (Parrish and Rykiel, 1979).

Examination of the geology underlying and adjacent to the Everglades indicates the way in which the Everglades trough was formed and the fickleness of sea level with respect to south Florida. The seas have inundated portions of south Florida at least five to seven times (Brooks, 1968; Perkins, 1977) since the beginning of deposition of the Tamiami Formation (Table 7.2). Freshwater limestone layers within the marine strata of the Fort Thompson Formation appear similar to marl sediments currently being deposited in southern parts of the Everglades. These occur widely at or near the bedrock surface in the northern Everglades (Parker and Cooke, 1944; Brooks, 1974) and are buried deeper beneath the Atlantic Coastal Ridge in what is now the Biscayne Aquifer (up to eight such layers) (Schroeder et al., 1958). Thus, the present Everglades is by no means the first occurrence of extensive shallow freshwater marsh environments in south Florida.

A generalized geologic map of south Florida and the Everglades is shown in Figure 7.2. The floor of the Everglades basin is made up of both the Fort Thompson and the Anastasia Formations in the northern half and mainly the Miami Limestone Formation in the southern half. Sediments comprising the eastern edge of the trough consist of the Anastasia Formation to the north and the Miami Limestone Formation to the south. The western side of the northern Everglades is bordered by sandy flatlands underlain by the Caloosahatchee Formation. The southern Everglades is bordered to the west mainly by Big Cypress Swamp, which is underlain by the more highly elevated rocks of the Tamiami Formation, along with some erosional remnants or outliers of Miami Limestone (Meeder, as cited in Duever et al., 1979; Duever, 1984).

The Tamiami Formation, deposited in Pliocene time, constitutes the oldest surface rock of the Everglades trough. Sandy limestone to even calcareous sandstone prevails and is widely fossiliferous. The top of the Tamiami Formation dips underground to the east from Big Cypress Swamp, occurring beneath younger limestones.

The Caloosahatchee Formation, found near the Caloosahatchee River and around Lake Okeechobee, is characterized by a wide variety of sediment types including sands and shell beds and a variety of well-preserved fossil shells, many taxa of which are now extinct. The formation is thin, from about 3 m to a maximum of about 13 m, and is buried beneath the Fort Thompson Formation.

The Fort Thompson Formation occupies the axis of the current Everglades

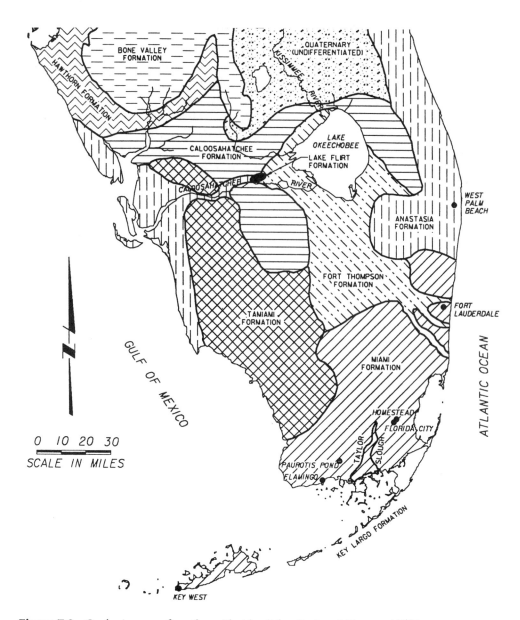

Figure 7.2 Geologic map of southern Florida. (After Puri and Vernon, 1964.)

trough in the northern portion. The Fort Thompson Formation has been redefined by geologists numerous times, and discussion continues. The exact number of recognized units within this formation is debated. Sellards (1919) and Parker and Cooke (1944) describe it as having alternating marine and freshwater limestones and shell beds, with oscillations of sea level therefore indicated. The formation was deposited on a formerly exposed erosional (solutioned) surface, and two additional such surfaces are found within and at the top of the formation (Parker and Cooke, 1944).

Very recent freshwater calcitic muds, forming the surface in the southern Everglades, are deposited near older limestones that are dissolving and in environments where the length of seasonal flooding (hydroperiod) is less than required for marsh peat deposition and accumulation. If peats were also originally deposited in the Fort Thompson Formation, they have all been eroded away by subsequent transgressions of the sea or oxidized by prolonged unflooded exposure. We speculate that during the time of deposition of the freshwater limestones of the Fort Thompson Formation, the present-day northern Everglades and adjacent Lake Okeechobee and coastal ridge were probably the sites of marl prairies and marshes similar to those which currently exist around the pinelands of Everglades National Park (but with less of a pinnacle-rock nature) and south and southeast of Homestead. (The Fresh Creek area at Andros, Bahama Islands, may be a better modern analog.)

The Miami Limestone consists of two distinct units: an upper unit called the oolitic facies, with spherical grains, and a lower unit called the bryozoan facies, after the principal fossil type (Hoffmeister et al., 1967). The bryozoan facies covers the greater part of Dade County and extends into adjoining counties. It forms the bedrock of the southern Everglades. To the east, it is covered by an elongated mound of cross-bedded oolitic limestone which forms the Atlantic Coastal Ridge. The Miami Limestone ends at the eastern edge of Big Cypress Swamp, except for the scattered relict deposits mentioned above. The Anastasia Formation has shellrock facies in the lower lying areas to the west of the Atlantic Coastal Ridge, and there may have been a shallow lagoon behind a bar deposit (Lovejoy, 1987).

The Key Largo Limestone is a fossil coral reef and forms the upper Florida Keys (Hoffmeister, 1974). The Tamiami Formation, at the western boundary of the Everglades, contains the oldest surface rocks. The Caloosahatchee Formation, Miami Limestone, and Fort Thompson Formation in part lie upon the Tamiami Formation and therefore are younger. The Miami Limestone, Anastasia Formation, and Key Largo Limestone are in part contemporaneous. The Fort Thompson Formation also underlies and predates the Miami Limestone. The general time correlation of the various formations either underlying or forming the boundary of the Everglades is shown in Table 7.1.

The Caloosahatchee Formation appears to be transitional in age between the Pliocene and Pleistocene periods and contains marine and terrestrial fossils and even freshwater fish remains. Bender (1973) dated corals from the Caloosahatchee Formation at 1.8 and 1.9 million years old, which puts the upper part of the formation in the early Pleistocene. The position of sea level and the paleogeography of Florida during the late Pliocene are shown on Figure 7.3.

For the Fort Thompson Formation, using a method based on degradation of organic chemicals, Mitterer (1975) determined the ages of three zones described by Perkins (1977) and correlated by Brooks (1968). The three zones are dated at 180,000, 236,000, and 324,000 YBP, moving deeper, respectively (Table 7.1).

The Miami Limestone and Key Largo Limestone date from the Sangamon interglacial period of higher sea level. The Anastasia Formation is also believed to be of Sangamon age by inference. Broecker and Thurber (1965) dated Key Largo Formation corals from the quarry at Windley Key at between 95,000 and 140,000 YBP (isotope disequilibrium dating). Osmond et al. (1965) similarly dated Miami

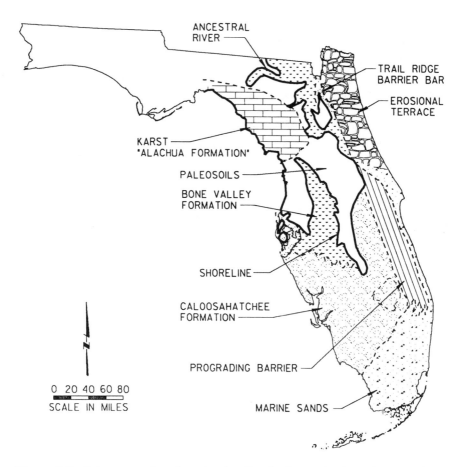

Figure 7.3 Paleogeography of peninsular Florida during the late Pliocene. (From Brooks, 1968.)

Limestone oolites at 130,000 YBP. The paleogeography of Florida and the position of the shoreline during the Sangamon interglacial are shown in Figure 7.4

These dates correspond generally with the Sangamon interglacial high sea level stand worldwide, although some may be too young (Figure 7.5). Additionally, the oolite date is subject to some additional uncertainty due to recrystallization (although it agrees well). Well-preserved fossil corals in late Pleistocene wave-cut or depositional terraces at 21 locations throughout the world have been dated and the results evaluated by Moore (1982). At almost all localities a high stand of sea level, higher than current sea level, occurred about 120,000–140,000 YBP.

A Pleistocene vertebrate fauna has been found in the Anastasia Formation at West Palm Beach. Most of the fossils are similar to faunas found near Vero Beach and Melbourne and reflect both freshwater and terrestrial communities. The assemblage suggests a wet savanna environment near the edge of a long-absent river or lake. Bison, mastodons and mammoths, land tortoises, horses, and camels were present. Tapirs, capybara, alligators, garfish, bowfin, and catfish represent the freshwater area. Carbon-14 dating of a section of mastodon rib from the site indicated 21,150 YBP (Converse, 1973). This date, having been determined from

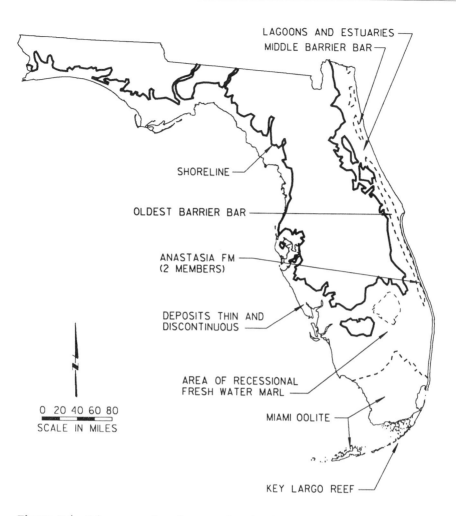

Figure 7.4 Paleogeography of peninsular Florida during Sangamon interglacial time. (From Brooks, 1968.)

bone (and possibly from below a marine shell bed) (Converse, personal communication), may very well not be reliable.

Climatic and Vegetational Changes

The last great glaciation, which culminated far to the north in Canada and the northern United States, began its retreat around 18,000–16,000 YBP (hereafter all noncalendar dates are determined by the radiocarbon method). New England, the Great Lakes region, and Canada all were covered by the massive thick Laurentide ice sheet. At that time, south Florida was very different in appearance and climate from what it is now. The sea is estimated to have been more than 100 m below its current level (Figure 7.5), and Florida's coastline was out on the present continental shelf or margin. By about 12,000 YBP most of the ice in the northeastern United States had melted, although final remnants of the Laurentide ice sheet

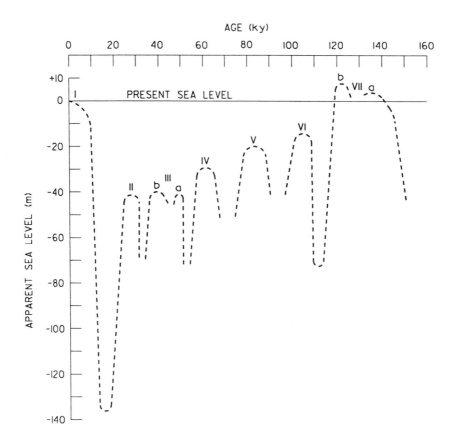

Figure 7.5 Late Pleistocene sea level curve relative to present sea level. (From Moore, 1982.)

in Canada were not dispersed until 6000 years ago (Prest, 1973). All this ice represented water removed from and much later returned to the sea.

Changes in ocean temperature must have strongly affected climatic conditions on the land. Ocean temperatures off the coast of Florida during the glacial maximum are estimated by some methods to have been 2–4°C colder than present temperatures (CLIMAP Project Members, 1981). Summer climate in the southeastern United States is dominated by the flow of moist marine air from the tropics, creating a summer rainy season. Lower ocean temperatures during the last glacial maximum likely reduced evaporation over the ocean and resulted in drier summers in Florida (Porter, 1988). Tropical low-pressure storms spawned by warm water, including hurricanes, now provide considerable rainfall on average to south Florida but were probably repressed in glacial times. Conditions in the southeastern United States during full glacial times are still subject to considerable variation in interpretation: from much colder in the Carolinas and Georgia (Watts, 1980; Watts and Hansen, 1988; Delcourt and Delcourt, 1981) to even warmer than at present, at least in winter (Carbone, 1983).

Little is directly known of terrestrial conditions in the basin prior to the development of the Everglades. This is because wetlands form the sediments that provide the record of conditions. In south Florida, the period during the last glaciation was one of erosion and solution rather than deposition. Pollen records from lakes in the central and southern Florida peninsula and from several small older wetland depressions yield information on the Holocene period somewhat prior to Everglades development, but only one evidences the glacial period preceding. The absence of glacial-age sediment in the Everglades area indicates that this long period of low sea level was without widespread persistent wetlands and probably without the southern peninsula's abundant present rainfall. Some thin calcareous crusts may date from this time period, but no substantial freshwater calcitic mud or limestone layer is yet strongly suggested to do so. Crusts do not appear to require inundation to form. The interpretative problem is complicated by the notorious unreliability of radiocarbon dates from carbonates greater than 20,000–25,000 YBP (and sometimes for dates even younger) (Morner, 1971; Stapor and Tanner, 1973). Deposits of fossil vertebrates that may include a record of these glacial-age times also are not firmly dated and may mix glacial and interglacial faunas (thus giving rise to the interpretations of cooler summers, but warmer winters, in glacial times).

Vegetational changes in the Everglades basin during the period of glacial shrinking are inferred from the changes shown through study of pollen in deep lake sediments from uplands of the central and southern peninsula, rather than by direct evidence of such changes in south Florida lowlands including the Everglades. Many of the lakes in Florida were dry in the last glacial time, although one lake has left a seemingly complete record (Watts, 1969, 1971, 1975, 1980; Watts and Hansen, 1988). A reasonable explanation for the drainage of shallow lakes in glacial times is the lowering of the water table due to reduced rainfall and a much lower sea level. Independent evidence arguing against any extreme aridity during the last and most extreme glacial period is found in the dating of Floridan Aquifer ground water in central and northern Florida, with recharge covering at least the last 30,000 years (Hanshaw et al., 1965; Navoy and Bradner, 1987). As late as about 12,000 YBP, after the entry of humans into south Florida, water levels in the few existing deep sinkholes were about 25 m below their present levels (Warm Mineral Spring and Little Salt Spring, Sarasota County) (Clausen et al., 1975a, 1975b, 1979; Cockrel and Murphy, 1978).

Glacial-age conditions eventually were replaced by open oak communities, possibly forming savanna, and then by southern pine forest on the Florida peninsula; the cypress swamps and Everglades marshes of south Florida were added finally (see below). Generalized regional vegetational maps covering the time period 14,000–200 YBP are shown in Figure 7.6 (Delcourt and Delcourt, 1981).

Lake Annie, located near Lake Placid on a southern projection on one of Florida's main sandy ridges, lies not far northwest of the Everglades and continuously records conditions as far back as 13,000 YBP (Watts, 1975, 1980; Watts and Hansen, 1988). Lake Tulane at Avon Park, about 50 km further north on the sand ridge, appears to have a continuous record including the last glacial period, when most other present Florida lakes apparently were dry (Watts and Hansen, 1988). These are the only sites known from near south Florida where the record of the regional vegetation is revealed and where sedimentation has been continuous from

Figure 7.6 Vegetational maps of the southeastern United States for (A) 18,000, (B) 10,000, (C) 5000, and (D) 200 YBP. (From Delcourt and Delcourt, 1981.) Subsequent work at Lake Tulane in south Florida by Watts and Hansen (1988) suggests that conditions there at 18,000 YBP were more like dry open pinewoods than very dry sand dune scrub (see A).

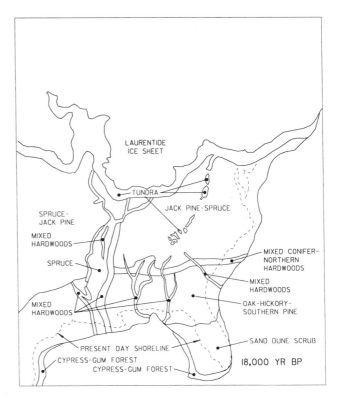

A

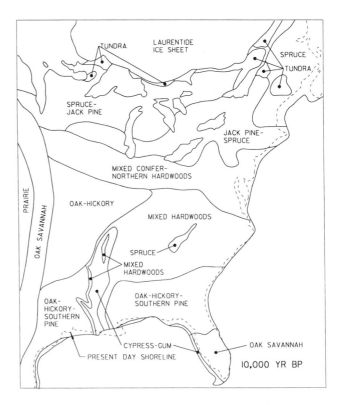

B

Figure 7.6 (continued)

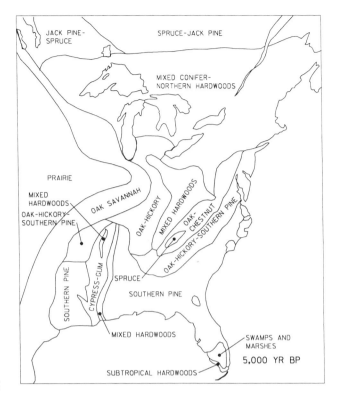

C

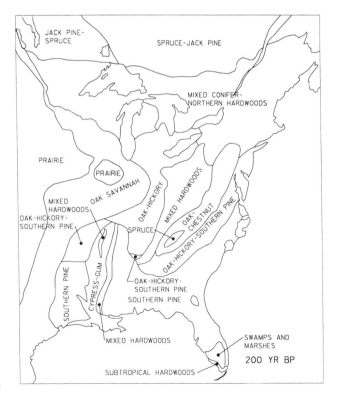

D

late or terminal glacial times to the present. At Lake Tulane, sediments from full and late glacial times (dating greater than 14,000 YBP) show pollen assemblages that suggest that the sandy uplands were then vegetated by pine woodlands with abundant grasses and herbaceous plants, either in prairie-like openings or beneath a sparse forest stand (Watts and Hansen, 1988). Pine varied widely in proportional representation. Some shrubs and herbs adapted to dry conditions were more prevalent than in near-modern times. Acidic sands surround the lake basin, and coexisting conditions on exposed hard limestone rock in the Everglades basin are unknown.

For early postglacial times at Lake Annie, near the Everglades but surrounded by acidic sandhills, Watts (1975) found an oak (*Quercus* spp.) and ragweed (*Ambrosia* spp.) pollen zone dating from 13,010 to 4715 YBP. (Actual dates such as these should be thought of as "about 13,000 YBP" or "about 4700 YBP" [or even better, "about 4500 YBP" or "about 5000 YBP," using the latter date as an example]). Pines (*Pinus* spp.) were rare to absent locally and presumably also uncommon in the regional vegetation because pine pollen is abundantly produced and wind-dispersed. An upland vegetation of oak savanna or oak scrub with prairie does not have a modern Florida example, but would be indicative of conditions drier than under the present climate. Because sea level and freshwater base level would have little hydrologic effect on sandhill vegetation, there is a strong suggestion of much less precipitation compared to modern conditions. At Lake Annie, modern vegetation and climate have prevailed since 4715 YBP. The pollen in this interval is dominated by pine, along with wax myrtle (*Myrica cerifera* L.), cypress (*Taxodium distichum* L.), holly (*Ilex* spp.), and white bay (*Gordonia lasianthus* L.), indicating pine forest associated with cypress swamps and bayheads.

Pollen studies further north in central and northern Florida and southeastern Georgia suggest that climatic conditions were somewhat drier during 8500–5000 YBP. Watts (1969, 1971) also found predominantly oak pollen at three lake sites in southeastern Georgia and peninsular Florida from 8500 to 5000 YBP, probably representing sclerophyllous or dry adapted forests, scrub, or savanna. About 5000 YBP, pine forest became predominant on upland sites throughout the coastal plain of the southeastern United States, and this may suggest a final change to an overall climate much like today's. A wet climate certainly was needed for the origin of the Everglades in south Florida. Cypress and bayhead vegetation, now common and widespread throughout Florida, became significant after 5000 YBP. Thus, the time of origin of the Everglades peatland was also one of widespread environmental change toward the present pine forest on the nonflooded upland sites in the region, while other wetlands were beginning to form on lowlands. "Upland" versus "lowland," of course, can be a matter of far less than a meter elevation difference locally in flat south Florida.

Peat deposition began somewhat earlier in the Okefenokee Swamp in southern Georgia than in the Everglades, but both occurred in mid-Holocene times. Older basal peats from Okefenokee Swamp date from about 6500 YBP (Cohen, 1973; Spackman et al., 1976) or about 1000 to 1500 radiocarbon years prior to those from the Everglades. Because Okefenokee Swamp is located at an elevation of about 35 m above present mean sea level (MSL) and the Everglades ranges from an elevation of sea level to about 6 m above MSL, it is unlikely that the onset of peat deposition in Okefenokee Swamp was caused by rising sea level. If a rising regional hydrologic base level was the most significant factor, then peats in the Everglades should have originated first, as evidenced by the much lower elevation. Thus, an increase in rainfall is likely a dominant factor in the development of the wetlands of the region, including the Everglades.

Large-scale peat deposition in south Florida began around 5700–4800 YBP, as indicated by roughly similar, older basal peat dates for the Everglades, Corkscrew Swamp (Collier County), and the savannas (St. Lucie County). However, smaller, isolated peat deposits associated with archeological sites have shown some considerably older dates. Other small deposits, and the peripheries of the large peat deposits, can have basal dates thousands of years younger.

Examples of the older peat deposit dates at archeological sites are a small marshy pond at the Windover site (Brevard County) with a basal date of 10,750 YBP (water lily peat) (Doran and Dickel, 1988), a marshy slough at Little Salt Spring (Sarasota County) with a near-basal date of 9350 YBP (Clausen et al., 1979), and a cypress strand pond at the Bay West site (Collier County) with a basal date of 7550 YBP (Beriault et al., 1981). These sites are important inasmuch as they indicate that while central and southern Florida may have been generally drier at the surface prior to about 5000 YBP, it was experiencing sufficient rainfall to give rise to some scattered wetland deposits of peat or muck sediments. This seeming anomaly may relate to upland sites responding more to climatic factors, particularly rainfall (but factors such as wind or insolation may be involved), and depression-focused sites at low elevation responding more to hydrologic base level or sea level.

Sea Level Rise

The global drop in sea level at the last glacial maximum, caused by build-up of ice in glaciers on land and cooling and contraction of sea water, is estimated to range between about 100 to 150 m below current sea level, based on worldwide studies (Shepard, 1973; Moore, 1982; Bloom, 1983). The time of sea level minimum is not known with certainty but best estimates range from 20,000 to 16,000 YBP, with 18,000 YBP widely accepted. The shoreline in south Florida at that time had shifted out to the Straits of Florida along the east coast and out toward the edge of the Floridian Plateau to the west. The melting of the continental ice sheets caused a period of rapid sea level rise between about 16,000 and 10,000 YBP (Bloom, 1983). Final disappearance of the remnants of the Laurentide ice sheet covering Canada occurred about 6000 years ago (Prest, 1973). As ocean waters flooded the south Florida land mass, the coastline moved inland and rose progressively higher. (Note, though, that many coastal land features, such as Cape Sable and various barrier islands, formed when the rise in sea level slowed, with an actual seaward or outward movement of the shoreline) (Smith, 1968; Roberts et al., 1977; Stapor et al., 1991).

Two primary schools of thought have developed worldwide with respect to the rise in sea level. The first, espoused by Fairbridge (1974), believes that sea level mainly rose until about 6000 YBP and since then has been oscillating above and below current sea level, in minor (to several meters) transgressions and regressions.

Table 7.3 Variation in the Rate of Sea Level Rise

7000–2000 YBP	2000 YBP–present
(7–0.75 m below MSL)	(0.75 m below MSL–MSL)
12 cm/100 years	4 cm/100 years

From Robbin, 1984.

The second school, espoused by Shepard (1973), Scholl et al. (1969), Robbin (1984), and many others in the region, posits that sea level mainly continued to rise with time, but the rate of this rise gradually slowed, especially after 6000 YBP. Most shallow sediments examined from south Florida seem to evidence the latter continuous gradual rise as more credible locally, but some coastal beach features offer serious challenges (Missimer, 1973, 1980; Stapor, 1983; Stapor et al., 1991). Scholl et al. (1969) based their sea level curve on radiocarbon dates of submerged peats from the south Florida coast, near shore, and mainland. Robbin similarly dated peat (mainly mangrove peat) plus three laminated limestone crust (caliche) samples from along the Florida reef tract from east of Miami to southwest of the Marquesas Keys. The locally derived sea level curves (Robbin, 1984; Scholl et al., 1969) agree closely between 6000 YBP and present.

Robbin's (1984) sea level curve at about 14,000 YBP is much shallower than other published sea level curves for the nearby Atlantic coast (Shepard, 1973; Bloom, 1983). The date is based on two submerged caliche crusts sampled from off the Florida Keys. Similar crusts form subaerially, and while many may form near sea level, not all are necessarily accurate indicators of proximity to sea level. Sea level may have been significantly below the elevation at which the crusts were forming. Scholl et al. (1969) and Robbin (1984) detected a decreasing rate of sea level rise during the past 5000–7000 years (Table 7.3).

Sea Level Elevation at the Time of Basal Peat Deposition

Robbin (1984) showed that at approximately 6000 YBP, about the time that Everglades calcitic muds were being laid down near Lake Okeechobee, sea level was only about 6 m below the current level. At 5000 YBP, about the time of initiation of the Everglades peatland, sea level was about 4.5 m below the current level.

The oldest freshwater basal peats in the Everglades Water Conservation Areas are 1–1.8 m above MSL, and thus widespread peat deposition began when sea level was about 5.5–6.3 m below these land surfaces, based on Robbin's (1984) curve at 5000 YBP.

Lake Flirt—An Enigmatic Pre-Everglades Wetland

No wetland sediments are known from the Everglades basin for around the time of the glacial maximum or termination. However, one such deposit of aquatic sediments is known as Lake Flirt which, until drainage, was a pond-like, freshwater body that straddled the headwaters of the Caloosahatchee River east of LaBelle toward Lake Okeechobee (Figure 7.1). It was described as a "shallow mud hole" in 1883 (Hanna and Hanna, 1948). Lake Flirt has since been drained by the Caloosahatchee Canal, which crosses it and exposes the sediments. Brooks (1968) recognized four principal mucky sand strata separated by three sandy calcitic mud layers. Brooks obtained a radiocarbon date of 20,900 YBP from plant material from above the lowest of the three calcitic mud beds and a date less than 7000 YBP from the uppermost mucky sand layer. The basal date roughly correlates with the glacial maximum period.

How could a glacial-age wetland have occurred at Lake Flirt but not within the Everglades? If the dates are correct (others of the same order have been obtained; P. Stone and R. Johnson, unpublished data), the conclusion can be made that at least moderate rainfall must have existed over the Lake Flirt basin in those early times, during the glacial extreme, and that base level position was not critical. The early origin and subsequent history of the deposit at Lake Flirt remains a dilemma.

FORMATION of the EVERGLADES MARSHLAND

The Calcitic Mud

The oldest postglacial wetland sediment dated from the Everglades is calcitic mud. Calcitic mud is a freshwater, frequently shelly, nonstratified, low-magnesium calcitic silt (Gleason et al., 1974). It is found underlying wide areas of the peat deposit in the northern Everglades (Davis, 1946). Within the modern surface Everglades, it is found in association with soft, thick algal mats called periphyton, which consist mostly of blue-green algae covered with calcium carbonate crystals. The periphyton grows luxuriantly in areas where the water chemistry is affected by nearby limestone exposures or canals that penetrate limestone. The growth of the algae precipitates the calcium carbonate found within the mats (Gleason, 1972; Gleason and Spackman, 1974). The algae are very sensitive to water quality and favor a slightly basic pH and relatively high calcium and bicarbonate concentrations (Browder et al., 1994).

Thicknesses of calcitic mud up to 2 m have been found southeast of Homestead, where they form the ground surface (Gallatin et al., 1958; Wanless, 1969). Calcitic muds are found underlying and interbedded with peats as far south as the saline tidal portions of Shark River Slough, Taylor Slough, Whitewater Bay, and Cape Sable (Scholl, 1964a, 1964b; Cohen, 1968; Smith, 1968; Gleason, 1972; Gleason et al., 1974) and occur in Florida Bay underneath Samphire and Russell Keys, south of Taylor Slough (Davies, 1980).

Freshwater calcitic muds have been found underlying open water in Lake Okeechobee, where samples dating $12,050 \pm 210$ and $13,160 \pm 190$ YBP (UM-190 and UM-561) (Gleason et al., 1974) indicate that seasonal wetland environments existed there during the terminal Pleistocene period of rapid glacial recession in the north. It is not yet known whether wetlands or aquatic conditions have existed continuously at Lake Okeechobee from that time.

The oldest postglacial-age sediment that has been dated from the Everglades area is a calcitic mud from below a muck-over-peat deposit on Kreamer Island in southern Lake Okeechobee. The lower layer of peat is contiguous with the adjacent Everglades onshore and represents a former extension of the Everglades. This calcitic mud is dated 6470 ± 120 YBP (UM-193) (Gleason and Stone, 1975), almost the same as a similar subpeat sample from the arm of the Everglades along the eastern shore of the lake (6320 YBP) (Brooks, 1974). The date from peat immediately above the calcitic mud beneath Kreamer Island was 5000 ± 90 YBP (UM-192). These calcitic mud dates agree well with another from the uppermost zone of calcitic mud beneath peat in Corkscrew Swamp in Collier County (6620 ± 105 YBP, UM-956) (Kropp, 1976).

The environmental significance of calcitic mud is that it indicates an annual period of flooding (hydroperiod) at the time of deposition, which was relatively short compared to those found in peat-forming marshes and was insufficient to generate a marsh peat deposit. (Here we interpret, by analogy, from the modern occurrences of calcitic mud formation and growth conditions of calcareous algal periphyton.) The environment of calcitic mud deposition is sparsely vegetated marsh, where the water surface is well lighted for the photosynthesizing algae. Peat-forming marshes tend to be denser. In interpreting quantified hydrological aspects of the depositional environment from modern areas where calcitic mud forms the marsh surface, one must remember that the regional hydrology has been altered in this century and that some present areas probably had a thin layer of overlying peat prior to the drainage (Gallatin et al., 1958). Development of a calcitic mud presently indicates relatively shallow water on average and considerable oxidation of organic material in the sediment throughout the year, but especially during the dry period. Although periphyton may grow in marsh waters more than 0.5 m deep, the average floodwater depth for all present marl-producing areas (or at least marl-floored areas) in the Everglades National Park vicinity ranges from 8.1 to 53 cm, averaging 21.6 cm (Tropical Bioindustries, 1987). The hydroperiod for those areas in Taylor Slough, located in Everglades National Park and covered by calcareous periphyton and marl, is estimated at 6–7 months (May–June through October–November), depending on when summer rainfall begins (Tropical Bioindustries, 1987). The association of calcitic mud "soil" with shallower shorter-flooded sites and marsh peats with slightly deeper, longer-flooded sites is clearly seen in Taylor Slough (Figures 7.6 and 7.7) (Gleason, 1972; Gleason et al., 1974; Olmsted et al., 1980).

The early postglacial onset of calcitic mud deposition at Corkscrew Swamp (10,600 ± 180 YBP, UM-958) (Kropp, 1976) and sites now forming the bed of Lake Okeechobee suggests that wetland environments of relatively short seasonal hydroperiod (compared to peat-forming marshes) were the first type to succeed upon previously terrestrial environments. However, most of the Everglades area was not accreting calcitic mud in early postglacial times. The area of the present Everglades may then have been completely terrestrial or experienced only very short, occasional, or local flooding, below the threshold for wetland calcitic mud formation (as in modern sandy "wet prairies" and the still drier pine flatwoods). Wetter environments and very local deposition must have occurred in various shallow solution pits and holes now buried beneath younger marl and peat in the area.

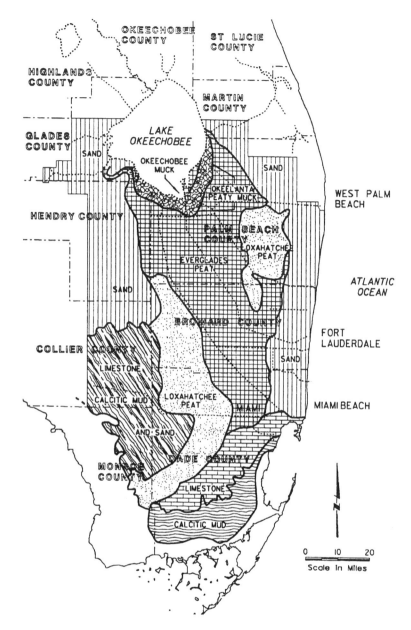

Figure 7.7 Map of surface sediments covering south Florida. (Adapted from Jones et al., 1948.)

Peat and Its Environmental Significance

Organic sediments, peats, and mucks are formed either by wetland plant communities in the case of the peats or in a nearshore environment adjacent to or in Lake Okeechobee in the case of an important muck deposit. The zonation of the peats, both vertically (and thus through time) and horizontally, is related to long-term developmental changes in the Everglades environment (i.e., hydroperiod, water flow, fire, floods, droughts, animal activity, formation of floating peat islands) and the adaptation of plant communities to those changes. The peats originate from the preservation of roots and rhizomes and, to a lesser extent, the stems and leaves of the Everglades plant communities. Distinct vertical peat-type zonation in cores reflects fairly long periods of deposition in one plant community, followed by changes to another through time. Muck consists of a mixture of fine peaty organic matter and fine sedimentary or detrital mineral material. Near Lake Okeechobee, muck was deposited on top of adjacent marshes and their accumulated peat layer.

The main areal distribution of marsh peat types is closely related to bedrock topography (discussed later). Distribution of tree island or hammock peats was also influenced by bedrock topography, in either a specific sense (certain tree islands occur on rock mounds or above depressions) or in a collective sense (for broad areas of tree island concentration). The shape of the many Everglades tree islands and their associated peat "ridges" was controlled by predrainage flow through the Everglades, as evidenced by their streamlined "cigar," "tadpole," or "sperm" shapes. The first is more typical of the deep peatlands discussed here (but which occur elsewhere in the Everglades) and the latter shapes are more typical of shallower peat areas, where bedrock features are mainly associated with the upstream "head" or high spot of the tree islands.

Davis (1946) and Jones et al. (1948) discuss the various peat types and their association with particular plant community types (Table 7.4). In their original meanings, these names referred to soil series that might include several different sediment types and horizons in a profile (i.e., the Okeechobee muck had a muck-over-peat sequence). In more recent times, the names have come to be associated more with the characteristic sediment (e.g., the muck layer itself). The geographic distribution of major peat types is shown in Figure 7.7.

The Okeechobee or custard apple muck refers to a sediment covering 130 km^2 (32,000 acres) around the southern and eastern shore of Lake Okeechobee (Jones et al., 1948). Harshberger (1914) describes the custard apple (*Annona*) swamp forest lining the south shore of Lake Okeechobee: "...the custard apple forest extends east and west, as far as the eye could see [from the intersection of the South New River Canal and Lake Okeechobee shoreline] and in a southward direction from the border of the Lake, a distance of about 4.8 km [3 mi]."

Table 7.4 Freshwater Organic Sediment Types and Associated Plant Communities

Sediment type	Plant community
Okeechobee muck or custard apple muck (muck-over-peat sequence)	Custard apple swamp[a]
Okeelanta peaty muck (peat-muck-peat sequence)	Elderberry–willow swamp[a]
Everglades peat	Sawgrass marsh
Loxahatchee peat	Water lily marsh or slough
Gandy peat	Tree island

[a] Not thought to be the vegetation that formed the organic sediment.

Allison and Dachnowski-Stokes (1932) describe the custard apple muck as follows:

> This series of profiles includes two-layered muck soils that appear in the position of a natural levee along the south and east shore of Lake Okeechobee. It is typical also on Ritta and Torry Islands and consists [in] the main of granular muck at the surface to varying depths that has been derived from a native formation of dense black to grayish black, finely divided, to colloidal sedimentary (allochthonous) peat varying in thickness from 6 to 60 inches. In locations where the surface material has the greatest depth this surface layer contains about midway one or more narrow layers of yellowish-brown, laminated, fibrous peat derived from fleshy rootstocks of semi-aquatic vegetation and grades into a lower layer of brown, fibrous, matted, poorly decomposed sawgrass peat ranging in thickness from 6 to 60 or 72 inches.

This muck is relatively high in mineral content. Clayton et al. (1942) indicate a content of 35–70% of dry mass. This fine mineral material presumably was washed in from Lake Okeechobee.

Allison and Dachnowski-Stokes (1932) suggest that the custard apple forest possibly had only recently established itself on a mud flat at the southern end of the lake because no woody peat was found below the trees and underlying sediments were of a fine transported nature. Heilprin (1887) visited Lake Okeechobee in 1866, but did not mention the custard apple forest, and it is possible that the forest developed during low water stages of the lake in following years. Observations in 1883 along the south shore of Lake Okeechobee indicated that custard apple was present at that time. Expansion of the custard apple forest may be related to drainage activities (Brooks, 1974).

The Okeelanta peaty muck is (was, unfortunately) characterized by a unique stratigraphy and in the past an unusual willow (*Salix* spp.) and elderberry (*Sambucus canadensis* L.) vegetation. Note that this name still refers to a profile sequence rather than a single sediment type. The willow–elderberry vegetation described by Davis (1943) as covering the Okeelanta peaty muck probably formed as a result of early drainage of the lake and Everglades by replacing sawgrass marsh. Harshberger (1914) did not recognize this as a distinct zone in his early botanical investigation of the Lake Okeechobee area. The distribution of this sediment sequence covers 105 km² (26,000 acres) (Jones et al., 1948). Allison and Dachnowski-Stokes (1932) describe it:

> Typically, the Okeelanta series is a three-layered peat profile and is probably the most extensive single group of peat soils to be found in the Upper Everglades. This series borders the Okeechobee muck to the south and east and represents that area or zone where the plastic layer of sedimentary muck has become covered with a surface layer of brown, fibrous peat derived for the most part from the roots and rhizomes of sawgrass. This surface material is porous, loose, light in weight, and varies in depth from 12 to 35 inches where a layer of dense, blackish, plastic material of sedimentary origin, 4 to 18 inches in thickness, is encountered.…This layer of sedimentary material passes into an underlying layer of fibrous sedge (sawgrass) peat that continues with minor variations to the rock or marl upon which the whole formation rests.

They indicate that the profile may be complicated by more than one layer of sedimentary material (muck). They also state that "…the position of the plastic or sedimentary layers in the profiles of the Okeechobee and Okeelanta series is well shown in continuous cross-section from the Lake at South Bay to a position 15 miles south along the North New River Canal...."

The presence of a sedimentary layer outward into the Everglades from the present shoreline of Lake Okeechobee suggests that at times in the past, open water conditions (or nearly so) capable of transporting suspended mineral clays prevailed over a considerably larger area (Dachnowski-Stokes, 1930). The paleoenvironmental significance of the sedimentary layer with transported mineral matter appears to have received virtually no attention since the early 1930s. The thinner, uppermost peat horizon no longer exists due to oxidation and plowing.

Everglades peat refers to sediments formed chiefly from sawgrass. This peat type is brown to black in color, fibrous to granular, and fairly low in mineral content (about 10% of the dry mass) (Clayton et al., 1942). These peats covered approximately 4420 km² (1,091,000 acres) of the Everglades marsh (plus underlying the mucky sediments) and are the most abundant of the organic sediments (Jones et al., 1948). The black color of certain of these peats has been attributed to an abundance of charcoalized material (fusinite) from fires which swept through the Everglades throughout their history (Smith, 1968). Much of the Everglades peat area has now been drained and developed or converted to agriculture.

Loxahatchee peat refers to peats derived primarily from the roots, rootlets, and rhizomes of the white water lily (*Nymphaea*). Loxahatchee peat is a light-colored, fibrous and spongy, reddish-brown peat. It covers 2950 km² (730,000 acres) within the Everglades basin and is second in abundance to sawgrass peat (Jones et al., 1948). Loxahatchee peat is related to water lily slough communities or water lily co-dominated wet prairie vegetation and is generally considered unsuitable for agriculture because of its high degree of shrinkage when drained, low mineral content, and easily degraded organic material. These peats form some of the thickest deposits, but also occur in shallow sediment profiles. Loxahatchee peat deposition currently prevails above the deeper bedrock subregional troughs of the Everglades which have superimposed wide swales in the peat. Such peat also underlies Everglades peat in parts of the sawgrass Everglades (see Figure 7.14) (Davis, 1946).

Gandy peat is a rooty and sometimes leafy peat formed by forest vegetation on Everglades tree islands. Davis (1946) found that Gandy peat overlaid Everglades and Loxahatchee peat on existing tree islands, but was rarely interbedded with sawgrass peat as well. Gandy peat, in stark contrast with marsh peat types, forms in the exposed oxidizing zone above surface water. Gandy peat covers only 77 km² (19,000 acres) in total among the thousands of widely scattered occurrences (Jones et al., 1948). This tree island peat, through an areal association with Loxahatchee peat, is concentrated in two large areas of deeper peat, lower bedrock elevation and where there is the greatest indication of broadly (subregionally) focused predrainage era surface water flow within the Everglades.

The Chronology of Peat Deposition

Peat deposition began in the Everglades at least as early as 5490 ± 90 YBP (UM-649), a date determined from basal peat in the extensive submerged peat deposit at the south end of Lake Okeechobee. This is the oldest freshwater peat yet dated from the Everglades (i.e., in this case, from one of the former extensions). A date of 5000 ± 90 YBP (UM-192) was obtained from a basal peat from Kreamer Island, also at the south end of Lake Okeechobee. Parker et al. (1955) report 5050 and 4900 YBP for near-basal peats in the northern Everglades. McDowell et al. (1969) dated basal peat east of Belle Glade at 4420 YBP. Gleason et al. (1975a) dated basal peats mainly from Water Conservation Areas 1 and 2. The oldest radiocarbon dates there were 4860 ± 170 YBP (GX-3282) and 4800 ± 100 YBP (UM-604) from the northern end of Water Conservation Area 2A and the center of Water Conservation Area 1, respectively. The middle Everglades area (southern Water Conservation Area 3) has given basal peat dates nearly as old: 4520 ± 160 YBP (UM-664, by Gage 3-26) and about 4800 YBP (Altschuler et al., 1983).

The contemporaneity of the onset of peat deposition in south Florida regionally is shown by basal dates of peats from Corkscrew Swamp and the Savannahs. Portions of the Corkscrew Swamp peat deposit seem slightly older than the main (central) Everglades peatland, 5685 ± 210 YBP (UM-955) (Kropp, 1976), but another site was about the same, 4720 ± 90 YBP (UM-635). The oldest of five basal peat dates from the savannas, near Fort Pierce, was 5022 ± 123 YBP (UM-961).

Thirteen basal peats from the northeastern Everglades were dated in order to develop a picture of the peatland expansion over time (Gleason et al., 1974, 1975a). The isochron pattern shown in Figure 7.8 suggests that peat deposition began nearly simultaneously from the southern shore of Lake Okeechobee to the current northern tip of Water Conservation Area 2. The anomalously older date at the south end of Water Conservation Area 2 may represent another focus of peat development located in a lower area west of the coastal ridge, possibly in response to ponding of water behind the ridge. Westward seepage discharge of ground water from beneath the ridge is also a possible factor.

The isochrons do not correlate well with bedrock elevations (Figure 7.9) presented by Jones et al. (1948). An expected correlation was of older dates with lower bedrock elevations, due to earlier prolonged flooding by the rising water table. One of the two oldest dates comes from an area of bedrock elevation about 0.7 m higher than the other, yet the dates are about the same. The bedrock surface drops significantly in elevation toward the south in Water Conservation Area 2; however, the dates there are not commensurately older, but rather are generally younger. In large part, this must result from a thickening wedge of sand over the bedrock as one approaches the ridge (see Figure 7.14) (Davis, 1946), whereby the surface of peatland initiation there was not the bedrock. There is undoubtedly some local relief on the limestone bedrock as well.

The possible role of other factors, such as the drainage into (recharging of) the underlying Biscayne Aquifer in certain areas and the original surface water drainage direction, indicates that more work is needed to understand the initial development of peat deposition. Gleason et al. (1974) note that the Biscayne Aquifer is found beneath the southeastern half of Water Conservation Area 2 and suggest that the more porous bedrock surfaces may have delayed both the ponding of water and the onset of peat deposition. They also note that a very rough correlation exists with predrainage era surface flow directions in that the oldest dates are generally in the upflow direction and the younger dates are from the downflow direction,

Figure 7.8 Ages of basal peats from the Everglades Water Conservation Areas.

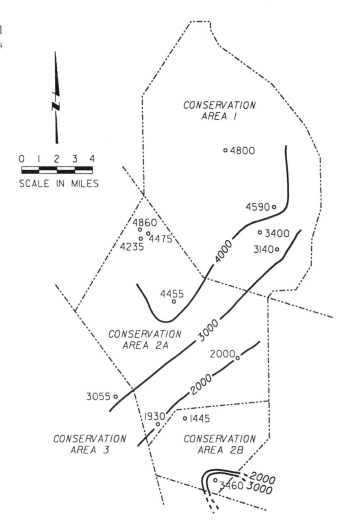

except for the one date at the south end of Water Conservation Area 2. Of course, thousands of years and much topographic alteration by peat deposition separate the onset of peat formation and the era for which we have good knowledge of the flow regime.

Relationship between Bedrock Topography and Vegetation

The configuration of the bedrock surface underlying the Everglades is shown in Figure 7.9. The bedrock surface is irregular, with broad topographic highs and lows. Gleason et al. (1974) named four main bedrock features within the Everglades basin, three of which are low areas (Loxahatchee Channel, Tamiami Basin, and Shark River Bedrock Slough) and one which is topographically high (West Everglades High). These topographic expressions of the bedrock surface appear to have had a profound effect on the type of vegetation that grew in various parts

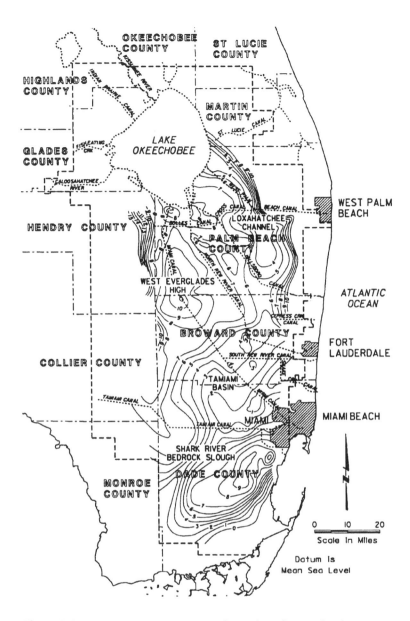

Figure 7.9   Approximate contours on the rock surface under the organic soils of the Everglades. (From Jones, 1948.)

of the Everglades, both through the prolonged development as revealed by peat type (compare Figures 7.7 and 7.9) and in late historical times (Davis, 1943) (see map included with this volume).

The Loxahatchee Channel, Tamiami Basin, and Shark River Bedrock Slough are depressions overlain mainly by Loxahatchee peats (water lily dominated) and topped by water lily sloughs and mixed wet prairies interspersed with thousands of tree islands. The bedrock high area is overlain mainly by Everglades peats (sawgrass dominated) and topped by sawgrass marshes of various densities along with sawgrass mixed with wax myrtle thickets. The bedrock low areas contain many elongated tree islands aligned with the predrainage era surface water flow direction. Thus, it appears that the bedrock topography and the imposed hydrology strongly influenced the Everglades plant communities through and after thousands of years of burial and topographic alteration by peat deposition. Harper (1909) similarly noted that the marsh and water lily "prairies" of Okefenokee Swamp, which are surrounded by the prevailing cypress swamp, are located in deeper peat areas that are topographically lower on the prepeat sand surface.

Areal Changes of the Peatland

The Everglades wetland and peatland, by means of peat deposition and a rising of the surface water level, simultaneously rose and expanded in its mineral sediment basin. So did Dismal Swamp in Virginia and North Carolina (Cocke et al., 1934; Whitehead and Oaks, 1979) and undoubtedly also Okefenokee Swamp in Georgia (evidenced by thick sequences of peat from shallow water vegetation) (Spackman et al., 1976). Eventually, the Everglades waters overtopped several low spots on the eastern margin (the Atlantic Coastal Ridge) and formed low rapids at the headwaters of small rivers (including New River and the Miami River) (Parker et al., 1955; Parker, 1974).

In some areas, the Everglades marshes contracted. Peat and calcitic mud sediments from former freshwater marshes now lie beneath Florida Bay to the south, nearby brackish and marine environments to the southwest, and Lake Okeechobee to the north of the adjacent present Everglades. These marshes were eliminated by the upslope migrations (transgressions) of the respective shorelines, caused by the rise of sea and lake level.

The drowning of the northern Everglades marshes by Lake Okeechobee may have proceeded slowly through time. Figure 7.10 shows the extent of Lake Okeechobee that is underlain by peats. No other sediments appear to overlie the peat and it is exposed on the bottom of the lake. The Everglades marsh at one time extended out into the lake. Prior to the dredging of canals to drain the Everglades (Stephens and Johnson, 1951), peats bordering Lake Okeechobee were 4–4.3 m (13–14 ft) thick adjacent to Lake Okeechobee, which, when added to the elevation of the bedrock (Figure 7.9), gave a surface elevation along the southern lakeshore in the range of 6–6.4 m (20–21 ft) MSL. Accumulation of peat and muck deposits in the Everglades, along the southern and eastern shore of the lake, in large part dammed the overflow from the lake and the lake level rose. Drowning of peat marshes by the lake suggests that water levels rose faster than peat could accumulate on the area that is now lake bottom, but the question remains as to how the organic sediment of the shoreline "dam" could rise more rapidly. Surface peats on the bottom of the lake were radiocarbon dated at 2670 ± 80 YBP (UM-650) and 4150 ± 90 YBP (UM-648), the latter from further offshore and in deeper water. These dates suggest two hypotheses: (1) some of the peat has been eroded, exposing older peats as or after the water level rose, or (2) the edge of the Everglades was pushed back over a time span of thousands of years, leaving a wedge of peat representing the rise in water

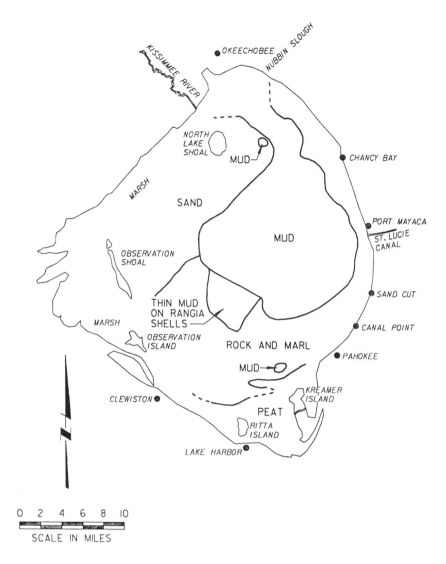

Figure 7.10 Bottom sediment types in Lake Okeechobee. (From Gleason et al., 1978.)

level, with older surface peats farther out and younger surface peats closer to the shore.

This long-term rise in water level within Lake Okeechobee and the Everglades does not depend upon an increase in precipitation after that associated with the initiation of peatland. Rather, the rise is a natural consequence of hydrologic feedback. Surface water levels increase from century to century as the ground (peat surface) elevation increases. In the long term, average water level rises with the peat surface. In the Everglades, the slowly rising and frequently overtopped peat and muck "dam" south of Lake Okeechobee eventually grew to 100 km long and up to 4 m high after about 5000 years. The contemporaneous rise of sea level tended toward the same result by raising the drainage base level.

Freshwater sediments underlying marine transgressive peats and aragonitic marls adjacent to Florida Bay are widespread and have been extensively investigated. Scholl (1964a, 1964b) showed that sedimentary sequences along the southwest coast of Florida were often of a transgressive nature, that is, resulting from an encroaching sea and shoreline; he found marine and brackish water (mangrove) peats lying above freshwater calcitic muds. Riegel (1965) and Spackman et al. (1966) found pollen evidence that marine sediments had deposited over freshwater peats. Using microscopic peat identifications, Cohen (1968) showed that Whitewater Bay, now a saline body of water, once was the site of a water lily–sawgrass marsh, and present marine environments along the Shark and Harney Rivers were also once freshwater marshes of a more extensive Everglades.

In a landmark investigation of Florida Bay, Davies (1980) found numerous occurrences of freshwater calcitic mud or freshwater peat underlying mud banks or islands in the bay. He found that in all cases, the basal peats from Florida Bay are of freshwater origin. The Ninemile Bank core, located far out in Florida Bay, evidenced a bayhead (or Gandy) type peat (willow, wax myrtle, and red bay) on top of bedrock. Figure 7.11A shows the sequence of events interpreted by Davies from 5000 YBP to present as sea level rose and Florida Bay was formed by the intrusion of open marine environments. Figure 7.11B shows the location of the cross section depicted in Figure 7.11A.

Davies (1980) found the oldest and deepest basal peats in the south and southwestern sections of the bay, where the bedrock surface is lowest. The deepest bedrock (4.9 m [16 ft] below MSL) occurred beneath Shell Key and had an overlying basal peat that dated 5685 YBP. The youngest basal peat was found in the shallow northeastern part of the bay beneath Eagle Key. Peats of freshwater origin occur deep beneath the islands and bays, but are absent beneath the open water areas. Peats occur in slight bedrock depressions 30 cm or more in depth (it appears by association that these pockets may have some control over the formation of the islands). Davies found that beneath the interwoven network of eastern banks and islands, peats are found in elongate "rill valleys." These may be Pleistocene interglacial (end of the Sangamon?) drainage channels which were subsequently filled in with freshwater peat and calcitic mud. This is similar to the distribution of swamp peats in nearby Big Cypress Swamp (J. Meeder, personal communication) and marsh peats in Taylor Slough (Gleason, 1972; Gleason et al., 1974; Olmsted et al., 1980).

Freshwater wetland environments thus appear to have migrated or retreated landward with rising sea level, suggesting an interdependence between the water table and sea level in areas close to shore. This differs in direct cause from interpreted conditions in the northern Everglades, where precipitation increases around 5000 YBP are suspected as a cause in peatland initiation and where the peatland expansion laterally up the slight basin slopes is thought to result from hydrologic feedback due to the thickening deposit. The authors believe that the difference lies in the northern Everglades—their interior location and the much less permeable nature of the near-surface Fort Thompson Limestone—both of which inhibit drainage. Florida Bay lies atop the highly permeable Miami Limestone in which groundwater drainage would be much more responsive to base level or sea level. Davies (1980) stated that the cover of peat was probably more complete in

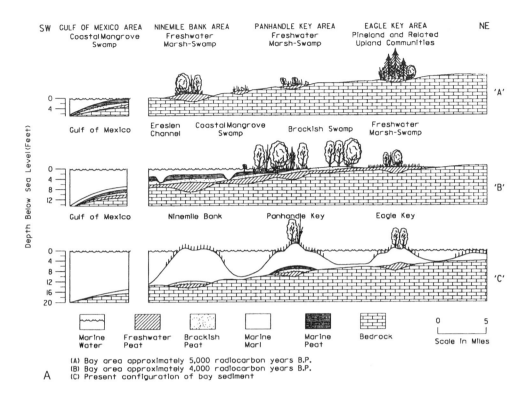

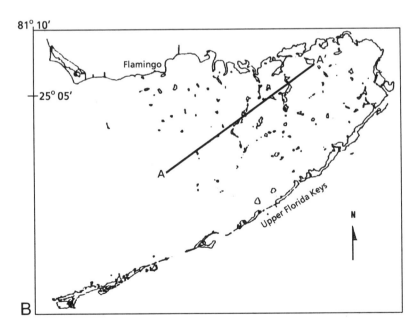

Figure 7.11 (A) Transgression of open marine environments over former freshwater marshes and swamps in Florida Bay. (From Davies, 1980). (B) Location of cross section.

the western half of the bay, while it was mainly restricted to isolated depressions or elongated slough areas in the east. Radiocarbon dates of basal peats by Davies indicate that freshwater marsh environments such as the Everglades existed in southwestern Florida Bay 5500–5000 YBP, in central Florida Bay 5000–4000 YBP, and in the more elevated northeastern part of the bay at least as late as 2500 YBP.

Origin of Red Bay (*Persea borbonia*) Tree Islands in the Northeastern Everglades

Two distinct types of tree islands are abundant on the deep peats of the large water lily-dominated portion of the northeastern Everglades (Loxahatchee peat area) (Figure 7.7). These tree island types are distinguished by length-to-width ratio, size, topography, and vegetation (Loveless, 1959). Small, roughly circular tree islands are dominated by red bay (*Persea borbonia*) and large elongated tree islands or strands are dominated by dahoon holly (*Ilex cassine*). Existing tree islands of both types in this area have formed relatively recently in the history of the Everglades. The formation of red bay tree islands appears mainly to be related to the development and evolution of floating peat islands.

Many other types of tree islands exist elsewhere in the Everglades and the south Florida region (and the world). Here, emphasis is on those deep peats resulting from marsh-swamp peatland processes and not from influences relict from the topography of the mineral sediment substrate. Some tree islands occupying peat lenses on freshwater calcitic mud terrains may similarly result exclusively from recent biogeomorphic and sedimentary processes (Figure 7.8) (Spackman et al., 1969; Egler, 1952). Elsewhere in the eastern, southwestern, and southern Everglades, limestone or sand mounds (or even depressions) underlie tree islands, and these features are often partially buried by peat (Craighead, 1971; Spackman et al., 1969).

Thousands of red bay tree islands dot Water Conservation Area 1 (Arthur R. Marshall Loxahatchee National Wildlife Refuge). Typical examples of these tree islands are 20–30 m in diameter. They are distinct topographic high spots with elevations of 0.5–1.2 m above the surrounding water lily slough or wet prairie peat surface. These peat mound tree islands are commonly zoned at their peripheries with different types of vegetation. Sawgrass often forms a narrow but distinct outermost sedge ring or fringe around the tree island. Shrubs such as cocoplum (*Crysobalanus icaco*) and wax myrtle grow on the perimeter and sometimes extend into the interior. The interior or higher ground supports large red bay trees. Ferns comprise a sparse ground cover.

Floating mats of peats are common in water lily sloughs (environments predominated by water lilies and somewhat more pond-like than typical marsh). The surface layer of slough bottoms is buoyant and will float when loosened, due to the air-filled, live water lily rhizomes, roots, and rootlets (and assisted by some gas bubbles in the peat). The dead peat itself is not buoyant and will not float when saturated. The floating peat islands are usually much less than 0.6 m thick and range in size from 2 to 15 m in longest dimension. They vary from round to elongated and irregular. The floating peat islands will often drift to the edge or out of the slough and become stranded in the surrounding marsh, eventually reattaching to the bottom. Therefore, they create both a topographic low spot in the slough and a nearby topographic high spot.

Bushy plant communities form a size–vegetational continuum from newly formed and newly colonized floating peat islands to the established red bay tree islands; in some cases, small but fruiting trees have been found on peat islands that were still fully afloat (Gleason et al., 1975a; Stone, 1978). This observational evidence, while not absolute confirmation as to the origin of the established red bay tree islands, strongly suggests a genetic relationship between the tree islands and floating peat islands.

Davis (1943) postulated that the change from deeper water slough conditions in the Everglades to islands of bushes and trees was due to the formation of "tussocks." He stated that at first this small group of plants may be floating, but felt that when it becomes fixed, the tussock expands laterally as other aquatic plants become attached to the edges and it slowly builds up in height to levels above the general water level. Eventually, in many instances, a tree island is thought to develop. Jones et al. (1948) explained that "thousands of small oval islands" result from plant succession on floating islands, followed by continued growth after the roots become attached to the underlying peat.

Loveless (1959) attributed the origin of small bayhead tree islands in Water Conservation Area 1 to succession upon floating peat masses. He did not consider the floating peat mass theory viable for the large elongated tree islands (such as the dahoon holly tree islands, discussed later) because of the large size of these tree islands and the thinness of the peat in some places (outside the deep peat area discussed here). He pointed out that many tree islands in the western and southern sections of the Water Conservation Areas (e.g., Water Conservation Area 3) were composed of woody peat down to the underlying rock and thus could not have formed upon buoyant patches of peat. They are mostly focused on bedrock mounds, which further argues against floating peat island involvement there.

Similar small, peat mound tree islands occur in the marsh "prairies" of Okefenokee Swamp. Cypert (1972) noted the similarity between the outlines of floating peat islands (called "batteries") and tree islands. Plant succession on the raised peat masses—from emergent marsh plants, to sparse trees, to forest patches—is suggested by apparent examples of all intermediate stages of succession existing in the modern marshes, from bare peat surfaces to old tree islands. However, Cypert also noted that after emergence for 12 years, herbaceous plants were still dominant on batteries he studied. Cohen (in Spackman et al., 1976) studied the peat stratigraphy of an Okefenokee tree island and found a layer of water loosestrife (*Decadon*) peat sandwiched between overlying forest peat and underlying water lily peat. He concluded that the tree island was derived from vegetational succession on a battery, because currently this plant occurs almost exclusively on batteries in the marshes. In a detailed study of tree islands in Okefenokee Swamp, Rich (1978; Rich and Spackman, 1979) found that microscopic pollen and peat analyses of 11 peat cores from 7 tree islands supported Cypert's theory. He found that the tree islands occur in areas that were originally inhabited by water lily-dominated marsh plants, subsequently colonized by grasses and sedges characteristic of floating or reattached batteries, and finally occupied by trees and shrubs which succeeded the earlier transitional species.

Gleason et al. (1980) tested the theory that tree islands in Water Conservation Area 1 developed from floating peat islands which, after arising, moved and reattached to an adjacent section of the marsh. If so, reversals with depth should

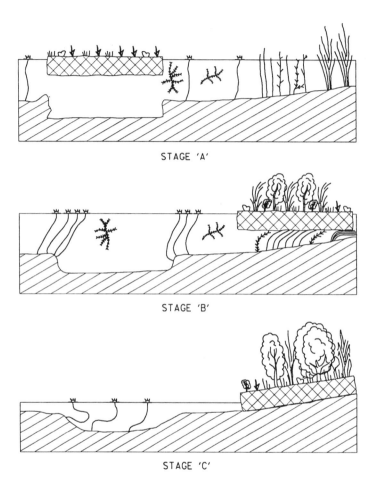

STAGE 'A'

STAGE 'B'

STAGE 'C'

Figure 7.12 Proposed stages in the development of small tree islands from a floating peat island. (A) Newly floated peat island in water lily slough. Shrunken water lily growth and seedings of colonizing plants are shown. (B) During high-energy events, such as periods of high winds, peat island drifts into denser vegetation toward edge of slough, corresponding to shallow water. High water levels facilitate this movement. (C) Local topographic elevation formed by stranded peat island, even when reattached, is emergent during low water stages and receives less seasonal inundation than surrounding sites. In the absence of local seed sources for woody plants (and perhaps for long seasonal hydroperiod), dense marsh rather than woody growth may succeed. Attached peat hummocks with steep edges, suggestive of this stage, occur in the Loxahatchee National Wildlife Refuge (Water Conservation Area 1). (From Gleason et al., 1980).

occur in radiocarbon dates due to superimposing an older peat (bottom of the floating island) on a then modern peat surface in the marsh. Finding such reversals is the difficulty. However, results from test cores taken through red bay tree islands have shown reversals. The sequence of events in converting a floating peat island to a tree island is graphically presented in Figure 7.12.

Parker and Hoy (1943) (site GS-12), Davis (1946, Figures 3 and 14), Cohen (1968; Figure 26 in Gleason et al., 1974), and Gleason et al. (1975a, cores 15–18) indicate that in the coring of a total of at least seven red bay tree islands in Water Conservation Area 1, the woody peats occur only toward the tops of the peat sequences. The tree island peats are on top of either water lily, sawgrass, or mixed water lily–sawgrass peats.

A large piece of wood (possibly red bay) taken from a depth of 90 cm on a red bay tree island was dated 780 ± 70 YBP (UM-557). Basal forest peat ("hammock" peat in the figures in Gleason et al., 1975a) in two cores rendered dates of 780 ± 80 YBP (UM-597) and 1280 ± 70 YBP (UM-665). These dates, along with the similar limited thicknesses of woody peats from other such tree islands, suggest that tree islands did not occur, at least not abundantly, throughout the early history of the Everglades within the area of their present concentration and superabundance in the northeastern Everglades; rather, they occurred only (or mainly) from about 1300 YBP to the present. Obviously, further dating could push this date back somewhat. The large number of red bay tree islands and the uniformity of their appearance in the prevailing established examples suggest that some stage or event may have initiated their formation. Speculation might invoke a stage of wetter conditions that promoted purer water lily vegetation which, in turn, promoted floating peat island formation. Alternatively, the emerged peat islands may have previously formed, but the resulting shallower microsites were colonized only by marsh emergents (e.g., sawgrass, arrowhead [*Sagittaria* spp.]) until seeds of woody plants were effectively dispersed around 1300 YBP or so, with no outside imposed hydrologic mechanism involved at all.

Origin of Dahoon Holly (*Ilex cassine* L.) Tree Islands in the Northeastern Everglades

Dahoon holly tree islands, which number more than 100 in the Loxahatchee Refuge vicinity (Water Conservation Area 1 and northern Water Conservation Area 2A), are the largest type of tree island in the northeastern Everglades. They range in size up to 1600 m long. These tree islands are distinctly elongated, are biconvex in outline, and show little distinction in shape, topography, and vegetation between northern (upstream) and southern ends. They are oriented parallel to the direction of predrainage era flow in the Everglades.

Dahoon holly tree islands are flooded throughout their general interior during high water, but the trees grow on a dense scattering of root-stump mounds or tussocks. The predominant vegetation is dahoon holly with an understory of ferns. These tree islands, in contrast to those south and southwest of this area, do not have "heads" or taller clumps of trees on an elevated, northern (upstream) apex of the tree island. Based on limited sampling, these dahoon holly tree islands appear to be on uniformly deep peats and bear no relationship to bedrock topography. This is in contrast to other types of elongated tree islands located elsewhere in the Everglades, which usually have heads located on distinct high spots of bedrock (or sand).

Single cores, each from five large dahoon holly tree islands (or their relicts) in the Loxahatchee Refuge vicinity (Gleason et al., 1974, 1975a), show that forest peat extends down 0.3–0.7 m below the general ground surface (cores were not taken through the narrow root-stump mounds). The forest peats immediately overlie sawgrass peat, except in one case where they overlie an arrowhead transitional peat. Deeper peats consist of water lily, water lily–sawgrass, and sawgrass peats.

This general trend of emergence from long-flooded marsh to only occasionally flooded bayhead forest is shown in greater detail by a single pollen and spore stratigraphic profile from one tree island (D. Nichols, in Gleason et al., 1975a). Water lily marsh, the dominant community in the area judging from both peats and the historic era environment, was succeeded first by sawgrass marsh (peat analysis) with abundant polypodiaceous ferns (spore identification), then by willows and wax myrtle, and finally, near the top of the peat profile, by dahoon holly, the modern dominant. This sequence weakly suggests that the elongation or shaping occurred during the sawgrass marsh stage (perhaps as a sawgrass "strand" or "ridge" as occurs today) and hints that the bush or tree colonization may have been facilitated by emerged microsites formed by fern root tussocks (Stone and Gleason, 1983; Stone and Brown, 1983). The dahoon holly pollen occurs only very close to the top of the peat profile (0–8 cm) and the wax myrtle pollen is only significant slightly deeper (8–15 cm). This shows that these tree islands are quite young and indicates that tree root intrusion has been responsible for the thicker deposit of forest peat.

Evidence for a Distinct Paleoclimatic Stage within the Everglades Era

Craighead (1969) and Gleason et al. (1974) have suggested that calcitic mud horizons within Everglades peat profiles may represent somewhat drier climatic periods. Hydroperiod requirements for the deposition of calcitic mud are relatively short compared to those for marsh peat accumulation. Gleason et al. (1974) indicate three reasons for the dominance of calcitic mud deposition rather than peats under short-hydroperiod marsh conditions. First, the less-wet conditions inhibit the luxuriant growth of peat-forming plants (additionally, sparse marsh vegetation facilitates a well-lighted water column for the growth of calcite-precipitating photosynthesizing algae). Second, seasonal drying favors the precipitation of calcium carbonate as a result of concentrating calcium and carbonate ions by evaporation. Third, a long season of dryness promotes the oxidation and breakdown of organic material that would form peat.

Peat sequences with interlayered calcitic mud are known from various parts of the middle and southern Everglades (Davis, 1946; Gallatin et al., 1958; Cohen, 1968; Craighead, 1969; Spackman et al., 1976; Gleason et al., 1974, 1975a; Altschuler et al., 1983). In order to determine the age relationships of a main calcitic mud layer at mid-level in the peat sequence, cores were taken in south Water Conservation Area 3A (north of Tamiami Trail) and to the west of Levee 67 (by the 3-26 Gage). Spackman et al. (1976) and Altschuler et al. (1983) have shown that the

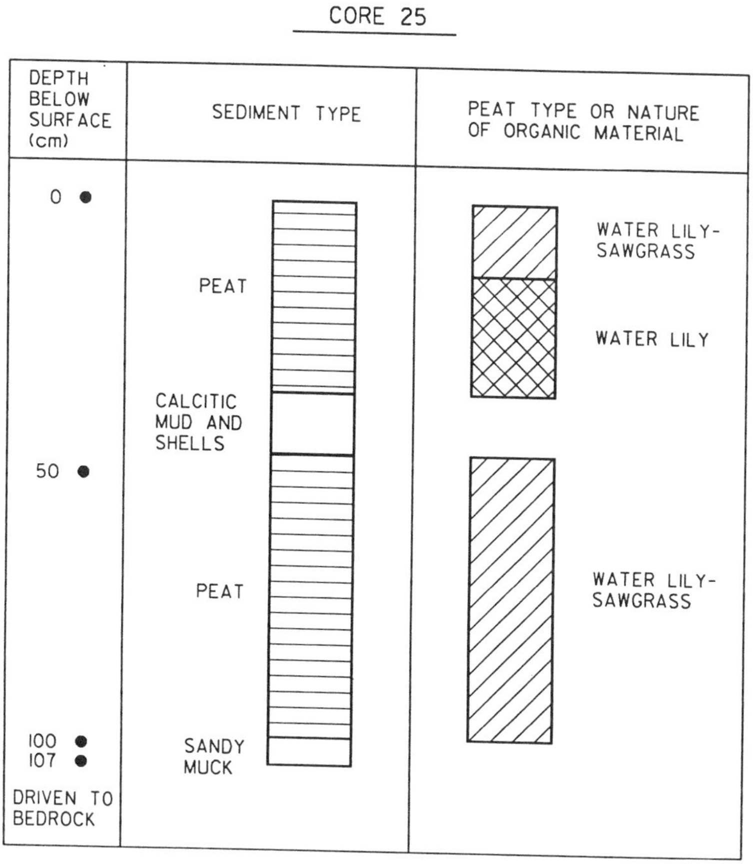

CORE 25

Figure 7.13 Stratigraphy of a peat and calcitic mud from near Gage 3-26 in Water Conservation Area 3A north of Tamiami Trail and west of Levee L-67.

thickness and position of the calcitic mud layer can vary over a very small area. The type of stratigraphy shown by three adjacent dated cores is shown in Figure 7.13 for one of the cores. Each core is about 1 m in length and exhibits a single calcitic mud layer. Analysis of one core revealed water lily–sawgrass peat below and water lily and water lily–sawgrass peat above the calcitic mud. Radiocarbon dates on the entire calcitic mud layer from the other two cores (a few meters apart) are 2432 ± 96 YBP (UM-698) and 2436 ± 99 YBP (UM-697). Peats immediately overlying the calcitic mud from the same two cores dated 2036 ± 76 YBP (UM-696) and 2082 ± 73 YBP (UM-700). The peats immediately underlying the calcitic mud dated 3171 ± 81 YBP (UM-695) and 3209 ± 77 YBP (UM-699). These dates strongly suggest that from about 3000 to 2000 YBP a stage occurred in the middle to southern Everglades that could most easily be interpreted as a millennium of reduced seasonal flooding. This agrees reasonably well with reductions in flood deposits between 3600 and 2200 YBP in the southwestern United Stated as a consequence of less frequent El Niño events (Ely et al., 1993). Elsewhere, dates as young as 1860 ± 90 YBP (UM-2370) have been obtained on the marl layer.

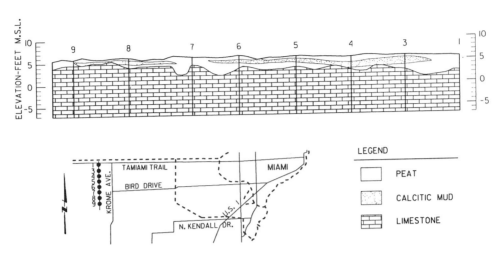

Figure 7.14 Stratigraphy of peripheral Everglades sediments west of Miami, showing lenses of calcitic mud in peat. (From Corps of Engineers, 1968.)

Curiously, the environments preceding and following the calcitic mud deposition appear to be of a fairly wet nature, with a typically long hydroperiod (i.e., water lily peat). Two not distant cores examined by Spackman et al. (1976) also show water lily peats above and below the calcitic mud horizons. Figure 7.14 shows the extent of interbedded calcitic mud in a transverse along Krome Avenue (U.S. 27) south of Tamiami Trail. Interbedded peat and calcitic mud are also known from Shark River Slough (Cohen, 1968) and Taylor Slough (Gleason, 1972).

Conditions during this time period in the northern Everglades are difficult to discern. Buried (but undated) archeological strata in peats here suggest water level fluctuations and some probable drier periods (Goggin, 1948). No interbedded calcitic mud layer occurs, but periods of higher and lower average water levels may in part be responsible for the fluctuation in marsh plant communities seen in peat core profiles (Gleason et al., 1975a). More significantly, a major transition from marsh peat to muck deposition occurred at the islands in southernmost Lake Okeechobee during this same time. Presumably, the same peat-to-muck sedimentary transition in the adjacent northern Everglades was at least roughly simultaneous. Uppermost peats beneath the muck at Kreamer and Torry Islands, respectively, dated 2500 ± 80 YBP (UM-666) and 2560 ± 80 YBP (UM-2030). However, the suggested change is not analogous. The shift from peat to muck deposition implies deeper water and, therefore, probably longer hydroperiods. It is postulated here that the onset of water conditions after 2200 YBP as a consequence of more frequent El Niño events (Ely et al., 1993) may have raised water levels in Lake Okeechobee, which terminated and eroded marsh in this area; this erosion in turn may have removed some peat, resulting in a slightly older date for the event. What is firmly demonstrated is the occurrence of distinct shifts and long-term environmental stages within the late Holocene period of essentially modern conditions and long after the establishment of the Everglades peatland.

Rate of Peat and Calcitic Mud Deposition

Radiocarbon dating of peats near Belle Glade (McDowell et al., 1969) suggests that the long-term average rate of deposition was 8.4 cm per century, but increased exponentially with time. Peat profile development was initially very slow, but proceeded more rapidly (7.3 cm/100 years [YBP]) from about 3500 to 1200 YBP. Deposition after 1200 YBP was approximately 16 cm/100 years (YBP). This increase in rate (corrected for postdrainage compaction) occurred while sea level rise was slowing and further suggests independent peatland processes and feedback mechanisms.

Scholl et al. (1969) used radiocarbon dates to estimate the rate of calcitic mud deposition in present coastal areas and found evidence for a decrease with time. The rate averaged 2.8 cm/100 years from 5000 to 4000 years ago (with minor corrections to calendar years from YBP measurements), 1.9 cm/100 years from 4000 to 3000 years ago, 1.7 cm/100 years from 3000 to 2000 years ago, and 1.2 cm/100 years from 1000 years ago to present. Scholl et al. suggest that this decreasing rate of deposition is attributable to the decreasing rate of sea level rise, but this trend may instead reflect the decrease in subaerial exposures of limestone as the Everglades basin became filled with peat and the dissolving limestone in progressively higher areas was covered with peat and calcitic mud (Gleason et al., 1974).

Absence of Wet Prairie Environments in the Peat Record

Wet prairie, or sparse marsh of grass-like (graminoid) vegetation, is a common vegetational type in the Everglades. In the northern Everglades peatland, it is characterized by beakrush (*Rhynchospora tracyi*), maidencane (*Panicum hemitomon*), or spikerush (*Eleocharis cellulosa*) and generally occurs in the depth and hydroperiod zone between sawgrass marsh and the deeper water lily sloughs, in areas that frequently are dry during the spring (Goodrick, 1974; Loveless, 1959). Wet prairie was the second most common plant community in Water Conservation Areas 2 and 3 at the time of the Loveless study.

Microscopic examination of peat profiles from numerous Everglades sites (Cohen, 1968; Cohen and Spackman, 1970; Spackman et al., 1976; Gleason, 1972; Gleason et al., 1974, 1975a) has not revealed wet prairie peats. A number of hypotheses may explain this. First, the wet prairie environment may have become much more abundant in the past century or so due to naturally lower or artificially lowered water levels. Second, the low biomass of wet prairie communities may not generate a sufficient abundance of plant material to form much peat. Third, roots and stems of wet prairie plants may not be as amenable to preservation as those of sawgrass and water lily and thus may be diluted by remains of succeeding vegetation or intermixed plants. Water lily commonly is co-dominant in mixed wet prairie in the Loxahatchee Refuge Area today.

Evidence for Cattail in the Prehistoric Everglades

Cattails (*Typha* spp.) are native plants and are occasionally encountered in undisturbed interior marshes of the Everglades, but the large and expanding dense

modern cattail marshes are located in areas of chemical and hydrological disturbance that relate mostly to receipt of agricultural or other impacted runoff waters
(Gleason et al., 1975b). Davis's (1943) vegetational map of the Everglades (included
with this volume) shows a significant area of cattail and sawgrass in what is now
the Everglades Agricultural Area, but this may well have represented disturbance
due to drainage or early agricultural runoff.

No cattail peats have been found in freshwater Everglades peat cores. Cattail
pollen was found deep in one core from near the northern tip of Water Conservation Area 2 (D. Nichols, in Gleason et al., 1975a). Therefore, cattail was common
at the initial stage of peatland development for that site, but as the peat
thickened, cattail pollen declined significantly. In cores possibly representing fresh
to brackish water conditions, Riegel (1965) and more recently Kuehn (1980) found
cattail pollen in buried peats at the coastline and offshore from the Broad and
Harney rivers on the southwest coast of Florida. Cattail has long been present
in and around the Everglades, but not in sufficient abundance to develop
organic deposits recognizable using petrographic methods (microscopic examination of thin sections). In many places today, a loose or "soupy" black mud, not
a peat, underlies cattails. The relative abundance of cattail early in the history of
the peat deposit (at least at one interior site) hints that isolation of the mineral
substrate from the rooting zone by peat accretion may somehow have been
responsible for the subsequent decrease in cattail abundance (or perhaps organic
preservation of cattail) by peat accretion becoming dominant over decomposition and recycling, resulting in a reduced availability of nutrients needed for
cattail growth.

FUTURE RATES of SEA LEVEL RISE

Tidal gauges around the state of Florida on both coasts confirm that relative sea
level is continuing to rise (National Oceanic and Atmospheric Administration,
1988). The period of record goes back to at least the 1930s except for one station.
Sea level is rising at a rate of 1.8–2.4 cm·yr^{-1}, which is significantly higher than
estimates of premodern ranges in Florida (0.3–1.3 cm·yr^{-1}), determined from
sediment dating (Robbin, 1984). These tidal gauge rates of sea level rise for Florida
are in good agreement with calculated average increases in sea level on a global
basis (Barnett, 1983; Revelle, 1983).

The greenhouse effect may increase the rate of sea level rise by further melting
of polar ice sheets and alpine glaciers, as well as by the thermal expansion of ocean
waters as they become warmer. The greenhouse effect causes increased atmospheric temperatures due to the accumulation of carbon dioxide and other gases
in the atmosphere, mainly from the burning of fossil fuels. Revelle (1983) indicates
that during the next century, sea level rise is likely to be on the order of 70 cm
if it is due only to melting of alpine glaciers, the Greenland ice cap, and heating
of the upper oceans. However, the melting of that portion of the West Antarctic
ice sheet which is above sea level would raise sea level worldwide by about 5 or
6 m. This almost would be equivalent to the recurrence of Sangamon interglacial
conditions. Any significant rise will mimic the mid-Holocene encroachment and

overtopping of freshwater marshes by mangroves or open salty water and the truncation of the southernmost Everglades.

NOTE on RADIOCARBON DATES

Radiocarbon dates are frequently assigned too much precision with respect to a particular event or boundary. This is especially so for thick samples of slow accreting, nonlaminated soft sediments such as peats and marsh muds produced in active wetland surface environments. The analysis is quite precise, but the reported uncertainty value is purely statistical and has no bearing on the integrity of the sample. Systematic biases from root intrusion in peat prior to deep burial or physical mixing of the uppermost zone of soft mud during accretion can exceed the counting uncertainties. Isotopically, peat and mud from the Everglades region are well suited for dating. Isotopic dilution from radiocarbon-depleted ancient limestone (a process which gives anomalously old dates elsewhere) is not grossly apparent in the sediments of the Everglades. The shallow calcitic-mud-forming marshes have good exchange with atmospheric carbon dioxide, which is also the direct source of carbon for the peats and the required reservoir for the dating method. A degree of physical mixing is the principal complication, and a sedimentological perspective should be maintained in interpretations using the dates. Nevertheless, radiocarbon dates are invaluable approximate estimates of actual ages for these sediments. Comparison from different sediment types should be regarded as more uncertain in interpretations because of the different biases (including some isotope fractionation effects). There are also systematic divergences between calendar (orbital) and radiocarbon years. Here, uncorrected radiocarbon dates are used, except in a few cases as noted. The uncertainty value and laboratory number are given for those dated samples from the authors' research.

ACKNOWLEDGMENTS

Dr. Jerry Stipp and his students of the radiocarbon laboratory, University of Miami, provided many of the critical dates used in this reconstruction through cooperative studies. Dr. William Spackman, The Pennsylvania State University, served as the inspiration for many of these investigations. James M. Montgomery, Consulting Engineers, Inc., actively supported the production of this manuscript.

LITERATURE CITED

Allison, R. V. and A. P. Dachnowski-Stokes. 1932. Physical and chemical studies upon important profiles of organic soils in the Florida Everglades. in Proc. and Papers of the 2nd Int. Congr. of Soil Science, Leningrad, pp. 223–245.

Altschuler, Z. S., M. M. Schnepfe, C. C. Silber, and F. O. Simon. 1983. Sulfur diagenesis in Everglades peat and origin of pyrite in coal. *Science,* 221:221–227.

Andrews, J. T. 1987. The Late Wisconsin glaciation and deglaciation of the Laurentide Ice Sheet. in *The Geology of North America,* Vol. K-3: North American and Adjacent Oceans during the Last Deglaciation, The Geological Society of America, Boulder, Colo., pp. 13–38.

Barnett, T. P. 1983. *Some Problems Associated with the Estimation of "Global" Sea Level Change,* Climate Research Group, Scripps Institution of Oceanography, University of California, San Diego.

Bender, M. L. 1973. Helium-uranium dating of corals. *Geochim. Cosmochim. Acta,* 37:1229–1247.

Beriault, J. G, R. S. Carr, J. J. Stipp, R. Johnson, and J. Meeder, 1981. The archaeological salvage of the Bay West site, Collier County, Florida. *Fla. Anthropol.,* 34:39–58.

Bloom, A. L. 1983. Sea level and coastal morphology of the United States through the late Wisconsin glacial maximum. in *Late Quaternary Environments of the United States,* Vol. 1: The Late Pleistocene, S. C. Porter (Ed.), University of Minnesota Press, pp. 215–229.

Broecker, W. S. and D. L. Thurber. 1965. Uranium-series dating of corals and oolites from Bahaman and Florida Key Limestones. *Science,* 149:58–60.

Brooks, H. K. 1968. The Plio-Pleistocene of Florida with special references to the strata outcropping on the Caloosahatchee River. in *Late Cenozoic Stratigraphy of Southern Florida— A Reappraisal, 2nd Annu. Field Trip of the Miami Geological Society,* R. D. Perkins (Ed.), Miami Geological Society, Miami, pp. 3–64, 103–110.

Brooks, H. K. 1974. Lake Okeechobee. in *Environments of South Florida: Present and Past,* P. J. Gleason (Ed.), Memoir No. 2, Miami Geological Society, Miami, pp. 256–286.

Browder, J. A., P. J. Gleason, and D. R. Swift. 1994. Periphyton in the Everglades: Spatial variation, environmental correlates, and ecological implications. in *Everglades: The Ecosystem and Its Restoration,* S. M. Davis and J. C. Ogden (Eds.), St. Lucie Press, Delray Beach, Fla., chap. 16.

Carbone, V. A. 1983. Late Quaternary environments in Florida and the southeast. *Fla. Anthropol.,* 36(1–2):3–17.

Carr, R. S. 1979. An Archaeological and Historical Survey of the Site 14 Replacement Airport and Its Proposed Access Corridors, Dade County, Florida, Report to the Federal Aviation Administration.

Carr, R. S. 1986. Preliminary report on the excavations at the Cutler Fossil Site (8Da2001) in southern Florida. *Fla. Anthropol.,* 39(3:2):231–232.

Carr, R. S. and J. G. Beriault. 1984. Prehistoric man in southern Florida. in *Environments of South Florida: Present and Past,* P. J. Gleason (Ed.), Miami Geological Society, Miami, pp. 1–14.

Carr, R. S. and D. Sandler. 1992. An Archaeological Survey and Assessment of the Westridge Property, Broward County, Florida, Tech. Rep. 18, Archaeological and Historical Conservancy, Miami.

Clausen, C. J., H. K. Brooks, and A. B. Wesolowsky. 1975a. The early man site at Warm Mineral Springs, Florida. *J. Field Archaeol.,* 2(3):191–213.

Clausen, C. J., H. K. Brooks, and A. B. Wesolowsky. 1975b. Florida springs confirmed as 10,000 year old early man site. *Fla. Anthropol.,* 28(3:2):1–38.

Clausen, C. J., A. D. Cohen, C. Emiliani, J. A. Holman, and J. J. Stipp. 1979. Little Salt Spring, Florida: A unique underwater site. *Science,* 203:609–614.

Clayton, B. S., J. R. Neller, and R. V. Allison. 1942. Water Control in the Peat and Muck Soils of the Florida Everglades, Bull. 378, University of Florida Agricultural Experiment Station.

CLIMAP Project Members. 1976. The surface of the ice-age earth. *Science,* 191:1131–1136.

Cocke, E. C., I. F. Lewis, and R. Patrick. 1934. A further study of Dismal Swamp peat. *Am. J. Bot.,* 21:374–396.

Cockrell, W. A. and L. Murphy. 1978. Pleistocene man in Florida. *Archaeol. East. North Am.,* 6: 1–13.

Cohen, A. D. 1968. The Petrology of Some Peats of Southern Florida (with Special Reference to the Origin of Coal), Ph.D. dissertation, The Pennsylvania State University, University Park.

Cohen, A. D. 1973. Possible influences of subpeat topography and sediment type upon the development of the Okefenokee swamp-marsh complex of Georgia. *Southeast. Geol.,* 15:141–151.

Cohen, A. D. and W. Spackman. 1970. Methods in peat petrology and their application to reconstruction of paleoenvironments. *Geol. Soc. Am. Bull.,* 83:129–142.

Cohen A. D. and W. Spackman. 1977. Phytogenic organic sediments and sedimentary environments in the Everglades–mangrove complex. Part II. The origin, description and classification of the peats of southern Florida. *Palaeontographica,* 162(Abt. B):71–114.

Converse, H. H., Jr. 1973. A Pleistocene vertebrate fauna from Palm Beach County, Florida. *Plaster Jacket* (The Florida State Museum), 21:1–14.

Corps of Engineers. 1968. Geology and soils. Appendix I. in Water Resources for Central and Southern Florida, Appendix IV, Survey-Review Report on Central and Southern Florida Project, Serial Number 58, Department of the Army, Jacksonville District, Jacksonville.

Craighead, F. C., Sr. 1969. Vegetation and recent sedimentation in Everglades National Park. *Fla. Nat.,* 42:157–66.

Craighead, F. C., Sr. 1971. *Trees of South Florida,* Vol. 1: The Natural Environments and Their Succession, University of Miami Press, Coral Gables, Fla.

Cypert, E. 1972. The origin of houses in the Okefenokee prairies. *Am. Midland Nat.,* 87:448–458.

Dachnowski-Stokes, A. P. 1930. Peat profiles of the Everglades of Florida: The stratigraphic features of the "upper" Everglades and correlation with environmental changes. *J. Wash. Acad. Sci.,* 20:89–107.

Davies, T. D. 1980. Peat Formation in Florida Bay and Its Significance in Interpreting the Recent Vegetational and Geological History of the Bay Area, Ph.D. dissertation, The Pennsylvania State University, University Park.

Davis, J. H., Jr. 1943. The natural features of southern Florida, especially the vegetation, and the Everglades. *Fla. Geol. Surv. Bull.,* 25:311.

Davis, J. H. 1946. The peat deposits of Florida. *Fla. Geol. Surv. Bull.,* No. 30.

Delcourt, P. A. and H. R. Delcourt. 1981. Vegetation maps for eastern North America: 40,000 yr b.p. to the present. in *Geobotany II,* R. C. Romans (Ed.), Plenum Press, New York., pp. 123–165.

Doran, G. H. and D. N. Dickel. 1988. Multidisciplinary investigations at the Windover site. in *Wet Site Archaeology,* B. A. Purdy (Ed.), Telford Press, Caldwell, N.J., pp. 263–289.

Duever, M. J. 1984. Environmental factors controlling plant communities of the Big Cypress Swamp. in *Environments of South Florida: Present and Past II,* P. J. Gleason (Ed.), Memoir No. 2, Miami Geological Society, Miami, pp. 127–137.

Duever, M. J., J. E. Carlson, J. F. Meeder, L. C. Duever, L. H. Gunderson, L. A. Riopelle, T. R. Alexander, R. F. Myers, and D. P. Spangler. 1979. Resource Inventory and Analysis of the Big Cypress National Preserve, Vol. 1: Natural Resources, U.S. National Park Service.

Egler, F. E. 1952. Southeast saline Everglades vegetation, Florida, and its management. *Vegetatio Acta Geobot.,* 3:213–265.

Ely, L. L., Y. Enzel, V. R. Baker, and D. R. Cayan. 1993. A 5,000-year record of extreme floods and climate change in the southwestern United States. *Science,* 262:410–412.

Fairbridge, R. W. 1968. Holocene, postglacial or recent epoch. in *The Encyclopedia of Geomorphology,* R. W. Fairbridge (Ed.), Reinhold, New York, pp. 525–536.

Fairbridge, R. W. 1974. The Holocene sea-level record in South Florida. in *Environments of South Florida, Present and Past,* P. J. Gleason (Ed.), Memoir No. 2, Miami Geological Society, Miami, pp. 223–232.

Gallatin, M. H., J. K. Ballard, C. B. Evans, H. S. Galberry, J. J. Hinton, D. P. Powell, E. Truett, W. L. Watts, and G. C. Willson. 1958. Soil survey (detailed-reconnaissance) of Dade County, Florida. *U.S. Soil Conserv. Serv. Ser.,* 1947(4):56.

Gleason, P. J. 1972. The Origin, Sedimentation and Stratigraphy of a Calcitic Mud Located in the Southern Freshwater Everglades, Ph.D. dissertation, The Pennsylvania State University, University Park.

Gleason, P. J. (Ed.). 1974. *Environments of South Florida: Present and Past* (2nd ed., 1984), Memoir No. 2, Miami Geological Society, Miami.

Gleason, P. J. and W. Spackman. 1974. Calcareous periphyton and water chemistry in the

Everglades. in *Environments of South Florida: Present and Past,* P. J. Gleason (Ed.), Memoir No. 2, Miami Geological Society, Miami, pp. 146–181.

Gleason, P. J. and P. A. Stone. 1975. Prehistoric Trophic Level Status and Possible Cultural Influences on the Enrichment of Lake Okeechobee, unpublished report, South Florida Water Management District, West Palm Beach.

Gleason, P. J., A. D. Cohen, W. G. Smith, H. K. Brooks, P. A. Stone, R. L. Goodrick, and W. Spackman. 1974. The environmental significance of Holocene sediments from the Everglades and saline tidal plain. in *Environments of South Florida: Present and Past,* P. J. Gleason (Ed.), Memoir No. 2, Miami Geological Society, Miami, pp. 287–341.

Gleason, P. J., P. A. Stone, R. Goodrick, G. Guerin, and L. Harris. 1975a. The Significance of Paleofloral Studies and Ecological Aspects of Floating Peat Islands to Water Management in the Everglades Conservation Areas, unpublished report, South Florida Water Management District, West Palm Beach.

Gleason, P. J., P. A. Stone, P. Rhoads, S. M. Davis, M. Zaffke, and L. Harris. 1975b. The Impact of Agricultural Runoff on the Everglades Marsh in the Conservation Areas of the Central and Southern Florida Flood Control District, unpublished report, South Florida Water Management District, West Palm Beach.

Gleason, P. J., P. A. Stone, and D. Benson. 1978. Bottom sediments of Lake Okeechobee. in *Hydrogeology of South-Central Florida,* M. P. Brown (Ed.), Publ. No. 20, Southeastern Geological Society, Tallahassee, Fla., pp. 44–60.

Gleason, P. J., D. Piepgras, P. A. Stone, and J. J. Stipp. 1980. Radiometric evidence for involvement of floating islands in the formation of Florida Everglades tree islands. *Geology,* 8:195–199.

Goggin, J. M. 1948. Florida archeology and recent ecological changes. *J. Wash. Acad. Sci.,* 38:225–233.

Goodrick, R. L. 1974. The wet prairies of the northern Everglades. in *Environments of South Florida: Present and Past,* P. J. Gleason (Ed.), Memoir No. 2, Miami Geological Survey, Miami, pp. 287–341.

Hanna, A. J. and K. A. Hanna. 1948. *Lake Okeechobee,* Bobbs-Merrill, New York.

Hanshaw, B. B., W. Back, and M. Rubin. 1965. Radiocarbon determinations for estimating groundwater flow velocities in central Florida. *Science,* 148:494–495.

Harper, R. M. 1909. Okefenokee Swamp. *Pop. Sci. Mon.,* 74:596–614.

Harshberger, J. W. 1914. The vegetation of South Florida. *Trans. Wagner Free Inst. Sci. Philadelphia,* 7(3):51–187.

Heilprin, A. 1887. *Explorations on the West Coast of Florida and in the Okeechobee Wilderness,* Wagner Free Institute of Science, Philadelphia.

Hoffmeister, J. E. 1974. *Land from the Sea,* University of Miami Press, Coral Gables, Fla.

Hoffmeister, J. E., K. W. Stockman, and H. G. Multer. 1967. Miami Limestone of Florida and its recent Bahamian counterpart. *Geol. Soc. Am. Bull.,* 78:175–190.

Jones, L. A. et al. 1948. Soils, geology, and water control in the Everglades region. *Univ. Fla. Agric. Exp. Stn. Bull.,* No. 442.

Kropp, W. 1976. Geochronology of Corkscrew Swamp Sanctuary. in Cypress Wetlands for Water Management, Recycling and Conservation, 3rd Annu. Rep., Center for Wetlands, University of Florida, Gainesville, pp. 772–785.

Kuehn, D. W. 1980. Offshore Transgressive Peat Deposits of Southwest Florida: Evidence for a Late Holocene Rise of Sea Level, M.S. thesis, The Pennsylvania State University, University Park.

Lovejoy, D. W. 1987. The Anastasia Formation in Palm Beach and Martin Counties, Florida. in *Symp. on South Florida Geology,* F. Maurrasse (Ed.), Memoir No. 3, Miami Geological Society, Miami.

Loveless, C. M. 1959. A study of the vegetation in the Florida Everglades. *Ecology,* 40:1–9.

Lundelius, E. L., Jr. 1988. What happened to the mammoth? The climatic model. in *Americans before Columbus: Ice-Age Origins,* R. C. Carlisle (Ed.), Department of Anthropology, University of Pittsburgh; *Ethnol. Monogr.,* 12:75–82.

Masson, M., R. S. Carr, and D. Goldman. 1988. The Taylor's head site (8BD74): Sampling a prehistoric midden on an Everglades tree island. *Fla. Anthropol.,* 41:336–350.

McDowell, L. L., J. C. Stephens, and E. H. Stewart. 1969. Radiocarbon chronology of the Florida Everglades peat. *Soil Sci. Soc. Am. Proc.,* 33:743–745.

Milanich J. T. and C. H. Fairbanks. 1980. *Florida Archaeology,* Academic Press, Orlando.

Missimer, T. M. 1973. Growth rates of beach ridges on Sanibel Island, Florida. *Trans. Gulf Coast Assoc. Geol. Soc.,* 23:383–388.

Missimer, T. M. 1980. Holocene sea level changes in the Gulf of Mexico: An unresolved controversy. in *Holocene Geology and Man in Pinellas and Hillsborough Counties Florida,* S. B. Upchurch (Ed.), Southeastern Geological Society Guidebook, Vol. 22, Southeastern Geological Society, Tallahassee, Fla.

Mitterer, R. M. 1975. Ages and digenetic temperatures of Pleistocene deposits of Florida based on isoleucine epimerization in *Mercenaria. Earth Planet. Sci. Lett.,* 28:275–282.

Moore, W. S. 1982. Late Pleistocene sea-level history. in *Uranium Series Disequilibrium: Applications to Environmental Problems,* M. Ivanovich and R. S. Harmon (Eds.), Oxford University Press, Oxford, pp. 481–496.

Morner, N.-A. 1971. The position of the ocean level during the interstadial at about 30,000 B. P.—A discussion from a climatic-glaciologic point of view. *Can. J. Earth Sci.,* 8:132–143.

Natural Oceanic and Atmospheric Administration. 1988. Sea Level Variations for the United States 1855–1986, U.S. Department of Commerce, Washington, D.C.

Navoy, A. S. and L. A. Bradner. 1987. Ground water resources of Flagler County, Florida. *U.S. Geol. Surv. Water-Res. Invest. Rep.,* No. 87-4021.

Olmstead, I. C., L. L. Loope, and R. E. Rintz. 1980. A Survey and Baseline Analysis of Aspects of the Vegetation of Taylor Slough, South Florida Research Center Rep. T-586, Everglades National Park, Homestead, Fla.

Osmond, J. K., J. R. Carpenter, and H. L. Windom. 1965. The age of the Pleistocene corals and oolites of south Florida. *J. Geophys. Res.,* 70:1943.

Parker, G. G. 1974. Hydrology of the pre-drainage system of the Everglades in southern Florida. in *Environments of South Florida: Present and Past,* P. J. Gleason (Ed.), Memoir No. 2, Miami Geological Society, Miami, pp. 18–27.

Parker, G. G. and C. W. Cooke. 1944. Late Cenozoic geology of southern Florida, with a discussion of the ground water. *Fla. Geol. Surv. Bull.,* No. 27.

Parker, G. G. and N. C. Hoy, 1943. Additional notes on the geology and ground water of southern Florida. *Soil Sci. Soc. Fla. Proc.,* 5-A:33–55, 77–94.

Parker, G. G., G. E. Ferguson, S. K. Love, et al. 1955. Water resources of southeastern Florida. *U.S. Geol. Surv. Water-Supply Pap.,* No. 1255.

Parrish, F. K. and E. J. Rykiel. 1979. Okefenokee Swamp origin: Review and reconsideration. *J. Elisha Mitchell Sci. Soc.,* 95:17–31.

Perkins, R. D. 1977. Depositional framework of Pleistocene rocks in south Florida. in *Quaternary Sedimentation in South Florida,* P. Enos and R. D. Perkins (Eds.), *Geol. Soc. Am. Mem.,* 147:131–198.

Petuch, E. J. 1986. The Pliocene reefs of Miami: Their geomorphological significance in the evolution of the Atlantic Coastal Ridge, Southeastern Florida, USA. *J. Coastal Res.,* 2:391–408.

Petuch, E. J. 1987. The Florida Everglades: A buried pseudoatoll? *J. Coastal Res.,* 3:189–200.

Porter, S. C. 1988. Landscapes of the last ice age in North America. in *Americans before Columbus: Ice Age Origins,* R. C. Carlisle (Ed.), Department of Anthropology, University of Pittsburgh; *Ethnol. Monogr.,* No. 12,

Prest, V. K. 1973. Retreat of the last ice sheet. in *The National Atlas of Canada,* G. Fremlin (Ed.), Department of Energy, Mines, and Resources, Ottawa, Ontario, Canada, pp. 31–32.

Puri, H. S. and R. O. Vernon. 1964. Summary of the geology of Florida and guidebook to the classic exposures. *Fla. Geol. Surv. Spec. Publ.,* No. 5, 312 pp.

Ralph, E. K. and H. N. Michael. 1974. Twenty-five years of radiocarbon dating. *Am. Sci.,* 62:553–560.

Revelle, R. R. 1983. Probable future changes in sea level resulting from increased atmospheric carbon dioxide. in *Changing Climate,* Report of the Carbon Dioxide Assessment Committee, National Academy of Sciences, National Academy Press, Washington, D.C.

Rich, F. J. 1978. The Origin and Development of Tree Islands in the Okefenokee Swamp as Determined by Peat Petrography and Pollen Stratigraphy, Ph.D. dissertation, The Pennsylvania State University, University Park.

Rich, F. J. and W. Spackman. 1979. Modern and ancient pollen sedimentation around tree islands in the Okefenokee Swamp. *Palynology,* 3:219–226.

Riegel, W. L. 1965. Palynology of Environments of Peat Formation in Southwestern Florida, Ph.D. dissertation, The Pennsylvania State University, University Park.

Robbin, D. M. 1984. A new Holocene sea-level curve for the upper Florida Keys and Florida reef tract, in *Environments of South Florida: Present and Past II,* P. J. Gleason (Ed.), Memoir No. 2, Miami Geological Society, Miami, pp. 437–458.

Roberts, H. H., T. Whelan, and W. G. Smith. 1977. Holocene sedimentation at Cape Sable, South Florida. *Sediment. Geol.,* 18:25–60.

Scholl, D. W. 1964a. Recent sedimentary record in mangrove swamps and rise in sea level over the southwestern coast of Florida. Part I. *Mar. Geol.,* 1:344–366.

Scholl, D. W. 1964b. Recent sedimentary record in mangrove swamps and rise in sea level over the southwestern coast of Florida. Part 2. *Mar. Geol.,* 2:343–364.

Scholl, D. W., F. C. Craighead, Sr., and M. Stuiver. 1969. Florida submergence curve revised: Its relation to sedimentation rates. *Science,* 163:562–564.

Schroeder, M. C. and H. Klein. 1954. Geology of the western Everglades area, south Florida. *U.S. Geol. Surv. Circ.,* No. 314.

Schroeder, M. C., H. Klein, and N. D. Hay. 1958. Biscayne aquifer of Dade and Broward Counties, Florida. *Fla. Geol. Surv. Rep. Invest.,* No. 17.

Sears, E. and W. Sears. 1976. Preliminary report on prehistoric corn pollen from Fort Center, Florida. *Southeast. Archaeol. Conf. Bull.,* 19:53–56.

Sears, W. 1982. Fort Center. *Ripley P. Bullen Monographs in Anthropology and History,* No. 4, The Florida State Museum, Gainesville.

Sellards, E. H. 1919. Geologic Sections across the Everglades of Florida, Fla. Geol. Survey 12th Annu. Rep., pp. 67–76.

Shepard, F. P. 1973. *Submarine Geology,* 3rd ed., Harper and Row, New York.

Smith, W. G. 1968. Sedimentary Environments and Environmental Change in the Peat Forming Area of South Florida, Ph.D. dissertation, The Pennsylvania State University, University Park.

Spackman, W., C. P. Dolson, and W. Riegel. 1966. Phytogenic organic sediments and sedimentary environments in the Everglades-Mangrove complex. Part 1. Evidence of a transgressing sea and its effects on environments of the Shark River area of southwestern Florida. *Palaeontographica,* 117(8):135–152.

Spackman, W., W. L. Riegel, and C. P. Dolson. 1969. Geological and biological interactions in the swamp-marsh complex of southern Florida. in *Environments of Coal Formation,* E. C. Dapples and M. E. Hopkins (Eds.), Geological Society of America Special Paper No. 114, pp. 1–35.

Spackman, W., A. D. Cohen, P. H. Given, and D. J. Casagrande. 1976. A Field Guidebook to Aid in the Comparative Study of the Okefenokee Swamp and the Everglades–Mangrove Swamp–Marsh Complex of Southern Florida, Coal Research Section, The Pennsylvania State University, University Park.

Stapor, F. W. 1983. Control of Sea Level and Sediment Supply on Depositional Sequences: The Holocene Example, Exploration Research Rep. EPR-24 EX-83, Exxon Production Research Company, Houston.

Stapor, F. W. and W. F. Tanner. 1973. Errors in the pre-Holocene C-14 scale. *Trans. Gulf Coast Assoc. Geol. Soc.,* 23:351–354.

Stapor, F. W, T. D. Matthews, and F. E. Lindfors-Kearns. 1991. Barrier-island progradation and Holocene sea-level history in southwest Florida. *J. Coastal Res.,* 7:815–838.

Stephens, J. C. and L. Johnson. 1951. Subsidence of organic soils in the upper Everglades region of Florida. *Soil Sci. Soc. Fla. Proc.,* XI:191–237.

Stewart, J. T. 1907. Report on Everglades Drainage Project in Lee and Dade Counties, Office of Experiment Stations, U.S. Department of Agriculture.

Stone, P. A. 1978. Floating Islands—Biogeomorphic Features of Hillsboro Marsh, Northeastern Everglades, Florida, M.A. thesis, Florida Atlantic University, Boca Raton, Fla.

Stone, P. A. and J. G. Brown. 1983. The pollen record of Pleistocene and Holocene paleoenvironmental conditions in southeastern United States. in *Variation in Sea Level on the South Carolina Coastal Plain,* D. J. Colquhoun (Ed.), Department of Geology, University of South Carolina, Columbia, pp. 169–206.

Stone, P. A. and P. J. Gleason. 1983. Environmental and paleoenvironmental significance of organic sediments (peats) in southeastern United States. in *Variation in Sea Level on the South Carolina Coastal Plain,* D. J. Colquhoun (Ed.), Department of Geology, University of South Carolina, Columbia, pp. 121–141.

Tropical Bioindustries. 1987. Key Environmental Indicators of Ecological Well-Being, Everglades National Park and the East Everglades, Dade County, Florida, partial draft report to the U.S. Army Corps of Engineers, Environmental Resources Planning Branch, Planning Division, Jacksonville, Fla., Contract DACW 17-84-C-0031.

Wanless, H. R. 1969. Sediments of Biscayne Bay—Distribution and Depositional History, University of Miami, Inst. of Marine Sciences Tech. Rep. 69:2, 260 pp.

Watts, W. A. 1969. A pollen diagram from Mud Lake, Marion County, north/central Florida. *Geol. Soc. Am. Bull.,* 80:631–642.

Watts, W. A. 1971. Postglacial and interglacial vegetation history of southern Georgia and central Florida. *Ecology,* 52:676–690.

Watts, W. A. 1975. A late Quaternary record of vegetation from Lake Annie, south-central Florida. *Geology,* 3:344–346.

Watts, W. A. 1980. The late Quaternary vegetation history of the southeastern United States. *Annu. Rev. Ecol. Syst.,* 11:387–409.

Watts, W. A. and B. C. S. Hansen. 1988. Environments of Florida in the Late Wisconsin and Holocene. in *Wet Site Archaeology,* B. A. Purdy (Ed.), Telford Press. Caldwell, N.J., pp. 307–323.

Webb, S. D. (Ed.). 1974. *Pleistocene Mammals of Florida,* University Presses of Florida, Gainesville.

Webb, S. D., J. T. Milanich, R. Alexon, and J. S. Dunbar. 1984. A *Bison antiquus* kill site, Wacissa River, Jefferson County, Florida. *Am. Antiq.,* 49:384–392.

Whitehead, D. R. and R. Q. Oaks. 1979. Developmental history of the Dismal Swamp. in *The Great Dismal Swamp,* P. W. Kirk (Ed.), University Press of Virginia, Charlottesville, pp. 25–43.

Willey, G. R. 1949. Excavations in southeast Florida. *Yale Univ. Publ. Anthropol.,* No. 42.

8

SEA LEVEL CONTROL on STABILITY of EVERGLADES WETLANDS

Harold R. Wanless

Randall W. Parkinson

Lenore P. Tedesco

ABSTRACT

The expansive coastal wetlands and freshwater marsh of south Florida are a result of the very slow relative rise of sea level during the past 3200 years (average rate of 4 cm/100 years). Prior to that time, relative sea level was rising at a rate of 23–50 cm/100 years—too fast for coastal swamp, marl, or sand environments to stabilize along south Florida's coastlines. The establishment of a broad, coastal wetland during the past 3200 years has provided a natural barrier to marine waters and permitted freshwater environments of the Everglades to expand.

Tide gauges throughout the United States record a dramatic increase in the rate of relative sea level rise beginning about 1930. During the following 60 years, the relative rise in sea level for south Florida has averaged 3–4 mm $\cdot$ yr^{-1} (equivalent to 20–40 cm/100 years). This rate is 6–10 times that of the past 3200 years and is triggering dramatic changes in the coastal wetland communities, including accelerated erosion of shore margins, landward encroachment of marine wetlands, and saltwater encroachment of surficial and ground waters.

Continuation at these rates or acceleration, as expected due to global warming, will cause dramatic to catastrophic modifications of both the coastal and freshwater wetlands of south Florida. Major hurricanes will cause dramatic steps of erosion as well as overstepping of coastal wetland margins.

INTRODUCTION

The mid to late Holocene (6500 years before present [YBP] to present) has been a time of continued relative sea level rise along the Gulf and southern Atlantic coasts of the United States. Curiously, a large number of coastal sediment bodies in south Florida were initiated or stabilized between 3300 and 3000 YBP and again between approximately 2300 to 2500 YBP. The origination and development of those coastal and shallow-marine carbonate and clastic sediment bodies is dependent on the interactive roles and timing of sea level changes, climatic fluctuations, storm influences, vegetative colonization, pre-existing topography, and provision of detrital (input or recycled) silicic–clastic and carbonate sediment. Of these, the rate of sea level rise appears to have been the primary control on the clastic, carbonate, and organic sediment bodies of south Florida. This chapter summarizes the general response of coastal sediment bodies in south Florida to Holocene sea level history and points out uncertainties that remain to be solved, before the Holocene record can provide hindsight guidance for specific modeling of future coastal evolution in response to future sea level changes.

This hindsight is especially important because beginning about 1930, relative sea level has been rising along Florida's shorelines at a rate of 20–40 cm/100 years. This is nearly ten times the rate recorded by earlier tide gauge records and by the past several thousand years of geologic history. This increased rate of rise is already exhibiting a significant influence on Florida shorelines and wetlands and will have dramatic consequences within the next century if continued. Further increases in this rate of rise are widely predicted due to global warming, resulting from increases in atmospheric carbon dioxide, methane, and other anthropogenic gases. Higher rates of relative sea level rise will further heighten the implications of future (1) rapid retreat of south Florida shorelines on sandy beaches, mangrove swamps, salt marshes, and marl levees; (2) inundation of low-lying inland areas; and (3) loss of freshwater resources through saline intrusion into ground water. A historical perspective is helpful in appreciating the reality of these conclusions.

HOLOCENE SEA LEVEL HISTORY

The latter stage of Holocene sea level history is commonly graphed as a smoothed curve. In the case of the Atlantic and Gulf coasts of the United States, this is a progressively decreasing rate of rise in relative sea level (Scholl et al., 1969; Bloom, 1977). Smoothed curves have also been used to describe the history of relative sea level for the Pacific coast (Curray et al., 1969). In a sense, this smoothed curve approach has become necessary because of limitations in the number of radiocarbon data points and the uncertainty of the exact level with respect to sea level that each point represents (Kidson, 1986).

Some authors have suggested that this curve of sea level rise (transgression) is not smooth, but rather contains a series of hinge points (Figure 8.1) (Neumann, 1972; Bloom, 1977; Wanless, 1982). These hinge points have become attractive to some coastal sedimentologists because they roughly coincide with either changes in the style of coastal and shallow marine sedimentation or the appearance or stabilization (end of landward transgression) of sediment bodies (Parkinson, 1987).

This chapter uses a late Holocene sea level curve with hinge points. The form

Figure 8.1 History of relative sea level for south Florida during the late Holocene, incorporating data from Scholl et al. (1969), Neumann (1972), and Wanless (1976). The increased rate of sea level rise during the past 60 years is based on tide gauge measurements.

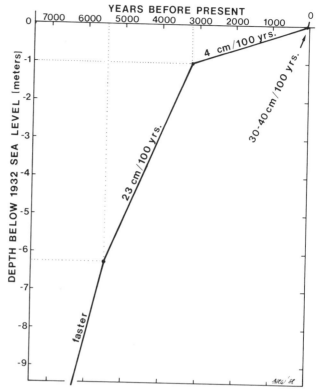

of the curve is derived from the curve of Neumann (1972) for Bermuda, which closely parallels Florida. The specific nature of the curve is based on data points from south Florida by Scholl et al. (1969) and Wanless (1969 and unpublished). As can be seen in Figure 8.1, the latter stages of the Holocene sea level rise over south Florida can be divided into three stages:

Time (YBP)	Relative rate of rise (cm/100 years)	Elevation at end of period (from present sea level) (m)
0–3200	4	0
3200–5500	23	–1.0
7500–5500	>50	–6.2

Figure 8.2 contours the intersection of sea level with the Pleistocene limestone surface at 5500 YBP (–6.2 m), 3500 YBP (–1.5 m), and the present (0 m). These contours represent the shoreline positions had there been no accumulation of Holocene sediment over Pleistocene limestone. The shaded areas between these time-depth lines are the zones of the shelf potentially transgressed. This map of potential transgression will be evaluated more closely after the sedimentary environments have been described.

Figure 8.2 Map of south Florida showing present coastline (line with dots adjacent) and the topography/ bathymetry of the underlying Pleistocene limestone surface. The –6-, –1.5-, and 0-m contours represent the potential position of the coastline had there been no Holocene sediment 5500 YBP, 3500 YBP, and at present. Also shown are the +1- and +2-m contours of bedrock elevation which are useful in considering future projections of sea level rise.

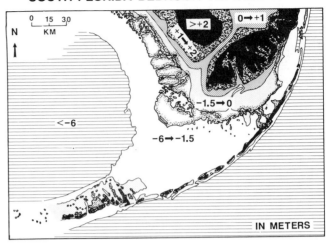

SOUTH FLORIDA BEDROCK TOPOGRAPHY

SOUTH FLORIDA ENVIRONMENTS

Sedimentary Influences

The following critical and fundamental attributes of the south Florida environment influence sedimentation:

1. A subtropical setting subjected to occasional freezes and periods of cold inshore waters
2. The warm Florida Current, which provides a tropical influence on the outer shelf waters and a stabilizing influence on air temperatures
3. The presence of seasonally strong freshwater discharge through channels, sheet flow, and ground water from mainland to coastal environments and catastrophic discharge of fresh water during some hurricanes
4. A seasonally rainy/humid climate, which receives an annual rainfall of 165 cm along the southeast mainland, decreasing to 90 cm in the southwestern Florida Keys
5. The influence of quartz sand as southward drifting barrier islands and shoals on both the southeastern and southwestern coasts
6. A pre-existing limestone topography with four low limestone ridges that have been partially inundated to create an outer and inner line of lagoons and a peat-choked freshwater embayment
7. Protection from oceanic waves and swells by the Bahama Banks to the east, by Cuba to the south, and by a broad, gently sloping continental shelf to the west
8. Semi-diurnal tides on the southeast coast, which average about 60 cm but are dampened to less than 15 cm in the interior bays, and mixed tides on the southwest coast, which reach 250 cm in a narrow zone in the eastern Ten Thousand Islands

Physiography and Coastal Landforms

Southern Florida is a partially inundated limestone platform on which Holocene carbonate, organic, and siliceous sediments of biogenic origin are forming and accumulating and to which quartzose sands are being provided by littoral transport, as well as by dissolution from limestones along the eastern and western coasts. Three shallowly inundated limestone depressions are defined by four topographically high limestone physiographic features. The depressions from west to east are the Everglades, the coastal bays, and the outer shelf depressions; the highs are the Big Cypress Ridge, the Atlantic Coastal Ridge, the Key Largo Limestone Ridge, and the shelf-edge reefal ridge (Figure 8.3), with the shelf-edge reef occurring just landward of the 20-m contour at the edge of the Straits of Florida.

The Everglades depression extends southward from Lake Okeechobee down the center of south Florida and is bounded to the west by exposed Pliocene limestone of the Big Cypress Ridge and to the east by the Atlantic Coastal Ridge, a late Pleistocene quartz sand and oolitic limestone ridge. The broad Everglades depression carries freshwater flow southward from Lake Okeechobee to the sea just west of Florida Bay. The depression is filled with Holocene freshwater peat and calcitic mud deposits. A natural coastal dam of mangrove peat and storm-levee marl limits saline intrusion in the depression to along the axis of channels such as Shark

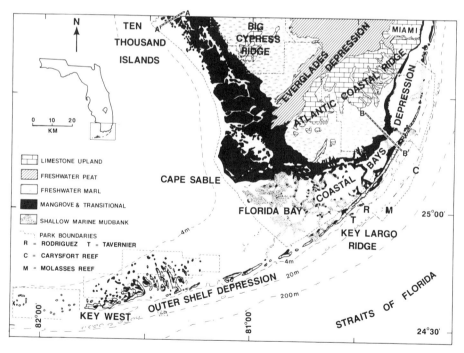

Figure 8.3 Map of south Florida showing the major physiographic features; the −4-, −20-, and −200-m depth contours; park boundaries (small dashes); and the location of cross sections A–A′ and B–B′ across the Ten Thousand Islands and across Card Sound (Figures 8.4 and 8.5). Mainland soil types are based on an unpublished map by Meeder and Wanless. (Adapted from Wanless et al., 1989.)

River Slough. The oolitic limestone ridge crests 2–6 m above sea level and extends north to south from north Miami to Homestead, where it then swings westward and is cross-cut by numerous swales, which are fossil tidal channels. These channels, largely filled with freshwater peat and quartz sand, served as conduits for freshwater flow through the ridge prior to drawdown of Everglades water levels in the early 1900s. These are still pathways for catastrophic storm discharge of fresh water.

The Coastal Bay Depression is bounded to the landward side by the Atlantic Coastal Ridge and to the seaward side by the Key Largo Limestone Ridge, a ridge of Pleistocene coral limestone of which the emergent portions form the present northern and central Florida Keys. The limestone surface beneath the coastal bays is generally less than 4 m and is shallower toward the mainland. Freshwater marshes, mangrove swamps, and coastal storm levees and flats partially fill the Coastal Bay Depression and isolate the various bay units. Florida Bay and Biscayne Bay are further subdivided by marine carbonate mud banks. Exposed limestone floors much of the bay bottom of southern Biscayne Bay, eastern Florida Bay, and the bays between. The Key Largo Limestone Ridge is now submerged along the eastern margin of central and northern Biscayne Bay. The seaward margin is defined by quartzose barrier islands to the north and a carbonate mud-bar belt (Safety Valve) seaward of central Biscayne Bay. Both cap the submerged limestone ridge.

Between the Florida Keys (Key Largo Limestone Ridge) and the shelf margin is a 5- to 10-km-wide outer shelf depression (or lagoon), commonly known as Hawk Channel, whose seaward margin is a Pleistocene reefal ridge. The shelf margin limestone ridge is now well submerged and capped with as much as 20 m of Holocene reefal sediment (Enos, 1977). The present reefal rim is well developed where seaward of an effective island barrier (the Florida Keys), but is poorly and sporadically developed where adjacent to breaks in the Florida Keys and north of Key Biscayne. In addition to the physiographic units defined by Pliocene and Pleistocene ridges, build-up of Holocene sediment generates coastlines and marine banks which define and subdivide the coastal bays and form the effective coastlines to southern and southwestern Florida.

The northern shore of Florida Bay is comprised of transgressive (landward eroding) to regressive (seaward expanding) coastal storm levees that are separated from the mainland mangrove coastline by a line of brackish lagoons (Figure 8.4). Depths to the limestone surface are about 1.5 m at the mainland mangrove fringe (Figure 8.5) and increase seaward to 2–4 m beneath the coastal storm levee. Cape Sable, at the northwestern corner of Florida Bay, is a complex of shell and sand beach ridges built in front of the relict (former) mouth of Everglades discharge. The coastal bays themselves, especially Florida Bay, are dissected by an anastomosing maze of carbonate mud and mangrove peat banks that have built nearly to sea level. Mud banks are dotted with intertidal to supratidal islands, some of which are relics from an earlier phase of mud bank nucleation on coastal storm levees and some of which are recent caps to marine mud bank deposits (Wanless and Tagett, 1989).

Along the southwestern coast of Florida, the Ten Thousand Islands form a 2- to 4-km-wide complex of mangrove islands, oyster banks, and shallow lagoons seaward of a continuous mangrove belt along the mainland shore. Depths to the

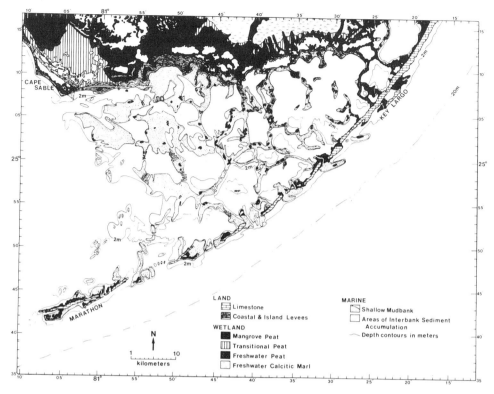

Figure 8.4   Map of Florida Bay showing major sedimentary environments including Holocene mud banks, islands, and coastal levees. The Gulf of Mexico is to the west. The Pleistocene Florida Keys form the southeastern border, and coastal levees adjacent to the freshwater Everglades define the northern margin. The coastal levees are erosional and somewhat sediment starved to the east, but to the west they become broad progradational features. The landward side of the bays northward of the coastal levees is about the contour of –1.5 m to limestone. (Adapted from Wanless and Tagett, 1989.)

limestone surface increase from about 1.5 m at the mainland shore to more than 5 m beneath the seaward margin of the Ten Thousand Islands (Figure 8.6).

Holocene Stratigraphy—Generalizations

Coastal swamp environments are transgressive on the eastern, mainland shore of Biscayne Bay and Barnes Sound. Cores from nearshore areas record a sequence from basal freshwater peat or marl, grading upward to coastal red mangrove (*Rhizophora mangle*) peat and capped by transgressive marine sands or muds (Wanless, 1976). The mangrove swamp coastlines on the eastern side of the northern Florida Keys are similarly erosionally transgressive, but cored sequences exhibit mangrove peat throughout their entire length.

Some mangrove peat or coastal levee deposits have accreted vertically from their inception, either as a persisting coastal environment (e.g., west shore of Card Sound; Figure 8.5) or as an island or peninsular remnant of such a coastline (e.g.,

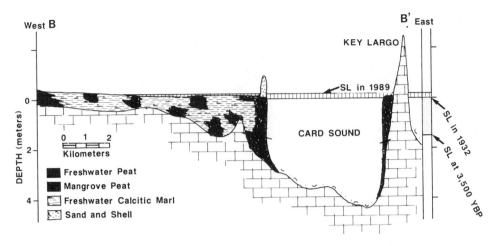

Figure 8.5   Cross section across the mainland mangrove fringe, Card Sound, and Key Largo in southeast Florida. Here, interaction of late Holocene sea level history and pre-existing topography caused a mangrove peat fringe barrier to form and persist during the past 3200 years. On the leeward east side of Card Sound, mangrove peats have persisted from earlier dates. See Figure 8.3 for location. (Adapted from Wanless, 1976.)

the mangrove peat neck separating Card Sound and Barnes Sound; the marl levee peninsulas and the high islands in northern Florida Bay).

The sedimentary sequences beneath the northwestern coastline of Florida Bay and the Ten Thousand Islands of southwest Florida record shallowing upward sequences during the past 3200 years as rise of relative sea level slowed. In both areas, the shallowing sequences are capped by supratidal deposits. On the northern coast of Florida Bay, this is a linear emergent storm levee of carbonate marl; in the

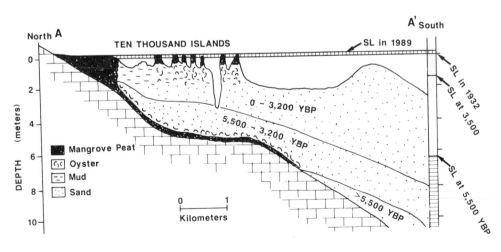

Figure 8.6   Cross section across the Ten Thousand Islands showing sediment packages deposited and preserved from >5500, 5500 to 3200, and <3200 YBP. The coastal sequence was transgressive until about 3200 YBP and then regressive since. Sea level positions at hinge points are shown at right. See Figure 8.3 for location. (Adapted from Parkinson, 1987.)

Ten Thousand Islands, this is a complex of mangrove islands capping oyster banks (Figure 8.6) (Parkinson, 1987; Perlmutter, 1982). In western Florida Bay there have also been steps of seaward progradation as storms formed a new storm levee shoreline bayward of the previous position.

The coral–algal banks on the landward part of the shelf lagoon are coarsening upward sequences, the shallower parts of which initiated 2500–3000 YBP. In the Pleistocene deep depression between the shelf margin reef tract and the Florida Keys, some sedimentation began before 5500 YBP, but the bulk of the sediment infill occurred following 5500 YBP. The seaward or outer shelf lagoon sequence is also coarsening upward, with carbonate mudstones in the base of the trough and *Halimeda*/bryozoan packstone at the top of the lagoonal sequence. Coral–algal packstones and grainstones dominate White Bank, a reef-derived skeletal sand bank extending into the lagoon from the reefal margin (Enos, 1977; Craig, 1983).

COASTAL and STRATIGRAPHIC RESPONSE to HOLOCENE SEA LEVEL RISE

A simple way to assess the influence of rate of relative sea level rise on coastal and shallow marine sedimentation is to (1) compare the potential inundation (intersection of sea level and the limestone bedrock surface) at the end of a period with the actual and (2) use information of the Holocene stratigraphy to time the origin and the transgressive (landward eroding and migrating) and/or regressive (seaward accreting) nature of sedimentary sequences.

15,000–5500 Years Before Present

About 15,000 YBP, the end of the last major ice age was near, and sea level stood about 130 m below present level. At this time the shoreline was about 10 km seaward of the present reefal margin of southeastern Florida. In the distance to the east across the Straits of Florida were the dramatic white cliffs of the limestone plateaus of the Bahamas (now inundated). The shoreline of the Gulf of Mexico was 100–200 km westward (across the present continental shelf) from the southwest coast of Florida.

The great continental ice sheets began to rapidly melt 14,000–15,000 years ago. Melt waters poured back into the oceans, causing a rapid eustatic (i.e., real input of water) rise in sea level. Between about 14,000 and 9000 YBP, sea level rose from −130 m (−400 ft) (Fairbanks, 1989) to near about −20 m, at a rate of 2 m/100 years. During this extreme rate of rise, all shorelines and shallow marine deposits either eroded landward or were left behind on the rapidly deepening marine shelf. Imagine developing shoreline property when the sea was rising 30–45 cm (1–1.5 ft) every 25 years and the coastline was migrating landward at a rate of over a kilometer (0.6 mi) every hundred years.

This eustatic rate slowed somewhat between 9000 and 5500 YBP. By 5500 years ago, the rapid input of melt water dramatically slowed, as did the eustatic sea level rise. After 5500 years ago, differences in land subsidence caused important differences in relative sea level change along the coastlines (Neuman et al., 1980;

Barnett, 1990). In much of the central Pacific, for example, sea level has remained relatively steady during the past 5500 years. Along the Atlantic coast of Brazil, however, sea level was nearly 2 m higher 5500 years ago than it is today. This relative lowering of sea level has been most dramatic along coastlines that were previously covered by thick glacial ice. As this ice load was removed, the land has gradually rebounded. The coastlines of Alaska, Maine, and the maritime provinces of Canada and Scandinavia are examples of places where rebounding has resulted in relative lowering of sea level.

In contrast, along the Gulf and Atlantic coasts of the United States south of Boston, 5500 years ago sea level was 5–8 m below present level. Using south Florida as an example, this relative rise in sea level appears to have occurred at a decreasing rate, with a hinge at about 3200 years ago. It is important to look at this rise more closely because it has an important bearing on forecasting the future of coastal response to sea level rise.

5500 Years Before Present

At 5500 YBP, sea level was at about –6.2 m. It had been rising at rates greater than 50 cm/100 years, but the rate of relative rise was slowing significantly. Areas that were potentially transgressed prior to 5500 YBP were, in fact, transgressed because little modern sediment build-up occurred (Figures 8.1 and 8.2). Shoreline deposits were thin, narrow, and ephemeral. Significant marine sediment bodies were accumulating only in areas of reefal growth or reef-derived sand influx on and adjacent to the southeastern shelf margin.

5500–3200 Years Before Present

Between about 5500 and 3200 years ago, relative sea level in south Florida rose from about –6 to –1 m at an average rate of about 23 cm/100 years. This rate of relative rise, although much slower than before, was still sufficient to cause rapid shoreline retreat. As seen in Figure 8.2, areas that were potentially transgressed between 5500 and 3200 YBP were, in fact, mostly transgressed because no modern sediment build-up occurred.

The coastal lagoons and estuaries of Florida were inundated during this time by repeated drowning inundation of coastal levees and the persistent retreat of mangrove swamp and cordgrass (*Spartina* spp.) marsh coastlines. Throughout this period these important coastal wetland communities, where present, formed only a narrow coastal band. Significantly, the character of coastal lagoons and estuaries was rapidly changing as seaward barriers were flooded or eroded, as bay depths increased, and as freshwater input and bay circulation evolved.

Biscayne Bay, a narrow channel prior to 5500 YBP, was largely inundated during this time by persistent erosional backstepping of mangrove peat and quartz sand shorelines (Figure 8.2) (Wanless, 1976). The Key Largo Limestone Ridge, which defines the seaward margin of Biscayne Bay, was extensively flooded, with only Elliott Key remaining as a significant emergent barrier to southern Biscayne Bay.

Florida Bay, emergent at 5500 YBP, was inundated to near its present configuration by 3200 YBP. This occurred through repeated erosion and overstepping of storm levee marl and peat shorelines (Figure 8.2) (Cottrell, 1987; Wanless and Tagett, 1989). Large volumes of carbonate mud and sand, recycled across outer Florida Bay with the transgressing sea, deposited several lines of coastal levees as Florida Bay was inundated. Most of these were overstepped and partially eroded but served as bathymetric features from which carbonate mud banks were subsequently initiated (Figures 8.3 and 8.4).

The Key Largo Limestone Ridge, defining the seaward margin of the coastal bays on the southeastern coast, was flooded in low areas, thus providing inlet passes between Florida Bay and the reefal shelf to the east and opening up much of the eastern margin of central Biscayne Bay.

On the southwestern coast (now the Ten Thousand Islands), Scholl (1966) and Parkinson (1987) have documented that there was only a narrow paralic mangrove band at the coast and that this was retreating landward between 5500 and 3200 YBP. Parkinson (1987) concluded that at about 3200 YBP this narrow, mangrove swamp shoreline was about at the position of the landward edge of the interior bays of the present Ten Thousand Islands (Figures 8.2 and 8.6). The limestone surface is about 1.5 m below present sea level at that position, which was sea level about 3500 years ago. Parkinson's (1987) stratigraphic and paleoecological study of core borings taken seaward of the paleoshoreline concluded that the transgressed adjacent marine environment was deepening and becoming more and more open marine between 5500 and 3200 YBP (Figure 8.6).

Significant persisting shoreline deposits were initiating only in areas such as (a) Key Biscayne on the southeast coast, where southerly littoral drift of sands from the Miami Beach area was focused along the seaward face of a Pleistocene ridge of Key Largo Limestone and Anastasia Limestone, and (b) in areas such as Sanibel Island on the southwest coast, where pre-existing topography was similarly focusing longshore drift. In some areas of southeast Florida, prolific coral reef growth was initiated and produced enough skeletal debris to keep up with relative sea level rise, but in many areas reefs were not growing upward as fast as sea level was rising (Shinn et al., 1989).

Along most of the southern Florida coast, the coastline was several kilometers landward of its present position by 3200 years ago. Had this rate of rise continued, the broad mangrove and *Spartina* coastal wetlands and the shallow reefal, barrier island, oyster bar, and mud bank environments, so characteristic of today's Atlantic and Gulf coastal system, would not have formed in such an extensive manner and the coastline now would have been tens of kilometers inland of its present position.

3200–60 Years Before Present

About 3200 YBP, relative sea level rise appears to have slowed dramatically; for south Florida the rise averaged only about 4 cm/100 years (equal to less than 2 in./100 years) between 3200 years ago and near the present. This much slower rate of relative rise during the last meter of inundation permitted many coastlines to stabilize or begin expanding seaward and many shallow marine environments to build upward to sea level. The rate of landward migration of barrier islands dramatically slowed, and some stabilized or began growing seaward.

The retreating mangrove, *Spartina,* and marl levees at the margins of coastal bays stabilized at about 3200 YBP and subsequently have been prograding seaward, especially in Florida Bay and the Ten Thousand Islands (Figures 8.2, 8.3, and 8.4). The low-energy southern and southwestern coastline of Florida prograded several kilometers seaward as mud banks and oyster bars shallowed the marine bottom to sea level, permitting mangroves to take hold. The maze of mud banks within and marginal to Florida Bay also has mostly built up to sea level and set the stage for mangrove colonization.

The marl levees forming the northern margin of Florida Bay stabilized and, in the vicinity of Flamingo and Cape Sable, have been accreting seaward. The shorelines of Card Sound were stabilized by mangroves and, with the continued very gradual rise of sea level, have accreted upward to produce a marginal mangrove peat "dam." This dam has permitted a landward-expanding freshwater marsh to form behind (Figure 8.5). Although this dam is cut by natural channels and sloughs (such as Shark River Slough and Taylor Slough), there is sufficient freshwater head during much of the year to inhibit saline intrusion, and during the dry season saline intrusion is limited to areas near these sloughs.

In the Ten Thousand Islands, the mainland mangrove shoreline stabilized about 3200 YBP and has accreted vertically and expanded seaward since. In addition, the shallow nearshore environment became colonized by oyster banks, which in turn were colonized by mangroves. This process has gradually extended this mangrove coastline 2–3 km seaward (Figure 8.6).

Importantly, this period of slowed relative sea level rise permitted coral reefs seaward of the Keys to catch up to sea level, except where water exchange with coastal bays through newly flooded inlets caused reef demise (Ball, 1967). Had the pre-3200 YBP rate of rise of 23 cm/100 years continued, the shallow reefs, sand and mud banks, barrier islands, oyster bars, and paralic (marginal marine) swamp environments and deposits seen today would not have formed in such an extensive manner. These broad coastal environments are all the product of the very slow sea level rise (4 cm/100 years) during the past 3200 years. Only along the sediment-starved shorelines of southwestern Biscayne Bay and northeastern Florida Bay did mangrove shorelines continue to gradually retreat.

All in all, over the last 3200 years the Gulf and Atlantic coastal and nearshore marine environments have been shallowing and the coastlines stable or expanding. Around the southern tip of Florida, an extensive natural coastal dam was built during this time by the coastal mangrove peats and storm levee marls. This wetland "dam" separates the landward freshwater environments from the sea.

Hindsight Uncertainties

Recent research from regressive coastlines (Brazil) and from intensely studied transgressive coastlines (Netherlands) indicates that the late Holocene contains a series of high-frequency sea level oscillations (either a rapid lowering followed by a rapid rise or stillstand followed by a rapid rise, 200–400 years in duration and 0.5–2 m in amplitude) superimposed on the overall trend of sea level.

In several parts of the world, relative sea level has been dropping for the past 5000–6000 years. Such a relative lowering has occurred in Brazil and has produced broad regressive strandplains of prograding beach ridges, emergent lagoons, and elevated skeletal encrustations of intertidal and subtidal marine organisms on rocky coasts (Martin et al., 1985; Dominguez, 1987). Radiocarbon dating of these lagoonal peats, shells, and intertidal skeletal encrustations indicates that there have been two or more high-frequency sea level oscillations during the past 5000 years (Dominguez et al., 1987). In Brazil, high-frequency sea level oscillations occurred about 4100–3800 YBP and 2900–2500 YBP (Martin et al., 1985; Dominguez, 1987). They appear to have had a duration of 200–400 years and an amplitude of 1–2 m. The oscillations consist of a rapid lowering followed by a rapid rise of sea level.

Morphostratigraphic reconstructions of Brazilian strandplain deposits indicate that high-frequency sea level oscillations have caused direct and predictable responses in the coastal system. The response on the strandplain coasts (coasts with seaward prograding sets of beach ridges) is strandplain progradation followed by a transgressive barrier island migration phase (Dominguez, 1987; Dominguez et al., 1987). These high-frequency oscillations occur within an overall gradual lowering of relative sea level between 5500 YBP and the present. The high-frequency sea level oscillations are visible in the Brazilian situation because the overall regression is leaving behind a visible, exposed, prograding strandplain record of the late Holocene.

In contrast, the late Holocene sea level of the Gulf and Atlantic coasts of the United States has continued to have a relative rise: transgressing, inundating, eroding, and obscuring the previous deposits. Deposits that may have been initiated or formed during brief stillstands or sea level lowerings are mostly modified and their origin at least somewhat obscured by the following rise in sea level and associated transgression. To date, most coastal and shallow-marine sediment bodies along the Atlantic, Gulf, and Pacific coasts of the United States have been thought to initiate (1) in response to a gradually decelerating rate of sea level rise or (2) at hinges in the decreasing rate of sea level rise. DePratter and Howard (1981), studying the Georgia coast, offered geological and archeological evidence that there was at least one significant lowering of sea level within the late Holocene transgression sometime between 3000 and 2400 YBP. They use radiocarbon dates of in-place tree stumps to suggest that sea level dropped from –1.5 m to about –4 m below present level during this period.

Detailed work in northwestern Europe (Morner, 1980) and the Netherlands (van de Plassche, 1982) showed that the late Holocene has been a period of continued relative sea level rise at a decelerating rate. Their research, however, also suggested that there are high-frequency sea level oscillations superimposed on this decelerating rise. In some curves, these oscillations are illustrated as relative sea level lowerings (Morner, 1980) and in others as periods of slowed rise or stillstand (van de Plassche, 1982). Some authors, however, caution that it is presently extremely difficult to recognize minor cycles within the overall Holocene (Kidson, 1986).

The coastal and associated environments of northwestern Florida Bay are individually and collectively very sensitive recorders of environmental change. There is a suggestion that a regression in sea level occurred between 3000 and 2400 YBP. The rapid coastal progradation following this period demonstrates that the coastal system had large volumes of sediment toward the end of the potential sea level oscillation.

The core borings and radiocarbon dates that have been obtained from the coastal complex adjacent to northwestern Florida Bay and in the Cape Sable area indicate that the surficial strandplain on Cape Sable was well into its regressive phase of sedimentation at 2200 YBP and had completed forming (as a quite rapid burst of prograding beach ridge sedimentation) by about 1200 YBP (Roberts et al., 1977). This burst of regressive sedimentation is what would be expected following a rapid transgression, as sediment bodies in disequilibrium following sea level rise are reworked and sediment is transferred to more stable sites. Cross sections by Roberts et al. (1977) across Cape Sable and across the seaward shore levee on northwest Florida Bay suggest that this progradation took place across a surface 1.5–2 m below present sea level.

The stratigraphic sequences of Roberts et al. (1977) through Cape Sable and those of Spackman et al. (1964) through environments adjacent to Whitewater Bay are shown as shallowing upward sequences, but the stratigraphic sequences and radiocarbon dates provided are very compatible with a regression between 2900 and 2500 YBP. In fact, freshwater calcitic mud dates in that interval occur well below the smoothed sea level curve of Scholl et al. (Scholl et al., 1969; Scholl, 1964a, 1964b).

The cross sections of Roberts et al. (1977) are very similar to those of Parkinson (1987) for the Ten Thousand Islands area farther west on the southwest Florida coast. Both record a halt in the landward migration of coastal environments before 3000 years ago, but find that most of the coastal progradation and marine shallowing has occurred since 2500 YBP.

HISTORICAL RECORD of SEA LEVEL

Tide gauge records from south Florida document a dramatic increase in the rate of relative sea level rise beginning in about 1930 (Wanless, 1982, 1989). For Key West, this increase is from a prior rate equivalent to about 3 cm/100 years to a post-1932 rate equivalent to about 38 cm/100 years (Figure 8.7). The tide gauge records for Miami and Naples (which begin in the 1930s and the 1950s, respectively) show comparable trends and rates during the years since 1932.

Similar increases beginning in about 1930 are recorded in tide gauges along the Atlantic, Gulf, and Pacific coasts of the United States (Wanless, 1989). Analyzing tide gauge records throughout the world, Peltier and Tushingham (1989) concluded that there has been a global eustatic rise in sea level of 2.4 mm·yr^{-1} since 1920 and that at least 75% of that rise must be explained by glacier and ice sheet melting.

This six- to tenfold increase in the rate of relative sea level rise in southern Florida has persisted for 60 years and has already resulted in a relative rise of over 20 cm. This has initiated dramatic changes in the character of coastal wetlands and estuaries and has set the stage for dramatic (storm-driven) future modifications in coastal deposits.

The rate of relative sea level rise since 1930 is faster than the rise that occurred between 5500 and 3200 YBP. The lesson from that period is that, with few exceptions, all types of coastlines steadily eroded and retreated landward. Because this landward backstepping tends to occur in storm-driven steps, an immediate

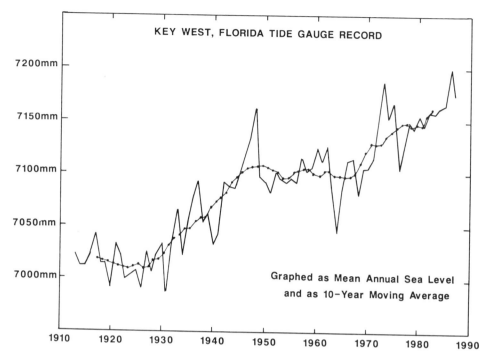

Figure 8.7 Plot of tidal gauge data from Key West, Florida, plotted as mean annual sea level and as a 10-year running average. Since 1932, relative sea level has been rising at a dramatically increased rate (equivalent to 3.8 mm/year). Pulses of rise occurred in the 1940s–50s and again in the 1970s–80s. Data were provided to the author by National Ocean Survey. Vertical scale is in 5-cm intervals; absolute scale is not pertinent.

response to this recent increase should not be expected. Thus, it should not be surprising that this recent rise is accurately expressed by the upward migration of oysters and barnacles on the seawalls, but is not yet extensively shown by dramatic shore retreat.

IMPLICATIONS

Future Coastal Response

The simple conclusion is that if relative sea level rise continues at its present rate, the sandy, mangrove, and levee coasts of south Florida will erode at an accelerated rate, low-lying freshwater wetlands will become saline, and increased saltwater intrusion will diminish freshwater resources. The lesson from the geologic record is that any prolonged rate at or faster than 23 cm/100 years will lead to rapid and complete coastal erosion. This erosion includes landward migration and storm dissection of sandy barrier islands, erosion of coastal wetlands at a faster rate than they are expanding landward, and deepening and increased marine influence in coastal bays and estuaries. Coastlines along northeastern Florida Bay, which have demonstrated instability during the past 3000 years, should respond first and most dramatically.

The tenfold increase in the rate of sea level rise during the past 60 years and the projection for continuation of that rise must be considered a minimum. All indications are that there will be a further dramatic increase in the rate of sea level rise within the next century in response to global warming, caused by the build-up of carbon dioxide and other gasses in the atmosphere. The U.S. Environmental Protection Agency, for example, forecasts that by the year 2100, there will be an additional eustatic rise in sea level of 55–335 cm (average 150–210 cm).

For south Florida, all this could be turned into a simple forecast for the future, except for two problems: the limestone topography of south Florida is complicated and the capability for rapid upward growth of its natural coastal dam of mangrove peat and levee marl is not certain. The Everglades depression is just that—a low-lying swale between the limestone and quartz sand Atlantic Coastal Ridge (on which Fort Lauderdale, Miami, and Homestead are situated) and the limestone ridge associated with the Big Cypress Swamp. The limestone surface is less than 1 m above sea level in much of the southern Everglades drainage basin. Addition-ally, both the Atlantic Coastal Ridge and the Big Cypress Swamp Ridge are dissected by numerous peat-filled swales or channels. If the coastal dam is eroded back or overstepped, these swales and the Everglades depression itself may become saline. As this happens, remaining fresh groundwater resources will be further stressed, and undoubtedly there will be accentuated saltwater intrusion beneath the remain-ing upland areas. The swales and channels dissecting the ridges will encourage saltwater intrusion, probably by landward jumps as individual ridge segments are effectively breached.

An important difference between the present phase of accelerated sea level rise and that of 5500–3200 YBP is that the present follows a prolonged period of growth of coastal sediment bodies creating large coastal and nearshore sediment reservoirs. For example, the volume of mangrove peat, levee marls, and lagoonal muds spread across a coastal complex several kilometers wide at the northern margin of Florida Bay, even if unstable, will take time to rework.

Using the knowledge available, the following forecasts can be made:

1. With a continued rate of relative rise of sea level of about 30 cm/100 years, nearly all south Florida coastlines will become erosional. The mangrove coastline should continue to build upward, maintaining a coastal dam, but the seaward mangrove margin will be eroded. The landward margin of the saline wetlands will migrate landward; coastal bays will become deeper and more saline, and breaching of new inlets through the Florida Keys will further diminish reef growth.

2. With the rate of relative sea level rise increased to 60 cm/100 years, the coastal marl and wetland dam will become overstepped in numerous places, causing rapid landward encroachment of the sea into the Everglades wet-lands. In association, there will be rapid loss of transitional and freshwater habitats, extensive saltwater encroachment, and increased water levels in the adjacent freshwater wetlands. Physical erosion of the transgressed coastal peats and marls will increase turbidity and nutrient levels to the coastal bays.

Erosion of mud banks within and opening of inlets into the coastal bays will occur. Bay waters will become less restricted but more turbid, and the bay bottoms will become deeper, with significant changes resulting in benthic marine communities. Recall that minimum U.S. Environmental Protection Agency forecasts are for 55 cm in the next 110 years in addition to the current rate (30–40 cm/100 years).

3. Rates of relative sea level rise greater than 90 cm/100 years will produce catastrophic inundation of southern Florida, loss of coastal wetlands, and loss of freshwater resources.

Role of Catastrophic Events

Large catastrophic hurricanes cause the real changes in shoreline configuration and loss of coastal wetlands; sea level rise just sets the stage.

Hurricanes

As south Florida had not been subjected to a hurricane event between 1965 and 1991, the opportunity for major storm modification had not occurred until very recently (see discussion of Hurricane Andrew). Each area in south Florida should be affected by hurricane-force winds once every 7.5 years (Neumann et al., 1978). Hurricanes can affect shorelines of all exposures and can significantly affect interior lakes and ponds.

The Great Labor Day Hurricane of 1935 and Hurricane Donna in 1960 swept across Florida Bay and the Everglades, causing major modifications (Craighead, 1964). These storms caused significant loss of exposed mangrove shoreline, erosion of the margins of interior lagoons and ponds, and major erosion of mangrove flats adjacent to deeply penetrating channel systems (Figures 8.8 and 8.9). Much of these eroded areas are now permanently marine. With higher sea levels, these steps of storm erosion will become greater and greater as storm frequency, intensity and level of inundation increase.

Significantly, with global warming the tropical Atlantic should warm, encouraging hurricanes to become more frequent and more intense (Barron, 1988).

Effect of Hurricane Andrew on Coastal Evolution

Hurricane Andrew swept across south Florida on August 24, 1992. The storm moved east to west across south Florida, with the center of the eye crossing Elliott Key, Turkey Point nuclear power plant, Homestead, and the Everglades at Lostman's River. The eye contracted to about 20 km (12 mi) in diameter as the storm approached shore, and the general winds in the eye wall increased to about 235 $km \cdot hr^{-1}$ (145 mph). There were, however, numerous vortices within the eye wall that generated winds at or above 320 $km \cdot hr^{-1}$ (200 mph). The storm moved onto and across south Florida extremely rapidly, at a forward velocity of 35 $km \cdot hr^{-1}$ (22 mph). As a result, the coastal environment was subjected to intense but very brief wind and storm waves, currents, and surge.

This was the first hurricane to impact south Florida in 26 years and affords an

Figure 8.8   Low-level aerial photograph of a the mangrove swamp complex in a portion of Little Sable Creek, north of Cape Sable. The barren areas were eroded by storm surges during the hurricanes of 1935 and 1960. This erosion occurred adjacent to tidal channels well back in the mangrove swamp. The photograph, taken in 1983, shows no trend toward regrowth.

opportunity to consider some of the predictions of storm-generated modifications during a time of rising sea level. Two environments will be considered—beaches and mangrove coastlines.

Sandy Coastlines.   Key Biscayne (on the east coast) and Highland Beach (on the west coast) were the only beaches which received the full force of the storm. Initial overflights of the area led some scientists to conclude that there was little erosion of the Key Biscayne shoreline. Key Biscayne received a building north-to-south longshore current followed by a brief strong onshore surge which increased southward from 2 to 4 m (6 to 12 ft). The longshore current first swept a large volume of nearshore sand to the south. The onshore surge then swept a large volume of sand landward from the beach onto the island and reprofiled the beach into a very gently sloping storm ramp which intersected sea level at about the same position as before the storm. Since the storm, prevailing easterlies, tropical waves, and winter storm waves and swells have dramatically steepened the beach face and resulted in 20–30 m (65–100 ft) of erosion to an initially 60- to 80-m-wide (195- to 260-ft) beach (renourished in 1986). This is a dramatic stepping back of this shoreline.

The storm exited Florida along the broad mangrove forest between Cape Sable

Figure 8.9   Ground photograph of mangrove swamp area in Figure 8.8 that was destroyed by storm surges during the hurricanes of 1935 and 1960. The downed red mangrove trees in the distance record destruction during flood surge. The intertidal red mangrove peat surface remains barren in this 1979 photograph. Rising sea level and marine erosion processes are transforming the subtidal environments of these areas.

and Everglades City in Everglades National Park. Here, the south side eye wall, although moving offshore at more than 33 km·hr^{-1} (>20 mph), generated a brief but intense onshore surge that reached +4 m (13 ft). This surge generated 5–15 m (15–50 ft) of landward migration of the narrow quartz sand and coarse shell beach ridge on Highland Beach and left a very gently sloping beach profile similar to Key Biscayne. The onshore surge swept a carbonate storm layer through the broad mangrove swamp and washed an organic and leaf-litter layer into the interior bays. The swamp storm layer was commonly capped with stranded dead fish, frequently in densities of more than one fish per square meter.

As the storm passed, the impounded surge water, which had built up in the broad mangrove swamp, surged back seaward, dissecting Highland Beach with more than 20 ebb channels, each with an ebb-tidal delta spread across the shallow seaward platform. This breaching has severely weakened the beach ridge as an effective shoreline barrier, separating the interior mangrove swamp from wave and tidal processes of the marine environment. (Hurricane Donna caused similar flooding in 1960, but not as extensive channel breaching.) The ebb surge also created thick mud deltas from side creeks feeding into the large penetrating tidal channels.

Mangrove Coastlines.   Over 28,329 ha (70,000 acres) of mangrove forest was destroyed by Hurricane Andrew. Most of this destruction was by wind stress on the taller forests. The onshore surge, reaching 5.2 m (16.9 ft) on the east coast and about 4 m (13 ft) on the west coast, also flattened many trees (Figure 8.10). The storm surge and waves, however, caused less than 15 m (50 ft) of erosion at the shoreline. Again, at first glance, the hurricane appeared to have caused little erosion to the shoreline environment.

The 28,000 ha of flattened tall forest was mostly adjacent to the coastline or tidal channels. In these areas, 90–100% of the trees (20–30 m [65–100 ft] in height) were either uprooted or broken at the lower trunk and in either case killed (Figure 8.10). Although former storm layers are covered with coarse charcoal (the next step in the taphonomy of this forest), in times of rising sea level, the widespread flattening of mangrove forests can be recognized as a step in transgression. It is very likely that much of this downed forest will convert to a marine environment. The once flat swamp surface is now tortuously rugged with over 1.5 m (5 ft) of relief from widespread uprooting. This has resulted in stagnant ponds and supratidal patches. Over the next 5–10 years, marine processes will modify portions of the downed mangrove community substrate and may inhibit recolonization by mangrove seedlings.

Figure 8.10   The flattened mangrove forest at Highland Beach following Hurricane Andrew. Over 90% of the mangrove forest was destroyed by uprooting and trunk breakage, and the former level peat surface is now a highly irregular topography (foreground). Photograph taken September 13, 1992.

Figure 8.11 A low-level oblique aerial photograph looking into one of the branches of Little Sable Creek penetrating north Cape Sable at a low tide. Previous hurricanes flattened the mangrove forest adjacent to these creeks. These flattened mangrove forests did not recover, but rather have largely evolved to a deepening intertidal environment. The decaying oriented trees can still be seen in areas (arrow). Photograph taken August 21, 1992.

Similar to Andrew, Hurricane Donna in 1960 caused some coastline erosion, but the main damage was destruction of the interior mangrove forest. Hurricane Donna generated an onshore surge that penetrated far into many of the tidal channel complexes of northern Cape Sable. This surge flattened the mangroves in a 50- to 100-m-wide band (165- to 325-ft) adjacent to the channels and extending 1–2 km (0.6–1.2 mi) into the swamp. Since that storm, these flattened mangrove forests have evolved into a deepening intertidal to subtidal environment in which mangrove seedlings have not been able to recolonize (Figure 8.11), and marine burrowing and erosion processes are gradually removing and deepening the peat substrate. The present rapid rise of sea level is helping to assure that the storm-eroded areas are not recolonized as mangrove swamp, but evolve into deepening, expanding bays.

Hurricane Andrew and Hurricane Donna thus show that the historical loss of wetland habitat and evolution to marine conditions is primarily due to the loss of interior wetland damaged by hurricane winds and surges. By watching only for coastline setback, the initiation of a major transgression may be overlooked.

Hurricanes are causing significant changes in coastal environments. With an increase in the rate of sea level rise, these steps can be expected to become more dramatic and less reversible.

Winter Storms

Each winter, 30 to 40 cold fronts pass through south Florida. Effective erosional wind waves are from the north and northeast following frontal passage (Wanless et al., 1989). Winter storms thus cause persistent erosion of north- and east-facing shorelines if there is not an adequate supply of sediment. The north- and east-facing shores of islands in Florida Bay, of bay margins, and of the smaller brackish lakes in the southern Everglades complex all tend to be erosional. With increased rates of sea level rise, the rates of erosion will increase.

Response

There is little that can be done at a local, state, or national level to prevent future relative rises of sea level. Proper education and legislation are necessary to adopt wetland and coastal management strategies that will accommodate the inevitable modifications to the environment.

A dike cannot be built around south Florida to keep the rising water level at bay, because the sand and limestone substrate is much too porous. Planning is the only way to minimize impact. Some of the key planning elements are as follows.

First and foremost is the need for planning and programs to protect freshwater surface and groundwater resources—in terms of both availability and quality. What effect will a 0.5-, 1-, or 2-m sea level rise, with various scenarios for coastal retreat and wetland evolution, have on freshwater resources? Are these areas presently being zoned and managed to assure future availability and quality of water?

Regional and local zoning and policy for residential and commercial development, waste dumps, wetland management, and coastal modification should be completely re-evaluated in light of various 50- to 200-year scenarios for relative sea level rise and coastal retreat. Building codes, stormwater discharge, and flooding management should be similarly re-evaluated. It must be emphasized that many low-lying inland areas of south Florida will be as seriously impacted by a 1-m sea level rise as will the coastal zone.

It will be necessary to carefully evaluate projected changes in coastal circulation, salinity, environments, and marine habitats that will result from various projected rates of sea level rise and then to re-evaluate the economic future of the various marine resources and development programs. Any forecast planning must be based on coastal response forecasts rooted in firm knowledge of how coastal environments will, in fact, respond and interact. This is presently inadequately understood, as are the details of sea level for the past several thousand years, on which hindsight knowledge for forecasting will be based.

The cost effectiveness of the various shore protection/management programs for sandy shorelines needs careful re-evaluation in light of the present increased rates of relative sea level rise, especially beach nourishment versus relocation.

It is important not to overreact to short-term (less than decade scale) fluctuations in relative sea level or climate. Most importantly, however, it is time to begin obtaining the necessary background information from which to make useful forecasts and to begin serious re-evaluation of coastal regulations, management, and policy. This is not yet a crisis, and planning can be done in a logical progression and rational atmosphere. In 30 years, that may not be possible.

SUMMARY

The Past

The broad coastal wetlands and the broad freshwater marsh of south Florida are a result of the very slow relative rise of sea level during the past 3200 years. Prior to that time, relative sea level was rising at a rate of 30–50 cm/100 years (3–5 mm·yr^{-1})—too fast for coastal swamp communities to dominate the south Florida coastal environment. Beginning about 3200 years ago, this relative sea level rise slowed to less than 4 cm/100 years (<0.4 mm·yr^{-1}) This slowed rate permitted shallow marine sediments and organic coastlines to build upward. As shallow marine sediments caught up with sea level, coastal mangrove swamps prograded seaward across them. The resulting broad, low-gradient coastal swamp has provided a natural barrier to marine waters and has permitted the freshwater environments of the Everglades to spread seaward behind and on top of the coastal swamp deposits. Swamp coastlines continued to erode only in inner Florida Bay and on the southwest coast of Biscayne Bay.

The Present

Tide gauges throughout the United States record a dramatic increase in the rate of relative sea level rise beginning about 1930. In the following 60 years, the relative sea level rise for south Florida has averaged 30–40 cm/100 years (3–4 mm·yr^{-1}). This rate is nearly 10 times that of the past 3200 years. As a result of this increased rate of rise, the following changes are taking place in the coastal wetland environments: (1) relative sea level has risen 18–24 cm; (2) marine waters have encroached significantly landward, setting back freshwater marsh communities; (3) surficial sea water has encroached further landward seasonally and during storms, affecting freshwater communities and soils; (4) a setting has been created for major hurricanes to cause increased erosion and saltwater invasion; and (5) marine waters can penetrate further landward in natural and artificial channels, causing increased saltwater intrusion into the ground waters.

The Future

If the historical rate of sea level rise for the past 60 years continues (and all forecasts are that the relative rate of rise will be much greater), the coastal mangrove swamp will erode at an increasing rate, both along the shoreface and along penetrating channels and creeks. The erosion will occur in dramatic hurricane pulses. This will narrow and dissect the protective coastal swamp that presently serves as a barrier between marine waters and the freshwater marshes.

If future rates of relative sea level rise are much faster than at present, marine waters can be expected to penetrate far into freshwater environments as tidal sheet flow. In addition, storm erosion will be increased further. Because the limestone substrate to the main axis of the Everglades (Shark River Slough) lies near or below present sea level as far north as Alligator Alley, there is the strong potential for catastrophic loss of freshwater wetlands and diminished fresh groundwater resources during the next 100–200 years.

ACKNOWLEDGMENTS

This chapter is adapted from articles by Wanless and Parkinson (1989) and Wanless (1989). Research funding is in part from National Science Foundation Grants EAR-77-13707 and EAR-92-24480 and National Park Service Grant 5280-2-0990. The authors thank National Ocean Survey for providing tide gauge data.

LITERATURE CITED

Ball, M. M. 1967. Carbonate sand bodies of Florida and the Bahamas. *J. Sediment. Petrol.,* 37: 556–591.

Barnett, T. P. 1990. Recent changes in sea level: A summary. in *Sea-Level Change,* National Research Council, Geophysics Study Committee, National Academy Press, Washington, D.C., pp. 37–51.

Barron, E. A. 1988. Severe storms during Earth history. *Geol. Soc. Am. Bull.,* 101:731–741.

Bloom, A. L. 1977. *Atlas of Sea-Level Curves,* International Geological Correlation Program 61, Sea-Level Project, Cornell University, Ithaca, N.Y.

Cottrell, D. J. 1987. Holocene evolution of the northeastern coast and islands of Florida Bay (abstract). A symposium on Florida Bay, a subtropical lagoon. in Proc. and Abstr. U.S. Natl. Park Service/ENP and University of Miami/RSMAS, Coral Gables, Fla.

Craig, G. S., 1983. Holocene Sedimentation in a Pleistocene Depression, South Florida, Master's thesis, University of Miami, Coral Gables, Fla.

Craighead, F. C. 1964. Land, mangroves and hurricanes. *Fairchild Trop. Gard. Bull.,* 19(4):1–28.

Curray, J. R., F. J. Emmel, and P. J. Crampton. 1969. Holocene history of a strand plain lagoon coast, Nayarit, Mexico. Proc. Int. Symp. Coastal Lagoons, Mexico City, 1967, pp. 63–100.

DePratter, C. B. and J. D. Howard. 1981. Evidence for a sea level lowstand between 4500 and 2400 B.P. on the southeast coast of the United States. *J. Sediment. Petrol.,* 51:1287–1295.

Dominguez, J. M. L. 1987. Quaternary Sea Level Changes and the Depositional Architecture of Beach-Ridge Strand Plains along the East Coast of Brazil, Ph.D. dissertation, University of Miami, Coral Gables, Fla.

Dominguez, J. M. L., L. Martin, and A. C. S. P. Bittencourt. 1987. Sea level history and the Quaternary evolution of river mouth-associated beach-ridge plains along the east/southeast coast of Brazil—a summary. in *Sea Level Fluctuation and Coastal Evolution,* D. Nummedal, O. H. Pilkey, and J. D. Howard (Eds.), *Soc. Econ. Paleontol. Mineral. Spec. Publ.,* No. 41, pp. 115–128.

Enos, P. 1977. Quaternary sedimentation in south Florida. Part 1. Holocene sediment accumulations of south Florida shelf margin. *Geol. Soc. Am. Mem.,* 147:1–130.

Fairbanks, R. G. 1989. A 17,000-year glacio-eustatic sea level record: Influence of glacial melting rates on the Younger Dryas event and deep-ocean circulation. *Nature,* 342:637–642.

Kidson, C. 1986. Sea-level changes in the Holocene. in *Sea-Level Research: A Manual for the*

Collection and Evaluation of Data, O van de Plassche (Ed.), Galliard Printers Ltd., Great Yarmouth, pp. 27–64.

Martin, L., J. M. Flexor, D. Blitzkow, and K. Suguio. 1985. Geoid change indications along the Brazilian coast during the last 7,000 years. *Proc. 5th Int. Coral Reef Congr.*, 3:85–90.

Morner, N. A. 1980. The northwest European "sea-level laboratory" and regional Holocene eustacy. *Paleogeogr. Paleoclimatol. Paleoecol.*, 29:281–300.

Neuman, W., L. J. Cinquemani, R. R. Pardi, and L. F. Marcus, 1980. Eustacy and deformation of the geoid: 1000–6000 radiocarbon years BP. in *Earth Rheology, Isostacy and Eustacy*, N. A. Morner (Ed.), Wiley, New York, pp. 555–567.

Neumann, A. C. 1972. Quaternary sea-level history of Bermuda and the Bahamas. in 2nd Natl. Conf. Abstr., American Quaternary Association, pp. 41–44.

Neumann, C. J., G. W. Cry, E. L. Caso, and J. D. Kerr. 1978. Tropical Cyclones of the North Atlantic Ocean, 1871–1977, Stock #003-17-0042502, National Climatic Center, U.S. Government Printing Office, Asheville, N.C.

Parkinson, R. W. 1987. Holocene Sedimentation and Coastal Response to Rising Sea Level along a Subtropical Low Energy Coast, Ten Thousand Islands, Southwest Florida, Ph.D. dissertation, University of Miami, Coral Gables, Fla.

Peltier, W. R. and A. M. Tushingham. 1989. Global sea level rise and the greenhouse effect: Might they be connected? *Science*, 244(4906):806–810.

Perlmutter, M. A. 1982. The Recognition and Reconstruction of Storm Sedimentation in the Nearshore, Southwest Florida, Ph.D. dissertation, University of Miami, Coral Gables, Fla.

Roberts, H. H., T. Whelan, and W. G. Smith. 1977. Holocene sedimentation at Cape Sable, south Florida. *Sediment. Geol.*, 18:25–60.

Scholl, D. W. 1964a. Recent sedimentary record in mangrove swamps and rise in sea level over the southwestern coast of Florida. Part 1. *Mar. Geol.*, 1:344–366.

Scholl, D. W. 1964b. Recent sedimentary record in mangrove swamps and rise in sea level over the southwestern coast of Florida. Part 2. *Mar. Geol.*, 2:343–364.

Scholl, D. W. 1966, Florida Bay: A modern site of limestone formation. in *Encyclopedia of Oceanography*, R. W. Fairbridge (Ed.), Reinhold, New York, pp. 282–288.

Scholl, D. W., F. C. Craighead, Sr., and M. Stuiver. 1969. Florida submergence curve revised— Its relation to coastal sedimentation rates. *Science*, 163:562–564.

Shinn, E. A., B. H. Lidz, J. L. Kindinger, J. H. Hudson, and R. B. Halley. 1989. Reefs of Florida and the Dry Tortugas: A Guide to the Modern Carbonate Environments of the Florida Keys and the Dry Tortugas, U.S. Geological Survey, St. Petersburg, Fla.

Spackman, W., D. W. Scholl, and W. H. Taft. 1964. Field Guidebook to Environments of Coal Formation in Southern Florida, Geological Society of America Field Trip.

van de Plassche, O. 1982. Sea-level change and water-level movements in the Netherlands during the Holocene. *Meded. Rijks Geol. Dienst*, 36:1–93.

Wanless, H. R. 1969. Sediments of Biscayne Bay—Distribution and Depositional History, Tech. Rep. 69-2, Institute of Marine Science, University of Miami, Coral Gables, Fla.

Wanless, H. R. 1976. Geologic setting and recent sediments of the Biscayne Bay region, Florida. in *Biscayne Bay: Past/Present/Future*, A. Thorhaug and A. Volker (Eds.), University of Miami Sea Grant Publ., Coral Gables, Fla., pp. 1–31.

Wanless, H. R. 1982. Sea level is rising—So what? *J. Sediment. Petrol.*, 52:1051–1054.

Wanless, H. R. 1989. The inundation of our coastlines: Past, present and future with a focus on south Florida. *Sea Frontiers*, 35(5):264–271.

Wanless, H. R. and R. W. Parkinson. 1989. Late Holocene sea level history of southern Florida: Control on coastal stability. in Proc. 8th Symp. on Coastal Sedimentology: Coastal Sediment Mobility, Florida State University, Tallahassee, pp. 197–213.

Wanless, H. R. and M. G. Tagett. 1989. Origin and dynamic evolution of carbonate mudbanks in Florida Bay. *Bull. Mar. Sci. Fla. Bay Symp.*, 44(1):454–489.

Wanless, H. R., L. P. Tedesco, V. Rossinsky, Jr., and J. Dravis. 1989. Environments and sequences of Caicos platform with an introductory evaluation of south Florida. in 28th Int. Geol. Congr. Field Trip Guidebook T374, American Geophysical Union, Washington, D.C.

9

The CLIMATE of SOUTH FLORIDA and Its ROLE in SHAPING the EVERGLADES ECOSYSTEM

M. J. Duever

J. F. Meeder

L. C. Meeder

J. M. McCollom

ABSTRACT

South Florida has a tropical climate, with a summer wet season and a dry season from mid fall through late spring. Average temperatures are warm all year, with only occasional freezes associated with winter cold fronts. Thunderstorms are the major source of rainfall, although erratically occurring tropical cyclones and winter frontal systems can contribute significantly in some years. Besides the annual cycle, rainfall patterns are associated with a minor bimodal peak during the wet season and a 5- to 6-year cycle associated with global climate cycles. The long-term trend in total annual precipitation has been essentially constant over the past 100 years. Evapotranspiration is lowest during the cool winter months and highest in late spring, after which it declines only slightly during the summer months. Freezes play a large role in controlling the distribution of tropical elements of the fauna and flora of south Florida. In general, they are more severe farther north or inland from the ocean. There is no clear evidence of any change in frequency or severity of freezes over the last 40 years. Droughts can significantly alter composition and structure of aquatic animal communities, provide opportunities for germination of wetland vegetation, and set the stage for fire. Individual droughts may affect all of south Florida, but they are frequently restricted only to portions of it. Tropical cyclones are major late summer/early fall storms that can severely affect coastal areas. However, they have relatively minor long-term effects on biota of the more interior portions of south Florida. The erratic distribution in space and time of these climatic disturbance events contributes to the maintenance of the structural and biological diversity of the overall Everglades landscape.

INTRODUCTION

This chapter has two objectives. The first is to provide a general introduction to the climatic setting of the Everglades. The second objective is to document some of the more significant influences of climate on the biological characteristics of this ecosystem. Available climatic summaries have been the primary source of data, and thus information from a variety of stations in south Florida has been utilized, depending on where information on a particular topic could be found.

Six weather stations within or near the Everglades had long-term and reasonably continuous temperature and precipitation records that could be compared over similar time periods (National Oceanic and Atmospheric Administration, 1987) (Figure 9.1). These include Clewiston U.S. Engineers, Belle Glade Experiment Station, and Loxahatchee (located west to east along the northern Everglades); Tamiami Trail 40 Mile Bend and Homestead Experiment Station (located west to east in the southern Everglades); and Flamingo Ranger Station (located just below the Everglades along Florida Bay).

Data from these as well as other National Oceanic and Atmospheric Administration weather stations were used to document certain aspects of the south Florida climate. Fort Myers FAA/AP (in southwest Florida) had the longest continuous precipitation records available in south Florida and provided information on long-term precipitation trends. Information on seasonal thunderstorm frequency was

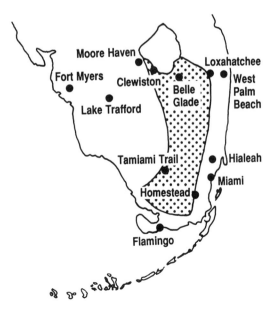

Figure 9.1 Map of south Florida showing location of the Everglades relative to the weather stations mentioned in this chapter.

available from Davis and Sakamoto (1976) for only three south Florida stations, which they listed as Miami, West Palm Beach, and Fort Myers. Evaporation pan data were available from five National Oceanic and Atmospheric Administration weather stations in south Florida: Clewiston U.S. Engineers, Belle Glade Experiment Station, Moore Haven Lock 1, Tamiami Trail 40 Mile Bend, and Hialeah. Because long-term averages were not available, only data for calendar 1988 year were used to illustrate the annual pattern. Duever et al. (1986) reported on the most severe droughts in south Florida recorded at three stations: Miami, Fort Myers FAA/AP, and Lake Trafford near Immokalee. These data provided insights on the spatial continuity of major droughts across the southern tip of the Florida peninsula.

SOUTH FLORIDA CLIMATE

The climate of south Florida has more in common with tropical regions than with any other area within the continental United States, according to the Köppen classification (Hammond, Inc., 1967) or the Trewartha classification (Espenshade and Morrison, 1978). As in many tropical regions, it has a more distinctive annual pattern of wet and dry seasons than it does a pattern of winter and summer seasons, such as characterizes much of North America. Hela (1953) described the area occupied by the Everglades as having a tropical savanna climate, which "has a relatively long and severe dry season, and the rainfall of the wet period is insufficient to compensate for the drought."

Temperature

Average daily temperatures show relatively little seasonal change (Figure 9.2). Average daily maximum temperatures are consistently above 27°C from April through October at the more northern stations and from March through November at the more southern stations (National Oceanic and Atmospheric Administration, 1985). Winters are also warm, with daytime temperatures frequently above 25°C. Highest recorded temperatures are only about 5–10°C higher than average maximum temperatures during the winter and 3–5°C higher during the summer. Average daily minimum temperatures rarely fall below 15°C from May through October and are above 10°C even through the winter months. However, temperatures can drop below freezing following the passage of cold fronts, which periodically move through the area. During some winters, only a few cold fronts reach south Florida, while they can be frequent in other years.

Precipitation

Precipitation is the main route by which water enters the Everglades ecosystem. Average annual precipitation at five of the six Everglades stations (1951–80) ranged from 119 to 157 cm (National Oceanic and Atmospheric Administration, 1987), while individual years can be as low as 86 cm or as high as 224 cm (Sculley, 1986). Approximately 60% falls from June through September, and only about 25% falls from November through April at the five Everglades stations (Figure 9.3). Rainfall

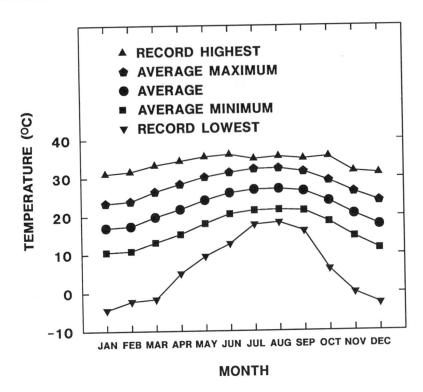

Figure 9.2 Average daily and extreme maximum temperatures at Belle Glade Experiment Station (1951–80) illustrate the small seasonal variation in these parameters in south Florida. Extreme minimum temperatures were considerably more variable. (From National Oceanic and Atmospheric Administration, 1985.)

in May and October tends to be transitional and variable from one year to the next, depending on when the wet season begins and ends. Wet season precipitation is produced primarily by localized thunderstorms and, as a result, is erratically distributed in both space and time. Tropical cyclones can significantly increase wet season precipitation over large areas in those years when they cross or come close to south Florida. Dry season precipitation results primarily from the passage of warm maritime or cold continental air masses. These storms affect large areas and can drop large amounts of rain. However, the frequency with which they reach south Florida is variable from one year to the next.

Thomas (1974) described the physical basis for a bimodal pattern of wet season precipitation in south Florida, which shows up as two minor maxima in early and late summer. He conducted power spectral density analyses of monthly rainfall data from 127 stations south of latitude 29° N to investigate the occurrence of the pattern. These analyses indicated the presence of the bimodal pattern throughout the region, although it was "barely above the background 'noise'" in the higher rainfall areas within about 25 km of the coast from Miami to West Palm Beach (Figure 9.4).

The bimodal pattern of wet season precipitation might be explained by rainfall

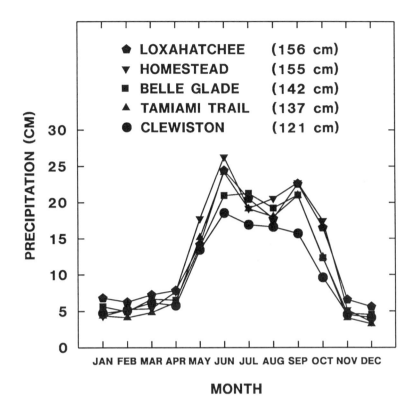

Figure 9.3 Average monthly rainfall at five of the six weather stations in the Everglades (National Oceanic and Atmospheric Administration, 1987). Numbers following the station names are average annual rainfall for 1951–80.

associated with tropical cyclones. To study the influence of thunderstorms, the monthly distribution of thunderstorm days as reported for Fort Myers, West Palm Beach, and Miami by Davis and Sakamoto (1976) was examined. All three stations showed relatively high thunderstorm activity from June through September, with a peak in July and August (Figure 9.5). This seemed like a fair match with the rainfall patterns for the five Everglades stations for which long-term averages were available (Figure 9.3), except for the slightly lower rainfall period during July and August as compared to June and September. To study the influence of tropical cyclones on the annual rainfall pattern, records were compiled on monthly numbers of tropical storms and hurricanes that passed over or near south Florida (Figure 9.6). Outside of a small peak in June, almost all of these events were spread more or less evenly from August through October, while wet season rainfall was more or less evenly distributed from June through September. October was a transitional month between the wet and dry seasons, with relatively little rainfall compared to the summer wet season months. Thus, the long-term seasonal pattern of rainfall appears to be most strongly influenced by thunderstorm activity, particularly during the summer months (Figure 9.7). It is less influenced by tropical cyclones and winter storms, although rainfall associated with these events can play

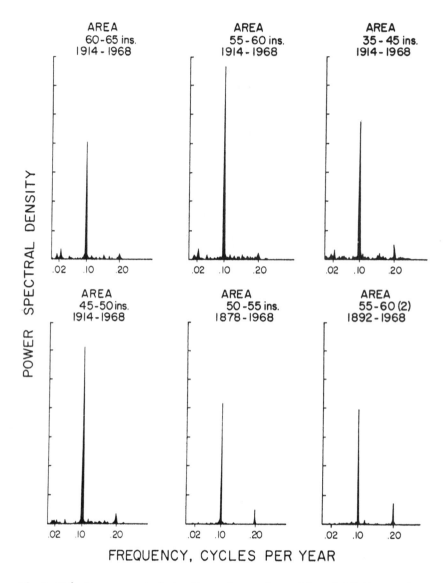

Figure 9.4 Power spectral density analyses of monthly rainfall time series for 127 stations in 6 areas of the Florida peninsula south of latitude 29° N (Thomas, 1974). Return frequencies of cyclic behavior in rainfall: 0.20 = 2 cycles/year, 0.10 = 1 cycle/year, 0.02 = 1 cycle/5 years.

other important roles, such as greatly increasing overland flows through the system and moderating severity of dry season droughts.

Another type of cycle is associated with a lower frequency of occurrence. As part of his power spectral density analyses, Thomas (1974) reported a slight peak at about 5 years (Figure 9.4). It was present, however, primarily in areas east of the Everglades and in the Florida Keys. Several more recent studies have suggested the occurrence of a cycle with a similar frequency. Working in the Kissimmee River basin, Huber et al. (1982) found a periodicity of 5–6 years. Pielke and Gray (1988)

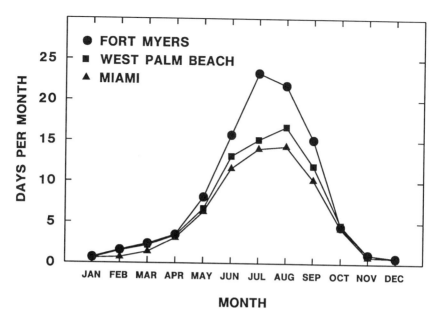

Figure 9.5 Mean monthly number of thunderstorm days at three weather stations bordering the Everglades. (From Davis and Sakamoto, 1976.)

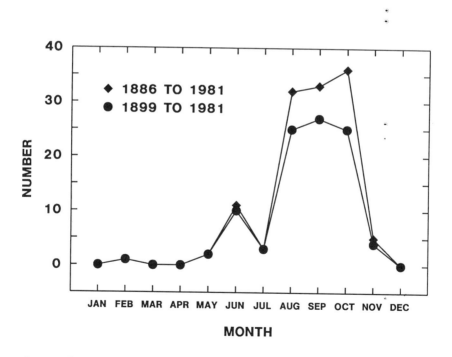

Figure 9.6 Number of tropical storms and hurricanes that passed over or close to south Florida during each month. While the data available prior to 1899 were considered "less accurate," their inclusion did not change the seasonal pattern. (Compiled from Neumann et al., 1981.)

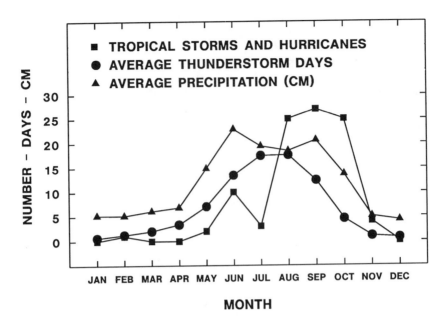

Figure 9.7   Comparison of monthly distributions of average monthly precipitation for the five Everglades weather stations in Figure 9.3, average monthly thunderstorm days for the three stations in Figure 9.5, and number of tropical storms and hurricanes (1899–1981) in Figure 9.6.

Figure 9.8 Cumulative annual rainfall reported for the Fort Myers weather station from 1892–1988 indicates that there has been no major change in rainfall patterns over this period. (Compiled from a variety of published and unpublished weather records from the National Oceanic and Atmospheric Administration weather station in Fort Myers.)

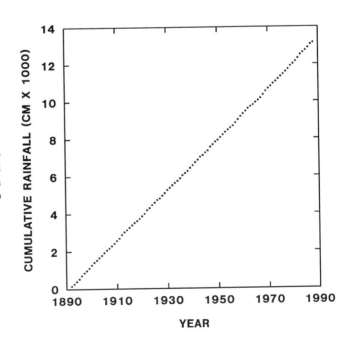

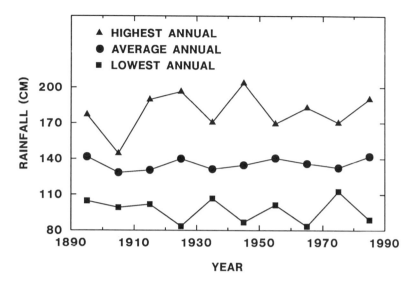

Figure 9.9 Variability in average annual rainfall for each 10-year period from 1892 to 1988 (actually 9 years for 1892–1900 and 8 years for 1981–88) at the Fort Myers weather station. (Compiled from a variety of published and unpublished weather records from the National Oceanic and Atmospheric Administration weather station in Fort Myers.)

suggested a similar cycle for south Florida dry season rainfall, which appeared to be associated with the interaction of El Niño events, the Quasi-Biennial Oscillation, and solar sunspot activity.

Despite the possible occurrence of several types of cyclic patterns, the longest rainfall record in south Florida at Fort Myers indicates that the long-term trend in total annual precipitation has not changed since the late 1800s (Figure 9.8). In addition, the variability in mean annual precipitation for each 10-year interval has not changed over this period (Figure 9.9).

Evapotranspiration

Evapotranspiration is a particularly important aspect of the Everglades climate, because it is the primary mechanism by which water leaves the ecosystem. In undisturbed south Florida wetlands, it has been estimated to export on the order of 70–90% of the rainfall entering these systems (Black, Crow and Eidsness, Inc., 1974; Dohrenwend, 1977; Kenner, 1966). The annual pattern of evaporation from an open water surface peaks in late spring, when temperatures and wind speeds are high and relative humidities are low (Figure 9.10). It is lowest during the winter, when temperatures and wind speeds are low. However, actual evapotranspiration from natural ecosystems is greatest during the summer wet season months, when water is readily available for surface evaporation or vegetative transpiration and temperatures are high. While climatic conditions are more favorable in late spring, the relatively low water table at this time of year normally limits actual evapotranspiration rates.

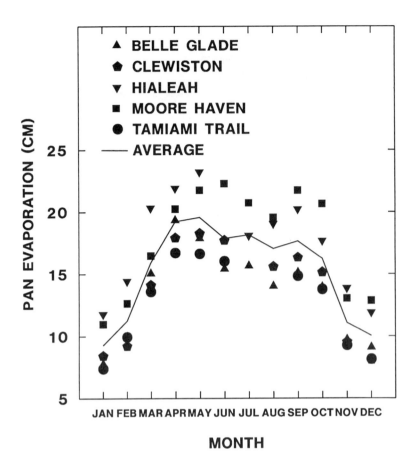

Figure 9.10 Monthly pan evaporation at weather stations in south Florida during 1988 illustrates the general seasonal pattern of evaporation from an open water surface. (From National Oceanic and Atmospheric Administration, 1988.)

EXTREME CLIMATIC EVENTS

Average climatic conditions play a major role in determining the character of the Everglades ecosystem. However, extreme events, which are often perceived as destructive, are also important in shaping the natural character and distribution of its communities. Three types of extreme climatic events that significantly influence the south Florida ecosystem will be discussed: freezes, droughts, and tropical cyclones.

Freezes

The relatively mild climate of south Florida has facilitated its colonization by a large number of tropical plant and animal species. However, occasional freezing temperatures following the passage of cold fronts has limited the distribution, both regionally and locally, of these populations. The primary factor influencing the

Figure 9.11 Mean January isotherms for south and central Florida illustrate the strong moderating influence of the ocean on cold winter temperature along the southern tip of the peninsula. (From Thomas, 1974.)

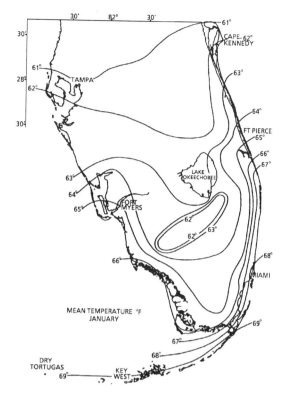

distribution of tropical species in south Florida is the presence of large bodies of surface water (particularly the ocean) which moderate freezing temperatures associated with winter cold fronts. This effect is clearly seen in the pattern of decreasing January mean temperatures as one moves inland from the coast (Figure 9.11). In some years, the Everglades itself has sufficiently large areas of standing water during the winter months to moderate cold temperatures. A secondary pattern of declining temperatures occurs as one moves away from major wetland areas or lakes in the interior of the peninsula.

The structural characteristics of tropical woody species are also influenced by periodic freezes. As the regional frequency and severity of freezes increase, their growth form increasingly becomes that of a shrub (Myers, 1986). Thus, while many species may be able to survive in south Florida north of the Florida Keys, farther north and/or inland they are increasingly pruned back by freezes and survive only as stunted multi-stemmed plants, their size being largely a function of time since the last freeze.

Freezes are erratic in their year-to-year and seasonal occurrence, as well as in the severity of individual freeze events in different parts of south Florida. The authors assessed the occurrence and severity of freezes (temperatures 0°C or lower) at six weather stations in and near the Everglades for their common period of record from winter 1949–50 to winter 1987–88 (Table 9.1). While there was an expected pattern of more frequent and severe freezes as one moved from south to north, there were a number of exceptions to this pattern.

Table 9.1 Summary of 39 Years (1949–87) of Freeze Data for Six Weather Stations in or near the Everglades

	Weather Stations					
	Belle Glade	Loxahatchee	Clewiston	Homestead	Tamiami Trail	Flamingo
Lowest temperature °F (°C)	21 (–6.1)	24 (–4.4)	26 (–3.3)	27 (–2.8)	28 (–2.2)	27 (–2.8)
Number of days						
≤32°F (0°C)	84	84 {1}	36	39 {5}	14 {3}	13 {4}
≤30°F (–1.1°C)	38	46 {1}	14	18 {1}	4 {2}	8 {1}
≤28°F (–2.2°C)	18	23 {1}	3	4 {1}	1	1
≤26°F (–3.3°C)	7	9	1	0	0	0
≤24°F (–4.4°C)	3	2	0	0	0	0
≤22°F (–5.6°C)	1	0	0	0	0	0
Number of years						
≤32°F (0°C)	29	28	19	20 {1}	8	8 {2}
≤30°F (–1.1°C)	20	21	7	11 {1}	3	5 {1}
≤28°F (–2.2°C)	15	19	2	4 {1}	1	1
≤26°F (–3.3°C)	6	8	1	0	0	0
≤24°F (–4.4°C)	3	2	0	0	0	0
≤22°F (–5.6°C)	1	0	0	0	0	0

Note: The numbers in brackets are possible additional freeze events at a particular station. In these instances, minimum temperatures had not been recorded at a station on a particular date when freezing temperatures were noted at other stations in south Florida. The estimated severity of the freeze for missing data was based on the relative severity of freezes at this station compared with other stations.

In general, Belle Glade and Loxahatchee, both in the northeastern drained portion of the Everglades, had similarly more frequent and severe freezes than the other four stations (Table 9.1). Also, the warmer temperatures at Flamingo and Tamiami Trail 40 Mile Bend were similar. This was not expected, because Flamingo is the southernmost station and is surrounded by Florida Bay or extensive coastal wetlands, while Tamiami Trail is an interior site in the central Everglades along its border with the Big Cypress Swamp. A possible explanation is that the Tamiami Trail site is located adjacent to the undrained Big Cypress Swamp and at the bottom of the Water Conservation Areas, which tend to maintain standing water along their downstream boundaries later into the dry season than does most of the Everglades. Clewiston and Homestead had a similar frequency and severity of freezes which were intermediate between the other two pairs of stations. This was unexpected, because Clewiston is among the more northern and Homestead among the more southern of the six stations. Homestead is not far from the coast but is in a generally drained agricultural area, while Clewiston is along the southern edge of Lake Okeechobee but is otherwise surrounded by a generally drained agricultural area.

The number of freeze events per year and the lowest annual temperatures over the period 1949–88 suggested a pattern of increasing frequency and severity of freezes at all stations except Homestead, which exhibited the reverse pattern. Linear regression analyses were conducted for each station to evaluate the significance of these trends. In the first analysis, daily minimum temperatures at each station were compared over time to test for a trend in the severity of freeze events. Of the six stations, only Tamiami Trail showed a significant (p = 0.05) relationship in these analyses, which suggests that there has not been a regionwide increase in the severity of freezes over the past 40 years. In the second analysis, the number of freezes per year was compared over time to test for a trend in the frequency of these events. Of the six stations, Clewiston and Flamingo showed a significant positive relationship (p = 0.05) and Homestead a significant negative relationship (p = 0.05) in these analyses. Thus, it was not possible to detect a consistent trend in the frequency of freezes for the period from 1949 to 1988 in the overall Everglades region.

However, even though there were no long-term trends, there were periods when freezes were less frequent and severe (1948–54 and 1970–75), as well as periods when they were more frequent and severe (1955–65 and 1976–85) (Figure 9.12).

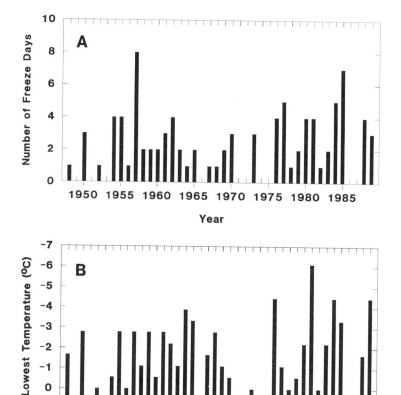

Figure 9.12 Freezes at Belle Glade Experiment Station for 1948–89. Winters (e.g., 1948–49 = 1948) with no data did not have a daily minimum temperature ≤0°C. (A) Number of days during a winter when minimum temperatures were ≤0°C. (B) The lowest temperature that occurred during a winter. (Compiled from monthly National Oceanic and Atmospheric Administration climatic summaries for Florida.)

For the purposes of this chapter, analysis is focused on temperatures 0°C and below, because this would clearly indicate freezing temperatures at a site. It is also the only type of information available for the entire Everglades region. Temperatures above freezing can have significant effects on plant and animal communities, because damaging frosts can occur at measured temperatures above 0°C. Many tropical species are undoubtedly affected by temperatures above 0°C, and some temperate species require below-freezing temperatures to complete portions of their life cycles. Thus, while many temperate and tropical species could survive in south Florida, they are unable to successfully reproduce and survive there over the long run.

Over the years the authors have worked in south Florida, they have made numerous qualitative observations on the ways that freezing temperatures affect plant communities. Aspects of apparent significance include minimum temperature, duration of temperatures below freezing, frequency of freezes within a single winter, and condition of vegetation at the time of a freeze. This mix of factors is at least part of the reason for the occurrence of spatially different effects of a single freeze event in south Florida, as well as different effects of different freeze events at the same site.

How low the temperature drops is obviously important in terms of the severity of its effect on vegetation. However, if circumstances are such that even a relatively low temperature is of short duration, it can have less of an effect than a higher freezing temperature that persists for a longer period. This suggests that for vegetation to be adversely affected by a freeze, its temperature must be lowered to some level, which is a result of a mix of freeze severity and duration.

The vegetative canopy of a plant tends to provide an insulating blanket over itself, so that a freeze generally affects only the outer leaves and branches. However, once these parts of a plant have been pruned by a freeze, a subsequent freeze will affect more interior portions of the plant, pruning it back much more severely than would a single freeze event during a winter. Thus, multiple freeze events in a single winter, even if relatively mild, can have a much greater effect than a more severe single freeze during that winter.

Another aspect of freeze impacts is the timing of a freeze. If vegetation is not "hardened" for winter conditions, it is much more susceptible to freeze damage. In late February 1989, a severe freeze occurred in the northern portion of the Big Cypress Swamp near the end of what had been a mild winter. Many temperate species had begun new growth and proved to be vulnerable to the freeze. Among these, many wax myrtle (*Myrica cerifera*) were killed or severely pruned back to larger branches, while new growth of slash pine (*Pinus elliottii*) and cypress (*Taxodium distichum*) were killed.

Another important but indirect effect of freezes on plant and animal communities is the increased fuel loads that develop as a result of the death of herbaceous vegetation during these events.

Droughts

South Florida wetlands, including the Everglades, exist where the water table is at times above and at other times below the ground surface for extended periods during an average annual cycle. Precipitation and evapotranspiration are the major factors affecting the timing and extent of this fluctuation on sites with a natural hydrologic regime. These processes result in a distinctive pattern of heavy rainfall and high water levels during the summer months, followed by a slow decline in the water table during the winter and a much more rapid decline during spring. Climatic conditions are such that wet season rainfall more than compensates for high summer evapotranspiration rates. When the wet season ends, fall and winter temperatures are relatively low and vegetative use of water is reduced; thus, while rainfall is low, evapotranspiration is also low and the water table declines slowly. During the spring, however, temperatures increase and vegetative use of water increases, while rainfall does not, resulting in a rapid decline in the water table.

An analysis of precipitation and water level relationships at Corkscrew Swamp in southwest Florida indicated that surface water levels in any particular wet or dry season reflected the amount of rain that had fallen during the respective wet or dry season (Duever et al., 1975). Thus, a wet season with abundant rainfall had high water levels for extended periods, but water levels in the subsequent dry season were high or low based solely on the amount of rain that fell during the dry season. The same applied to a dry season following a summer wet season when little rain fell and water levels remained low. Compared to the Everglades, Corkscrew Swamp is a relatively small wetland. Its size and its location near the top of its watershed could combine to make this site relatively responsive to rainfall inputs onto its upstream watershed. In contrast, the vastness of the Everglades and the relative remoteness of many of its rainfall sources could be expected to result in a more variable and delayed response to any individual wet or dry season rainfall inputs. This could be particularly true in the more downstream portions of the system, such as Everglades National Park.

Extended periods of little or no rainfall represent a particularly important aspect of precipitation patterns. The low water tables associated with severe droughts can influence a variety of components of the Everglades ecosystem. Wading birds and other predators may temporarily find abundant concentrations of aquatic prey in drying wetlands. However, populations of aquatic organisms, both predators and prey, can be severely reduced as less and less habitat is available during a drought; some of the longer lived species can take years to recover following a severe drought (Loftus and Eklund, 1994). Also, as the water table drops during a drought, seeds of many wetland plants are able to germinate and survive on the newly exposed soil surface (Duever et al., 1986; Gunderson, 1984). Survival of these seedlings is enhanced by a drought that lasts long enough to allow them to attain a size sufficient to avoid complete inundation during subsequent wet seasons. However, a severe drought can lower the water table below their roots, resulting in death. Low water tables, even those associated with normal dry season conditions, can reduce protection from freezes for sensitive tropical species.

Droughts also set the stage for fire. Even relatively short periods without rain during the hot summer months have dried out pinelands and some marshes sufficiently to permit the occurrence of fires. In dense marshes that have not burned for some time and have accumulated large amounts of standing dead litter, these fires can even burn over water after several days without rain (Wade et al., 1980). At the other extreme, a long drought can lower water tables enough to permit the burning of organic soils and the conversion of associated wetland communities to aquatic environments.

Droughts can be measured in a variety of ways: periods (days, weeks, months) with no measurable rainfall or with total rainfall less than some specified amount, periods when rainfall is some percentage less than normal during that time of year, or water levels (ground water, lakes, rivers) that fall below specified levels. The perceived significance of these measures can vary tremendously, depending on when they occur during the year and how they affect human interests.

The great spatial and temporal variability in south Florida rainfall (MacVicar and Lin, 1984) results in great spatial and temporal variability in droughts or even normal dry season water levels. This variability is beneficial to mobile predators, such as wading birds, which depend on a falling water table to concentrate their prey, because it produces a mosaic pattern of drydown that provides food over a much longer period of time than would be the case if the entire system dried down simultaneously. It also results in the occurrence of refuges for aquatic organisms, which can then recolonize more severely drought-impacted sites with the return of wet season rains.

In an effort to evaluate how widespread drought conditions are in south Florida, Duever et al. (1986) compiled data on the worst droughts recorded at Fort Myers, Lake Trafford (near Immokalee), and Miami (Figure 9.1). These data have been reanalyzed to allow more accurate comparisons between stations by utilizing data from similar time periods. While the accuracy of the analyses has been substantially improved, the basic conclusions did not change. Because these analyses were conducted using available data, droughts at each site were based on different types of information. Drought in Miami was based on unpublished data from the National Hurricane Center. These data document extended time periods over which minimal amounts of rainfall occurred between 1912 and 1965. Lake Trafford drought is based on minimum water levels during each year between 1941 and 1977 as recorded by the U.S. Geological Survey. Fort Myers drought is based on data for rainfall in January through May for the years 1892–1977 provided by the National Oceanic and Atmospheric Administration weather station in Fort Myers.

Data were available for both Miami and Fort Myers for the period from 1912 to 1965 (Figure 9.13). Among the ten worst droughts at each of these stations during this period, four occurred at both stations, and all of these occurred prior to 1941, when data from Lake Trafford were first recorded. Data were available for both Lake Trafford and Miami for the period 1941–65 (Figure 9.13). While continuous records were available for Lake Trafford, data for Miami were only available for the ten worst droughts between 1912 and 1965. Four of the Miami droughts occurred during the period 1941–65, but only one of these matched the droughts recorded during this period at Lake Trafford. Data were available for both Fort Myers and Lake Trafford for the period 1941–77 (Figure 9.13). Among the ten worst droughts at each of these stations during this period, seven occurred at both stations.

Because Fort Myers is only 50 km from Lake Trafford, the large number of coincident severe droughts recorded at these two stations is not surprising. However, Fort Myers is approximately 200 km from Miami, and almost half of the

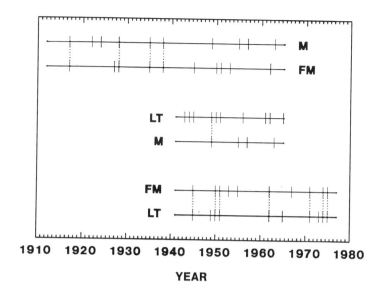

1910 1920 1930 1940 1950 1960 1970 1980

YEAR

Figure 9.13 Comparison of timing of most severe droughts (indicated by short vertical lines) among three stations in south Florida. Abbreviations: M = Miami, FM = Fort Myers, LT = Lake Trafford.

severe droughts at these stations were still coincident. This indicates that some severe droughts, at least in the early part of this century, were fairly widespread. These data also indicate that a number of these events were localized, even over distances of only 50 km.

Tropical Cyclones

Tropical cyclones are classified on the basis of their maximum sustained wind speeds, which does not consider the stronger wind gusts or tornadoes that frequently accompany them. Tropical depressions have wind speeds of less than 63 km·hr^{-1}. These events have relatively little impact on human interests and as a result have been poorly documented over the years. Tropical storms (with wind speeds of 63–119 km·hr^{-1}) and hurricanes (with wind speeds in excess of 119 km·hr^{-1}) have been much better documented because of their importance to man's interests. All three are primarily late summer and early fall weather patterns (Figure 9.14) which move west across the Atlantic Ocean just north of the equator. They may continue west and dissipate over Mexico or northern Central America, or they may turn north at some point and move into North America or the open Atlantic (Figure 9.15). They typically range from 185 to 1110 km in diameter at maturity (Neumann et al., 1987). Winds increase toward the center, reaching 242 km·hr^{-1} in severe storms, until the calm central "eye" is reached.

There are no readily available long-term records documenting the occurrence of tropical depressions, although some contribute significantly to wet season rainfall in south Florida. Therefore, the rest of this discussion of tropical cyclones will focus on North Atlantic tropical storms and hurricanes. Neumann et al. (1987)

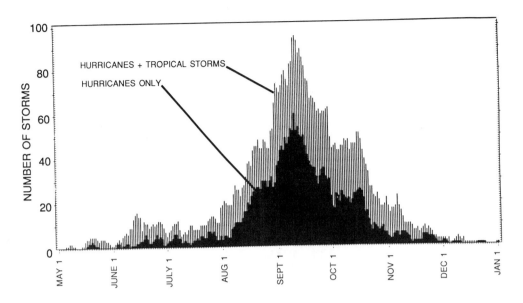

Figure 9.14 Seasonal variation in the 100-year frequency of North Atlantic tropical storms and hurricanes based on the 1886–1986 period of record. (From Neumann et al., 1987.)

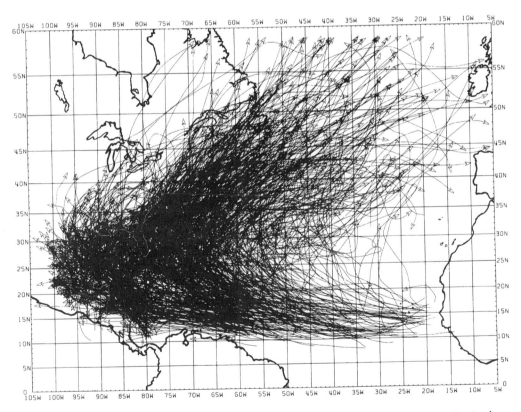

Figure 9.15 Tracks of 845 tropical cyclones reaching at least tropical storm intensity in the North Atlantic over the 1886–1986 period of record. (From Neumann et al., 1987.)

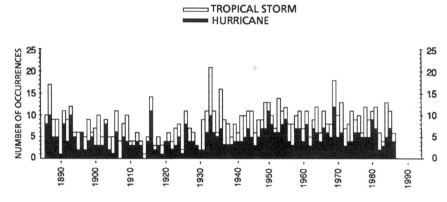

Figure 9.16 Annual distribution of 845 North Atlantic tropical cyclones from 1886 through 1986. (From Neumann et al., 1987.)

summarized much of the information available on these 2 more intense events, 934 of which occurred during the period 1871–1986. As many as 21 tropical storms or hurricanes occurred during a single year (1933), and every year had at least one tropical storm or hurricane. Since 1886, when tropical storms and hurricanes began to be distinguished, only two years did not have a hurricane somewhere in the North Atlantic Ocean (Figure 9.16). The period 1910–30 had smaller numbers of these more intense storms each year than did earlier or later periods. The increase after 1930 could be a result of an improved detection and monitoring capability, but the higher numbers prior to 1910 suggest an actual decrease in frequency from 1910-1930 (Gentry, 1974).

Based on maps of storm paths between 1871 and 1981 (Neumann et al., 1981), 138 tropical storms and hurricanes crossed or came close enough to south Florida that they are likely to have affected the Everglades ecosystem (Figure 9.17). Hurricanes accounted for 78 of these events and tropical storms for 44. Storm type for 16 events that occurred prior to 1886 was not distinguished. The paths of these storms are more or less uniformly distributed over the area, indicating that nowhere in south Florida has been spared, while at the same time no particular area has been affected by a disproportionately large number of events.

The main effects of tropical storms and hurricanes on natural communities are felt in coastal areas from storm surges, which can severely erode some areas and bury others. Noncoastal sites are affected primarily by heavy rains and strong winds.

Most south Florida plants are not adversely affected by flooding associated with intense storms, but terrestrial and nesting animals (such as alligators) may be affected. However, these high rainfall events can be important in initiating a rapid and early rise in the wet season water table when they occur in June or in prolonging a high wet season water table when they occur in late October or November, after evapotranspiration rates have substantially declined.

Even the strong winds associated with hurricanes appear to have few long-term effects on plant and animal communities within the interior of south Florida.

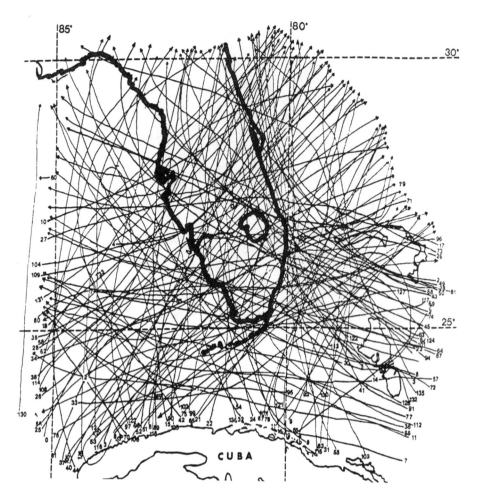

Figure 9.17 Tracks of tropical storms and hurricanes that passed over or close to south Florida from 1871 through 1981, as traced from Neumann et al. (1981). Numbers are in order of occurrence and indicate the start of the track as shown in this figure.

Woody vegetation tends to be more severely affected than herbaceous communities. Trees may be topped or lose major branches, and some may be killed. However, the basic character of the community is unchanged, and within a few years the canopy fills in again. On some sites, almost all trees can be killed, requiring decades for the structure of the forest to be restored. A few types of plants appear to be more vulnerable to these high winds, including strangler figs (*Ficus aurea*) and epiphytes, which may be directly affected by the storm or merely lost when trunks or branches supporting them are broken off. Although large numbers of animals may be killed by hurricane winds, no species have been more than temporarily set back by these events.

Hurricane Andrew

The following discussion is based on information contained in the National Park Service Hurricane Andrew Resource Assessment of Park Service lands in south

Florida (Davis et al., 1992). The assessment was conducted 22–30 days after the storm.

Andrew was a small but intense hurricane. When it made landfall in south Florida on August 24, 1992, it was a Category 4 hurricane (5 is the most severe). It had maximum sustained winds of 242 km·hr^{-1}, nearly the highest on record in Florida. However, it was a relatively dry storm, with only about 5 cm of precipitation recorded at several stations in the interior Everglades.

The storm severely affected a swath of vegetation about 50 km wide as it moved from east to west across the Everglades. The greatest damage was in or near the eye of the hurricane. Severity of damage decreased farther away from the eye, as evidenced by fewer effected trees, loss of only branches rather than stem breakage or uprooting, and a reduced frequency of patches of severe damage associated with isolated strong wind gusts.

Loss of woody biomass was most severe in hardwood communities. When located near the eye of the storm, virtually every canopy or subcanopy tree sustained major damage. Many were uprooted or had their main stems broken, and most of the remainder lost substantial numbers of larger branches. Smaller trees and shrubs were buried under the fallen canopy of larger trees. Both cypress and pine communities were much less severely affected. Impacts on cypress were minor, with only 1–2% of the trees in the center of the storm suffering major damage. Pine stands in the hurricane eye lost 25–40% of the canopy trees, when their trunks snapped or they were uprooted. Trees not downed sustained relatively minor damage, primarily consisting of loss of some branches and needles.

There was a virtually complete loss of leaves from trees along the central storm track, grading to only a general thinning near the margins of the storm-affected area. Again, hardwood forests were most severely affected, in terms of both loss of branches supporting leaves and defoliation of most remaining branches. Cypress retained much of their foliage, although many of the leaves were killed. Pines undoubtedly lost some needles, but still had a substantial amount of live foliage after the storm. Varying degrees of leafout occurred in the weeks following the storm, with a number of tropical hardwood species showing the fastest recovery.

Everglades marshes appeared to have been largely unaffected by the storm. The most severe impacts were on the periphyton mat, which was completely eliminated in some areas. This may have been caused by strong directional winds pushing water out of some areas and then allowing it to "slosh" back when the eye passed and the winds reversed direction.

There was very little evidence of effects on water quality in the Everglades three weeks after the hurricane. This was somewhat surprising given the amounts of living vegetation that were dumped into the system.

As expected, there was little indication of significant direct impacts on Everglades wildlife populations. Greatest concerns were associated with damage to wildlife habitat, such as nest trees (e.g., eagles, woodpeckers), roosting sites (e.g., wading birds), and feeding areas (e.g., deer, butterflies), which may result in more subtle, long-term impacts.

One of the most significant concerns regarding the recovery of Everglades plant communities from Hurricane Andrew is the potentially rapid invasion of exotic vegetation into disturbed areas.

CONCLUSIONS

It is relatively simple to present average monthly and annual climatic conditions for the Everglades. This information is useful in providing a general understanding of the climate of the area and how it compares with other regions.

However, when relating climatic factors to the spatial and temporal patterns of biological populations and communities, these simple and readily available data can be misleading. Some of the reasons for this have been illustrated in this chapter. Of all the measures of temperature in south Florida, probably the most important to the ecosystem is the occurrence of extreme low temperatures. Another aspect is the variability of weather data, in both space and time. Use of average annual rainfall data from a single or even several weather stations provides little information on how wet the wet seasons were, how dry the dry seasons were, or how these conditions were spatially distributed. Climatic events often considered to have major impacts on an ecosystem may be a more accurate reflection of impacts on human interests rather than on natural communities. Hurricanes could be considered in this light, at least as they affect noncoastal portions of south Florida. This is because their main effect appears to be on individual organisms rather than on the long-term viability of populations or communities.

On a different scale, the potential synergistic role of hurricanes in combination with sea level rise in the erosion of natural coastal levees in the southern Everglades represents a major concern of ecosystem-level impacts of hurricanes during the next century.

Thus, it is important when trying to explain changing population or community patterns to clearly identify the relevant climatic parameters and be aware of how they vary in space and time. This variability in the system is likely a crucial aspect of the Everglades that allows mobile species to survive and more sedentary species to recolonize a site from which they have been extirpated.

While there is clear evidence of short-term cycles and variability in climatic patterns, the long-term climatic trend for south Florida appears to have been fairly stable over the past 50–100 years. The most predictable and obvious short-term climatic pattern of interest in this chapter is the annual cycle, with a wet season during the warm summer months and a dry season through the rest of the year. Twice-a-year and once-in-five-year rainfall cycles have also been reported, but they are much less predictable. Freezes and tropical cyclones are likely every year in south Florida, but it is not possible to predict their actual occurrence, frequency, or severity in any particular year. Droughts, depending on how they are defined, can occur a number of times every year or can occur at intervals of a decade or more. This is an excellent example of a climatic parameter whose definition should be based on the requirements of each particular study in which it is considered important.

Although temperatures become warmer as one moves south, they are also less extreme toward the coast. This is a fairly subtle change in terms of average temperature, but an important one in terms of freezing temperature. The effects of severe cold fronts on the Everglades, however, can also be influenced by local conditions, such as the amount of water standing above the ground surface in an area or the previous history of cold weather during that winter. Rainfall is highest along the southeast coastal ridge and declines to the west across the Everglades. However, beyond this general statement, seasonal and year-to-year variability can produce very erratic and unpredictable rainfall, and therefore drought patterns, throughout the Everglades. The size of tropical cyclones dictates that their effects will be felt throughout the Everglades to some degree.

LITERATURE CITED

Black, Crow and Eidsness, Inc. 1974. Hydrologic Study of the G.A.C. Canal Network, Collier County, Florida, Black, Crow and Eidsness, Inc. Gainesville, Fla., 71 pp.

Davis, G. E., L. Loope, C. Roman, G. Smith, and J. Tilmant. 1992. Hurricane Andrew Resources Assessment of Big Cypress National Preserve, Biscayne National Park, and Everglades National Park, Draft report, National Park Service, Homestead, Fla., September 15–23, 1992, 131 pp.

Davis, J. M. and C. M. Sakamoto. 1976. An atlas and tables of thunderstorm and hail day probabilities in the southeastern United States. *Auburn Univ. Agric. Exp. Stn. Bull.,* No. 477, 75 pp.

Dohrenwend, R. E. 1977. Evapotranspiration patterns in Florida. *Fla. Sci.,* 40:184–192.

Duever, M. J., J. E. Carlson, and L. A. Riopelle. 1975. Ecosystem analyses at Corkscrew Swamp. in *Cypress Wetlands for Water Management, Recycling and Conservation,* H. T. Odum, K. C. Ewel, J. W. Ordway, and M. K. Johnston (Eds.), 2nd Annual Report to National Science Foundation and Rockefeller Foundation, Center for Wetlands, University of Florida, Gainesville, pp. 627–725.

Duever, M. J., J. E. Carlson, J. F. Meeder, L. C. Duever, L. H. Gunderson, L. A. Riopelle, T. R. Alexander, R. L. Myers, and D. P. Spangler. 1986. *The Big Cypress National Preserve,* Research Report No. 8, National Audubon Society, New York, 455 pp.

Espenshade, E. B., Jr. and J. L. Morrison. 1978. *Goode's World Atlas,* Rand McNally, Chicago, 372 pp.

Gentry, R. C. 1974. Hurricanes in South Florida. in *Environments of South Florida, Present and Past. Memoir 2,* P. J. Gleason (Ed.), Miami Geological Society, Miami, pp. 73–81.

Gunderson, L. H. 1984. Regeneration of cypress in logged and burned strands at Corkscrew Swamp Sanctuary, Florida. in *Cypress Swamps,* K. C. Ewel and H. T. Odum (Eds.), University Presses of Florida, Gainesville, pp. 349–357.

Hammond, Inc. 1967. *Hammond World Atlas and Gazetteer,* Hammond, Inc., Maplewood, N.J., 48 pp.

Hela, I. 1953. Remarks on the climate of southern Florida. *Bull. Mar. Sci. Gulf Carib.,* 2:438–447.

Huber, W. C., K. Maalel, E. Foufoula, and J. P. Heaney. 1982. Long-Term Rainfall/Runoff Relationships in the Kissimmee River Basin, Final Report to South Florida Water Management District, University of Florida Department of Environmental Engineering Sciences and Florida Water Resources Research Center, Gainesville, 63 pp.

Kenner, W. E. 1966. Runoff in Florida, Map Series No. 22, Florida Board of Conservation, Division of Geology, Tallahassee, 1 p.

Loftus, W. F. and A.-M. Eklund. 1994. Long-term dynamics of an Everglades small-fish assemblage. in *Everglades: The Ecosystem and Its Restoration,* S. M. Davis and J. C. Ogden (Eds.), St. Lucie Press, Delray Beach, Fla., chap. 19.

MacVicar, T. K. and S. S. T. Lin. 1984. Historical rainfall activity in central and southern Florida: Average, return period estimates and selected extremes. in *Environments of South Florida, Present and Past, Memoir 2,* P. J. Gleason (Ed.), Miami Geological Society, Miami, pp. 477–509.

Myers, R. L. 1986. Florida's freezes: An analog of short-duration nuclear winter events in the tropics. *Fla. Sci.,* 49:104–115.

National Oceanic and Atmospheric Administration. 1985. Climatography of the United States No. 20, Climatic Summaries for Selected Sites, 1951–80. Florida, National Climatic Data Center, Asheville, N.C., 89 pp.

National Oceanic and Atmospheric Administration. 1987. Climatological Data No. 91, Annual Summary. Florida. 1987, National Climatic Data Center, Asheville, N.C., 36 pp.

National Oceanic and Atmospheric Administration. 1988. Climatological Data No. 92, Annual Summary. Florida. 1988, National Climatic Data Center, Asheville, N.C., 36 pp.

Neumann, C. J., B. R. Jarvinen, and A. C. Pike. 1981. Tropical Cyclones of the North Atlantic Ocean 1871–1981, Historical Climatology Series 6-2, National Climatic Data Center, Asheville, N.C., 174 pp.

Neumann, C. J., B. R. Jarvinen, A. C. Pike, and J. D. Elms. 1987. Tropical Cyclones of the North Atlantic Ocean 1871–1986, Historical Climatology Series 6-2, National Climatic Data Center, Asheville, N.C., 186 pp.

Pielke, R. A. and W. Gray. 1988. Current Level of Knowledge Regarding the Predictability of Changes in South Florida Weather Due to Major Natural Large Scale Atmospheric Perturbations, Aster, Inc., Boulder, Colo., 109 pp.

Sculley, S. P. 1986. Frequency Analysis of SFWMD Rainfall, Tech. Publ. 86-6, South Florida Water Management District, West Palm Beach, 167 pp.

Thomas, T. M. 1974. A detailed analysis of climatological and hydrological records of south Florida with reference to man's influence upon ecosystem evolution. in *Environments of South Florida, Present and Past, Memoir 2,* P. J. Gleason (Ed.), Miami Geological Society, Miami, pp. 82–122.

Wade, D. D., J. J. Ewel, and R. H. Hofstetter. 1980. Fire in South Florida Ecosystems, Forest Service General Tech. Rep. SE-17, Southeastern Forest Experiment Station, Asheville, N.C., 125 pp.

10

A COMPUTER MODEL
to SIMULATE NATURAL
EVERGLADES HYDROLOGY

Robert J. Fennema

Calvin J. Neidrauer

Robert A. Johnson

Thomas K. MacVicar

William A. Perkins

ABSTRACT

This chapter presents the first attempt to use a computer simulation model to describe the hydrology of the Everglades under a more natural landscape. This simulation is accomplished by eliminating the man-made modifications to the landscape (canals, structures, pumps, etc.) from an existing calibrated and verified regional hydrologic model of today's managed system, the South Florida Water Management Model. All the dominant hydrologic processes are incorporated in this two-dimensional, integrated ground and surface water model. The model domain covers an area of 19,736 km² (7620 mi²) south of Lake Okeechobee and runs on a daily time step using historical meteorological conditions for the period 1965–89.

The Natural System Model was developed to estimate surface water flows and stages in the remnant Everglades areas, such as Everglades National Park and the Water Conservation Areas. Comparisons were made between the results of the natural and managed system models to provide some insight in evaluating alternatives for future restoration initiatives. However, the lack of available historical information on past hydrological and ecological conditions prevented sufficient modifications of the data sets to reflect what is generally perceived as predrainage or preproject conditions in the northern and western portions of the modeled area.

In evaluating the output of the Natural System Model, the South Florida Water Management Model, and recorded flows, the hydrologic changes in flow patterns and timing that have occurred in the Everglades become apparent. Greater spatial and temporal extent of surface water depths occurred under natural conditions, as well as a relocation of some of the deeper pools from the west side of the coastal ridge to the present-day Water Conservation Areas. A comparison of ground and surface water flow volumes in all parts of the modeled area shows significant differences in the quantity and timing of flows throughout the Everglades basin.

INTRODUCTION

The settlement and subsequent development of south Florida has occurred during the short period between the establishment of western man's first foothold in the late 19th century and his total dominance of the landscape less than a century later. Even the earliest settlers had an impact on the ecosystem far in excess of their numbers, from the wanton massacre of wading birds to the diversion of Lake Okeechobee water to the coast. However, by the time the Everglades water system was completely subdued, public sentiment had begun to shift in favor of preserving and restoring the remaining environmental elements of the system. The fact that the early human settlements could only succeed by dramatically altering the natural environment assured future scientists that there would be little, if any, systematic documentation of basic natural phenomena with which to reconstruct a semblance of the natural landscape. This presents unique challenges to those now involved in the protection and restoration of the regional ecosystem. This chapter summarizes a first attempt to use modern computer simulation techniques to establish detailed data sets for estimating the hydrology of the Everglades prior to its alteration by man.

The computer model, called the Natural System Model, was developed by the South Florida Water Management District using the calibrated algorithms and parameters from the South Florida Water Management Model. The South Florida Water Management Model (MacVicar et al., 1984) is a large-scale mathematical model of the present-day network of canals, structures, levees, pumps, and well fields that make up the highly managed water system around which south Florida has developed. The Natural System Model attempts to apply the knowledge gained from using the South Florida Water Management Model for modern south Florida conditions to a hypothetical, more natural south Florida which has been little influenced by man. The Natural System Model uses two-dimensional numerical techniques to simulate the integrated surface and ground water hydrology of a 19,736-km² (7620-mi²) area south of Lake Okeechobee. There are no canals, structures, pumps, levees, wells, or other appurtenances of today's system—only the overland flow, groundwater, evapotranspiration, and rainfall processes that drove the original Everglades. The primary use of the model to date has been to estimate the seasonal and annual fluctuations in overland flow rates and water levels in the areas of the remaining Everglades.

To develop the Natural System Model it was necessary to make several gross assumptions about important elements of the hydrology on the periphery of the system. These assumptions will limit the utility of the current version of the model to making assessments on the gross regional scale of the domain. The Natural System Model, in its current form (and largely due to data limitations), does not simulate the natural Everglades, defined as the hydrological and ecological functions prior to the influence of man, but rather tries to simulate the natural hydrologic response of the system as it would function under current meteorological conditions and without the existence of man's influence in the form of the structural improvements of the Central and Southern Florida Project. Specifically, the current version of the model does not contain a reliable estimate of the overflow from Lake Okeechobee. The overflow function used for this version of the model was chosen only to show the sensitivity of the Everglades system to potential overflow from the lake, rather than to provide actual estimates of the quantity of water moving south. Similarly, the eastern flow from the Everglades through the transverse glades to the ocean has not been quantified, but a crude system was included in the model so that the possible influence of this feature could be evaluated. Most of the data sets used in the model have not yet been modified to more accurately reflect predrainage conditions; additional work is being performed to document estimates of predrainage topography, land use patterns, and other pertinent data sets and parameters.

In its present form, labeled NSM36 (for version 3.6), the model can be used as a general approximation of the hydrologic behavior of an unmanaged Everglades. It would not be appropriate to attempt to replicate the flow rates from this model to remaining pieces of the system both because the rates themselves may be inaccurate and because the application of predevelopment flow rates to a much smaller, and in other ways modified, system would not necessarily produce a positive ecological response.

Cooperative efforts between the staffs of Everglades National Park and the South Florida Water Management District are underway to improve the algorithms and data sets of NSM36. The goal is to develop a set of models which will mimic the hydrologic response of the Everglades as accurately as possible, within the limits of recorded history and thus ultimately accurately reflect the hydrologic history of a changing Everglades ecosystem.

METHODOLOGY

The current release of the Natural System Model (NSM36) is a two-dimensional coupled surface/ground water model, incorporating the dominant physical processes affecting the hydrologic conditions in south Florida. Using almost the identical parameters, data, and algorithms of a calibrated and verified South Florida Water Management Model (version 1.1), originally developed by MacVicar et al. (1984), the NSM36 is a preliminary attempt to simulate the hydrology under more natural conditions. For purposes of this model, the influence of man is evidenced by the changes in land use and the impacts of construction and operation of the Central and Southern Florida Project. This influence was removed in the NSM36 primarily by eliminating the Central and Southern Florida Project features and changing some of the input data (e.g., land use descriptions and ground surface elevations). Although much additional work is needed on the input data, the NSM36 is instructive in describing regional flow patterns and water depths; however, flow volumes and water depths at particular sites should be used with caution, as these are likely to be of insufficient accuracy for site-specific hydrologic or ecological assessments.

For a more detailed description of the algorithms used in the NSM36 and the changes which have taken place from earlier versions, the documentation report by MacVicar et al. (1984), the draft report by Perkins and MacVicar (1991), the evaluation report by Fennema (1992), and the documentation memo by Neidrauer (1992) should be consulted. This section will deal only with a general description of the major processes, data sets, and parameters and will discuss the fundamental differences between the South Florida Water Management Model and NSM36. The domain of the NSM36 includes all of the South Florida Water Management Model area, as well as part of Hendry County; the entire area encompasses 19,736 km^2 (7620 mi^2) south of Lake Okeechobee (Figure 10.1). The area is divided into 6.4-km^2 (4-mi^2) grid cells, for a maximum of 65 rows and 41 columns, with a total of 1905 grid cells. Limited historical data on rainfall and evapotranspiration are available for input to the NSM36. By using the same input data as the South Florida Water Management Model, direct regional comparisons between the two models become possible.

The hydrologic processes which are simulated in the NSM36 and the South Florida Water Management Model are driven by two primary input data sets: rainfall and pan evaporation. The response processes include evapotranspiration, infiltration, overland flow, groundwater flow, and channel flow. Rainfall is added to the model by increasing the surface water depth in each grid cell according to the location of the cell in 1 of 16 rainfall basins and the precipitation value for the particular day. Lake Okeechobee plays an important role in adding water to the northern portion of the domain. The average daily lake water level is read by the model at every time step, and water levels in the cells adjacent to the lake are explicitly set to these elevations. A description of the remaining governing processes, which distribute water in the model, is provided in the next section, followed by an overview of the input data.

Governing Processes

Overland Flow

After water is added to the domain (by rainfall and/or lake inflows), the model simulates overland flow in each grid cell by using a solution to Manning's equation between adjacent grid cells to the east and south. The outflow from each grid cell is thus computed twice, once for west-east flow and then a second time for south-north flow, as follows:

$$V_{of} = \frac{1.49}{n} * h_m^{5/3} \left(\frac{h_u - h_d}{L} \right)^{1/2} * 43,200 \, \Delta t \qquad (10.1)$$

where V_{of} is the outflow from a cell in cubic feet (0.028 m^3) per half day, n is Manning's coefficient, W is grid cell width, h_m is the average of the ponding depths,

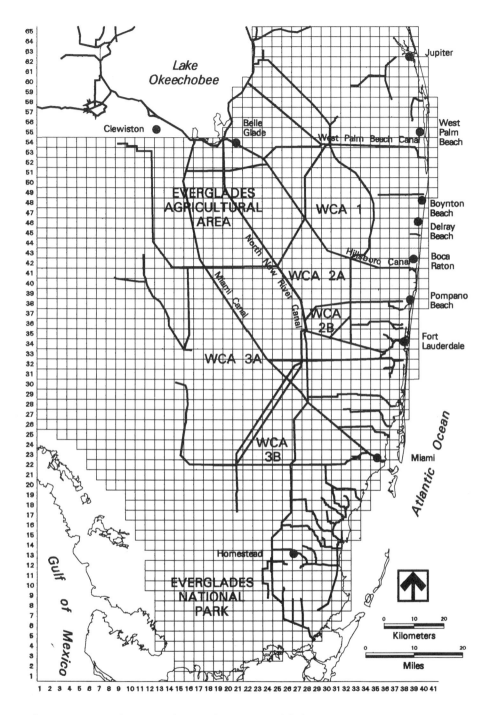

Figure 10.1   Study area and Natural System Model domain.

Figure 10.2 Variation of Manning's n with surface water depth.

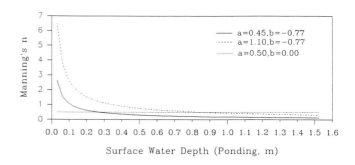

b_u is the upstream stage, b_d is the downstream stage, L is grid cell length, and Δt is the time step increment. Ponding depth is the depth of surface water above the ground surface elevation of the grid cell, while stage refers to the elevation (in feet NGVD) of the water. Overland flow is limited by a surface detention depth (below which no flow is allowed) to simulate the depressions on the surface. The algorithm is a kinematic solution whose primary limitations are loss of the acceleration and pressure terms and the assumption of uniform flow. The friction caused by the vegetation and the smoothness of the surface is computed by using a variable Manning's coefficient (Chow, 1959), which in this model is based on a power equation of the form

$$n = a\,p^b \qquad (10.2)$$

where p is the ponding depth, and a and b are coefficients based on one of five land use categories. These coefficients are identical, except for land uses 2 (slough) and 4 (coastal ridge). The range of Manning's n for a sequence of relevant ponding depths is shown in Figure 10.2 for the three sets of coefficients used in NSM36. The increase of Manning's n with lower ponding depth is an attempt to account for the increased resistance by the vegetation. The stages in the grid cells along the marine perimeter of the model domain are set to explicit values at each time step, and no flow is allowed to take place across the boundary and along the northern side and most of the western side.

Infiltration

Exchange between ground and surface water is by infiltration. Each grid cell with ponded water above the detention depth adds to the ground water if the level of the ground water is below land surface. This process is controlled by an infiltration rate coefficient which may be adjusted based on the rainfall basin (Figure 10.3) of the grid cell. Thus, the amount of infiltration to the ground water can be reduced by lowering the rate to reflect, for example, surficial peat layers. The coefficients used for each of the rainfall basins are

Rainfall basin	1	2	3	4	5	6	7	8
Infiltration (cm/day)	30	30	30	8	30	15	8	30
Rainfall basin	9	10	11	12	13	14	15	16
Infiltration (cm/day)	15	15	15	8	15	30	30	15

Figure 10.3 Rainfall basins used in the NSM36 and the South Florida Water Management Model.

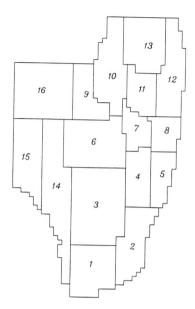

If the groundwater depth exceeds the ground surface elevation during later computations, the excess water is added to the surface water directly.

Evapotranspiration

The next step in the computational scheme is the evapotranspiration (ET) process, which reflects the return of moisture back to the atmosphere by evaporation from surface and ground water and transpiration from vegetation. Because of the significance of this process in the NSM36, a new algorithm, based on the CREAMS model (Knisel, 1980), was developed. To maintain the compatibility between models, this algorithm was subsequently incorporated in the South Florida Water Management Model. The basic parameters are simulated by computing a daily pan evaporation value (P) for each rainfall basin, which is modified by a daily crop coefficient (K), reflecting the type of vegetation cover (Figure 10.4) and its seasonal variation.

If there is no surface water, ET is computed from ground water by adjusting the value based on a deep (DRZ) and shallow (SRZ) root zone for each of the five land use types:

Root zone	Land use (cm)				
	1—Sawgrass	2—Slough	3—Glades	4—Ridge	5—Rim
Shallow	0	0	0	61	0
Deep	122	46	274	305	122

The computation of ET from ground water then becomes

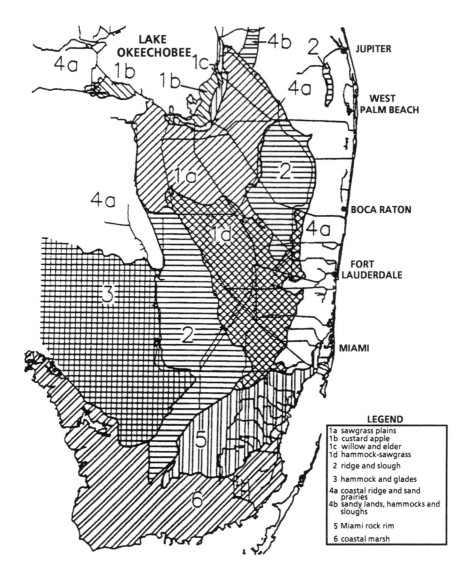

Figure 10.4 Vegetation cover information used in NSM36. (Reproduced from Jones, 1948.)

$$ET = 0 \qquad\qquad GWD > DRZ$$

$$ET = K * P * \frac{(DRZ - GWD)}{(DRZ - SRZ)} \qquad DRZ > GWD > SRZ$$

$$ET = K * P \qquad\qquad GWD < SRZ \qquad (10.3)$$

where GWD is the depth to ground water. If ET is from surface water, a value is computed if ponding is less than 91 cm (3 ft). If the value of ponding is greater than 91 cm (3 ft), the ET rate is explicitly set to 81% of the pan evaporation value,

Figure 10.5 Schematic of evapotranspiration computation in the NSM36 and the South Florida Water Management Model.

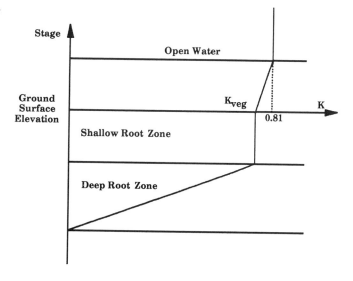

to resemble open water evaporation. Otherwise, the value is computed by adjustment based on the crop coefficient:

$$ET = 0.81 * P \qquad\qquad SWD > 3.0'$$

$$ET = P * \left[\frac{SWD}{3.0} * (0.81) - K) + K \right] \quad SWD \leq 3.0' \qquad (10.4)$$

where SWD is the ponding in the grid cell. A graphical representation of the ET computation for both ground and surface water is shown in Figure 10.5. The monthly crop coefficients for each of the five land use types in the model are given in Table 10.1.

Table 10.1   Monthly Crop Coefficients for the Five Land Use Types

Month	Land use				
	1—Sawgrass	2—Slough	3—Glades	4—Ridge	5—Rim
January	0.721	0.751	0.753	0.590	0.721
February	0.817	0.802	0.762	0.581	0.817
March	0.895	0.829	0.757	0.585	0.895
April	0.903	0.821	0.756	0.595	0.903
May	0.692	0.643	0.763	0.583	0.692
June	0.627	0.611	0.758	0.579	0.627
July	0.722	0.701	0.787	0.594	0.722
August	0.784	0.771	0.825	0.630	0.784
September	0.798	0.785	0.799	0.622	0.798
October	0.754	0.742	0.854	0.631	0.754
November	0.703	0.694	0.856	0.642	0.703
December	0.763	0.751	0.804	0.601	0.763

Ground Water

Ground water is computed using an explicit finite difference solution of the two-dimensional transient groundwater equation:

$$\frac{\partial}{\partial x} T_{xx} \frac{\partial h}{\partial x} + \frac{\partial}{\partial y} T_{yy} \frac{\partial h}{\partial y} = S \frac{\partial h}{\partial t} + R \qquad (10.5)$$

where T_{xx} and T_{yy} are the aquifer transmissivity values in the east-west and north-south direction, respectively, h is the groundwater stage, S is the aquifer storage coefficient, and R is the recharge. The recharge term accounts for the infiltration, evapotranspiration, and seepage interaction with channel flow. Transmissivity values are computed at each time step following the unconfined aquifer theory and thus involve hydraulic conductivity, aquifer depth, and groundwater stage.

No-flow boundaries exist along most of the northern and western limits of the domain, with the southern and eastern marine boundaries set to the explicit value of 12 cm (0.4 ft), which reflects the level of Florida Bay and the Atlantic Ocean. No attempt has been made to allow for the spatial and seasonal fluctuations of marine water levels, which are particularly important along the Florida Bay boundary, where fluctuations can be up to 30 cm (1 ft).

Channel Flow

An important contribution of surface flow to the Atlantic coast was through shallow sloughs and channels (the transverse glades), which flowed unregulated through the coastal ridge, where several of the channels formed rapids along their course seaward. Historical accounts of these rapids (the forks of the Miami River and Arch Creek, for example) had vertical drops of 1.8 m (6 ft) over a length of 1.4 km (0.87 mi), which are an indication of very high water levels in the sandy flatlands west of the coastal ridge. The NSM36 contains 21 channels which course through the coastal ridge and discharge to the Atlantic. Channel width is kept constant at 30 m (100 ft), and the total length of an individual channel ranges from 6.4 km (4 mi) to 19.3 km (12 mi). The basic channel routine in the NSM36 processes channel flow using a weir equation and averages the channel depth calculated in each reach to compute ground and surface water seepage.

Grid Cell Static Data

The NSM36 input data that do not change during the entire run are referred to as static data. The static data are defined based on (1) grid cell location, (2) grid cell land use, or (3) grid cell rainfall basin. Evapotranspiration parameters, infiltration rates, and Manning's coefficients (discussed previously) are based on items 2 and 3. Each grid cell is also defined by a set of five static data, providing the ground surface elevation, rainfall basin, land use, hydraulic conductivity, and aquifer depth.

Elevations

The topography of the NSM36 was adapted from the South Florida Water Management Model which in turn is based on topographic data from a variety of sources ranging from the early 1950s to the 1980s. For the NSM36, subsidence near the major canals was "smoothed," which resulted in increased elevations of up to 0.61 m (2 ft). For this preliminary model no attempt was made to raise land elevations in the Everglades Agricultural Area or to further raise the elevations between the major canals. Both of these areas are known to have undergone substantial subsidence. Especially vulnerable to errors in the topography are the areas which historically contained some depth of muck or peat. Early survey information along the major canals indicates that elevations in these peatlands were as much as 1.5 m (5 ft) higher than found in more recent surveys (Fennema, 1992).

Rainfall Basins

The spatial distribution of rainfall is defined by 16 rainfall basins, 15 of which are used in the South Florida Water Management Model. Unfortunately, scant information is available for predrainage years, and little is known about the climatic changes which may have occurred over the past 150 years. For these reasons, no changes were made in the NSM36, except for the additional basin added because of the addition of Hendry County. Alteration in the timing and distribution of rainfall over this time period would undoubtedly influence the behavior of the model, because rainfall is the principal driving mechanism of water distribution (Obeysekera and Loftin, 1990). To offer a valid comparison between the NSM36 and the South Florida Water Management Model, the spatial and temporal distribution of rainfall was kept the same in both models.

Land Use

Several parameters used by the NSM36 (e.g., the infiltration and the evapotranspiration routines) depend on the type of surface cover that predominates in a particular grid cell. The South Florida Water Management Model uses distinctions based on agriculture, urban, natural, etc. Because the NSM36 attempts to reflect a natural landscape, the five land uses types are based on natural vegetative cover and soil types. To establish the spatial composition of these types, the vegetation map by Jones (1948) (Figure 10.4) was adapted to divide the modeling area into five classifications as follows:

Land use no.	Classification
1	Sawgrass plains and coastal marsh
2	Hammock-sawgrass, slough, and ridge
3	Hammock, glades, custard apple, willow, and elder
4	Coastal ridge and sand prairies
5	Miami rock rim

Further refinement of these classifications is possible based on historical accounts of soil types, soil depths, and vegetative densities, but they have not been investigated for this preliminary version of the Natural System Model.

Hydraulic Conductivity and Aquifer Depth

These two static data sets are used exclusively by the groundwater routine to compute transmissivity values of the aquifer. Both data sets are fairly well estimated in the coastal counties, but additional work is needed in the areas to the west, where drilling information is scarce. Values of hydraulic conductivity range from 3 m/day (10 ft/day) to 11,350 m/day (37,238 ft/day), with the largest values found in Dade County. Coupled with the large storage coefficients ($S = 0.2$) in these areas, these data sets form the bases for accurate representation of groundwater flow volumes to the lower Everglades system. Groundwater flows are especially important in the dry season, when flow volumes in several areas are of the same order of magnitude as surface water flow. The South Florida Water Management Model and NSM36 contain the same values for both data sets, and incorporation of current research in the western areas (Water Conservation Areas, Everglades Agricultural Area, and Big Cypress basin) is needed to refine this component.

SENSITIVITY ANALYSIS

The response of NSM36 to changes in certain input data sets or parameters was measured by the variation in stages at selected locations and flows across selected grid cells. This was done to demonstrate the importance of those input data to the model output. The majority of the input to the NSM36 is from the calibrated and verified version of the South Florida Water Management Model (version 1.1). Thus, the algorithms have been calibrated to the current managed system. The behavior of a natural system, as compared to today's managed Everglades, may be different. By identifying the sensitivity of the model to the magnitude of the variation in the key data, informed decisions can be made regarding areas needing further work and the reliability of the current version of the model.

The parameters selected for this analysis were those identified as being most important to the sensitivity of overland flow, evapotranspiration, and the variation of Lake Okeechobee stages. The first analysis presented is one of the most important and, unfortunately, one of the least understood processes: evapotranspiration. The NSM36 uses pan evaporation data and computes evapotranspiration based on several parameters. Rather than determining the sensitivity of the model to changes in each of these parameters, the pan evaporation data were adjusted by multiplication with a constant, and the response was recorded by a time series of stages and flows in the model. The second analysis concerns overland flow. The variation in stages and flows as a result of adjusting the parameter in Manning's equation (n) again by multiplication with a constant was used to illustrate the importance of using a correct resistance coefficient. Third, the response of the model to increased or decreased overflows of Lake Okeechobee was tested by adding a constant to the lake's historical daily stage data.

The NSM36 was run for the period 1966–89, with the specific changes made to the data or code. Average weekly stages at three gauges were recorded: 1–7 in what is now Water Conservation Area 1, 3A–4 in what is now Water Conservation Area 3A, and P-33 in what is now Everglades National Park (Figure 10.6). Percent

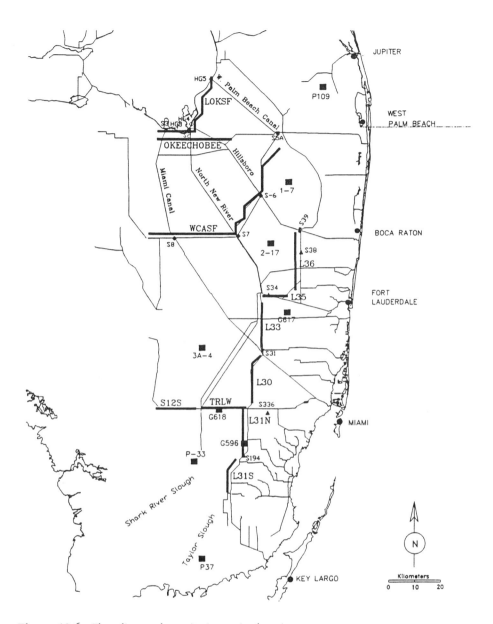

Figure 10.6 Flow line and monitoring point locations.

exceedance curves were generated for each of the gauges to illustrate the variation in stage for the different sensitivity runs. The range of the vertical axis of these curves was kept the same to facilitate comparison between plots. In addition to stage output, the total annual flow across selected flow lines was computed to document the variation in surface flows for each of the runs. The annual flow across a flow line is calculated by adding the flows out of a preset number of grid cells. The model computes flows in a south-north and west-east direction, either one of which can be specified for flow line computation. For the two selected flow lines (Tamiami Trail and Okeechobee) (Figure 10.6), only south-north flows were added to obtain the total flow. The Tamiami Trail flow line was used to illustrate the changes in flows across a natural Shark River Slough flow way and is located across what is now the northern boundary of Everglades National Park. The Okeechobee flow line was used to compare the discharges from Lake Okeechobee to the downstream Everglades.

Sensitivity to Changes in Pan Evaporation

Evapotranspiration is dependent on many factors, including weather, vegetation, soil type, and water depth. Most hydrologic models, including the NSM36, have to simplify the evapotranspiration process for lack of data. The most important parameters (pan evaporation, land cover, and water depth) are used in the NSM36 to adjust the evapotranspiration values. To illustrate the importance of this process on the resulting stages and flows, the pan evaporation data were changed by multiplying the data by 0.8 and 1.2, for a 20% reduction and a 20% increase in evapotranspiration, respectively.

The stage record for gauge 1–7 (Figure 10.7) shows a small variation in stage exceedance, except in dry years, when there is no surface water and evapotranspiration comes entirely from ground water. At gauge 3A–4, the variation is much larger. Again, the water level drops much faster when evapotranspiration comes from ground water, but even during wet years the difference in stage is larger, and an identical stage is exceeded 30–40% less often with increasing evapotranspiration. At P-33 the variation increases with reduced surface water, but this area stays inundated, except for 5% of the time, at the 120% evapotranspiration level. This appears to indicate that a sharp reduction in inflow from the north occurs, as evidenced by the reduction at 3A–4, and local rainfall is not sufficient to maintain stages with the increase in evapotranspiration. The decrease in flows with increasing evapotranspiration is illustrated by the surface flow line across Tamiami Trail in the bottom plot of Figure 10.7. A 20% decrease in evapotranspiration causes an average increase in inflow of 73% into Shark River Slough, while a 20% increase reduces flow by 60%, an average annual reduction of 247 million m^3 (200,000 acre-ft) over many years.

Sensitivity to Changes in Manning's *n*

Changes in the overland flow parameter (Manning's *n*) were evaluated by multiplying the computed *n* value by 0.05 or 2.00, a 95% reduction or 100% increase, respectively. Even with these large shifts in values, the NSM36 response to the stages and flows (Figure 10.8) is less than the evapotranspiration shifts reported above. Increasing Manning's *n* has much less of an effect on the stage exceedance curves, especially at P-33, where decreasing *n* results in up to 15% lower stages. On the average, 64% more flow went across Shark River Slough when the resistance to flow, as measured by the parameter *n*, was decreased to 95% of the value, while a 20% decrease was noted when the *n* value was increased by 100%.

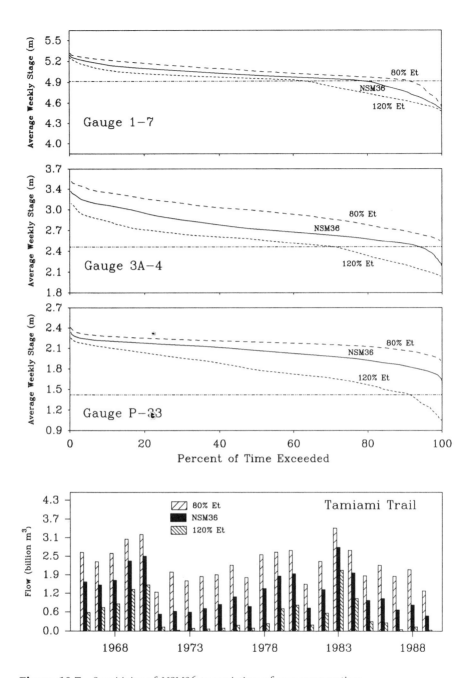

Figure 10.7 Sensitivity of NSM36 to variation of pan evaporation.

Sensitivity to Changes in Lake Okeechobee Stages

The effects of Lake Okeechobee stages on the model were tested by adding an offset to the daily lake stage ranging from 0 to 152 cm (0 to 5 ft). Because no records exist for the variation of predrainage lake stages, the model uses the

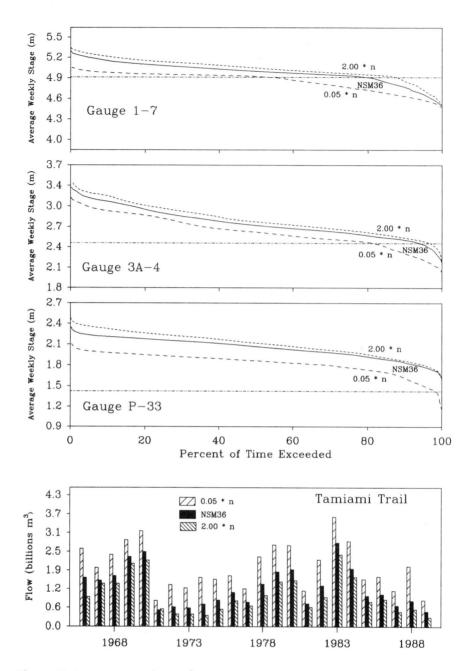

Figure 10.8 Sensitivity of NSM36 to variation of Manning's *n*.

managed historical (1965–89) stages under the assumption that the effects of management do not overshadow the rainfall stage relationship which would exist in the natural landscape (see E. G. Stewart, 1907). The normal NSM36 simulation uses historical daily average lake stages with an offset of +46 cm (1.5 ft). The water thus supplied from Lake Okeechobee to the domain of the model is handled by setting the stages in adjacent grid cells equal to the lake stages at each time step.

Note that one limitation of the assumed lake stage simulation is that it does not provide a mechanism for adjusting lake stages during a spill from the lake. Consequently, NSM36 could overestimate the volume of flow from the lake during a spill.

The effects of altered lake stages are shown in Figure 10.9. By not adjusting the historical values (offset is equal to zero), the lake stage exceedence curves are

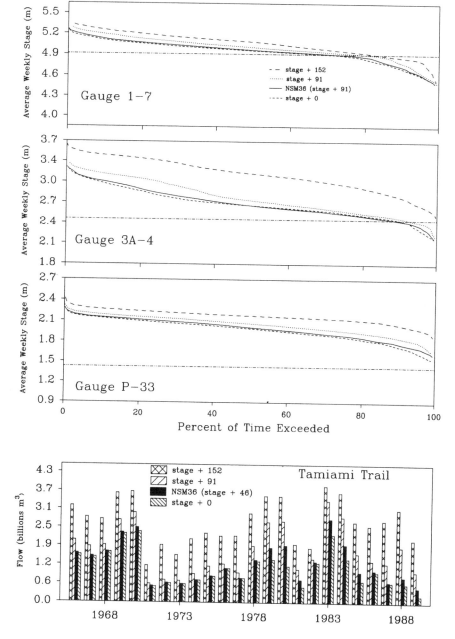

Figure 10.9   Sensitivity of NSM36 to variation of Lake Okeechobee stages.

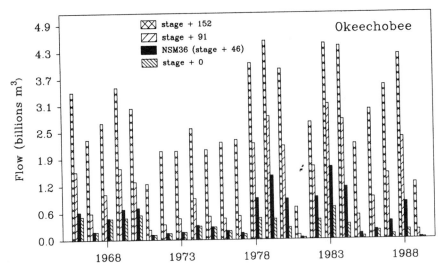

Figure 10.10 Sensitivity of NSM36 to the variation in lake stages at the Lake Okeechobee flow line.

nearly identical to the normal NSM36 run. The effect of increasing lake stages is felt throughout the model domain, even as far south as P-33 in the southern end of the Everglades. The increase in annual flow across the Tamiami Trail flow line using no offset as compared to adding 46 cm (1.5 ft) accounts for an average of 10% more flow during the simulation, while a stage addition of 91 cm (3 ft) increases flow by 34%. A lake stage increase of 152 cm (5 ft) increases flow over 100%. The flows across the Okeechobee flow line (Figure 10.10) were summarized and compared, in a relative sense, to those across the Tamiami Trail line. Under the normal NSM36 run, 130% more flow occurs across the Tamiami Trail line than the Okeechobee line, while if there is no offset, the flow is almost 300% of the Okeechobee flows. Average annual flows across Tamiami Trail at 152 cm (5 ft) (the largest offset used) are 9% lower than at the Okeechobee line. Increasing the offset to 91 cm (3 ft) and then to 152 cm (5 ft) doubles the flows each time across the Okeechobee line.

It is important to note that the sensitivity analyses presented herein were made to demonstrate the high degree of sensitivity of the NSM36 output to changes in some of the more uncertain parameter values. The parameter values selected for the sensitivity runs do not necessarily reflect reasonable ranges of the uncertain parameters. By using the calibrated parameter values from the South Florida Water Management Model for NSM36, the uncertainty in the parameter values was somewhat reduced. Still, additional research efforts are needed to further reduce this uncertainty.

RESULTS

In this section, the results of the NSM36 are compared to recent historical data and to the output from the South Florida Water Management Model (version 1.1—

calibration/verification version). The results presented include (1) flow vector comparisons, (2) inundation and hydroperiod comparisons, (3) flow line comparisons, and (4) stage hydrograph and duration comparisons. These comparisons are made primarily to illustrate the NSM36 results in the context of changes in the regional-scale hydrologic response of the Everglades that are estimated to have occurred from the predrained, or natural, system to the current managed system.

Flow Vector Comparisons

Surface Water Flow Patterns

Figure 10.11 presents a comparison of surface water flow patterns for the natural and managed systems for a condition following above normal rainfall (September 30, 1988). Only surface water flow (overland flow or sheet flow) is shown on these maps. Structure discharges or canal flows are not shown for the managed system. Note that the size of the vectors on these maps is proportional to the flow volumes. Although the Water Conservation Area levees L-31N and C-111 do not exist in the natural system, they are shown on the map for reference.

These maps indicate that water management produced a significant change to the natural overland flow patterns. Overland flow in the natural system appears

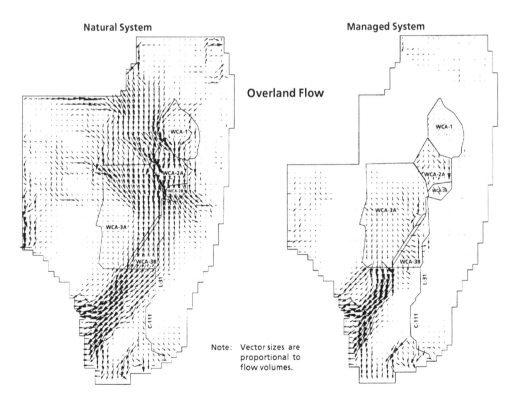

Figure 10.11 Comparison of natural and managed system overland flow for September 30, 1988 (condition following above normal rainfall).

continuous from the south shore of Lake Okeechobee through what is now the Everglades Agricultural Area and Water Conservation Areas into Shark River Slough. Overland flow in the managed system has been significantly reduced by the impoundment of Lake Okeechobee, construction of the Water Conservation Area levees, and irrigation/drainage canals in the Everglades Agricultural Area. Under managed conditions, the largest overland flow occurs in Shark River Slough within Everglades National Park.

Groundwater Flow Patterns

Figure 10.12 provides a comparison of groundwater flow patterns for the natural and managed systems for the same date (September 30, 1988). Note that the size of the vectors is proportional to the flow volumes but cannot be directly compared to the size of the vectors in Figure 10.11, because they are shown at different scales. That is, for a vector in Figure 10.11 and a vector in Figure 10.12 of the same length, the surface flow volume is roughly about ten times larger than the groundwater flow volume.

Under natural conditions the largest hydraulic gradients occurred along the coastal ridge. Consequently, the largest groundwater flow occurred at that location. Excess water from the ridge flowed both east and west. With the construction of

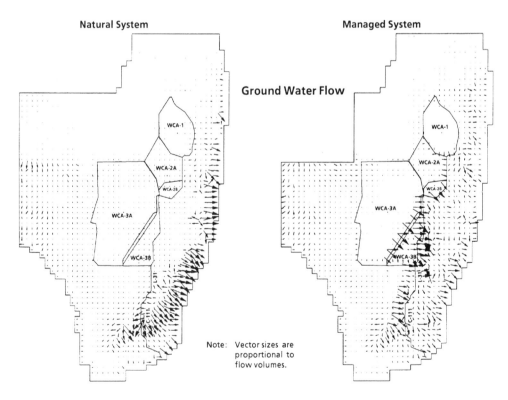

Figure 10.12   Comparison of natural and managed system groundwater flow for September 30, 1988 (condition following above normal rainfall).

the Water Conservation Area levees and the drainage of the lower east coast developed areas, the largest groundwater gradients were shifted westward to the Water Conservation Area levees. Thus, the largest groundwater flow under managed conditions occurs at the levees. The lowering of groundwater levels in the developed areas of the lower east coast reduced the water table gradients along the coast that occurred under natural conditions. This has increased the potential for saltwater intrusion (Leach et al., 1972). Groundwater flows are now predominantly to the east, increasing the drainage of the remnant Everglades, especially along the L-31N and C-111 canals, as shown in Figure 10.12.

Inundation Depth and Hydroperiod Comparisons

The depth of water above the land surface is defined as the inundation depth. Natural and managed system inundation depth patterns are presented for four different historical meteorologic conditions: end of a wet (or above normal rainfall) wet season, end of a dry (or below normal rainfall) wet season, end of a wet dry season, and end of a dry dry season. Also presented is a comparison of the natural and managed system hydroperiod patterns for an average year. These comparisons are made in order to (1) illustrate the various inundation depth patterns that result from different seasonal conditions and (2) compare the natural system inundation depth and hydroperiod patterns to those of the managed system.

End of a Wet Wet Season

Plate 2* presents a comparison of natural and managed system inundation depths at the end of a wet wet season. The 5-month wet season ending October 31, 1985 produced 90.2 cm (35.5 in.) of rainfall over the region, the highest wet season amount during the 1980s. Normal wet season rainfall for the region based on the period 1980–89 was 78.5 cm (30.9 in.). For the natural system as represented by the NSM36, there appears to be 30 cm (1 ft) or more of inundation in most of the original Everglades. The central portion of the deepest water was located at the northern end of what is now Water Conservation Area 3B and was up to 120 cm (4 ft) deep. Under these conditions, only the highest parts of the coastal ridge experienced no inundation. For the managed system, as represented by the South Florida Water Management Model, inundation depths greater than 30 cm were restricted to the Water Conservation Areas and Shark River Slough. Note that little or no inundation occurred in the Everglades Agricultural Area or the lower east coast developed areas—a consequence of the efficient drainage system that was developed to drain those areas for agricultural production and urban development. Also note that the Water Conservation Areas generally impound more water at their southern portions. Inundation depths up to 60 cm (2 ft) occurred in Water Conservation Area 1 and depths up to 120 cm (4 ft) were evident along L-67A in Water Conservation Area 3A.

* Plate 2 appears following page 432.

End of a Dry Wet Season

Plate 3* shows a comparison of natural and managed system inundation depths at the end of a dry wet season. The 5-month wet season ending October 31, 1980 was the driest wet season during the 1980s and produced 66.5 cm (26.2 in.) of rainfall over the region, which is 11.9 cm (4.7 in.) below normal. For the natural system, inundation depths greater than 30 cm (1 ft) still occurred in most of the original Everglades, even under dry wet season conditions. Note that there are no significant differences between the natural system inundation depth patterns for the wet wet season conditions (Plate 2) as compared to the dry wet season conditions shown in Plate 3. As with the wet wet season condition, no inundation occurred in the Everglades Agricultural Area or in the lower east coast developed areas under the managed system. However, as compared to Plate 2, a much larger part of Everglades National Park and Big Cypress basin experienced no inundation. The deepest water occurred in the south end of Water Conservation Area 2A (up to 150 cm or 5 ft) and along L-67A in Water Conservation Area 3A (up to 90 cm or 3 ft).

End of a Wet Dry Season

Plate 4* shows a comparison of natural and managed system inundation depths at the end of a wet dry season. The 7-month dry season ending May 31, 1983 produced 72.6 cm (28.6 in.) of rainfall over the region, the highest amount during the 1980s. Normal dry season rainfall for the region based on the period 1980–89 was 50.3 cm (19.8 in.). The first four months of the 1983 dry season (November 1, 1982 to February 28, 1983) brought record high rainfall to the region—50.5 cm (19.9 in.)—which is 26.9 cm (10.6 in.) more than normal. In the natural system simulation, Lake Okeechobee stages were at a maximum and a spill out of the lake occurred during this period. Plate 4 shows that up to 120 cm (4 ft) of inundation occurred in the heart of the original Everglades, and a large portion of the original Everglades had depths up to 60 cm (2 ft). These deep water areas extended as much as 32 km (20 mi) east of the current location of the Water Conservation Area 3B eastern levees. For the managed system, only parts of the Water Conservation Areas, Shark River Slough, and the Holeyland showed water depths up to 60 cm (2 ft). The rest of the area was predominantly dry. Again, the effects of the drainage system and Water Conservation Area levees were evident.

End of a Dry Dry Season

Plate 5* shows a comparison of natural and managed system inundation depths at the end of a dry dry season. The 7-month dry season ending May 31, 1989 was the driest dry season during the 1980s and produced only 31.0 cm (12.2 in.) of rainfall over the region, which is 19.3 cm (7.6 in.) below normal. For the natural system, only a relatively small part of the original Everglades experienced any inundation. Depths greater than 30 cm (1 ft) occurred only in areas south of the current location of highway I-75, and depths up to 60 cm (2 ft) occurred only in

---

* Plates 3–5 appear following page 432.

scattered parts of what is now Water Conservation Area 3B, Northeast Shark River Slough, and in Shark River Slough. For the managed system, almost the entire region was dry on this date. Only small areas at the south ends of Water Conservation Areas 1 and 2A, along L-67A in Water Conservation Area 3A, and in the southern C-111 basin experienced water depths up to 30 cm (1 ft).

Hydroperiod Pattern Comparison

A hydroperiod is defined as the length of time a specified area is inundated over a 12-month period. Plate 6* illustrates a comparison of the natural and managed system hydroperiod patterns (number of days each cell was inundated) during the 1986 calendar year. The region experienced a near normal rainfall of 129 cm (50.8 in.) during 1986. Normal rainfall based on the 10-year period from 1980 through 1989 was 135 cm/year (53.0 in./year). Also, the seasonal rainfall amounts during this time were near normal. For the wet season (June–October 1986), rainfall was 77.5 cm (30.5 in.), which is only 1 cm (0.4 in.) below the 78.5-cm (30.9-in.) normal. For the dry season (November 1986–May 1987), rainfall was normal at 50.3 cm (19.8 in.).

Plate 6 shows that under natural conditions, a significant portion of the original Everglades had between 10- and 12-month hydroperiods. Under managed conditions, the long-hydroperiod areas were limited to the Water Conservation Areas, the Corbett area, the Holeyland, central Shark River Slough, parts of Taylor Slough, and the southern C-111 basin. As with the inundation pattern comparisons, the effects of the drainage system are evident, as the majority of the Everglades Agricultural Area and lower east coast developed areas experienced zero hydroperiods. Clearly, the hydrologic response of the predrainage system, as estimated by the NSM36, shows that the natural Everglades was generally more of a flowing system with a greater spatial extent and longer periods of inundation than exist under today's managed conditions.

Surface Water Flow Line Comparisons

Monthly surface water flow patterns were analyzed for the period January 1980–December 1989 along four flow lines located at the intersections of the major water management basins in the Everglades watershed. Because each basin is separated by a man-made levee system, the simulated overland flow estimated by the NSM36 was compared to the surface water discharges from the appropriate water control structures corresponding with each particular flow line. In each case, the discharge data represent the total net structure flow (outflow minus inflow) from the upstream basin into the downstream basin, based on the published flow records of the U.S. Geological Survey or the discharge data maintained by the South Florida Water Management District (if the station is unpublished). Figure 10.6 shows the location of the four flow lines used in this analysis (LOKSF, WCASF, L36 + L35 + L33 + L30 & L31N + L31S, and S12S + TRLW) and their associated water control structures.

---

* Plate 6 appears following page 432.

Lake Okeechobee Outflows

The northernmost surface water flow line compares the estimated flows from Lake Okeechobee into the northern Everglades/Everglades Agricultural Area based on the NSM36 with the recorded historical structure discharges. The water control structures used for this analysis included the hurricane gates at the lake: HG-3/S-3 at the confluence with the Miami Canal, HG-4/S-2 at the confluence with the North New River Canal, and HG-5 at the confluence with the West Palm Beach Canal. Figure 10.13 is a comparison of the monthly total outflows from Lake Okeechobee into the downstream Everglades for the period 1980–89. The results of the NSM36 suggest that surface water outflows from Lake Okeechobee into the Everglades were sporadic and were more closely tied to variations in rainfall. Periods of high flows occurred during years of above normal rainfall (such as 1982–83), and little or no flow occurred during below normal rainfall years (such as 1981, 1985, and 1989). The monthly historical structure flows from Lake Okeechobee into the northern Everglades/Everglades Agricultural Area displayed an irregular pattern throughout the study period. In contrast to the results of the NSM36 simulation, the historical inflows were highest during the drier years (such as 1981, 1985, and 1989). This suggests that the pattern of surface water outflows from Lake Okeechobee has shifted from a natural response to rainfall toward a response to the water supply needs of the downstream developed areas.

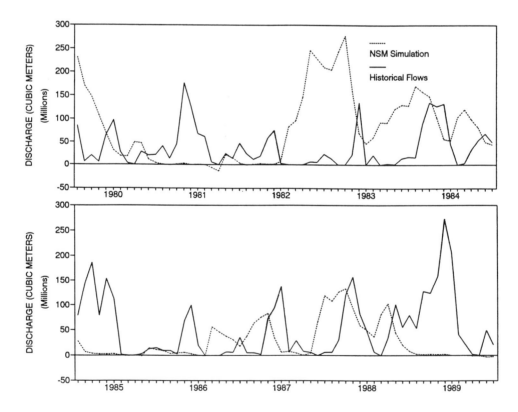

Figure 10.13 Comparison of natural system simulation and historical monthly flows from Lake Okeechobee into the northern Everglades.

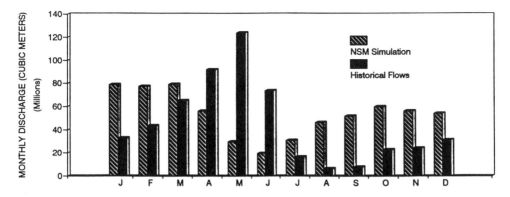

Figure 10.14   Comparison of natural system simulation and historical average monthly flows from Lake Okeechobee into the northern Everglades.

Figure 10.14 is a plot of the average monthly flows for the 10-year study period. Although NSM36-simulated outflows steadily increase throughout the wet season, and actually peak during the dry season months of January through March, this dry season flow pattern may be biased by the high flows during the 1983–84 dry seasons. Also, the use of historical managed lake stages in the NSM36 may not replicate the natural seasonal fluctuations of the lake. Furthermore, because NSM36 does not simulate the effects of lake spills on lake stages, the spill volumes from the lake could be considerably overestimated. The seasonal shift of northern Everglades/Everglades Agricultural Area inflows now steadily increases from the late wet season into the late dry season; on average, the historical dry season flows represent approximately 76% of the annual total flow, as compared to 67% for the NSM36 simulation.

Everglades Agricultural Area Outflows

The next surface water flow line compares the estimated flows from the Everglades Agricultural Area into the downstream Water Conservation Areas. The water control structures used in this analysis included S-5A1 which is a flow measurement point just upstream of the S-5A complex on the West Palm Beach Canal, S-6 on the Hillsboro Canal, S-7 on the North New River Canal, and S-8 on the Miami Canal (see Figure 10.6). Note that S-5A1 represents the total outflow from the Everglades Agricultural Area via the West Palm Beach Canal, including flows that enter Water Conservation Area 1 and those which pass eastward beyond this area into the lower east coast. Figure 10.15 is a comparison of the monthly total outflows from the Everglades Agricultural Area into the downstream basins based on the NSM36 simulation and historical flows. The NSM36 simulation shows that the timing of Everglades Agricultural Area outflows is very similar to that observed at the Lake Okeechobee flow line and that periods of high flows again coincide with years of above normal rainfall. The vertical scale for this figure has been doubled, and thus the peak Everglades Agricultural Area outflows are approximately twice as large as the corresponding inflows from Lake Okeechobee (shown in Figure 10.13). In contrast to the NSM36 results, the historical Everglades

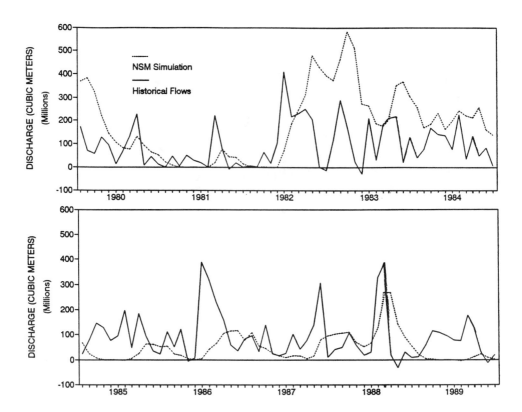

Figure 10.15 Comparison of natural system simulation and historical monthly flows from the Everglades Agricultural Area into the northern Everglades.

Agricultural Area outflows persisted slightly longer into the dry season, with rapid pulses scattered throughout the study period, and exhibited fewer periods of little or no flow during low rainfall years. Also in contrast to the NSM36 results, periods of peak Everglades Agricultural Area structure outflows frequently did not coincide with the periods of above normal rainfall, suggesting that the historical Everglades

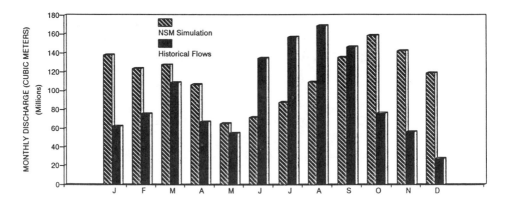

Figure 10.16 Comparison of natural system simulation and historical average monthly flows from the Everglades Agricultural Area into the northern Everglades.

Agricultural Area outflow pattern is a response to short-term pumping at the major water control structures. The short periods of negative flows represent a reversal of structure flows back into the Everglades Agricultural Area during the dry season, presumably for water supply.

Figure 10.16 is a plot of the average monthly Everglades Agricultural Area outflows for the 10-year period. The average monthly NSM36 outflows again show a similar pattern of steadily increasing flows throughout the wet season, with higher than expected flows in the early dry season. Again, the use of managed lake stages in NSM36 is a likely cause of the high dry season flows. The average monthly historical Everglades Agricultural Area outflows displayed a seasonal pattern of increasing outflows throughout the wet season, with the maximum outflows occurring during August.

Lower East Coast Inflows

The surface water patterns along the flow line that forms the intersection between the lower east coast and the Water Conservation Areas showed very different characteristics along its upper and lower reaches. For this reason, the flow line was divided into two segments. The northern reach of the lower east coast flow line includes the surface water flows from the Hillsboro Canal to the Miami Canal. The water control structures used in this analysis included S-39 in the Hillsboro Canal, S-38 in the C-14 canal, S-34 in the North New River Canal, and S-31 in the Miami Canal. The southern reach of the lower east coast flow line includes the surface water flows from the Miami Canal to the C-103 canal in southern Dade County. The water control structures used in this analysis included S-336 in the C-4 canal, S-338 in the C-1 canal, S-194 in the C-102 canal, and S-196 in the C-103 canal.

Figure 10.17 shows a comparison of the monthly total flows from the Water Conservation Areas into the northern portion of the lower east coast. The monthly historical structure flows into this portion of the lower east coast were again more erratic. The NSM36 simulation indicates that the pattern of lower east coast inflows was similar to the patterns observed at the upstream flow sections, with peak flows corresponding to periods of above normal rainfall. On average, the historical dry season outflows from the Water Conservation Areas into the northern portion of the lower east coast represent approximately 65% of the annual total flow, close to the 63% obtained from the NSM36 simulation. These historical surface water outflows are approximately 45% of the Natural System Model estimated flows, which represents an average annual flow reduction of approximately 295 million m^3 (240,000 acre-ft). Most of this difference is due to much higher NSM36 flows during an above normal rainfall period from late 1982 through early 1983.

Figure 10.18 shows that the NSM36 average monthly flows display a seasonal pattern, with inflows gradually increasing throughout the wet season and decreasing flows into the late dry season. The average monthly historical inflows display a pattern of decreasing flows throughout the wet season and peak flows during the late dry season. This suggests that wet season inflows to the lower east coast are controlled to limit the risk of flooding, while the inflows are increased during the dry season for water supply.

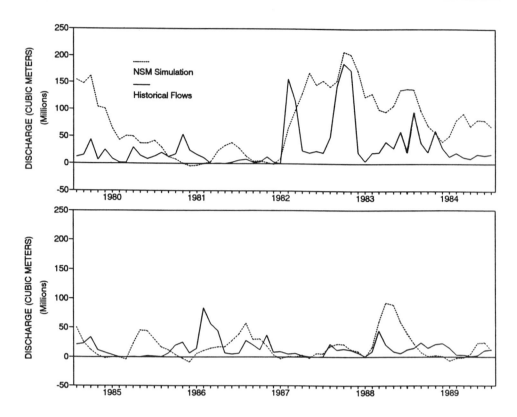

Figure 10.17 Comparison of natural system simulation and historical monthly flows from the Water Conservation Areas into the northern portion of the lower east coast.

Figure 10.19 shows a comparison of the monthly total flows from the Water Conservation Areas into the southern portion of the lower east coast flow line. The surface water flow patterns are markedly different in this area as a result of water management. The NSM36 simulation has the primary surface water flow direction from east to west (NSM36 flow convention assigns these flows a negative sign). This indicates that under predrainage conditions, surface water flowed from the western

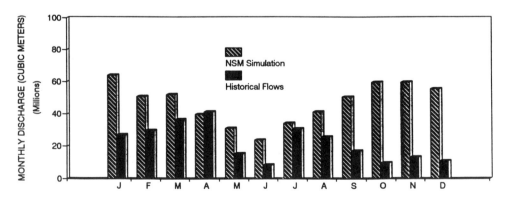

Figure 10.18 Comparison of natural system simulation and historical average monthly flows from the Water Conservation Areas into the northern portion of the lower east coast.

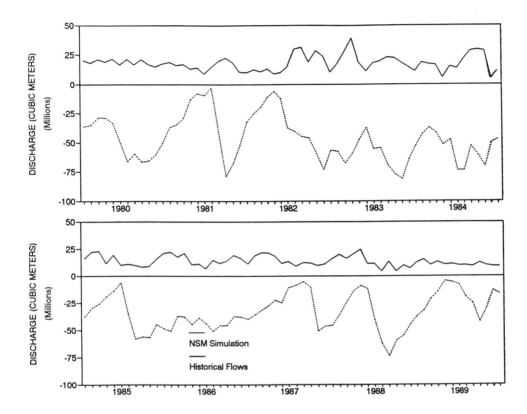

Figure 10.19 Comparison of natural system simulation and historical monthly flows from the Water Conservation Areas into the southern portion of the lower east coast.

portion of the coastal ridge westward into the Everglades. This flow pattern is a result of the topography, which shows a distinct depression oriented from the northeast to the southwest. The largest component of this east to west flow crosses the lower east coast flow line along the upper reach of the L-31N canal. This natural flow way is the headwater of the historical Shark River Slough basin, which originally captured surface flows from the northern wetlands and redirected them to the southwest, into Northeast Shark River Slough. After levee construction, this natural flow way was blocked, and the surface water east of these levees was captured by the adjacent canal system and routed south and east to the Atlantic. The estimated monthly NSM36 flows toward the west were more than twice the volume of the historical flows that now pass eastward to the Atlantic.

The average monthly NSM36 flows (Figure 10.20) persisted well into the dry season, supplying more uniform flows to the Shark River Slough basin. The seasonal fluctuation is smoother than the northern flow lines, probably due to the decline of the influence of Lake Okeechobee stages. On average, the historical dry season outflows from the Water Conservation Areas into the southern portion of the lower east coast represent 58% of the annual total flow, compared to 51% for the NSM36 simulation. The sum of the estimated NSM36 western inflows and the historical eastward diversions represents a reduction of annual flows to the downstream Everglades of approximately 665 million m^3 (540,000 acre-ft).

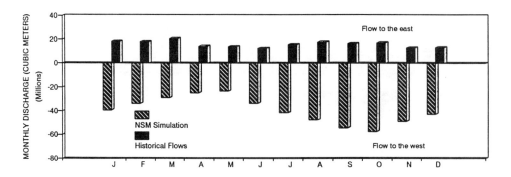

Figure 10.20   Comparison of natural system simulation and historical average monthly flows from the Water Conservation Areas into the southern portion of the lower east coast.

Shark River Slough Inflows

The last surface water flow line compares the estimated flows from the Water Conservation Areas into the Shark River Slough basin. The water control structures used for this analysis included the sum of the four S-12 structures (published as flows between L-67 and 40-Mile Bend) and the S-333 inflows minus S-334 outflows (published as flows between L-30 and L-67). At this location the monthly NSM36 simulations and the historical flows compare quite closely, particularly after mid-1985 (Figure 10.21). The major differences in flows are the result of sharp pulses of regulatory releases in response to high water levels in Water Conservation Area 3A. Also, in the early period (prior to mid-1983), there was strict adherence to a minimum delivery schedule during periods of normal or low water conditions in Water Conservation Area 3A. Since 1985, Shark River Slough inflows have been delivered according to a rainfall-based water delivery formula developed by the South Florida Water Management District (Neidrauer and Cooper, 1989).

The average monthly flows suggest that on average, both the NSM36 simulation and historical flows (Figure 10.22) produce a gradual increase in wet season flows and a smooth transition into the dry season. The historical inflows, however, tend to drop off much more rapidly following the end of wet season rainfall. On average, the historical dry season flows represent approximately 45% of the annual total flows, compared to 56% for the NSM36 simulation. Based on the 10-year comparison period, the historical inflows from the Water Conservation Areas into Shark River Slough are approximately 35% lower than the NSM36 estimates, which represents an estimated average annual flow reduction of approximately 340 million m³ (280,000 acre-ft). Most of this difference occurred during 1983, when much of the historical Water Conservation Area 3A outflow was diverted eastward to the Miami Canal via S-31 (see Figure 10.17).

Groundwater Flow Line Comparisons

A comparison was also made of the monthly groundwater flows between the NSM36 simulation and the results of flow estimates based on the South Florida Water Management Model for the same 10-year period. The most significant groundwater flows occur along the levee system separating the Water Conservation

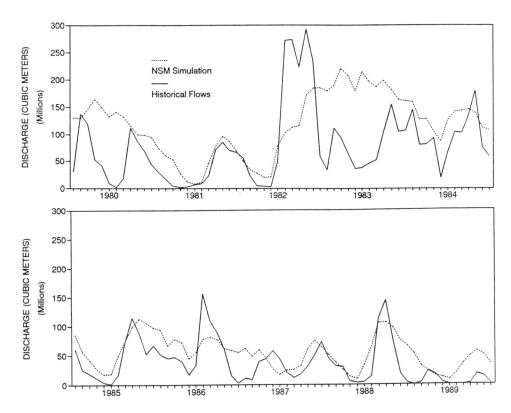

Figure 10.21   Comparison of natural system simulation and historical monthly flows from the Water Conservation Areas into Shark River Slough.

Areas and the lower east coast. For this reason, the groundwater flow comparisons will only be presented for this area. The same flow lines were used to perform this analysis as was described in the preceding section on surface water flows. For the Natural System Model simulation, there is a single subsurface flow component. The South Florida Water Management Model simulation computes the groundwater flows and the seepage losses through the levee system separately; however, they were combined for this analysis.

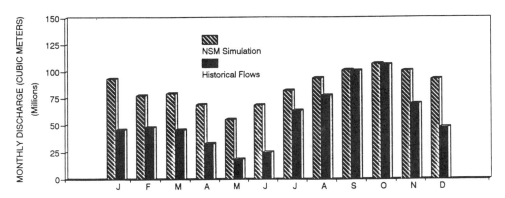

Figure 10.22   Comparison of natural system simulation and historical average monthly flows from the Water Conservation Areas into Shark River Slough.

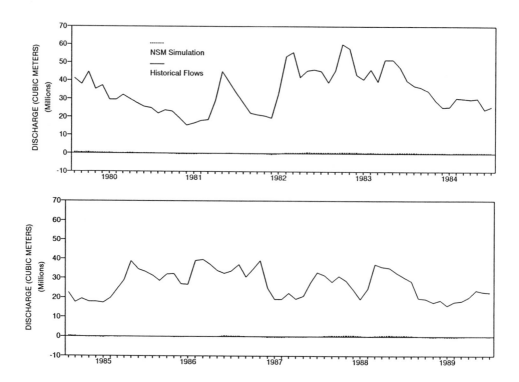

Figure 10.23 Comparison of the monthly groundwater flows from the Water Conservation Areas into the northern portion of the lower east coast based on the natural system and managed system simulations.

Lower East Coast Groundwater Flows

Figure 10.23 shows a comparison of the monthly total groundwater flows from the Water Conservation Areas into the northern portion of the lower east coast. This northern reach includes the groundwater flow/levee seepage along the L-36, L-35, and L-37/L-33 levees. The groundwater flows under the NSM36 simulation were

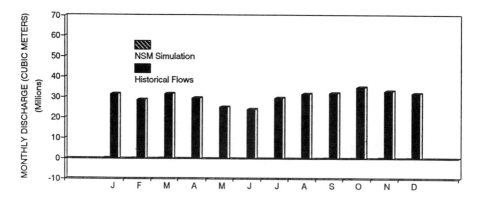

Figure 10.24 Comparison of the average monthly groundwater flows from the Water Conservation Areas into the northern portion of the lower east coast based on the natural system and managed system simulations.

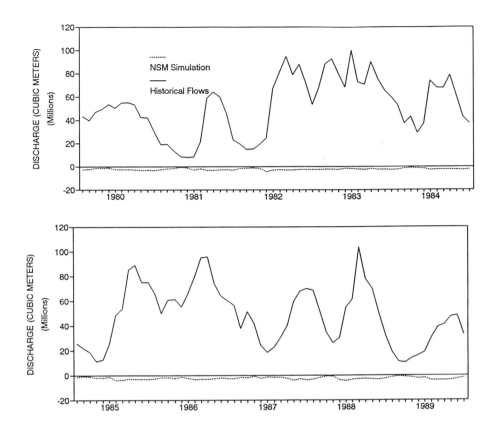

Figure 10.25 Comparison of the monthly groundwater flows from the Water Conservation Areas into the southern portion of the lower east coast based on the natural system and managed system simulations.

small throughout the study period, as compared to the South Florida Water Management Model which estimated large subsurface flows to the east. Maximum groundwater/levee seepage flows coincided with periods of high Water Conservation Area water levels resulting from above normal rainfall. Figure 10.24 shows that the average monthly groundwater flows peaked at the end of the wet season and gradually fell as stages dropped throughout the dry season. Results indicate that peak monthly groundwater flows frequently exceeded 40 million m^3, and the average annual groundwater flows exceeded 365 million m^3 (295,000 acre-ft).

Figure 10.25 shows the same comparison of the monthly total groundwater flows from the Water Conservation Areas into the southern portion of the lower east coast. This southern reach included groundwater flow/levee seepage along the L-30 and L-31N levees. The NSM36 simulation shows that the groundwater flows were again small, although the primary groundwater flow direction was clearly toward the west, or off the coastal ridge toward the Everglades. The South Florida Water Management Model simulation shows that the groundwater flow is extremely high and is from the Everglades to the lower east coast. Figure 10.26 again shows that the groundwater flows peak in the late wet season and gradually decrease into the dry season. Results show that peak monthly groundwater flows from the

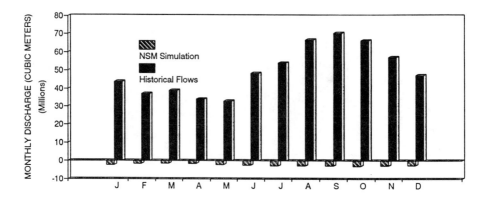

Figure 10.26 Comparison of the average monthly groundwater flows from the Water Conservation Areas into the southern portion of the lower east coast based on the natural system and managed system simulations.

managed system simulation frequently exceeded 80 million m^3, and the average annual groundwater flows were approximately 600 million m^3 (485,000 acre-ft). Note that these groundwater flows are only 13% lower than the historical average annual deliveries made to the Shark River Slough basin via the S-333 and S-12 structures.

Water Depth/Hydroperiod Comparisons at Selected Gauging Locations

In addition to the set of inundation depth and hydroperiod maps that were generated to examine the differences in spatial patterns between the natural and the managed system simulations, a comparison was also made of the temporal differences between the water depths and hydroperiods at nine sites selected to represent the major water management basins in the modeled area. In each case, the water levels estimated by the NSM36 simulation for a particular grid cell were compared with historical stages for key long-term monitoring stations located in the grid cell. All water depth and hydroperiod comparisons were based on analyses of monthly average values (computed from daily stage output) using the ground surface elevation for each grid cell, as represented in the simulation model.

Eastern Everglades Sites

Three sites were selected along the eastern edge of the historical Everglades watershed, in areas that have experienced a reduction of surface water inflows and/ or water level reductions resulting from canal drainage projects. All three sites are currently published by the U.S. Geological Survey as groundwater stations, although all experienced routine surface water flooding in their early records. The northernmost site (PB-109) is in northwest Palm Beach County adjacent to Loxahatchee Slough (see Figure 10.6). The central site (G-617) is in western Broward County adjacent to the North New River Canal. The southern site (G-596) is in western Dade County adjacent to the L-31N canal. Note that the NSM36

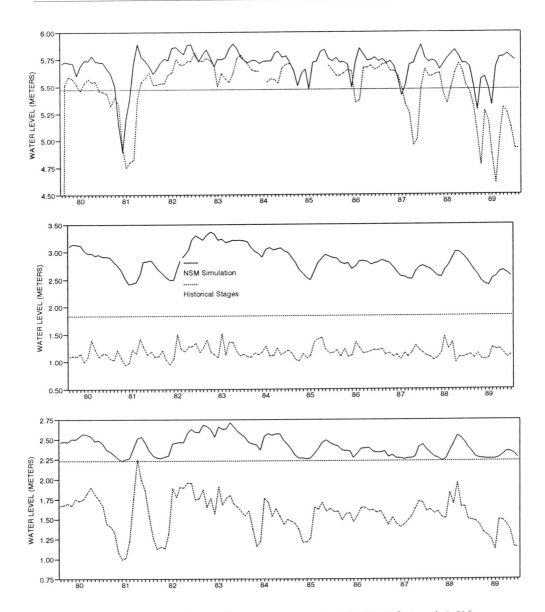

Figure 10.27   Comparison of monthly average stages for PB-109, G-617, and G-596.

probably overestimates water depths along the eastern edge of the Everglades, which is related to the poor topographic detail in these areas.

Figure 10.27 is a comparison of the monthly average stages for the three sites, based on the NSM36 simulation and published records for the 1980–89 study period. The NSM36 suggests that all three of these stations would have been inundated throughout much of the study period. Note that the NSM36 results for PB-109 and G-596 show that average water depths were rarely more than 0.15 m (0.50 ft) above the land surface, except during periods of high rainfall. Average water depths at G-617 were approximately 0.65 m (2.10 ft) above the land surface throughout the 1980–89 period. The NSM36 results also suggest that PB-109 and G-596 maintained 12-month hydroperiods during normal to wet years, but experienced drydowns during lower rainfall years (1981, 1985, and 1989). In contrast, historical stages at all of these sites are markedly lower. Historical average water depths are approximately 0.20, 1.65, and 0.85 m (0.65, 5.40, and 2.80 ft) lower than the NSM36 results for sites PB-109, G-617, and G-596, respectively. These results suggest that much of the eastern Everglades in Broward and Dade counties (as represented by sites G-617 and G-596) experienced a complete loss of surface water as a result of water management. In northern Palm Beach County (as represented by PB-109), the frequency of surface flooding was reduced by approximately 50%, and hydroperiods of 2 months or less occurred during all but six of the study years.

Water Conservation Area Sites

The next three sites were selected to examine the water depth and hydroperiod differences in the Water Conservation Areas. The northern site (1–7) is in the central portion of the Arthur R. Marshall Loxahatchee National Wildlife Refuge (Water Conservation Area 1). The central site (2–17) is in the central portion of Water Conservation Area 2A. The southern site (3A–4) is in the south-central portion of Water Conservation Area 3A (see Figure 10.6 for site locations). Figure 10.28 is a comparison of the monthly average stages for the three sites based on the NSM36 simulation and the published record. The Natural System Model results suggest that all three sites were completely inundated during all but the driest years. Average water depths at 2–17 and 3A–4 were approximately 0.30 m (1.00 ft) above the land surface. In contrast, average water depths at 1–7 were approximately 0.10 m (0.30 ft) above the land surface. This again points out a possible problem with the ground surface elevation data used in the NSM36, because several other sources of information suggest that Water Conservation Area 1 had deeper water than the downstream Water Conservation Areas.

Historical average water depths were 0.10–0.25 m (0.30–0.80 ft) lower than the NSM36 estimates at sites 1–7 and 2–17, respectively. In contrast, historical average water depths were approximately 0.15 m (0.50 ft) higher at site 3A–4. On average, historical hydroperiods were 2–4 months shorter at sites 1–7 and 2–17 than the NSM36 estimates, while historical hydroperiods at 3A–4 were slightly longer, particularly after 1985. These results suggest that the northern Everglades marshes in Water Conservation Areas 1 and 2A are drier (particularly in the dry season) than the Natural System Model estimates. This most likely is a result of the removal of surface water to meet the dry season water supply needs of the Everglades Agricultural Area and the lower east coast developed areas. The increased water depth and hydroperiod changes in Water Conservation Area 3A are most likely a result of the detention of dry season flows into the downstream basins.

Everglades National Park Sites

The last three sites are located within the Shark River Slough and Taylor Slough wetlands of Everglades National Park. P-33 is located in the western portion of the Shark River Slough watershed, downstream of the S-12 structures (see Figure 10.6).

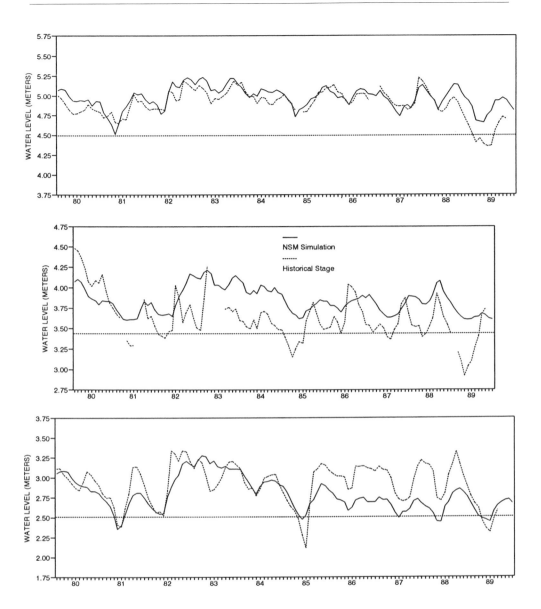

Figure 10.28 Comparison of monthly average stages for 1–7, 2–17, and 3–4.

G-618 is located in the central flow way of Shark River Slough just south of Tamiami Trail. P-37 is located in the downstream portion of the Taylor Slough watershed. Figure 10.29 is a comparison of the monthly average stage for the three sites, based on the NSM36 simulation and the historical record. The NSM36 simulation suggests that the two Shark River Slough sites would have been inundated during the entire 10-year period. In contrast, the stage pattern at P-37 indicates that the Taylor Slough watershed experienced a seasonal drydown. A review of the topography of the Everglades National Park area suggests that the P-33 and P-37 cells are located in slight depressions, which are approximately 0.1–0.2 m lower than the surrounding terrain. Average NSM36 water depths at sites P-33 and G-618 were approximately

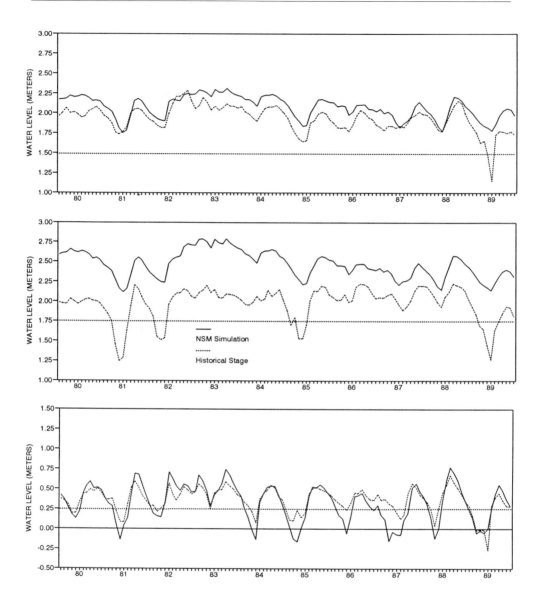

Figure 10.29 Comparison of monthly averages stages for P-33, G-618, and P-37.

0.60 m (2.00 ft) above the land surface. In contrast, the NSM36 water depths at P-37 were only approximately 0.20 m (0.65 ft) above the land surface.

Historical average water depths at sites P-33 and G-618 were 0.15–0.50 m (0.50–1.65 ft) lower, respectively, than the NSM36 results. The historical average hydroperiod for P-33 was only slightly shorter than the Natural System Model estimate, while the historical hydroperiod at G-618 was reduced by 4–5 months during the low rainfall years. These results suggest that the reduction in surface water inflows has had a significant impact on the Shark River Slough watershed, leading to lower water depths and substantially shorter periods of continuous inundation in the central slough. The historical hydroperiod at P-37 was approximately 1 month longer than the Natural System Model estimate as a result of slightly higher dry season flows, even though the average wet season stages dropped by approximately 0.05 m (0.15 ft). Note that the results of the Natural System Model simulation are less reliable for areas close to the coastal boundary of the model, such as P-37, because of problems associated with the boundary condition.

CONCLUSIONS

The development of the Natural System Model is an important first step toward establishing a tool which mimics natural and, eventually, predrainage hydrology, within the limitations of recorded history. The most important conclusions from this chapter are listed below in the same order in which the topics were presented.

Methods

NSM36 is one of the first steps toward estimating the hydrology of the natural system. Several gross assumptions were made for this preliminary model, including:

1. The simulation of natural Lake Okeechobee stages was crudely approximated by adding a fixed offset to the historical data. This approximation assumes that natural lake stages follow a temporal pattern similar to the historical/managed lake stages. A limitation of this approximation is that the effects of lake spills on lake stages are not simulated. This could produce an overestimate of the flows from the lake to the Everglades.
2. Flows through the transverse glades to the coastal areas were modeled by a crude channel system to allow evaluation of this feature.
3. Elevation data were only roughly adjusted to eliminate subsidence along the canals.
4. Rainfall and evaporation data from a recent time period (1965–89) were used.
5. Land use patterns were incorporated from fairly recent data.

Sensitivity

The sensitivity of the model to changes in pan evaporation, Manning's n, and lake stages was investigated and the following observations were made:

1. The model is very sensitive to evapotranspiration. Relatively small changes in evapotranspiration affect the stages and flows throughout the model domain.
2. NSM36 is far more sensitive to changes in Manning's n than the South Florida Water Management Model, as expected, because overland flow rather than canal flow is the principal surface water component in the NSM36. Decreases in n values affect the stages and flows more than corresponding increases.
3. Variation in lake stages affects the central portion of the model domain, but significant changes are observed as far south as Everglades National Park.

Results

1. The hydrologic response of the natural system shows that the natural Everglades was generally more of a flowing system with a greater spatial extent and longer periods of inundation than exists today under managed conditions.

2. Drainage efforts and the lowering of the water table in the lower east coast developed areas resulted in a diversion of significant flow volumes to tide and also caused a westward shifting of the largest groundwater gradients from the areas near the coastal ridge to the east coast protective levee system. Under managed conditions the high groundwater gradients and flows from the Everglades Water Conservation Areas and Everglades National Park are an important source of recharge to the Biscayne Aquifer, thereby providing a significant amount of water supply to the lower east coast developed areas.

3. The results of the Shark River Slough flow comparison, coupled with the major alterations at the southern lower east coast flow line, indicate that past water management changes (primarily the impoundment of water in Water Conservation Areas 3A and 3B and the diversions of surface water flows to the east) have had a major impact on flow reductions into the southern Everglades.

LITERATURE CITED

Chow, V. T. 1959. *Open Channel Hydraulics,* McGraw-Hill, New York.

Fennema, R. J. 1992. Natural System Model. An Evaluation of Version 3.4, Report to the National Park Service, Everglades National Park, Homestead, Fla.

Jones, L. A. (Ed.). 1948. Soils, Geology, and Water Control in the Everglades Region, Bull. 442, University of Florida Agricultural Experiment Station, Gainesville.

Knisel, W. G. (Ed.). 1980. CREAMS: A Field Scale Model for Chemicals, Runoff, and Erosion from Agricultural Management Systems, Conservation Res. Rep. No. 26, Science and Education Administration, U.S. Department of Agriculture.

Leach, S. D., H. Klein, and E. R. Hampton. 1972. Hydrologic Effects of Water Control and Management of Southeast Florida, Report of Investigation No. 60, Bureau of Geology, Florida Department of Natural Resources, Tallahassee, 115 pp.

MacVicar, T. K., T. Van Lent, and A. Castro. 1984. South Florida Water Management Model Documentation Report, Tech. Publ. 84-3, Resource Planning Department, South Florida Water Management District, West Palm Beach.

Neidrauer, C. J. 1992. Documentation of Modifications Performed to Establish Version 3.6 of the Natural System Model, Memorandum, Lower District Planning Division, South Florida Water Management District, West Palm Beach.

Neidrauer, C. J. and R. M. Cooper. 1989. A Two-Year Field Test of the Rainfall Plan—A Management Plan for Water Deliveries to Everglades National Park, Tech. Publ. 89-3, South Florida Water Management District, West Palm Beach.

Obeysekera, J. T. B. and M. K. Loftin. 1990. Hydrology of the Kissimmee River Basin—Influence of man-made and natural changes. in Kissimmee River Restoration Alternative Plan Evaluation and Preliminary Design Report, M. K. Loftin, L. A. Toth, and J. T. B. Obeysekera (Eds.), South Florida Water Management District, West Palm Beach.

Perkins, W. A. and T. K. MacVicar. 1991. A Computer Model to Simulate Natural South Florida Hydrology, Draft report, South Florida Water Management District, West Palm Beach.

South Florida Water Management District. 1993. Draft Working Document in Support of the Lower East Coast Regional Water Supply Plan, South Florida Water Management District, West Palm Beach.

Stewart, J. T. 1907. Report on Everglades Drainage Project in Lee and Dade Counties, Florida, Office of Experiment Stations, Irrigation and Drainage Investigations, United States Department of Agriculture, Washington, D.C., 110 pp.

11

FIRE PATTERNS in the SOUTHERN EVERGLADES

L. H. Gunderson

J. R. Snyder

ABSTRACT

Fire records from freshwater wetlands in Everglades National Park (1948–92) and Water Conservation Areas 2 and 3 (1980–90) were analyzed for temporal and spatial patterns. During the 45 years of record in Everglades National Park, 752 fires were registered, with sizes ranging up to ~75,000 ha; over the 11 years of record in the Water Conservation Areas, 127 fires burned, the largest of which was ~34,000 ha. Rank order pattern of fire sizes follows a log-normal distribution, with an anomalous clustering of fires ranging from 8000 to 15,000 ha. Fourier analyses of the two data sets revealed dominant cycles with frequencies of 7 months, 1 year, and 10–14 years. The annual and monthly frequencies occur at the same scale as seasonal variation in both drying patterns (rainfall and surface moisture) and in nonhuman ignition sources. The factor influencing the longer cycles appears to be climatic variation, although the causal mechanisms are unclear. Human-caused fires account for most fires and area burned, although more lightning fires have been recorded in recent years.

INTRODUCTION

This chapter describes the spatial and temporal patterns of fires in the Everglades ecosystem. The introduction begins with a brief review of the role of fire in the ecosystem, followed by descriptions of fire patterns and a review of the primary factors that influence them.

Role of Fire in the Everglades Ecosystem

In a review of south Florida fire history up to the middle of this century, Robertson (1953) pointed out the casual attitude held by the local populace toward the frequent and widespread fires. The prevailing attitude was that fire was a natural, or at least benign, part of the south Florida environment because of the rapid regrowth of the vegetation. A series of severe fires during the 1940s led to concerns that water management practices were producing disastrous results. This was summed up by Egler (1952) in what has become the most quoted statement on fire in the Everglades: "the herbaceous Everglades and the surrounding pinelands were born in fires...they can survive only with fires...they are dying today because of fires." Egler's bleak prophesy has not yet materialized, but has probably helped to focus attention on the need to understand the role of fire in the ecosystem.

The ecological effects of fire on the southern Everglades have been qualitatively described by Egler (1952), Robertson (1953), and Craighead (1971) and summarized most recently by Wade et al. (1980). The vegetation in general responds to fire by resprouting from below-ground parts, rapidly reaching prefire species composition and biomass. The major exception to this can occur in areas where there is an organic substrate (such as dense sawgrass strands or tree islands) during a severe drought. Here, if water levels are below the soil surface and the soil moisture is low enough to allow combustion, the substrate will burn, killing the vegetation and lowering the soil surface; this usually results in a conversion to a different type of vegetation type. In this manner dense sawgrass marshes that have heavy fuel loads can be converted into a spikerush marsh, which will not carry fire for many years (Craighead, 1971; Hoffman et al., 1994), and tree islands can turn into rock platforms sparsely covered by herbaceous plants. A major uncertainty still remains as to whether these severe conditions have occurred more frequently and become more extreme since drainage of the Everglades began (e.g., Robertson, 1953; Alexander and Crook, 1984) or whether extreme fire events have always occurred in the ecosystem. Even fires under less severe conditions can have significant effects, such as killing hardwoods and maintaining herbaceous dominance in marsh communities. The effects of fire are to some degree dependent on the time of year at which they occur (Robbins and Myers, 1992). For example, fire that occurs just before a rapid rise in water level can result in elimination of sawgrass without loss of substrate (Herndon et al., 1991).

Previous Work on Fire Patterns

The earliest works describing fire patterns in the Everglades were largely qualitative. Robertson (1953) was able to piece together records from newspaper reports and identify the "severe" fire years in the system; this was the first documentation of interannual variation in fire patterns. Other workers such as Egler (1952), Craighead (1971), and Alexander and Crook (1984) inferred return periods and sizes of fires from vegetation patterns. The advent of public land management agencies in the late 1940s necessitated keeping fire records, primarily to track control costs.

Taylor (1981) compiled one of the first quantitative assessments of fire patterns in the Everglades by examining 30 years of fire records from Everglades National Park. He quantified the range of fire sizes, calculated fire return intervals, and attempted correlate fire patterns and hydrologic variables. Taylor (1981) estimated an interval of 5.8–7.5 years between severe fire years, defined as total area burn greater than 8094 ha (20,000 acres). A shorter interval (3.3 years) was estimated between moderate fire years, defined as total area burned between 4000 and 8000 ha (10,000 to 20,000 acres). Little correlation was found between fire measures (total area burned and number of fires) and hydrologic measures of rain or water level (Taylor, 1981; Snyder, 1991).

Factors Influencing Fire Patterns

The factors influencing space and time patterns of fire in the landscape can be grouped into three classes: (1) processes involved with ignition, (2) biotic variables that influence fuel type and structure, and (3) physical variables such as wind and moisture regime that affect the patterns of spread.

Fire ignition comes primarily from two sources: lightning and humans. Lightning is generally associated with convective thunderstorms, most common during the summer months. Southern Florida is well known for thunderstorm activity and averages about 90 thunderstorm days per year (Chen and Gerber, 1990). People light fires for a variety of reasons, including criminal intent (arson) and habitat improvement. Land management agencies prescribe fire to reduce the impact of wildfire and protect specific resources such as hammocks and tree islands (Bancroft, 1976; Taylor, 1980). Both humans (Carr and Beriault, 1984) and fires (Cohen 1984) have been present in the ecosystem since it formed thousands of years ago; however, the magnitude of human influence on fire patterns continues to be a topic of debate (Egler, 1952; Robertson, 1953; Snyder, 1991).

Vegetation characteristics influence both spatial and temporal aspects of fires. The regrowth rate and fuel properties of the vegetation structure can influence fire frequency. For example, the biomass in the wet prairie association reaches a steady-state fuel load approximately 3 years postburn (Herndon and Taylor, 1986), and stands of sawgrass can recover to preburn height and biomass within 2 years (Loveless, 1959; Forthman, 1973; Yates, 1974) yet continue to accumulate dead material for years. Nevertheless, the interaction between vegetation and fire return interval is not restricted to a single stable state, and indeed multiple stable states have been observed. As noted above, peat-consuming fires can transform a site to a very different vegetation type, with a very different fuel structure and fire return period. The spatial patterns of fire are related in part to the distribution of plant associations. Fires that burn under moist conditions are constrained by contiguous strands of sawgrass and may not consume wet prairie or tree island vegetation types (Wade et al., 1980).

The physical variables that influence fire spread include surface water and soil moisture conditions, wind direction, wind speed, and humidity, as shown by the prescription conditions outlined for different management actions (Everglades National Park, 1991; Wade et al., 1980).

The preceding arguments indicate that the factors influencing fire patterns in the landscape occur at different space and time scales. For example, lightning strikes vary annually, associated with thunderstorms. Vegetation regrowth rates can influence the number of years between fires. Longer term climatic variation in rainfall and drying conditions influence the severity of the impact of fire. These examples of factors operating at different time scales all interact to influence temporal patterns of fires. Likewise, cross-scale spatial factors such as topography, vegetation, and other physical factors all influence the spatial patterns of fires.

The cross-scale interactions of these important factors shape the spatial and temporal patterns of fires in the landscape. These processes interact to create fire patterns that are either random or distinguishable. If patterns are present, a derivation of emerging cross-scale theory (Holling, 1992; Gunderson, 1992) would postulate that they are entrained by a small number of variables (such as ignition sources, vegetation characteristics, and moisture regimes). This entrainment by variables that operate over a range of scales should result in fire patterns that have a few dominant cycles over time and a discontinuous distribution of fire sizes. The sources of data and methods used to determine whether the fire records of the southern Everglades reveal such patterns are the subject of the next section.

METHODS

Data Sources

Long-term data on fire patterns were obtained from published reports and unpublished records from various land management agencies. Fire data from Everglades National Park for 1948–82 were reported by Taylor (1981) and Doren and Rochefort (1984). Data for the period 1982–92 were obtained from records kept by the fire management staff at Everglades National Park. Fire data from Water Conservation Areas 2 and 3 during the period 1980–90 were provided by the Florida Game and Fresh Water Fish Commission. Because the data cover different lengths of time and areas, they are referred to separately as Water Conservation Area and Everglades National Park (Figure 11.1).

Dates of burn, location of burn, size of the fire, and source of ignition were recorded for each fire within the sample regions. The date of ignition was used to determine the month of burn for the time series analysis. For the few fires that burned longer than one month, this represents a source of error. The locations of the fires were resolved to coordinates of township and range, as early reports listed locations in these units; locations of fires reported as latitude and longitude were translated to corresponding township and range. For fires that burned areas larger than a township, the initial location listed in the fire report was used. All fires from Long Pine Key and mangrove areas were removed from the data set, so that the data set reflected patterns from the freshwater wetland complex of Everglades National Park and the Water Conservation Areas. However, the criteria were based on township and range location, resulting in the irregular-shaped sample area (Figure 11.1) which may include some fires in the pineland and mangroves and exclude some wetlands complex. Three categories of causes of ignition were recorded: human (including incendiary or arson fires), lightning, and prescribed.

Temporal Patterns

Total area burned and number of fires were summarized for each year and each month over the period of record for both Everglades National Park and the Water Conservation Area. These summaries provide a graphical depiction of fire patterns over time. Time series data on fire size were also analyzed using Fourier techniques to detect cyclical patterns. The Fourier analysis was done on two data sets: a 512-

Figure 11.1 Areas of Everglades ecosystem used in the analysis of fire patterns. Two sample areas were used: the Water Conservation Areas and most of the freshwater wetland complex of Everglades National Park.

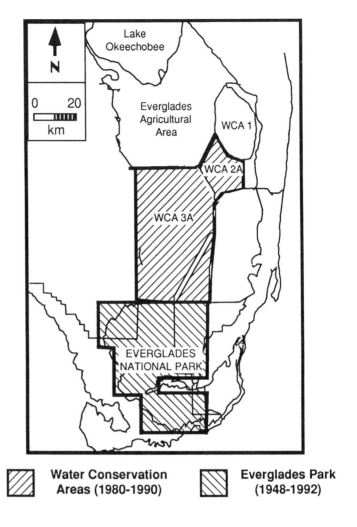

Water Conservation Areas (1980-1990)   Everglades Park (1948-1992)

month record from Everglades National Park and a 128-month record from the Water Conservation Area. A monthly time series of fire size (expressed as a logarithm to base 10 of total area burned) was then used in the Fourier analysis. For months with no fires, a value of zero was assigned.

The analyses were done with the fast Fourier algorithm in the SYSTAT software for the Macintosh microcomputer. The technique decomposes the time series into a finite set of sine waves and then partitions the total sample variance to each of the frequencies (Platt and Denman, 1975). The fast Fourier technique is a modification which utilizes data sets with windows that are 2^n units (hence the 128- and 512-month records). For each data set, the mean was subtracted from every value and the data detrended, so that the values varied above and below zero, with no overall change or trend in the mean.

Spatial Patterns

Because many factors influence the size of fires, the distribution of fire sizes may be multimodal, in which case statistics such as mean fire size or variance may not be meaningful. A number of techniques were used to investigate the distribution of fire sizes. A rank order distribution of the observed fire sizes was used, rather than frequency histograms, so that the observed data generate the pattern rather than the patterns being determined by a subjectively chosen category size. One method of searching for multiple modes in the rank order data used a statistic that looks for bumps or "stair-steps" in the distribution (Silverman, 1986). Another statistic used was the "gap detector" or relative difference between succeeding ranked observations (Holling, 1992). The final approach was to compare the observed data set with standard distributions such as normal and log-normal; if the observed distribution was unimodal, then it should fit one of these theoretical distributions.

RESULTS

Annual and Seasonal Burn Patterns

The annual totals of area burned and number of fires indicate interannual (year-to-year) cycles of fires in Everglades National Park and the Water Conservation Area sites. The pattern can be characterized by large areas burned during some

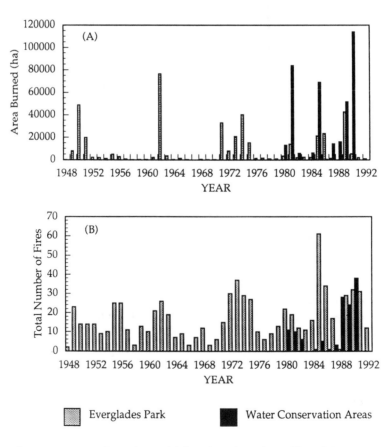

Figure 11.2   Total area burned (A) and total number of fires (B) per year over the time period 1948–92 (Everglades National Park) and 1980–90 (Water Conservation Areas).

years, followed by a series of years when a much smaller area burned. The total number of fires followed a similar pattern, as shown by a positive significant correlation between annual log area burned and annual number of fires. In Everglades National Park, large areas (defined as greater than the annual geometric mean of 93 km^2) burned in 1950, 1951, 1962, 1971, 1973, 1974, 1975, 1981, 1985, 1986, and 1989 (Figure 11.2). In Everglades National Park, the most severe fire year was 1962, when about 764 km^2 burned. In the Water Conservation Area (and in Everglades National Park), large areas burned in 1981, 1985, 1989, and 1990 (Figure 11.2).

While the interannual patterns were similar in both Everglades National Park and the Water Conservation Areas, the seasonal (month-to-month) patterns of fires in the two areas were different. During the 44 years of record in the park, the largest areas burned were ignited during the late dry season months of April and May (Figure 11.3). The total number of fires in Everglades National Park followed a sinusoidal pattern over the months, with peaks in May and July. The highest number of fires and area burned in the Water Conservation Areas occurred earlier in the dry season, during December and January (Figure 11.3). Most of this was

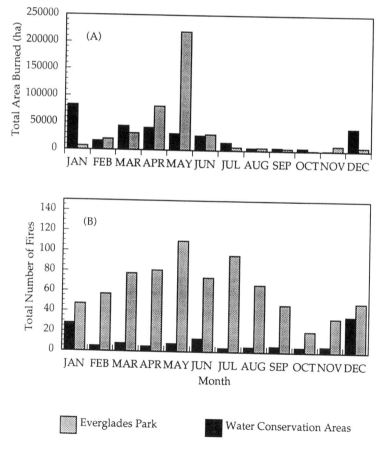

Figure 11.3   Total area burned (A) and total number of fires (B) recorded by month in the southern Everglades.

Figure 11.4 Area burned and number of fires, by ignition source, 1948–92, Everglades National Park.

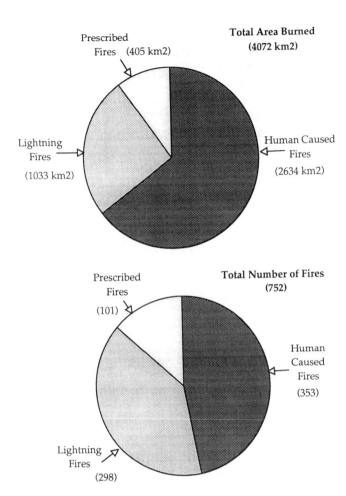

prescribed burning of sawgrass to protect tree islands for deer habitat (S. Coughlin, personal communication).

Some of the interannual patterns can be assigned to the source of ignition of fires. Fires caused by humans account for most of the fires and area burned in Everglades National Park (Figure 11.4). The total area burned and number of fires assigned to either lightning or prescribed fires has dramatically increased since the mid-1970s, so that the reported ignitions prior to this time may reflect a tendency to assign cause to humans rather than other sources. There was a reluctance to attribute fires to lightning before 1951, when observations from two fire towers confirmed numerous fires from this source (Robertson, 1953). The number of lightning fires and area burned by them have apparently increased since 1972, probably due to increased surveillance (Taylor, 1981). Substantial prescribed burning in sawgrass areas began in 1971 (Klukas, 1973), but this has not accounted for much of the area burned in Everglades National Park since that time. It appears that fewer human-caused wildfires have burned less area in Everglades National Park since 1975.

Human-caused fires were ignited from December through May, with the number of fires peaking in April and area burned peaking in May (Figure 11.5).

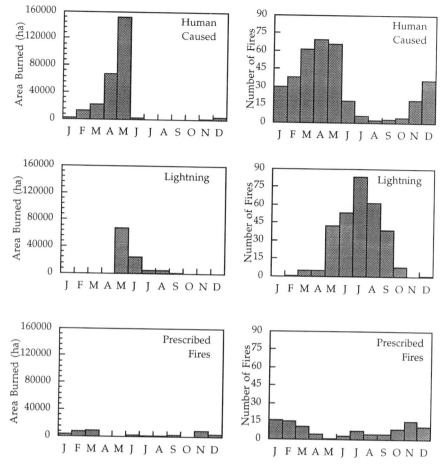

Figure 11.5 Monthly time course of total area burned and total number of fires by ignition source. Data are from Everglades National Park, 1948–92.

The number of lightning fires follows the seasonal thunderstorm pattern, with a peak in July. Because the ecosystem generally becomes wet during June, the largest area burned results from lightning ignitions in May and declines through the summer months. Most prescribed fires are ignited during October through March.

Spectral Analyses

Fourier analyses indicate the presence of at least three cycles in the monthly fire data sets from both Everglades National Park and the Water Conservation Areas. These cycles represent return frequencies with a period of 6–7 months, 1 year, and 10–14 years. One way of interpreting this analysis is to consider the time series plot of the log-transformed fire sizes (Figures 11.6 and 11.7) to be a composite of three sine waves, with frequencies of 6–7 months, 1 year, and 10–14 years; each cycle adds to the composite signal. The dominant (accounting for most of the variation) cycle is the annual one, followed by the 10- to 14-year cycle, followed

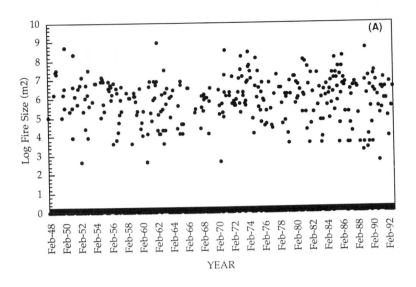

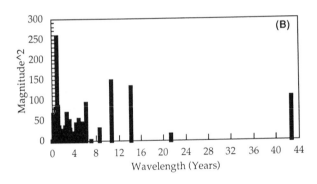

Figure 11.6 (A) Time course of log-transformed area burned by month, for 45 years of record in Everglades National Park. (B) Spectral plot of same data set. Height of bars indicates magnitude of component frequency.

by the 6- to 7-month frequency. The long-term data from the park suggest two other periods: 40+ years and a cycle of about 6 years. These cycles are not exact and are constrained by the frequency of the data and the number of data points. However, a number of pieces of evidence indicate that the frequencies detected with this method may indeed reflect composite processes. For example, the interannual data indicated that large areas burn approximately every 10–15 years. Taylor (1981) reports return frequencies of 6–7 years for severe fires (>8000 ha). Furthermore, the summaries by month (Figure 11.3) indicate the presence of an annual cycle.

Fire Sizes

During the period from 1948 through 1992, 752 fires were registered in the freshwater wetlands sample area, including Everglades National Park. These fires

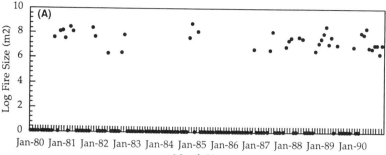

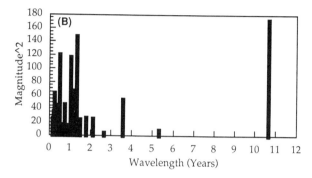

Figure 11.7   (A) Time course of log-transformed area burned by month, for 10 years of record in Water Conservation Areas. (B) Spectral plot of same data set. Height of bars in spectral plot indicates magnitude of component frequency.

ranged in size from 4 m^2 to 74,860 ha, with a geometric mean fire size of 551 ha. The largest fire recorded is close to the size of the largest patches of sawgrass in this mosaic (Gunderson, 1992). During the time period from 1980 through 1990, 127 fires were recorded in the Water Conservation Areas. These fires ranged from 0.81 to 34,200 ha, with a geometric mean of 2420 ha. These statistics provide an estimate of the range of fire sizes and perhaps indicate a central tendency. The next analysis will examine the characteristics of the distribution of fire sizes. The rank order of fire sizes roughly followed a log-normal distribution. Deviations from the log-normal are apparent at the upper and lower tails of the distribution as "bumps" or "stair-steps" in the rank order plot (Figure 11.8). These deviations were also noted in quantile plots of the log-transformed data and in the extreme values of the difference index (Figure 11.8). The bumps at the lower end of the distribution all occur at fire sizes less than 1 ha and probably reflect the rounding off associated with estimating sizes of small fires. That is, many small fires were estimated at sizes of 1 or 2 acres, resulting in many observations of this size. The bumps at the upper end of the distribution reflect a clustering of fires of between 8000 and 15,000 ha, more than would be expected if the fire sizes follow a strictly log-normal relationship.

Figure 11.8 Rank order plot (A) and difference index as a function of fire size (B) for Everglades National Park data.

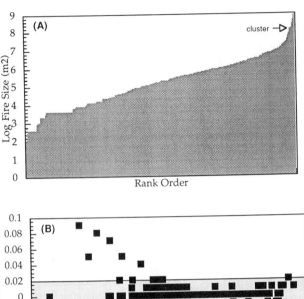

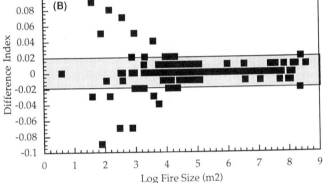

DISCUSSION

The time series analysis indicates the presence of at least three dominant cycles that correspond to documented phenomena related to fluctuations in moisture conditions. The annual cycle is on the same frequency as the wet and dry seasons of southern Florida. Most of the rain occurs during the months of June through September in association with convective thunderstorms. During the dry season, rainfall is mainly associated with the passage of cold fronts. The increase in insolation and air temperatures, coupled with the lack of rain, creates the "driest" conditions in April and May. These are the months with the largest areas burned in Everglades National Park. The largest burns in the Water Conservation Areas have occurred during the months of December and January, earlier in the dry season. The longer term cycle (some 10–14 years long) occurs at similar time ranges as other environmental variables in the system; water levels, water flows, and pan evaporation all fluctuate on cycles of similar time frames (Gunderson, 1992). These variables indicate longer term fluctuations in wet or dry conditions and, hence, susceptibility to burning. The shortest term cycle (6–7 months) is intriguing. Analyses of rainfall data indicate a similar periodicity (Gunderson, 1992), so that this cycle may reflect a shorter term fluctuation in moisture/dryness conditions. Even though there is some evidence for multiple cycles of moisture and drought regimes, other factors influence the temporal patterns of fires.

Ignition sources exhibit seasonal patterns, while longer term cycles are less well documented. The number of lightning fires show seasonal patterns (Figure 11.5), similar to the seasonal pattern of thunderstorm days (Chen and Gerber, 1990; Davis and Sakamoto, 1976). Longer term lightning cycles are not well documented or understood, but may be related to sunspot activity (H. T. Odum, personal communication). Human-caused fires exhibit a seasonal pattern, yet prescribed fires do not (Figure 11.5).

Fire as a Landscape Process

One way of summarizing the spatial and temporal domain of fire is to map the observed results in a space–time plot. The spatial domain of observed fires can be estimated by the smallest and largest fires (4 m² and 74,000 ha) during the period of record. The square roots of the smallest and largest observed fire sizes result in linear dimensions of 2 m and 8.6 km and define the spatial bounds in Figure 11.9. The temporal domain is established by estimates of the fire cycles from Fourier analyses. For the Everglades National Park data set, these bounds are 6 months and

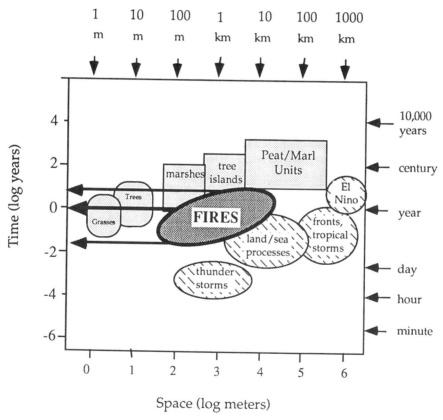

Figure 11.9 Domain of Everglades fires in a space–time plot. Spatial bounds of the ellipse were determined by smallest and largest recorded fire over 45-year period of record. Temporal bounds were determined by dominant cycles from Fourier analyses. Vegetation and atmospheric hierarchies from Gunderson (1992).

a little more than a decade. Fire is a landscape-level process, operating over a broad spatial domain and on multiple time scales, resulting from the interaction of factors that occur at multiple scales. Fire occupies an intermediate position in space or time dynamics between the vegetation hierarchy and the atmosphere. The structure and dynamics of vegetation occur at "slower" scales over broader regions and are influenced by the "faster" dynamics of fire. Atmospheric dynamics (wind, rain, drying conditions, etc.) generally occur over broader areas and much more rapidly.

SUMMARY

Examination of the multiple decades of fire records indicates at least three cycle periods: seasonal, annual, and decadal. The cycles indicate that fires occur at multiple time scales, at least one order of magnitude between the annual and longer term period. Other physical factors (water level, rainfall, evaporation, water flow) exhibit cycling at similar frequencies. These analyses also suggest an interaction between spatial and temporal patterns: large areas burn at approximately decadal cycles, while smaller fires occur more frequently. Recognition of the temporal cycles and the associated spatial extent of fires will hopefully aid in the efforts to restore the remnant Everglades ecosystem.

ACKNOWLEDGMENTS

The authors gratefully acknowledge the assistance of the key people who provided data, provided access to data, or created computer files within a very short period of time. Dave Lentz of Everglades National Park was most helpful in providing access to fire records for the park for the last decade. Steven P. Coughlin provided data for the Water Conservation Areas from the files of the Florida Game and Fresh Water Fish Commission. Special thanks go to Holly Belles, Bev Gunderson, and Keiley Pilotto for their work in entering and correcting the computerized data. Jack Ewel and Bill Platt provided useful comments on earlier drafts of the manuscript.

LITERATURE CITED

Alexander, T. R. and A. G. Crook. 1984. Recent vegetational changes in southern Florida. in *Environments of South Florida: Present and Past, Memoir 2*, P. J. Gleason (Ed.), Miami Geological Society, Coral Gables, Fla., pp. 199–210.

Bancroft, L. 1976. Fire management in Everglades National Park. *Fire Manage.*, 37(1):18–21.

Carr, R. S. and Beriault. 1984. Prehistoric man in southern Florida. in *Environments of Florida: Present and Past, Memoir 2*, P. J. Gleason (Ed.), Miami Geological Society, Coral Gables, Fla.

Chen, E. and J. F. Gerber. 1990. Climate. in *Ecosystems of Florida*, R. L. Myers and J. J. Ewel (Eds.), University Presses of Florida, Gainesville.

Cohen, A. D. 1984. Evidence of fires in the ancient Everglades and coastal swamps of southern Florida. in *Environments of South Florida: Present and Past, Memoir 2*, P. J. Gleason (Ed.), Miami Geological Society, Coral Gables, Fla.

Craighead, F. C., Sr. 1971. *The Trees of South Florida,* Vol. I: The Natural Environments and Their Succession, University of Miami Press, Coral Gables, Fla.

Davis, J. H., Jr. 1943. The Natural Features of Southern Florida, Especially the Vegetation, and the Everglades, Florida Geological Survey, Tallahassee.

Davis, J. M. and S. M. Sakamoto. 1976. An atlas and tables of thunderstorm and hail day probabilities in the southeastern United States. *Auburn Univ. Agric. Exp. Stn. Bull.,* No. 477.

Doren, R. F. and R. M. Rochefort. 1984. Summary of Fires in Everglades National Park and Big Cypress National Preserve, 1981, Report SFRC-84/01, South Florida Research Center, National Park Service, U.S. Department of the Interior, Homestead, Fla.

Egler, F. E. 1952. Southeast saline Everglades vegetation, Florida, and its management. *Vegetatio,* 3(4–5):213–265.

Everglades National Park Staff. 1991. Fire Management Plan, Everglades National Park, Homestead, Fla.

Forthman, C. A. 1973. The Effects of Prescribed Burning on Sawgrass, *Cladium jamaicense* Crantz, in South Florida, M.S. thesis, University of Miami, Coral Gables, Fla.

Gunderson, L. H. 1992. Spatial and Temporal Hierarchies in the Everglades Ecosystem, with Implications to Water Deliveries to Everglades National Park, Ph.D. dissertation, University of Florida, Gainesville.

Herndon, A. K. and D. Taylor. 1986. Response of a *Muhlenbergia* Prairie to Repeated Burning: Changes in Above-Ground Biomass, South Florida Research Center, National Park Service, U.S. Department of the Interior, Homestead, Fla.

Herndon, A. K., L. H. Gunderson, and J. R. Stenberg. 1991. Sawgrass, *Cladium jamaicense,* survival in a regime of fire and flooding. *Wetlands,* 11:17–27.

Hoffman, W., G. T. Bancroft, and R. J. Sawicki. 1994. Foraging habitat of wading birds in the Water Conservation Areas of the Everglades. in *Everglades: The Ecosystem and Its Restoration,* S. M. Davis and J. C. Ogden (Eds.), St. Lucie Press, Delray Beach, Fla., chap. 24.

Hofstetter, R. H. 1976. Effects of Fire in the Ecosystem. Part I. Appendix K in South Florida Environmental Project, University of Miami, Coral Gables, Fla.

Holling, C. S. 1992. Cross-scale morphology, geometry and dynamics of ecosystems. *Ecol. Monogr.,* 62(4):447–502.

Klukas, R. W. 1973. Control burn activities in Everglades National Park [Florida]. *Proc. Annu. Tall Timber Fire Ecol. Conf.,* 12:397–425.

Loveless, C. M. 1959. A study of the vegetation of the Florida Everglades. *Ecology,* 40(1):1–9.

Platt, T. and K. L. Denman. 1975. Spectral analysis in ecology. *Annu. Rev. Ecol. Syst.,* 6:189–210.

Robbins, L. E. and R. L. Myers. 1992. Seasonal Effects of Prescribed Burning in Florida: A Review, Misc. Publ. No. 8, Tall Timbers Research, Inc., Tallahassee, Fla., 96 pp.

Robertson, W. B., Jr. 1953. A Survey of the Effects of Fire in Everglades National Park, Everglades National Park, National Park Service, U.S. Department of the Interior, Homestead, Fla.

Silverman, B. W. 1986. in *Density Estimation for Statistics and Data Analysis. Monographs on Statistics and Applied Probability,* D. R. Cox, D. V. Hinckley, D. Rubin, and B. W. Silverman (Eds.), Chapman and Hall, London.

Snyder, J. R. 1991. Fire regimes in subtropical south Florida. *Proc. 17th Tall Timbers Fire Ecol. Conf.,* 17:303–319.

Taylor, D. L. 1980. Summary of Fires in Everglades National Park and Big Cypress National Preserve, 1979, Report T-595, South Florida Research Center, National Park Service, U.S. Department of the Interior, Homestead, Fla.

Taylor, D. L. 1981. Fire History and Fire Records for Everglades National Park, 1948–1979, Report T-619, South Florida Research Center, National Park Service, U.S. Department of the Interior, Homestead, Fla.

Wade, D. D., J. J. Ewel, and R. H. Hofstetter. 1980. Fire in South Florida Ecosystems, General Technical Report SE-17, SE Forest Experimental Station, Forest Service, U.S. Department of Agriculture, Asheville, N.C., 125 pp.

Yates, S. A. 1974. An Autecological Study of Sawgrass, *Cladium jamaicense,* in Southern Florida, Masters thesis, University of Miami, Coral Gables, Fla., 93 pp.

12

SYNTHESIS: SPATIAL and TEMPORAL CHARACTERISTICS of the ENVIRONMENT

Donald L. DeAngelis

ABSTRACT

The chapters in this section described the abiotic driving forces that shape the Everglades. In this synthesis, the driving forces are classified into three general types: gradually changing characteristics (e.g., sea level and climate), disturbances (e.g., storms, fires, freezes, floods, and droughts), and natural periodicities (e.g., annual hydropattern). Every structural aspect of the Everglades, from tree islands on the microscale to the shape of the shoreline and the major vegetational regions on the macroscale, is the product of processes controlled by these driving forces. The effects of the driving forces vary across the Everglades landscape and contribute to the heterogeneity of this landscape. It is this landscape heterogeneity that provides the diversity of habitats and environmental conditions that make a diversity of biota possible.

INTRODUCTION

The objective of the chapters in this section has been to describe the abiotic (geologic, climatic, and hydrologic) driving forces acting in south Florida to form the Everglades ecosystem. By emphasizing the variation in the temporal and spatial scales at which these forces influence the Everglades landscape, the volume has taken a significant step beyond the snapshot-in-time view typical of such percep-

© St. Lucie Press CCC 0-9634030-2-8 1/94/$100/$.50

tions of ecosystems. The picture of the Everglades that emerges is of an ecosystem that has been shaped by temporally and spatially varying abiotic forces and which may continue to change as those forces change over time. The hope for successful restoration of the Everglades rests on the understanding of both natural and human-induced changes.

In the theoretical overview in Chapter 2 by DeAngelis and White (1994), variations in the abiotic driving forces were divided into three general classes: gradual changes in the environment, temporally discrete events or disturbances, and natural periodicities. First, the understanding of conditions and driving forces that shaped the Everglades as presented in this section will be summarized. Next, synthesis of the chapters will be addressed in terms of the classification of abiotic driving forces and the interactions among these forces. Finally, findings on spatial variability presented in this section, as well as the implications of these findings for the Everglades ecosystem, will be discussed.

DRIVING FORCES SHAPING the EVERGLADES

The development of the Everglades ecosystem illustrates the actions of abiotic forces driving ecosystem processes on a variety of temporal scales, which DeAngelis and White (1994) described as a general feature of ecosystems. As outlined by Gleason and Stone (1994), geological processes of alternating erosion and deposition have acted, particularly during the past five million years or so, to form much of the bedrock topography that includes the Everglades trough. The ridge bounding the Everglades to the east originated during one of the Pleistocene interglacials (the Sangomon period, 128,000–67,000 years before present [YBP]). Everglades formation began about five millennia ago when rising sea level, together with the basic topography, retarded water runoff sufficiently to produce a freshwater wetland.

On a geological time scale, the Everglades ecosystem is relatively young, and on this time scale the bedrock topography can be considered approximately unchanging. Yet during the about 5000 years of existence of the Everglades, the forces of climate and sea level change have continued to shape and reshape many of its basic physical structures or features (the shoreline, the soil and subsoil, the spatial pattern of water flow, tree islands, and nutrient trophic status) by altering the rates of the main processes that control these structures (see Chapter 2, Table 2.1).

The Everglades, like other ecosystems, is partly a product of climatic driving forces. Also, like other ecosystems, it derives many of its important unique features from topography and other aspects of local geography. According to the Holdridge diagram (see Chapter 2, Figure 2.2), which takes into account only average annual climatic factors, a region characterized by the average yearly temperature (about 20°C) and average yearly precipitation (about 1175–1550 mm) of the Everglades should be occupied by moist forest. Other factors must play a role in creating the conditions leading to the formation of the Everglades. In Chapter 7, Gleason and Stone (1994) point out that among the most important of these are topographic features and the seasonality of rainfall. Principal topographic features include the bedrock depression underlying the Everglades and ridges to the east that limit the potential drainage of water into the Atlantic, causing the water table to rise above the surface for part of the year. The primary feature of seasonality is the distinct occurrence of rainy and dry seasons.

Every structure in a natural landscape is the result of processes caused by abiotic and biotic forces. Often there are two different processes in opposition or "competition." The deposition of peat and marl soils under conditions of inundation and, alternatively, biotic oxidation under dry conditions are important, competing geological processes, which result in either a net build-up or a net degradation of organic matter. For example, near Belle Glade peat deposition in the Everglades has taken place at about 16 cm per century during the last 1200 years (Gleason and Stone, 1994). The reverse process of oxidation can occur at about 30 cm per decade under favorable conditions. This asymmetry in the potential rates of these processes of soil build-up and reduction makes the Everglades ecosystem highly unstable with respect to disturbances in the hydrologic cycle, as noted in Chapter 3 by Maltby and Dugan (1994).

Despite these continual changes, the bedrock topography has continued to exert a strong influence on vegetation through its influence on the patterns of water depth and flow. The lower areas, such as the Loxahatchee Channel, Tamiami Basin, and Shark River Bedrock, are occupied by water lily sloughs and tree islands. Higher areas are overlain by sawgrass and sawgrass mixed with wax myrtle thickets (Gleason and Stone, 1994). An interesting point made by Gleason and Stone is that, despite peat deposition, areas of low bedrock have tended to remain habitats of aquatic plants. This differs from expectations based on the classical view of autogenic succession that low, wet areas should eventually fill in and become terrestrial habitats. The competing allogenic process of organic matter degradation may prevent this classical succession from occurring.

A further effect of the particular type of bedrock is that the high calcium of the marine and freshwater limestone underlying the Everglades favors periphyton growth, a key component at the base of the Everglades food web (Gleason and Stone, 1994; Browder et al., 1994).

The seasonal hydrologic pattern in the Everglades is one of the key factors causing its uniqueness and divergence from the Holdridge diagram prediction mentioned earlier. The pattern of water flow in the Everglades may be thought of as a dynamic physical structure in some sense, determined by the differences between precipitation and evaporation, as mediated by topography. Because the pattern of flow is strongly seasonal, it will be considered in detail later in the section on natural cycles.

On a smaller scale, the maintenance of tree islands may also be thought of as the steady state of a competition between environmental processes. As mentioned by Gleason and Stone (1994), some tree islands in the Everglades may be as old as 1200 years. According to one hypothesis, the initiation of certain types of tree islands, those dominated by red bay (*Persea borbonia*), requires a discrete disturbance, such as the floating up of a mass of peat. After the mat has dislodged, however, physical processes of erosion, on the one hand, and accretion of matter through formation of a reverse eddy behind the mat, on the other hand, compete to determine whether the mat grows or shrinks. (This is analogous to the long-term maintenance of barchan sand dunes in the desert, which maintain integrity over many years through the simultaneous wind erosion on one end and growth on the other end from sand deposition in the reverse eddies [Barndorff-Nielsen et al., 1985].) Autogenic processes of vegetative growth eventually occur if the island survives long enough and can stabilize them for a long period of time, perhaps until environmental conditions change.

As the chapters thus illustrate, the Everglades consists of processes occurring over a wide range of temporal and spatial scales. Gunderson (1994, Table 13.2) has ordered processes influencing vegetation according to three spatial scales (1–3 m, 100–300 m, and >100–300 m) and three temporal scales (fast, 1–10 years; intermediate, 10–100 years; and slow, >100 years). To encompass processes beyond those directly associated with vegetation, a broader set of scales would be necessary.

The long-term natural driving forces of climate and geology determined what sort of ecosystem now exists in south Florida; they explain the formation of the Everglades and its primary features. Driving forces on shorter time scales (such as the seasonal variations in rainfall and the frequencies of droughts, storms, fires, and freezes) determined, and continue to determine, which of the many potential colonizing species have appropriate life cycles to survive and successfully compete with other species within the Everglades. These will be considered in later sections. First, we return to consider how gradual abiotic changes may continue to be an important factor in the Everglades.

GRADUALLY CHANGING ENVIRONMENTAL CONDITIONS

Major examples of the three general types of variations in driving forces (gradual changes, disturbances, and natural periodicities) are shown in Figure 12.1. Normally, on time scales of interest to humans (decades or, at most, centuries), changes that are very gradual, such as those of climate and sea level, can be regarded as relatively constant. Applied ecologists and resource managers tend to focus their attention on much shorter term phenomena than changes in climate and sea level, for example, the effects of wet and dry cycles on the order of years or decades.

Wanless et al. (1994) described sea level as generally monotonically rising during the late Holocene, although the rate of change during the last 3200 years has been about 4 cm/100 years, clearly less than the rates during the period 5500–3200 YBP (about 23 cm/100 years) and the period 7500–5500 YBP (>50 cm/100 years). An important consequence of the deceleration over the past three millennia has been that processes opposing marine transgressions have roughly kept pace with sea level. The shorelines have been stabilized by mangroves and by marl levees produced by storms. The build-up of mud banks has led to some seaward coastline expansion. Only in some places, probably due to high exposure to storms, has coastal erosion occurred during this last period. This recent stabilization of shoreline contrasts with the earlier time periods, when saltwater transgressions were common. Thus, the competing processes of erosion and sedimentation have shifted in dominance, with approximate dynamic equilibrium between the two occurring when the rates are nearly equal.

Figure 12.1 Three general types of alterations in the environmental driving forces can occur: gradual continuous change, discrete disturbances, and natural periodicities. These types of alterations are not independent and can mutually influence each other.

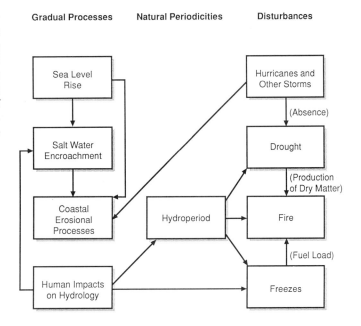

Gradual Processes   Natural Periodicities   Disturbances

If the rate of sea level change were to remain 4 cm/100 years, then it could safely be excluded from the list of current major concerns for the Everglades. However, considerable evidence is presented by Wanless et al. (1994) to indicate that sea level rise has returned to a much higher rate during the past five decades and is currently rising relative to the coast of southern Florida at a rate of 30–40 cm per century. This is a rate that is fast enough to present problems on the time scale of decades. In addition, global climate models predict that the current rate of increase in atmospheric carbon dioxide could lead to both a significant global warming (average temperature increase of greater than 1°C) as well as to an even faster rise in sea level within the next several decades. If the current rate continues long enough, and especially if it increases, it will lead to marine encroachment into freshwater areas, although the sedimentation that occurred during the last few thousand years will slow this encroachment. Major transgressions, if they occur, will probably be aided by storms. For these reasons, there is high interest in the future effects of rapid changes in climate and sea level.

Unlike climate change, which poses an unknown challenge in the future of the Everglades, more direct human activities are already changing the Everglades ecosystem. Primary among these changes are alterations in the original hydrologic regime of the Everglades through a system of levees, canals, and water impoundments (Fennema et al., 1994; Light and Dineen, 1994). The hydrologic regime of a system is a property that would normally change on geologic time scales, but the Central and Southern Florida Project of water regulation has accelerated this process. The changes are complex, affecting the magnitude, the seasonal timing, and the spatial distribution of annual flows of water. For example, the levee system along the Tamiami Trail (L-29) has increased the water depth in Water Conservation Area 3A, disrupting the seasonal pattern of drying and flooding, while the diversion of surface water flow into Northeast Shark River Slough has caused the loss of persistent surface water (Fennema et al., 1994). These and other changes, as noted by the authors, may be affecting species whose life cycles are attuned to the natural hydrologic pattern. Fennema et al. (1994) have taken the important step of reconstructing the natural hydrologic regime through computer modeling. Better knowledge of the natural hydrologic regime will allow one to make sharply defined hypotheses on how the current hydrologic regime is affecting the Everglades and its species populations.

Another human impact is the change in nutrient inputs to the Everglades from agricultural areas to the north, particularly phosphorus, which is most likely to limit primary production (Davis, 1994). The nutrient input to the Everglades comes primarily from this upstream flow and from precipitation. Nutrient input is episodic, increasing during periods of high water flow and high precipitation. The average input of nutrients to the Everglades has been low, so that the Everglades is relatively oligotrophic. However, the spread of cattails, a species characteristic of eutrophic conditions, into the northern Water Conservation Area 2A suggests a change in nutrient status on a shorter time scale.

DISTURBANCES in the EVERGLADES

Disturbances were defined by DeAngelis and White (1994) as relatively discrete events in time, although such phenomena as prolonged droughts, which are certainly not instantaneous, are often classified as disturbances. Among the primary types of disturbances affecting the Everglades are storms, freezes, droughts, and fires.

The term *disturbance* often has a negative connotation that can be misleading as far as its role in the ecosystem is concerned. Disturbances that occur with some degree of regularity are "integrated" by the ecosystem so that they are to some extent indispensable to the maintenance of the existing biotic community. For example, ecosystems that are vulnerable to frequent fires typically consist of plant species that are either fire resistant or have life cycles shorter than the average periodicity of fire occurrence and may even require fire for their regeneration. Also, essential processes such as nutrient recycling are often accelerated by fires and other disturbances. Because of this, some ecologists do not like to use the term disturbance for such cases, although it will be used here with the understanding that disturbances play an essential role as much as they do a disruptive role in an ecosystem.

Examples of this important role are hurricanes and other large storms, which "contribute significantly to the wet season rainfall in south Florida" (Duever et al., 1994). Along with this possibly essential role of providing water, the damage that severe storms can do to plant communities must also be considered, particularly in coastal areas, as demonstrated by the Okeechobee storms of 1926 and 1928 (W. B. Robertson, Jr., personal communication) and Hurricane Andrew.

Fire disturbances are also a natural part of south Florida ecosystems, although some part of the pre-Columbian fire regime may be attributable to burning by native Americans (Snyder, 1991). The importance of fire to the plant community is supported by observations of changes in the community following fire exclusion. When fire is excluded from the Miami rock ridge pinelands, these succeed to closed hardwood hammocks that contain none of the more than 15 species of herbs endemic to the area (Robertson, 1953).

However, like hurricanes, fires occupy a dual position of being both an integral part of as well as a danger to the Everglades. Egler (1952, cited by Gunderson and Snyder, 1994), stated that, "The herbaceous everglades were born in fires; that they can survive only with fires; that they are dying today because of fires." Frequency of intense fires is probably the most important factor when discussing the possible negative effect on the Everglades ecosystem. Intense fires can have effects that last 3 to 10 decades, and a high frequency of such fires could have strong effects on inland plant communities and organic substrates (W. B. Robertson, Jr., personal communication).

Other types of disturbances may also have pervasive impacts. Anywhere from zero to several freeze events may occur in the Everglades during the winter. These are generally mild and of short duration. However, these disturbance events may play a major role in limiting the abundances of tropical invaders which would otherwise be favored by the average climatic conditions (Duever et al., 1994). Freeze events exemplify other aspects of disturbances in general that were mentioned by DeAngelis and White (1994); that is, that the particular characteristics of the event determine its effect. How a freeze event will affect the biota in an ecosystem depends on the duration of the freeze, its absolute temperature, when it occurs (whether the organisms are cold-hardened), and whether there have been previous freezes that winter (Duever et al., 1994).

A "hierarchical pattern" of disturbances discussed by DeAngelis and White (1994) is exemplified by patterns in the Everglades. Disturbances that affect large spatial areas tend to occur less frequently than those that affect small areas. The ranges of frequencies and spatial extent for several types of disturbances important to the Everglades are plotted in Figure 12.2. These include storms, floods, droughts, freezes, and fires.

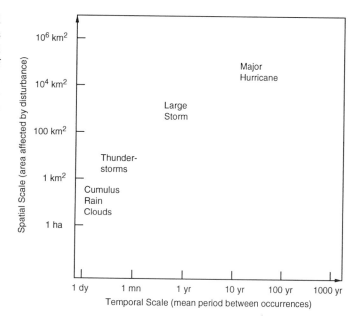

Figure 12.2 The spatial scale of a disturbance tends to be positively correlated with the mean period of time between occurrences. Some typical storm disturbances that can affect the Everglades are plotted on a rough scale of spatial extent and mean time between events.

Disturbances are usually thought of as stochastic, or randomly occurring, events. This is not completely true, however. Because storms, droughts, fires, and some other disturbances tend to be seasonal phenomena, there are biases in the probabilities of when they occur within a given year. In addition, many weather phenomena in the Everglades, such as long droughts, recur on fairly regular cycles of several years. Precipitation records indicate that periods of drought, or alternatively, of high precipitation, recur on periods of 4–6 years. The periodicity in precipitation results in other periodicities, such as fire frequency and flood frequency. The latter has the additional consequence of leading to a similar period in high freshwater runoff into Florida Bay, as the periodicity in fluorescent banding in corals noted by Smith et al. (1989) shows. Spectral analysis of fire frequency data indicates the occurrence of a cycle of 10–14 years in fire frequency (Gunderson and Snyder, 1994). As Gunderson and Snyder remark, the similarity of this period to that of the sunspot cycle may be coincidental, but it may also reflect increased incidence of electrical storms. Weaker signals in the spectral pattern show possible periods of 40+ years and 6 years. The latter may be related to the periodicity of precipitation.

Finally, the spatial pattern of disturbances in the Everglades evokes some of the landscape ideas discussed by DeAngelis and White (1994). These are discussed further in a later section of this synthesis.

NATURAL PERIODICITIES in the EVERGLADES

The primary natural periodicity in the Everglades is the alternation of wet and dry seasons within the year. The periodicity in temperature is much less pronounced (Duever et al., 1994). The wet-dry cycle of precipitation is of particular importance in the Everglades, as it results in a spatially varying annual hydropattern that influences the floristic and faunal communities across the Everglades. Normally, the annual hydrologic cycle includes a wet season, during which there is a rapid rise of water level, followed by a relatively dry season, during which the water level generally falls. Rainfall and evaporation dominate over long-distance water flow in the input-output budget in most places in the Everglades. Thus, while there may be some smoothing out of the effects of rainfall events on the water level pattern due to surface inflows and outflows, individual local rainfall events are important as direct causes of local water level rises.

Related to the annual hydrologic cycle, there is a pronounced period in fire frequency of one year, due to the high prevalence of human-induced fires late in the dry season and lightning-caused fires in the rainy season. This is indicated by a sharp peak in the fire frequency spectral plot (Gunderson and Snyder, 1994).

Although the species that inhabit the Everglades evolved long before the Everglades was formed, they exist there because their life cycles are well adapted to such characteristics of the Everglades as its hydrologic regime. As emphasized in the case history studies of the following vegetation and fauna sections, the details of the temporal pattern of falling water level (drydown) during the dry season may greatly affect breeding success of some species. The dependence of reproductive success in populations on the details of annual abiotic periods is not unique to

Figure 12.3 Schematic diagram of a drying front that moves down an elevational gradient in the Everglades during the dry season. Large spatial areas of low water level favorable to wading bird foraging are created along the front.

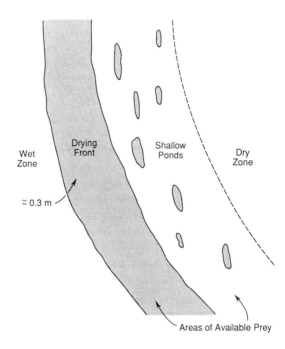

the Everglades. Many freshwater temperate fish species, for example, are dependent on the details of the scenario of water temperature change through the spawning season. Reversals in the steady increase in water temperature can lead to nest abandonment.

DeAngelis and White (1994) noted that annual periodicities often lead to annual occurrences of transient spatial features called *fronts,* or moving transition zones, between two environmental conditions. In the Everglades an annual drying front moves from higher to lower elevations as the water level drops through the dry season (Figure 12.3). This is important to Everglades wading birds, because the lowering of water levels during the dry season concentrates forage fish and aquatic invertebrates, enabling wading birds to obtain an adequate supply of food for nestlings. However, this requires that the drop in water level proceed continuously at a rate that provides new areas of shallow water at some minimum rate. The period of water level decrease must also last long enough for the offspring to be fledged.

INTERACTIONS between DRIVING FORCES

Figure 12.1 illustrates some of the primary interrelationships between changes in the different abiotic driving forces in the Everglades that were discussed in the chapters of this section. From left to right are gradual environmental changes, changes in natural periodicities, and disturbances.

Sea level rise and human-induced changes in the Everglades drainage are classified here as gradual changes imposed exogenously. Hurricanes and other storms are seen as exogenous disturbances. The principal natural periodicity of the system is the hydropattern. Sea level rise and drainage can both have long-term effects on the hydropattern, the former lengthening it by causing increasing retardation in water movement, and the latter reducing or eliminating it. Changes in the frequency of hurricanes and other storms can modify the normal hydropattern by adding pulses of rains at various times.

Sea level rise has a direct effect on the gradual changes of coastal erosion and saltwater intrusion. Hurricanes and other storms may greatly increase the progress of these latter processes by causing boundaries to be overstepped (Wanless et al., 1994). Storms may be specific events in which sea level transgressions occur during a time of rising sea level.

The hydropattern affects the probability of droughts and fires. Because water acts as a buffer to temperature changes, an increase in the hydroperiod would be expected to decrease the probability of freezes. Drainage and changes in land use can have a direct effect on the frequency of freezes and the frequency and size of fires. Both drought and freeze disturbances increase fire probabilities because they increase the amount of dead organic matter that is available to burn.

SPATIAL VARIABILITY in the EVERGLADES

It is clear from the chapters in this section that the Everglades is a system of enormous spatial variability. The types of variability may be classified on the macroscale, mesoscale, and microscale. The macroscale, for present purposes, would be of the order of tens to hundreds of kilometers (and corresponding areas of hundreds to thousands of square kilometers), the mesoscale of the order of 1–10 km (1–100 km^2), and the microscale of the order of particular features such as tree islands (less than 1 km^2). Table 12.1 lists examples of each type that were discussed. Like any arbitrary classification, this representation is imperfect, but it is useful as an organizational tool.

The geological bedrock of south Florida varies on a macroscale not only in elevation but also in type (Figure 7.2, Gleason and Stone, 1994). Bedrock topography directly affects the pattern of water flow (Figure 7.9, Gleason and Stone, 1994), and thus of vegetation (Plates 7–9,* Davis et al., 1994), as well as of surface sediment (Figure 7.7, Gleason and Stone, 1994). These effects have led to the evolution of a number of distinct regions in south Florida.

The features that are more or less permanent on a human time scale (e.g., landscape topography and distance from the ocean) affect the more transient phenomena of climate-related influences, which vary in time and space: floods, droughts, freezes, and fires. Along landward lines perpendicular to the coast are a decreasing gradient in salinity and an increasing gradient in the frequency of temperature extremes.

Droughts show great spatial variability. The fact that Fort Myers and Miami had only one drought year in common over a period of about 50 years is an indication of how localized such disturbances can be (Duever et al., 1994). Hurricanes and

---

* Plates 7–9 appear following page 432.

Table 12.1 Summary of Features of the Everglades and Their Characteristic Spatial Scales (Lineal and Areal): Microscale (10–1000 m or 100–10^4 m^2), Mesoscale (1–10 km or 1–100 km^2), Macroscale (10–>100 km or 100–10,000 km^2)

Feature	Spatial scale
Major geological features (e.g., bedrock type)	Macroscale
Tree islands	Microscale
Ponds, alligator holes	Microscale
Microtopography of bedrock	Microscale
General climate of south Florida	Macroscale
Droughts	Mesoscale to macroscale
Freezes	Mesoscale to macroscale
Floods	Mesoscale
Hurricanes	Macroscale
Storm effects: lightning, tornadoes, downdrafts	Microscale to mesoscale
Major hydrologic features (e.g., sloughs)	Macroscale
Drying fronts	Mesoscale to macroscale
Levees and canals	Microscale to macroscale
Fires	Microscale to macroscale

other storms exhibit at least as much variability, usually affecting only parts of the Everglades at any given time. Duever et al. (1994) refer to this as producing a mosaic pattern in the Everglades that is favorable to the stability of populations, because the entire population of any species in the Everglades will not be affected by a particular drought or other disturbance. It is a central idea of the discipline of landscape ecology that, far from being spatially uniform, most ecosystems are mosaics resulting from the differential effects of disturbances.

Fires are influenced by the landscape pattern of the Everglades, and fires themselves contribute to the formation of the landscape pattern. For example, although the factors that determine fire sizes are still poorly understood, "the largest fire recorded in the tree island/wet prairie/sawgrass mosaic is close to the size of the largest patches of sawgrass in this mosaic" (Gunderson, 1992). Over the period from 1948 to 1992, the 992 registered fires varied in size from 4 m^2 to 74,860 ha, with a clumping between 8000 and 15,000 ha. A single fire can destroy the vegetation where it occurs. Following this event, succession through a series of vegetation types will occur on that site until the prefire vegetation type is restored. The pattern of fires maintains a mosaic of different vegetation types. Thus, although a single fire is destructive to the vegetation at a local site, the effect of a fire regime is usually to maintain a relatively constant spatial mosaic of vegetation types.

The implications of this spatial variability are profound. As has been amply shown, there is great variability in the seasonal rainfall pattern, especially in the occurrence of large rainfall events (which may be termed *disturbances*) during the normal dry season. These affect the hydropattern and may cause local reversals in the falling water levels, thereby adding a risk to wading bird nesting success (Karr and Freemark, 1985; Kushlan, 1979). However, spatial nonuniformities, such as microscale and mesoscale elevational irregularities, tend to even out the interannual rainfall variability. If there is sufficient spatial heterogeneity, then, regardless of how the rainfall pattern varies from year to year, there will be a variety of hydrologic patterns represented in the foraging range of the individual wading bird, or at least the population as a whole.

On the macroscale, as Fennema et al. (1994) observed, the persistent water in the deeper portions of the downstream wetlands under natural conditions acted as a dry season refuge for animals. The existence of early dry season shallow water (eastern transitional zone) provided food that allowed wading birds to start feeding early (Loftus et al., 1992). The effect, as Duever et al. (1994) point out, is of a mosaic pattern of drydown, which provides food for mobile consumers, such as wading birds, over a long period.

Spatial extent is the crucial factor here, particularly the spatial extent relative to the size of potential disturbances. If spatial extent is great enough, the diversity at the macroscale level means that the ecosystem will provide sufficient habitat of the right types over the entire region, even if disturbances affect local habitats. The normal range of disturbances in this case not only do not threaten populations, but they impart a diversity of conditions that maintains populations and communities.

DISCUSSION and CONCLUSIONS

The chapters in this section set the stage for the biotic case studies of the following sections. A key fact brought out here is that the age of the Everglades is much smaller than that of probably all of the species that inhabit it. These species have, therefore, colonized from other places. Successful species were preadapted in most ways.

Is this biological community stable? Stability is a relative concept and depends on spatial and temporal scales. Stability is the tendency for species to persist and not fluctuate wildly. Communities occupying small, spatially homogeneous areas are subject to two basic types of disruption: (1) environmental stresses or disturbances causing extinctions because of high mortalities or lack of successful reproduction in a population and (2) overly strong biotic interactions, including competition from invading exotic species, resulting in drastic population oscillations and competitive exclusions.

The chapters here are of relevance to the first type of disruption, that from abiotic effects. It is clear that spatial extent and heterogeneity can decrease the total effect of these disruptions. If the spatial extent of a system is large compared to the spatial scale on which disruptive forces act, then the ecological community will be subjected to destabilizing forces only in some areas of the overall domain of the community at any given time. Depending on their degree of mobility and local habitat constraints, animals may be able to move to areas not disrupted or, if local extinctions occur, the areas where they occur may be recolonized from other areas after the disruptive event. This was probably always the case in the Everglades prior to the last several decades. Today, as changes in the Everglades populations indicate, this may no longer be true.

ACKNOWLEDGMENTS

Research was sponsored by the National Science Foundation Ecosystem Studies Program under Interagency Agreement No. 40-689-78 with the U.S. Department of Energy under Contract DE-AC05-84OR21400 with Martin Marietta Energy Systems, Inc.

LITERATURE CITED

Barndorff-Nielsen, O. E., P. Blaesild, J. L. Jensen, and M. Sorensen. 1985. The fascination of sand. in *A Celebration of Statistics,* A. C. Atkinson and S. E. Feinberg (Eds.), Springer-Verlag, New York, pp. 57–87.

Browder, J. A., P. J. Gleason, and D. R. Swift. 1994. Periphyton in the Everglades: Spatial variation, environmental correlates, and ecological implications. in *Everglades: The Ecosystem and Its Restoration,* S. M. Davis and J. C. Ogden (Eds.), St. Lucie Press, Delray Beach, Fla., chap. 16.

Davis, S. M. 1994. Phosphorus inputs and vegetation sensitivity. in *Everglades: The Ecosystem and Its Restoration,* S. M. Davis and J. C. Ogden (Eds.), St. Lucie Press, Delray Beach, Fla., chap. 15.

Davis, S. M., L. H. Gunderson, W. A. Park, J. Richardson, and J. Mattson. 1994. Landscape dimension, composition, and function in a changing Everglades ecosystem. in *Everglades: The Ecosystem and Its Restoration,* S. M. Davis and J. C. Ogden (Eds.), St. Lucie Press, Delray Beach, Fla., chap. 17.

DeAngelis, D. L. and P. S. White. 1994. Ecosystems as products of spatially and temporally varying driving forces, ecological processes, and landscapes: A theoretical perspective. in *Everglades: The Ecosystem and Its Restoration,* S. M. Davis and J. C. Ogden (Eds.), St. Lucie Press, Delray Beach, Fla., chap. 2.

Duever, M. J., J. F. Meeder, L. C. Meeder, and J. M. McCollom. 1994. The climate of south Florida and its role in shaping the Everglades ecosystem. in *Everglades: The Ecosystem and Its Restoration,* S. M. Davis and J. C. Ogden (Eds.), St. Lucie Press, Delray Beach, Fla., chap. 9.

Egler, F. E. 1952. Southeast saline Everglades vegetation, Florida, and its management. *Vegetatio,* 3:213–265.

Fennema, R. J., C. J. Neidrauer, R. A. Johnson, T. K. MacVicar, and W. A. Perkins. 1994. A computer model to simulate natural Everglades hydrology. in *Everglades: The Ecosystem and Its Restoration,* S. M. Davis and J. C. Ogden (Eds.), St. Lucie Press, Delray Beach, Fla., chap. 10.

Gleason, P. J. and P. A. Stone. 1994. Age, origin, and landscape evolution of the Everglades peatland. in *Everglades: The Ecosystem and Its Restoration,* S. M. Davis and J. C. Ogden (Eds.), St. Lucie Press, Delray Beach, Fla., chap. 7.

Gunderson, L. H. 1992. Spatial and Temporal Hierarchies in the Everglades Ecosystem, with Implications to Water Deliveries to Everglades National Park, Ph.D. dissertation, University of Florida, Gainesville.

Gunderson, L. H. 1994. Vegetation of the Everglades: Determinants of community composition. in *Everglades: The Ecosystem and Its Restoration,* S. M. Davis and J. C. Ogden (Eds.), St. Lucie Press, Delray Beach, Fla., chap. 11.

Gunderson, L. H. and J. R. Snyder. 1994. Fire patterns in the southern Everglades. in *Everglades: The Ecosystem and Its Restoration,* S. M. Davis and J. C. Ogden (Eds.), St. Lucie Press, Delray Beach, Fla., chap. 11.

Karr, J. R. and K. E. Freemark. 1985. Disturbance and vertebrates: An integrative perspective. in *The Ecology of Natural Disturbance and Patch Dynamics,* S. T. A. Pickett and P. S. White (Eds.), Academic Press, New York, pp. 153–168.

Kushlan, J. A. 1979. Design and management of continental wildlife reserves: Lessons from the Everglades. *Biol. Conserv.,* 15:281–290.

Light, S. S. and J. W. Dineen. 1994. Water control in the Everglades: A historical perspective. in *Everglades: The Ecosystem and Its Restoration,* S. M. Davis and J. C. Ogden (Eds.), St. Lucie Press, Delray Beach, Fla., chap. 4.

Loftus, W. F., R. A. Johnson, and G. Anderson. 1992. Ecological impacts of the reduction of groundwater levels in short hydroperiod marshes of the Everglades. in 1st Int. Conf. on Ground Water Ecology, American Water Resources Association Proc., pp. 199–208.

Maltby, E. and P. J. Dugan. 1994. Wetland ecosystem protection, management, and restoration: An international perspective. in *Everglades: The Ecosystem and Its Restoration,* S. M. Davis and J. C. Ogden (Eds.), St. Lucie Press, Delray Beach, Fla., chap. 3.

Pickett, S. T. A. and P. S. White. 1985. Natural disturbance and patch dynamics: An introduction. in *The Ecology of Natural Disturbance and Patch Dynamics,* S. T. A. Pickett and P. S. White (Eds.), Academic Press, New York, pp. 3–13.

Robertson, W. B., Jr. 1953. A Survey of the Effects of Fire in Everglades National Park, National Park Service, U.S. Department of the Interior, Homestead, Fla.

Smith, T. J., III, J. H. Hudson, M. B. Robblee, G. V. N. Powell, and P. J. Isdale. 1989. Freshwater flow from the Everglades to Florida Bay: A historical reconstruction based on fluorescent banding in the coral *Solenastrea bournoni. Bull. Mar. Sci.,* 44:274–282.

Snyder, J. R. 1991. Fires regimes in subtropical South Florida. in High Intensity Fire in Wildlands: Management Challenges and Options, Proc. 17th Tall Timbers Fire Ecology Conf., E. V. Komerek (Ed.), Tall Timbers Research Station, Tallahassee, Fla.

Wanless, H. R., R. W. Parkinson, and L. P. Tedesco. 1994. Sea level control on stability of Everglades wetlands. in *Everglades: The Ecosystem and Its Restoration,* S. M. Davis and J. C. Ogden (Eds.), St. Lucie Press, Delray Beach, Fla., chap. 8.

III

VEGETATION COMPONENTS
and PROCESSES

13

VEGETATION of the EVERGLADES: DETERMINANTS of COMMUNITY COMPOSITION

Lance H. Gunderson

ABSTRACT

The apparent tranquillity of the Everglades landscape masks the dynamics of processes that affect vegetation. The composition of Everglades plant communities has been changing since the ecosystem began to form some 5000 years ago. Change has accelerated during the past 100 years, in both subtle and dramatic ways, due to natural and anthropogenic factors. By virtue of its geographic location on a peninsula extending from a temperate continent into the subtropics, the Everglades has a flora comprised of tropical, temperate, and endemic taxa. Human activity during this century has increased the rate of species introduction, resulting in the naturalization of many alien plants, some of which have transformed the character of plant associations. Vegetation patterns are also influenced by a set of environmental processes that operate at distinct spatial scales (from meters to hundreds of kilometers) and distinct temporal rates (from days to centuries). Of these processes, disturbances (such as severe fires and drought/flood cycles, which operate at intermediate scales of tens of kilometers and decades) account for much of the variation in spatial distribution of major types of vegetation communities.

INTRODUCTION

The Everglades region has long contained many mysteries and wonders. At first glance, the landscape appears monotonous—expanses of open water and gray-green marshes speckled with islands of dark green trees. The tranquil landscape, however, belies the dynamic nature of the biota, especially the vegetation. The vegetative components of the Everglades ecosystem are constantly changing, as a result of both internally induced processes and the frequent disturbances described in other chapters of this volume. The unique floral assemblages occupy pivotal positions in the ecosystem, controlling and modifying many aspects of the structure (e.g., soil elevation via peat accretion) and processes (e.g., flow of water and nutrients) of the Everglades ecosystem. The vegetation also responds to a set of fluctuating environmental variables that operate at different spatial and temporal scales. The vegetation communities, then, can be regarded as integrators of changing environmental conditions, responding to natural and anthropogenic changes in both dynamic and evolutionary ways.

For the purposes of this chapter, the Everglades is defined as the freshwater wetland complex that is bordered by Lake Okeechobee to the north, the mangrove forests along Florida Bay to the south, the Big Cypress area to the west, and the coastal ridge to the east, as outlined by Davis (1943) and Davis et al. (1994). The vegetation of Long Pine Key is included in this chapter, because it is intricately linked to the nearby wetlands and it is the only remaining large section of rockland vegetation in Florida (Snyder et al., 1990).

The purpose of this chapter is twofold. The first objective is to describe the vegetation of the Everglades in terms of the plant species now present in the region and their organization into associations. These associations form hierarchies which are structured by environmental processes that control the spatial and temporal distributions of communities. The second purpose is to discuss the interplay between plant associations and environmental determinants. One striking pattern of Everglades vegetation is the spatial mosaic of plant communities, which is a result of the interaction between the plant communities and temporal environmental variation. Although descriptions of the environmental fluctuations are presented elsewhere in this volume, this chapter will concentrate on how these fluctuations interact with organizational patterns of vegetation.

FLORA of the EVERGLADES

Just as the human population of southern Florida represents a melting pot of different ethnic backgrounds, the flora of the Everglades is also a melting pot or vegetation soup representing a multiplicity of source regions. The Everglades region is located in the transition zone between temperate and tropical areas. The flora of the Everglades has many representatives from these two areas, but it also has elements from regions further away (aliens or exotics), as well as taxa that are endemic to the region (Long, 1974) (Figure 13.1). The native tropical species are primarily from the Antillean–Caribbean region of the neotropics. Propagules of these species are thought to have crossed the saltwater barrier to Florida carried either by hurricanes, migrating birds, or (for those able to tolerate a period of exposure to salt water) ocean currents. The temperate species range northward into the coastal plain of the southeastern United States. The endemic taxa consist of species and subspecies which have evolved in unique habitats of southern Florida and are found nowhere else in the world. The exotic flora was, for the most part,

Figure 13.1 Four source regions for the flora of the Everglades.

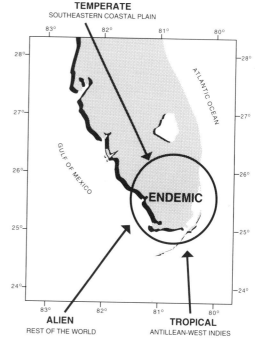

introduced into southern Florida for ornamental and horticultural purposes. Many of these introduced species were so well adapted to conditions that they have become aggressive invaders.

The flora of the Everglades has been included in volumes that cover larger areas, such as the southeastern United States (Small, 1932) and southern Florida (Long and Lakela, 1971). These authors describe a flora represented by species both tropical and temperate in origin, but do not quantify the percentage of taxa that come from the tropics, probably because of the confounding effects of exotics from other tropical regions. The most recent detailed plant listing from an area within the Everglades was prepared by Avery and Loope (1983), who enumerated some 830 species within Everglades National Park. Although this represents only the southern fifth of the Everglades region, Everglades National Park includes all of the major habitats of the Everglades region. Avery and Loope (1983) identified 141 exotic species or about 17% of the total number. The endemic taxa comprise the smallest percentage of the flora and may be relicts from previous climatic regimes. Robertson (1953) counted 122 taxa that were considered endemic, but later work by Avery and Loope (1980) refined the number to 65 taxa or about 8% of the total plant species in Everglades National Park. The flora of the area can be described as primarily herbaceous. The most speciose family in the Everglades National Park inventory was the Poaceae, with 101 taxa. Other species-rich families include the Asteraceae (77 taxa), Fabaceae (54), Cyperaceae (45), Orchidaceae (35), and Euphorbiaceae (35). Here, nomenclature follows Avery and Loope (1983).

The composition of the flora is in part a result of variable rates of species introduction and extinction. The temperate elements probably arrived in a relatively slow progression, via a gradual continuous movement over the land connections with the southeastern United States. The arrival of native tropical taxa was probably also a relatively slow but intermittent process. The speciation of endemic taxa probably occurred long before the Everglades as it is now known was formed some 5000 years ago (Gleason et al., 1984), and these taxa are perhaps relics of former conditions. Within this century, however, the rates of species introduction and extinction have accelerated tremendously. As humans became more mobile, they increased the rates of plant movement, introducing many nonnative or exotic organisms, especially plants, into southern Florida. A highly aggressive few of these introduced species are now conspicuous components in the landscape of the Everglades.

FLORAL ASSEMBLAGES

The approximately 850 plant species found in and around the Everglades region aggregate into fewer than 20 associations. The plant communities described in the works of Harshberger (1914), Harper (1927), and Davis (1943) do not include exotic plants. As of the 1950s (Egler, 1952; Robertson, 1953), exotic taxa were recognized as becoming established in the native flora. More recent descriptions (Craighead, 1971; Wade et al., 1980; Olmsted and Loope, 1984; Gunderson and Loftus, 1993) are evidence of the increasing importance of exotic plants in the vegetation assemblages, yet all native types described earlier are present today.

A classification scheme (Table 13.1) based on major environmental dimensions indicates one hierarchy of plant communities. The initial major division in the grouping scheme is based on the hydro-edaphic conditions of a site, that is, whether the soil is saturated for at least part of the year (wetland) or not (upland). Secondary divisions within the wetland groups are based on water chemistry (primarily soil salinity) and growth form of the vegetation (forested versus graminoid).

Upland Complex

The upland complex includes pine forests, tropical hardwood hammocks, and areas of newly created or disturbed land. The uplands are found along the eastern edge of the Everglades on the Atlantic Coastal Ridge and extend into Everglades National Park as Long Pine Key. Tropical hardwood hammocks occur within the Everglades, but cover relatively little area. Distributions of these two forest types are due not only to the effects of fire on species composition and structure, but also to substrate differences.

Pine forests once dominated the Atlantic Coastal Ridge. Communities of sand pine (*Pinus clausa*) were found over the deep, white sands in the northern areas of the ridge (in and around West Palm Beach). From northern Dade County and extending south into the area now in Everglades National Park, outcrops of oolitic limestone are the dominant surficial feature, which historically supported a rockland vegetation comprised of a matrix of pine forests enclosing hardwood hammocks. Approximately 10% of the original rockland forests is conserved as park area, most of which is in Everglades National Park (Snyder, 1986). Small outcrops of bedrock

Table 13.1 Plant Communities of Everglades, Grouped by Major Ecological Classes

---

I. Upland Communities—Long Pine Key
   A. Rockland pine forests
      1. Low-stature hardwood understory
      2. Tall hardwood understory
   B. Tropical hardwood hammocks
      1. Mature phase
      2. Successional phase
II. Wetland communities
   A. Freshwater wetlands
      1. Forested communities (tree islands)
         a. Bayheads
         b. Willow heads
         c. Cypress forests
      2. Graminoid associations (marshes and prairies)
         a. Sawgrass marshes
            i. Tall stature
            ii. Intermediate stature
         b. Wet prairies (peat)
            i.   *Eleocharis* flats
            ii. *Rhynchospora* flats
            iii. Maidencane marshes
         c. Wet prairies (marl)
      3. Little or no emergent vegetation
         a. Ponds and creeks
         b. Sloughs

Note: Groups are derived from classes of Davis (1943), Loveless (1959), and Gunderson and Loftus (1993).

support areas of hardwood hammocks and are scattered throughout the wetland areas of the Everglades proper.

Rockland Pine Forests

The remnant rockland pine forests are dominated by a monotypic overstory of south Florida slash pine (*Pinus elliotti* var. *densa*). Olmsted et al. (1983) and Snyder et al. (1990) recognize two types of pine forests: those with low-stature understory and those with well-developed hardwood understory. Stands with open, low-stature understory are regularly burned, whereas hardwood abundance increases rapidly in the absence of fire.

The floristically significant feature of the rockland pine forests is the species richness of the understory strata. Loope et al. (1979) counted 186 species in this association, making it the most diverse community in Everglades National Park. At least 50 to 75 species of hardwoods (primarily West Indian in origin) are found in the understory (Taylor and Herndon, 1981). The understory plants in these forests attain heights of 2 m within an average fire return interval (i.e., 5–10 years). Dominant hardwoods include rough velvetseed (*Guettarda scabra*), indigo berry (*Randia aculeata*), varnish leaf (*Dodonea viscosa*), myrsine (*Myrsine floridana*), and willow bustic (*Bumelia salicifolia*) (Olmsted et al., 1983; Snyder, 1986; Taylor and Herndon, 1981). Important palm species are cabbage palm (*Sabal palmetto*) and saw palmetto (*Serenoa repens*). Grasses and forbs cover a small percentage of the ground, but account for the bulk of species diversity. Common graminoid species are Florida bluestem (*Schizachyrium rhizomatum*) and silver bluestem (*Andropogon cabanisii*) (Olmsted et al., 1983).

The flora of the pinelands contains 51 taxa which are species of special concern. Many of these are endemic taxa, such as pineland dyschoriste (*Dyschoriste oblongifolia*), Florida five-petaled leaf flower *(Phyllanthus pentaphyllus)*, *Borreria terminalis*, Florida Keys noseburn (*Tragia saxicola*), Porter's hairy-podded spurge (*Chamaesyce porteriana* var. *porteriana*), pineland clustervine (*Jacquemontia curtissii*), small-leaved cat tongue (*Melanthera parvifolia*), and pine wood privet (*Forestiera segregata* var. *pinetorum*). One plant species that is classified as federally endangered is Garber's spurge (*Chamaesyce garberii*). Taxa that are rare, and therefore of concern, include alvaradoa (*Alvaradoa amorphoides*), Blodgett's wild mercury (*Argythamnia blodgettii*), rhacoma (*Crossopetalum rhacoma*), white ironwood (*Hypelate trifoliata*), Cuba colubrina (*Colubrina cubensis* var. *floridana*), *Passiflora sexflora*, smooth strongbark (*Bourreria cassinifolia*), *Verbena maritima*, and *Ernodea littoralis* var. *angusta*.

The areas of tall stature hardwoods contain the same hardwoods and palm species found in the understory of lower stature pinelands. The hardwoods attain heights of 8–10 m. The taller hardwoods also shade out the ground cover, resulting in few if any herbs and forbs found in these associations.

Common exotic plants found in the rockland forests include Brazilian pepper (*Schinus terebinthifolius*) and plume grass (*Neyraudia reynaudiana*). *Schinus* behaves much like native hardwoods, in that with repeated burning it is relegated to a shrub-sized plant. However, with increased fire intervals both *Schinus* and *Neyraudia* quickly dominate pinelands (Wade et al., 1980).

Tropical Hardwood Forests

Tropical hardwood hammocks appear on rock substrates and occur more often in the southern Everglades. These forests are comprised of broad-leaved evergreen species primarily from the West Indian–Antillean region. Hammocks are found interspersed throughout the pine forests (Johnson et al., 1983; Olmsted et al., 1983): at the interface between pine forests and sloughs (e.g., Royal Palm Hammock in Everglades National Park), on elevated outcrops on the upstream side of some tree islands, and scattered throughout mangrove forests (Craighead, 1971; Russell et al., 1980). Because hammocks are found on the highest, driest sites, most show evidence of disturbance by humans.

Three tree species, live oak (*Quercus virginiana*), wild tamarind (*Lysiloma latisiliquum*), and gumbo limbo (*Bursera simaruba*), tend to be the largest, tallest trees in hammocks (Robertson, 1953; Olmsted et al., 1981). Other less common trees which are found in the overstory include sugarberry (*Celtis laevigata),* mastic (*Mastichodendron foetidissimum*), and mahogany (*Swietenia mahogani*) in the southern hammocks. One suite of species does not exceed subcanopy stature but accounts for the tremendous stem density found in hammocks; these commonly found smaller trees include willow bustic (*Bumelia salicifolia*), lancewood (*Nectandra coriacea*), several species of stoppers (*Eugenia* spp.), pigeon plum (*Coccoloba diversifolia*), and marlberry (*Ardisia escallonioides*). Few if any plants are found on the ground due to the dense canopy. Most of the herbaceous flora are epiphytes, including vines, orchids, and bromeliads.

Tropical hammocks support the greatest number (59 taxa) of rare and threat-ened plants. Most of these are epiphytes from the bromeliad (Bromeliaceae), orchid (Orchidaceae), and fern families. Solution holes in hammocks support at least ten of the rare plant species in this association. The species which are rarest in this association are hand fern (*Ophioglossum palmatum*), Floridian royal palm (*Roystonea elata*), bromeliads (*Catopsis floribunda, Guzmania monostachia, Encyclia boothiana*), cowhorn orchid (*Cyrtopodium punctatum*) and other orchids (such as *Brassia caudata, Oncidium floridanum, Macradenia lutescens*), and myrtle-of-the-river (*Calyptranthes zuzygium*). Some of these taxa, such as *B. caudata,* may be recently extirpated.

Hammocks which have recently burned or have histories of frequent past fires are in a successional, often scrubby, phase. Hammocks with recurrent or severe fire damage are often colonized by native weedy species such as Florida trema (*Trema floridana*) and bracken fern (*Pteridium aquilinum)* or the exotic tree Australian pine (*Casuarina* spp.).

Many nonnative plants are found in hammocks as a result of human cultivation. Fruit trees, such as citrus (*Citrus* spp., e.g., orange, lime, and grapefruit) and banana (*Musa × paradisicum*), are common.

Freshwater Wetland Complex

The native freshwater wetlands of the Everglades are comprised of both forested and nonforested communities. All communities are described by the dominant species or groups. The forested wetlands include bayheads, willow heads, and cypress forests. Ponds, open water sloughs, sawgrass marshes, and wet prairies comprise the nonforested or graminoid wetlands. The communities are typically arrayed in a type of mosaic, as shown in vegetation maps (Rintz and Loope, 1979; Gunderson et al., 1986).

Bayhead/Swamp Forests

The broad-leaved hydrophytic hardwood associations of the Everglades are also referred to as tree islands. These clumps of hardwood trees are taller than the surrounding marsh, and hence the name tree island. Tree islands are mostly swamp forests, although some small areas of higher elevation support tropical hardwood vegetation. The larger tree islands within the Everglades are in the shape of an elongated teardrop, generally oriented with the main axis parallel to the main axis of flow. The teardrop shape may be a result of a bedrock bump near the upstream end of the island, which interacts with the flow pattern to form this shape. The smaller tree islands found in Water Conservation Area 1 have a circular shape (Gleason et al., 1984; Gleason and Stone, 1994) which may be related to their mode of formation, where chunks of peat break loose and form floating islands or "batteries" and are colonized by hardwoods.

Bay trees dominate the swamp forest/tree islands; hence, the name bayhead is often applied (Davis, 1943; Craighead, 1971). Canopy species include red bay (*Persea borbonia*) and swamp bay (*Magnolia virginiana*), as well as dahoon holly (*Ilex cassine*), pond apple (*Annona glabra*), and wax myrtle (*Myrica cerifera*). Other less common species include willow (*Salix caroliniana*) and strangler fig (*Ficus aurea*). A dense shrub layer is generally found beneath the canopy, composed primarily of cocoplum (*Chrysobalanus icaco*), but it can also include smaller individuals of the above-mentioned overstory species. Other plants in the shrub stratum include buttonbush (*Cephalanthus occidentalis*) and the large leather fern (*Acrostichum danaeifolium*). Some areas of the understory are devoid of ground cover due to the dense shade of the overstory.

Willow Heads

Stands of willow (*Salix caroliniana*) are also called willow heads (Davis, 1943; Loveless, 1959; Craighead, 1971). These stands are generally monotypic, with willow the abundantly dominant woody plant. Other less common associates include phragmites (*Phragmites australis*), sawgrass (*Cladium jamaicense*), and alligator flag (*Thalia geniculata*). Herbaceous vines such as whitevine (*Sarcostemma clausum*), climbing hempweed (*Mikania scandens)*, and Everglades morning glory (*Ipomoea sagittata)* are commonly found.

Willow heads are found on sites with a history of severe soil disturbance, such as following a peat fire, lumbering, farming, or alligator excavation (Craighead, 1971; Alexander and Crook, 1973, 1975; Gunderson et al., 1986; Olmsted et al., 1980). As a consequence, willow is commonly found around alligator holes or ponds. Craighead (1971) reports that willow has become much more widespread due to changes in hydrology and the associated impacts of dry season fires.

Cypress Forests

Cypress forests are a relatively minor feature of the Everglades, occurring in the southern Everglades and along the western border with the Big Cypress area. Two types of forests occur—domes and cypress prairie—and both are dominated by pond cypress (*Taxodium ascendens*).

The domes are characterized by dense, tall pond cypress. Those with lowest soil surface elevations have understories comprised of a few aquatic species. Domes with higher soil elevations support many of the hardwood species found in bayheads (red bay, swamp bay, wax myrtle, and cocoplum) and have been called cypress heads (Olmsted et al., 1980).

Cypress prairies of the Everglades were first described by Small (1933). Also called dwarf or hatrack cypress, the cypress trees have a stunted growth form and are widely spaced, with a grassy or prairie-like understory composed of sawgrass, muhly grass (*Muhlenbergia filipes*), and other herbs and grasses.

Pond Apple Forests

Prior to conversion to agriculture, a large forest dominated by pond apple, with some willow and elderberry (*Sambucus simpsonii*), was located on the southern rim of Lake Okeechobee (Davis, 1943). Although this forest is gone, small stands of pond apple still occur throughout the Everglades.

Sawgrass Marshes

Sawgrass (*Cladium jamaicense*) is the common, characteristic plant species of the freshwater Everglades system. Sawgrass is a rhizomatous, perennial sedge, rather than a grass, as its common name implies. The plant is well adapted to the conditions of flooding and burning which occur in the Everglades. Although capable of surviving variable water depths, from dry soil to flooding of the lower portions of the plant, sawgrass is killed if high water levels are prolonged (Davis, 1990; Hofstetter and Parsons, 1979; Herndon et al., 1991). Sawgrass also has low nutrient requirements (Steward and Ornes, 1975), a characteristic which promotes its dominance in the oligotrophic waters of the Everglades. The adaptation of sawgrass to fire has been well documented (Craighead, 1971; Davis, 1943; Forthman, 1973; Hofstetter and Parsons, 1979; Wade et al., 1980; Yates, 1974). Its leaves are extremely flammable (Wade, 1980). Plants normally survive burning, as the meristem is protected by spongy tissue (Conway, 1938) which is inflammable except under extreme drought conditions. With the meristem intact, regrowth is rapid following a fire (Forthman, 1973; Tilmant, 1975) and preburn structure (height and biomass) is attained within 2 years (Loveless, 1959).

Two types of sawgrass marsh occur in the Everglades: tall stature or dense marsh and short stature or intermediate sawgrass marsh. Apparently, soil depth accounts for much of the observed difference in size. At sites with deeper accumulations of peat (over 1 m), sawgrass attains heights of 3 m. At sites with less organic soil, sawgrass normally attains heights of 80–150 cm. Both types of marsh are dominated by sawgrass in terms of biomass and density. Few other species occur in the tall, dense marshes (Craighead, 1971), except for some woody plants, such as willow or pond apple, which establish in openings in or on the border of dense marshes. Only 14 other species have been found in association with sawgrass in the sparse sawgrass marshes (Olmsted and Loope, 1984). Other plants found include spikerush (*Eleocharis cellulosa*), water hyssop (*Bacopa caroliniana*), marsh mermaid weed (*Proserpinaca palustris*), and *Ipomoea sagitatta.*

Sawgrass marshes in the northern and eastern Everglades have been invaded by woody exotic trees, resulting in a shift from a grassland to a wooded wetland. *Melaleuca* seed was sown from aircraft over the Everglades in the late 1920s by a well-intentioned forester, J. C. Gifford (Meskimen, 1962). Many of these trees survive today and along with other intentional plantings (for windbreaks) form the nuclei for rapidly spreading populations (Myers, 1975).

Wet Prairies (Peat)

Wet prairies include a group of low stature, graminoid marshes. They are found over peat and marl, and each soil type supports distinct communities. Wet prairies over peat occur in the central, wetter areas of the Everglades. Those over marl are found in the southern Everglades on higher, drier sites (Craighead, 1971; Davis, 1943).

Wet prairies are generally described by the dominant plant found in association with them. Loveless (1959) described these associations over peat as *Eleocharis, Rhynchospora,* and *Panicum* flats. Craighead (1971) described them as spikerush and sedge flats. Common emergent aquatic plants which are found in wet prairies include spikerush (*Eleocharis cellulosa*), beakrush (*Rhynchospora tracyi),* maidencane (*Panicum hemitomon*), arrowhead (*Sagittaria lancifolia*), and pickerel weed (*Pontederia lanceolata*). At least 25 taxa occur in these associations, but generally spikerush, beakrush, and maidencane dominate. Submerged aquatics include ludwigia (*Ludwigia* spp.) and bladderworts (*Utricularia* spp.). Periphyton is a conspicuous and important feature of these associations (Browder et al., 1994). As with sawgrass marshes, wet prairies have been invaded by *Melaleuca* within the last 50 years.

Sloughs

Slough communities encompass all associations of floating aquatic plants and are found on the lowest, wettest sites in the Everglades. Dominant macrophytes include white water lily (*Nymphaea odorata*), floating hearts (*N. aquatica*), and spatterdock (*Nuphar advena*). The remainder of the flora of these associations is comprised of submerged aquatics and periphyton. The common submerged aquatics are the bladderworts—leafy bladderwort (*Utriculara foliosa*) and two-flower bladderwort (*U. biflora*). The submerged aquatics provide structure for periphyton and are a key component in what is described as the periphyton mat complex (Browder et al., 1981, 1982, 1994).

Wet Prairies (Marl)

Wet prairies over marl substrates occur in the southern Everglades on the east and west margins of the Shark River Slough and Taylor Slough, where bedrock elevations are slightly higher and hydroperiods shorter. These communities have been called the southern coastal marsh prairies (Davis, 1943), southeast saline Everglades (Egler, 1952), marl Everglades, marl prairies (Harper, 1927; Werner, 1975; Olmsted and Loope, 1984), *Muhlenbergia* prairies (Olmsted et al., 1980), and southern marl marshes (Davis et al., 1994).

Most of the marl prairies in Everglades National Park are dominated by two species: muhly grass and sawgrass. Other species that can be locally dominant include blackrush (*Schoenus nigricans*), arrowfeather (*Aristida purpurascens*), Florida bluestem (*Schizachyrium rhizomatum*), and Elliot's lovegrass (*Eragrostis elliottii*). Beakrush is common in the lower, wetter areas of the marl prairies. The association is diverse. Olmsted and Loope (1984) reported over 100 species, the majority of which are herbaceous plants; yet these comprise less than 1% of the ground cover (Olmsted et al., 1980). Typically, the graminoid vegetation is less than 1 m tall. Olmsted et al. (1980) found from 9 to 12 species per square meter. Above-ground biomass ranges from 50 to 500 g·m^{-2}, depending on soil depth, hydroperiod, and time since fire (Herndon and Taylor, 1986).

The marl prairies have undergone changes in hydrologic and fire regime, with changes in vegetation resulting. One such shift has been the increasing abundance of muhly grass (Atwater, 1954). Other changes involve the establishment of both native and nonnative woody hardwoods in some prairie areas. The southeastern coastal prairies have been invaded by *Casuarina* since the 1950s (Egler, 1952). Areas to the north (known as the East Everglades) show significant invasion by *Melaleuca* and *Casuarina* (Craighead, 1971; Wade et al., 1980; Bodle et al., 1994).

SCALES of ENVIRONMENTAL PROCESSES

The remainder of this chapter provides a conceptual cross-scale framework for comparison of the environmental factors influencing the distribution and abundance of vegetation communities. These factors can be grouped into two broad categories: (1) hydrologic-edaphic conditions and (2) disturbances. The variables associated with both categories operate on varying spatial and temporal scales, but can be hypothetically grouped into three categories (i.e., fast, medium, and slow) (Holling, 1986) of different space/time scales. Fast variables operate on spatial scales less than 100 m^2 and within time frames of months and years. Intermediate variables operate on scales of tens of years, with spatial domains on the order of tens of square kilometers The slow variables change over centuries and hundreds of square kilometers. The component variables of each of the categories (such as soil type and hydroperiod) interact within these groupings, as well as between classes (such as fire effects and hydroperiod). The interactions of these categories of variables are theorized to account for much of the spatial and temporal variation in the vegetation communities (Table 13.2).

Hydrologic-Edaphic Processes

One of the important hydrologic processes that affect vegetation is the physical hydrology of a site. It is most readily characterized by measures of depth of flooding and duration of soil saturation, or hydroperiod. Physical hydrology is a

Table 13.2 Hypothetical Spatial and Temporal Domains of Environmental Processes Influencing Vegetation in the Everglades

Temporal scale (years)	Spatial scale (m)		
	<1–3	<100–300	>100–300
Fast (1–10)	Grassy vegetation regrowth to steady-state biomass	Nonsevere fires	Hydropattern Water quality
Intermediate (10–100)	Residual impact of alligator activity (holes and nest mounds) Woody vegetation regrowth to steady-state biomass		Severe fires Hurricanes Freezes Drought/flood cycle
Slow (>100)	Soil accretion to stable depth Microbedrock topography		State of sea level Macrobedrock topography

Note: Groupings are approximate, based on qualitative estimates of range and central tendency.

result of the interaction between the topography (soil surface elevation) and the net integration of all phases of the hydrologic cycle. The interplay between water chemistry and vegetation is also an important determinant of vegetation patterns (Davis, 1994).

Topography

The soil surface elevation is a function of the underlying bedrock topography and accumulations of organic soils. Modifications (primarily dissolution) of the bedrock topography operate in the regime of slow variables. Rates of soil accretion are also slow variables, dependent primarily upon the extant vegetation and physiochemical hydrologic regime. Exceptions to these rates are found in the battery formation process (Gleason and Stone, 1994) and effects of severe fire.

Soil type reflects a long-term integration of topographic and hydrologic regimes and vegetation type (Gleason and Stone, 1994). Peat accretion occurs in wetland vegetation types—primarily slough, sawgrass, and tree islands—on sites with the longest hydroperiods. The correlation between the histosols of the Everglades and vegetation type is shown by the soil nomenclature. Everglades peat is comprised of sawgrass remains, Loxahatchee peat is made up of water lily and other emergent plants, and Gandy peat occurs beneath tree islands (Davis, 1943; Craighead, 1971). Marl (Perrine and Flamingo marls) is a calcitic mud generated by reprecipitation of calcium carbonate by the blue-green algal mat in the seasonally inundated, short-hydroperiod marshes (Browder et al., 1994). The rockland soils are outcrops of Miami or oolitic limestone. Sand, organic matter, and marl are found in the depressions of the pitted rock substrate, while variable thicknesses of leaf litter accumulate in the time periods between fires.

Little spatial variation of soil surface elevations reflects the flat bedrock topography and the slow rates of modification. The overall topographic gradient is on the order of $2 \text{ cm} \cdot \text{km}^{-1}$ in a north-south direction, but larger variations occur on smaller spatial scales. The range of relative elevation between the lowest and highest vegetation communities is on the order of 1.5 m. Distinct, quantitative differences in soil surface elevations among communities have not been established because the intracommunity variations exceed the differences among community types. Elevational differences interact with temporally variable rainfall, resulting in differences in the hydrologic regime. The plant community types can be qualitatively ranked from lowest to highest: ponds, sloughs, wet prairies (peat), sawgrass marshes, tree islands, marl prairies, pine forests, and tropical hammocks (Davis, 1943; Gunderson and Loftus, 1993).

Hydrologic Regime

The hydrologic regimes that affect the vegetation communities are influenced by three variables, each operating in a characteristic spatial/temporal domain. The slow variable is the state of sea level. The oldest Everglades peats date back 5000 years before present (Gleason et al., 1984), a time that loosely correlates to the phase in which a relatively rapid rise in sea level occurred (Fairbridge, 1984;

Robbin; 1984, Gleason and Stone, 1994; Wanless et al., 1994). This rise created the current hydrologic conditions for the Everglades, including the current mosaic of vegetation communities (Olmsted and Loope, 1984). The intermediate hydrologic variable is measured by the return periods of droughts and floods. The flood/drought cycle is determined by macro- and mesoscale weather phenomena, such as El Niño. Dramatic shifts in type of vegetation community and composition can occur during floods and droughts. Prolonged flooding modifies forested communities by increasing mortality of hardwood species (Craighead, 1971; Gunderson et al., 1988; Worth, 1987). Extended flooding may also kill sawgrass and other sensitive graminoid species. Drought effects are discussed later in relation to fire regimes.

The fast hydrologic variable is the annual hydrologic regime. Annual hydropatterns in plant communities are driven by seasonal variation in rainfall. The summer rains cause water levels to rise and reach annual maxima by September and October in most years. As rainfall decreases through the fall and winter months, water levels decline and reach annual minima during the spring months. Davis (1943) first related the influence of hydroperiod to the distribution of these communities. As long-term hydrologic data became available, quantification of the hydrologic regime within some of these plant communities was possible. Gunderson (1990) studied water level records in five wetland communities of Everglades National Park and found that historical hydropatterns did not differ statistically among community types, due to the high year-to-year variability. The wettest associations are the ponds and slough communities, with year-round inundation and mean annual water depths of 30 cm (Gunderson, 1990). All remaining wetland communities are inundated for at least some period during the year. The driest graminoid wetland is marl prairie, where inundation averages 3–7 months/year and mean depths average 10 cm (Olmsted et al., 1980). The hardwood tree islands are inundated for shorter periods, averaging 4–5 months/year (Olmsted et al., 1980). The pine forests can become locally inundated during extremely rare rain events for periods of approximately 1 month. Tropical hammocks are rarely, if ever, inundated (Gunderson and Loope, 1981). The loose correlation between plant community type and annual hydrologic regime may be explained in part by the hierarchy of hydrologic variables. The intermediate variable (i.e., drought and flood interval) appears to influence the composition of the community (Worth, 1987; Gunderson et al., 1988), while the fast variable (i.e., annual hydropattern) affects processes such as primary productivity, decomposition, water flux, and phenologic/reproductive activity of the vegetation.

Two aspects of water chemistry influence the distribution of vegetation: salinity and nutrients. The border between the freshwater and saline wetlands is a large, dynamic zone determined by the equilibrium between the freshwater overland flow and the downstream tidal influence. This zone is related to the slow and intermediate hydrologic variables (e.g., sea level, coastal storm tides, and frequency of drought). Most of the native vegetation of the Everglades is adapted to low nutrient levels, including both wetland vegetation such as sawgrass (Steward and Ornes, 1975; Davis, 1994) and upland pineland vegetation (Snyder, 1986).

Disturbances

The disturbances which influence the communities include fire, hurricanes, frosts, and animal activity (Davis, 1943; Loveless, 1959; Craighead, 1971). Fire plays a significant role in the ecology of all of the rockland and freshwater wetland associations. Lightning and hurricanes have significant effects on the forested communities and less significant impacts on other vegetation types. Freezes influence sensitive species in all communities. Alligators are the most important animal affecting the freshwater wetland associations (Craighead, 1971). As with hydrologic variables, the effects of disturbance variables have characteristic spatial and temporal domains; the scales of most (except animals) appear to fall into the intermediate category.

Fire is inferred to have two qualitatively different impacts on vegetation. More frequent fires appear to burn smaller areas and have short-term impacts on vegetation, whereas fires with severe impact have longer return periods, burn larger areas, and can result in dramatic changes in vegetation. These severe fires have been a part of the ecology of the Everglades throughout the existence of current vegetation patterns, as evidenced by the presence of charcoal through-out the basal peats of the wetlands (Cohen, 1984; Gleason and Stone, 1994). The impact of severe fires is related to the consumption and alteration of organic soil. Subtle changes in species composition can occur in the high-frequency fire regimes.

The floristic integrity of the pine forests is maintained by high-frequency, low-impact fires. These fires burn pine forests on the average every 3–7 years (Robertson, 1953; Hofstetter, 1976; Taylor, 1981; Wade et al., 1980) and consume fuels, including pine needle litter, standing grasses, palms, and hardwood leaves and stems (Snyder, 1986). Plants top-killed by the fire resprout readily from protected meristems. Frequent fires prevent the understory hardwoods from domi-nating a site, a conversion that has been documented to occur within a 25-year absence of fire (Robertson, 1953; Alexander, 1967; Hofstetter, 1976). Survival of the rich endemic flora is dependent upon maintenance of an open canopy associated with frequent fires (Robertson, 1953; Wade et al., 1980; Loope et al., 1979). While the pine forests are maintained by the fast fire variable, hammocks appear better adapted to a more intermediate fire regime. Most of the hammocks of Long Pine Key have been burned within the last 40 years (Olmsted et al., 1983), yet have recovered rapidly via two mechanisms: resprouting from undamaged root systems and regeneration via external seed sources (Robertson, 1953).

The same two groups of fire regimes and their associated effects influence the communities of the wetland complex. One critical factor that determines impacts is the ambient moisture condition. Frequent fires burn sawgrass communities (Robertson, 1953; Forthman, 1973; Yates, 1974; Hofstetter, 1976; Werner, 1975; Wade et al., 1980), resulting in rapid sawgrass regrowth (Loveless, 1959; Forthman, 1973; Yates, 1974) and little or no change in species composition. Less frequent, severe fires, which occur during drought conditions, can remove the organic soils, lowering soil elevation and altering plant communities. Peat-consuming fires can result in the conversion of sawgrass marsh to *Eleocharis* flat (Craighead, 1971) and bayhead tree island communities to willow (Wade et al., 1980).

Impacts associated with hurricanes occur as a result of strong winds, heavy rainfall, storm surge, and increased wave action (Craighead and Gilbert, 1962). Mangrove forests are affected by all of these processes, whereas inland types are affected by wind and rain. High winds defoliate the vegetation and uproot trees of the woody associations, such as large mahogany trees (Craighead, 1971). Heavy rains can produce some flood mortality. Massive mortality occurred in the mangrove forests of Everglades National Park following Hurricane Donna in 1960, after the storm surge and waves deposited a fine mud which suffocated the root systems and the high winds defoliated the trees (Craighead, 1971). The woody plant communities in Everglades National Park were hardest hit by Hurricane Andrew in 1992. Andrew caused uprooting and defoliation of broad-leaved trees and thinning of pine forests when high winds blew over trunks. Five months after the storm, there still was no well-developed canopy structure, although there was major regrowth from stumps and seedlings. Nonforested wetlands did not appear to be affected by Andrew in terms of structural impacts.

The influence of animals on vegetation patterns is related to both local and regional patterns. Alligators influence soil elevations by the physical sculpting associated with their movement activity. Alligator nests may also be incipient tree islands, because they serve as foci for colonization by swamp hardwoods, which then may cascade in size to larger tree island. Both of these are small-scale (<10 m^2) effects. More mobile species, such as birds, can influence post-disturbance recovery by acting as vectors of seed movement over a broader range of spatial scales.

LITERATURE CITED

Alexander, T. R. 1967. A tropical hammock on the Miami limestone: A twenty-five year study. *Ecology,* 48:863–867.

Alexander, T. R. and A. G. Crook. 1973. Recent and Long-Term Vegetation Changes and Patterns in South Florida. Part I: Preliminary Report, Appendix G, South Florida Ecological Study, Report to the National Park Service, 215 pp.

Alexander, T. R. and A. G. Crook. 1975. Recent and Long-Term Vegetation Changes and Patterns in South Florida. Part II: Final Report, South Florida Ecological Study, Report to the National Park Service, 827 pp.

Atwater, W. G. 1954. Hair grass takes over. *Everglades Nat. Hist.,* 2:43.

Avery, G. N. and L. L. Loope. 1980. Endemic Taxa in the Flora of South Florida, Tech. Rep. T-558, South Florida Research Center, National Park Service, U.S. Department of the Interior, Homestead, Fla.

Avery, G. N. and L. L. Loope. 1983. Plants of the Everglades National Park: A Preliminary Checklist of Vascular Plants, 2nd ed., Tech. Rep. T-574, South Florida Research Center, National Park Service, U.S. Department of the Interior, Homestead, Fla.

Bodle, M. J., A. P. Ferriter, and D. D. Thayer. 1994. The biology, distribution, and environmental consequences of *Melaleuca quinquenervia* in the Everglades. in *Everglades: The Ecosystem and Its Restoration,* S. M. Davis and J. C. Ogden (Eds.), St. Lucie Press, Delray Beach, Fla., chap. 14.

Browder, J. A., S. Black, P. Schroeder, M. Brown, M. Newman, D. Cottrell, D. Black, R. Pope, and B. Pope. 1981. Perspective on the Ecological Causes and Effects of the Variable Algal

Composition of Southern Everglades Periphyton, Tech. Rep. T-643, South Florida Research Center, National Park Service, U.S. Department of the Interior, Homestead, Fla.

Browder, J. A., S. Black, and D. Black. 1982. Comparison of Laboratory Growth of *Hyla squirrella* Tadpoles on Everglades Periphyton of Three Different Taxonomic Compositions, Draft Report to Everglades National Park, Rosenstiel School of Marine and Atmospheric Science, University of Miami, Miami.

Browder, J. A., P. J. Gleason, and D. R. Swift. 1994. Periphyton in the Everglades: Spatial variation, environmental correlates, and ecological implications. in *Everglades: The Ecosystem and Its Restoration,* S. M. Davis and J. C. Ogden (Eds.), St. Lucie Press, Delray Beach, Fla., chap. 16.

Cohen, A. D. 1984. Evidence of fires in the ancient Everglades and coastal swamps of southern Florida. in *Environments of South Florida: Present and Past, Memoir 2,* P. J. Gleason (Ed.), Miami Geological Society, Coral Gables, Fla.

Conway, V. C. 1938. Studies on the autecology of *Cladium mariscus* R. Br., the growth of leaves. *New Phytol.,* 37:254–278.

Craighead, F. C., Sr. 1971. *The Trees of South Florida,* Vol. I: The Natural Environments and Their Succession, University of Miami Press, Coral Gables, Fla.

Craighead, F. C. and V. C. Gilbert. 1962. The effects of Hurricane Donna on the vegetation of southern Florida. *Q. J. Fla. Acad. Sci.,* 25:1–28.

Davis, J. H., Jr. 1943. The natural features of southern Florida, especially the vegetation, and the Everglades. *Fla. Geol. Surv. Bull.,* No. 25.

Davis, S. M. 1990. Sawgrass and Cattail Production in Relation to Nutrient Supply in the Everglades, Fresh Water Wetlands and Wildlife 9th Annu. Symp., R. R. Sharitz and J. W. Gibbons (Eds.), Savannah River Ecology Laboratory, Charleston, S.C., March 1986, pp. 24–27.

Davis, S. M. 1994. Phosphorus inputs and vegetation sensitivity in the Everglades. in *Everglades: The Ecosystem and Its Restoration,* S. M. Davis and J. C. Ogden (Eds.), St. Lucie Press, Delray Beach, Fla., chap. 15.

Davis, S. M., L. H. Gunderson, W. A. Park, J. Richardson, and J. Mattson. 1994. Landscape dimension, composition, and function in a changing Everglades ecosystem. in *Everglades: The Ecosystem and Its Restoration,* S. M. Davis and J. C. Ogden (Eds.), St. Lucie Press, Delray Beach, Fla., chap. 17.

Egler, F. E. 1952. Southeast saline Everglades vegetation, Florida, and its management. *Vegetatio Acta Geobot.,* 3:213–265.

Fairbridge, R. 1984. The Holocene sea level record in south Florida. in *Environments of South Florida: Present and Past, Memoir 2,* P. J. Gleason (Ed.), Miami Geological Society, Coral Gables, Fla., pp. 427–436.

Forthman, C. A. 1973. The Effects of Prescribed Burning on Sawgrass, *Cladium jamaicense* Crantz, in South Florida, M.S. thesis, University of Miami, Coral Gables, Fla.

Gleason, P. J. and P. A. Stone. 1994. Age, origin, and landscape evolution of the Everglades peatland. in *Everglades: The Ecosystem and Its Restoration,* S. M. Davis and J. C. Ogden (Eds.), St. Lucie Press, Delray Beach, Fla., chap. 7.

Gleason, P. J., A. D. Cohen, W. G. Smith, H. K. Brooks, P. A. Stone, R. L. Goodrick, and W. Spackman, Jr. 1984. The environmental significance of Holocene sediments from the Everglades and saline tidal plain. in *Environments of South Florida: Present and Past, Memoir 2,* P. J. Gleason (Ed.), Miami Geological Society, Coral Gables, Fla., pp. 297–351.

Gunderson, L. H. 1990. Historical Hydropatterns in Vegetation Communities of Everglades National Park. Freshwater Wetlands and Wildlife, Charleston, SC, Savannah River Ecology Laboratory, Aiken, S.C.

Gunderson, L. H. and W. F. Loftus. 1993. The Everglades: Competing land uses imperil the biotic communities of a vast wetland. in *Biotic Communities of the Southeastern United States,* W. H. Martin, S. C. Boyce, and A. C. Echternacht (Eds.), John Wiley & Sons, New York.

Gunderson, L. H. and L. L. Loope. 1981. Plant Communities of the Pinecrest Area, Big Cypress National Preserve, Report T-666, South Florida Research Center, National Park Service, U.S. Department of the Interior, Homestead, Fla.

Gunderson, L. H., D. P. Brannon, and G. Irish. 1986. Vegetation Cover Types of Shark River

Slough, Everglades National Park, Derived from LANDSAT Thematic Mapper Data, Tech. Rep. SFRC-86/03, South Florida Research Center, National Park Service, U.S. Department of the Interior, Homestead, Fla.

Gunderson, L. H., J. R. Stenberg, and A. K. Herndon. 1988. Tolerance of five hardwood species to flooding regimes in south Florida. in *Interdisciplinary Approaches to Freshwater Wetlands Research,* D. J. Wilcox (Ed.), Michigan State University Press, Ann Arbor.

Harper, R. M. 1927. Natural Resources of Southern Florida, 18th Annu. Rep., Florida Geological Survey, Tallahassee.

Harshberger, J. W. 1914. The vegetation of south Florida, south of 27 degrees 30′ north, exclusive of the Florida Keys. *Trans. Wagner Free Inst. Sci.,* 7:49–189.

Herndon, A. K. and D. Taylor. 1986. Response of a *Muhlenbergia* Prairie to Repeated Burning: Changes in Above-Ground Biomass, Tech. Rep. SFRC-86/05, South Florida Research Center, National Park Service, U.S. Department of the Interior, Homestead, Fla.

Herndon, A. K., L. H. Gunderson, and J. R. Stenberg. 1991. Sawgrass (*Cladium jamaicense*) survival in a regime of fire and flooding. *Wetlands,* 11:17–27.

Hofstetter, R. H. 1976. Effects of Fire in the Ecosystem, Part I. Appendix K in South Florida Environmental Project, University of Miami, Coral Gables, Fla.

Hofstetter, R. H. and F. Parsons. 1979. The Ecology of Sawgrass in the Everglades of Southern Florida, Proc. 1st Conf. on Scientific Research in the National Parks, R. M Linn (Ed.), National Park Service Transaction and Proceedings Series, No. 5, U.S. Department of the Interior, Washington, D.C., pp. 165–170.

Holling, C. S. 1986. The resilience of terrestrial ecosystems: Local surprise and global change. in *Sustainable Development of the Biosphere,* W. C. Clark and R. E. Munn (Eds.), IIASA, Laxenburg, Austria.

Johnson, J. M., I. C. Olmsted, and O. L. Bass, Jr. 1983. Vegetation Map of Long Pine Key, South Florida Research Center, Everglades National Park, U.S. Department of the Interior, Homestead, Fla.

Long, R. W. 1974. The vegetation of southern Florida. *Fla. Sci.,* 37(1):33–45.

Long, R. W. and O. Lakela. 1971. *Flora of Tropical Florida,* University of Miami Press, Coral Gables, Fla., 962 pp.; Banyon Books, Miami.

Loope, L. L., D. W. Black, S. Black, and G. N. Avery. 1979. Distribution and Abundance of Flora in Limestone Rockland Pine Forests of Southeastern Florida, Tech. Rep. T-547, South Florida Research Center, National Park Service, U.S. Department of the Interior, Homestead, Fla.

Loveless, C. M. 1959. A study of the vegetation of the Florida Everglades. *Ecology,* 40:1–9.

Meskimen, G. F. 1962. A Silvical Study of the Melaleuca Tree in South Florida, M.S. thesis, University of Florida, Gainesville.

Myers, R. L. 1975. The Relationship of Site Conditions to the Invading Capability of *Melaleuca quinquenervia* in Southwest Florida, M.S. thesis, University of Florida, Gainesville.

Olmsted, I. C. and L. L. Loope. 1984. Plant communities of Everglades National Park. in *Environments of South Florida: Present and Past II,* P. J. Gleason (Ed.), Miami Geological Society, Coral Gables, Fla.

Olmsted, I. C., L. L. Loope, and R. E. Rintz. 1980. A Survey and Baseline Analysis of Aspects of the Vegetation of Taylor Slough, Everglades National Park, Report T-586, South Florida Research Center, National Park Service, U.S. Department of the Interior, Homestead, Fla.

Olmsted, I. C., L. L. Loope, and C. E. Hilsenbeck. 1981. Tropical Hardwood Hammocks of the Interior of Everglades National Park and Big Cypress National Preserve, Report T-604, South Florida Research Center, National Park Service, U.S. Department of the Interior, Homestead, Fla.

Olmsted, I. C., W. B. Robertson, Jr., J. Johnson, and O. L. Bass, Jr. 1983. Vegetation of Long Pine Key, Everglades National Park, Tech. Rep. SFRC-83/05, South Florida Research Center, National Park Service, U.S. Department of the Interior, Homestead, Fla.

Rintz, R. E. and L. L. Loope. 1979. Vegetation Map of Taylor Slough, Everglades National Park, Everglades National Park, National Park Service, U.S. Department of the Interior, Homestead, Fla.

Robbin, D. M. 1984. A new Holocene sea level curve for the Upper Keys and Florida reef tract. in *Environments of South Florida: Present and Past, Memoir 2,* P. J. Gleason (Ed.), Miami Geological Society, Coral Gables, Fla., pp. 437–458.

Robertson, W. B., Jr. 1953. A Survey of the Effects of Fire in Everglades National Park, Mimeo, Everglades National Park, National Park Service, U.S. Department of the Interior, Homestead, Fla.

Russell, R. P., L. L. Loope, and I. C. Olmsted. 1980. Vegetation Map of the Coastal Region Between Flamingo and Joe Bay of Everglades National Park, Everglades National Park, National Park Service, U.S. Department of the Interior, Homestead, Fla.

Small, J. K. 1932. *Manual of the Southeastern Flora,* University of North Carolina Press, Chapel Hill.

Small, J. K. 1933. An Everglades cypress swamp. *J. N.Y. Bot. Gard.,* 34:261–267.

Snyder, J. R. 1986. Impact of Wet Season and Dry Season Prescribed Fires on Miami Rock Ridge Pineland, Everglades National Park, Tech. Rep. SFRC-86/06, South Florida Research Center, National Park Service, U.S. Department of the Interior, Homestead, Fla.

Snyder, J. R., A. K. Herndon, and W. B. Robertson, Jr. 1990. Rockland forests. in *Ecosystems of Florida,* R. Myers and J. J. Ewel (Eds.), University Presses of Florida, Gainesville.

Steward, K. K. and W. H. Ornes. 1975. The autecology of sawgrass in the Florida Everglades. *Ecology,* 56:162–171.

Taylor, D. L. 1981. Fire History and Fire Records for Everglades National Park, 1948–1979, Tech. Rep. T-619, South Florida Research Center, National Park Service, U.S. Department of the Interior, Homestead, Fla.

Taylor, D. L. and A. K. Herndon. 1981. Impact of 22 Years of Fire on Understory Hardwood Shrubs in Slash Pine Communities within Everglades National Park, Tech. Rep. T-640, South Florida Research Center, National Park Service, U.S. Department of the Interior, Homestead, Fla.

Tilmant, J. T. 1975. Habitat Utilization by Round-Tailed Muskrats, *Neofiber alleni,* in Everglades National Park, M.S. thesis, Humboldt St. University, Arcata, Calif.

Wade, D. D., J. J. Ewel, and R. Hofstetter. 1980. Fire in South Florida Ecosystems, General Tech. Rep. SE-17, SE Forest Experimental Station, Forest Service, U.S. Department of Agriculture, Asheville, N.C., 125 pp.

Wanless, H. R., R. W. Parkinson, and L. P. Tedesco. 1994. Sea level control on stability of Everglades wetlands. in *Everglades: The Ecosystem and Its Restoration,* S. M. Davis and J. C. Ogden (Eds.), St. Lucie Press, Delray Beach, Fla., chap. 8.

Werner, H. W. 1975. The Biology of the Cape Sable Sparrow, M.S. thesis, University of South Florida, Tampa.

Worth, D. 1987. Environmental Responses of Water Conservation Area 2A to Reduction in Regulation Schedule and Marsh Drawdown, Tech. Rep. 87-5, South Florida Water Management District, West Palm Beach.

Yates, S. A. 1974. An Autecological Study of Sawgrass, *Cladium jamaicense,* in Southern Florida, Master's thesis, University of Miami, Coral Gables, Fla., 93 pp.

14

The BIOLOGY, DISTRIBUTION, and ECOLOGICAL CONSEQUENCES of *MELALEUCA QUINQUENERVIA* in the EVERGLADES

Michael J. Bodle

Amy P. Ferriter

Daniel D. Thayer

ABSTRACT

The spread of the Australian melaleuca tree (*Melaleuca quinquenervia*) across the Everglades began with its introduction into south Florida early in the 20th century. Because the tree has tremendous reproductive, invasive, and survival capabilities, it has overtaken large areas.

Melaleuca primarily infests the Florida peninsula south of Lake Okeechobee. This area comprises about 7,500,000 total acres (3,035,000 ha), of which 500,000–1.5 million acres (202,000–607,000 ha) is estimated to be melaleuca infested. The degree of infestation varies from single trees to thousands of trees per acre. Past experience and a rate of expansion study indicate that the uncontrolled trees could overtake most of the region's remaining natural land within 30 years. Several mapping projects have been conducted or proposed to establish its range. Many public land management agencies are striving to eliminate the plant from their area of responsibility.

Melaleuca control techniques include herbicidal, biological, mechanical, and physical methods. Herbicidal control is currently the most practical and economically feasible technique. Biological control may result from the planned winter 1994 release of Australian melaleuca-feeding insects. Hopefully, these insects will halt the plant's spread by consuming new shoots, seedlings, flowers, and seeds. Mechanical removal is very expensive and can scar sensitive wetlands.

Melaleuca has been declared both a Federal Noxious Weed and a Florida Prohibited Aquatic Plant. These designations have prohibited its import into the United States and its transport within Florida, but have not financially supported ongoing control work. Also, Everglades restoration initiatives have not mandated continual melaleuca management. Therefore, only fitful progress has been made on management of the plant and research initiatives into its control. Thorough melaleuca management will require federal and state dedication and establishment of effective biological control.

INTRODUCTION

Early in this century, man introduced the melaleuca tree, *Melaleuca quinquenervia,* into Florida as a fast-growing ornamental tree and into the Everglades as a means to "drain the swamp." This was during a time when people such as Florida's Governor Napoleon Bonaparte Broward were proclaiming the greatness of drainage projects to promote growth in the state. Indeed, he won the governor's race with a "drain the Everglades" plank in his platform.

The silviculturists who championed the introduction of melaleuca as a means of Everglades "reclamation" succeeded only too well. However, casual observers of today's version of the Everglades usually presume that the ecosystem is still pristine and intact—in spite of the federal flood control project that bisects it and the development slowly gnawing at all its sides. Such presumptions are especially mistaken in light of 20th century water management practices in south Florida. Man's activities have permanently altered all remaining Everglades areas, and invasions of exotic plants figure prominently in these alterations (Figure 14.1).

This hardy, fast-growing tree, which was imported from Australia, has papery white bark resembling the white birch. These characteristics spurred its use as an

Figure 14.1 Melaleuca in the Everglades.

ornamental landscape tree, as agricultural windrows, and as protective living "guard rails" and soil stabilizers along canals. Melaleuca readily invades canal banks, buffer zones between pinelands and cypress areas, and uninterrupted sawgrass prairies (Myers, 1975, 1976; Austin, 1978; Duever et al., 1979; Wade et al., 1980; Woodall, 1981a, 1983). The potential spread of melaleuca in south Florida is variously held to be unlimited, ultimately encroaching upon all open land (Hofstetter, 1991), or limited to underutilized niches in the relatively young Florida landscape (Myers, 1975). However, acknowledgment of such alternative views embraces a common thread: melaleuca needs to be controlled—regardless of whether it could ultimately cover the peninsula.

Soil types fail to limit the ability of melaleuca to take hold. The tree grows equally well in the deep peat soil of the Arthur R. Marshall Loxahatchee National Wildlife Refuge (Water Conservation Area 1) in the northern portion of the remnant Everglades and in the inorganic, calcareous soil of western Dade County (Wade et al., 1980) (Figure 14.2). The incredible resilience of melaleuca and its capacity to invade all of south Florida's natural areas underscore the ecological imperative to control its spread.

BIOLOGICAL CHARACTERISTICS

Melaleuca quinquenervia is a member of the Australian Myrtaceae, myrtle family, which also includes the *Eucalyptus* (gum) and *Callistemon* (bottlebrush) genera. Many of the more than 3000 species of Myrtaceae are cultivated worldwide as sources of fruits, spices, aromatic oils, or timber. The flowers of *M. quinquenervia*, like most Myrtaceae, have numerous stamens on a cup-shaped receptacle with the ovary below. The leaves of Myrtaceae are simple and entire, and the plants are usually aromatic. The genus *Melaleuca* is native to Australia and includes more than 200 species. *M. quinquenervia* is the only representative of the genus in Florida. Australian common names include cajeput tree, punk tree, and broadleaf paperbark tree, its official common name. It is easily recognized by its white, spongy flaking bark, lanceolate longitudinally parallel-veined leaves, and clusters of woody seed capsules along the stems.

Saplings usually grow with a single main stem, although a damaged terminal bud is readily substituted by an auxiliary bud, resulting in new shoot growth along a new axis. Damage may also induce new shoot growth from the roots, resulting in multi-trunked trees or coppices. Because of these damage responses, older trees are often multi-stemmed.

Physical characters of *Melaleuca quinquenervia* are illustrated in Figure 14.3. Branching occurs at irregular intervals along the trunk. Vegetative and reproductive sections along each branch are distinct and alternate. Leaves are spirally arranged along the vegetative branch segments. Reproductive and vegetative segments are separated by overlapping scars of resting terminal bud scales or *scale-zones* (Tomlinson, 1980). New elongating branch growth occurs sporadically and varies from branch to branch. Periods of branch growth are separated by brief or lengthy periods of rest (Tomlinson, 1980). The relative length of the scale-zone increases with increased resting time between periods of shoot elongation. Reproduction

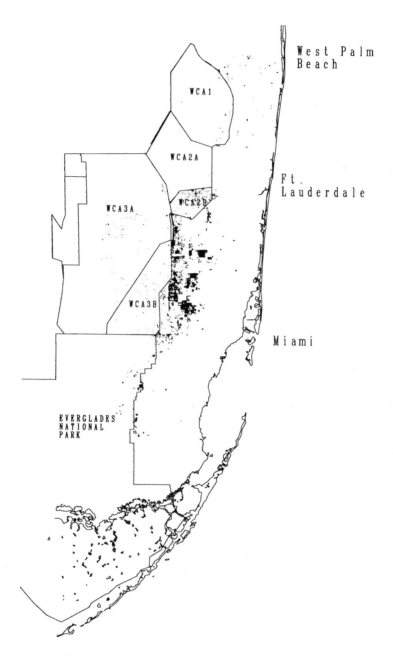

Figure 14.2 Mature melaleuca forests in southeast Florida. Shaded areas represent pure melaleuca stands 2.5 acres (1 ha) or larger.

occurs along flower-bearing branch segments. Previous inflorescences are apparent because of the persistent woody seed capsules. Different trees and different shoots on the same tree are nonsynchronous in their flowering.

An individual melaleuca tree may flower within 3 years of germination and as many as five times per year (Meskimen, 1962). Flowering occurs as vegetative

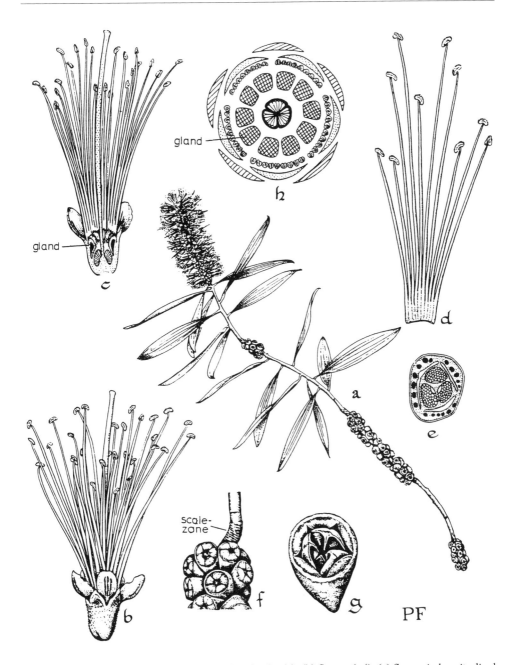

Figure 14.3 *Melaleuca quinquenervia:* (a) Habit (×1/2); (b) flower (×4); (c) flower in longitudinal section (×4); (d) stamen group (×6); (e) ovary in transverse section (×8); (f) unripe fruits (×2); (g) dehisced fruit (×4); (h) floral diagram. (Reprinted with permission from P. B. Tomlinson. 1980. *The Biology of Trees Native to Tropical Florida,* Harvard University Printing Office, Allston, Mass., p. 285. Illustration by Priscilla Fawcett.)

branch segments convert directly into flowering spikes. This conversion is preceded and succeeded by a resting phase indicated by the scale-zone. In Florida, main flowering periods are fall and summer, yet some flowering occurs all year. Flowers are clustered in trios, and each ovary is tripartite and surrounded by the calyx tube and ten hairy glandular outgrowths of the tube. These glands secrete nectar, which collects within the basal floral cavity. Six to ten stamens are opposite each of five sepals and five white petals. Flowers are pollinated by insects, and seed fertility is high. After ripening, the persistent woody capsules hold the mature seeds until desiccation induces dehiscence and the release of 200–300 very small seeds from each capsule. Capsules may remain on a branch for at least 10 years (Meskimen, 1962). A tree, or branches of a tree, will desiccate when stressed by freezing, drought, fire, herbicide treatment, or breakage. At such times, a single tree may release as many as 20 million seeds (Woodall, 1981b).

Melaleuca seeds germinate upon moist soils, usually within a few days of wetting, and may remain viable for up to 6 months under water or in wet soils (Meskimen, 1962; Myers 1975). Meskimen (1962) found that germination occurs more frequently in sun than in shade. The rarity of seedlings within dense stands of melaleuca may be from shading or allelopathic effects of melaleuca litter (DiStefano and Fisher, 1983). Seedlings less than several weeks or months old may die from fire or dry soils (Myers, 1975, 1983) but are soon able to withstand extreme conditions ranging from fire to total immersion for up to 6 months (Meskimen, 1962). Therefore, timing of fire, freeze, drought, herbicide application, breakage, and flooding events can determine whether melaleuca seedfall events lead to successful establishment and spread.

One computer model predicted that 99% of seeds released from "one tree during an ordinary year" (Browder and Schroeder, 1981) would disperse no further than 170 m, and of these only 5% would be viable. Hurricane (100-knot wind) events are predicted to cause maximum seed dispersal of 7 km. Because of the small dispersal zones predicted, this study suggested that control of outlier trees would be an effective control mechanism.

INTRODUCTION, ESTABLISHMENT, and POTENTIAL RANGE in SOUTH FLORIDA

In the early 20th century, several independent introductions of melaleuca were made in Florida, but John C. Gifford is commonly recognized as the first to introduce the species to southeast Florida in 1906 (Meskimen, 1962). Seedlings were planted at his home on Biscayne Bay in Coconut Grove and at the Frank Stirling and Sons Nursery in Davie, Broward County. Austin (1978) reported that "shortly thereafter" (i.e., after introduction) trees originating from these plantings in Davie began to invade the wet prairies and marshes of the Everglades. On the southwest coast, melaleuca was introduced through a similar, but independent, operation in Lee County, near Estero. In the early 1940s, trees were planted along the rim canal of Lake Okeechobee and at Monroe Station within today's Big Cypress National Preserve.

The subsequent spread of melaleuca has been described as explosive, with the rate of spread accelerating (Hofstetter, 1991; Cost and Craver, 1980). A time-series examination of melaleuca (Laroche and Ferriter, 1992) documents its invasion capacity. Rates for invasion of a sawgrass- (*Cladium jamaicense*) dominated community are described as exponential.

The potential range of melaleuca in Florida is typically noted as the entire peninsula south of Lake Okeechobee, excluding the saline zone. This tree is seemingly kept in check by frost, although Woodall (1981b) noted that most, if not all, mature melaleuca trees had survived severe frost damage during a record-breaking freeze in Florida in 1977. A few saplings were killed all the way to the ground line. However, sprouts were soon noted growing from the root collars, and planted specimens have survived far north of the range of naturalized populations in Florida. This suggests that the potential range of melaleuca may in fact stretch farther north than is currently accepted. Climatological models developed in Australia postulate that the entire Gulf coast of the United States provides conditions similar to those in the native Australian range of the tree (Center, 1993).

In south Florida, melaleuca grows in a variety of areas: along road-sides, on ditch banks, along rights-of-way, and in lake margins, pastures, pine flatwoods, mesic prairies, sawgrass marshes, and stands of cypress trees. Although salinity is thought to limit melaleuca encroachment, it also grows in mangrove areas (Center and Dray, 1986). Habits in its native range indicate a degree of salt tolerance, as frequent reports term melaleuca a "backmangrove species," growing directly behind mangroves and surviving both fresh and brackish water inundations (Myers, 1975).

The known growth characteristics and habitat requirements of melaleuca do not adequately explain the observed differences in its pattern of spread in Florida (Myers, 1975; Hofstetter, 1991). In southeastern Florida, melaleuca does not only invade areas where soil has been disturbed; rather, it has invaded essentially every existing community, including those in which native vegetation is seemingly healthy and vigorous. For example, melaleuca invades the ecotone between pine and cypress, successfully displacing cypress (Ewel et al., 1976). Because hydrologic patterns have been greatly altered in southeast Florida, and no freshwater wetlands remain in their natural (historical) state, virtually all of these communities could be considered indirectly disturbed to a significant degree and perhaps increasingly susceptible to exotic invasion.

Invasion potential of melaleuca tends to differ among plant communities. Wetter areas are more prone to melaleuca invasion than drier areas. In general, xeric communities such as scrub tend to be resistant, but not immune, to melaleuca invasion. When drier communities are wetter than normal, however, they may become more susceptible to invasion. Conversely, if a site is continually inundated for extended periods, seeds may be unable to germinate. Although prolonged inundation inhibits seed germination, and thus seedling establishment, it does not seem to affect established trees.

DISTRIBUTION and MAPPING

To date, no successful detailed inventory of melaleuca has been made in south Florida. Although a number of surveys have been conducted, each suffers from limitations (Hofstetter, 1991; Cost and Craver, 1981; Capehart et al., 1977; Arvanitis and Newburne, 1984). The earliest large-scale attempt to survey the species in south Florida was conducted by Capehart et al. in 1977. This effort used digital image analysis of LANDSAT data to determine the areal extent of melaleuca. The study aimed to provide baseline data which could be used in comparative studies with later surveys. However, "insurmountable obstacles" were encountered which prevented the production of a regionwide distribution map. These difficulties arose from the significant spectral reflectance variability found within melaleuca stands. Additionally, similar reflectance properties of other common south Florida vegetation (such as mangrove) caused erroneous classifications.

Cost and Craver (1981) conducted a large-scale survey of melaleuca for the U.S. Forestry Service and estimated that there were 40,300 acres (16,309 ha) of "pure" melaleuca in south Florida. In this study, observers registered the occurrence of melaleuca in over 7.6 million acres (3.08 million ha), as seen from fixed wing aircraft. Flight lines were spaced approximately 5 mi (8 km) apart and were flown in an east-west pattern. Pure melaleuca was defined as "stands that are at least 16.7 percent stocked wherein melaleuca comprises at least 50 percent of the stocking." The result was a dot map, with each mark roughly depicting the location of a melaleuca occurrence along the flight line. Melaleuca acreage was then statistically interpolated. This survey was repeated in 1987. A total of 47,000 acres (19,021 ha) was classified as "pure" melaleuca in the 1987 survey.

Arvanitis and Newburne (1984) evaluated the feasibility of using photographic remote sensing to define melaleuca within the region. This feasibility study was based on trials conducted on two test sites on both the southeast and southwest coasts of Florida. To detect melaleuca, including both individual trees and larger populations, analysis of color infrared photography at a scale of 1:12,000 was suggested. Because of the need for a high degree of contrast, winter photography was recommended for accurate interpretation. The recommendations of this feasibility study have never been applied.

The South Florida Water Management District (1992) recently completed an inventory of mature melaleuca monocultures within south Florida. This survey analyzed false color infrared photography at a scale of 1:40,000 and delineated mature, monotypic stands 2.5 acres (1 ha) or larger within south Florida. The estimated coverage for stands dominated by monotypical, mature melaleuca is 26,000 acres (10,522 ha). However, this survey did not include any infestations which were not pure or nearly pure and fully mature. Single trees, outlying small heads, and mixed plant community infestations were not included in this inventory. The Everglades portion of this regionwide map is shown in Figure 14.2.

Several factors can explain the discrepancies in these estimates of melaleuca area. First, the definition of "pure" melaleuca differs among the studies. Cost's 1980 Forestry Service survey defined "pure" as areas with greater than 50% melaleuca, while the South Florida Water Management District (1992) survey considered monocultures to be greater than 95% melaleuca. Also, the Forestry Service survey statistically estimated the area covered from observations made along flight lines flown at 5-mi (8-km) intervals. The South Florida Water Management District estimates were derived from the direct delineation and digitization of melaleuca areas from aerial photographs.

Both surveys indicate that the general distribution of melaleuca in south Florida predominantly centers around the areas of original introduction. The largest, most abundant monotypic stands of mature melaleuca in southeast Florida are concentrated west of urban southern Broward and northern Dade counties. Large, mature monotypic stands dominate the undeveloped land west of Pembroke Pines, Miramar, Cooper City, and Hialeah. Water Conservation Area 2B, located west of Fort Lauderdale, is the most severely infested area of the remnant Everglades. This area contains large, mature monocultures and substantial amounts of seedlings, saplings, and pioneering or "outlier" melaleuca. Moderate to heavy infestations (many outliers, small and large heads) occur in the Arthur R. Marshall Loxahatchee National Wildlife Refuge (Water Conservation Area 1), Big Cypress National Preserve, and the eastern half of the East Everglades Acquisition Area. Light to moderate infestations (widely scattered outliers and small heads) occur in Water Conservation Area 3 and the western half of the East Everglades Acquisition Area. Very few outlier melaleuca are found in the Holey Land, Everglades National Park, and Water Conservation Area 2A.

Detailed, large-scale outlier coverage data are not currently available for the entire area. However, as control operations proceed within the Water Conservation Areas, Loxahatchee National Wildlife Refuge, Everglades National Park, and Big Cypress National Preserve, data are collected on geographic position and tree densities. This is, in effect, an "after the fact" mapping process because mapped trees are also treated trees which, presumably, will die. These data could, however, be used to supplement other less detailed mapping projects. This would allow an estimate of outlier coverage and a more realistic depiction of the total area actually infested with melaleuca. For example, although the infestation level in Water Conservation Area 3 is described as light to moderate, 500,000 mature trees and 1.5 million seedlings have been treated by the South Florida Water Management District within a 200,000-acre (80,939-ha) section of the area (Figure 14.4). This helps to illustrate that even where infestations are relatively low, significant numbers of melaleuca can persist and continue to spread.

CONTROL METHODS

Integration of all available control techniques will be required to effectively eliminate melaleuca from the Everglades. These control techniques include biological, mechanical, physical, and herbicidal methods. The current melaleuca management program in Florida is not truly integrated. Biological control is not yet available and physical controls, such as fire and flooding, are not completely understood or easily managed. Use of mechanical control has been limited to manual removal of seedlings in natural areas. Consequently, herbicidal control currently offers the only practical and economically feasible method of limiting the further expansion of melaleuca in the Everglades.

Managing melaleuca requires a knowledge of its biology. The reproductive potential of melaleuca is tremendous. A mature tree may retain millions of seeds, all of which may be released from their protective capsules following a stressful event. This stress may be caused by desiccation, fire, frost, physical damage, or herbicide application (Meskimen, 1962). Once released, the exceptionally small seeds do not remain viable for longer than 1 year (Woodall, 1983). This relatively short period of seed viability limits the time for new seedling establishment, which also limits the monitoring period at a site following herbicidal control.

Figure 14.4   Treated melaleuca in Water Conservation Area 3. Each cross represents a treatment site; 500,000 trees and 1.5 million seedlings have been treated in this area.

Hypothetically, melaleuca can be eliminated from an area in 2 years. The first year of control would target all existing trees and seedlings in a given area. Using LORAN-C or Geographic Positioning System coordinates as a guide, control crews would return to the same site in the second year and remove any seedlings resulting from the previous year's work. Realistically, several years are required to eliminate melaleuca from a particular site in order to ensure that all trees and/or new seedlings are controlled.

Biological Control

The cornerstone of an effective and long-term management program for melaleuca is the introduction of biological control agents. Successful establishment of biological controls will be important in reducing the exponential rate of melaleuca expansion now occurring in south Florida wetlands (Laroche and Ferriter, 1992). In Australia, melaleuca seedlings are much less common than in the melaleuca-infested areas of Florida. The U.S. Department of Agriculture (USDA) has been investigating insects which feed on melaleuca seedlings, saplings, and new shoot growth in Australia. The USDA proposes to introduce these insects into south Florida and anticipates a reduction in the rate of spread of these trees. The hope is to reduce the distribution and rate of spread in Florida to that currently seen in Australia.

Since the USDA began Australian melaleuca biocontrol field work in 1986, more than 150 insects have been found feeding, at least occasionally, on melaleuca. The melaleuca-dependent insects comprise the candidates for biocontrol by insects. Tests, conducted in quarantine, eliminate any candidate found to feed, lay eggs, or pupate on any plant other than melaleuca.

To bring Australian insects into the United States, quarantine, application, and approval must be granted by the federal Technical Advisory Group on biological control. Typically, this testing takes several years and must be completed before final application is made to the Technical Advisory Group for release approval. The first insect candidate, the melaleuca weevil (*Oxyops vitiosa*), was approved for quarantine study in the United States in May 1992. If all studies now underway in Gainesville, Florida, confirm that the melaleuca weevil is a suitable candidate, field releases could begin in early 1994. Another insect, the melaleuca sawfly (*Lophyrotoma zonalis*), is expected to be approved for quarantine study in the United States soon.

No single insect species is likely to provide effective control. A complement of several insects that feed on melaleuca will likely be required to effectively reduce its reproductive potential. Investigations in Australia continue on insects which affect melaleuca reproduction and growth through stem boring, leaf feeding, or gall forming. Despite excellent progress, future funding is in doubt, in which case Australian research will have provided only two of the half dozen candidate insects that will likely be required to contain this pest plant.

Mechanical and Physical Control

Mechanical removal using heavy equipment is not appropriate in natural areas because of disturbances to soils and nontarget vegetation. However, this method of control can be applied along canal and utility rights-of-way and other similar areas adjacent to infested wetlands. Stumps left after any mechanical operation would require an herbicide application to prevent root sprouts and regrowth from cut surfaces. Increasing interest in uses of exotic pest trees for commercial purposes, such as mulch and boiler fuel, may lead to partnerships between the government and private industry. Innovative methods of removing trees from wetlands are being developed in timber-growing regions of the southeastern United States. This may make commercial exploitation profitable without environmental degradation. Currently, manual removal of melaleuca seedlings is the only form of mechanical control being used in the natural areas of south Florida. Hand pulling of trees is generally restricted to trees less than 2 m in height.

Woody vegetation can be stressed, or sometimes killed, by environmental alterations such as water level manipulation or fire. Constraints often limiting the usefulness of these methods include the ability to maintain required water levels, the effects of these manipulations on desirable vegetation, adequate conditions for prescribed burns, and the liability involved with burning. However, research which incorporates physical stresses into other methods of melaleuca control is important and ongoing.

Herbicidal Control

Exotic woody vegetation is most commonly managed by herbicide application. Early investigations (mid-1970s) into the control of melaleuca in the Everglades using herbicides were conducted by the Florida Game and Fresh Water Fish Commission (1975–80). Many noncrop and forestry herbicides were tested to examine various application methods and application rates. Herbicides were often applied at rates far beyond those that would normally kill woody vegetation. Effectiveness of these high-rate applications varied. Results of these tests showed that melaleuca would be very difficult to control with available herbicides and conventional application techniques. These field trials were not monitored frequently and no quantitative data were collected; however, they did provide the foundation for future studies.

In 1979, the U.S. Army Corps of Engineers began herbicide effectiveness trials on mature melaleuca trees and on seedlings along the Herbert Hoover Dike around Lake Okeechobee, near the towns of Okeechobee and Clewiston, respectively (Stocker and Sanders, 1980). Prior to treatment, randomly selected melaleuca trees in each plot were tagged with identification numbers. At each evaluation, vitality (live versus dead) was determined for each tagged individual. Results of herbicide treatments to seedlings indicated that 100% control could be obtained after 1 year with applications of 4.5 kg·ha^{-1} active ingredient (a.i.) HYVAR X (bromacil), 4.5 kg·ha^{-1} a.i. SPIKE 80W, 13.4 kg·ha^{-1} a.i. SPIKE 40P (tebuthiuron), and 2.2 kg·ha^{-1} a.i. VELPAR L (hexazinone). Mature trees proved more difficult to control; no treatment gave 100% control after an evaluation period of 1 year. Only the application of VELPAR L at a rate of 4.5 kg·ha^{-1} a.i. and SPIKE 40P at a rate of 11.2 kg·ha^{-1} a.i. produced greater than 80% control. However, effectiveness ratings may have risen if treatment evaluations had continued for longer than 1 year.

More recent South Florida Water Management District herbicide trials are investigating foliar-active herbicides. Foliar-active herbicides are generally more selective and have shorter soil residence times than soil-active herbicides. Foliar applications are usually made by diluting the herbicide in water and applying it to leaves with ground or aerial application equipment. Dilution is usually about 20:1 for aerial applications and from 50 to 400:1 for applications from the ground. Foliar applications can be directed to either minimize damage to nontarget vegetation or broadcast. Broadcast applications are used where damage to nontarget vegetation is not a concern or where an herbicide which selects for melaleuca is used. Soil applications may be diluted or concentrated sprays which are applied directly to the base of individual trees or granular herbicide formulations which are applied directly to soils. Investigations are also under way to evaluate the effectiveness of aerial application with foliar-active and soil-active herbicides. These investigations will evaluate herbicide effectiveness on melaleuca, effects on nontarget vegetation, and environmental fate. The herbicides being evaluated include VELPAR L, SPIKE 40P, ARSENAL (imazapyr), GARLON 3A (triclopyr), and RODEO (glyphosate). Various application techniques, herbicide combinations, and nonherbicidal spray additives are also being evaluated. Until these studies are completed, individual tree application methods—frill (or girdle) and cut/stump—will likely be the control methods of choice. Individual tree application methods expose the cambium layer, and herbicide (concentrated or diluted) is applied directly to the cut surface at rates recommended by the manufacturer. Recommended rates of herbicide application are those which have been specified on the label included with each product. Individual tree applications are labor intensive and therefore costly. However, these applications have no adverse effect on nontarget vegetation and generally completely kill the target plant. Soil applications are labor conservative and cost effective, but negatively affect surrounding vegetation. Studies of effects on non-target vegetation continue.

To date, the "girdle" technique for applying herbicide to cambium exposed around the entire circumference of a tree has been the method of choice by the U.S. National Park Service in Everglades National Park, the U.S. Fish and Wildlife Service in the Arthur R. Marshall Loxahatchee National Wildlife Refuge in Water Conservation Area 1, and the South Florida Water Management District in Water Conservation Areas 2 and 3. ARSENAL has proven to be the most consistently effective herbicide and is presently the most frequently used herbicide in these areas. ARSENAL also has a "Special Local Need" label, allowing this manner of use in flooded areas. South Florida Water Management District costs for controlling melaleuca in the Water Conservation Areas from October 1992 to the present have averaged $1.70 per tree. This cost includes labor, equipment, and herbicide. Transport of crews to each treatment site, usually by helicopter, has been the largest single cost.

Labor-intensive, manual herbicide applications to individual "outlier" trees and small tree groups continue to be the primary melaleuca control method throughout south Florida. Several south Florida agencies and herbicide manufacturers are cooperating to complete small-scale studies with herbicides that are broadcast onto dense melaleuca stands. Results of these tests may suggest that large-scale broadcast herbicide treatments onto melaleuca forests can be made safely. These labor-conservative methods will be adopted only if test results indicate acceptable herbicide persistence times and effects on native vegetation. Such affordable large-scale herbicide applications could, if deemed environmentally acceptable, further melaleuca management in south Florida. However, perpetual control hinges on establishment of effective biological control agents.

MELALEUCA MANAGEMENT POLICY and MANDATES

Congressional and Florida legislative actions have termed melaleuca both a Federal Noxious Weed and a Florida Prohibited Plant. These designations primarily prohibit commercial melaleuca trade, but indicate the widespread recognition of the threat posed by its invasion and may spur increased financial support for its management. Major public land management agencies (U.S. National Park Service, U.S. Fish and Wildlife Service, Florida Game and Fresh Water Fish Commission, South Florida Water Management District) are all attacking the plant in their respective areas of responsibility. Most south Florida counties and municipalities require melaleuca removal during real estate development. Melaleuca was extirpated within the city limits of Sanibel Island, when its removal was required on all land, public or private, at the owner's expense. However, vast stands remain on private lands abutting most of south Florida's public land. These seed sources will continually threaten preserved areas until the public will demands their control or introduced biocontrol insects curtail seed production.

CONCLUSIONS

Work progresses to control melaleuca throughout south Florida, especially on state and federal lands. In the past 20 years, many scientists and informal coalitions of concerned citizens have highlighted the severe threat posed by the uncontrolled spread of melaleuca. During the 1980s, members of the newly formed Exotic Pest Plant Council worked to assist with invasive plant problems. In 1990, the South Florida Water Management District helped to form a state and federal interagency melaleuca task force to draw upon the experience of the area's scientists, resource managers, the Exotic Pest Plant Council, and others to develop a regional melaleuca management plan. This task force brought much-needed coordination to melaleuca management and focused attention on funding, public awareness, and legislative needs. This coordination of effort must continue if there is to be long-term control of melaleuca in south Florida.

ACKNOWLEDGMENT

The authors thank Dr. Ron Hofstetter, University of Miami Biology Department, for his significant contributions to this chapter.

REFERENCES

Arvanitis, L. G. and R. Newburne. 1984. Detecting *Melaleuca* trees and stands in south Florida. *Photo. Eng. Remote Sensing*, Jan.:95–98.
Austin, D. F. 1978. Exotic plants and their effects in southeastern Florida. *Environ. Conserv.*, 5(1):25–34.
Browder, J. A. and P. B. Schroeder. 1981. Melaleuca seed dispersal and perspectives on control. in Proc. Melaleuca Symp., R. K. Geiger (Ed.), Division of Forestry, Florida Department of Agriculture and Consumer Services, Tallahassee, Sept. 23–24, 1980, pp. 17–21.
Capehart, B. L., J. Ewel, B. Sedlik, and R. Myers. 1977. Remote sensing survey of *Melaleuca*. *Photo. Eng. Remote Sensing*, 43(2):197–206.
Center, T. D. and F. A. Dray, Jr. 1986. *Melaleuca quinquenervia*, a Serious Threat to South Florida

Wetlands, Report of the U.S. Department of Agriculture, Agricultural Research Service, Aquatic Plant Laboratories, Fort Lauderdale.

Center, T. 1993. Personal communication, Agricultural Research Service, U.S. Department of Agriculture, March 25, 1993.

Cost, N. D. and G. C. Craver. 1981. Distribution of *Melaleuca* in south Florida measured from the air. in Proc. Melaleuca Symp., R. K. Geiger (Ed.), Division of Forestry, Florida Department of Agriculture and Consumer Services, Tallahassee, Sept. 23–24, 1980, pp. 1–8.

DiStefano, J. F. and R. F. Fisher. 1983. Invasion potential of *Melaleuca quinquenervia* in southern Florida, U.S.A. *For. Ecol. Manage.*, 7:133–141.

Duever, M. J. et al. 1979. Resource Inventory and Analysis of the Big Cypress National Preserve, Center for Wetlands, University of Florida, Gainesville, and Environmental Research Unit, National Audubon Society, Naples, Fla., 1225 pp.

Ewel, J. J. et al. 1976. Studies of Vegetation Changes in South Florida, Report to U.S. Forest Service on Research Agreement 18-492, University of Florida, Gainesville.

Hofstetter, R. H., 1991. The current status of *Melaleuca quinquenervia* in southern Florida. in Proc. Symp. on Exotic Pest Plants, R. F. Doren (Ed.), Nov. 2–4, 1988, Miami, pp. 159–176.

Laroche, F. B. and A. P. Ferriter. 1992. Estimating expansion rates of melaleuca in south Florida. *J. Aquat. Plant Manage.*, 30:62–65.

Light, S. S. and J. W. Dineen. 1994. Water control in the Everglades: A historical perspective. in *Everglades: The Ecosystem and Its Restoration,* S. M. Davis and J. C. Ogden, St. Lucie Press, Delray Beach, Fla., chap. 4.

Meskimen, G. F. 1962. A Silvical Study of the Melaleuca Tree in Florida, M.S. thesis, University of Florida, Gainesville, 177 pp.

Myers, R. L. 1975. The Relationship of Site Conditions to the Invading Capability of *Melaleuca quinquenervia* in Southwest Florida, M.S. thesis, University of Florida, Gainesville, 151 pp.

Myers, R. L. 1976. Melaleuca field studies. in Studies of Vegetation Changes on South Florida, J. Ewel et. al. (Eds.), U.S. Forest Service Research Agreement 18-492, pp. 4–24.

Myers, R. L. 1983. Site susceptibility to invasion by the exotic tree *Melaleuca quinquenervia* in southern Florida. *J. Appl. Ecol.*, 20:645–658.

South Florida Water Management District, Operations and Maintenance Department. 1992. Inventory of Melaleuca Monocultures in South Florida, unpublished study.

Stocker, R. and D. R. Sanders. 1980. Chemical control of *Melaleuca quinquenervia*. in Proc. Melaleuca Symp., R. K. Geiger (Ed.), Division of Forestry, Florida Department of Agriculture and Consumer Services, Sept. 23–24, 1980, pp. 129–134.

Tomlinson, P. B. 1980. *The Biology of Trees Native to Tropical Florida,* Harvard University Printing Office, Allston, Mass., pp. 271–274.

Wade, D. D, J. J. Ewel, and R. H. Hofstetter. 1980. Fire in South Florida ecosystems. *U.S. For. Serv. Gen. Tech. Rep. SE,* No. SE-17, 125 pp.

Woodall, S. L. 1979. Spatial and Temporal Aspects of Wind Dispersal of Melaleuca Seed, paper presented to Florida Field Biologists Annu. Meet., March 16, 1979, Orlando, 13 pp.

Woodall, S. L. 1981a. Integrated methods for melaleuca control. in Proc. of Melaleuca Symp., R. K. Geiger (Ed.), Division of Forestry, Florida Department of Agriculture and Consumer Services, Tallahassee, Sept. 23–24, 1980, pp. 135–140.

Woodall, S. L. 1981b. Site requirements for Melaleuca seedling establishment. in Proc. Melaleuca Symp., R. K. Geiger (Ed.), Division of Forestry, Florida Department of Agriculture and Consumer Services, Tallahassee, Sept. 23–24, 1980, pp. 9–15.

Woodall, S. L. 1983. Establishment of *Melaleuca quinquenervia* seedlings in the pine-cypress ecotone of southwest Florida. *Fla. Sci.*, 46(2):65–71.

15

PHOSPHORUS INPUTS and VEGETATION SENSITIVITY in the EVERGLADES

Steven M. Davis

ABSTRACT

Changes in the quantity, timing, and distribution of phosphorus supply to the Everglades during the 20th century are examined, and data are reviewed concerning the sensitivity of sawgrass and cattail communities to these changes.

The Everglades ecosystem apparently evolved under conditions of relatively low phosphorus inputs, mostly from direct rainfall, which contributed approximately 196 tonnes annually to the 500,000-ha freshwater marsh within the Water Conservation Areas and Everglades National Park. Phosphorus inputs to the Water Conservation Areas have increased nearly threefold (from 129 tonnes under predrainage conditions to 376 tonnes currently) due to inflows of agricultural runoff water. Everglades National Park appears to have experienced smaller increases in annual phosphorus input, from 88 tonnes predrainage to 89 tonnes currently. The spatial distribution of surface water phosphorus within Everglades marshes takes the form of gradients from high concentrations at or near inflow points to nearly nondetectable concentrations within interior marsh areas.

Cattail has become established in sawgrass in Everglades marsh areas near anthropogenic phosphorus inputs. The differing responses of cattail and sawgrass to phosphorus enrichment provide a plausible hypothesis to explain the occurrence of persistent cattail stands in eutrophic areas. Sawgrass nutrient requirements appear to be low relative to cattail. Both species increase annual production and annual phosphorus allocation to living biomass per unit area of marsh in response to phosphorus enrichment, although cattail phosphorus allocation is nearly three times that of sawgrass. Cattail possesses more temporal flexibility than sawgrass in its ability to vary annual production and phosphorus allocation with annual variations in surface water enrichment. Contrasting responses of the two species to nutrient enrichment characterize sawgrass as a low nutrient status species that is competitive in infertile habitats, in contrast to cattail as a high nutrient status species that is competitive when nutrient supply increases. Cattail density has been documented to increase, relative to sawgrass density, at eutrophic Everglades sites. Evidence is reviewed which suggests that the successional status of cattail may change from an early colonizer of naturally disturbed oligotrophic sites to a persistent dominant of eutrophic sites in areas where sawgrass was previously persistent.

Conversion to cattail is one of many ecosystem changes that accompany eutrophication of an Everglades marsh. Relatively fine, flocculent cattail sediment covers the marsh floor at eutrophic sites, in contrast to relatively compact, fibrous sawgrass sediment at background sites. Calcareous blue-green, green, diatom-rich, and desmid-rich periphyton communities of the Everglades are either eliminated or replaced by a community dominated by the filamentous algae *Microcoleus lyngbyaceus* and other pollution indicator species, apparently with adverse effects on food webs and oxygen budgets. Diurnal fluctuation in dissolved oxygen is dampened so that mid-morning concentrations remain below $0.2 \text{ m} \cdot \text{L}^{-1}$, in comparison to levels of ~2.0 $\text{mg} \cdot \text{L}^{-1}$ at oligotrophic sites. Colony counts of facultative bacteria and fungi that colonize leaf litter are almost an order of magnitude lower at eutrophic sites in comparison to oligotrophic sites. The number of macroinvertebrate taxa that colonize leaf litter is reduced, Diptera numbers are reduced, snails are nearly eliminated, isopods are eliminated, and density of annelid worms is more than doubled at eutrophic compared to oligotrophic sites. Total phosphorus concentrations in surface soil are elevated to 1200–1600 $\text{mg} \cdot \text{kg}^{-1}$ at eutrophic sites compared to 400 $\text{mg} \cdot \text{kg}^{-1}$ at oligotrophic sites. Transpiration and stomatal conductance rates during the dry season are higher for cattail than sawgrass and at eutrophic compared to oligotrophic sites.

INTRODUCTION

This chapter examines 20th century changes in the quantity, timing, and distribution of phosphorus supply to the Everglades and reviews data concerning the sensitivity of vegetation communities to these changes. Patterns of precipitation and hydrology are interpreted in order to construct a generalized representation of phosphorus inputs and dynamics in the predrainage Everglades. Based on this historical framework, broad changes in phosphorus inputs that appear to have resulted from 20th century changes in hydrology and land use are examined. The significance of these changes is reviewed as they relate to surface water quality and vegetation community structure, function, and resilience.

Vegetation community shift as a result of phosphorus enrichment was documented by Walker et al. (in preparation) in Everglades National Park, where a wet prairie/slough community dominated by *Eleocharis* spp., *Utricularia* spp., and periphyton changed to a community dominated by *Panicum hemitomon, Pontederia lanceolata,* and *Sagittaria lancifolia.* Community change that resulted from nitrogen enrichment alone was substantially less than that due to phosphorus, while change resulting from additions of both phosphorus and nitrogen was greater than that due to phosphorus alone. Steward and Ornes (1983) observed that high levels of phosphorus inhibited dry matter production, shoot elongation, and shoot production in sawgrass seedlings, indicating potential adverse impacts. These results suggest that phosphorus may be more important than nitrogen as a limiting macronutrient for Everglades vegetation. High total N/P ratios of greater than 100 reported by Walker (1991) in surface water inflows into Everglades National Park also support phosphorus limitation; hence the focus on phosphorus here.

PREDRAINAGE PERSPECTIVE

The predevelopment Everglades has been characterized as an oligotrophic ecosystem that received external water and nutrient inputs primarily through direct rainfall. Rainfall was the major external source of phosphorus to the parts of the ecosystem that now remain in the Water Conservation Areas and Everglades National Park, partly because of the dominance of direct rainfall in Everglades hydrologic inputs (Parker, 1974; Fennema et al., 1994) and partly because phosphorus inputs from Lake Okeechobee overflows would likely be assimilated by the once extensive sawgrass (*Cladium jamaicense*) marshes between the lake and the Water Conservation Areas. The mean rainfall total phosphorus (total P) concentration of 0.029 mg·L^{-1} (wetfall + dryfall, $\bar{X}$ 1979–88) at West Palm Beach (South Florida Water Management District, 1992) is reduced to levels near detection limits of 0.01 mg·L^{-1} total P and 0.002 mg·L^{-1} soluble reactive phosphorus (SRP), which are typical in the surface water of interior Everglades marshes that remain primarily rainfall fed (Davis, 1989, 1991; Swift and Nicholas, 1987; Urban et al., in press; South Florida Water Management District, 1992). Therefore, the Everglades ecosystem apparently evolved under conditions of low phosphorus inputs.

Historic atmospheric phosphorus inputs (wetfall + dryfall) are derived from rainfall quantity multiplied by the total P concentration of 0.029 mg·L^{-1} from the West Palm Beach collection site. This value is lower than mean annual total P concentrations of 0.05–0.11 mg·L^{-1} in rainfall at sites within the Everglades Agricultural Area (South Florida Water Management District, 1992). Factors contributing to higher measured atmospheric phosphorus inputs over the Everglades Agricultural Area have not been quantified, but elevated rainfall concentrations there coincide with the annual burning of cane fields and the dust of agricultural operations, high atmospheric phosphorus deposition rates primarily due to elevated dryfall deposition (Hendry et al., 1981), and a higher frequency of contamination effects by insects, bird feces, etc. (Walker, 1989). Because the Everglades Agricultural Area was Everglades marsh prior to its drainage and conversion to agriculture, the nonagricultural, coastal West Palm Beach collection station probably approximates predrainage Everglades phosphorus inputs better than agricultural sites at this point in time.

The 0.029-mg/L value used here should be considered a maximum estimate (and likely an overestimate) of rainfall phosphorus inputs to the natural system. Limitations of the measurement methodology make it difficult to separate the above potential sampling errors, and the use of measured (wet, dry, or bulk) deposition rates from any station to formulate external mass balances for the Everglades likely overestimates true atmospheric loads (particularly predevelopment values). Furthermore, rainfall total P concentrations that were used in loading calculations are based on arithmetic means rather than volume-weighted means because only an arithmetic mean was available for the 10-year database that is used here from the West Palm Beach site. Use of arithmetic mean may lead to overestimation of rainfall phosphorus. Everglades National Park unpublished data (D. Scheidt, personal communication) indicate that the arithmetic mean total P concentration exceeded the volume-weighted mean by a factor of 2.6 at the park.

Rainfall quantity used in phosphorus loading calculations is based on Water Conservation Area and Everglades National Park rainfall records for the period 1963–87. Rainfall exhibited north to south spatial variability during that period. Mean annual rainfall decreased southward, from 1.25 m in Water Conservation Area 1, to 1.19 m in Water Conservation Area 2, to 1.13 m in Water Conservation Area 3, but then increased to 1.40 m in Everglades National Park (Table 15.1) (Cooper and Roy, 1991). Although rainfall exhibited considerable seasonal and interannual variation, long-term unidirectional trends were not apparent (Duever et al., 1994). Everglades National Park annual rainfall averaged 1.39 m during the period 1941–62 and, correspondingly, averaged 1.40 m during 1963–87. The 1963–87 period of record is used as an approximation of long-term mean annual rainfall.

Based on the above calculations, the mean annual atmospheric input of phosphorus (wetfall + dryfall) to the 500,000-ha freshwater marsh within the Water Conservation Areas and Everglades National Park is estimated to amount to 196 tonnes, or approximately 36 $mg \cdot m^{-2} \cdot yr^{-1}$ (Table 15.1). This estimate compares favorably with the 35-$mg \cdot m^{-2} \cdot yr^{-1}$ phosphorus deposition rate derived from peat accretion data in remote areas of Water Conservation Area 2A (Walker, 1993). An advantage of backcalculating deposition rate using peat accretion is that it reflects net deposition, whereas direct measurements of atmospheric inputs may be biased because they reflect the sum of net deposition, local recycling (i.e., ash and dust),

Table 15.1 Estimated Mean Annual Phosphorus Inputs from Rainfall into the Water Conservation Areas and Everglades National Park

	Surface area (ha)	Rainfall[a] (m)	Rainfall (m³)	Estimated rainfall P input (tonnes) based on concentration of 0.029 mg·L⁻¹ in West Palm Beach rainfall
Water Conservation Areas				
1	57,552	1.25	7.19×10^8	20.9
2A	42,706	1.19	5.08×10^8	14.7
2B	11,338	1.19	1.35×10^9	3.9
3A	198,726	1.13	2.24×10^9	65.1
3B	41,318	1.13	4.67×10^8	13.5
Everglades National Park	191,636[b]	1.40	2.68×10^9	77.8

[a] 1963–87.
[b] Everglades freshwater wetlands, as delineated by Davis et al. (1994) within the boundaries of Everglades National Park.

and sample contamination, as discussed earlier. Despite limitations of the measurement methodology for atmospheric inputs, the similarity of the two estimates supports the 0.029-mg/L phosphorus concentration used here and the resulting estimates of atmospheric loadings to the Everglades in Table 15.1.

In addition to spatial patterns of rainfall phosphorus input, temporal patterns include interannual variation in climate, wet season phosphorus input, and dry season recycling. The wet/dry season rainfall pattern characteristic of southern Florida (Duever et al., 1994) presumably concentrated atmospheric phosphorus inputs during the June–October wet season in most years. The November–May dry season appeared to be characterized by lower rates of phosphorus inputs but accelerated rates of phosphorus regeneration within the marsh. Temporary high phosphorus concentrations in surface water reported by Swift and Nicholas (1987) after marsh drying and reflooding and by Forthman (1973) after marsh burning likely resulted from remineralization of phosphorus in oxidized organic sediments. Such pulses of available phosphorus appear to be short lived, with durations of days or weeks.

The annual influx of wading birds to the Everglades during the winter-spring dry season also contributes to phosphorus regeneration and transport due to birds ingesting aquatic organisms in areas of receding water and then defecating elsewhere, particularly at rookeries and nesting colonies. Kushlan (1977) estimated that 23,800 pairs of white ibis nesting in the Everglades consumed ~882 tonnes (wet mass) of food during the 1972 nesting season and 15,500 pairs consumed ~575 tonnes during the 1973 nesting season. Stinner (1983) reported that an average addition of 4.6 $g \cdot m^{-2} \cdot yr^{-1}$ phosphorus to an Okefenokee Swamp colony site by 20,000–30,000 wading birds (mostly white ibis) resulted in significant increases in phosphorus in aquatic macrophytes during nesting. Oliver and Legovic (1988) estimated that a normal nesting of 8000 white ibis in an Okefenokee Swamp colony increased mean daily phosphorus input from ~1.5 $mg \cdot m^{-2} \cdot d^{-1}$ due to precipitation alone to ~40 $mg \cdot m^{-2} \cdot d^{-1}$ due to the birds. This would amount to a contribution of ~9.7 $g \cdot m^{-2} \cdot yr^{-1}$ by the birds over the nesting season.

Despite an apparently high level of phosphorus enrichment at wading bird colony sites, the phosphorus mass transported to colonies may have been small relative to historic atmospheric phosphorus inputs to the Everglades. Frederick and Powell (1994) estimated that phosphorus transport due to ingestion and excretion by Water Conservation Area, Everglades National Park, and East Everglades wading bird populations amounted to ~3.3 tonnes annual phosphorus deposition at colonies before 1890, which equaled ~1.7% of the estimated 196 tonnes atmospheric phosphorus input at that time. However, colony phosphorus deposition could have important local effects on Everglades vegetation, seagrass communities, and probably food webs (Powell et al., 1989, 1991).

TWENTIETH CENTURY CHANGES

The Everglades ecosystem has experienced changes in phosphorus inputs, distribution, and regeneration patterns since drainage and development began in the early 1900s. Increases in phosphorus supply to the Everglades have resulted

Table 15.2 Estimated Historic and Current Phosphorus Inputs
into the Water Conservation Areas and Everglades National Park

	Historic P input (tonnes)	Current P input (tonnes)
Water Conservation Areas		
Rainfall input	118	118
Overland or pumped input	11[a]	258
Total input	129	376
Everglades National Park		
Rainfall input	78	78
Overland or pumped input	10[b]	11
Total input	88	89

[a] Based on an estimated 1.07×10^9 m^3 (869,875 acre-ft) mean annual
overland flow into the Water Conservation Areas (Fennema et al.,
1994) with a mean annual total P concentration of 0.010 mg·L^{-1}
(South Florida Water Management District, 1992).

[b] Based on an estimated 1.04×10^9 m^3 (842,179 acre-ft) mean annual
overland flow into Everglades National Park (Fennema et al., 1994)
with a mean annual total P concentration of 0.010 mg·L^{-1}.

from point source inflows of runoff water from drained agricultural lands into the
Water Conservation Areas (Table 15.2). Four major pump stations (S-5A, S-6, S-7,
and S-8) (Figure 15.1) plus several smaller inflows provide a mean annual
phosphorus input of 258 tonnes ($\overline{X}$ 1979–88) to the Water Conservation Areas
(South Florida Water Management District, 1992), compared to an estimated
overland flow phosphorus input of ~11 tonnes that entered the Water Conservation Area marsh from the north and west under predrainage conditions. Including
atmospheric phosphorus input, the mean annual supply of phosphorus to the
Water Conservation Areas appears to have increased nearly threefold, from
levels of ~129 tonnes under predrainage conditions to ~376 tonnes in the 1980s
(Table 15.2).

The annual phosphorus input to Everglades National Park appears to have
increased less since predrainage conditions. Current phosphorus inputs of 11
tonnes per year through water control structures to the park appear to represent
a 1-tonne increase over the estimated predrainage, overland flow input of 10
tonnes (Table 15.2). However, a trend of increasing phosphorus loads to the park
is evident. Walker (1991) demonstrated that Everglades National Park surface
water inflow concentrations increased at an average rate of 5% per year between
1978 and 1989. As discussed later, areawide phosphorus budgets may be misleading in that they underestimate local effects of surface water flows and loadings on
areas immediately downstream of inflow structures. Thus, the relatively small
increases in phosphorus loads to the park do not preclude locally high loads and
impacts.

Calculations of current phosphorus input are based on the assumption that
atmospheric input has not increased due to drainage and development in south

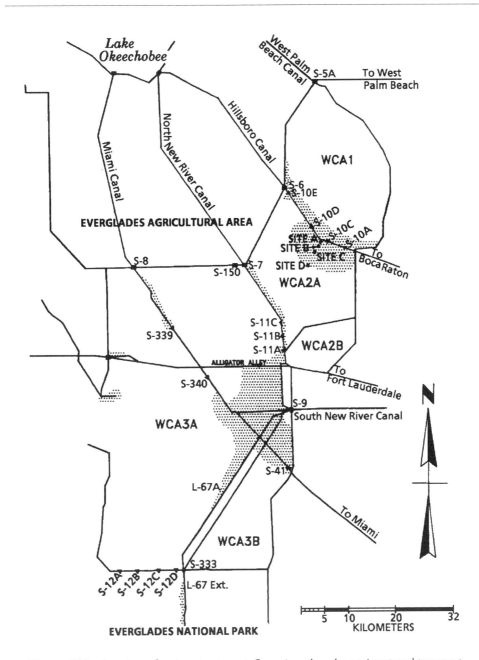

Figure 15.1 Location of major structures influencing phosphorus input and transport in the Everglades. Shading represents approximate distribution of major cattail stands in the Water Conservation Areas and Everglades National Park. Sites A–D indicate locations of sawgrass and cattail sampling locations referenced in the text.

Florida and that the estimates of 0.029 mg·L^{-1} and 36 mg·m^{-2}·yr^{-1} apply to current as well as historic conditions. Although measured atmospheric inputs of $0.037–0.050$ mg·L^{-1} over the Water Conservation Areas are currently higher (South Florida Water Management District, 1992), the methodology limitations noted earlier make it difficult to separate sampling problems and local recycling as factors contributing to higher measurements. There is no direct evidence of a temporal increase in atmospheric phosphorus loads.

The spatial penetration of surface water phosphorus within each Water Conservation Area and Everglades National Park (Waller and Earle, 1975; Millar, 1981; Waller, 1982; Urban et al., in press) is influenced by both the extent of overland flow across the marsh and phosphorus retention within the marsh. Water and phosphorus inflows into Water Conservation Area 1 through pump stations S-5A and S-6 (Figure 15.1) mostly bypass the marsh and follow perimeter canals. As a result, the interior marsh is characterized by low surface water phosphorus concentrations and conductivities (Swift and Nicholas, 1987; South Florida Water Management District, 1992). Much of the pumped water and phosphorus that enter Water Conservation Area 1 empty into Water Conservation Area 2A through the gated S-10 structures (Figure 15.1), although the 60 tonnes of S-10 phosphorus input into Water Conservation Area 2A (South Florida Water Management District, 1992) is not reflected in Table 15.2 as an external input to the Water Conservation Areas because the water and phosphorus first entered Water Conservation Area 1. Richardson et al. (1992) reported: "Over the past three decades nutrient inputs of N and P from the EAA into WCA-2A have increased 12.4 times and 10.0 times, respectively." The absence of north to south interior canals within Water Conservation Area 2A forces the S-10 water and phosphorus inputs to flow overland. Here, surface water phosphorus is carried deeper into the marsh, to distances of ~1.6 km from inflows during the 1970s and nearly 7.0 km during the 1980s (South Florida Water Management District, 1992; Urban et al., in press). Water and phosphorus inputs into the southwest corner of Water Conservation Area 2A through the S-7 pump station mostly bypass the marsh to the north and flow directly through the three S-11 gates into Water Conservation Area 3A (Figure 15.1).

Phosphorus transport in the marsh and in canal systems to the south of Water Conservation Area 2A is less well documented. The S-11 gates which drain Water Conservation Area 2A and the S-9 pump station which drains eastern urban areas (Figure 15.1) contribute 31 tonnes of phosphorus annually to the northeast corner of Water Conservation Area 3A (South Florida Water Management District, 1992). However, the proportions of these inputs that penetrate the marsh have not been quantified. Similarly, the fate of the 57 tonnes of phosphorus input annually into Water Conservation Area 3A from pump station S-8 (South Florida Water Management District, 1992) is poorly understood. Much of that input probably is contained within the Miami Canal, which flows diagonally through Water Conservation Area 3A. Two gated plugs in the Miami Canal (S-339 and S-340) force water flows around the canal and into the marsh (Figure 15.1). The extent of overland flow and re-entry of diverted water into the canal also lacks quantification. Much of the Miami Canal water that flows from Water Conservation Area 3A into Water Conservation Area 3B is forced southward across the Water Conservation Area 3B marsh because the S-31 outflow gate from this area is usually closed (Figure 15.1); again, however, the split of water and phosphorus between the outflow and the marsh has not been fully evaluated.

Everglades National Park, at the downstream end of the system, receives water and phosphorus directly from the Water Conservation Area 3A marsh and from canal L-67A along the eastern border of Water Conservation Area 3A (Figure 15.1). Canal L-67A collects water and phosphorus from the Miami Canal (S-8), Water Conservation Area 2A (S-11A, B, and C), the North New River Canal (S-150), and the South New River Canal (S-9). The multitude of upstream water and phosphorus sources to L-67 potentially could contribute 115 tonnes of phosphorus annually to Everglades National Park. However, current annual phosphorus inputs of 11 tonnes to the park via the S-12 gates and pump station S-333 (Figure 15.1) amount to only one-tenth of the phosphorus that could potentially be input from the canal systems that feed the park (South Florida Water Management District, 1992).

The spatial distribution of surface water phosphorus within the Everglades marshes takes the form of concentration gradients from high phosphorus concentrations at or near inflow points to nearly nondetectable concentrations within interior marsh areas. Surface water phosphorus gradients have been examined most intensively in Water Conservation Area 2A by Urban et al. (in press) (Figure 15.2). These gradients indicate that the marsh effectively reduces surface water phosphorus concentrations to levels near detection limits within a 7.0-km distance from points where inflows enter the marsh. Marshes adjacent to inflow structures and canals continue to function in this capacity three decades after phosphorus inputs through water control structures into the system began in the late 1950s and early 1960s. The prolonged assimilation capacity of Everglades marshes provides an explanation for the current relatively low phosphorus inputs into Everglades National Park despite much higher inputs into marsh and canal systems that ultimately drain into the park.

Phosphorus concentration gradients in the Everglades marsh also indicate that annual rates of input or uptake per unit area cannot be calculated when considering each Water Conservation Area or Everglades National Park as a whole. Anthropogenic phosphorus supply currently reaches only a small percentage of the marshlands within each area, as evidenced by the prevalence of near-detection-limit surface water concentrations beyond the immediately affected areas. Retention of anthropogenic phosphorus supply also occurs only in the gradient zones, and areas beyond appear to contribute little to that retention. Thus, phosphorus mass balances for each Water Conservation Area and for Everglades National Park do not reflect rates of supply or uptake per unit area of the entire marsh, but rather reflect a more intensive supply and uptake within limited portions of each area.

Higher mean annual surface water phosphorus concentrations in eutrophic

Figure 15.2 Surface water phosphorus concentration gradient in Water Conservation Area 2A. Values represent means of bi-weekly samples for the 6-year period 1986–91 ± SE. (From N. H. Urban.)

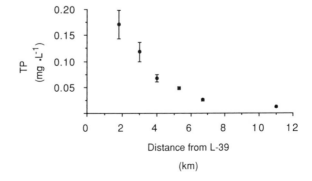

areas of the Everglades actually reflect increased temporal variability in phosphorus concentrations at sites near inflow structures (Figure 15.2) (Urban et al., in press). A mean surface water phosphorus concentration over one or more years at a eutrophic site represents a high temporal variation from near detection limits between discharges to peak values during discharges. In contrast, a mean concentration at an interior oligotrophic marsh site during the same time period represents low temporal variation, with values only slightly exceeding detection limits throughout the period. At eutrophic sites, a high temporal variability in surface water phosphorus concentrations, probably due to variation in anthropogenic inputs and regeneration, is superimposed on a low temporal variability due to rainfall inputs. This suggests that the assimilative capacity of the Everglades sawgrass marsh can accommodate the temporal variation in rainfall phosphorus inputs, but that assimilative capacity may be exceeded by the higher peaks due to the wider temporal variability of anthropogenic phosphorus inputs. This concept will be explored further as vegetation sensitivity to phosphorus enrichment.

In addition to changes in phosphorus inputs, regeneration patterns in present-day Everglades marshes appear to have changed since predrainage conditions. Two changes that would affect regeneration patterns are hydrology and wading bird abundance. Shortened hydroperiods in Water Conservation Areas 2B and 3B and the East Everglades (Fennema et al., 1994) may have accelerated phosphorus regeneration because of greater sediment oxidation and mineralization during extended dry periods. Conversely, deeper water and lengthened hydroperiods in the southern pooled area of Water Conservation Area 3A and in Water Conservation Area 2A during the 1960s and 1970s (Fennema et al., 1994) may have slowed rates of phosphorus regeneration from organic sediments. The magnitude of phosphorus recycling and transport by Everglades wading birds probably has decreased as populations have declined. The estimated 3.3 tonnes of annual phosphorus transport by wading birds to colony sites before 1890 may have been reduced to ~2.0 tonnes by 1934 and ~0.2 tonnes by 1980 (Frederick and Powell, 1994). With relocation and species changes of major wading bird colonies, Frederick and Powell also noted that the location of phosphorus deposition in colonies has shifted northward from the mangrove-freshwater interface in Everglades National Park to the freshwater marsh in the Water Conservation Areas. Thus, phosphorus deposition due to wading bird defecation represented a phosphorus transfer from freshwater to brackish systems historically, in contrast to a relocation within the freshwater system today.

VEGETATION SENSITIVITY: SAWGRASS and CATTAIL COMMUNITIES

Spatial Patterns of Cattail Spread

Everglades marsh areas which are near anthropogenic phosphorus inputs have been invaded by cattail (*Typha domingensis*). The distribution of major cattail stands in the Everglades depicted in Figure 15.1 is based on a 2-day helicopter reconnaissance of the Water Conservation Areas and the eastern border of Everglades National Park in September 1990 by S. M. Davis and K. A. Rutchey. Cattail dominates ~2000 ha of the perimeter of Water Conservation Area 1 in areas of former slough/wet prairie and sawgrass (South Florida Water Management District, 1992). In Water Conservation Area 2A, where surface water phosphorus has penetrated further into the marsh, cattail has also penetrated further. A monospecific cattail stand of ~2400 ha, located below the S-10 inflow gates, invaded an additional ~5700 ha during the 1980s to produce a mixed cattail and sawgrass community in an area that was formerly sawgrass (S. M. Davis, personal observation). Cattail has become established in southwest Water Conservation Area 2A downstream of pump station S-7. Cattail grows in smaller stands along the western border of Water Conservation Area 3A, near inflows from canals that drain agricultural lands to the west and adjacent to structure S-339, which was built to divert Miami Canal water into the marsh. Downstream of these structures, an extensive cattail stand dominates the marsh near the confluence of the Miami Canal and canal L-67A. The band of cattail continues southwest along canal L-67A, where it has invaded gaps in the spoil bank, and from there moved into the Water Conservation Area 3A marsh. To the south, a band of cattail has been observed along the western border of L-67 extension in Everglades National Park.

Factors Other than Nutrients Contributing to Cattail Distribution

It would be a misinterpretation to conclude that nutrient supply alone explains the occurrence of cattail in the Everglades. Cattail is native to the Everglades and widely distributed in small stands in oligotrophic Everglades areas (Davis et al., 1994). Water depth and site disturbance history provide two plausible hypotheses for factors affecting cattail distribution in oligotrophic areas.

T. domingensis represents a deep-water species of cattail (Grace, 1989) which is enhanced by nutrient additions in medium to deep waters (>58 to <105 cm) but not in shallow water (<58 cm) in experimental ponds (Grace, 1988). Eh data suggest that *T. domingensis* may have a high capacity to aerate its roots and rhizosphere, which corresponds to its ability to thrive in the above depths (Grace, 1988).

The occurrence of cattail in oligotrophic Everglades areas often coincides with colonization after disturbance. Canal banks are usually bordered by cattail, and it is a common (and sometimes dominant) community on tree islands which were burned prior to construction of the Water Conservation Areas and flooded thereafter (Alexander and Crook, 1973). Davis (1943) documented a mixed sawgrass, fern, and cattail community in an elevated area ~4.9 m (16 ft) above mean sea level which is now within the Everglades Agricultural Area; he attributed the presence of that community to severe fire stress compared to lower elevation marshes during years of overdrainage prior to construction of the Water Conservation Areas. Disturbance also contributed to the 2400-ha cattail stand below the S-10 structures in the north end of Water Conservation Area 2A. This stand occupies a depression adjacent to the Hillsboro Canal that apparently was created through accelerated soil subsidence around the canal before Water Conservation Area 2A was constructed. This willow (*Salix caroliniana*) dominated subsidence valley in the 1950s (Loveless, 1959) mostly converted to cattail upon lengthening and deepening of hydroperiod after construction of Water Conservation Area 2A. Cattail expansion beyond the 2400-ha disturbed area during the 1980s corresponded to higher annual input and penetration of nutrients into the marsh, with the diversion of agricultural runoff water from Lake Okeechobee to the Water Conservation Areas and with the accelerated growth of agriculture and sugarcane production in the Everglades Agricultural Area beginning in the early 1970s (Light and Dineen, 1994; Snyder and Davidson, 1994).

Based on 6 years of monitoring sawgrass and cattail densities in Water Conservation Area 2A, Urban et al. (in press) reported that deep water, fire, and nutrient enrichment worked synergistically to stimulate cattail expansion into sawgrass communities. Nutrient enrichment appeared to support an expanding distribution and increasing density of cattail in sawgrass stands during wet years and after fires, but was only temporarily set back by droughts. Cattail increased at both nutrient-enriched and low nutrient sites following fire. The postburn increases in cattail density were consistent with the concept of cattail as an early colonizer following disturbance. Yet nutrient enrichment appeared to increase the rate of postburn cattail development. Total P loading best explained plant density fluctuations at most sites close to the levee, while hydrology best explained plant density fluctuations at the site most distant from the levee.

It is noteworthy that cattail remains a minor component of oligotrophic Everglades vegetation communities, despite the chronic fire history (Gunderson and Snyder, 1994) and widely fluctuating hydrologic conditions (Fennema et al., 1994) under which the ecosystem developed. Thus, water depth and fire history do not appear to represent sufficient hypotheses to completely explain the occurrence of extensive cattail stands adjacent to nutrient inflows into the Everglades.

Differential Response of Sawgrass and Cattail to Nutrients

Sawgrass covered much of the Everglades marsh historically and continues to be prevalent in the remaining Everglades (Gunderson, 1994; Davis et al., 1994). Much of the Everglades landscape consists of a mosaic of vast, nearly monospecific stands of this large sedge broken by tree islands on higher elevations and ponded slough and wet prairie areas at lower elevations.

An early survey of nutrient levels in Everglades macrophytes indicated that sawgrass tissue phosphorus concentrations were low compared to other species, but varied significantly from site to site (Volk et al., 1975). Steward and Ornes (1975a, 1975b) further concluded that sawgrass phosphorus requirements were relatively low based on nutrient additions into enclosures and tissue nutrient levels. These conclusions suggested that Everglades sawgrass marshes had limited potential for phosphorus uptake. The occurrence of a cattail zone, which in the early 1970s extended approximately 1.6 km below the S-10 inflows into Water Conservation Area 2A, also suggested that Everglades nutrient inputs might contribute to a shift in plant communities. This prompted a series of studies of nutrient impacts and uptake potentials for sawgrass and cattail in the Everglades marsh beginning in the mid-1970s. Interim reports of sawgrass and cattail production and nutrient uptake were provided by Davis and Harris (1978) and Davis (1984).

Upon completion of these studies, Davis (1989) concluded that both sawgrass and cattail increased annual production in response to phosphorus enrichment, but the two species differed in how they responded. Sawgrass production reflected general site characteristics of surface water enrichment, but did not vary with interannual changes in phosphorus concentrations. Cattail differed from sawgrass in that it was able to take advantage of interannual variations in surface water phosphorus by increasing production during years of high concentrations, suggesting that cattail was more opportunistic than sawgrass in utilizing temporal variability in surface water phosphorus enrichment. Davis (1991) estimated the phosphorus uptake accompanying leaf production and reported that cattail annual allocation of phosphorus to growing leaves was nearly three times that of sawgrass. However, dying cattail tissue leached or translocated proportionately larger amounts of phosphorus than sawgrass, and the amount of phosphorus remaining in standing dead leaves was comparable for both species.

Greater phosphorus allocation to leaf tissue and increased responsiveness to temporal variations in surface water phosphorus inputs in cattail corresponded to a shorter plant longevity and a more rapid leaf turnover rate relative to sawgrass (Figure 15.3). Cattail longevity was approximately one-fifth that of sawgrass; individually measured cattail plants lived a maximum of 88 weeks, while sawgrass plants survived up to nearly 400 weeks (Davis, 1989). Comparison of leaf turnover rates (ratio of annual production to mean annual biomass) indicated that cattail leaf turnover for plants of a given age was approximately twice that of sawgrass. Shorter longevity and more rapid leaf turnover resulted in a rapid formation of new leaf tissue, which allowed cattail to vary allocation of phosphorus to leaves in response to temporal variation in surface water phosphorus.

The correlation of cattail phosphorus allocation to temporal variability in surface water phosphorus concentrations also corresponded to the role of adventitious roots as uptake sites. Tracer experiments utilizing phosphorus-32 indicated that, compared to other living plant parts, roots growing upward into the water column between dead leaf sheaths of sawgrass and cattail yielded highest scintillation counts 10 days after introduction of the tracer into the surface water of enclosures (Davis, 1982). Scintillation counts of 1285 $pCi \cdot g^{-1}$ in cattail adventitious roots at eutrophic site A in Water Conservation Area 2A were approximately five times higher than counts of 241 $pCi \cdot g^{-1}$ at oligotrophic site D (Figure 15.1). In contrast, counts of 861 $pCi \cdot g^{-1}$ in sawgrass adventitious roots at the eutrophic site were only a third the value of 2434 $pCi \cdot g^{-1}$ for the oligotrophic site. Adventitious roots appeared to represent a direct pathway for phosphorus uptake by the living plants of both species; however, utilization of this pathway by cattail appeared to be higher at the enriched site, while its utilization by sawgrass appeared to be greater at the background site.

Contrasting responses of the two species to nutrient enrichment indicated that sawgrass was a low nutrient status species that was competitive only in infertile habitats, but cattail was a high nutrient status species that was competitive when nutrient supply increased (Table 15.3). In eutrophic habitat, such as the north end of Water Conservation Area 2A, relatively high rates of growth and uptake predictably allowed cattail to dominate available space, light, and nutrient re-

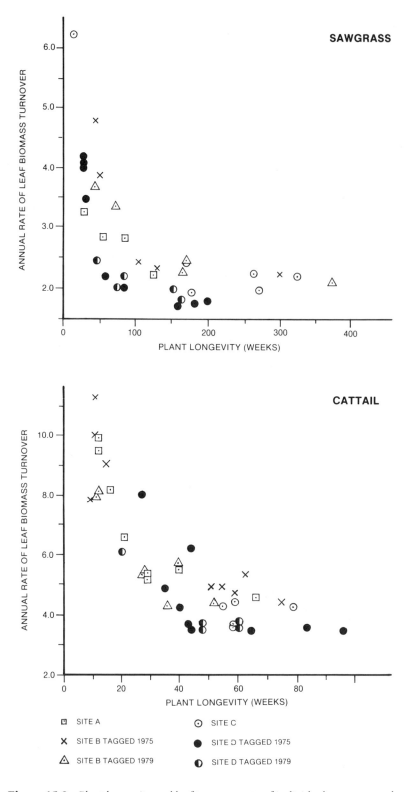

Figure 15.3 Plant longevity and leaf turnover rate of individual sawgrass and cattail plants in Water Conservation Area 2A.

Table 15.3 Contrasting Responses of Sawgrass and Cattail to Phosphorus Enrichment (Davis, 1991) Grouped According to Competitive Strategies (Chapin, 1980; Grime, 1977)

Low nutrient status, stress-tolerant species from infertile habitats	High nutrient status, competitive species from fertile habitats
Sawgrass	**Cattail**
Relatively small leaf growth response to temporal variations in phosphorus enrichment	Relatively large leaf growth response to temporal variations in phosphorus enrichment
Relatively low uptake capacity under phosphorus-enriched conditions	Relatively high uptake capacity under phosphorus-enriched conditions
Lower rate of phosphorus loss from senescing leaves	Higher rate of phosphorus loss from senescing leaves
Longer leaf longevity	Shorter leaf longevity
Lower leaf growth rate	Higher leaf growth rate
Slower leaf turnover rate	More rapid leaf turnover rate
Well-developed leaf cuticle	Poorly developed leaf cuticle

sources. Urban et al. (in press) reported that phosphorus loading into Water Conservation Area 2A best explained cattail density fluctuations at eutrophic sites during a 6-year monitoring study.

Davis (1991) reported that sawgrass and cattail sequestered phosphorus in detritus resulting from plant growth, death, and 2 years of decomposition. Phosphorus retention by both species increased with enrichment. Increases in detritus phosphorus content over 2 years of decomposition at eutrophic sites (Davis, 1991) corresponded to findings that leaf litter assimilated and concentrated phosphorus-32 added to surface water at a eutrophic site (Davis, 1982). Under highly enriched conditions in the north end of Water Conservation Area 2A, an upper limit for phosphorus retention through growth and 2 years of decomposition was estimated to be 1.4 $g \cdot m^{-2} \cdot yr^{-1}$ for cattail and 1.1–1.3 $g \cdot m^{-2} \cdot yr^{-1}$ for sawgrass, compared to estimates of 0.30 and 0.07 $g \cdot m^{-2} \cdot yr^{-1}$ for cattail and sawgrass, respectively, at an interior site. These estimates were based solely on above-ground production and decomposition and did not include values for below-ground roots and rhizomes. Corresponding estimates by Toth (1987, 1988) for roots and rhizomes indicated that below-ground phosphorus uptake accounted for ~14% of the total annual uptake by cattail and ~11% of the uptake by sawgrass. Richardson et al. (1992) estimated a rate of approximately 0.9 $g \cdot m^{-2} \cdot yr^{-1}$ long-term phosphorus accumulation in accreted peat at a highly enriched site in Water Conservation Area 2A. Walker (1993) estimated a rate of 0.035 $g \cdot m^{-2} \cdot yr^{-1}$ phosphorus deposition in accreted peat in a relatively oligotrophic, remote area in Water Conservation Area 2A. Comparison of the estimates of Davis (1991), Richardson et al. (1992), and Walker (1993) suggests that long-term phosphorus deposition in accreted peat amounts to half (or slightly more) of the phosphorus that is sequestered by sawgrass and cattail through above-ground growth, death, and 2 years of decomposition.

VEGETATION SENSITIVITY: COMMUNITY FUNCTION

Accompanying the shift from sawgrass to cattail at eutrophic sites are changes in organic sediment texture, periphyton communities, dissolved oxygen, microbial communities, macroinvertebrate communities, and plant transpiration rates. The texture of organic sediments on the marsh floor changes from relatively compact, fibrous sawgrass sediment at oligotrophic locations to relatively fine, flocculent cattail sediment at eutrophic sites (Davis, 1991).

Calcareous blue-green, green, diatom-rich, and desmid-rich periphyton communities found in low-nutrient waters of the Everglades are replaced by a community dominated by the filamentous algae *Microcoleus lyngbyaceus* and other pollution indicator species at nutrient-enriched sites characterized by high nutrient and major ion concentrations in surface water (Swift and Nicholas, 1987; Browder et al., 1994). Browder et al. (1994) reviewed evidence that periphyton communities dominated by *Microcoleus* in nutrient-enriched areas may provide poor quality food to animal consumers and thus adversely affect periphyton-based aquatic food webs in the Everglades. Periphyton communities appear to be excluded from nutrient-enriched areas of cattail monoculture and mixed dense sawgrass and cattail in Water Conservation Area 2A, which offers an explanation for depressed dissolved oxygen concentrations in these areas (Swift and Nicholas, 1987).

Dissolved oxygen concentrations in surface water tend to fluctuate diurnally in the Everglades marsh, except at eutrophic sites, where daytime increases are dampened (Browder et al., 1994; Belanger and Platko, 1986). As a result, mid-morning dissolved oxygen concentrations of ~2.0 mg·L^{-1} at background sites were depressed to <0.2 mg·L^{-1} at eutrophic sites in Water Conservation Area 2A during the microbiological studies of Reeder and Davis (1983). Anoxic conditions at eutrophic sites appeared to have caused a shift in leaf litter microbial communities toward a reduction in densities of facultative bacteria and fungi (Reeder and Davis, 1983). Colony counts of these organisms were almost an order of magnitude lower at eutrophic sites compared to oligotrophic sites and lower in cattail litter compared to sawgrass. Ramifications of such a shift on Everglades food webs are poorly understood but may be significant, because detrital food webs play important roles in many aquatic and wetland ecosystems.

Colonization of sawgrass and cattail detritus by macroinvertebrates at enriched and oligotrophic sites was monitored by N. H. Urban (unpublished data) over a 2-year decomposition period in Water Conservation Area 2A. Macroinvertebrates were collected from litterbags that were loosely tied to contain the litter while allowing free access of macroinvertebrates into the bags. (See Reeder and Davis [1983] for the litterbag methodology.) Bags containing sawgrass and cattail freshly dead leaf material were placed in sawgrass and cattail stands, respectively, at sites A and D (Figure 15.1) in July 1982, and six replicate bags were retrieved periodically over 2 years. Retrieved bags were preserved in 10% formalin solution with Rose Bengal, after which the detrital contents of each bag were washed and scraped and organisms were removed. Macroinvertebrate numbers are reported as the total number of organisms of each taxon collected from sawgrass or cattail litter at each site over the course of the 2 years (Table 15.4). Differences in communities

Table 15.4 Numbers of Macroinvertebrates Colonizing
Sawgrass and Cattail Leaves in Litterbags during 2 Years
of Decomposition at Nutrient-Enriched and Background Sites
in Water Conservation Area 2A (N. H. Urban, unpublished data)

	Number of individuals			
	Background		Enriched	
	Sawgrass	Cattail	Sawgrass	Cattail
Annelida[a]	210	372	579	662
Gastropoda[b]	129	105	4	0
Isopoda[c]	89	408	0	0
Amphipoda[d]	45	73	64	101
Diptera[e]	110	126	197	10
Coleoptera	6	47	40	34
Other insects	5	10	2	0
Total	594	1141	886	807
	Number of taxa			
	24	36	25	18

[a] Primarily *Dero pectinata* and *D. nivea.*
[b] *Ferrissia, Gyraulus, Helisoma, Pyrgophorus,* and *Physa* at background site; *Physa* at enriched sites.
[c] *Asellus obtusus.*
[d] *Hyalella aztecus.*
[e] Primarily Chironomidae.

of macroinvertebrates that colonized both sawgrass and cattail leaf litter at the eutrophic site compared to the oligotrophic site included a near elimination of snails, elimination of isopods (*Asellus obtusus*), and more than doubling the number of annelid worms (*Dero* spp.). Cattail litter at the eutrophic sites supported half the number of macroinvertebrate taxa compared to the oligotrophic site. Conflicting results were reported by Rader and Richardson (1992), who found higher macroinvertebrate diversity in sweep-net samples from a nutrient-enriched cattail location compared to a more interior wet prairie/slough location in Water Conservation Area 2A, although habitat types and sampling methods are not comparable with those of Urban.

Transpiration measurements by Koch and Rawlik (1993) indicate that nutrient enrichment, as well as replacement of sawgrass by cattail, could alter the water budget of the Everglades. During dry season months, cattail transpiration and stomatal conductance rates per unit area of marsh significantly exceeded those of sawgrass. Average annual transpiration rates for both species were greater at nutrient-enriched compared to transitional and oligotrophic sites in Water Conservation Area 2A. Increased transpiration water loss resulting from eutrophication and cattail spread is of concern, in that it exacerbates the trends of diminished water flows and hydroperiods within the remaining ecosystem (Fennema et al., 1994).

IMPLICATIONS CONCERNING COMMUNITY PERSISTENCE

The vegetation mosaic of the oligotrophic Everglades ecosystem developed under chronic perturbations of climate (Duever et al., 1994), hydrology (Fennema et al., 1994), and fire (Gunderson and Snyder, 1994). Persistence in an environment subject to such variations would necessarily be a strong selection factor governing the development of Everglades vegetation in areas free of anthropogenic influences, and this appears to be the case today (Gunderson, 1994; Davis et al., 1994). Tolerance to low-nutrient stress also appears to have been a selection factor governing community development, as evidenced by sawgrass production and nutrient allocation traits characteristic of low-nutrient stress tolerant plants (Davis, 1991). It is noteworthy that cattail remains a small and widely scattered component of the vegetation mosaic throughout oligotrophic Everglades marshes today (Davis, 1943; Loveless, 1959; Gunderson, 1994; Davis et al., 1994), despite historic disturbance regimes of climate, hydrology, and fire.

Cattail invades oligotrophic Everglades marshes after severe disturbance (Alexander and Crook, 1973), but there it must be viewed as an early colonizer that eventually is replaced by another vegetation type. Otherwise, cattail would be much more prevalent in the Everglades, considering the disturbance history of the ecosystem. Anthropogenic nutrient enrichment might be viewed as yet another type of disturbance which, like natural perturbations, results in cattail colonization of formerly oligotrophic marsh communities. However, the role of cattail appears to change from an early colonizer to a long-term dominant in formerly oligotrophic marshes that have become eutrophic. Where it has become established in such areas in the Water Conservation Areas, cattail has been observed to persist and spread, rather than undergo succession to other vegetation types, over a time scale of decades (S. M. Davis, personal observation).

A prolonged drought that occurred between 1988 and 1990 caused cattail dieback throughout the Everglades. These reversals appeared to be temporary in eutrophic areas, as evidenced by the re-emergence of cattail shoots upon marsh reflooding (Urban et al., in press). A survey of major cattail stands in the Everglades (Figure 15.1) in September 1990 indicated a dense groundcover of *Polygonum* sp., with scattered new cattail shoots emerging in areas which were observed to be dense cattail prior to the drought.

The distribution of soil phosphorus in Water Conservation Area 2A provides one explanation for cattail persistence in nutrient-impacted marshes. Reddy et al. (1991) reported elevated total P concentrations of approximately 1200–1600 $mg \cdot kg^{-1}$ in the top 10 cm of soil cores collected from eutrophic areas near inflow structures compared to 400 $mg \cdot kg^{-1}$ from oligotrophic areas of the interior marsh (Figure 15.4). Anthropogenic surface water nutrient supply through the S-10 gates has created a pool of soluble phosphorus in the peat interstitial water, which represents a chronic source for the high phosphorus requirements of cattail despite intermittent S-10 inputs. This, combined with the opportunistic ability of cattail to allocate temporally variable surface water phosphorus inputs into growing tissue (Davis, 1991), offers a plausible explanation for the ability of cattail to persist in nutrient-impacted areas compared to its role as an early colonizer in oligotrophic marshes

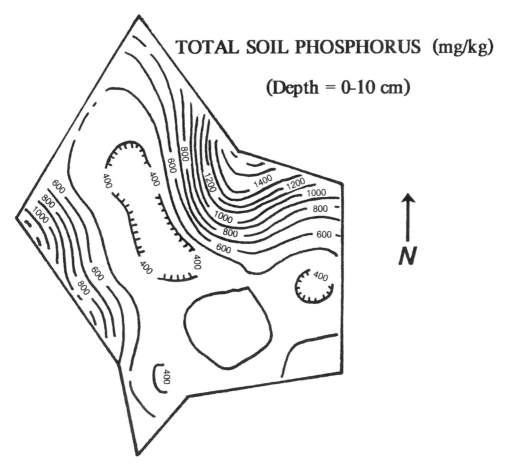

Figure 15.4 Spatial distribution of total phosphorus in the surface 10-cm soil layer in Water Conservation Area 2A. Contour line values are expressed as mg $P \cdot kg^{-1}$ dry soil. (From Reddy et al., 1991.)

beyond zones of nutrient enrichment. The apparent change in the role of cattail in the Everglades from an early colonizer of oligotrophic disturbed sites to a long-term dominant of eutrophic sites may represent a loss of persistence of sawgrass and wet prairie/slough communities within the ecosystem.

ACKNOWLEDGMENTS

Scores of researchers contributed to the work on Everglades phosphorus-vegetation relationships reviewed in this chapter. In particular, credit goes to J. W. Dineen for his encouragement and support of long-term ecosystem research in the Everglades. W. A. Park assisted with the preparation and editing of the manuscript. J. B. Grace, K. K. Steward, W. W. Walker, and M. S. Koch provided reviews that strengthened the manuscript considerably.

LITERATURE CITED

Alexander, T. R. and A. G. Crook. 1973. South Florida Ecological Study. Part 1. Recent and Long-Term Vegetation Changes and Patterns in South Florida, Rep. NTIS PB 231-939, University of Miami, Coral Gables, Fla., 148 pp

Belanger, T. V. and J. R. Platko II. 1986. Dissolved Oxygen Budgets in the Everglades Water Conservation Area 2A, Florida Institute of Technology, Report to the South Florida Water Management District, West Palm Beach, 112 pp.

Browder, J. A., P. J. Gleason, and D. R. Swift. 1994. Periphyton in the Everglades: Spatial variation, environmental correlates, and ecological implications. in *Everglades: The Ecosystem and Its Restoration,* S. M. Davis and J. C. Ogden (Eds.), St. Lucie Press, Delray Beach, Fla., chap. 16.

Chapin, F. S. 1980. The mineral nutrition of wild plants. *Annu. Rev. Ecol. Syst.,* 11:223–260.

Cooper, R. M. and J. Roy. 1991. An Atlas of Surface Water Management Basins in the Everglades: The Conservation Areas and Everglades National Park, Tech. Memo., South Florida Water Management District, West Palm Beach, September 1991, 87 pp.

Davis, J. H. 1943. The natural features of southern Florida. *Fla. Geol. Soc. Geol. Bull.,* No. 25, 311 pp.

Davis, S. M. 1982. Patterns of Radiophosphorus Accumulation in the Everglades after its Introduction into Surface Water, Tech. Publ. 82-2, South Florida Water Management District, West Palm Beach, 28 pp.

Davis, S. M. 1984. Cattail Leaf Production, Mortality, and Nutrient Flux in Water Conservation Area 2A, Tech. Publ. 84-8, South Florida Water Management District, West Palm Beach, 40 pp.

Davis, S. M. 1989. Sawgrass and cattail production in relation to nutrient supply in the Everglades. in Freshwater Wetlands and Wildlife, R. R. Sharitz and J. W. Gibbons (Eds.), Office of Scientific and Technical Information, U.S. Department of Energy, Oak Ridge, Tenn., pp. 325–341.

Davis, S. M. 1991. Growth, decomposition, and nutrient retention of *Cladium jamaicense* Crantz and *Typha domingensis* Pers. in the Florida Everglades. *Aquat. Bot.,* 40:203–224.

Davis, S. M. and L. A. Harris. 1978. Marsh plant production and phosphorus flux in Everglades Conservation Area 2. in Environmental Quality through Wetlands Utilization, M. Drew (Ed.), Council on the Restoration of the Kissimmee River Valley and Taylor Creek–Nubbin Slough Basin, Tallahassee, pp. 105–131.

Davis, S. M., L. H. Gunderson, W. A. Park, J. Richardson, and J. Mattson. 1994. Landscape dimension, composition, and function in a changing Everglades ecosystem. in *Everglades: The Ecosystem and Its Restoration,* S. M. Davis and J. C. Ogden (Eds.), St. Lucie Press, Delray Beach, Fla., chap. 17.

Duever, M. J., J. F. Meeder, L. B. Meeder, and J. M. McCollum. 1994. The climate of south Florida and its role in shaping the Everglades ecosystem. in *Everglades: The Ecosystem and Its Restoration,* S. M. Davis and J. C. Ogden (Eds.), St. Lucie Press, Delray Beach, Fla., chap. 9.

Fennema, R. J., C. J. Neidrauer, R. A. Johnson, T. K. MacVicar, and W. A. Perkins. 1994. A computer model to simulate natural south Florida hydrology. in *Everglades: The Ecosystem and Its Restoration,* S. M. Davis and J. C. Ogden (Eds.), St. Lucie Press, Delray Beach, Fla., chap. 10.

Forthman, C. A. 1973. The Effects of Prescribed Burning on Sawgrass, *Cladium jamaicense* Crantz, in South Florida, M.S. thesis, University of Miami, Coral Gables, Fla., 83 pp.

Frederick, P. C. and G. V. N. Powell. 1994. Nutrient transport by wading birds in the Everglades. in *Everglades: The Ecosystem and Its Restoration,* S. M. Davis and J. C. Ogden (Eds.), St. Lucie Press, Delray Beach, Fla., chap. 23.

Grace, J. B. 1988. The effects of nutrient additions on mixtures of *Typha latifolia* L. and *Typha domingensis* Pers. along a water-depth gradient. *Aquat. Bot.,* 31:83–92.

Grace, J. B. 1989. Effects of water depth on *Typha latifolia* and *Typha domingensis. Am. J. Bot.,* 76:762–768.

Grime, J. P. 1977. Evidence for the existence of three primary strategies in plants and its relevance to ecological and evolutionary theory. *Am. Nat.,* 111:1169–1194.

Gunderson, L. H. 1994. Vegetation of the Everglades: Determinants of community composition. in *Everglades: The Ecosystem and Its Restoration*, S. M. Davis and J. C. Ogden (Eds.), St. Lucie Press, Delray Beach, Fla., chap. 13.

Gunderson, L. H. and J. R. Snyder. 1994. Fire patterns in the southern Everglades. in *Everglades: The Ecosystem and Its Restoration*, S. M. Davis and J. C. Ogden (Eds.), St. Lucie Press, Delray Beach, Fla., chap. 11.

Hendry, C. D., P. L. Brezonik, and E. S. Edgerton. 1981. Atmospheric deposition of nitrogen and phosphorus in Florida. in *Atmospheric Pollutants in Natural Waters*, S. J. Eisenreich (Ed.), Ann Arbor Science, Ann Arbor, Mich., pp. 199–215.

Koch, M. S. and P. S. Rawlik. 1993. Transpiration and stomatal conductance of two wetland macrophytes (*Cladium jamaicense* and *Typha domingensis*) in the subtropical Everglades. *Am. J. Bot.*, 80(10):1146–1154.

Kushlan, J. A. 1977. Energetics of the white ibis. *Auk*, 94:114–122.

Light, S. S. and J. W. Dineen. 1994. Water control in the Everglades: A historical perspective. in *Everglades: The Ecosystem and Its Restoration*, S. M. Davis and J. C. Ogden (Eds.), St. Lucie Press, Delray Beach, Fla., chap. 4.

Loveless, C. M. 1959. A study of the vegetation in the Florida Everglades. *Ecology*, 40:1–9.

Millar, P. S. 1981. Water Quality Analysis in the Water Conservation Areas 1978 and 1979, Interim Progress Report, Tech. Memo., South Florida Water Management District, West Palm Beach, May 1981, 63 pp. plus appendix.

Oliver, J. D. and T. Legovic. 1988. Okefenokee marshlands before, during and after nutrient enrichment by a bird rookery. *Ecol. Model.*, 43:195–223.

Parker, G. G. 1974. Hydrology of the pre-drainage system of the Everglades in southern Florida. in *Environments of South Florida: Present and Past, Memoir No. 2*, P. J. Gleason (Ed.), Miami Geological Society, Coral Gables, Fla., pp. 18–27.

Powell, G. V. N., W. J. Kenworthy, and J. W. Fourqurean. 1989. Experimental evidence of nutrient limitations of seagrass growth in a tropical estuary with restricted circulation. *Bull. Mar. Sci.*, 44:324–340.

Powell, G. V. N., J. W. Fourqurean, W. J. Kenworthy, and J. C. Zieman. 1991. Bird colonies cause seagrass enrichment in a subtropical estuary: Observational and experimental evidence. *Mar. Coastal Shelf Sci.*, 32:567–579.

Rader, R. B. and C. J. Richardson. 1992. The effects of nutrient enrichment on algae and macroinvertebrates in the Everglades: A review. *Wetlands*, 12:121–135.

Reddy, K. R., W. F. DeBusk, Y. Wang, R. DeLaune, and M. Koch. 1991. Physico-Chemical Properties of Soils in the Water Conservation Area 2A of the Everglades, Final Report, Soil Science Department, Institute of Food and Agricultural Sciences, University of Florida, Gainesville, submitted to South Florida Water Management District, West Palm Beach, 118 pp. plus appendix.

Reeder, P. B. and S. M. Davis. 1983. Decomposition, Nutrient Uptake and Microbial Colonization of Sawgrass and Cattail Leaves in Water Conservation Area 2A, Tech. Publ. 83-4, South Florida Water Management District, West Palm Beach, 24 pp.

Richardson, C. J., C. B. Craft, R. R. Johnson, R. G. Qualls, R. B. Rader, L. Sitter, and J. Vymazal. 1992. Effects of Nutrient Loadings and Hydroperiod Alterations on Control of Cattail Expansion, Community Structure and Nutrient Retention in the Water Conservation Areas of South Florida, Duke Wetland Center Publ. 92-11, School of the Environment, Duke University, Durham, N.C., 441 pp.

Snyder, G. H. and J. M. Davidson. 1994. Everglades agriculture: Past, present, and future. in *Everglades: The Ecosystem and Its Restoration*, S. M. Davis and J. C. Ogden (Eds.), St. Lucie Press, Delray Beach, Fla., chap. 5.

South Florida Water Management District. 1992. Draft Surface Water Improvement and Management Plan for the Everglades, Supporting Information Document, South Florida Water Management District, West Palm Beach, 472 pp.

Steward, K. K. and H. Ornes. 1975a. The autecology of sawgrass in the Florida Everglades. *Ecology*, 56:162–171.

Steward, K. K. and H. Ornes. 1975b. Assessing a marsh environment for wastewater renovation. *J. Water Pollut. Control Fed.*, 47:1880–1891.

Steward, K. K. and W. H. Ornes. 1983. Mineral nutrition of sawgrass (*Cladium jamaicense* Crantz) in relation to nutrient supply. *Aquat. Bot.*, 16:349–359.

Stinner, D. H. 1983. Colonial Wading Birds and Nutrient Cycling in the Okefenokee Swamp Ecosystem, Ph.D. dissertation, University of Georgia, Athens.

Swift, D. R. and R. B. Nicholas. 1987. Periphyton and Water Quality Relationships in the Everglades Water Conservation Areas, Tech. Publ. 87-2, South Florida Water Management District, West Palm Beach, 44 pp.

Toth, L. A. 1987. Effects of Hydrologic Regimes on Lifetime Production and Nutrient Dynamics of Sawgrass, Tech. Publ. 87-6, South Florida Water Management District, West Palm Beach, 31 pp.

Toth, L. A. 1988. Effects of Hydrologic Regimes on Lifetime Production and Nutrient Dynamics of Cattail, Tech. Publ. 88-6, South Florida Water Management District, West Palm Beach, 26 pp.

Urban, N. H., S. M. Davis, and N. G. Aumen. In press. Fluctuations in sawgrass and cattail densities in Everglades Water Conservation Area 2A under varying nutrient, hydrologic and fire regimes. *Aquat. Bot.*

Volk, B. G., S. D. Schemnitz, J. F. Gamble, and J. B. Sartain. 1975. Baseline data on Everglades soil-plant systems: Elemental composition, biomass, and soil depth. in Mineral Cycling in Southeastern Ecosystems, CONF 740513, National Technical Information Service, U.S. Department of Commerce, Springfield, Va., pp. 658–672.

Walker, W. W. 1989. Rainfall Total Phosphorus Concentrations and Loadings in Everglades National Park, Report to U.S. Department of Justice, August 1989.

Walker, W. W. 1991. Water quality trends at inflows to Everglades National Park. *Water Resour. Bull.*, 27(1):59–72.

Walker, W. W. 1993. A Mass-Balance Model for Estimating Phosphorus Settling Rate in Everglades Water Conservation Area 2A, Report to U.S. Department of Justice, March 8, 1993.

Walker, D. R., M. D. Flora, R. G. Rice, and D. J. Scheidt. In preparation. Response of the Everglades Marsh to Increased Nitrogen and Phosphorus Loading. Part II: Macrophyte Community Structure and Chemical Composition, Tech. Rep., South Florida Research Center, Everglades National Park, Homestead, Fla.

Waller, B. G. 1982. Water Quality Characteristics of Everglades National Park, 1959–1977, with Reference to the Effects of Water Management, Water Resour. Invest. 82-34, U.S. Geological Survey, Miami, 51 pp.

Waller, B. G. and J. E. Earl. 1975. Chemical and Biological Quality of Water in Part of the Everglades, Southeastern Florida, Water Resour. Invest. 56-75, U.S. Geological Survey, Miami, 157 pp.

16

PERIPHYTON in the EVERGLADES: SPATIAL VARIATION, ENVIRONMENTAL CORRELATES, and ECOLOGICAL IMPLICATIONS

Joan A. Browder

Patrick J. Gleason

David R. Swift

ABSTRACT

The periphyton community, made up of many taxa of microalgae, covers much of the Everglades, serving as a food web base, as well as building calcitic mud sediment and oxygenating the water column. An overview of the literature on Everglades periphyton is provided, as well as a proposed classification system based on general taxonomy and extent of calcite encrustation. Specific community types that are extremes in gradually changing sets of dominant taxa along environmental gradients are described. At least three environmental gradients—hydroperiod/water depth, phosphorus concentration, and aspects of water chemistry involving the major ions, especially calcium—affect the taxonomic composition, growth characteristics, structure, and extent of calcite encrustation of Everglades periphyton. Degree of saturation with respect to $CaCO_3$, rather than $CaCO_3$ concentration alone, determines calcite encrustation. The authors propose that hydrologic conditions influence $CaCO_3$ saturometry through their influence on the release of carbon dioxide into, and its diffusion from, the water column.

Water quality at various Everglades locations is determined largely by whether the main source of surface water is local rainfall or canal discharge. Hydroperiod and water depth at many Everglades locations have been changed by canals, levees, and other water control structures. Because they affect both water quality and hydrologic conditions, water management activities influence spatial distributions of various periphyton communities.

Dissolved oxygen and diet studies with small aquatic animals suggest that changes in the periphyton community caused by either nutrient enrichment or shortened hydroperiod may reduce habitat quality and carrying capacity. In restoration efforts, periphyton communities can be used as sensitive, site-specific indicators of environmental quality.

INTRODUCTION

The smallest but most widely distributed plant community in the Florida Everglades is an assemblage of microalgae that live on shallow, submerged substrates and are referred to collectively as periphyton, aufwuchs, or the algal mat. This algal complex covers the submerged portions of most aquatic macrophytes in the Everglades and forms a thick carpet covering the sediment in many locations.

Everglades periphyton is composed of many different species of microalgae. Most belong to one of three major divisions, or groups, of algae: Myxophyceae (blue-green algae), Bacilliophyceae (diatoms), and Chlorophyceae (green algae). Green algae in the families Mesotaeniaceae and Desmidiaceae are separated from the other green algae in most discussions of Everglades periphyton. Called *desmids,* they consist of many species that tend to occur together under environmental conditions different from those where other major types of green algae in Everglades periphyton are found.

The periphyton community is significant for several reasons. Converting light and CO_2 to organic matter and fed upon by many aquatic organisms, it is a primary producer. Along with the detritus of aquatic macrophytes, periphyton forms a base of the Everglades food web.

Calcareous periphyton has a pronounced effect on the chemistry of surrounding and overlying waters, particularly pH, partial pressures of dissolved oxygen and carbon dioxide, and concentrations of calcium and bicarbonate. Certain periphyton taxa fix nitrogen. Precipitation of calcite may influence the bioavailability of phosphorus and may be responsible for the low natural concentrations of phosphorus in Everglades waters.

The large quantities of calcite precipitated as a result of periphyton photosynthesis form a calcitic mud that is one of the two major soil types in the Everglades. Calcitic sediment promotes water retention and provides habitat for burrowing aquatic organisms and infauna as well. The remains of algae and other microorganisms associated with periphyton may also contribute substantially to sawgrass peat deposits (Hatcher et al., 1986).

The taxonomic composition of periphyton reflects local water chemistry and hydrologic conditions. Therefore, periphyton can be used as an environmental indicator. From a geologic perspective, the calcitic mud deposited by the calcareous periphyton is an important indicator of both periphyton occurrence in the geologic past and past hydrologic and water quality conditions.

Everglades periphyton communities are classified and characterized according to the paradigm that periphyton occurs along at least three major environmental gradients which determine taxonomic composition, structure, appearance, and extent of calcite encrustation. At environmental extremes of their occurrence, periphyton communities differ. Community types merge, and the hypothetical point on each environmental gradient where a change of type occurs is fuzzy. Because environmental variables seldom differ one at a time outside of laboratories, this characterization is imperfect. Gaps in research coverage prevent complete understanding; however, when viewed together, existing studies provide a new perspective.

This proposed classification and characterization will be used to describe the current spatial distribution of each periphyton type, its probable historical distribution, and probable reasons for distribution changes. The possible ecosystem-scale consequences of these changes will then be explained. Other chapters in this volume will provide descriptions of regional climate, hydrology, soils, and aquatic animals which are essential to understanding the Everglades periphyton communities.

All freshwater wetlands south of the Tamiami Trail will be referred to as the southern Everglades (Figure 16.1). This includes Shark River Slough, Taylor Slough, and their flanking wet prairies, as well as wet prairies inside the Buttonwood Embankment, which separates the Everglades from the coast. The latter includes the C-111 area, named for the major canal bisecting it. Although referred to by Egler (1952) as the southeast "saline" Everglades, this area consists primarily of freshwater wetlands (Tabb et al., 1967). The wetlands that lie south of Tamiami Trail and east of Everglades National Park, including northeast Shark River Slough and upper Taylor Slough, are called the East Everglades. Hydrologic conditions differ to some extent from historic conditions in most of these areas. The differences will be described in a later section.

HISTORY of RESEARCH

The south Florida periphyton community was first described by Dachnowski-Stokes (1928). Hunt (1952) showed the importance of periphyton in south Florida aquatic food webs. Van Meter-Kasanof (1973) provided the first comprehensive phycological and ecological description. Wood and Maynard (1974) described the periphyton and phytoplankton at several Everglades sites. Browder et al. (1981, 1982) performed a quantitative comparison of Everglades periphyton at sites subjected to a range of environmental conditions. Several investigators, including Hunt (undated), Belanger and Platko (1986), and Browder et al. (1982), estimated primary production of Everglades periphyton. Gleason (1972) and Gleason and Spackman (1974) examined the effect of the physiological activity of periphyton on water chemistry, which varied diurnally and was related to photosynthesis and respiration. Gleason and Spackman (1974) described diurnal changes in water quality in relation to periphyton photosynthetic activity. Gleason (1972) examined Everglades periphyton from a geological and sedimentological perspective. Gleason and Spackman (1974) examined the calcium carbonate encrusting

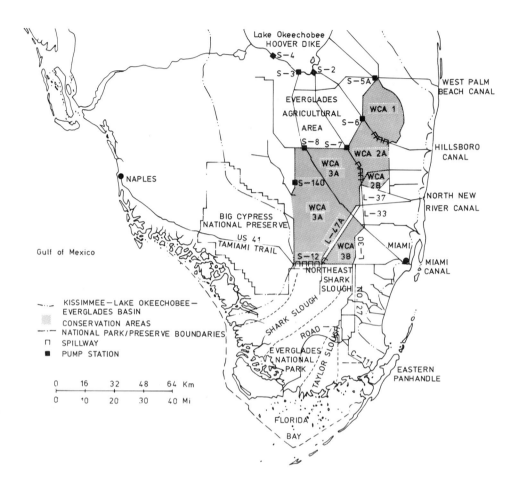

Figure 16.1 Map of Everglades and southeast Florida, showing present wetland hydrologic units: Everglades Agricultural Area; Water Conservation Areas 1, 2A, 2B, 3A, and 3B; Everglades National Park; East Everglades, and C-111 area.

characteristics of periphyton in relation to taxonomic composition and water chemistry. Swift (1981, 1984) and Swift and Nicholas (1987) described periphyton communities in the Water Conservation Areas and identified taxonomic indicators of water quality. The contributions of these authors to an understanding of Everglades periphyton ecology will evolve from the development of an ecological synthesis.

The variation in Everglades periphyton is briefly described in this review. Calcite encrustation of periphyton, relation of periphyton to macrophytes, repopulation mechanisms, biomass, aquatic primary productivity, and the diurnal variation in water chemistry associated with the metabolism of the periphyton community are described next. The chapter concludes with a proposed classification of periphyton types in relation to environmental variables and a hypothesis regarding the interacting effects of certain environmental factors on periphyton taxonomic composition, chemical composition, and growth habit.

Variation and Spatial/Temporal Patterns in Everglades Periphyton Communities

Everglades periphyton communities vary in three major ways: (1) extent of encrustation with calcite, (2) taxonomic composition, and (3) growth habit. Variation in primary production and biomass is related to growth habit. At some locations, periphyton cover all available surfaces, whereas at others they cover only the leaves and stems of submerged macrophytes.

Blue-Green, Green, and Diatom-Rich Periphyton

Several authors have described the taxonomy of Everglades periphyton. Descriptions from representative studies are given in Table 16.1. The descriptions can roughly be separated into those made in the Everglades Water Conservation Areas and those made in the southern Everglades, which includes Everglades National Park and the East Everglades (Figure 16.1). The Water Conservation Area studies were conducted by Gleason and Spackman (1974), Swift (1981), and Swift and Nicholas (1987). The southern Everglades studies were conducted by Gleason (1972), Van Meter-Kasanof (1973), Wood and Maynard (1974), and Browder et al. (1981). Van Meter-Kasanof (1973) described pioneers of two periphyton communities: blue-green and green. Swift (1981) and Swift and Nicholas (1987) described distinct periphyton communities associated with three types of water chemistry. Browder et al. (1981) described a broad continuum of periphyton types occurring along a hydroperiod gradient. Gleason and Spackman (1974) distinguished between high desmid and low desmid periphyton whose presence was determined by water quality. Further discussion of this topic will be deferred until a later section in which what is known about the effect of environmental variables on periphyton taxonomic composition is more fully described.

Calcareous and Noncalcareous Periphyton

Gleason and Spackman (1974) observed a visual difference between periphyton in different parts of Water Conservation Area 1. At sites near the canals that form the perimeter of Water Conservation Area 1, they found a "calcareous" periphyton, which exhibited a thick, white, calcareous appearance with an annular ring structure. At interior Water Conservation Area 1 sites, they found a "noncalcareous" periphyton which formed a thin, brown coating on plant stems. Measured ash mass, which provides a rough indication of calcite content, was about 12% of total dry mass in noncalcareous, desmid-rich periphyton in interior Water Conservation Area 1. In contrast, ash constituted from 30 to 50% of the dry mass of calcareous periphyton in peripheral Water Conservation Area 1 (taken from a figure in Gleason and Spackman, 1974). The calcareous periphyton at the peripheral Water Conservation Area 1 location did not form an algal mat, unlike the calcareous periphyton observed by Gleason (1972) at Paurotis Pond and by others elsewhere in the southern Everglades.

Ash mass measured by Browder et al. (1982) in the southern Everglades (Figure 16.1) was higher than that reported by Gleason and Spackman in peripheral Water Conservation Area 1. Southern Everglades ash mass represented from 49 to 81%

Table 16.1   Dominant or Representative Taxa in Everglades Periphyton,
as Described by Various Investigators

Van Meter (1965) and Van Meter-Kasanof (1973)

This investigator defined two types of Everglades periphyton, green and blue-green. Succession in green periphyton began with green algae (especially desmids), progressed to the addition of filamentous green and some blue-green algae, and finally to the addition of more filamentous blue-green algae. Green algae and diatoms were present throughout succession. The composition of the blue-green periphyton was blue-green algae from beginning to end, with other algal classes rare.

Most important taxa in terms of number of individuals and effect on the ecosystem		"Most widely distributed" taxa	
Filamentous blue-greens		Desmids	Filamentous blue-greens
Scytonema	*Phormidium*	*Cosmarium*	*Oscillatoria*
Lyngbya	*Plectonema*	*Staurastrum*	*Chroococcus*
Microchaete	*Leptobasis*	Diatoms	*Lyngbya*
Oscillatoria	*Schizothrix*	*Cymbella*	*Phormidium*
		Penium	
		Mastogloia	
		Synedra	

Wood and Maynard (1974)

These investigators identified a sequence in the formation of periphyton that began with a sessile diatom flora and continued into an increasing blue-green algal extension of the community.

Most frequent blue-green algae	Abundant in upper part of Everglades National
Oscillatoria (dry algal mats consist almost entirely of this taxon)	Park, particularly sinkholes, and possibly indicative of eutrophication
Microcystis (next most frequent)	Filmentous greens
Chroococcus (next in frequency)	*Cladophora* *Oedogonium*
	Bulbochaete *Spirogyra*

Gleason and Spackman (1974)

These investigators referred to two distinct communities, one prevalent in waters supersaturated with respect to calcium carbonate and the other prevalent in waters undersaturated with respect to calcium carbonate. A high diversity of desmid species and certain diatom species was characteristic of the undersaturated waters. Several species of calcareous filamentous blue-green algae, other diatom species, and one desmid genus were characteristic of supersaturated waters.

Most important blue-green genera		Filamentous greens at low-conductivity stations	
Scytonema	*Oscillatoria*	*Bulbochaete* sp. *Oedogonium* spp.	
Lyngbya	*Plectonema*	*Mougeotia* sp. *Spirogyra* spp.	
Microchaete	*Leptobasis*	Others	
Phormidium	*Schizothrix*	Diatoms	
Desmids important at high-conductivity stations		*Cymbella* spp.	
Cosmarium spp.		*Nitzchia* spp.	
Desmids at low-conductivity stations		*Navicula* spp.	
78 species		*Mastogloia smithii* var. *lacustris*	
		Others	

Table 16.1 (continued)   Dominant or Representative Taxa in Everglades Periphyton, as Described by Various Investigators

Browder et al. (1981)

These investigators examined algal taxonomic composition in terms of relative cell volumes of the major taxonomic units: blue-green algae, green algae, and diatoms. They found significant differences among sites and related these differences to hydroperiod. Blue-greens dominated the periphyton in terms of volume at most sites, but diatoms and green algae were substantial periphyton components at sites with longer hydroperiods.

Dominant genera
  Filamentous blue-greens
    Scytonema (highest in volume, >50% at
      most stations)
    Schizothrix (second in volume at stations
      where *Scytonema* made up >50% of
      volume)

Green algae of major volumetric importance
  at some stations
  Spirogyra (filamentous green)
  Bulbochaete (filamentous green)
Most important diatoms in terms of volume
  Cymbella   *Amphora*
  Gomphorema   *Navicula*
  Mastogloia   *Synedra*

Swift and Nicholas (1987)

These investigators distinguished three periphyton communities and related them to water chemistry.

Dominant taxa in waters of high mineral
($CaCO_2$) content and low nutrient content
  Filamentous blue-greens
    Schizothrix calcicola (the most frequently
      encountered taxon on glass slides)
    Scytonema hofmannii (50–80% of algal
      volume in mat cores)
  Diatoms
    Mastogloia smithii var. *lacustris*
    Cymbella ruttneri
    Anomoeneis vitrea
    Synedra pahokeensis sp. nov.
  Other diatoms characteristic of highly miner-
    alized ($CaCO_2$) waters although not dom-
    inant
    Annomoeneis vitrea
    Cymbella ruttneri
    Mastogloia smithii var. *lacustris*
    Cymbella minuta var. *psuedogracilis*
    Synedra pahokeensis sp. nov.
    Gomphonema cf. *affine* var. *insigne*
    Nitzschia storchii sp. nov.
    Amphora venta
Dominant taxa in waters of high nitrogen and
phosphorus content
  Early summer
    Filamentous greens
      Oedogonium spp.
      Stigeoclonium sp.
  Late summer and fall
    Filamentous blue-greens
      Microcoleus lyngbyaceus (previously
        Oscillatoria tenius)

Winter
  Diatoms
    Gomphonema parvulum
    Nitzschia amphibia
    Navicula disputans
    Navicula confervaceae
    Nitzschia palea
  Other diatoms characteristic of high
  nutrient waters, but not dominant
    Nitzschia   *Navicula*
      tarda   *confervacea*
      sigmoidea   *disputans*
      sp. 7   *cuspidata*
Taxa characteristic of low nutrient, low mineral
waters
  Desmids of many species
  Filamentous greens
    Mougeotia sp. and others
  Diatoms (typical of low nutrient, low mineral
  waters, but generally not dominant)
    Cymbella amphioxys
    Anomoeoneis serians var. *brachysira*
    Anomoeoneis serians
    Anomoenoneis vitrea
    Frustulia rhomboides var. *saxonica*
    Frustulia rhomboides var. *crassinervia*
    Cymbella minuta var. *silesiaca*
    Nitzchia sp. 7 sp. nov.
    Eunotia naegelii
    Synedra tenera
    Pinnularia biceps
    Navicula subtilissima
    Stenopterobia intermedia

(mean = 67%) of total periphyton dry mass in the Browder et al. (1982) study and varied little between seasons.

Calcite content of periphyton in the southern Everglades studied by Van Meter-Kasanof (1973) on glass slides ranged from 10% in 1-week growths to 73% in the heaviest and oldest growths. The ash component of the periphyton samples studied by Wood and Maynard (1974) varied between roughly 10% at two interior Shark River Slough sites and from 80 to 90% in Taylor Slough and near Paurotis Pond (Figure 16.1). The low ash content of interior Shark River Slough samples studied by Wood and Maynard (1974) is similar to that of the noncalcareous, desmid-rich periphyton studied by Gleason and Spackman (1974).

According to Van Meter-Kasanof (1973), cylinder thickness on stems in the southern Everglades ranged from 0 to 6 mm in diameter ("light" periphyton) and from 19 to 50 mm in diameter ("heavy" periphyton). According to Gleason and Spackman (1974), the diameter of periphyton cylinders around stems can be as great as 63 mm. Wood and Maynard (1974) reported algal mats as thick as 100 mm in parts of the southern Everglades.

Gleason and Spackman (1974) identified two filamentous blue-green algal species as the principal encrusting taxa: *Scytonema hofmannii* and *Schizothrix calcicola. Microcoleus lyngbaceous,* another filamentous blue-green alga frequently found in Everglades periphyton, also encrusts (Gleason and Spackman, 1974) but is not as important a component of calcareous periphyton as the other two species. Other filamentous blue-green algae such as *Johannesbaptistia pellucida,* unicellular blue-green algae, green algae, and diatoms do not encrust.

Physical Structure

As described by Gleason and Spackman (1974), *Scytonema* and *Schizothrix* form the basic physical structure of the periphyton mat or cylinder. *Schizothrix* occurs as felts of filaments wrapped around and between larger *Scytonema* filaments. *Schizothrix* felts form horizontal layers, and vertical strands of *Scytonema* connect the layers.

Filamentous blue-green algae have three components: a trichome which is a linear arrangement of cells, a noncellular and nonliving sheath surrounding the trichome that has been secreted by the cells, and a mucilage coating on the sheath which has the appearance of a clear, amorphous gel (Figure 16.2). The calcite precipitates associated with blue-green algae form crystals that vary in characteristics by species (Gleason and Spackman, 1974). Gleason and Spackman (1974) found the dendritic crystal associated with *Schizothrix calcicola* unusual and possibly indicative of supersaturated conditions. The highly structured vertical microstratigraphy of calcareous periphyton consists of many layers (Gleason and Spackman, 1974). Monty (1965) described the general organization of a similar mat in Andros Island as an alternation between calcified layers and hyaline layers. Van Meter-Kasanof (1973) found that her "green" periphyton was amorphous and did not exhibit the concentric ring structure associated with cylinders of blue-green periphyton. Gleason and Spackman (1974) noted that the layered structure of calcareous periphyton was not present in noncalcareous periphyton. The filamentous blue-green algae are the only algae in the periphyton that have the capacity to precipitate calcite. Therefore, a positive correlation might

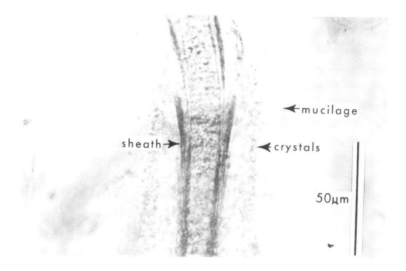

Figure 16.2 Filament structure of the blue-green algae *Scytonema hofmannii* showing sheath, mucilage, and crystals. (From Gleason and Spackman, 1974.)

be expected between the relative cell volume of blue-green algae and the relative weight of calcite. Browder et al. (1981) tested this hypothesis with regression analysis but found no support.

Association with Macrophytes

Calcareous periphyton shows a preference for some plant species over others (Van Meter, 1965; Gleason and Spackman, 1974). Water hyssop (*Bacopa* sp.), sawgrass (*Cladium jamaicensis*), spikerush (*Eleocharis cellulosa* and *E. elongata*), beakrush (*Rhynchospora traceyi*), purple-flowered bladderwort (*Utricularia purpurea*), spider lily (*Hymenocallis* sp.), arrowhead (*Sagittaria* sp.), string lily (*Crinum americanum*), and red mangrove (*Rhizophora mangle*) are readily coated by periphyton. On the other hand, yellow-flowered bladderwort (*Utricularia* sp.), pipewort (*Eriocaulon compressum*), and white water lily (*Nymphaea odorata*) are not normally encrusted. Possible reasons for differential colonization of macrophytes by periphyton are discussed by several authors in a volume edited by Wetzel (1983).

The association of periphyton with *Utricularia* is particularly strong and widespread. Floating mats in Water Conservation Areas 2A and 3A and the peripheral marsh of Water Conservation Area 1 are made up almost entirely of periphyton-encrusted *Utricularia* (D. R. Swift, personal communication). Bosserman (1981) observed an association of periphyton with bladderwort in the Okefenokee Swamp.

Repopulation after Drydown

Differential abilities to either survive on site or to recolonize after drying may be a major basis for the variation in taxonomic composition resulting from different hydroperiods. Wood and Maynard (1974) listed several possible ways in which periphyton communities in the Everglades may be restored after drying. Some algal taxa may grow faster than others under newly reflooded conditions, which would make them more volumetrically important in areas of frequent and prolonged drying.

Survival within a dry, even burned, area could be achieved by means of desiccation- and fire-resistant spores, protection within the interior of the algal mat, or protection within peat sediments. The mucus coating of some filamentous blue-green algae allows vegetative cells to survive dry conditions. In addition, some species of filamentous blue-green algae have desiccation-resistant resting hetero-cysts and akenetes that can survive long droughts. Diatoms form auxospores (i.e., desiccation-resistant resting spores) which may help them become re-established in seasonally drying areas. Wood and Maynard (1974) found viable algal spores and vegetative filaments in algal mats, even after fire. They reported that algae taken 35 cm below the surface in peat cores proved viable after 8 months in a core tube.

Wood and Maynard (1974) proposed that recolonization could result from the wind transport of surface foams containing diatoms, as observations by Maynard (1968a, 1968b) suggest. Some filamentous green algae may also colonize this way. Continuously flooded source areas are necessary for recolonization to occur by this mechanism.

Some algae may survive dry periods as animals do—in sinkholes and deeper ponds. Certain filamentous green algae such as *Spirogyra, Bulbochaetae,* and *Oedogonium* tolerate nutrient-rich conditions (Swift and Nicholas, 1987) which often exist in small ponds during the dry season.

Algae that can withstand high water temperatures may have some advantage in shallow Everglades marshes, where Browder et al. (1981) observed water tempera-tures as high as 36°C. The occurrence of blue-green algae in thermal streams documents their adaptation to high temperatures (Ruttner, 1972). *Schizothrix calcicola,* an Everglades periphyton species, is found in the Ohanapecosh Hot Springs of Mount Rainier National Park, Washington, where temperatures range from 35 to 48°C (Stockner, 1968).

Biomass

Periphyton forms a substantial part of the vegetative biomass of the Everglades. Wood and Maynard (1974) determined that organic periphyton biomass (excluding ash) often represents more than 50% of total dry vegetative biomass at Shark River Slough sites (Table 16.2). Sites in Shark River Slough, accessible only by air boat, had the heaviest periphyton growth.

Organic periphyton averaged 33% of total dry plant biomass at sites investigated by Browder et al. (1982). Maximum organic periphyton biomass reached roughly 43% of total dry plant biomass between October and December, the end of the periphyton growing season at annually drying sites.

Biomass measurements taken by Browder et al. (1982) were lower than those by Wood and Maynard (1974), but Browder et al. had no stations in central Shark River Slough, where Wood and Maynard found their highest periphyton biomass. The two studies shared one general study location, the Taylor Slough bridge (SR

Table 16.2   Everglades Periphyton Biomass ($g \cdot m^{-2}$) Measured by Various Investigators

Description	Organic (ash free)	Total (incl. ash)	Author
Taylor Slough minimum	T	1	Browder et al. (1982)
Taylor Slough maximum[a]	542	2682	
Taylor Slough bridge	293	1492	
Shark River Slough minimum	1	2	
Shark River Slough maximum	419	1430	
Taylor Slough bridge	950	2620	Wood and Maynard (1974)
Shark River Slough minimum	0	0	
Shark River Slough maximum	2550	5960	
Water Conservation Area 1 minimum (interior)	40	44	Gleason and Spackman (1974)
Water Conservation Area 1 maximum (peripheral)	225	445	

[a] Maxima for organic periphyton and total periphyton in Taylor Slough are separate cases.

27) area. Periphyton biomass measurements by Wood and Maynard (1974) at that site were roughly three times those of Browder et al. (1982).

Browder et al. (1982) found that the bulk of periphyton on standing macrophytes was associated with dead rather than live stems. In four samples, 99% of total periphyton dry mass on standing macrophytes occurred on dead stems. Observations during collecting suggested that a large proportion of the algal mat was associated with the fallen dead stems of macrophytes.

Biomass measurements by Gleason and Spackman (1974) (Table 16.2) illustrate the difference between calcareous (peripheral) and noncalcareous (interior) periphyton in terms of both ash content and total biomass. If their samples were representative, the biomass of calcareous periphyton in Water Conservation Area 1 is lower than that in the southern Everglades. There was no algal mat at peripheral Water Conservation Area 1 sites studied by Gleason and Spackman, which may explain the lower biomass of calcareous periphyton there.

Aquatic Primary Production and Associated Water Chemistry

Large diurnal changes in dissolved oxygen (DO) and the partial pressure of carbon dioxide (pCO_2) demonstrate the high biological activity of the periphyton community (Hunt, undated; Van Meter, 1965; Van Meter-Kasanof, 1973; Browder et al., 1982; Belanger and Platko, 1986; Belanger et al., 1989). DO commonly varied from 0 to 3 $mg \cdot L^{-1}$ at dawn to supersaturation (as much as 15 $mg \cdot L^{-1}$) by midday at periphyton-rich sites investigated by Van Meter-Kasanof (1973). Diurnal DO fluctuations were much lower where periphyton was not a major community component (Van Meter-Kasanof, 1973; Belanger and Platko, 1986).

Oxygen produced in photosynthesis is sometimes trapped in periphyton as conspicuous air bubbles (Van Meter-Kasanof, 1973). The trapped bubbles create a buoyancy that causes portions of the periphyton mat to break away from the substrate and rise to the surface. The resulting periphyton rafts create a shaded microhabitat utilized by largemouth bass and other aquatic animals.

Gleason (1972), Gleason and Spackman (1974), and others described the diurnal fluctuations in calcium, pCO_2, alkalinity, and pH in the water column over periphyton communities (Figure 16.3) and ascribed them to photosynthesis and respiration. Calcite precipitation is influenced by photosynthesis and respiration through their effect on pCO_2, which affects $CaCO_3$ saturometry (Figure 16.3). $CaCO_3$ concentrations required for saturation increase with increasing pCO_2.

High rates of gross primary production (GPP) and low rates of net primary production (NPP) are typical of shallow Everglades periphyton areas (Table 16.3). The low NPP despite high GPP is caused by a high net community metabolism. The highest GPP and lowest net community metabolism measured by Belanger and Platko (1986) was in Water Conservation Area 2A. They found positive net community metabolism only at their sites with periphyton. They view periphyton

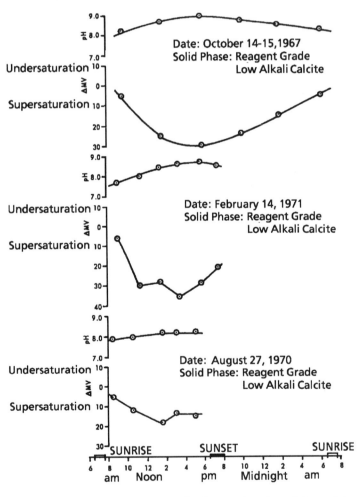

Figure 16.3 pH and change in millivolts as a function of time of day in water overlaying an algal mat. (From Gleason and Spackman, 1974.)

Table 16.3 Estimates of Aquatic Primary Productivity,
Net Community Metabolism, and P/R Ratios in the Everglades

Description		Gross primary production (g·m⁻²)	Net community metabolism (g·m⁻²)	P/R ratio	Citation
		24-hour			
Summer minimum	SS	1.41	−0.649	0.684	Browder et al. (1982)
Summer maximum	TS	4.49	2.01	1.808	Browder et al. (1982)
Winter minimum	SS	1.16	−0.015	0.988	Browder et al. (1982)
Winter maximum	TS	2.00	−0.042	0.999	Browder et al. (1982)
June minimum	TS		0.98		Hunt (unpublished)
June maximum	TS		3.88		Hunt (unpublished)
September	WCA 2A	10.7	−2.42		Belanger and Platko (1986)
November	WCA 2A	3.28	1.28		Belanger and Platko (1986)
Summer minimum	Park	3.4			Goldstein et al. (1980)
Summer maximum	Park	14.0			Goldstein et al. (1980)
Average	Park			1.1	Goldstein et al. (1980)
Maximum (Oct.)	SWEVER2	3.102	−0.158	0.90	Swift (1989)
Minimum (Mar.)	SWEVER2	0.436	−0.018	0.92	Swift (1989)
Average	SWEVER2	1.611	−0.147	0.892	Swift (1989)
Minimum (Feb.)	C-111:7	0.587	−0.075	0.77	Swift (1989)
Maximum (Jul.)	C-111:7	6.144	0.182	1.06	Swift (1989)
Average	C-111:7	2.371	−0.147	0.89	Swift (1989)

Note: Maxima and minima are based on GPP. P/R = GPP/night respiration.

photosynthesis as a major contributor to the DO content of Everglades waters. Both Browder et al. (1982) and Hunt (undated) found positive net community metabolism during the summer at Taylor Slough sites with conspicuous periphyton communities. Swift (1989) found positive net community production at a C-111 site with periphyton (Table 16.3).

Browder et al. (1982) estimated annual GPP and net community metabolism at two sites based on summer and winter measurements in a flow-through chamber. An annual GPP of 1184 g O_2·m⁻² and net community metabolism of 359 g O_2·m⁻² were estimated for a long-hydroperiod site in Taylor Slough. An annual GPP of 469 g O_2·m⁻² and net community metabolism of −121 g O_2·m⁻² were estimated at a long-hydroperiod site in Shark River Slough.

The annual estimate by Browder et al. (1982) at the Taylor Slough sites roughly translates into a net production of 359 g·m⁻²·yr⁻¹ dry organic matter (assuming every positive gram of O_2 in community metabolism is roughly equivalent to production of 1 g of dry organic matter [Odum and Hoskins, 1958]). This is similar to the estimate of 381 g·m⁻²·yr⁻¹ (1043 mg·m⁻²·d⁻¹) by Van Meter-Kasanof (1973) for Shark River Slough, based on harvesting periphyton grown on glass slides.

Two other estimates of annual GPP and net community production are based on 24-hour oxygen diurnals for each month calculated by Swift (1989) and are summarized in Table 16.3. Working from Swift's original table and substituting values from adjacent months for months with no data, the 12 values were summed and multiplied by 30.4 (average number of days per month). Calculated by this method, annual GPP and net community metabolism at his SWEVER2 station were 652 and -23 g $O_2 \cdot m^{-2} \cdot yr^{-1}$, respectively. At his C-111/7 station, they were 944 and -28 g $O_2 \cdot m^{-2} \cdot yr^{-1}$, respectively. The GPP estimates are higher than those of Browder et al. (1982), and the net community metabolism estimate is intermediate to that of Browder et al.

Wilson (1974) estimated aquatic GPP at 5.7 g $O_2 \cdot m^{-2} \cdot yr^{-1}$ and NPP at 4.29 g $O_2 \cdot m^{-2} \cdot yr^{-1}$ at periphyton sites in the C-111 area. She used a bell jar and measured oxygen pressure in the air above the water with a manometer. Her estimates seem exceptionally low compared to those of other investigators and may not have taken into account changes in the degree of saturation of the water column with oxygen during the measurement cycle.

By comparison, Stockner (1968) estimated a GPP equivalent to 1355 g·yr^{-1} dry mass and an NPP of 1150 g·yr^{-1} dry mass in a thermal spring in Washington. Respiration was much lower than that in the Everglades (equivalent to 205 g·yr^{-1} dry mass), probably because little organic material was added to that system from outside sources.

In general, the Everglades estimates indicate a system of moderately high aquatic GPP in which most of the production is used by the photosynthesizing organisms and decomposers. The rate of decomposition relative to photosynthesis is high due to the presence of emergent plants, which make up more than half the dry mass at periphyton-rich sites (Browder et al., 1982). They contribute little oxygen to the water column but deposit dead plant parts, whose decomposition consumes oxygen.

The production of emergent macrophytes should be viewed as part of the production of shallow marsh sites. Browder et al. (1982) estimated an annual net production of between 419 and 1744 g·m^{-2} by emergent plants at Everglades sites. This estimate of net production by emergents is lower than estimates for fertile marsh sites by Lieth and Whittaker (1975).

Correlations with Environmental Factors

Water chemistry, nutrient concentrations, hydrologic conditions, and sediment type have been shown to correlate with variation in taxonomic composition and degree of calcite encrustation of Everglades periphyton. Related aspects of water chemistry, such as total dissolved solids, pH, bicarbonate, and $CaCO_3$ concentration with respect to saturation, play a role in explaining variations. Phosphorus appears to be the important controlling nutrient. Data from both southern Everglades and Water Conservation Area sites have contributed to current understanding of effects of $CaCO_3$ saturometry on periphyton taxonomic composition. The most detailed studies of nutrient effects on taxonomic composition have been conducted in the Water Conservation Areas, which are exposed to agricultural drainage water as well as mineral-rich ground water. Studies implicating hydroperiod, water depth, and sediment type have been confined to the southern Everglades, where a full range of hydroperiods is found and sediments contain a range of percentages of calcite and organic matter.

Water Chemistry Relations

Gleason and Spackman (1974) found that the extent to which agricultural runoff influenced local water chemistry determined whether calcareous periphyton or a desmid-rich, noncalcareous periphyton was present (Table 16.1). Desmid species diversity, the number of green algal species, blue-green algal abundance, the calcareous character of the blue-green algae, periphyton biomass, and the ash content of the periphyton appeared to be affected by agricultural runoff. The calcareous periphyton, dominated by blue-green algae (particularly *Scytonema* and *Schizothrix*), was found in water that was more basic, higher in calcium and bicarbonate, higher in total dissolved solids, and closer to saturation with respect to calcite than that where desmid-rich periphyton was found. The desmid-rich periphyton, which also contained filamentous green algae and diatoms, was characteristic of the more pristine waters of the marsh interior, recharged primarily from rainfall.

Gleason and Spackman (1974) concluded that although the presence of calcareous periphyton correlated with conductance, the relationship was indirect. High periphyton biomass correlated well with stations having a basic pH and high alkalinity. Gleason and Spackman (1974) hypothesized that the extent of encrustation with calcite was a function of the degree of saturation of the water with respect to $CaCO_3$. The presence of blue-green calcareous periphyton in Water Conservation Area 1 corresponded with water saturated with respect to $CaCO_3$, whereas the presence of green algae, particularly numbers of desmid taxa, correlated with water undersaturated with respect to $CaCO_3$.

Swift and Nicholas (1987) compared the taxonomic composition of periphyton at sites exposed to three types of water: low nutrient/low dissolved minerals, low nutrient/high dissolved minerals, and high nutrient/high dissolved minerals. The low nutrient/low dissolved minerals water occurred in areas that appeared influenced mainly by rainfall, particularly interior Water Conservation Area 1. The high nutrient/high dissolved minerals waters occurred in areas fed by canals carrying drainage water from the Everglades Agricultural Area. The low nutrient/high dissolved minerals water occurred in the interior part of Water Conservation Areas 2A and 3A and the peripheral marsh of Water Conservation Area 1.

Principal taxa of the periphyton communities growing in the various areas are summarized in Table 16.1. Desmids such as *Mougeotia* sp. and filamentous green algae were the principal species in low nutrient/soft water (i.e., low mineral) habitats in interior Water Conservation Area 1. Calcareous blue-green algae (*Schizothrix calcicola* and *Scytonema hofmannii*) and a group of hard water (i.e., high mineral) diatoms were well represented in low nutrient/hard water habitats. Peripheral areas of Water Conservation Areas 1 and 2 receiving high phosphorus and nitrogen loading supported a specialized community of pollution-tolerant algae dominated by *Microcoleus lyngbyaceus*.

Swift and Nicholas (1987) distinguished three distinct diatom assemblages whose presence or absence reflected water quality (Table 16.1). Low nutrient/hard water sites harbored one group of diatom species. Acidic, low mineral waters with moderately low nutrient levels contained another group of diatoms. Waters high in both nutrients and minerals supported a third diatom group.

Browder et al. (1981) did not find the correlation between taxonomy and high levels of dissolved solids in southern Everglades periphyton that Gleason and Spackman (1974) and Swift and Nicholas (1987) found in the Water Conservation Areas. Lower conductivity sites (generally 200–400 umhos·cm^{-1}) studied by Browder et al. (1981) were located in Taylor Slough where, in general, the lowest relative cell volumes of diatoms and green algae were found. The high conductivity sites (generally 450–900 umhos·cm^{-1}) were in Shark River Slough where, in general, the highest relative cell volumes of diatoms and green algae were found. High conductivity at some Shark River Slough sites was caused by the L-67 canal (Flora and Rosendahl, 1982), which delivers water to Everglades National Park.

Hydrologic conditions may override the effect of high levels of dissolved solids on southern Everglades periphyton. Most Shark River Slough stations had higher water depths and longer hydroperiods than most Taylor Slough stations. Browder et al. found that the highest conductivity sites had the longest hydroperiods, and the lowest conductivity sites had the shortest hydroperiods.

Inorganic Phosphorus (Ortho P) Relations

Swift and Nicholas (1987) determined by factor analysis that *Microcoleus* predominated at sites with higher phosphorus availability in the water column and higher phosphorus concentrations in periphyton tissue. *Microcoleus* abundance also appeared to correspond to high conductivity and high calcium concentrations. Filamentous green algae (*Oedogonium* sp. and *Stigeoclonium* sp.) were also correlated with waters containing high mineral and nutrient concentrations. *Scytonema* and *Schizothrix,* two blue-green algae that dominated the periphyton at low nutrient/high mineral sites, were replaced by the blue-green *Microcoleus* at high nutrient/high mineral sites.

Experimental addition of nutrients to parts of Water Conservation Area 3B eliminated the algal mat. Stewart and Ornes (1975) added nutrients to marsh enclosures containing periphyton, sawgrass, and bladderwort to evaluate the ability of Everglades marsh to renovate waste water. Natural background dissolved phosphorus levels ranged from undetectable to 0.022 mg·L^{-1}, and total phosphorus ranged from 0.004 to 0.041·L^{-1} in control enclosures. In the treated enclosures, concentrations were increased gradually to 0.8 mg·L^{-1} of dissolved phosphorus (1.2 mg·L^{-1} total P). The periphyton was eliminated and replaced by dense and continuous algal blooms (the algal taxa responsible for the blooms were not documented).

Nutrient dosing experiments in Everglades National Park (Scheidt et al., 1987) demonstrated that inorganic P concentrations of 0.033 mg·L^{-1} (as compared to the natural level of roughly 0.004 mg·L^{-1}) eliminated the periphyton mat and changed taxonomic composition to more pollution-tolerant species. In the Water Conservation Areas, inorganic P concentrations during the wet season exist near or below detectable limits (<0.004 mg·L^{-1}), while total P averages near 0.01 mg·L^{-1} (Swift and Nicholas, 1987).

Browder et al. (1981) found a statistically significant negative relationship between inorganic P concentration and desmid percent cell volume. They reported water inorganic P concentrations from 0.006 to 0.039 mg·L^{-1} at southern Everglades stations. The highest values (>0.015 mg·L^{-1}) were found immediately after reflooding

(June 1 and 2, 1978) at the Taylor Slough stations, but were widespread and corresponded to similar levels in the nearby C-31 canal. Samples were not filtered prior to analysis for inorganic P, which may prevent their valid comparison with other reported concentrations.

Inorganic Nitrogen Relations

Scheidt et al. (1987) found nitrate levels of 0.096 mg·L^{-1} as nitrogen (N) in an environment where concentrations were ordinarily 0.01 mg·L^{-1}. This led to elimination of the periphyton mat within months and a significant decrease in periphyton biomass, which remained only on submerged leaves and stems.

Goldstein (1980) concluded that the blue-green algal component of periphyton, particularly *Scytonema*, was a potential source of nitrogen in the Everglades. Using the acetylene reduction technique, they estimated daylight nitrogen fixation of 0.0961–0.593 mg N_2·m^{-2}·h^{-1} at Everglades sites. Assuming a 12-hour day, results were 1.2–7.1 N_2·m^{-2}·d^{-1}. The higher rate is roughly one-half the 12.1-mg N_2·m^{-2}·d^{-1} average reported by Bautista and Paerl (1985) for a blue-green algal mat on an intertidal flat. Other than the student project by Goldstein (1980), no investigations into the role of periphyton in Everglades nitrogen cycling have been made.

Hydroperiod and Water Depth Relations

Van Meter-Kasanof (1973) identified two distinct periphyton communities based on taxonomic composition of pioneer and climax stages on glass slides and the hydroperiod of the location (Table 16.1). She concluded that green periphyton (which had a larger component of green algae, particularly desmids) required year-round flooding. Hydroperiods of 5–7 months promoted the occurrence of blue-green periphyton. She considered the blue-green periphyton to be adapted for exposure to rigorous environmental extremes. She found blue-green periphyton mainly in Taylor Slough, but also found that it occurred following a year of drought at a Shark River Slough site that had burned.

Browder et al. (1981) found a statistically significant positive relationship ($p <$ 0.1) between the percentage volume of both green algae and diatoms in periphyton and hydroperiod-related variables. Sites representing both hydroperiod extremes were located in both Shark River Slough and Taylor Slough, but the majority of the short-hydroperiod sites were in Taylor Slough.

The genus *Scytonema*, which dominates the periphyton at short-hydroperiod sites, is a soil algae (Monty, 1965) and may be better adapted to desiccation than the other dominant algal taxa occurring in Everglades periphyton. Observations by Browder et al. (1981) suggest that frequent and prolonged drying may arrest succession in periphyton communities, promoting a near monoculture of blue-green algae by preventing the development of a diverse flora and limiting the diatoms and green algae.

Swift and Nicholas (1987) found a statistically significant negative relationship between water depth and the percent cell volume of *Scytonema* and *Schizothrix* grown on glass slides at Water Conservation Area sites. They found a significant positive relationship between water depth and the percent cell volume of *Microcoleus*. These relationships seemed secondary to the effect of water chemistry on taxonomic composition in the Water Conservation Areas.

Observations by Gleason and Spackman (1974) suggest that water depth affects the formation of calcareous periphyton. Calcareous periphyton develops on plants and the substrate at variable water depths. However, the best development is in the upper 0.67 m of the water column. The periphyton appears to lose its calcite crystals with depth and takes on a deep green appearance. The algal mat, or epipelic calcareous periphyton, is sensitive to changes in depth and shading and will degenerate to a crumbly mass or a thin coating of algae at depths greater than roughly 60 cm. At roughly 20–30 cm deeper than the surrounding wet prairie environment, shallow ponds on the western side of Taylor Slough lacked any epipelic calcareous periphyton, although the wet prairie exhibited a luxurious calcareous periphyton. These observations suggest that calcareous periphyton mats probably will not be found at depths greater than approximately 60 cm in open water. Where macrophytes shade the water and retard photosynthesis at the bottom the depth allowing algal mat formation may be less.

Calcareous periphyton and the formation of an algal mat are typical of Taylor Slough, and general conditions there provide an indication of conditions favorable to the formation of calcareous periphyton. Tropical BioIndustries (1987) estimated that areas in Taylor Slough covered by calcareous periphyton and marl had a 6- to 7-month hydroperiod. The average floodwater depth for all marl-producing areas along the southeastern boundary of Everglades National Park, from Shark River Slough to C-111, ranged from 8.1 to 53.1 cm (mean equal to 21.6 cm) during the early 1960s (Tropical BioIndustries, 1987).

Bottom Sediment Relations

Several investigators have found correlations between bottom sediment and characteristics of periphyton communities. All believed that the relationship with bottom sediment was secondary and other factors were controlling. Browder et al. (1981) found a significant positive correlation between sediment organic matter and the percent cell volume of green algae and diatoms in calcareous periphyton. They concluded that hydroperiod determined taxonomic composition and that long-term hydrologic conditions determined the organic content of the sediment. Swift (1984) found that peat sediment depth was positively related to filamentous green algal percent cell volume and negatively related to blue-green algal percent cell volume. His main conclusion was that water chemistry determined taxonomic composition. Van Meter-Kasanof (1973) found that green periphyton (actually a calcareous periphyton dominated by blue-green algae) was typically associated with peat substrates with year-round flooding.

The documented occurrence of the desmid-rich periphyton community described by Gleason and Spackman (1974) and Swift (1984) is in areas underlain by peat, but the coincidence with peat is believed to be secondary. Gleason and Spackman (1974) concluded that the desmid-rich periphyton found in Water Conservation Area 1 occurs where water chemistry dictates that calcareous periphyton will not flourish and nutrient concentrations are low. A peat substrate will not necessarily give rise to a desmid-rich periphyton unless the overlying water is undersaturated with respect to $CaCO_3$ and the water is low in nutrients.

A calcitic mud substrate probably will give rise to calcareous periphyton under low nutrient conditions because of the effect of the calcitic mud on overlying water chemistry. Gleason (1972) and Gleason and Spackman (1974) found that under natural water chemistry conditions in Everglades National Park, calcareous periphyton flourished in areas near limestone outcroppings, such as northern Taylor Slough and Paurotis Pond. Both locations possessed a calcitic mud substrate, a short hydroperiod, and a water chemistry which, at that time, appeared largely controlled by rainwater seeping from adjacent pinelands or uplands and carrying dissolved limestone.

Calcareous periphyton is not confined to calcite substrates. It grows over peat in the Water Conservation Areas. Gleason and Spackman (1974) described the calcareous periphyton growing over peat at peripheral marsh sites in Water Conservation Area 1 that are exposed to canal water containing high concentrations of $CaCO_3$. Calcareous periphyton was also growing over peat in parts of Water Conservation Area 3A not directly exposed to canal water (D. R. Swift, personal observation). A decrease in hydroperiod may be the reason for the growth of calcite-forming periphyton in areas where the peat substrate suggests no previous calcite deposition.

Calcareous periphyton can grow on a variety of substrates including calcitic mud, peat, and rock, as well as submerged vegetation. Thin peats and peats up to 9 ft (2.7 m) thick in Water Conservation Area 1 are covered by calcareous periphyton. Although substrate type plays a major role in other periphyton habitats, current understanding of Everglades algae suggests that substrate alone does not dictate the presence or absence of Everglades periphyton, its calcite content, or its taxonomic composition. Effects of substrate can be overridden by water chemistry and hydrologic conditions.

Factors Influencing Primary Productivity

Swift and Nicholas (1987) proposed that phosphorus was limiting to periphyton primary productivity in the Everglades. They examined phosphorus and nitrogen concentrations in periphyton communities in relation to phosphorus concentrations in marsh water and found that N:P mass ratios decreased exponentially with increases in water phosphorus concentrations. Maximum periphyton growth rates in the Water Conservation Areas occurred at N:P ratios less than 15:1. N:P ratios were highest at interior marsh sites unaffected by canal water inflows, where ratios ranged from 50:1 to 200:1. Such areas are characterized by low concentrations of phosphorus within algal cell tissue and in surface waters, soils, and macrovegetation, suggesting nutrient-limiting conditions.

Swift and Nicholas (1987) suggested that the availability of phosphorus for plant uptake may be closely tied to the calcite precipitation rate. Inorganic P rapidly coprecipitates with $CaCO_3$ at pH 9 and greater, which commonly occurs under high rates of photosynthesis. The process is enhanced by temperature (Otsuki and Wetzel, 1972) and could be expected to be intense in the Everglades. The interaction between $CaCO_3$ and phosphorus may be an important factor influencing phosphorus availability in the Everglades.

Data by Hunt (undated) suggest that shortages of CO_2 may sometimes limit photosynthesis in the periphyton community. Maximum rates of oxygen production occurred in the morning, before the CO_2 in the water column was depleted. Diurnal variation in inorganic P, caused by both phosphorus uptake in photosynthesis and interaction of phosphorus with $CaCO_3$, may also have played a role in the variable rate of photosynthesis observed by Hunt.

Periphyton Utilization

Van Meter-Kasanof (1973) identified the aquatic microinvertebrates of shallow water marshes and open ponds of the Everglades. The organisms she identified in 79 samples came from 6 phyla: Protozoa, Coelenterata (class Hydrozoa), Nematoda (class Phasmidia), Rotatoria (class Monogonta), Arthropoda, and Mollusca. Three protozoan classes were included: Mastigophora, Sarcodina (including amoebas), and Ciliata (including the paramecia). Arthropods were represented by crustaceans (such as cladocerans, daphnids, cyclops, and amphipods) and arachnids (such as Hydracarina). Mollusks were represented by gastropods and pelecypods. Although food habits were not examined, it seems likely that many were feeding either directly on the algae of the periphyton or on each other.

Hunt (1952) examined the food relationships of larger aquatic animals in the Tamiami Canal, which bisects the Everglades. He identified the stomach contents of the Florida spotted gar (*Lepisosteus platyrhincus*), nine fish species consumed by gar, and the freshwater shrimp *Palaemonetes paludosa*. The shrimp was abundant in the canal and figured prominently in the stomach contents of gar and fish that gar consume. Four fish species were carnivorous, with diets consisting of invertebrates alone: bluegill (*Lepomis microlophus*), spotted sunfish (*Lepomis punctatus*), warmouth (*Chaenobryttus coronarius*), and golden topminnow (*Fundulus chrysotus*). Three were omnivorous, consuming both algae and invertebrates: redfin killfish (*Chriopeops goodei*), eastern mosquitofish (*Gambusia holbrooki*), and least killifish (*Heterandria formosa*). Two fish species were herbivorous, their stomach contents consisting primarily of algae and plant fragments: sailfin molly (*Poecilia latipinnis*) and flagfish (*Jordanella floridae*). Algae and fragments of rooted vegetation made up the diets of the shrimp and most other invertebrates.

Hunt (1952) thought the net plankton in the canal too sparse to provide sufficient food for fish that fed on algae or microinvertebrates. The periphyton, "a heterogeneous assemblage of algae and minute animals which covers all solid submerged objects with a slimy coating," seemed a more likely food source.

Cell counts on stems indicated maximum densities of diatoms and green algae several times greater than Young (1945) found in Douglas Lake, Michigan. The number of rotifers per square centimeter averaged 450. Hunt (1952) found oligochaetes, chironomid larvae, copepods, cladocerans, ostracods, water mites, and snails within the canal periphyton. All algal species and most small animal taxa identified in the intestinal tracts of fishes and invertebrates were also present in the periphyton.

Based on available literature, Hunt (1952) listed the following as herbivorous, aquatic invertebrates that may have been feeding directly on the periphyton algae: oligochaetes, some rotifers, copepods, cladocereans, freshwater shrimp, mayfly nymphs, caddis larvae, some Hemiptera (Corixa), some beetles (Haliplidae), and snails. Omnivorous taxa that may have been acting as both primary and secondary consumers included some rotifers, ostracods, scuds (Hyalella), and crayfish. Carnivorous taxa included some rotifers, dragonfly and damselfly nymphs, most Hemiptera, most beetles, and water mites.

Based on frequency of occurrence in stomachs or intestinal tracts, Hunt (1952) concluded that periphyton was the primary food source of the freshwater shrimp. Stomachs of mayfly nymphs (*Caenis diminuta* and *Callibaetis floridana*) contained green and blue-green algae, diatoms, and higher plant debris. Stomachs of chironomid larvae contained diatoms and masses of fine plant debris. The intestinal tracts of the damselfly nymphs contained only animals, including mayfly nymphs and chironomid larvae.

Hunt (1952) roughly calculated that the canal produced up to a tonne of gar per hectare per year. He concluded that the great production of plants, particularly the algae in the periphyton, was one factor responsible for the extremely large gar population.

Browder et al. (in review) examined the food habits of some small fish and macroinvertebrates in shallow Everglades marshes. They determined that roughly 50% of the diet of crayfish was algae, the remainder consisting of higher plant detritus. Apple snails also consumed considerable quantities of algae. Although small invertebrates were the main dietary item of mosquitofish samples in August, algae were their main food source in December. There may have been fewer insects in the winter, making periphyton a critical winter food source for mosquitofish. Current understanding of the Everglades aquatic food web, based on periphyton and detritus from macrophytes, is summarized in Figure 16.4.

Periphyton grazers may consume more diatoms and green algae than blue-green algae in relation to their relative cell volumes in the environment. Browder

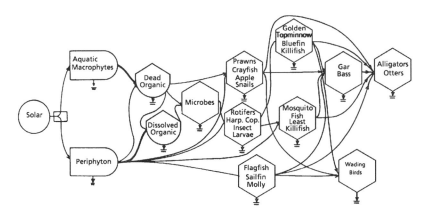

Figure 16.4 Everglades food web, showing energy flow from periphyton to upper trophic level consumers.

et al. (in review) examined differential consumption of periphyton taxa in several species of fish and macroinvertebrates at two different sites and on two different dates. Sampled digestive tracts had lower cell volume proportions of blue-green algae and higher proportions of other types of algae than were found in periphyton sampled from the local environment. Despite small sample sizes, differences were statistically significant in several cases and for the animals as a whole.

Growth rates of squirrel tree frog (*Hyla squirella*) tadpoles, a common Everglades amphibian, were shown to differ depending upon the type of periphyton they were fed in a laboratory experiment. Separate rations were fed to four groups of tadpoles for 12 days by Browder et al. (undated). The first ration was from periphyton rich in diatoms, the second from periphyton rich in filamentous green algae, and the third from periphyton consisting almost entirely of filamentous blue-green algae. The fourth ration was a commercial tadpole food. Although the most rapid growth was achieved with the commercial food, tadpoles fed diatom-rich periphyton grew significantly faster than those fed either green-rich or blue-green-rich periphyton. Growth on green-rich periphyton was significantly greater than growth on blue-green periphyton. The tadpoles fed the blue-green periphyton experienced no significant growth during the 12-day feeding experiment.

The two animal studies suggest that the diatoms and green algae in periphyton may be higher quality dietary elements than the blue-green algae. Diatoms are known to be a rich source of fatty acids, some of which are essential to growth. The sheath material of filamentous blue-green algae is difficult for most animals to digest, which may make this group a less suitable food.

Reark (1961) estimated the standing crop of aquatic animals at Everglades sites, from dense sawgrass to light rush with bladderwort. He found the highest biomass (2.5–200 kg·ha^{-1} wet mass) on the light rush with bladderwort. The lowest biomass was in dense sawgrass (<0.5 kg·ha^{-1} wet mass) and light rush in open water (0.25–0.5 kg·ha^{-1} wet mass). Standing stocks were from 0.5 to 10 kg·ha^{-1} wet mass in the other vegetation types. Highest standing stocks coincided with areas of greatest diurnal physicochemical variation (DO, pH, etc.), presumably due to the activity of periphyton and submerged plants. Estimates by Reark (1961) seem extremely low in comparison to those by Lagler et al. (1971), who estimated fish biomass of 64 kg·ha^{-1} wet mass in a "grass marsh" in Zambia (Lowe-McConnell, 1975).

DISCUSSION

Some differences among study results and conclusions and possible reasons for them will be discussed. Then, both commonalties and differences from these studies will be used to develop three new paradigms for viewing periphyton ecology in the Everglades. The first is an explanation of how hydrologic conditions influence water chemistry and, therefore, calcite precipitation. The second explains the influence of hydrology on bottom sediment. The third is a classification scheme for Everglades periphyton communities, relating periphyton community types to environmental variables. This synthesis will be followed by a rough description of distributional changes in periphyton communities and their probable causes.

Finally, possible uses for periphyton in environmental assessment, monitoring, and restoration will be discussed.

Reasons for Differences in Conclusions among Previous Studies

Descriptions of south Florida periphyton communities and influencing environmental conditions have been provided by five main investigators or investigative teams: (1) Van Meter (1965) and Van Meter-Kasanof (1973), (2) Gleason (1972) and Gleason and Spackman (1974), (3) Wood and Maynard (1974), (4) Browder et al. (1981, 1982), and (5) Swift (1981, 1984) and Swift and Nicholas (1987). Differences in their conclusions may be due to four main factors: (1) differences in geographical area (southern Everglades versus Water Conservation Areas in the central to northern Everglades), (2) differences in methodology (e.g., slides versus natural substrate), (3) differences in authorities used for nomenclature, and (4) differences in field work years.

Investigations conducted in the Water Conservation Areas and those conducted in the southern Everglades resulted in different conclusions about the primary factors governing the taxonomic composition of Everglades periphyton communities. Studies in the Water Conservation Area emphasized the importance of nutrient concentrations and other aspects of water chemistry, whereas studies in Everglades National Park emphasized the importance of hydrologic conditions, particularly hydroperiod. Clearly, ranges in variation in the environmental conditions in the two areas helped shape the respective studies and facilitated their conclusions. Variations in nutrient concentrations and other aspects of water chemistry are more pronounced in the Water Conservation Areas than in Everglades National Park because of the direct inflow of agricultural runoff to the Water Conservation Areas. Variation in hydroperiod is greater in Everglades National Park and the East Everglades because natural water flows have been replaced by scheduled water releases influenced by other water management priorities. The wide variations in water chemistry in the northern Everglades create dramatic differences in algal taxonomic composition and extent of calcite encrustation. Wide variation in hydroperiod in the southern Everglades causes substantial changes in periphyton taxonomic composition, but variants are not as conspicuously different as in the northern Everglades because southern Everglades periphyton communities are all calcareous.

Some differences in results among quantitative studies can be explained by differences in degree of taxonomic detail. Browder et al. (1981) distinguished only four taxonomic groups in their quantitative analysis: blue-greens, desmids, other greens, and diatoms. Swift and Nicholas (1987) measured cell volume at finer taxonomic detail. This was useful because there are subsets of species or genera within each major group that respond differently to variation in water chemistry and nutrient concentrations and thus represent reliable water quality "indicators." For example, the blue-greens *Scytonema* and *Schizothrix* grow best in low nutrient conditions and the blue-green *Microcoleus* grows most prolifically in high nutrient, particularly high phosphorus, conditions (Swift and Nicholas, 1987). Some of the filamentous green algae (*Spirogyra* and *Oedogonium*) also respond favorably to higher nutrient conditions. Groups of diatom species grow well in hard water, eutrophic conditions, while others require low nutrient, calcium-rich waters. Similarly, desmid species can be separated into those that thrive in soft waters and those that prefer hard waters. Taxonomic groupings by Browder et al. (1981) worked well in distinguishing differences in hydrologic conditions, but were not suitable for detecting water chemistry differences.

Van Meter (1965), Van Meter-Kasanof (1973), Swift (1981), and Swift and Nicholas (1987) all used glass slides, whereas the work by Browder et al. (1981) was done entirely with periphyton growing on natural substrates. Observations by Swift and Nicholas (1987) indicated that taxonomic composition on glass slides and natural substrates differs. The results of Browder et al. (1981) did not support the observation of Van Meter-Kasanof (1973) that climax stages of both green periphyton and blue-green periphyton were composed largely of blue-green algae. Browder et al. (1981) found that diatoms and/or filamentous green algae consistently made up more than 20% of cell volume in well-established periphyton communities at longer hydroperiod sites.

The study by Wood and Maynard (1974) stands apart from the others in its description of the main blue-green algal components of Everglades periphyton. Their departure may be because they used different nomenclature authorities from the other investigators. Two major works affecting the nomenclature of blue-green algae in the group Oscillatoriaceae were published in the 1960s (Drouet, 1963, 1968). Identification and classification of some of the blue-greens was confusing up until publication and acceptance of Drouet's work. Drouet (1963) showed that *Schizothrix calcicola* could be identified as 54 separate species, depending upon growth type, which varied according to environmental conditions. Similarly, Drouet (1968) considered specimens identified as *Tolopthrix* (mentioned by Wood and Maynard [1974] as one of the blue-greens generally present) to be a growth form of *Scytonema*. Drouet (1968) grouped hundreds of filamentous blue-green algal species into 6 genera and 23 species. Investigators using earlier references could easily have identified *Microcoleus lyngbyaceus* as one of five different species, including both *Oscillatoria* spp. and *Lyngbya* spp. Identification of *Scytonema* as *Tolopthrix* could partly explain why Wood and Maynard (1974) thought neither was as important as *Oscillatoria*. (See Swift and Nicholas [1987] for a more complete discussion of nomenclature problems as related to Everglades periphyton.) All Everglades periphyton authors other than Wood and Maynard (1974) appear to follow Drouet's (1968) revision of the Oscillatoriaceae. Although Wood served as an advisor to Van Meter-Kasanof (1973), her nomenclature usage seems more consistent with that of the other investigators than with Wood and Maynard (1974).

Study differences that may have resulted from the passage of time and alterations in environmental conditions in the study area will be discussed later. Setting aside other differences, the major theses of the various reports are accepted here and paradigms based upon them are offered.

Proposed Mechanism for Hydrologic Effect on Calcite Encrustation

Hydrologic conditions may have an effect on water chemistry and, therefore, on calcite precipitation, which has not previously been discussed in the south

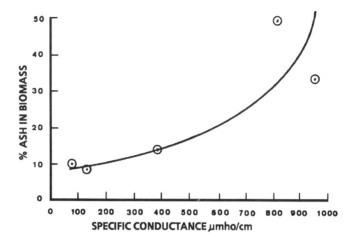

Figure 16.5 Ash component of periphyton as percent total weight versus specific conductance in Water Conservation Area 1. (The ash component is almost entirely calcite.) (From Gleason and Spack-man, 1974.)

Florida periphyton literature. Gleason (1972) and Gleason and Spackman (1974) hypothesized that degree of saturation of the water column with respect to CaCO₃, rather than calcium concentration alone, determines the extent of periphyton encrustation with calcite and the rate of accumulation of calcitic mud. Data of Gleason and Spackman (1974) show an increase in percentage calcite in periphyton with conductivity (Figure 16.5). However, data of Browder et al. (1982), re-examined for this report, do not show that relationship (Figure 16.6). The range of specific conductance represented in the Browder et al. (1981, 1982) data is almost as great as that in Gleason and Spackman (1974). At the Browder et al. (1981, 1982) stations, specific conductance ranged from 202 to 910 umhos·cm⁻¹.

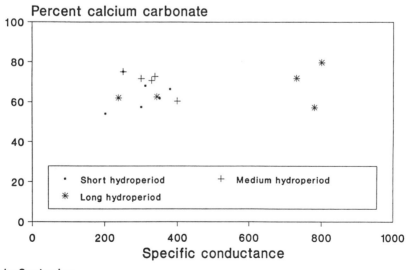

July-September

Figure 16.6 Ash component of periphyton as percent total weight versus specific conductance in Everglades National Park and the East Everglades (from values presented by Browder et al., 1981, 1982). (Ash component is almost entirely calcite.)

At the Gleason and Spackman (1974) stations, specific conductance ranged roughly from 100 to 875 umhos·cm^{-1} (taken from graphs). Gleason and Spackman (1974) found the noncalcareous, desmid-rich periphyton community between specific conductance of 100–300 umhos·cm^{-1}. Browder et al. (1981, 1982) found only calcareous periphyton at stations where specific conductance ranged from 200 to 400 umhos·cm^{-1} over the year and also at stations where specific conductance ranged from 400 to 910 umhos·cm^{-1} over the year. The Browder et al. (1981, 1982) data support the hypothesis by Gleason and Spackman (1974) that degree of saturation with respect to $CaCO_3$, rather than $CaCO_3$ concentration alone, determines the extent of calcite encrustation of periphyton communities. It is proposed here that hydrologic conditions influence degree of saturation with $CaCO_3$.

The Browder et al. (1981, 1982) data were collected along a gradient in hydroperiod and water depth. Their shortest hydroperiod/shallowest water stations were in Taylor Slough and coincided with their lowest conductivities, ranging roughly from 200 to 400 umhos·cm^{-1}. Their longest hydroperiods and deepest waters were at Shark River Slough stations, several of which had high conductivities (400–910 umhos·cm^{-1}) because of nearness to the canal 67 extension. It is proposed here that the lack of correlation between periphyton calcite content and water column-specific conductance in the Browder et al. (1981, 1982) data was due to an effect of hydrologic conditions on the carbonate equilibrium, which determines the $CaCO_3$ concentration required to saturate the water column.

Whether a given concentration of $CaCO_3$ results in saturation is determined by both pH and the partial pressure of CO_2. The higher the pCO_2, the greater the $CaCO_3$ concentration required for saturation. Conversely, the lower the pCO_2, the lower the $CaCO_3$ concentration required for saturation. Precipitation occurs only after saturation is reached. For an in-depth discussion of $CaCO_3$ precipitation and the influence of carbon dioxide, refer to Ruttner (1972), Gleason and Spackman (1974), and Hutchinson (1975).

Hydrology influences the pCO_2 in the water column in several ways. Carbon dioxide is produced by respiration and the decomposition of aquatic plants and then is released into the water column. Consistently higher partial pressures are found in the water column than in the overlying air, and frequently the water column pCO_2 is 30 times greater (Gleason and Spackman, 1974). Therefore, the outward rate of diffusion is greater than the inward rate. The outward diffusion rate is an inverse function of water depth. For this reason, deeper waters contain higher pCO_2, particularly at the bottom. Thus, at the bottom of deeper waters, a higher $CaCO_3$ concentration is required to reach saturation. Hence, deeper waters are less likely to have a calcareous algal mat. This was confirmed by Gleason and Spackman (1974), who noted the absence of calcareous periphyton in ponds adjacent to wet prairies containing rich growths of calcareous periphyton.

Hydroperiod may also affect water column pCO_2. Decomposition of dead aquatic plant biomass can be a major source of CO_2 to the water column. Decomposition of dead plant material on exposure to air is more rapid than decomposition under water. Furthermore, material that decomposes during a drydown will not be present to provide a source of CO_2 to the water column once the area is reflooded. A short hydroperiod reduces the contribution of CO_2 to the water column from decaying plants. Therefore, a lower concentration of $CaCO_3$

may be required to precipitate calcite in a short-hydroperiod environment than in a long-hydroperiod environment. This may explain why Browder et al. (1981, 1982) found that the extent of calcite encrustation of periphyton at shorter hydroperiod/ lower conductivity sites was as great as the calcite encrustation at longer hydroperiod/ higher conductivity sites. This may also be the reason for the lack of an algal mat reported by Gleason and Spackman (1974) at peripheral Water Conservation Area 1 stations, despite high $CaCO_3$ concentrations in the water column.

In addition to the effect of depth on the rate of diffusion of CO_2 from the water column and the effect of hydroperiod on the proportion of CO_2 released from dead aquatic plant material into the water column, there may be at least one other reason why higher CO_2 concentrations may be associated with longer hydroperiods. Cursory observations suggest that, in general, emergent plant production increases with hydroperiod until, at one end, continuous high water depths exclude macrophytes and, at the other end, lack of flooding creates conditions more suitable for terrestrial plants. Higher emergent plant net production rates will result in higher rates of litter fall, causing more CO_2 to be released into the water column through decomposition. The effect of plant production on pCO_2 and its subsequent effect on calcite precipitation may be reflected in the "tails" downstream from tree islands in the C-111 area, where D. Swift (personal communication) noted the conspicuous absence of the calcareous periphyton mat. It is proposed here that CO_2 from decomposing material produced by the tree island is released into the water flowing past the island. The high pCO_2 raises the saturation point with respect to $CaCO_3$ above the concentration of $CaCO_3$ in the water column. Therefore, calcite does not precipitate along a line directly downstream from the tree island.

Influence of Hydrology on Bottom Sediment

Calcareous periphyton can give rise to a calcitic mud in areas of short hydroperiods and shallow water. In order for a calcitic mud to be deposited, the rate of deposition of organic material must be low in comparison with the rate of deposition of algally precipitated calcite. For reasons that were proposed in the previous section, hydroperiod and water depth can affect (1) the rate of organic matter production by aquatic plants and (2) the rate of decomposition of organic matter. Organic matter will decay faster where there is a short hydroperiod because exposure to air promotes rapid oxidation. Thus, there will be less accumulation of organic material in bottom sediments at short-hydroperiod sites. Van Meter (1965) estimated that the hydroperiod at blue-green periphyton areas in Taylor Slough was 4–7 months. Hydroperiods in much of this area were shorter during the Browder et al. (1981, 1982) study.

Small, deep water ponds within the calcitic mud prairies of Taylor Slough probably have a surface water chemistry similar, if not identical, to the calcitic mud prairies, but examination of bottom sediments indicates a much more organic-rich sediment with no epipelic periphyton in the center of the pond. Hydroperiod, water depth, shading by emergent vegetation, rate of production and deposition of vegetation growing in the pond, and preservation of that material under more reducing conditions than on the adjacent prairie all appear to play a role in determining the nature of the soil being deposited.

Figure 16.7 Sediment core from Water Conservation Area 3A showing calcite layer embedded between two peat layers. (From P. J. Gleason, personal communication.)

Epipelic periphyton is eliminated when the water is more than roughly 60 cm deep. This prevents coverage by periphyton of more than the stems of plants in deeper water areas such as the central, deeper parts of Shark River Slough (Gleason and Spackman, 1974). The rate of deposition of calcitic mud must be much lower in areas where periphyton coverage is limited to the stems of plants, even when the periphyton is calcareous.

In the geologic past, calcitic mud replaced peat in various areas of the Everglades north and south of Tamiami Trail and in Taylor Slough (Gleason and Stone, 1994). Coring has revealed multiple oscillations of calcitic mud and peat (Gleason et al., 1984), indicating that at times in the past, calcitic mud was deposited on top of peat, created a distinctive band of sediment, and then was overlain by peat deposition. The example in Figure 16.7 is from interior Water Conservation Area 3A. Cores with alternating layers of peat and calcite may reflect prehistoric climate changes, with periods of low rainfall and short hydroperiods alternating with periods of high rainfall and long hydroperiods. In recent times, calcite overlying peat or vice versa may reflect water management changes in hydrologic conditions.

When calcareous periphyton forms over peat and photosynthetic activity leads to supersaturated DO conditions in the water column, it is possible that oxidation of the peat substrate is promoted. Presumably, this would be a temporary condition until calcite deposition separates the peat from the water column.

Classification of Periphyton Types

Taxonomic composition and degree of calcite encrustation are the two major structural characteristics of Everglades periphyton, and both are quantifiable. These characteristics respond to certain environmental factors and vary as they do. The state of the environment at any one location determines the type of periphyton

Figure 16.8 Periphyton community types, shown in relation to principal environmental gradients.

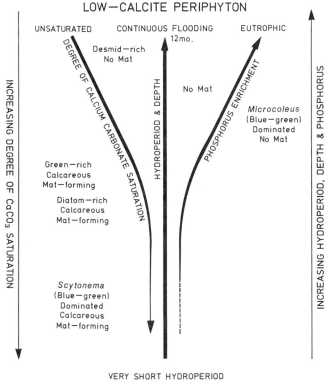

LOW—CALCITE PERIPHYTON

UNSATURATED   CONTINUOUS FLOODING   EUTROPHIC
12mo.

DEGREE OF CALCIUM CARBONATE SATURATION

Desmid—rich
No Mat

No Mat

PHOSPHORUS ENRICHMENT

Microcoleus
(Blue—green)
Dominated
No Mat

HYDROPERIOD & DEPTH

INCREASING DEGREE OF CaCO₃ SATURATION

INCREASING HYDROPERIOD, DEPTH & PHOSPHORUS

Green—rich
Calcareous
Mat—forming

Diatom—rich
Calcareous
Mat—forming

Scytonema
(Blue—green)
Dominated
Calcareous
Mat—forming

VERY SHORT HYDROPERIOD

CALCAREOUS PERIPHYTON

found there, although there may be some change in both taxonomic composition and degree of encrustation over time. Here the types of periphyton communities that can be found at the locally occurring extremes of the environmental gradients are described, and a conceptual model is summarized in Figure 16.8. The general layout of the model reflects the preceding discussion concerning the interrelated effects of hydroperiod-water depth, phosphorus enrichment, and degree of $CaCO_3$ saturation and their effects on taxonomic composition, degree of calcification, and mat formation of periphyton.

Water chemistry and hydroperiod are major factors influencing the taxonomic composition of Everglades periphyton. Major ion concentration has separate effects from the concentration of nutrients, particularly phosphorus. Therefore, there are at least three environmental variables whose geographic gradients modulate the taxonomic expression of Everglades periphyton.

Degree of calcite encrustation is another structural variable of periphyton. Water chemistry, in particular the degree of saturation with respect to $CaCO_3$, is a major factor influencing the degree of encrustation. Water depth and hydroperiod affect encrustation by influencing water chemistry.

Calcareous Periphyton versus Desmid-Rich Periphyton

The two most disparate types of periphyton identified are calcareous periphyton and noncalcareous, desmid-rich periphyton, which occur in low nutrient waters of opposite extremes in major-ion-related chemistry, as depicted by positioning these two periphyton types at the top and bottom of the conceptual model in Figure 16.8. Calcareous periphyton is defined as periphyton that has a substantial (possibly greater than 15%) calcite component and a recognizable structure created by planar felts of *Schizothrix calcicola* held together by filaments of *Scytonema hofmannii*. These species occur primarily in water containing 0.01 mg·L^{-1} total P and 50–70 mg·L^{-1} calcium and exhibiting a pH range from 6.9 to 7.5.

Desmid-rich periphyton contains a large percent cell volume of desmids, possibly greater than 70%. The noncalcareous, desmid-filamentous green algal periphyton community occurs in low nutrient, poorly buffered, acidic (pH 5.7) water of low mineral content (calcium <5 mg·L^{-1}). This community also includes a specific group of acid/soft water indicator diatom species such as *Cymbella amphioxys, Anomoeoneis serians* var. *brachysira,* and others. This type of periphyton has only a minor ash component, probably no more than 15%.

Calcareous and desmid-rich periphyton communities are strikingly different in appearance. The calcareous periphyton is a thick, cream-colored to slightly yellowish-brown coating of primarily blue-green algae mixed with calcite crystals, whereas the desmid-rich periphyton is a thin, hairy, and sometimes slimy coating of green to brown algae.

Desmid-rich periphyton forms an algal mat only when associated with *Utricularia.* Its principal substrate is submerged plants. This form of periphyton is known to occur in interior Water Conservation Area 1 and in other small wetland sites in the Water Conservation Areas isolated from canal inflows. Desmid-rich periphyton may have been more widespread in the past. For example, it may have occurred in Shark River Slough in Everglades National Park, when waters were deeper and flooding more continuous than currently.

Calcareous periphyton is found throughout wetlands of the Everglades, parts of the Big Cypress Swamp, Taylor Slough, and the prairie south and southeast of Homestead, extending all the way to brackish water. Within the Everglades, it is found on the peripheral canal-bordered fringes of Water Conservation Area 1, throughout Water Conservation Areas 2 and 3, and throughout Shark River Slough. It also is the main periphyton community found in Big Cypress Swamp and in the extensive western marsh of Lake Okeechobee.

Calcareous periphyton coats all submerged surfaces wherever water depth and canopy cover allow sufficient light penetration. Generally, the periphyton coats vegetation and bottom sediments in slough and wet prairie environments. The submerged margins of tree hammocks and sawgrass communities are not a favored site, although periphyton will cover sawgrass in ecotonal situations (Gleason and Spackman, 1974).

Intermediates between desmid-rich and calcareous periphyton community types are well represented in low nutrient environments of varying degrees of saturation with respect to $CaCO_3$.

Calcareous Blue-Green, Green, and Diatom-Rich Periphyton

Within the southern Everglades, three types of calcareous periphyton are recognized: calcareous blue-green, calcareous diatom-rich, and calcareous green.

All are calcareous periphyton because of their high inorganic component, no less than 49% by mass (Browder et al., 1981, 1982). They are positioned along the hydroperiod–degree CaCO₃ saturation scale in the conceptual model (bottom to center). The first and third types correspond to the blue-green and green periphyton described by Van Meter-Kasanof (1973). Based on a statistical comparison by Browder et al. (1981), blue-green periphyton are considered to be any calcareous periphyton comprised of greater than 80% blue-green algae, by cell volume, during the latter part of the wet season.

The term *diatom-rich* periphyton is used to label calcareous periphyton communities with a substantial diatom component. Periphyton communities are considered to be diatom rich if their cell volume consists of less than 80% blue-green algae and diatoms make up a greater proportion of cell volume than greens.

Green periphyton contains less than 80% cell volume as blue-green algae and a larger volume of green algae and desmids than diatoms. To conform to the definition given by Van Meter-Kasanof (1973), the green algal component includes desmids, particularly *Pleurotaenium*. The desmids in calcareous green periphyton are not the same acid water species as those described by Gleason and Spackman (1974) and Swift and Nicholas (1987). Rather, they are a less speciose group that tolerates hard water. Sufficient information is not available to correctly position this community relative to the diatom-rich community, but it is tentatively placed at slightly greater depths and lower degree of CaCO₃ saturation than the diatom-rich community. The contrast in green communities described by Van Meter-Kasanof (1973) and Browder et al. (1981) suggests that there may be more than one calcareous green community.

Eutrophic or Pollution-Tolerant Periphyton

Distinct from the calcareous blue-green, green, and diatom-rich periphyton communities and the desmid-rich periphyton communities, which are found in low nutrient waters at the two extremes of major-ion-related chemistry, a specialized periphyton community dominated by *Microcoleus lyngbyaceus* consists of filamentous blue-green algae that occurs, along with a number of pollution-indicator diatom species and several filamentous green algae, in waters containing both high major ion and high nutrient concentrations. This community was placed along the phosphorus enrichment scale of the conceptual model in Figure 16.8. This type of periphyton may precipitate calcite under some conditions, but does not form a distinct algal mat under the high nutrient conditions where it develops on glass slides in the Water Conservation Areas. The occurrence of this type of periphyton community may be transitional. If water levels are deep (>2 m) and stable, it may be replaced by a phytoplankton bloom, as in a nutrient enrichment experiment by Steward and Orne (1975). If water levels are lower and variable, cattails might become established and both periphyton and phytoplankton will be eliminated by shading. Knowledge of the eutrophic or pollution-tolerant periphyton community is limited primarily to growth on glass slides in areas of open water.

Water Management Effects on Spatial Distributions

Studies relating periphyton community types to environmental variables suggest that water management activities in the Everglades have changed the spatial distribution of the various types of periphyton communities.

Water Chemistry Related Distributions

The data collected by Gleason and Spackman (1974), Swift (1981, 1984), and Swift and Nicholas (1987) indicate that variations in taxonomic composition of periphyton within the Water Conservation Areas are due to water chemistry variations rather than water depth and hydroperiod. Variations in species composition among the calcareous periphyton, the desmid-rich periphyton, and the pollution-tolerant algal community appear related to $CaCO_3$ equilibria, pH, calcium concentration, bicarbonate concentration, and nutrient concentrations in the water. In order for the calcareous periphyton to occur, the water in which the periphyton is growing must be at, above, or very close to saturation with respect to $CaCO_3$. However, if the water is characterized by high nutrient concentrations in addition to the above condition, then *Microcoleus,* filamentous green algae (such as *Stigeoclonium, Spirogyra,* and *Oedogonium*), and certain pollution-tolerant diatom species will develop rather than the calcareous periphyton. At the opposite extreme, if the water is acidic, low in calcium concentration, low in bicarbonate concentration, and undersaturated with respect to $CaCO_3$ and if nutrient concentrations are low, then a predominant desmid and filamentous green (*Mougeotia* sp.) algal flora will develop.

The area of Water Conservation Area 2A influenced by agricultural drainage water has expanded (Davis, 1994). Cattails have become established in the area of nutrient enrichment. A mixed community of dense cattails and sawgrass extends south of the cattail monoculture. Periphyton communities appear to be excluded from much of this area, except in open water and at edges. Expansion of cattails and dense sawgrass leads to a decrease in the coverage of periphyton.

The coverage of calcareous periphyton in the Everglades may be increasing, except in eutrophic areas. A source of $CaCO_3$ is required for calcareous periphyton to proliferate. As explained earlier, hydrologic conditions influence the concentration of $CaCO_3$ required for precipitation. Canals, which are cut through limestone and are also in contact with mineralized ground water, introduced $CaCO_3$ to parts of the Everglades not previously exposed to high major ion concentrations, particularly peripheral parts of Water Conservation Areas 1 and 2. Calcareous periphyton now is proliferating in areas with a geologic history of peat deposition.

Today, water originating in the Everglades Agricultural Area is discharged to Everglades National Park through structures at the Tamiami Trail. From 1970 to 1983, water from the Everglades Agricultural Area was released through canal C-67 extension into central Shark River Slough, changing the water quality of this area (Flora and Rosendahl, 1982). Although water no longer is delivered to Everglades National Park through the L-67 extension, central Shark River Slough still receives water of higher ionic content than was contained in historic sheet flow (Mattraw et al., 1987).

Hydroperiod-Related Distributions

The original observations of Van Meter-Kasanof (1973) and the statistical analysis by Browder et al. (1981) suggest that the higher taxonomic composition of periphyton in the southern Everglades is controlled by hydroperiod and water depth. Percent soil organic material may be a separate controlling factor, but more likely is correlated with hydroperiod. Short hydroperiods, extremely shallow waters, and low percentages of organic matter favor the overwhelming dominance of calcite-encrusting blue-green taxa, particularly *Scytonema* and, to a lesser extent, *Schizothrix*. Long hydroperiods, deeper waters, and high percentages of sediment organic matter allow the growth of diatoms and filamentous green algae, which together can sometimes comprise more than 98% of cell volume.

Water management structures and operations have shortened hydroperiods in some areas and increased water depths in others (Fennema et al., 1994; Johnson and Conner, in preparation). Hydrologic conditions in the Water Conservation Areas have been greatly altered, with deeper water and extended hydroperiods in southern Water Conservation Areas 1, 2A, and 3A and with shallow water and shortened hydroperiods in northern Water Conservation Areas 1 and 3A, northwestern Water Conservation Area 2A, and Water Conservation Areas 2B and 3B. Coverage of the hydrologically altered Water Conservation Areas by periphyton is probably less today than historically. South of the Tamiami Trail, northeast Shark River Slough and upper Taylor Slough experience shortened hydroperiods because of diversion of former inflows.

The statistical relationship between algal taxonomic composition and hydroperiod found by Browder et al. (1981) suggests that the shortened hydroperiod caused a reduction in the volume proportion of diatoms and green algae, including desmids, in periphyton. The blue-green algae *Scytonema* appears favored by drier conditions.

Today, calcareous blue-green periphyton, poor in diatoms and green algae, covers most of Taylor Slough, except for the deeper part of the slough south of SR 27, Flamingo Road in Everglades National Park. The biomass of the calcareous periphyton community in Taylor Slough may have been greater in the 1960s than in the late 1970s. Deeper parts of Taylor Slough contain calcareous diatom-rich periphyton. The best development found by Browder et al. (1981) in the Taylor Slough area was immediately south of the Anhinga Trail and east of Royal Palm Hammock. In the late 1970s, calcareous blue-green periphyton was found in upper northeast Shark River Slough near C-67 extension. Calcareous blue-green periphyton was also found in the northern part of Everglades National Park along the Shark River Tower Loop Road at a site of oolitic outcropping. In 1979, calcareous diatom-rich algae occupied the deeper northern part of Shark River Slough, both inside and outside Everglades National Park, near the south end of L-67 extension. Periphyton with a low calcite content occupied the southern part of Shark River Slough in the mid-1960s (the time of the Wood and Maynard [1974] study). Observations by D. Swift suggest that calcareous periphyton occupies the area today. Calcareous diatom-rich algae occurs in the C-111 area south of the canal. This type of periphyton probably was present historically throughout the low-lying wet prairies landward of the Buttonwood Embankment (a rise of roughly 1 m that partially separates the wet prairies from the mangrove-lined coast). No information is available on the type of periphyton present in that area today.

Sensitivity of Periphyton and Its Use as an Indicator

Periphyton is a sensitive indicator of water chemistry, including nutrient concentrations and hydroperiod. The algae in periphyton respond almost immediately to changed hydrologic as well as nutrient conditions. The presence of calcareous periphyton, made up at least in part of the filamentous blue-green algae *Scytonema* and *Schizothrix,* is an indication of water with a low nutrient concentration, basic pH, and relatively high calcium and bicarbonate concentration. This water will be saturated or supersaturated with respect to $CaCO_3$. The presence of a desmid-rich periphyton, as described by Gleason and Spackman (1974) and Swift (1984), indicates water with a low nutrient concentration, an acid pH, low concentration of calcium and bicarbonate, and low total dissolved solids. This water will be undersaturated with respect to $CaCO_3$.

The presence of the filamentous blue-green algae *Microcoleus* indicates high water concentrations of both minerals and nutrients. *Microcoleus* growth on slides was used by Swift (1984) to determine the distance into the Water Conservation Area 2A marsh that agricultural drainage water affected nutrient levels (Figure 16.9).

Diatoms should be used in conjunction with filamentous blue-greens to evaluate water quality. Three distinct ecological groups of diatom species are found within the southern Everglades periphyton, and each group indicates one of the

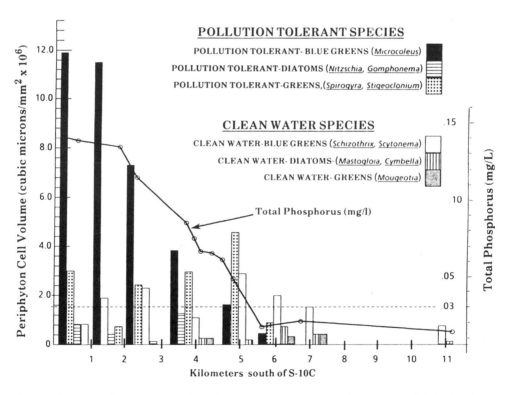

Figure 16.9 Periphyton relative abundance and species composition south of S-10C, where nutrient-rich agricultural drainage water is released into Water Conservation Area 2A. (From D. R. Swift, personal communication.)

following water quality conditions: (1) low nutrient/low mineral (acidic water), (2) low nutrient/high mineral (basic water), and (3) and high nutrient/high mineral (basic water) (Table 16.1) (Swift and Nicholas, 1987).

Periphyton composition is also an indicator of hydroperiod (Van Meter-Kasanof, 1973; Browder et al., 1981). Calcareous periphyton consisting of more than 50% *Scytonema,* by cell volume, indicates a short hydroperiod, probably less than 6 months. Calcareous periphyton with a strong representation of diatoms (at least 20% by cell volume in the latter part of the wet season) probably represents a hydroperiod of more than 6 months. Calcareous periphyton with a diatom and green component making up more than 50% of cell volume probably represents an area that is flooded throughout wetter-than-average years. Flooded areas where there is no algal mat probably are covered by more than 60 cm of water throughout the year. A constantly deep water area may be difficult to differentiate from a flooded area that is eutrophic on the basis of lack of periphyton alone; however, a low level of phytoplankton in the former and a high level in the latter should serve to distinguish these two situations. If there is no emergent or submergent vegetation, then water depth probably is the reason for lack of periphyton. The calcitic mud which periphyton deposits is a sensitive indicator of hydroperiod and water depth (Tropical BioIndustries, 1987). The geologic history of hydrologic conditions at a site is recorded in sediment cores.

The use of periphyton as an indicator should include quantitative measurements. At present, the most practical and reliable method is volumetric analysis, as used by Browder et al. (1981), Swift (1984), and Swift and Nicholas (1987). As can be seen from the foregoing discussion, it is difficult to compare periphyton from area to area and time to time without the quantitative data obtained with this approach.

Dominant organisms should be identified to species when possible and quantified separately from their main taxonomic group. Ecological groups of species, once recognized, also should be quantified separately. For example, diatoms can be used as water quality indicators only at the species level. Ecological groups of diatoms make valuable water quality indicators. Within the blue-green algae, *Scytonema* and *Schizothrix* are indicative of entirely different environmental conditions than *Microcoleus*. Similarly, certain species of filamentous green algae thrive under high nutrient conditions, whereas others may not. While statistical analysis is more tractable when dealing with higher taxonomic groups, quantitative measurements of dominant species and subsets of species are needed for meaningful examination of periphyton communities along some environmental gradients. Quantitative descriptions should not neglect estimation of calcite content, an important varying characteristic of Everglades periphyton.

Ecosystem-Level Implications of Periphyton Changes

Periphyton appears to be an important food web base. Many microscopic animals live in or on the surface of periphyton, presumably feeding on the algae or each other. The periphyton complex, which includes this animal component, provides direct nourishment to herbivores and omnivores, including some thought to be mainly detritovores. These, in turn, are a source of food to carnivores, ranging from fish to wading birds and alligators. Animals that forage on periphyton may select diatoms, desmids, and other green algae in preference to blue-green algae. The food value of the blue-green algal species dominating Everglades periphyton may be lower than that of other algal taxa.

Shortening the hydroperiod appears to result in a decrease in the diatoms, desmids, and other green algae associated with periphyton. This may decrease the food value of periphyton and affect the overall productivity of the Everglades system. Wetlands in which the periphyton has changed from rich in diatoms, desmids, and other green algae to consisting only of blue-green algae may exhibit lower secondary productivity. A loss of the diatom-desmid-green algal component of the periphyton, while not obvious to a wetlands manager, could represent a major functional change in the ecosystem. Nutrient-enriched aquatic systems dominated by *Microcoleus* may provide poor quality animal food. A related taxon, *Lyngbea* (current taxonomists include most *Lyngbea* species in the genus *Microcoleus*), was avoided by amphipods in a study by Moore (1974).

Aquatic secondary and tertiary production depends not only on net production, but also on habitat. Sufficient DO is an important habitat requirement. Emergent plants contribute to aquatic productivity, but contribute little, if any, DO to the water column with their photosynthesis. At the same time, the submerged decomposition of emergent plant material removes oxygen from the water column. As Belanger and Platko (1986) pointed out, the periphyton community makes an important DO contribution to shallow Everglades waters. Periphyton may be as important for its role in providing an oxygenated habitat for aquatic animals as for its role in the food web. Even though periphyton algae consume but do not produce oxygen at night, 24-hour measurements by D. Swift show that DO levels were higher throughout the 24-hour period in the water column at a low nutrient Everglades site than at a nutrient-enriched site (Figure 16.10) (Davis, 1994; South Florida Water Management District, 1992). Periphyton was well represented at the low nutrient site.

Because of the effect of plant metabolism and decomposition on water quality, the relationship between primary production and animal carrying capacity and

Figure 16.10   Diurnal oxygen curves at nutrient-enriched and nonnutrient-enriched Water Conservation Area 2A marsh sites, June 1979. (From D. R. Swift, personal communication.)

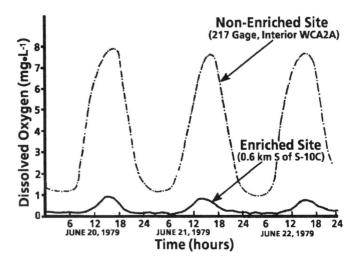

production may not be linear. High biological oxygen demand and other problems associated with nutrient enrichment make it unlikely that the increased nutrient concentrations will result in higher aquatic secondary and tertiary production in the Everglades. In fact, quite the opposite could be the case. DO shortages caused by prolific growth of emergent macrophytes could be more of a concern in the Everglades than in temperate wetlands because of the year-round higher temperatures that increase the rate of decomposition.

The higher level production of south Florida wetlands is magnified by the expansion and contraction of water area (Browder, 1976). The periphyton community, at the base of the food chain, appears designed by nature to maintain itself under spatially and interannually varying conditions of hydroperiod, mineral content, and nutrient availability. Within the periphyton community, a group of species capable of maintaining community activities is available for all conditions, although total biomass, productivity, and food quality may be greater under some conditions than others.

Potential Role of Periphyton and Its Use as an Indicator

The observations of Swift and Nicholas (1987) and Gleason and Spackman (1974) suggest that calcareous periphyton does not exist in areas of nutrient enrichment. This may be because of the high production and subsequent decomposition of organic material in these areas, which promotes a high partial pressure of carbon dioxide and retards calcite precipitation.

The previous discussion on the effect of hydrologic conditions on the partial pressure of carbon dioxide suggests that nutrients may have a stronger effect on periphyton encrustation in continuously flooded, deeper areas than in shallow areas that dry periodically. Water management procedures that promote shallower water and shorter hydroperiods in cells of proposed wetland water quality treatment areas (South Florida Water Management District, 1992) might induce calcite precipitation. The tendency of phosphorus to coprecipitate with calcite could cause a reduction in dissolved phosphorus concentrations and the related eutrophication problem. In hydroperiod manipulation, care might have to be taken to avoid promoting oxidation of organic sediments, which could release phosphorus that presently is bound to peats. The extent to which water manipulation could control phosphorus concentrations would depend upon the availability of $CaCO_3$; however, this is an abundant commodity in the nutrient-charged canal waters that are causing the eutrophication problems. It might be possible to induce calcite and phosphorus precipitation, even under continuously flooded conditions, with calcite supplementation.

LITERATURE CITED

Bautista, M. F. and H. W. Paerl. 1985. Diel N fixation in an intertidal marine cyanobacterial mat community. *Mar. Chem.,* 16:369–377.

Belanger, T. V. and J. R. Platko II. 1986. Dissolved Oxygen Budgets in the Everglades Water

Conservation Area 2A, Report to the South Florida Water Management District, West Palm Beach, 107 pp.

Belanger, T. V., D. J. Scheidt, and J. R. Platko II. 1989. Effects of nutrient enrichment in the Florida Everglades. *Lake Reserv. Manage.,* 5:101–111.

Bosserman, R. W. 1981. Elemental composition of *Utricularia*-periphyton ecosystems from Okefenokee Swamp. *Ecology,* 64:1637–1645.

Browder, J. A. 1976. Water, Wetlands, and Wood Storks, Ph.D. dissertation, University of Florida, Gainesville, 405 pp.

Browder, J. A., S. Black, P. Schroeder, M. Brown, M. Newman, D. Cottrell, D. Black, R. Pope, and P. Pope. 1981. Perspective on the Ecological Causes and Effects of the Variable Algal Composition of Southern Everglades Periphyton, Report T-643, South Florida Research Center, Homestead, Fla., 110 pp.

Browder, J. A., D. Cottrell, M. Brown, M. Newman, R. Edwards, J. Yuska, M. Browder, and J. Krakoski. 1982. Biomass and Primary Production of Microphytes and Macrophytes in Periphyton Habitats of the Southern Everglades, Report T-662, South Florida Research Center, Homestead, Fla., 49 pp.

Browder, J. A., A. Black, and D. Black. Undated. Comparison of Laboratory Growth of *Hyla squirella* Tadpoles on Everglades Periphyton of Three Taxonomic Compositions, Unpublished report to Everglades National Park, Homestead, Fla., 12 pp.

Browder, J. A., R. L. Pope, and P. B. Schroeder. In review. Quantitative comparison of periphyton as food for aquatic animals in the southern Everglades.

Dachnowski-Stokes, A. P. 1928. A preliminary note on blue-green algal marl in southern Florida in relation to the problem of coastal subsidence. *J. Wash. Acad. Sci.,* 18:476–480.

Davis, S. M. 1994. Phosphorus inputs and vegetation sensitivity in the Everglades. in *Everglades: The Ecosystem and Its Restoration,* S. M. Davis and J. C. Ogden (Eds.), St. Lucie Press, Delray Beach, Fla., chap. 15.

Drouet, F. 1963. Ecophenes of *Schizothrix calcicola* (Oscillatoriaceae). *Proc. Acad. Nat. Sci. Philadelphia,* 115:261–277.

Drouet, F. 1968. Revision of the classification of the Oscillatoriaceae. *Monogr. Acad. Nat. Sci. Philadelphia,* No. 15.

Egler, F. E. 1952. Southeast saline Everglades vegetation, Florida, and its management. *Vegetatio Acta Geobot.,* 3(4–5):213–265, 1950.

Fennema, R. J., C. J. Neidrauer, R. A. Johnson, T. K. MacVicar, and W. A. Perkins. 1994. A computer model to simulate natural Everglades hydrology. in *Everglades: The Ecosystem and Its Restoration,* S. M. Davis and J. C. Ogden (Eds.), St. Lucie Press, Delray Beach, Fla., chap. 10.

Flora, M. D. and P. C. Rosendahl. 1982. An analysis of surface water nutrient concentrations in the Shark River Slough, 1972–1980, Report T-653, South Florida Research Center, Homestead, Fla., 40 pp.

Gleason, P. J. 1972. The Origin, Sedimentation, and Stratigraphy of a Calcitic Mud Located in the Southern Fresh-Water Everglades, Ph.D. thesis, The Pennsylvania State University, University Park.

Gleason, P. J. and W. Spackman, Jr. 1974. Calcareous periphyton and water chemistry in the Everglades. in *Environments of South Florida: Present and Past, Memoir No. 2,* P. J. Gleason (Ed.), Miami Geological Society, Coral Gables, Fla., pp. 146–181.

Gleason, P. J. and P. A. Stone. 1994. Age, origin, and landscape evolution of the Everglades peatland. in *Everglades: The Ecosystem and Its Restoration,* S. M. Davis and J. C. Ogden (Eds.), St. Lucie Press, Delray Beach, Fla., chap. 7.

Gleason, P. J., A. D. Cohen, H. K. Brooks, P. Stone, R. Goodrick, W. G. Smith, and W. Spackman, Jr. 1984. The environmental significance of Holocene sediments from the Everglades and saline tidal plain. in *Environments of South Florida: Present and Past, Memoir No. 2,* P. J. Gleason (Ed.), Miami Geological Society, Coral Gables, Fla., pp. 297–351.

Goldstein, J. D. 1980. Effects of Water Management on Primary Production in the Florida

Everglades, Summary of completed project, Student-oriented studies, National Science Foundation, Washington, D.C., 1 p. (abstract).

Gunderson, L. H. Undated. Historical Hydropatterns in Wetland Communities of Everglades National Park, Unpublished report, South Florida Research Center, Everglades National Park, Homestead, Fla.

Hatcher, P. G., E. C. Spiker, and W. H. Orem. 1986. Organic geochemical studies of the humification process in low-moor peat. in *Peat and Water,* C. H. Fuchsman (Ed.), Elsevier, New York, pp. 195–213.

Hunt, B. P. 1952. Food relationships between Florida spotted gar and other organisms in the Tamiami Canal, Dade County, Florida. *Trans. Am. Fish. Soc.,* 82:13–33.

Hunt, B. P. Undated. A Preliminary Survey of the Physico-chemical Characteristics of Taylor Slough with Estimates of Primary Productivity, Mimeographed report, University of Miami, Coral Gables, Fla., 26 pp.

Hutchinson, G. E. 1975. *A Treatise on Limnology,* Vol. 1, Part 2: Chemistry of Lakes, John Wiley & Sons, New York.

Johnson, R. A. and Conner, S. C. In preparation. A review of the hydrologic changes in the Everglades basin: 1940–1992.

Kahl, M. P., Jr. 1964. Food ecology of the wood stork (*Mycteria americana*) in Florida. *Ecol. Monogr.,* 34:97–117.

Kolipinski, M. C. and A. L. Higer. 1969. Some Aspects of the Effects of the Quantity and Quality of Water on Biological Communities in Everglades National Park, Open File Report 69007, U.S. Geological Survey, Tallahassee, Fla.

Kushlan, J. A., J. C. Ogden, and A. L. Higer. 1975. Relation of Water Level and Fish Availability to Wood Stork Reproduction in the Southern Everglades, Florida, Open File Report 75-434, U.S. Geological Survey, National Park Service. U.S. Department of the Interior, Homestead, Fla.

Lagler, K. F., J. M. Kapetsky, and D. J. Steward. 1971. The Fisheries of the Kafue Flats, Zambia, in Relation to the Kafue Gorge Dam, University of Michigan Tech. Rep., FAO, Rome, No. FI:SF/ZAM II Tech. Rep., I, pp. 1–161.

Lieth, H. and R. H. Whittaker. 1975. *Primary Productivity of the Biosphere,* Springer-Verlag, New York, 339 pp.

Lowe-McConnell, R. H. 1975. *Fish Communities in Tropical Waters,* Longman, New York, 337 pp.

Mattraw, H. C., D. J. Scheidt, and A. C. Federico. 1987. Analysis of Trends in Water Quality Data for Water Conservation Area 3A, the Everglades, Florida, Water Resources Investigations Report 86-4142, U.S. Geological Survey, Tallahassee, Fla., 52 pp.

Maynard, N. G. 1968a. Aquatic foams as an exological habitat. *Z. Allg. Mikrobiol.,* 8:119–126.

Maynard, N. G. 1968b. Significance of airborne algae. *Z. Allg. Mikrobiol.,* 8:225–226.

Monty, C. L. U. 1965. Geological and Environmental Significance of Cyanophyta, Ph.D. dissertation, Princeton University, Princeton, N.J.

Monty, C. L. U. 1967. Distribution and structure of recent stromatolitic algal mats, eastern Andros Island, Bahamas, *Ann. Soc. Geol. Belg.,* T90:55–100.

Moore, J. W. 1974. The role of algae in the diet of *Asellus aquaticus* L. and *Gammarus pulex* L. *J. Anim. Ecol.,* 44:719–729.

Odum, H. T. and C. M. Hoskin. 1958. Comparative studies on the metabolism of marine waters. *Publ. Inst. Mar. Sci. Univ. Tex.,* 5:16–46.

Otsuki, A. and R. G. Wetzel. 1972. Coprecipitation of phosphate with carbonates in a marl lake. *Limnol. Oceanogr.,* 17:763–767.

Reark, J. B. 1961. Ecological Investigations in the Everglades, 2nd Annual Report, Biology Department, University of Miami, Coral Gables, Fla.

Ruttner, F. 1972. *Fundamentals of Limnology,* University of Toronto Press, Toronto, Canada, 295 pp.

Scheidt, D. J., M. D. Flora, and D. R. Walker. 1987. Water quality management for Everglades National Park, paper presented at the North American Lakes Management Society 75th Int. Symp., Orlando, Nov. 4, 1987, 26 pp.

Scholl, D. W. 1964a. Recent sedimentary record in mangrove swamps and rise in sea level over the southwestern coast of Florida: Part 1. *Mar. Geol.,* 1:344–366.

Scholl, D. W. 1964b. Recent sedimentary record in mangrove swamps and rise in sea level over the southwestern coast of Florida: Part 2. *Mar. Geol.,* 2:343–364.

Smith, W. G. 1968. Sedimentary Environments and Environmental Change in the Peat Forming Area of South Florida, Ph.D. dissertation, The Pennsylvania State University, University Park.

Steward, K. K. and W. H. Ornes. 1975. Assessing a marsh environment for wastewater renovation. *J. Water Pollut. Control Fed.,* 47(7):1880–1891.

Stockner, J. G. 1968. Algal growth and primary productivity in a thermal stream. *J. Fish. Res. Board Can.,* 25:2037–2058.

Swift, D. R. 1981. Preliminary Investigation of Periphyton and Water Quality Relationships in the Everglades Water Conservation Areas, Tech. Publ. 81-5, South Florida Water Management District, West Palm Beach, 83 pp.

Swift, D. R. 1984. Periphyton and water quality relationships in the Everglades Water Conservation Areas. in *Environments of South Florida: Present and Past, Memoir No. 2,* P. J. Gleason (Ed.), Miami Geological Society, Coral Gables, Fla., pp. 97–117.

Swift, D. R. 1989. Water quality section. in Vegetation, Water Quality, Estuarine Salinity, and Productivity in C-111 Basin, Unpublished report, South Florida Water Management District, West Palm Beach.

Swift, D. R. and R. B. Nicholas. 1987. Periphyton and Water Quality Relationships in the Everglades Water Conservation Areas, 1978–1982, Tech. Publ. 87-2, South Florida Water Management District, West Palm Beach, 44 pp.

Tabb, D. C., T. A. Alexander, T. M. Thomas, and N. Maynard. 1967. The Physical, Biological and Geological Character of the Area South of C-111 Canal in Extreme Southeastern Everglades National Park, Florida, Unpublished report to the U.S. National Park Service, University of Miami Institute of Marine Science, Miami, 55 pp.

Tropical BioIndustries, Inc. 1987. Key Environmental Indicators of Ecological Well Being, Everglades National Park and East Everglades, Dade County, Florida, Partial draft report to the Planning Division, Environmental Resources Planning Branch, U.S. Army Corps of Engineers, Jacksonville, Fla., Contract No. DACW 17-84-C-0031.

Van Meter, N. 1965. Some Quantitative and Qualitative Aspects of Periphyton in the Everglades, Master's thesis, University of Miami, Coral Gables, Fla.

Van Meter-Kasanof, N. 1973. Ecology of the microalgae of the Florida Everglades. Part I. Environment and some aspects of freshwater periphyton, 1959 to 1963. *Nova Hedwigia,* 24:619–664.

Waller, B. G. and J. E. Earle. 1975. Chemical and Biological Quality of Water in Part of the Everglades, Southeastern Florida, Water Resources Investigations 56–75, U.S. Geological Survey, Tallahassee, Fla., 157 pp.

Wetzel, R. G. (Ed.). 1983. *Periphyton of Freshwater Ecosystems,* W. Junk Publishers, Boston.

Wilson, S. U. 1974. Metabolism and Biology of a Blue-Green Algal Mat, M.S. thesis, University of Miami, Coral Gables, Fla.

Wood, E. J. F. and N. G. Maynard. 1974. Ecology of the micro-algae of the Florida Everglades. in *Environments of South Florida: Present and Past, Memoir No. 2,* P. J. Gleason (Ed.), Miami Geological Society, Coral Gables, Fla., pp. 123–145.

Young, O. W. 1945. A limnological investigation of periphyton in Douglas Lake, Michigan. *Trans. Am. Fish. Soc.,* 64:1–20.

17

LANDSCAPE DIMENSION, COMPOSITION, and FUNCTION in a CHANGING EVERGLADES ECOSYSTEM

Steven M. Davis

Lance H. Gunderson

Winifred A. Park

John R. Richardson

Jennifer E. Mattson

ABSTRACT

Vegetation change in the Everglades is assessed at two spatial and temporal scales. At the broadest scale of physiographic landscapes, changes are evaluated by comparing the spatial extent of landscape units on a reconstructed map of the predrainage Everglades with the coverage of these landscapes today. At a smaller spatial and shorter temporal scale, change in percent cover of plant communities within the remaining physiographic landscapes is measured in 25 square-mile (259-ha) plots during a time interval of 15–21 years of recent management practices.

Since the turn of this century, approximately one-half of the 1.2 million ha once covered by Everglades wetlands has been converted for human uses of agriculture and development. Three of seven predrainage landscapes, a custard apple forest (dominated by *Annona glabra* L.), peripheral wet prairie, and cypress forest (*Taxodium distichum* (L.) Rich.), have disappeared completely. Three-fourths of a dense, monotypic sawgrass plain (*Cladium jamaicense* Crantz) that once covered the northern Everglades has been replaced by agricultural crops. The other landscape units, including the wet prairie-slough-sawgrass-tree island mosaic, the sawgrass-dominated mosaic, and the southern marl marshes, have decreased in spatial extent to a lesser degree.

A community shift from wet prairie and slough to sawgrass occurred in the square-mile plots during 15–21 years of recent management practices throughout the extensive mosaic of sawgrass, wet prairies, sloughs, and tree islands. Mean percent cover of wet prairie and slough decreased significantly ($p = 0.05$) from 48% to 35%, while sawgrass increased significantly from 39% to 50%. Plausible hypotheses to explain the decrease in cover by wet prairie and slough include lowered water levels due to drainage activities that began during the first half of this century and suppression of severe fires due to fire management practices.

Functional losses to the Everglades ecosystem that have accompanied landscape and plant community change appear to include (1) reduction in spatial extent of system aquatic productivity with loss of half of the ecosystem, (2) reduction in aquatic productivity of the southern Everglades due to shortened hydroperiods and interrupted flows as a result of development of marshes upstream, (3) reduction in cover of wet prairie and slough and related aquatic productivity in the remaining system, (4) loss of habitat diversity at the landscape and community level, and (5) reduction in early dry season feeding habitat of wading birds. Extended hydroperiods and flows during most years, punctuated by severe fires during periodic drought years, appear to be temporal patterns that supported Everglades landscape patterns and functions prior to drainage and development. Allowing such natural disturbance patterns to operate at full impact is suggested as a fundamental guideline for Everglades ecosystem restoration.

INTRODUCTION

Everglades landscape scenes consisting of sawgrass (*Cladium jamaicense* Crantz) strands interlaced with lily-pad-covered sloughs and dotted with forested islands (Figure 17.1) give little indication of the changes that have occurred during 5000 years of ecosystem evolution and a century of human development. The dynamic nature of vegetation in the Everglades is evident throughout the geological history of this young ecosystem. Gleason and Stone (1994) describe many fluctuations in the recent history of the Everglades: shifts between marl-forming and peat-forming marshes in the southern Everglades, repeated conversions between sawgrass marshes and white water lily (*Nymphaea odorata* Ait.) sloughs in areas of peat substrate, the appearance of smaller bayhead tree islands 900–1000 years ago and larger tree islands more recently, and an ever-present fire regime. Variables that have contributed to vegetation change in the Everglades include climate, sea level, topography, hydroperiod, fire, alligator activity, and, more recently, anthropogenic nutrient input (Davis, 1943; Craighead, 1971; Olmsted and Loope, 1984; Davis, 1994). Gunderson (1994) identified and ranked these variables according to slow, intermediate, and fast temporal scales of environmental fluctuations that result, respectively, in large, intermediate, and small spatial scales of vegetation change.

These processes have created a tapestry of vegetation patterns, with mosaics of plant clusters nested within larger mosaics of plant communities and landscapes at expanding spatial scales. When viewed over time, this tapestry becomes

Figure 17.1 Everglades landscape scenes (1990): plot 75 in central Water Conservation Area 1 (top) and plot 58 in the interior of Everglades National Park (bottom).

analogous to a kaleidoscope of changing patterns, with the larger scale mosaics changing at a slower rate than the finer patterns. Change not only creates these patterns, but is also an integral part of the pattern. An Everglades landscape scene can be viewed as a snapshot of this kaleidoscope in time and space.

The dynamic nature of Everglades landscapes poses a problem of how changing patterns can be used as guidelines for ecosystem restoration. The approach that is taken in this chapter is to identify changes during this century that cannot be explained by natural spatial/temporal relationships as they are understood and that appear to be primarily due to human influence. These changes are then examined as to their impacts on the ecosystem function of the Everglades. Restoration guidelines are suggested to stop or reverse human-induced changes which appear deleterious to ecosystem function.

Landscape change in the Everglades is assessed at two spatial and temporal scales. At the broadest scale of physiographic landscapes covering areas on the order of hundreds of square kilometers, human-induced during changes this century involve the conversion of portions of each landscape to agri-urban development. Predrainage landscape patterns are reconstructed based on early accounts by explorers and botanists, which provide snapshots of what Everglades vegetation looked like at the turn of the century. Changes are evaluated by comparing the spatial extent of landscape units on a reconstructed map of predrainage Everglades landscapes with the coverage of these landscapes today. At a smaller spatial scale and shorter temporal scale, change in coverage of plant communities is measured in 25 square-mile (259-ha) plots within the remaining physiographic landscapes during a time period of 15–21 years of recent management practices.

EARLY VEGETATION ACCOUNTS (1841–1929)

Due to its impenetrable nature, the Everglades remained virtually unexplored territory prior to the middle of the 19th century. Consequently, the physiography of the region was left to speculation. When explorer Hugh Willoughby (1898) came across a French map of the Florida peninsula dated 1750, he was amused to find that, "According to this map, Florida was wedge-shaped and a wonderfully mountainous country, the high peaks extending almost to the extreme south." Early Spanish maps were of greater accuracy, particularly in their coastal detail; however, they were kept highly classified and were then "eventually buried somewhere in Spain's official archives" (Tebeau, 1968), contributing little to public knowledge of the interior.

Crisscrossing the peninsula in dugout canoes, Indians were essentially the only community with significant knowledge of the region until the mid-1800s, when the Seminole Wars (1835–42,1855–59), realization of the potential for commercial enterprise, and scientific curiosity stimulated the demand for intelligence gathering and initiated an era of Everglades enlightenment. Reports from military reconnaissance missions provide some of the earliest accounts of Everglades vegetation before major drainage efforts began in the early 1900s. Lieutenant J. C. Ives (1856) produced a map based on observations from several of these pioneer expeditions, which represents one of the earliest attempts to classify the flora of the Everglades into major plant communities (see inside front cover of this volume). Sawgrass, swamp, marsh, prairie (wet and dry), scrub, and hammock (wet and dry) were delineated.

C. F. Hopkins (1884) noted distinct landscape patterns during a surveying expedition that began at the southern shore of Lake Okeechobee. After crossing the custard apple swamp (dominated by *Annona glabra* L.) that fringed the southern border of the lake, the expedition plunged into the interior of the Everglades and then emerged at the headwaters of the Harney River (Figure 17.2). According to Hopkins, the route of a previous 1841 expedition of Major Childs in Ives (1856) (Figure 17.2) followed the dividing line between an upper and lower Everglades; "The upper or northern glades being a solid body of dense saw grass extending 30 miles east and west and 40 miles north and south, and which has hitherto been considered impenetrable. The lower Everglades is also saw grass but cut up by sloughs and dotted with innumerable small islands, varying in area from 1 to 20 acres" (0.4–8.1 ha).

Hugh Willoughby's record (1898) of his trek across the Everglades from Harney River to Miami (Figure 17.2) provides information pertaining to the vegetation of the region just prior to the turn of the century. He encountered dense strands of sawgrass divided by "leads" (wet prairies, sloughs, or open areas) of water during his initial venture into Shark River Slough. As the journey progressed eastward, his journal and photographs indicate a change in vegetation from a pattern of wet prairie-slough-tree island-dense sawgrass to one of less dense sawgrass where "The rock was more frequently near the surface, and could not support the heavier growth." He also noted that the tree islands became more numerous and closer together.

The Miami and North New River drainage canals had been dug by the time botanist John Harshberger (1914) conducted his phytogeographic survey of the Everglades. One of his routes followed the North New River Canal (Figure 17.2). Beginning at the southern shore of Lake Okeechobee, he inventoried the custard

Figure 17.2 Routes of explorers who provided early accounts of Everglades vegetation.

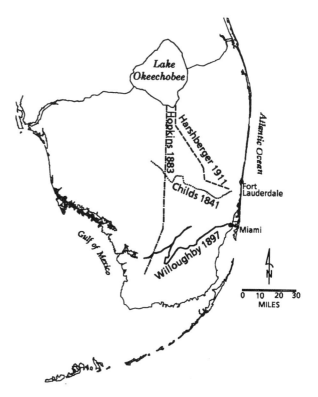

apple swamp. "It consists of an almost pure growth of the custard-apple, *Annona glabra* L., with an occasional cypress tree…As seen from the cupola of the New River Canal, the custard-apple forest extends east and west, as far as the eye could reach, and in a southward direction from the border of the lake, a distance of about 4.8 kilometers [3 miles]." (Timber estimator W. A. Roberts in Stewart [1907] determined the height of the custard apple trees to be 15–20 ft [4.5–6 m]). Proceeding south, Harshberger noted that "Before us, as far as the eye can reach, is a level prairie-like expanse of saw-grass." He considered this area the "upper Glades, because the vast saw-grass marsh is unbroken either by hammocks or channels." Landscape patterns of the "lower Glades" became apparent approximately 30 mi (48 km) south of Lake Okeechobee as "channels, lagoons and islands" began to break up the solid expanse of sawgrass. Clumps of bushes and tropical hammock vegetation were noted with increased frequency in the sawgrass-dominated mosaic as he traveled farther south toward Fort Lauderdale. Included in Harshberger's plant community classifications is a description of cypress swamps (*Taxodium distichum*) "found along the eastern border of the Everglades and the headwaters of the larger streams that flow eastward into the Atlantic."

In his report to the Florida Geological Society, Harper (1927) observed that "the south end of the Everglades has a limestone or marl substratum" and that "The vegetation…is similar in aspect to that of the northern part of the Everglades,…but differs in composition, on account of the calcareous soil." His inventory of plants associated with the marl marshes included wax myrtle (*Myrica cerifera* L.), sawgrass, and round grass (*Eleocharis cellulosa* Torr.). Small (1923, 1927) botanized throughout the southeastern Everglades marl marshes, which he described as "the front prairie on the Miami limestone." In his documentation of the southern prairie vegetation, he noted that in this region "the principal herbaceous growth is saw-grass" and recorded nearly 200 additional species, representative of 60 families (predominantly Poaceae, Cyperaceae, and Asteraceae).

The combined accounts of early explorers and scientists described a very generalized pattern of broad physiographic landscapes interconnected by transition zones. A custard apple swamp forest bordered the south shore of Lake Okeechobee. South and east of the forest was a vast plain of dense, nearly impenetrable monotypic sawgrass. Farther southeast, the sawgrass was broken first by sloughs and then by small islands of brush. Tree islands and sloughs became increasingly numerous and well developed eastward and southward, where they formed mosaics interwoven with sawgrass strands. The landscapes described to this point formed a vast peatland that extended from the shores of Lake Okeechobee to the mangroves bordering Florida Bay. The eastern edge of the peatland was bordered by a strand of cypress that began near Lake Okeechobee and extended south almost to Fort Lauderdale. Farther to the east and south of the cypress strand, sand-bottomed peripheral wet prairies of mixed grasses, sedges, and other macrophytes intermingled with higher elevation pine flatwoods and pine islands (Austin, 1976; Richardson, 1977). To the south, the peatland blended into marl and rock-bottomed marshes of mixed herbaceous plants, short sawgrass, and tree islands on both the eastern and western sides of the southern Everglades. The southeastern Everglades marl and rock-bottomed marshes were bisected diagonally by a rock ridge of pineland and tropical hammock that extended southwest from Miami.

Table 17.1 Interpretation of Vegetation Types from Davis (1943) into Predrainage Everglades Landscapes

Vegetation types from Davis (1943)	Predrainage landscapes						
	Swamp forest	Sawgrass plains	Sawgrass-dominated mosaic	Slough, tree island, sawgrass mosaic	Southern marl-forming marsh	Cypress strand	Peripheral wet prairie
Custard apple zone	X						
Willow and elderberry zone	X						
Sawgrass marshes (dnese)		X					
Sawgrass marshes (with abundant ferns and cattails)		X					
Sawgrass marshes (with wax myrtle thickets)		X	X				
Sawgrass marshes (medium dense to sparse)		X	X	X			
Sloughs, ponds, and lakes (with aquatic plants)			X	X			
Tree islands (bayhead forests)			X	X	X		
Tree islands (hammock forests)				X	X		
Southern coast marsh prairies					X		
Marsh prairies (southern Everglades)					X		
Cypress forests					X	X	
Wet prairies							X

Davis (1943) provided the first comprehensive map of Everglades vegetation; previous descriptions were made from observations along expedition routes. Davis identified 13 major wetland vegetation types in the Everglades (Table 17.1) and mapped their distribution based on aerial photography, reconnaissance flights, and ground truthing around 1940 (see map included with this volume).

LANDSCAPE-SCALE DYNAMICS

Predrainage Landscapes

The 13 vegetation types of Davis (1943) are grouped into 7 physiographic landscape categories (Table 17.1) and mapped as predrainage landscapes (Plate 7)* according to the following interpretations of the Davis map in combination with

---

* Plate 7 appears following page 432.

earlier accounts. Everglades boundaries are from Davis (1943) unless otherwise noted in the landscape descriptions. The seven groups were independently mapped for this study, but closely agree with the physiographic divisions mapped by Jones (1948).

The swamp forest landscape is a combination of the custard apple zone, which Davis (1943) described as bordering the south shore of Lake Okeechobee, and an elderberry (*Sambucus canadensis* L.) and willow (*Salix caroliniana* Michx.) zone, which Davis (1943) described as surrounding the custard apple zone to the south. Neither Hopkins (1884) nor Harshberger (1914) distinguished an elderberry and willow zone within this swamp forest. Davis (1943) grouped both zones in his calculations of swamp forest area.

The nearly monospecific sawgrass plain in the north-central Everglades is a landscape comprised of four vegetation types mapped by Davis (1943). His categories of dense sawgrass and medium dense to sparse sawgrass continue to be apparent today in remaining Everglades marshes over peat substrate. Davis (1943) attributed the difference in sawgrass density to drainage by man; soil subsidence and fires resulted in conversion from uniform dense sawgrass to patches of medium to sparse sawgrass. Similarly, Davis (1943) explained a patch of sawgrass with abundant ferns and cattails (*Typha* spp.) within the sawgrass plain as an area of previously dense sawgrass impacted by drainage, subsidence, and fire. The fourth Davis vegetation type included in the sawgrass plain landscape is a finger of sawgrass with wax myrtle thickets having a southeast to northwest orientation between the Miami and North New River canals. Davis categorized this area as part of the sawgrass marsh in a transition zone between the nearly uniform sawgrass and the wet prairie-slough and tree island area to the south. However, this portion of sawgrass mixed with myrtle thicket zone also may have represented an artifact of drainage by the two canals at the time Davis did his survey. Hopkins (1884) described only sawgrass as far as the eye could see along a route that passed adjacent to the northwest end of this finger before the expedition entered the wet prairie-slough and tree island region to the south. This finger of sawgrass with wax myrtle is interpreted as an area of previously monotypic sawgrass that was invaded by woody vegetation after drainage began.

A mosaic made up of wet prairies, sloughs, tree islands, and sawgrass strands is the dominant landscape feature in the Hillsborough Lake of the northeastern Everglades and the Shark River Slough of the southern Everglades. Sawgrass strands are interwoven with aquatic sloughs and wet prairies throughout this landscape. Bayhead tree islands are also common throughout, while tropical hammock islands appear in the southern portion of Shark River Slough. The landscape mosaics of Hillsborough Lake and Shark River Slough are delineated as they appear in the Davis map.

The marsh that lies between the Hillsborough Lake and Shark River Slough also appeared to form a mosaic of sawgrass, wet prairies, sloughs, and tree islands, but with sawgrass more prominent in larger unbroken stands and with wet prairies, sloughs, and tree islands generally smaller and more scattered. This landscape, designated the sawgrass-dominated mosaic, combines four of the Davis vegetation types. Much of this landscape was classified by Davis as medium dense to sparse sawgrass, although he did identify larger bayhead tree islands widely scattered among the sawgrass. Most of the remainder of this landscape was described by Davis as a band of sawgrass marsh with wax myrtle thickets that curved around the western and southern borders of the Hillsborough Lake and then extended southward along the eastern edge of the Everglades. Harshberger (1914) described his route along the North New River Canal as sawgrass with scattered sloughs and low clumps of trees and occasional large tree islands east of the bend in the canal, blending with the sawgrass plain in a transition zone approximately 6 mi (9.7 km) wide to the northwest of the bend. The transitional zone described by Harshberger corresponded to an area on the Davis map where the sawgrass with myrtle thickets converged to form a bottleneck of medium to dense sawgrass on either side of the North New River Canal. This zone is interpreted as the transition between the sawgrass plain to the northwest and the sawgrass-dominated mosaic to the southeast. The sawgrass-dominated landscape is a very broad interpretation of a sawgrass marsh, broken by occasional wet prairies, sloughs, and/or tree islands of varying size, which interconnected the Hillsborough Lake and Shark River Slough and formed a transition zone between the sawgrass plain and the western border of the Hillsborough Lake. This landscape delineation closely corresponds to the hammock-sawgrass physiographic division as mapped by Jones (1948).

Peripheral wet prairies and the eastern cypress strand are delineated as Everglades predrainage landscapes as they were mapped by Davis. Wet prairie areas along the east coast are interpreted to be a functional part of the ecosystem, although Davis's Everglades boundary excluded them. The southern marl marsh landscape combines five of the Davis vegetation categories. The southern marsh coast prairie, which Davis designated to be outside of the Everglades boundary, is interpreted to be a functional part of the ecosystem that is not clearly distinguishable floristically from the southern Everglades marsh prairie category of Davis. Both are combined into the southern marl marsh landscape. The marsh is dotted with bayhead tree islands and tropical hammock islands throughout. Scattered dwarf cypress trees and occasional cypress heads grow in this marsh in an area designated as cypress forest by Davis, but macrophytes, rather than cypress, prevail.

Spatial Changes in Predrainage Landscapes

The seven physiographic landscapes of the Everglades, as reconstructed in Plate 7, comprised an almost 1.2-million-ha wetland system before major drainage activities began early this century (Table 17.2). Most of the historic Everglades marsh was peatland (788,000 ha) and was covered by four landscape types. Approximately 311,000 ha of the peatland was in the wet prairie-slough-tree island-sawgrass mosaic. The sawgrass-dominated mosaic in the north-central Everglades covered 179,000 ha of the peat substrate area. The other landscape units in the Everglades peatland included the 238,000-ha sawgrass plain and the 60,000-ha swamp forest south of Lake Okeechobee. The peatland landscapes were bounded by a 12,000-ha cypress strand, 117,000 ha of peripheral wet prairie, and 249,000 ha of southern marl marshland. In combination, the mosaic of landscapes formed an immense subtropical macrophyte marsh that covered much of south Florida, was unique to North America, and was one of the largest in the world.

Drainage and human development during this century have reduced the

Table 17.2 Spatial Coverage of Everglades Landscapes

Landscape type	Predrainage (ha)	Current (ha)	Percent loss
Swamp forest	60,000	0	100
Sawgrass plains	238,000	63,000	74
Slough/tree island/sawgrass mosaic	311,000	271,000	13
Sawgrass-dominated mosaic	179,000	94,000	47
Peripheral wet prairie	117,000	0	100
Cypress strand	12,000	0	100
Southern marl-forming marshes	249,000	190,000	24
Total coverage	1,166,000	618,000	49

Everglades by half (Plate 8,* Table 17.2). Three of the seven physiographic landscapes have been eliminated entirely. The swamp forest along the south shore of Lake Okeechobee was cleared for agriculture by the 1920s (Harper, 1927). Since then, the peripheral wet prairies and the eastern cypress strand have undergone agricultural or urban development to the point that only isolated patches of a few hectares remain (Austin, 1976). The extent of development of the four remaining landscapes has been dependent on accessibility and soil fertility. The deep Everglades peat of the sawgrass plain (Gleason and Stone, 1994) provided an incentive for conversion of 175,000 ha (74%) of this landscape to the sugarcane, vegetable, and sod farms of the Everglades Agricultural Area (Snyder and Davidson, 1994). Undeveloped remnants of the sawgrass plain mostly lie within the Holeyland and Rottenburger tracts and northeast Water Conservation Area 3A. Urban encroachment from the east has consumed 85,000 ha (47%) of the sawgrass-dominated mosaic, which remains mostly in Water Conservation Areas 2A, 2B, 3A, and 3B. The deeper water mosaic of wet prairies, sloughs, tree islands, and sawgrass strands remains mostly undeveloped, with only a 40,000-ha (13%) loss in spatial extent. The Hillsborough Lake region has incurred most of this loss in areas outside the levees of Water Conservation Area 1. Most of the Shark River Slough remains undeveloped but is dissected by levees and canals into Water Conservation Areas 3A and 3B, the East Everglades, and Everglades National Park. Urban and agricultural encroachment has claimed 59,000 ha (24%), of the southern marl marshland, all on the eastern side of the Everglades. The undeveloped marl marshes cover most of the freshwater zones of Everglades National Park, the East Everglades, and the C-111 area.

Human utilization of the natural resources of the Everglades during this century has resulted in a dramatic reconfiguration of the historic Everglades landscape. One-half of the original wetland has been converted to alternative land uses. Landscape heterogeneity has been reduced due to the loss of three of seven predrainage landscapes, combined with the near elimination of a fourth. The remaining half is subjected to various internal management practices (such as fire and water management), as well as impacts associated with surrounding land uses. In order to assess the impacts of human activities and natural events on the

---

* Plate 8 appears following page 432.

vegetation composition of the remaining landscape units, an evaluation of vegetation dynamics was done at a finer spatial scale to discern changes in coverage of plant associations.

PLANT ASSOCIATION SCALE DYNAMICS

Vegetation dynamics at the plant community level were assessed by comparing vegetation patterns within 25 square-mile (259-ha) plots between two time periods. Changes in areal cover of vegetation types were determined by comparing sets of maps made from imagery taken in the period 1965–71 with maps made from photography taken in the period 1984–87. The plots sample the three major remaining Everglades landscapes (Figure 17.3) and were established by Alexander and Crook (1973, 1975) to qualitatively assess vegetation changes in southern Florida during the time period from the 1940s through the late 1960s. Coordinates for the plot locations are provided in Appendix 17.I. The Alexander and Crook maps were used as the first set of maps in this temporal comparison. Plots were chosen based on availability of imagery from the 1980s. The time periods between map sets ranged from 15 to 21 years.

Methods

The map series from the 1980s was made by tracing vegetation patterns apparent on aerial photographs onto transparent overlays. The square-mile boundary was determined by using the measured scale of the imagery to calculate the appropriate box dimensions. The scale of imagery ranged from 1:10,000 to 1:12,000, resulting in maps with dimensions from 5.4 × 5.4 in. (13.7 × 13.7 cm) to 6.4 × 6.4 in. (16.2 × 16.2 cm). The sample square was aligned and modified by visual comparison with the Alexander and Crook maps in order to achieve a precise overlap of mapped areas. Distinguishable vegetation units were outlined using a fine-tipped (00) pen. Outlined areas on both sets of maps were classified into 1 of 19 vegetation cover types (Table 17.3). The mapped cover types were groupings of plant associations that were identifiable on both the photographs and the published map set of Alexander and Crook (1975). The recent maps were ground truthed by helicopter overflights during the summer of 1989.

Separate maps were made of each cover type for each time period from the composite vegetation maps. The separation introduced a small error of overestimation of the size of each mapped unit, as borders were counted twice. Each of the separated maps was scanned, using an Apple Scanner, to a resolution of 150 dots per inch (59 dots per centimeter). The scanning rendered the maps into a raster data set, with each cell (or pixel) representing approximately 4 m^2 (~2 × 2 m) on the ground. The scanned maps were edited to remove irregularities introduced by the scanning process. All polygons were closed and filled. Borders of the mapped cover types were truncated at the edges of the square-mile quadrant. A series of programs (ERDAS™) with clumping and sorting algorithms were used to determine the pixel count of each patch within a mapped area. A patch or clump was defined as a discrete spatial unit of a given vegetation type, that is, an area enclosed or bounded by another type. The scale of each map was determined by dividing the

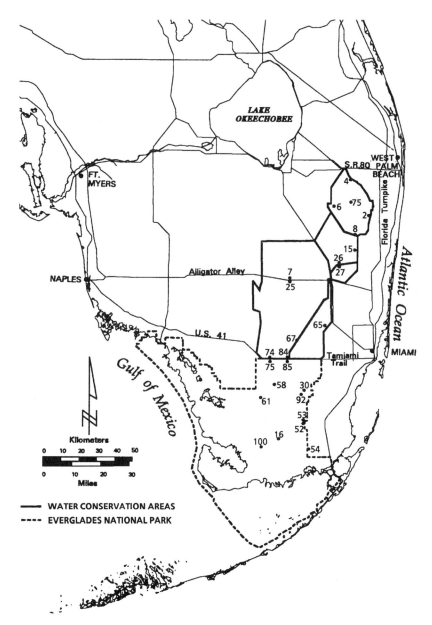

Figure 17.3 Everglades vegetation study sites: location of 25 square-mile plots. Plot numbers were designated by Alexander and Crook (1975).

linear distance of the map boundary (1609 m) by the number of pixels in both the *x* and *y* directions. The resulting values were multiplied to yield the ground area covered by each pixel. Individual scales were determined for each map, and the resulting area/pixel value was used to calculate the area of each clump.

Changes in percent cover were calculated as differences between the two time periods. Percent cover of a vegetation type was calculated by dividing the total area of all clumps of that type by the summed total area of all clumps over all vegetation

Table 17.3   Vegetation Cover Types Mapped in Square-Mile Plots

Sawgrass	Nearly monospecific *Cladium jamaicense* stand
With scattered shrubs	*Myrica cerifera* and *Baccharis halimifolia*
With scattered melaleuca	*Melaleuca quinquenervia*
With scattered cattail	*Typha domingensis*
Wet prairie/slough	A composite of two plant associations growing in deeper areas of peat plots and characterized by *Eleocharis cellulosa, E. elongata, Rhynochospora tracyi, R. inundata, Panicum hemitomon, Nymphaea odorata, Nymphoides aquatica, Nuphar luteum, Utricularia foliosa, U. biflora,* and periphyton
With scattered melaleuca	*Melaleuca quinquenervia*
Cattail	Nearly monospecific *Typha domingensis* stand
Willow	Seasonally flooded *Salix caroliniana* heads
Swamp forest tree island	Seasonally flooded tree islands with overstory of *Persea borbonia, Ilex cassine, Magnolia virginiana, Myrica cerifera,* and *Chrysobalanus icaco*
Tropical hammock	Upland tree islands with overstory of *Quercus virginiana, Lysiloma latisiliquum, Bursera simaruba, Celtis laevigata, Mastichodendron foetidissimum,* and *Swietenia mahogani*
Cypress head	Forested depression with *Taxodium ascendens* overstory
Marl prairie/sawgrass	Low-stature *Clamidium jamaicense, Muhlenbergia filipes,* and other macrophytes on marl or exposed limestone
With scattered shrubs	*Myrica cerifera and Baccharis halimifolia*
Slash pine forest	Upland islands with overstory of *Pinus elliotti*
Red mangrove	Scattered *Rhizophora mangle* along freshwater-saline transition zone
Melaleuca/casaurina	Nearly monospecific stands of *Melaleuca quinquenervia* or *Casaurina* sp.
Pond	Open water, no emergent or floating vegetation
Bare rock	
Levee canal	

types within a plot. A change was recorded if the cover of a given vegetation type varied by more than 1% of the sample plot area (~2.6 ha, or about 6400 pixels). Changes in percent cover represent a percentage of square-mile plot area, rather than percent change from the cover during the earliest time period. For example, an association type that covered 40% of the plot during the first time period and 50% during the second time period would be reported as having an increase in cover of 10%.

Vegetation Change

Percent cover of the 19 identifiable vegetation cover categories (Table 17.3) in the 25 plots during the 2 sampling periods is provided in Appendix 17.II. Vegetation cover is interpreted to include plant communities (such as sawgrass) that are clearly dominated by one or more species and variations in plant communities (such as sawgrass with scattered cattail) that result from invasion of plant species into a community where the dominant species remains the same. Similarly, change in vegetation cover is classified either as a community shift (which entails a change in dominants) or a community invasion (which involves the appearance of new species in a scattered pattern within communities where the dominant species do not change). Both types of vegetation change were recorded, and changes are analyzed for three groupings of plots which lie in the three major remnants of predrainage Everglades landscapes.

The wet prairie-slough-sawgrass-tree island mosaic of the Hillsborough Lake and Shark River Slough was represented by 12 plots (Table 17.4) that were distributed in Water Conservation Areas 1, 2A, 3A, and 3B and Everglades National Park (Figure 17.3). Plant communities that were widespread among these plots included sawgrass, wet prairie/slough, swamp forest tree island, willow, and cattail (Appendix 17.II). For the purpose of analyzing plant association scale dynamics, wet prairie and slough communities are combined into a single category referred

Table 17.4   Percent Cover of Selected Plant Communities
in Alexander and Crook Vegetation Plots during 1960s and 1980s

	1960s			1980s	
Wet prairie/slough-sawgrass-tree island mosaic (plots 2, 4, 7, 8, 15, 25, 58, 61, 67, 75, 84, 85)					
Wet prairie/slough	48%	(6.3)[a]	*[b]	35%	(4.1)
Sawgrass	39%	(6.5)	*	50%	(6.2)
Swamp forest tree island	4%	(1.2)	ns	4%	(1.4)
Willow	1%	(0.7)	ns	2%	(1.5)
Cattail	3%	(0.8)	ns	4%	(1.3)
Sawgrass-dominated mosaic (plots 6, 26, 27, 65)					
Wet prairie/slough	24%	(7.5)	ns	32%	(4.7)
Sawgrass	68%	(9.5)	ns	62%	(5.3)
Swamp forest tree island	4%	(3.8)	ns	3%	(2.6)
Willow	<1%	—	ns	1%	(0.8)
Cattail	3%	(3.0)	ns	<1%	—
Southern marl marshes (plots 16, 30, 52, 53, 54, 74, 76, 92, 100)					
Marl prairie/sawgrass	82%	(4.6)	ns	81%	(4.0)
Swamp forest tree island	5%	(2.1)	ns	5%	(1.5)
Tropical hammock	2%	(0.9)	ns	2%	(0.8)
Willow	2%	(0.9)	ns	2%	(1.2)

[a] Values represent mean (with standard errors in parentheses) for groupings of plots within physiographic landscapes.
[b] Asterisks indicate significant differences ($\propto$ = 0.05) between time periods based on paired t comparisons.

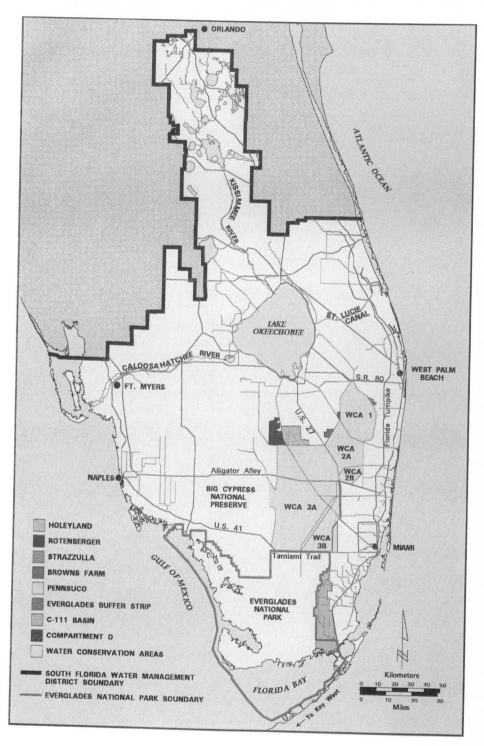

Plate 1 Undeveloped Everglades wetlands in southern Florida.

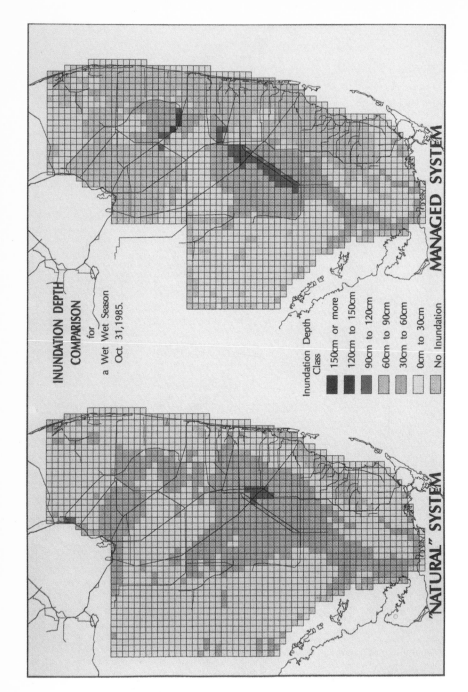

Plate 2 Comparison of natural system and managed system inundation depths at the end of a wet wet season (October 31, 1985).

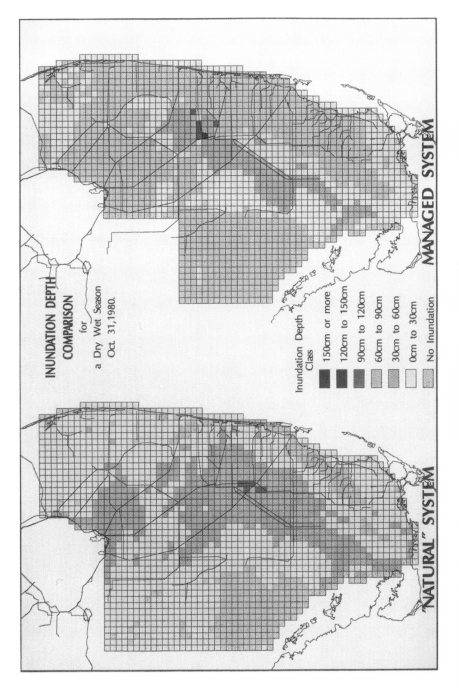

Plate 3 Comparison of natural system and managed system inundation depths at the end of a dry wet season (October 31, 1980).

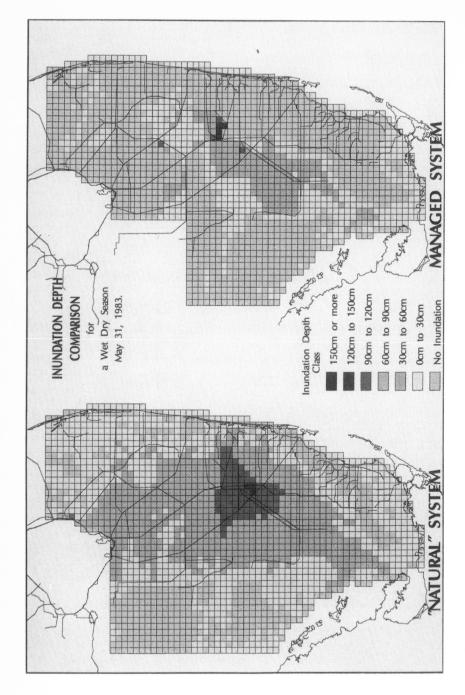

Plate 4 Comparison of natural system and managed system inundation depths at the end of a wet dry season (May 31, 1983).

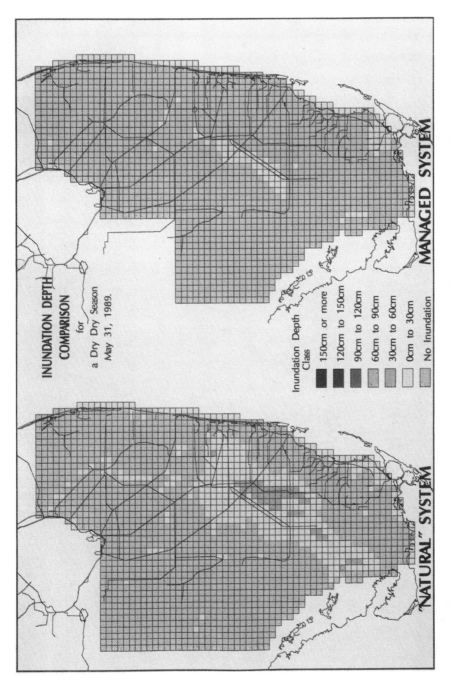

Plate 5 Comparison of natural system and managed system inundation depths at the end of a dry dry season (May 31, 1989).

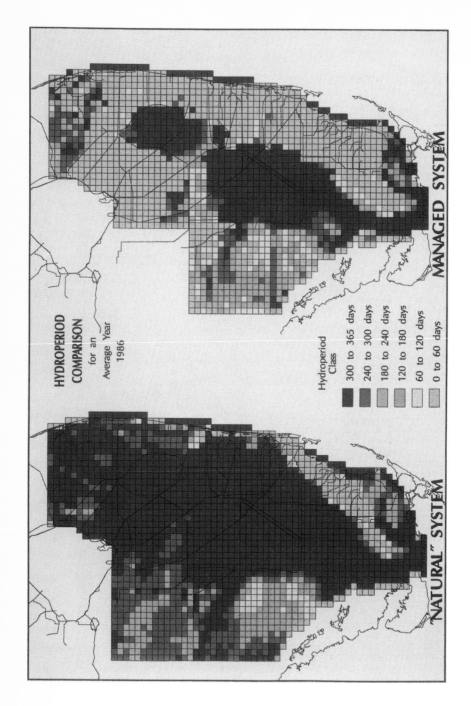

Plate 6 Comparison of natural system and managed system hydroperiod patterns for an average year (1986).

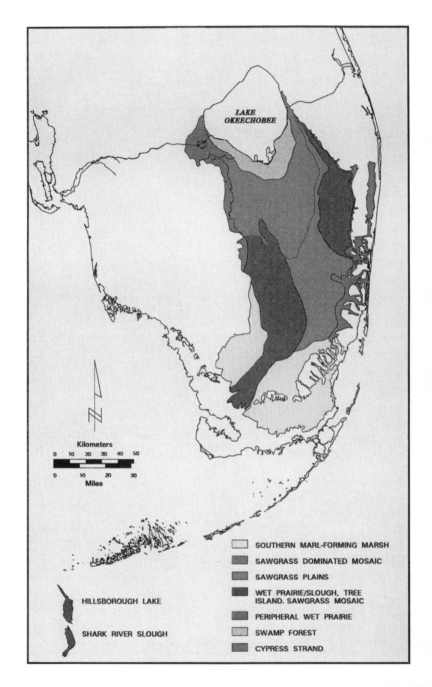

Plate 7 Historic Everglades vegetation: predrainage landscapes (around 1900).

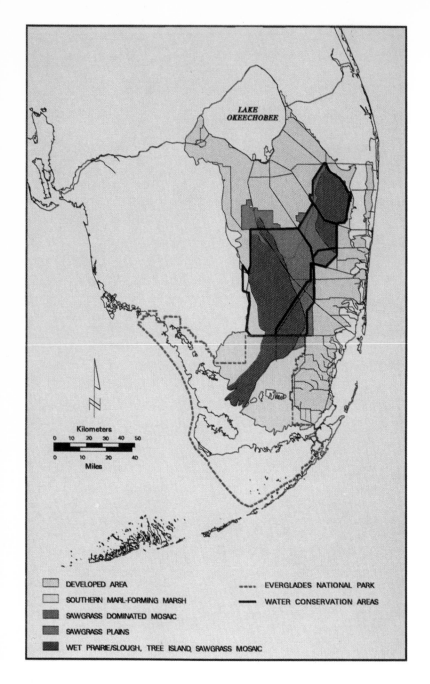

Plate 8 Current Everglades vegetation: remnants of predrainage landscapes (1990).

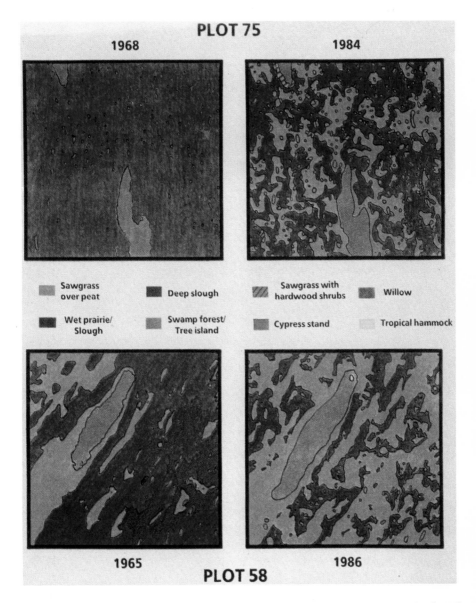

Plate 9 Vegetation cover in plot 75 in central Water Conservation Area 1 (top) and in plot 58 in the interior of Everglades National Park (bottom) during the two sampling periods.

Plate 10 Alligator pond surrounded by sawgrass (gray) and button bush (green). Photo extent ≈3 meters.

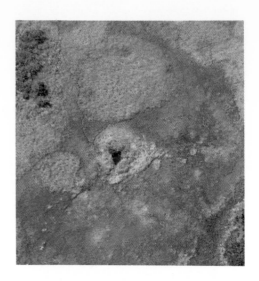

Plate 11 Plant communities of sawgrass marshes (gray), tree islands (green), and wet prairies. Note alligator trails. Photo extent ≈100 meters.

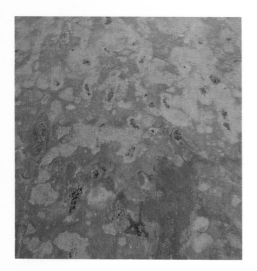

Plate 12 Mosaic of plant communities in landscape. Color scheme similar to previous plate. Photo extent ≈1 kilometer.

Plate 13 Satellite image showing plant communities: tree islands (red), sawgrass (gray), and wet prairie (black). Dark side of image is recent fire. Photo extent ≈10 kilometers.

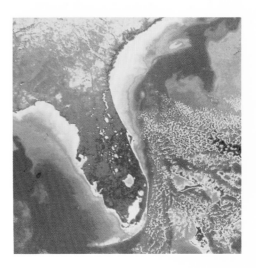

Plate 14 Satellite image of southern Florida. Geologic features are sloughs, marl wetlands, and urban coastal ridge. Photo extent ≈100 kilometers.

Plate 15 Satellite image of the southeastern United States. Features are Florida peninsula and ocean zones. Photo extent ≈1000 kilometers.

Plate 16 Photograph of the western hemisphere. Photo extent ≈5000 kilometers.

Plate 17 Photograph of Earth. Photo extent ≈15,000 kilometers

to as wet prairie/slough. These two deeper and more open water associations (Gunderson, 1994) were not readily distinguishable at the scale of resolution of the imagery. Variations in the sawgrass community, in addition to dense nearly monotypic sawgrass, were sawgrass with scattered hardwood shrub, melaleuca (*Melaleuca quinquenervia* [Cav.] Blake), or cattail. A variation in the wet prairie/slough community was wet prairie/slough with scattered melaleuca. Large expanses of composite sawgrass and wet prairie slough communities were the predominant vegetation, and each covered on the average between one-third and one-half of the plot areas (Table 17.4). The smaller tree islands, willow heads, and cattail stands that dotted the marsh each comprised 4% or less of the plot areas on the average.

A community shift from wet prairie/slough to sawgrass occurred throughout the wet prairie-slough-sawgrass-tree island mosaic during the 15- to 21-year measurement interval. Mean percent cover of wet prairie/slough decreased significantly ($p = 0.05$) from 48% to 35% (Table 17.4), while sawgrass increased significantly from 39% to 50%. Wet prairie/slough coverage decreased in 10 of 12 plots, while sawgrass increased in 9 of 12 plots. Surprisingly, two of the plots that appeared to be farthest removed from civilization and that appeared visually to represent the best examples of natural landscape scenes in the 1980s (Figure 17.1) had undergone the largest conversions from wet prairie/slough to sawgrass (Plate 9). Plot 75 in central Water Conservation Area 1 declined in wet prairie/slough from 90% of the plot in 1968 to 47% in 1984. Wet prairie/slough in plot 58 in Shark River Slough, within the interior of Everglades National Park, decreased from 65% in 1965 to 29% in 1986.

Water stages at plots 75 and 58 during the period of record from the 1950s to the 1980s are presented in Figure 17.4. The 1954–84 record of water stages at plot 75 included 6 years before the closure of levee L-39 and commencement of a water regulation schedule for Water Conservation Area 1 in 1960, 8 years under a 14- to 17-ft (4.3- to 5.2-m) regulation schedule prior to the 1968 vegetation map, and 17 years under water regulation schedules of 14–17 ft (4.3–5.2 m) and 15–17 ft (4.6–5.2 m) during the interval between the 1968 and 1984 vegetation maps. Mean water stages during the three intervals were 15.7, 15.9, and 16.0 ft (4.8, 4.8, and 4.9 m) mean sea level (MSL), respectively. The 1953–86 record of water stages near plot 58 included 10 years before the closure of L-29 (which regulated water flow into Everglades National Park beginning in 1963), water deliveries to the park with no minimum annual water release until 1970, a 260,000-acre-ft ($3207 \cdot 10^5$ m^3) minimum annual water delivery between 1970 and ~1981, and a 315,000-acre-ft ($3885 \cdot 10^5$ m^3) minimum delivery beginning in 1981 (when the South Dade Conveyance System became operational). Mean water stages at plot 58 during the four intervals were 6.0, 5.8, 6.1, and 6.4 ft (1.8, 1.8, 1.9, and 1.9 m) MSL, respectively. The interval between the 1965 and 1986 vegetation maps of plot 58 covered all but 2 years of the period since L-29 was completed. The loss of wet prairie/slough in plots 75 and 58 did not correspond to declining water levels during this period.

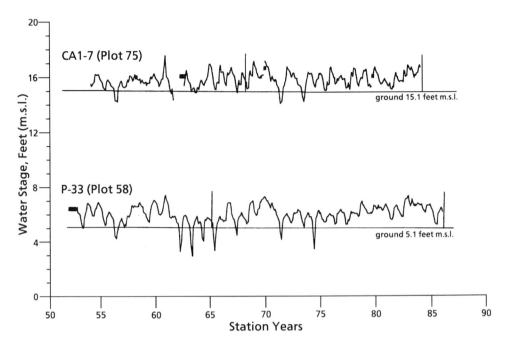

Figure 17.4 Hydrographs for gauge CA1–7 (in plot 75) and gauge P-33 (near plot 58). Vertical lines represent vegetation mapping years.

Within the sawgrass and wet prairie/slough communities of the wet prairie-slough-sawgrass-tree island mosaic, invasion by melaleuca and cattail was observed in plots 7 and 25 adjacent to Alligator Alley (Appendix 17.II). Scattered melaleuca replaced scattered hardwood shrubs in sawgrass in 34% of plot 7 north of the highway, while scattered cattail replaced scattered shrubs in sawgrass in 42% of plot 25 to the south. Slough/wet prairie was invaded by melaleuca on both sides of the highway in 47% of plot 7 and 41% of plot 25.

By comparison, tree islands, willow heads, and cattail stands remained consistent components of the wet prairie-slough-sawgrass-tree island landscape. Widespread net change in areal coverage of these communities was not significant, and decreases at particular sites were balanced for the most part by increases at other sites. Community variations in tree islands, willow heads, and cattail stands due to invasions were not detected at the scale of resolution of this study.

The sawgrass-dominated mosaic landscape was represented by four plots (Table 17.4) that were distributed in Water Conservation Areas 1, 2A, 2B, and 3B (Figure 17.3). Plant communities that were widespread among these plots were the same as those described for the wet prairie-slough-sawgrass-tree island mosaic, although sawgrass was, as expected, more abundant (62–64% mean cover) in the sawgrass-dominated mosaic and the other communities were reduced in coverage (Table 17.4). Statistically significant community shifts were not detected in the sawgrass-dominated mosaic; however, this may be an artifact of the small sample size. Invasions of the sawgrass community by melaleuca and cattail were documented (Appendix 17.II). Melaleuca invaded sawgrass in Water Conservation Area 2B and eastern Water Conservation Area 3B, replacing scattered hardwood shrubs to cover 53% of plot 27 and permeating dense sawgrass to cover 55% of plot 65. Cattail invaded sawgrass in Water Conservation Area 2A, covering 8% of plot 26.

The southern marl marsh landscape was represented by nine plots (Table 17.4) that were distributed in Everglades National Park, the East Everglades, and southeast Water Conservation Area 3A (Figure 17.3). By far, the most widespread plant community in this landscape was mixed marl prairie and sawgrass, which covered on the average 81–82% of the plots (Table 17.4). Variations of plant associations, which actually form a mosaic within the broad composite community designation of marl prairie, were not evident at the scale of resolution in this study. Tree islands, tropical hammocks, and willow heads dotted the expansive marl prairie and sawgrass community but covered a relatively small percent of the landscape (5% for tree islands and 2% each for tropical hammocks and willow heads). Vegetation coverage of the southern marl marshes appeared stable during the 15- to 21-year measurement interval at the scale of resolution of the images.

Mean percent cover of the marl prairie/sawgrass, tree islands, tropical hammocks, and willow remained constant in the southern marl marshes (Table 17.4), and only small changes were recorded in individual plots (Appendix 17.II). Mixed prairie and sawgrass, tree islands, and willow colonized an area of previously bare ground in 13% of plot 30 in the East Everglades, while bare ground replaced the same communities in 10% of the adjacent plot 92. Mixed prairie and sawgrass in combination with willow replaced tree islands in 10% of plot 74 in southeast Water Conservation Area 3A. Invasions appeared to have had negligible effects on community composition in the plots at the scale of resolution used. Vegetation change in the southern marl marshes progressed slowly relative to change in the peatland landscapes.

DISCUSSION

Causes and Ecosystem Significance of Changes at the Landscape Level

Due to agri-urban development of the northern and eastern borders of the ecosystem, four of the seven predrainage landscapes can be considered to be extinct, or nearly so, within the ecosystem. Remaining wetlands mostly fall within remnants of three predrainage landscapes. A corridor of peatland extends from the wet prairie-slough-sawgrass-tree island mosaic of the Hillsborough Lake, through the sawgrass-dominated mosaic, and into the wet prairie-slough-sawgrass-tree island mosaic of the Shark River Slough. Remnant marl marshes span much of the southern Everglades on either side of the peatland corridor.

The overall loss of half of the Everglades wetland system has resulted in a decline in the areal extent of aquatic productivity. The Everglades appears to represent a system with relatively low productivity per unit area compared to other wetlands in southeastern North America, as evidenced by periphyton production

(Browder et al., 1994), fish densities (Loftus and Eklund, 1994) alligator growth rates (Mazzotti and Brandt, 1994), and low nutrient status (Davis, 1994). This would suggest that the primary and secondary production that provided the support base for wading birds and other higher level consumers in the ecosystem was dependent on a wetland expansive in area, in compensation for a system low in productivity per unit area. Reduction in areal extent of the wetland by half may place a fundamental limitation on its support capacity for populations of wading birds that once utilized this ecosystem in much greater numbers (Ogden, 1994).

The direct contribution of the sawgrass plain to ecosystem function remains uncertain, but the conversion of most of the sawgrass plain in the northern Everglades to agricultural fields and drainage canals may have had a significant indirect effect on hydropattern and aquatic productivity in the marshes to the south. The dense monotypic sawgrass apparently slowed the southward flow of water from rainfall and Lake Okeechobee overflow. The result appears to have been a prolonged or continuous hydroperiod in the marshes currently in the Water Conservation Areas throughout normal dry seasons due to gradual inputs from the north of runoff from the previous wet season (Fennema et al., 1994; Walters et al., 1992). The replacement of the sawgrass plain by drained farmland, canals, and pump stations almost certainly accelerated the rate of southward flow of water and changed the timing of water supply (Fennema et al., 1994; Walters et al., 1992). Pumps and canals in the Everglades Agricultural Area were designed to drain water from the land within days rather than months. As a result, southward conveyance of water is presently sporadic and rapid, depending on rainfall and Water Conservation Area regulation schedules, in contrast to the steady conveyance that appeared to be characteristic of the predrainage system. Densities of the small fishes that provide an important forage base for wading birds are positively correlated to number of months of flooding (Loftus and Eklund, 1994). Thus, the sawgrass plain in the northern Everglades may have provided a support function for downstream parts of the ecosystem by prolonging hydroperiods.

The drainage of peripheral wet prairies that once framed both sides of the northern Everglades and the development of southern marl marshes in portions of the southern Everglades reduced the spatial coverage of shorter hydroperiod wetlands in the ecosystem (Fennema et al., 1994). Although these landscapes probably sustained lower densities of small fishes than longer hydroperiod marshes (Loftus and Eklund, 1994), the early wet season water recession patterns (Fennema et al., 1994) provided suitable water depths for wading bird foraging, which in turn may have allowed the birds to initiate nesting early in the dry season (Ogden, 1994).

Causes and Ecosystem Significance of Changes at the Community Level

The causal agents for the observed loss in wet prairie/slough are probably not slow variables, such as sea level or climate change. During this century sea level has risen (Wanless et al., 1994), which would tend to decrease hydrologic gradients, thereby creating additional impoundment and wetter conditions in the freshwater marshes and hence more wet prairies/sloughs. Declining rainfall also did not explain the wet prairie/slough loss. Mean annual rainfall during the measurement interval approximated long-term mean annual rainfall for the Everglades (Davis, 1994). Anthropogenic changes in hydrology and fire regimes are examined as more likely hypotheses to explain the plant community shift from wet prairie/slough to sawgrass.

Shortened hydroperiods, interrupted flows, and decreased flow velocity resulting from the development of the sawgrass plain landscape and construction of drainage canals during the early 1900s provide a possible explanation for loss of the deeper water wet prairie/slough communities. M. Maffei (personal communication) has suggested that canal construction and drainage activities during the first half of this century have interrupted and diverted the sheet flow of water that once sustained prolonged hydroperiods in Everglades marshes and that the relatively rapid encroachment of sawgrass into wet prairie/slough during the past two decades represents a continuing trend of sawgrass spread and wet prairie/slough decline that started with relatively slow change during previous decades, beginning with construction and drainage activities. Evidence to support this hypothesis includes the documentation by Kolipinski and Higer (1969) of a decrease in area of slough and wet prairie communities that was balanced by an increase in the area of sawgrass in Shark River Slough between 1940 and 1964. A parallel change during the same time interval was reported by Alexander and Crook (1973) in plot 75 of the Hillsborough Lake region, where slough areas were being encroached by sawgrass margins expanding out from bayheads during the period 1940–58. The hypothesis is further supported by the hydrologic models of Fennema et al. (1994) and Walters et al. (1992), which indicate that the remaining areas of the Everglades were generally wetter before development commenced. This hypothesis infers that hydrologic changes before construction of the flood control project set the stage for continuing vegetation changes that were measured during the mapping interval covering 15–21 years of recent management practices.

Modeling results of Walters et al. (1992) and other studies (Smith et al., 1989) further indicate that flows through the system may have been curtailed as a result of drainage activities early this century. Flow velocities may have been important to the maintenance of Everglades plant communities, as evidenced by the configuration and alignment of major wet prairie/slough systems and tree islands in relation to general flow directions in Shark River Slough (Gleason and Stone, 1994). Thus, decreased flow velocity may represent another causal factor contributing to sawgrass expansion into wet prairie/slough.

More recent changes in hydrology due to water management practices since construction of the Central and Southern Florida Flood Control Project might also have contributed to the conversion of wet prairie/slough to sawgrass, if water depth had declined at study plot locations. The hydrologic records for plots 75 and 58 (Figure 17.4), where the greatest losses of wet prairie/slough occurred, do not support this hypothesis.

A third hypothesis to explain the conversion of wet prairie/slough to sawgrass is the alteration of fire regimes. One model for wet prairie/slough formation involves fire that removes standing vegetation (primarily sawgrass) and lowers soil elevation, after which wet prairie/slough vegetation becomes established (Craighead,

1971). The observed conversion of wet prairie/slough to sawgrass may have occurred during an interdisturbance period between intense fires. This hypothesis is supported by reports by Hoffman et al. (1994) of extensive peat combustion in northern Water Conservation Area 3A resulting from fires during the 1989 drought. Sawgrass was killed in large patches and replaced by open-water habitats in depressions burned into the soil. Fire management in Water Conservation Areas 2 and 3 and Everglades National Park successfully reduced the frequency and extent of severe burns through programs of fire abatement and/or controlled burning during the last two decades (Gunderson and Snyder, 1994). However, fire control in Water Conservation Area 1 was negligible during that period (M. Maffei, personal communication). During the 1-in-100-year drought of 1989, when the water table was ~60 cm below the ground surface, an intense fire burned ~8100 ha in Water Conservation Area 1. This fire did not produce extensive peat burns, although it did kill woody vegetation (M. Maffei, personal communication; Hoffman et al., 1994). Thus, fires severe enough to lower soil elevation and promote establishment of wet prairie/slough in former sawgrass areas appear to be unpredictable and probably represent stochastic events of severe natural disturbance. Allowing natural disturbance patterns of infrequent severe fires to lower soil elevation, expand wet prairie/slough coverage, and offset sawgrass encroachment merits further consideration and experimentation as a restoration guideline in an ecosystem where wet prairie/slough appears to be disappearing.

The loss of wet prairie/slough exacerbates an overall loss of aquatic productivity due to reduction in areal extent of the ecosystem by half. The wet prairie/slough community is the primary location of periphyton, which is one of the primary pathways for energy flow in the Everglades (Browder et al., 1994). This community is also the major habitat for aquatic fauna, particularly crustaceans and fishes (Loftus and Eklund, 1994) and a primary foraging habitat for wading birds (Hoffman et al., 1994). The openness of macrophytic cover and the presence of an important food source add to the importance of these areas as wading bird feeding sites.

RESTORATION GUIDELINES

Functional losses to the Everglades ecosystem that have accompanied landscape change appear to include (1) reduction in spatial extent of aquatic productivity and total system productivity, (2) reduction in aquatic productivity of the southern Everglades due to shortened hydroperiods and interrupted flows, (3) reduction in areal extent of wet prairie/slough and related aquatic productivity throughout the remaining wet prairie/slough-sawgrass-tree island mosaic, (4) loss of habitat diversity at the landscape and community level, and (5) reduction in possible early dry season feeding habitat of wading birds.

The hydrologic contribution of the sawgrass plain landscape to prolonging flows and hydroperiods downstream provides a guideline that should be considered in rainfall-driven models of water supply and hydroperiod restoration for the remaining wetland system. Wet season rainfall over the Everglades Agricultural

Area should be included in a model of water supply to the Water Conservation Areas during the following dry season, in order to generate flows and hydroperiods as if the sawgrass strand continued to function in the capacity of slowly discharging water to the south. Similarly, the prolongation of water flow across the marsh that now lies within the Water Conservation Areas provides a guideline for refinement of rainfall-driven models of water supply to the southern Everglades within Everglades National Park.

The hypothesis that wet prairie/slough loss may also be a phenomenon of vegetation recovery during an interval between severe fires suggests a restoration guideline of experimentally allowing naturally occurring fires to burn unimpeded in the system, rather than attempting to control their severity through current prescribed burning programs. Hoffman et al. (1994) reported extensive wading bird use of flooded depressions burned into the soil as the result of intense fires in Water Conservation Area 3A during the 1989 drought. Perhaps it is not coincidental that highest densities of wading birds in the Everglades during the 1930s and 1940s coincided with decades of severe peat burns (W. B. Robertson, personal communication).

Extended hydroperiods and flows during most years, punctuated by severe fires during periodic drought years, appear to be temporal patterns under which the Everglades developed (Fennema et al., 1994; Gunderson and Snyder, 1994; Gleason and Stone, 1994). Restoration of these patterns is suggested as a fundamental guideline for recovery of aquatic productivity and habitat diversity in remaining Everglades landscapes. In this multi-equilibrium concept of Everglades vegetation dynamics, the emphasis for management to maintain diversity moves away from decreasing the impact of disturbances toward allowing disturbances to occur naturally, with the full breadth and scope of effect. Perhaps the Everglades landscape scene described at the beginning of this chapter can persist only through constant change, both in widely fluctuating climatic, hydrologic, and fire regimes and in vegetation response to these variables.

Human activities during the past century have changed the spatial and temporal relationships between environmental variables and their influence on landscapes and plant communities in the Everglades. Natural change in landscape-level features occurred over the time frame of thousands of years, as opposed to a century, as was the case with agri-urban development. Management actions such as water or fire management in the Everglades have attempted to dampen oscillations in hydrologic and fire disturbance regimes that naturally had return intervals of decades to centuries. Preventing further spatial reduction of Everglades landscapes, prolonging flows and hydroperiods as if predrainage landscapes were intact, and allowing natural disturbances to operate at full impact on the remaining segments of the historical ecosystem are crucial restoration actions suggested by this chapter to restore and maintain vegetation heterogeneity in a dynamic, viable Everglades ecosystem. A stepwise and adaptive approach to implementation of such guidelines is suggested, considering the uncertainty of response of an ecosystem already reduced in size by half.

Appendix 17.I   Vegetation Study Site Location[a]

	Coordinates[b]					
Plot	Latitude (deg/min/sec)			Longitude (deg/min/sec)		State plane (x, y)
02	26	27	09	80	14	15	749485	770865
04	26	38	05	80	20	19	716059	836916
06	26	30	34	80	25	54	685861	791234
08	26	21	35	80	18	48	724851	737003
75	26	31	21	80	20	04	717632	796132
15	26	17	14	80	18	32	726448	710659
26	26	13	15	80	24	14	695431	686374
27	26	12	17	80	24	15	695366	680517
07	26	09	08	80	40	42	605517	661118
25	26	08	01	80	40	25	607083	654358
74	25	45	54	80	47	01	571215	520315
76	25	45	06	80	46	51	572137	515471
84	25	46	13	80	41	04	603847	522299
85	25	45	06	80	41	04	603863	515535
58	25	38	25	80	45	05	581905	475006
61	25	34	32	80	49	55	555395	451443
30	25	36	26	80	35	26	634929	463125
92	25	35	50	80	36	27	629356	459473
52	25	26	58	80	36	03	631714	405774
53	25	27	56	80	35	54	632521	411632
54	25	19	26	80	34	12	642035	360177
16	25	22	11	80	43	52	588784	376695
65	25	55	20	80	28	35	672097	577740
67	25	51	01	80	38	01	620495	551417
100	25	19	28	80	50	14	553767	360184

[a] Plot numbers established by Alexander and Crook (1975).
[b] Coordinates locate a point within a plot area of 1 m².

Appendix 17.II   Percent Cover of Vegetation in Square-Mile Plots

Wet Prairie/Slough-Sawgrass-Tree Island Mosaic Landscape

		Sawgrass					Wet prairie/slough		
Plot	Year	Mono-spec.	Scat. shrub	Scat. mel.	Scat. cat.	Total	No mel.	Scat. mel.	Total
2	1968	34.0	—	—	—	34.0	59.9	—	59.9
	1985	45.0	—	—	—	45.0	48.1	—	48.1
4	1968	1.4	23.0	—	—	24.4	38.5	—	38.5
	1984	1.3	32.2	—	—	33.5	22.4	—	22.4
7	1971	—	34.5	—	—	34.5	64.7	—	64.7
	1986	—	—	44.9	4.5	49.5	—	46.9	46.9
8	1968	14.5	—	—	—	14.4	41.8	—	41.8
	1984	3.1	—	—	—	3.1	41.6	—	41.6
15	1971	46.1	—	—	—	46.1	35.6	—	35.6
	1986	63.3	—	—	—	63.3	31.8	—	31.8

Appendix 17.II (continued) Percent Cover of Vegetation in Square-Mile Plots

Wet Prairie/Slough-Sawgrass-Tree Island Mosaic Landscape

Plot	Year	Sawgrass					Wet prairie/slough		
		Mono-spec.	Scat. shrub	Scat. mel.	Scat. cat.	Total	No mel.	Scat. mel.	Total
25	1971	—	42.4	—	—	42.4	52.1	—	52.1
	1986	—	—	—	54.8	54.8	—	40.9	40.9
58	1965	28.6	—	—	—	28.6	65.4	—	65.4
	1986	62.8	—	—	—	62.8	29.2	—	29.2
61	1965	53.3	30.1	—	—	83.4	5.9	—	5.9
	1986	53.7	36.2	—	—	89.9	2.4	—	2.4
67	1971	33.4	—	—	—	33.4	52.3	—	52.3
	1987	33.0	—	—	—	33.0	54.4	—	54.4
75	1968	—	—	—	—	—	89.6	—	89.6
	1984	37.7	0.6	—	—	38.3	47.3	—	47.3
84	1971	58.5	—	—	—	58.5	29.5	—	29.5
	1986	56.7	—	—	—	56.7	29.9	—	29.9
85	1971	40.6	25.1	—	—	65.7	29.5	—	29.5
	1986	52.1	13.6	—	—	65.7	28.4	—	28.4

Plot	Year	Cat-tail	Wil-low	S.F. tree isl.	Trop. ham.	Mel./casa.	Pond	Cypr. head	Bare	Levee canal
2	1968	—	0.1	5.7	—	0.1	—	—	—	—
	1985	3.6	0.3	1.2	—	0.3	0.9	—	—	—
4	1968	12.4	5.4	11.5	—	—	7.7	—	—	—
	1984	14.9	18.1	11.1	—	0.1	—	—	—	—
7	1971	—	0.3	0.5	—	—	—	—	—	—
	1986	—	—	0.1	—	—	—	—	—	3.5
8	1968	2.3	7.2	1.7	—	0.1	—	—	—	32.6
	1984	29.5	5.4	0.1	—	—	—	—	—	20.5
15	1971	18.1	0.2	—	—	—	—	—	—	—
	1986	4.5	0.4	—	—	—	—	—	—	—
25	1971	—	0.1	0.1	—	0.3	—	—	—	4.3
	1986	—	—	—	—	—	—	—	—	4.3
58	1965	—	0.1	6.0	—	—	—	—	—	—
	1986	—	—	7.8	0.1	—	—	0.1	—	—
61	1965	—	0.1	8.2	2.4	—	—	—	—	—
	1986	—	—	4.9	2.8	—	—	—	—	—
67	1971	—	—	3.6	—	—	—	—	—	10.7
	1987	1.1	—	3.0	—	—	—	—	—	8.5
75	1968	—	—	10.4	—	—	—	—	—	—
	1984	—	—	14.5	—	—	—	—	—	—
84	1971	—	—	1.9	—	—	—	—	—	10.1
	1986	—	1.3	1.3	—	—	—	—	—	10.8
85	1971	—	0.6	2.1	0.2	0.8	—	—	1.1	0.1
	1986	—	1.1	3.5	0.2	1.1	—	—	—	0.1

Appendix 17.II (continued)   Percent Cover of Vegetation in Square-Mile Plots

Sawgrass-Dominated Mosaic Landscape

		Sawgrass					Wet prairie/slough		
Plot	Year	Mono-spec.	Scat. shrub	Scat. mel.	Scat. cat.	Total	No mel.	Scat. mel.	Total
6	1968	82.7	4.9	—	—	87.6	10.6	—	10.6
	1984	54.5	—	—	—	54.5	39.9	—	39.9
26	1971	42.3	—	—	—	42.3	44.7	—	44.7
	1986	53.9	—	—	8.7	62.6	36.4	—	36.4
27	1971	—	49.8	23.7	—	73.5	24.9	—	24.9
	1986	—	—	77.1	—	77.1	18.8	—	18.8
65	1971	68.2	—	—	—	68.2	16.1	—	16.1
	1986	—	—	54.6	—	54.6	34.3	—	34.3

Plot	Year	Cat-tail	Wil-low	S.F. tree isl.	Trop. ham.	Mel./ casa.	Pond	Cypr. head	Bare	Levee canal
6	1968	—	0.8	—	—	—	0.9	—	—	—
	1984	0.6	3.2	0.6	—	—	1.1	—	—	—
26	1971	12.2	—	—	—	—	—	—	—	0.8
	1986	—	—	—	—	—	—	—	—	1.0
27	1971	—	0.1	—	0.1	—	—	—	—	1.5
	1986	—	—	0.8	—	1.4	—	—	—	1.9
65	1971	—	—	15.5	0.1	—	—	—	—	—
	1986	—	—	10.8	0.2	—	—	—	—	—

Southern Marl Marsh Landscape

		Marl prairie/ sawgrass				S.F.							
Plot	Year	No shrub	Scat. shrub	Cat-tail	Wil-low	tree isl.	Trop. ham.	Cypr. head	Pine	Red mang.	Mel./ casa.	Bare	Levee canal
16	1971	89.8	—	—	0.1	—	1.4	—	8.6	—	—	—	—
	1986	81.2	—	—	0.3	—	5.1	0.7	12.7	—	—	—	—
30	1970	56.2	9.5	—	0.6	1.0	5.4	—	—	—	0.2	25.0	2.2
	1986	73.5	—	0.4	1.5	3.8	6.2	—	—	—	0.3	12.0	2.4
52	1970	95.9	—	—	1.3	1.7	—	—	—	—	—	—	1.6
	1986	93.7	—	—	1.8	3.3	—	—	—	—	—	—	1.2
53	1970	85.6	—	—	1.5	11.8	—	—	—	—	—	—	1.1
	1986	85.4	—	—	0.4	12.6	1.5	—	—	—	—	—	1.1
54	1970	95.5	—	—	—	0.6	0.1	1.2	—	—	—	—	2.7
	1986	94.8	—	—	0.6	—	—	1.2	—	—	0.5	—	2.9
74	1971	69.9	—	—	5.2	18.1	0.5	—	—	—	—	—	6.4
	1986	74.7	—	—	11.2	7.7	—	—	—	—	0.2	—	6.2
76	1971	81.9	—	—	7.5	8.1	—	—	—	—	—	—	2.6
	1986	80.8	—	—	5.4	9.3	—	—	—	—	—	—	4.6

Appendix 17.II   Percent Cover of Vegetation in Square-Mile Plots

Southern Marl Marsh Landscape

Plot	Year	Marl prairie/ sawgrass — No shrub	Scat. shrub	Cat-tail	Wil-low	S.F. tree isl.	Trop. ham.	Cypr. head	Pine	Red mang.	Mel./ casa.	Bare	Levee canal
92	1970	89.8	—	—	2.7	—	7.4	—	—	—	0.2	—	—
	1986	84.7	—	—	0.7	—	4.3	—	—	—	0.3	10.1	—
100	1965	58.2	—	—	—	6.1	1.6	—	3.1	30.2	—	—	—
	1986	56.0	—	—	—	5.3	2.8	—	3.1	28.1	—	3.8	—

LITERATURE CITED

Alexander, T. R. and A. G. Crook. 1973. Recent and Long Term Vegetation Changes and Patterns in South Florida, Part I, Preliminary Report, South Florida Ecological Study, Appendix G, University of Miami, Coral Gables, Fla.

Alexander, T. R. and A. G. Crook. 1975. Recent and Long Term Vegetation Changes and Patterns in South Florida, Part II, Final Report, South Florida Ecological Study, Appendix G, NTIS PB 264462, University of Miami, Coral Gables, Fla.

Austin, D. F. 1976. Vegetation of southeastern Florida. I: Pine Jog. *Fla. Sci.,* 39:230–235.

Browder, J. A., P. J. Gleason, and D. R. Swift. 1994. Periphyton in the Everglades: Spatial variation, environmental correlates, and ecological implications. in *Everglades: The Ecosystem and Its Restoration,* S. M. Davis and J. C. Ogden (Eds.), St. Lucie Press, Delray Beach, Fla., chap. 16.

Craighead, F. C., Sr. 1971. *The Trees of South Florida,* Vol. I: The Natural Environments and Their Succession, University of Miami Press, Coral Gables, Fla.

Davis, J. H., Jr. 1943. The Natural Features of Southern Florida, Especially the Vegetation and the Everglades, Bull. No. 25, Florida Geological Survey, Tallahassee.

Davis, S. M. 1994. Phosphorus inputs and vegetation sensitivity in the Everglades. in *Everglades: The Ecosystem and Its Restoration,* S. M. Davis and J. C. Ogden (Eds.), St. Lucie Press, Delray Beach, Fla., chap. 15.

Fennema, R. J., C. J. Neidrauer, R. A. Johnson, T. K. MacVicar, and W. A. Perkins. 1994. A computer model to simulate natural Everglades hydrology. in *Everglades: The Ecosystem and Its Restoration,* S. M. Davis and J. C. Ogden (Eds.), St. Lucie Press, Delray Beach, Fla., chap. 10.

Gleason, P. J. and P. A. Stone. 1994. Age, origin, and landscape evolution of the Everglades peatland. in *Everglades: The Ecosystem and Its Restoration,* S. M. Davis and J. C. Ogden (Eds.), St. Lucie Press, Delray Beach, Fla., chap. 7.

Gunderson, L. H. 1994. Vegetation of the Everglades: Determinants of community composition. in *Everglades: The Ecosystem and Its Restoration,* S. M. Davis and J. C. Ogden (Eds.), St. Lucie Press, Delray Beach, Fla., chap. 13.

Gunderson, L. H. and J. R. Snyder. 1994. Fire patterns in the southern Everglades. in *Everglades: The Ecosystem and Its Restoration,* S. M. Davis and J. C. Ogden (Eds.), St. Lucie Press, Delray Beach, Fla., chap. 11.

Harper, R. M. 1927. Natural Resources of Southern Florida, 18th Annu. Rep., Florida Geological Survey, Tallahassee.

Harshberger, J. W. 1914. The vegetation of South Florida, south of 27°30′ north, exclusive of the Florida Keys. *Trans. Wagner Free Inst. Sci. Philadelphia,* 7:49–189.

Hoffman, W., G. T. Bancroft, and R. J. Sawicki. 1994. Foraging habitat of wading birds in the Water Conservation Areas of the Everglades. in *Everglades: The Ecosystem and Its Restoration,* S. M. Davis and J. C. Ogden (Eds.), St. Lucie Press, Delray Beach, Fla., chap. 24.

Hopkins, C. F. 1884. Field notes of a reconnaissance made in Oct. and Nov. 1883. in Report on Everglades Drainage Project in Lee and Dade Counties, Florida, Jan. to May 1907, J. T. Stewart (Ed.), Office of Experiment Stations, Irrigation and Drainage Investigations, U.S. Department of Agriculture, Washington D.C.

Ives, J. C. 1856. *Memoir to Accompany a Military Map of the Peninsula of Florida South of Tampa Bay,* Wynkoop, Book and Job Printer, New York.

Jones, L. A. 1948. Soils, Geology, and Water Control in the Everglades Region, Bull. 442, University of Florida Agricultural Experiment Station, Gainesville.

Kolipinski, M. C. and A. L. Higer. 1969. Some Aspects of the Effects of Quantity and Quality of Water on Biological Communities in Everglades National Park, Open File Report, U.S. Geological Survey, Tallahassee, Fla., Oct. 1968 and Sept. 1969.

Loftus, W. F. and A.-M. Eklund. 1994. Long-term dynamics of an Everglades small-fish assemblage. in *Everglades: The Ecosystem and Its Restoration,* S. M. Davis and J. C. Ogden (Eds.), St. Lucie Press, Delray Beach, Fla., chap. 19.

Mazzotti, F. J. and L. A. Brandt. 1994. Ecology of the American alligator in a seasonally fluctuating environment. in *Everglades: The Ecosystem and Its Restoration,* S. M. Davis and J. C. Ogden (Eds.), St. Lucie Press, Delray Beach, Fla., chap. 20.

Ogden, J. C. 1994. A comparison of wading bird nesting colony dynamics (1931–1946 and 1974–1989) as an indication of ecosystem conditions in the southern Everglades. in *Everglades: The Ecosystem and Its Restoration,* S. M. Davis and J. C. Ogden (Eds.), St. Lucie Press, Delray Beach, Fla., chap. 22.

Olmsted, I. C. and L. L. Loope. 1984. Plant communities of Everglades National Park. in *Environments of South Florida: Present and Past II, Memoir No. 2,* P. J. Gleason (Ed.), Miami Geological Society, Coral Gables, Fla., pp. 167–184.

Richardson, D. R. 1977. Vegetation of the Atlantic Coastal Ridge of Palm Beach County, Florida. *Fla. Sci.,* 40:289–329.

Small, J. K. 1913. *Flora of Miami,* published by the author, New York.

Small, J. K. 1923. Green deserts and dead gardens. *J. N.Y. Bot. Gard.,* 24:237–239

Small, J. K. 1927. Among floral Aborigines: A record of exploration in Florida in the winter of 1922. *J. N.Y. Bot. Gard.,* 23:25–29.

Smith, T. J., H. Hudson, G. V. N. Powell, M. B. Robblee, and P. J. Isdale. 1989. Freshwater flow from the Everglades to Florida Bay: A historical reconstruction based on fluorescent banding in the coral *Solenastrea bournoni. Bull. Mar. Sci.,* 44:274–282.

Synder, G. H. and J. M. Davidson. 1994. Everglades agriculture: Past, present, and future. in *Everglades: The Ecosystem and Its Restoration,* S. M. Davis and J. C. Ogden (Eds.), St. Lucie Press, Delray Beach, Fla., chap. 5.

Stewart, J. T. 1907. Report on Everglades Drainage Project in Lee and Dade Counties, Florida, Office of Experiment Stations, Irrigation and Drainage Investigations, U.S. Department of Agriculture, Washington, D.C.

Tebeau, C. W. 1968. *Man in the Everglades: 2,000 Years of Human History in the Everglades National Park,* University of Miami Press, Coral Gables, Fla.

Walters, C. J., L. H. Gunderson, and C. S. Holling. 1992. Experimental policies for water management in the Everglades. *Appl. Ecol.,* 2:189–202.

Wanless, H. R., R. W. Parkinson, and L. P. Tedesco. 1994. Sea level control on stability of Everglades wetlands. in *Everglades: The Ecosystem and Its Restoration,* S. M. Davis and J. C. Ogden (Eds.), St. Lucie Press, Delray Beach, Fla., chap. 8.

Willoughby, H. L. 1898. *Across the Everglades, A Canoe Journey of Exploration,* J. M. Dent and Co., London.

18

SYNTHESIS: VEGETATION PATTERN and PROCESS in the EVERGLADES ECOSYSTEM

Peter S. White

ABSTRACT

An understanding of vegetation pattern and process is critical for the protection, restoration, and management of the Everglades ecosystem. Nine broadly defined physical driving forces and two additional biological processes, operating on a wide range of scales, contribute to this pattern and process. Two of the most critical are the hydrologic and fire regime, both of which are correlated with relative elevation and are the product of the interaction of a series of physical driving forces, as well as biological processes and feedback between biology and environment. Significant improvements in our understanding of these factors will require investigations of scale dependence and the incorporation of explicit reference to spatial position and configuration. The Everglades is oligotrophic; natural dynamics and periodicities are key parameters in aquatic productivity and in the concentration of prey items supporting birds and other large animal populations. The restoration of natural dynamic processes is critical, but the 50% reduction in the spatial extent of the ecosystem and several other human-induced changes (e.g., exotic species invasions) will represent significant challenges even when natural hydrologic and fire regimes are restored. A long-term commitment is needed for both proactive management and the research that will evaluate and refine that management.

INTRODUCTION

The plant communities of the Everglades exhibit a great deal of structural and compositional variation, from open water sloughs with sparse macrophytes to

---

sedge- and grass-dominated freshwater marshes, open pine stands, and dense broad-leaved evergreen forest. Coastal saline habitats vary from open flats of salt-tolerant herbs to dense mangrove forests. The substrates that support these communities are also diverse, ranging from deep accumulations of organically produced peat and marl to limestone rocklands and sands. While the north/south topographic gradient of absolute elevation is only 2 cm·km^{-1}, relative elevation is a critical variable, influencing hydroperiod, fire, and peat accumulation and, hence, ecosystem composition, structure, and function. Change in relative elevation produces large changes in vegetation over short distances and, hence, the mosaic-like appearance of the landscape as a whole (Gunderson, 1994). This mosaic is far from static; Davis et al. (1994) have proposed the appealing image of the kaleidoscope for the south Florida wetland landscape.

An understanding of Everglades vegetation is critical if this ecosystem is to be protected, restored, and managed. Primary production initiates energy flow through Everglades trophic webs, plant communities often contribute physical structure to habitats in the distribution of living and dead organic matter, and the vegetation not only reflects physical driving forces but also influences the effect of several critical forces to a considerable degree (DeAngelis and White, 1994; DeAngelis, 1994).

This chapter will present a synthesis of patterns and processes that are described in the vegetation section and the physical environment (described in the preceding chapters on driving forces) that influences them. The last 100 years, during which humans began to dominate the south Florida landscape, has seen dramatic and directional change, not only in terms of natural processes, but also in terms of the spatial extent of the ecosystem (Davis et al., 1994). Both kinds of change have critical influences on the overall productivity and condition of the Everglades ecosystem. This chapter thus addresses human impacts and management, as well as natural processes in this diverse landscape.

PATTERN, PROCESS, and PATCH DYNAMICS in VEGETATION

Pattern and process (Watt, 1947) describes an approach to the study of plant communities which focuses on spatial pattern within communities and the dynamic processes responsible for that pattern. In this context, pattern is the spatial distribution of measured vegetation properties. Such properties include structural characteristics (e.g., basal area, biomass, leaf area, density, and organic matter pools) and compositional characteristics (relative population size of component species). While pattern is relatively easy to study (it has been the subject of descriptive plant ecology for over 100 years), recent advances have brought new perspectives. In particular, there may be important scale dependencies in vegetation pattern: different relationships may emerge when the vegetation is studied at different spatial scales. Another important issue is whether pattern and process can be understood statistically without reference to spatial position or configuration or whether explicit spatial reference is required.

Process, which is both more challenging and more interesting to study, includes the mechanisms that give rise to pattern. Processes include those operating on time scales coincident with individual life history events (e.g., birth, dispersal, growth, senescence, mortality, disturbance, and build-up of allelochemics in soils) and those that operate on time scales of greater than single generation times (e.g., soil development and geomorphological processes).

Important questions about pattern and process include the relative roles of environment and biology, whether the pattern is stable (locally or as a statistical property across all sites), whether process rates are stationary or varying, whether there is feedback between patch state and process, and whether spatial configuration must be explicitly incorporated. All of these are key issues in any understanding of the Everglades ecosystem.

Patch dynamics (Pickett and White, 1985) has been applied to the study of pattern and process in biotic communities, with emphasis on individual patches within the pattern and the time course and interactions of these patches. Disturbance is one of the key processes by which patches are formed.

The pattern and process and patch dynamics views of vegetation are essential to any discussion of Everglades vegetation. What do we know about the pattern of plant communities in this landscape? What do we know about the processes that shape and change that pattern?

PHYSICAL DRIVING FORCES, ECOSYSTEM PROCESSES, and VEGETATION

Nine broadly defined physical driving forces and two additional biological processes influence Everglades vegetation (DeAngelis and White, 1994; DeAngelis, 1994; Gunderson, 1994). The driving forces range from those that show gradual change (e.g., climate change, sea level rise, and bedrock elevation, for which change is measured on time scales of $>10^2$ years) to natural periodicities and disturbances (e.g., hydrology, flood, drought, fire, storms, and freezing temperatures, for which change is measured on time scales of 10^{0-2} years). Biological processes contribute to both ecosystem dynamics and the playing out of these physical forces (e.g., productivity, decomposition, and species interactions, for which change is measured over times scales of 100^{-2} years).

What is the general character of environmental signals that the physical driving forces represent and what are typical vegetation responses to these forces? Figure 18.1 presents a conceptual model for this problem. Some physical forces are characterized by low variance and slow change over time (Figure 18.1A, the gradual driving forces of DeAngelis and White, 1994). If there is low within- and between-year variation when these forces are measured at a short time step (1–30 days), this model is a sufficient representation of the environmental signal influencing the vegetation (e.g., change in topography and climate). Such forces may be effectively regarded as constants by managers interested in 1- to 10-year time periods (DeAngelis and White, 1994). Vegetation responses include the adaptations of the species present to current conditions and population shifts as conditions and competitive relations change. However, when vegetation presents structural resistance to gradual change, change can be abrupt and constitute a threshold-type response.

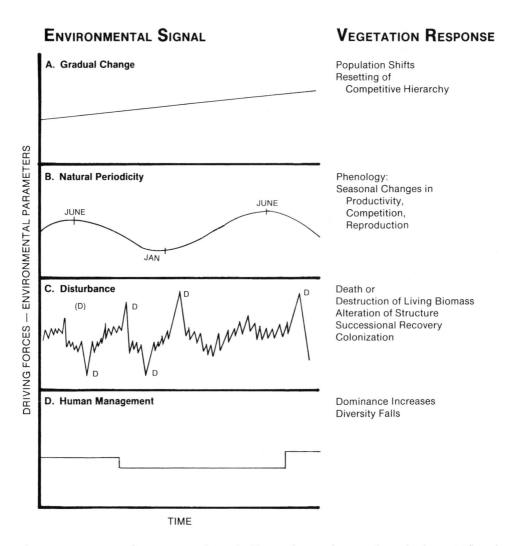

Figure 18.1   Forms of environmental signals (driving forces of DeAngelis and White, 1994) and types of vegetation response.

In addition to climate, sea level, and topography, inputs of nutrients from natural sources in the freshwater wetlands can be regarded as roughly constant and thus a part of the physical context of the Everglades. The Everglades is an oligotrophic system (Gunderson, 1994; Davis et al., 1994; Browder et al., 1994). While nutrients such as phosphorus may increase locally after fire or soil disturbance, the effect is quickly attenuated (Davis, 1994).

In contrast to climate, sea level, topography, and nutrient supply under natural conditions, many physical driving forces show substantial and predictable within-year (e.g., seasonal) variation. These forces can be depicted by averages computed at an intermediate time step (10–100 days) (Figure 2.1B, the "natural periodicities" of DeAngelis and White, 1994). For some seasonal phenomena with low year-to-year variation, this model will give an accurate and sufficient picture of the environmental signal (e.g., the seasonal progression of photoperiod and potential solar beam radiation). Vegetation response consists of adaptation to a seasonal environment as well as phenological changes such as rhythms of production, decomposition, mortality, and reproduction.

Hydroperiod and temperature are two driving forces which exhibit a natural periodicity and have a strong effect on vegetation composition and structure. For these and many other physical driving forces, however, there is not only substantial within-year variation, but also considerable between-year variation in intensity and spatial scale (Figure 18.1C). Extreme values of physical forces such as temperature, precipitation, and wind are encountered in the Everglades on a time period of 3–10 (or more) years (Gunderson, 1994).

When the structural resistance and physiological tolerance of the vegetation is exceeded, substantial and sudden destruction of living biomass occurs; hence the recognition of these events as disturbances (e.g., drought, flood, severe fire, damaging freezes, and wind). These often operate on intermediate to long time scales and intermediate to large spatial scales (Table 18.1) (Gunderson, 1994). These disturbances tend to be discrete events in time (i.e., their duration is measured in hours, days, or, at most, months and is thus many times shorter than the interval between events) and infrequent (i.e., greater than 10^{0-2} years for return time). Disturbance regime can be correlated with other aspects of the physical environment such as topographic position (Harmon et al., 1983).

Although initial regrowth or recolonization can be quite fast, full vegetation recovery (including recovery of soil organic matter and peat where it has been removed) tends to be slow (at least compared to the duration of the disturbance event), producing a basic asymmetry: destruction is fast, full recovery slow. Hence, disturbance has an effect out of proportion to its brief operation (White, 1979). Time lags become important and ensure a historic component to vegetation explanations (present structure and composition may reflect past, not present, conditions). This serves to decrease the strength of the predictions of vegetation composition and structure from physical variables (such as elevation) and current environment (such as hydroperiod) and may underlie the failure of researchers to find statistically significant differences in recently measured hydroperiods between plant community types (Gunderson, 1994). History becomes an important explanation of vegetation and may also influence current process.

Vegetation response to disturbance includes vegetative regrowth (e.g., low-severity disturbances) and recolonization from propagules. If the substrate is intact, disturbance may increase nutrient availability and stimulate a quick accumulation of biomass; where substrates and nutrient availability are decreased, recovery can be extremely slow. On sites with greater productivity, there may be a sorting of species, with the initial colonists being species with high reproductive output but low competitive ability and later dominants being species with low reproductive output and high competitive ability. Recolonization depends on the recolonists being present somewhere in the landscape; hence, spatial configuration and the relationship between disturbance patch size to landscape area becomes critical (White, 1987; Pickett and Thompson, 1978; DeAngelis and White, 1994).

All three of the driving force types (Figure 18.1A–C) can be found in the same environmental variable. There can be some overall gradual trajectory in the annual

Table 18.1 Major Plant Communities (Gunderson, 1994),
Major Driving Forces, and Human Influences of the Everglades Ecosystem

Vegetation	Relative hydrology	Dominant influences in recent time Natural	Dominant influences in recent time Human-caused
Upland communities			
Florida slash pine	7	Rockland soils	Development
		Fire	Fire suppression
Hardwood hammocks	8	Driest sites	Development
		Storms	Exotic species
		Freezes	
Freshwater wetland communities—forested			
Bayhead	5	Hydrology	Hydrologic changes
			Exotic species
Willow head	5	Hydrology	Hydrologic changes
		Soil disturbance	Exotic species
Cypress	5	Hydrology	Hydrologic changes
Pond apple	5	Hydrology	Development
Freshwater wetland communities—gramineous			
Sawgrass marsh	4	Hydrology	Hydrologic changes
		Fire	Fire suppression
			Exotic species
			Nutrient inputs
			Cattail invasion
			Development
Wet prairie—peat	3	Hydrology	Hydrologic changes
		Fire	
Wet prairie—marl	6	Hydrology	Hydrologic changes
		Fire	
Freshwater wetland communities—little or no emergent vegetation			
Ponds and creeks	1	Hydrology	Hydrologic changes
Sloughs	2	Hydrology	Hydrologic changes
Saline wetlands			
Mangrove forests		Storms	Development
		Lightning	
		Freezes	
Salt marsh		Storms	
Coastal prairie		Storms	

mean of a parameter that also has marked seasonal pattern in monthly means and considerable day-to-day or week-to-week fluctuation in actual values. For example, temperature has a slowly changing mean value, a marked seasonal pattern in monthly means, and daily and weekly variation, with significant extreme values every 3–10 years (e.g., damaging freezes). Vegetation responds to all of these forms of the temperature signal.

Four of the most important physical driving forces and ecological processes fall uniquely in the domain of human influence: fire, hydrology, vegetation structure, and exotic species invasions. For good or ill, humans are able to and have, in fact, altered these forces and processes. Humans often reduce natural variation and extremes in changeable and sometimes destructive phenomena such as hydroperiod and fire (Figure 18.1D). One effect of monotonous driving forces might be loss of diversity (Connell, 1978; Denslow, 1985): species tend to be uniquely adapted to different environmental conditions, and any regularization of physical driving forces will favor some species over others, creating a directional trajectory (White, 1987). There is evidence that much change in the Everglades in the last 100 years is the result of reductions in hydroperiod, severe fire, and spatial extent (Davis et al., 1994).

Unfortunately, we rarely have a full understanding of the nature of the driving forces in an ecosystem. Further, this understanding is made more challenging by the interaction among physical forces and by the way the biological components of the ecosystem contribute to overall dynamics. Finally, the driving forces and biological components may be constrained in their behavior by spatial and temporal context—the effect of driving forces may change depending on position in space and time, so that accurate models must deal with explicit spatial and temporal position. Each of these complications is discussed briefly.

Interaction of Driving Forces. There are many examples of the interaction of the driving forces in the Everglades (Figure 18.2). Drought lowers fuel moisture and increases the probability of high-intensity fire. Temperature regime and climate change influence evapotranspiration and hydroperiod. The effect of freezing temperatures depends on thermal buffering which, in turn, derives from the water level. Prior drought promotes freeze damage. Poor winter hardening, due to a lack of prior exposure to cold temperatures, increases freeze damage. Storms deposit debris, which changes hydrology and resists retreat of the coastline. Two of the

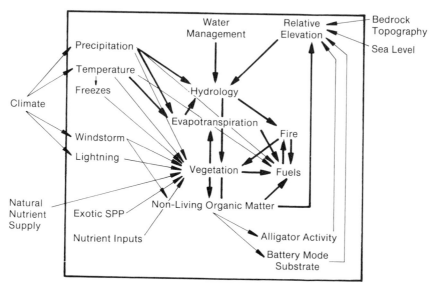

Figure 18.2 Major influences of driving forces on vegetation in the Everglades ecosystem. The heavier lines represent the most important interactions between driving forces and vegetation that determine vegetation composition and structure. Factors outside the box represent slowly changing parameters that are often regarded as constants in short-term management and analysis.

most important driving forces, hydrology and fire, are produced by the interactions of a number of these factors: precipitation, temperature, evapotranspiration, and biological processes.

Vegetation Influences on Driving Forces.   Vegetation reflects the physical driving forces, but it also influences these forces and plays a major role in two of the most important, hydrology and fire (Table 18.1). In effect, vegetation contributes inertia (the vegetation structure may reflect previous, rather than current, environmental conditions) and resistance (the vegetation structure physically resists some forces such as wave action and wind). Again, examples are numerous: living and dead organic matter are the fuels that burn in wild and managed fires (ecosystem state determines fuel loadings), evapotranspiration influences the water table, organic matter builds soil elevation and changes water flow and hydrology, and mangrove vegetation stabilizes shorelines against all but extreme storms. When there is a strong feedback between the successional age and the probability of a subsequent disturbance (e.g., when the likelihood of disturbance increases with time since the last disturbance), the vegetation may serve to regulate disturbances, a condition that, with large landscape size relative to disturbance patch size, can produce a dynamic equilibrium of disturbance patches (White and Pickett, 1985; Shugart, 1984).

Spatial and Temporal Position.   Several of the driving forces in the Everglades show a correlation with geographic position, as do some vegetation types (Table 18.1). Distance from the coast influences overall climate, the probability of hurricane disturbance, salinity encroachments, and damaging frosts. Probability of nutrient inputs from the Everglades Agricultural Area depends on distance from that area and its drainage canals (Davis, 1994). Many animal populations shift with changing hydroperiods; as a result, the spatial configuration (e.g., size and distance between deep water pools) of habitats may sometimes be important in the dynamics of population colonization. Deeper ponds serve as refuges for many large and small organisms and the size and number of these probably play a role in population dynamics. As discussed earlier, the effect of a given physical driving force may depend on temporal position (e.g., the time since the last disturbance).

A full model of vegetation and environment in the Everglades would include the interactions between driving forces, the feedback between ecosystem processes and environment, biological processes (production, decomposition, birth, and mortality, but including such phenomena as inertia and resistance), explicit reference to spatial and temporal position or context as an influence on process, and the issue of scale. Such a model could be examined for dominant influences in actual landscape patterns. A very simple outline of some of these factors (minus some of the most interesting and important, such as spatial position and scale) is presented in Figure 18.2.

Despite the complications in building a complete model of the vegetation environment relationship in the Everglades, two parameters stand out as dominant influences on the current condition of 11 of the 15 plant communities of the Everglades (Table 18.1): hydrology and fire. These two parameters are the product of other physical driving forces and the vegetation itself, are important short-term influences on vegetation structure, have been altered by past human activities, and represent variables at least partially under management control. Hence, they are central parameters in any future management schemes.

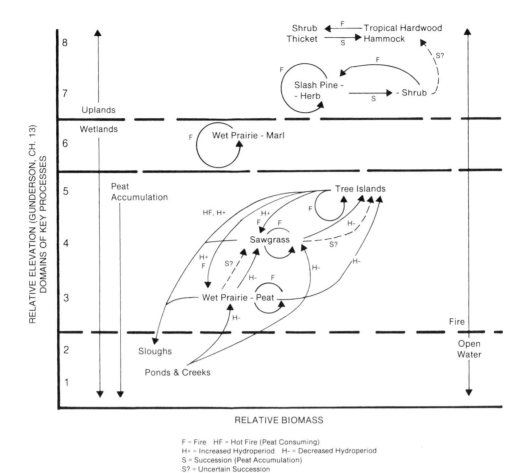

F = Fire   HF = Hot Fire (Peat Consuming)
H+ = Increased Hydroperiod   H- = Decreased Hydroperiod
S = Succession (Peat Accumulation)
S? = Uncertain Succession

Figure 18.3   Major plant communities of the central freshwater Everglades and the processes that produce them (Gunderson, 1994; Davis et al., 1994). This diagram does not show any explicit scale parameters and lacks information about spatial and temporal context.

Relative elevation can be used to rank the major plant communities of the Everglades (Table 18.1 and Figure 18.3) (Gunderson, 1994). Relative elevation, in turn, affects two of the most important driving forces, hydrologic and fire regime. The hydrologic regime is determined by relative elevation, rainfall, temperature, evapotranspiration, management, and such variables as ecosystem structure (e.g., vegetation physically slows water movement and affects relative elevation through peat deposits). Hydroperiod affects the important biologically mediated process of peat accumulation (Gunderson, 1994).

Because of their effect on vegetation structure, fuel loadings, fuel moisture, hydroperiod, and relative elevation contribute to fire occurrence. Fires range from frequent (10^{0-1} recurrence intervals) small area/low intensity to infrequent ($\sim10^2$ recurrence intervals) large area/high intensity. A critical phenomenon in high-intensity fires is whether peat is consumed; if so, there is an effect on hydrology as relative elevation is lowered. This, in turn, means that recovery times will be long. The lowering of relative elevation has been proposed as one of the few mechanisms (alligator activity is another, but more local, process) by which deeper waters and longer hydroperiods are generated (Davis et al., 1994).

The major plant communities of the Everglades freshwater ecosystems are ranked by relative elevation and relative biomass in Figure 18.3. Four elevation-related domains are indicated in the figure: (1) uplands with fire as an important factor (Florida slash pine and tropical hardwood hammocks), (2) wetlands with fire as an important factor and little peat deposition (wet prairie on marl), (3) wetlands with fire as an important factor and peat deposition as an important phenomenon (tree islands, sawgrass, and wet prairie on peat), and (4) wetlands without fire as an important factor (open water sloughs, ponds, and creeks).

Several important summary statements apply to Figure 18.3. First, the eight community types are weakly or not at all linked by a successional progression. A plausible exception to this is the progression from wet prairie on peat to sawgrass to tree island with extended peat deposition, but steady progression in this manner is probably rare; when these changes occur, it is more likely the result of a changed hydroperiod (Davis et al., 1994). Second, normal seasonal fluctuations in hydroperiod, with fire in the dry season and regrowth in the next wet season, are capable of maintaining all of the communities at a near stable composition and structure. In other words, in many years the vegetation is conditioned by a natural periodicity (hydroperiod) and a mild disturbance (low-intensity fire). Third, while fire can make conversions among the three central communities (tree islands, sawgrass, and wet prairie on peat), predicting the response to any given fire is difficult because of all of the interacting variables. Finally, among these three community types, peat and the processes affecting it (production, erosion, hydroperiod, and fire) are critical aspects of the ecosystem.

HUMAN INFLUENCES in the EVERGLADES ECOSYSTEM

Human influence has had two major effects in the Everglades: (1) a change in natural processes and dynamics, including a reduction in hydroperiod and the occurrence of severe fire, and (2) a reduction in the spatial extent of the Everglades ecosystem. The chapters reviewed here suggest that both of these human influences have lowered overall productivity in the Everglades ecosystem. Each of these topics is discussed below.

Change in Process

Hydrology and Fire

During the last 20 years, there has been a roughly 25% loss in aerial coverage of wet prairie and slough and a similar gain in sawgrass on selected study sites, probably as a result of lower water levels and interrupted water flow resulting from drainage structures built between 1900 and 1950 (Davis et al., 1994). Wet prairies and sloughs were major sites of periphyton productivity, as well as productivity of crustaceans and fish. The result of these changes has probably been reduced aquatic productivity, made all the more significant because the Everglades is a low-productivity ecosystem. Loss of aquatic productivity may, in turn, be linked to declines in wading bird populations (Ogden, 1994; Bancroft et al., 1994). It has also been inferred that there has been a reduction in fires severe enough to burn organic deposits and thus lower soil elevation, a phenomenon that would increase hydroperiod on long (>10^2 years) time scales.

Davis et al. (1994) suggest that the original Everglades ecosystem was characterized by extended hydroperiods during most years, with slow and continuous water flow. Water management has contributed to faster movement of water and episodic, rather than sustained, release. These wet conditions were punctuated by severe fires during extreme drought years.

The periphyton community appears to be controlled primarily by hydroperiod and water quality (Browder et al., 1994). This community forms the basis of important Everglades food chains. Shortened hydroperiods reduce overall productivity. Shorter hydroperiods, shallower waters, and lower percentages of organic matter in the sediment result in dominance by blue-green algae, which are less valuable in food chains. Higher values of these parameters result in dominance by diatoms and green algae. The sensitivity and rapid turnover rate of algae and other aquatic organisms may mean that they can be used in monitoring ecosystem function (Browder et al., 1994). The Everglades can be characterized by an oligotrophic, pulse system (Browder et al., 1994; Davis et al., 1994). The seasonal concentration of prey means that a larger consumer biomass is supported than in nonpulsing systems of the same primary productivity (Browder et al., 1994).

Phosphorus Additions

Significant and ongoing additions of phosphorus are currently confined to northern areas adjacent to the Everglades Agricultural Area. In the Water Conservation Areas, atmospheric phosphorus inputs are up by 70% due to agricultural dust; total inputs, from the atmosphere plus pumping, to the Water Conservation Areas have more than tripled (Davis, 1994). This increase in phosphorus results in conversion from sawgrass to cattail dominance, a change that brings with it a change in ecosystem structure, function, and productivity. Although phosphorus is quickly taken up and its influence decreases steeply from areas along canals and water control structures, some 300–400 km² of cattail has been created by this influence, thereby representing another source of decreased aquatic productivity (Davis, 1994; Browder et al., 1994).

Exotic Plant Invasion

The separations between the continents resulted in the evolution of biotas in isolation from one another (Elton, 1958). When species are transported purposefully or accidentally across these natural barriers, they may be preadapted to new environments or find environments similar to their native range. At the same time, they often leave behind natural enemies. Some species populations become invasive, with the pattern of invasion often described as explosive.

South Florida, perhaps because of its tropical climate, is extremely vulnerable to exotic species problems. Among these, *Melaleuca quinquenervia* is a severe threat to the interior wetlands that are the core of the Everglades ecosystem (Bodle et al., 1994): this species thrives in all but the wettest hydroperiods (e.g., sloughs and wet prairies). *Melaleuca* drastically changes ecosystem structure and dynamics—forest replaces gramineous marsh, thus changing animal use; leaf litter and woody debris change relative soil elevation and hence hydrology; tree weight can compress underlying peat deposits; organic matter results in heavy fuel loads of very combustible materials, leading to very hot fires; higher leaf areas increase evapotranspiration and lower water tables; and leaf litter may produce allelochemics and combined with dense evergreen shade may eliminate understory species (Bodle et al., 1994). *Melaleuca* benefits from many of the natural processes that support the original vegetation mosaic of the Everglades: mass seeding follows disturbances such as fire, and disturbance to native forests allows understory *Melaleuca* to invade the canopy.

Control of *Melaleuca* will be difficult. One program seeks to eliminate this species from successive bands east of Everglades National Park under the leadership of the Exotic Pest Plant Council. There is some evidence that the trees can be drought stressed and that hot fires following drought stress will kill established trees (Bodle et al., 1994), a solution made complicated by the desire to generally increase hydroperiod for ecosystem restoration. Some herbicides are effective in certain situations but are difficult to safely apply in this wetland ecosystem. A search is underway for natural enemies to import from its native range in Australia.

Change in Spatial Extent

In addition to the alteration of natural processes just described, man has drastically changed the spatial extent of the Everglades ecosystem, and this reduction in size has important consequences of its own for ecosystem function (Davis et al., 1994). As described by Davis et al. (1994), three of the seven predrainage Everglades landscape types have been lost entirely. Farming and urbanization have accounted for direct removal of 50% of the entire Everglades ecosystem. The Everglades Agricultural Area has itself replaced 75% of the original dense sawgrass marsh. Urban areas east and north of Everglades National Park have destroyed 90% of the Florida slash pine community that occurred on the oolitic limestone ridge to the south and east of the main area of Everglades marsh. In addition to this loss of spatial extent, there has been a shift in hydroperiod and invasion in some areas by *Melaleuca* and cattail, as described earlier.

This loss of spatial extent and the shortened and interrupted hydroperiods reduced the total productivity in aquatic communities (Browder et al., 1994) of the Everglades ecosystem, a generally oligotrophic ecosystem to begin with. Some component of changes in bird populations is due to this loss of productivity alone. There has also been loss of early dry season feeding habitat of wading birds in the peripheral marshes in the northern part of the Everglades ecosystem and the marl prairie areas in the south; these feeding areas were important in initiation of nesting in some species (Ogden, 1994; Bancroft et al., 1994).

Loss of spatial extent also has a consequence for vegetation dynamics. While there was always local change in the Everglades landscape, at broad spatial and temporal scales these local changes probably contributed to a rough stability in populations and ecosystem parameters (DeAngelis and White, 1994). However, such a dynamic equilibrium requires a landscape with a large spatial extent compared to disturbance patch size (White and Pickett, 1985; Shugart, 1984; Pickett and Thompson, 1978). Spatial extent is thus itself a buffer, so that patchwise population changes do not remove species from the area as a whole. The smaller size and changed hydrology of the Everglades ecosystem thus makes the area doubly vulnerable.

CONCLUSIONS

The vegetation of the Everglades is affected by processes acting on a wide range of scales. In the long term, bedrock, climate, and sea level represent background variables that set the context of the ecosystem. Imposed on this context are a series of dynamic processes, with hydrology and fire being particularly important. In seeking to conserve nature, we are protecting systems that must change (White and Bratton, 1980). Natural dynamics have become an important problem to understand in a wide variety of conservation areas (e.g., Romme and Knight, 1981). The dynamic processes which produced Everglades vegetation must be used as a basis for management for the long-term survival and productivity of the landscape. Scale becomes a critical issue in terms of scale dependence in the processes measured, as well as in terms of the reduction in the spatial extent (about 50%) of this ecosystem. A key issue becomes whether the reduction in the size of the ecosystem will itself affect pattern and process in the ecosystem once natural processes are restored.

LITERATURE CITED

Bancroft, G. T., A. M. Strong, R. J. Sawicki, W. Hoffman, and S. D. Jewell. 1994. Relationships among wading bird foraging patterns, colony locations, and hydrology in the Everglades. in *Everglades: The Ecosystem and Its Restoration,* S. M. Davis and J. C. Ogden (Eds.), St. Lucie Press, Delray Beach, Fla., chap. 25.

Bodle, M. J., A. R. Ferriter, and D. D. Thayer. 1994. The biology, distribution, and ecological consequences of *Melaleuca quinquenervia* in the Everglades. in *Everglades: The Ecosystem and Its Restoration,* S. M. Davis and J. C. Ogden (Eds.), St. Lucie Press, Delray Beach, Fla., chap. 14.

Browder, J. A., P. J. Gleason, and D. R. Swift. 1994. Periphyton in the Everglades: Spatial variation, environmental correlates, and ecological implications. in *Everglades: The Ecosystem and Its Restoration,* S. M. Davis and J. C. Ogden (Eds.), St. Lucie Press, Delray Beach, Fla., chap. 16.

Connell, J. H. 1978. Diversity in tropical rain forests and coral reefs. *Science,* 199:1302–1310.

Davis, S. M. 1994. Phosphorus inputs and vegetation sensitivity in the Everglades. in *Everglades: The Ecosystem and Its Restoration,* S. M. Davis and J. C. Ogden (Eds.), St. Lucie Press, Delray Beach, Fla., chap. 15.

Davis, S. M., L. H. Gunderson, W. A. Park, J. Richardson, and J. Mattson. 1994. Landscape dimension, composition, and function in a changing Everglades ecosystem. in *Everglades: The Ecosystem and Its Restoration,* S. M. Davis and J. C. Ogden (Eds.), St. Lucie Press, Delray Beach, Fla., chap. 17.

DeAngelis, D. L. 1994. Synthesis: Spatial and temporal characteristics of the environment. in *Everglades: The Ecosystem and Its Restoration,* S. M. Davis and J. C. Ogden (Eds.), St. Lucie Press, Delray Beach, Fla., chap. 12.

DeAngelis, D. L. and P. S. White. 1994. Ecosystems as products of spatially and temporally varying driving forces, ecological processes, and landscapes: A theoretical perspective. in *Everglades: The Ecosystem and Its Restoration,* S. M. Davis and J. C. Ogden (Eds.), St. Lucie Press, Delray Beach, Fla., chap. 2.

Denslow, J. S. 1985. Disturbance-mediated coexistence of species. in *The Ecology of Natural Disturbance and Patch Dynamics,* S. T. A. Pickett and P. S. White (Eds.), Academic Press, New York, pp. 307–323.

Elton, C. S. 1958. *The Ecology of Invasions by Animals and Plants,* Chapman and Hall, London.

Gunderson, L. H. 1994. Vegetation of the Everglades: Determinants of community composition. in *Everglades: The Ecosystem and Its Restoration,* S. M. Davis and J. C. Ogden (Eds.), St. Lucie Press, Delray Beach, Fla., chap. 13.

Harmon, M. E., S. P. Bratton, and P. S. White. 1983. Disturbance and vegetation response in relation to environmental gradients in the Great Smoky Mountains. *Vegetatio,* 55:129–139.

Ogden, J. C. 1994. A comparison of wading bird nesting colony dynamics (1931–1946 and 1974–1989) as an indication of ecosystem conditions in the southern Everglades. in *Everglades: The Ecosystem and Its Restoration,* S. M. Davis and J. C. Ogden (Eds.), St. Lucie Press, Delray Beach, Fla., chap. 22.

Pickett, S. T. A. and J. N. Thompson. 1978. Patch dynamics and the design of nature reserves. *Biol. Conserv.,* 13:27–37.

Pickett, S. T. A. and P. S. White. 1985. *The Ecology of Natural Disturbance and Patch Dynamics,* Academic Press, New York.

Romme, W. H. and D. H. Knight. 1981. Landscape diversity: The concept applied to Yellowstone National Park. *BioScience,* 32:664–670.

Shugart, H. H. 1984. *A Theory of Forest Dynamics,* Springer, New York.

Watt, A. S. 1947. Pattern and process in the plant community. *J. Ecol.,* 35:1–22.

White, P. S. 1979. Pattern, process, and natural disturbance in vegetation. *Bot. Rev.,* 45:229–299.

White, P. S., 1987. Natural disturbance, patch dynamics, and landscape pattern in natural areas. *Nat. Areas J.,* 7:14–22.

White, P. S. and S. P. Bratton. 1980. After preservation: The philosophical and practical problems of change. *Biol. Conserv.,* 18:241–255.

White, P. S. and S. T. A. Pickett. 1985. Natural disturbance and patch dynamics: An introduction. in *The Ecology of Natural Disturbance and Patch Dynamics,* S. T. A. Pickett and P. S. White (Eds.), Academic Press New York, pp. 3–13.

IV

FAUNAL COMPONENTS
and PROCESSES

19

LONG-TERM DYNAMICS of
an EVERGLADES
SMALL-FISH ASSEMBLAGE

William F. Loftus

Anne-Marie Eklund

ABSTRACT

Only two long-term, quantitative studies of Everglades freshwater fishes have been conducted, both in the *Eleocharis* marshes of the southern Everglades within Everglades National Park. Each study attempted to describe the dynamics of the fish community with relation to hydrological conditions in the marshes. The earlier study, using pull-trap data from 1966 to 1972, has been published and widely cited; these data have been reanalyzed for comparisons in this chapter. In the second study presented here, fish data obtained from the pull traps were compared with those from throw traps to re-evaluate the conclusions of the original pull-trap study. The two long-term studies produced contradictory conclusions about the responses of small fishes to periods of prolonged flooding. The throw-trap results showed that annual mean densities of small fishes increased across the 1977–85 study period when the marshes did not dry. There was no evidence that predation by large fishes caused a reduction in small-fish densities during high-water periods. It is likely that small fishes use the densely vegetated marshes as refuges from predation. Fish community composition in the marshes did not shift from small to large species but remained stable. Shifts in assemblage dominance in the Everglades marshes may occur coincidentally with long periods of water level stability, but not within the temporal scale of the authors' study. The comparisons of pull-trap and throw-trap data sets collected concurrently demonstrate that pull-trap data do not accurately reflect small-fish community dynamics. Biases associated with the pull-trap method explain the conclusions of the first long-term study. The impacts of repeated drydowns on marsh fish communities are discussed, and the implications for marsh restoration are presented.

INTRODUCTION

The Everglades marsh system of southern Florida may be the most famous wetland in the world, its importance signified by its designation as a United States National Park, a United Nations Biosphere Reserve, and a World Heritage Site. Waters from this wetland sustain millions of people along the southeastern coast of Florida, provide the basis for successful agribusiness, and, despite extensive hydrological modifications, continue to support rich and diverse communities of plants and animals. The spectacular wading bird displays first attracted the attention of naturalists and plume hunters to the Everglades at the turn of this century, and it was mainly to preserve the bird populations that the southern segment of the original system was preserved as park land. The dependence of wading birds on freshwater fishes was recognized early, but the fishes remained poorly studied until recently.

Freshwater fishes are important components of the marsh system, filling roles in the aquatic food web ranging from primary consumers of vegetation and detritus, through intermediate levels as predators on aquatic insects and crustacea, to top-level carnivores such as Florida gar (*Lepisosteus platyrhincus*) and other piscivores. The fishes, in turn, are prey for a myriad of predators and scavengers. Gunderson and Loftus (1993) discussed and illustrated the role of fishes in aquatic food webs of the Everglades. The rapid life cycles of small Everglades fishes (Haake and Dean, 1983) enable them to respond rapidly to environmental changes, so that fishes may serve as indicators of change in habitat quality. The Everglades fishes and their life histories also provide insights into the ecological, zoogeographical, and evolutionary processes that have shaped the Everglades marsh system. By studying the responses of the fishes to varying environmental conditions, and through experimental manipulation, scientists may begin to understand and model ecosystem function and suggest appropriate water management actions to land managers.

The major motivation for the study of fish dynamics in the southern Everglades has been the search for an explanation for the sharp decline in the numbers and success of nesting wading birds during recent decades. The decline in wading bird nesting coincided with man-made hydrological changes to the system (Kushlan et al., 1975; Ogden, 1994). Frederick and Collopy (1988) suggested that the decline might be due to a reduction in the standing stocks or availability of prey fishes. It is probable that the disruption of the natural pattern of water quantity, hydroperiod, and flow distribution has affected the marsh fish community negatively.

REVIEW of PREVIOUS WORK

Loftus and Kushlan (1987) described the fish communities of the Everglades. They found an assemblage of 30 species in the freshwater marshes, all of which occurred in the spikerush (*Eleocharis* spp.) or wet prairie habitat. Small species of killifishes (Cyprinodontidae), livebearers (Poeciliidae), and juvenile sunfishes (Centrarchidae) were the common inhabitants of spikerush and sawgrass (*Cladium jamaicense*) habitats. The killifishes and livebearers are short-lived, rapidly growing species (Haake and Dean, 1983), which respond to favorable conditions with rapid increases in population. The deeper open-water alligator holes were used by larger fishes such as Florida gar, yellow bullhead (*Ameiurus natalis*), and adult sunfishes (*Lepomis* spp.), although smaller species including mosquitofish (*Gambusia holbrooki*) and sailfin molly (*Poecilia latipinna*) were also common. Fish communities similar to those of the Everglades were described from the neighboring Big Cypress Swamp (Figure 19.1) by Duever et al. (1986).

Information on Everglades marsh fishes is regrettably recent and, prior to the 1960s, mainly qualitative. Hunt (1952) studied the food habits of large and small fishes from the Tamiami Canal along U.S. Highway 41 (Figure 19.1) and presented information on the abundance of Florida gar. Reark (1961) was the first to collect fish density and biomass data with relation to vegetation cover. He compiled the only database, albeit very limited, from Shark River Slough before the construction of the S-12 water control structures (Figure 19.1) in 1962. Those floodgates placed control of southern marsh hydropatterns in the hands of water managers, resulting in significant departures in the timing and quantity of natural water flow and its spatial distribution (Johnson and VonHatten, in preparation). All subsequent fish data from the southern Everglades south of U.S. Highway 41 were collected after marsh hydrology was disturbed by the construction of the Central and Southern Florida Project (Rose et al., 1981; U.S. Army Corps of Engineers, 1982), which included the present system of water control structures. Qualitative data on the

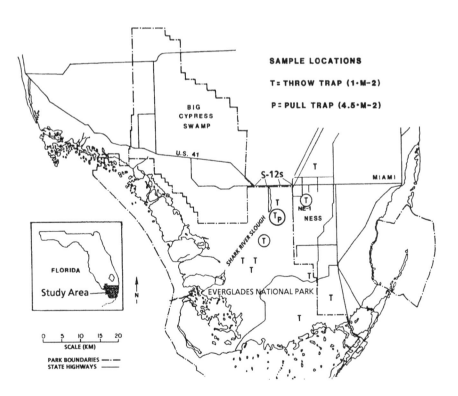

Figure 19.1 The southern Everglades, showing the locations of the pull-trap (P) and throw-trap (T) sites and water level recorders. Trap sites discussed in text are circled.

fishes of the northern Everglades, now impounded as the Water Conservation Areas, were presented by Dineen (1984); limited quantitative data have been published in annual reports of the Florida Game and Fresh Water Fish Commission (Dineen, 1968; Morello and Cook, 1984; Morello et al., 1985) and in recent studies from Water Conservation Area 1 (Wiechman, 1987; Chick et al., 1992). Comparable quantitative data on fish density and seasonal movements were collected from the Big Cypress Swamp by Carlson and Duever (1978).

From 1965 to 1972, the U.S. National Park Service contracted with the U.S. Geological Survey to design and implement a sampling program in Shark River Slough to relate changes in fish populations to hydrological conditions. The program, which used pull-trap gear, was described by Higer and Kolipinski (1967) and Kolipinski and Higer (1969). The results of the U.S. Geological Survey program were reported by Kushlan (1976, 1980) and Kushlan et al. (1975). In these papers, Kushlan concluded that, during a 27-month period without a drydown, fish densities decreased, but fish biomass, mean fish size, species diversity, and species richness increased. In examining the fish community composition, Kushlan (1976) described a functional shift in the trophic structure of the community during the wet period, from one dominated by small, omnivorous livebearers and killifishes to a large-fish community of carnivorous sunfishes and catfish. The wet period presumably fostered the survival of larger fishes. Kushlan (1976) concluded that the Everglades fish community shifts between opposing tendencies of species domination, depending on the degree of water level stability; the causal mechanism is increased predation by larger fishes when the environment is stable. The pull-trap data suggested that the greatest numbers of small prey fishes would be produced if the Everglades marsh were managed for frequent drydowns (Kushlan, 1987).

Following the U.S. Geological Survey study, the National Park Service supported an 8-year investigation of fish dynamics with relation to hydrology, using the more accurate 1-m^2 throw trap developed and tested by Kushlan (1981). The results of that 1977–85 study are reported in this chapter. The objective here is to compare simultaneously collected data from pull traps and throw traps during a high-water period to determine whether the two methods produce similar estimates of community dynamics. Data from the original pull-trap study (Kushlan, 1976, 1980) are re-examined to demonstrate how the biases of that method affected the conclusions reported in those papers. Also discussed is why management strategies derived from the pull-trap data (Kushlan, 1987) would result in marshes having reduced carrying capacities for aquatic animals. That scenario is demonstrated here by presenting fish assemblage data taken in an area subjected to frequent drying episodes.

METHODS

Study Sites

Sampling sites for this study and the earlier pull-trap program were located in the central region of Shark River Slough, the major drainage channel of the southern Everglades in Everglades National Park. Hydrological data for both studies were collected at gage P-33 (Figure 19.1), a continuous-recording instrumentation platform within 3 km of the biological stations. Hydrological data were also examined from short-hydroperiod marshes in Northeast Shark River Slough using continuous data from gage NE-1 (Figure 19.1).

The *Eleocharis* habitat comprising the wet prairies of the southern Everglades, although not as spatially extensive as sawgrass marshes, is used more by fishes and wading birds (Hoffman et al., 1994; Gunderson and Loftus, 1993). Emergent plant stem densities are lower than in sawgrass habitats, and the *Eleocharis* prairies retain standing water longer into the dry season. Major plant species in the wet prairies are spikerushes (mainly *Eleocharis cellulosa*), maidencane (*Panicum hemitomon*), beakrush (*Rhynchospora tracyi*), and arrowheads (*Sagittaria* spp.). Floating mats of bladderworts (*Utricularia* spp.), sometimes covered by thick periphyton, grow among the emergent stems. Organic peat soils are found in the long-hydroperiod wet prairies of the central Everglades, while marls or calcium carbonate sediments occur in marshes with shorter flooding periods (Gleason et al., 1984). The hydrology of the marsh system is complex, with rainfall and water levels varying greatly both seasonally and interannually. Water depths in wet prairies are usually shallow (<0.75 m), and the period of flooding each year depends on antecedent conditions, water management actions, and the local topography. During the dry season from November to May, some or all wet prairies may dry. This condition forces small fishes into deeper pockets of water, where they are vulnerable to resident piscivorous fishes or other predators. Those fishes remaining in the drying marshes are exposed to wading bird predation or desiccation. The severity of the dry season is the major abiotic factor determining the composition of the fish community and its abundance during the following year (Loftus and Kushlan, 1987). Because the wet prairies support large fish populations and attract feeding wading birds, and because the habitat is amenable to long-term quantitative sampling, studies of Everglades fishes have been concentrated within that habitat.

Sampling Regime

The first pull-trap study, from 1965 to 1972, utilized ten pull traps in northern Shark River Slough, sampled on two consecutive nights each month (*n* = 20 per month) (Figure 19.2). Unfortunately, the database contains many months toward the end of that period when samples were not collected. Each trap consisted of sheets of 3-mm bar nylon mesh affixed to a metal frame measuring 1.5 m × 3.0 m, sampling an area of 4.5 m². Operators at each end of the trap pulled on lines that lifted the net from the water, capturing animals swimming on or above the net. Two of the original ten pull traps (traps 3 and 4 in this chapter) were retained on site in Shark River Slough and were pulled monthly after 1972 to monitor the fish community. Because Kushlan (1974) discerned the biases inherent in the stationary pull traps, he devised and tested the 1-m² throw trap (Kushlan, 1981) that was employed from 1977 to 1985. The 70-cm-high metal frame trap was enclosed by 3.2-mm nylon netting and weighed approximately 9 kg.

Although fish community data have been collected with throw traps at 13 marsh sites throughout the Everglades, data are presented here only from the two sites in central Shark River Slough located nearest the original pull-trap sites (Figure 19.1). These sites, called upper slough and middle slough, were estimated to have had mean annual hydroperiods (length of time with surface water) greater than

	P	P	T
SAMPLE PERIOD	1965 – 1972	1972 – 1985	1977 – 1985
TIME OF DAY	NIGHT	DAY	DAY
FREQUENCY	TWICE MONTHLY	MONTHLY	MONTHLY
SUB–SAMPLES	20	2	7 – 15
AREA OF SAMPLE (SQUARE METERS)	4.5	4.5	1.0

Figure 19.2   Summary of the sampling regimes used in the two long-term fish studies. P = pull trap and T = throw trap.

11.5 months/year for several decades. Each month, between 7 and 15 traps (subsamples) were collected at each site; the sample size was calculated according to the procedure outlined by Kushlan (1981). Fishes were removed by using small-mesh (0.8-mm bar) and large-mesh (5.5-mm bar) dip nets until ten consecutive sweeps produced no additional specimens. Throw-trap sampling data collected from a shorter hydroperiod site in Northeast Shark River Slough between 1978 and 1985 are also presented to show how frequent, dry periods affect fish assemblages.

Data Analysis

Pull Traps

Analytical procedures used in the early pull-trap study were explained by Kushlan (1976, 1980). The original pull-trap database was accessed to calculate the monthly and annual means and variances for comparisons. Annual means were calculated for biological and hydrological data according to water years (which extend from June to May) and tested for differences among the annual means for density, wet biomass, and fish size (Kruskal–Wallis test, SPSS, Inc., 1988), followed by a post-hoc, multiple range test (Zar, 1984). The monthly data from pull traps 3 and 4, collected concurrently with throw-trap samples from 1977 to 1985, were averaged to provide a monthly estimate of mean density, wet biomass, and relative abundance.

Throw Traps

Fishes from the throw traps were measured and weighed individually for the first several years. The subsequent generation of length-mass equations (Kushlan et al., 1986) eliminated the need to weigh each specimen. Monthly density, wet biomass, and dry biomass estimates were calculated for both throw-trap sites, in addition to estimates of annual and seasonal densities and biomass. Correlation analysis (Spearman Correlation, SPSS, Inc., 1988) was used to test the stability of fish densities over time, and statistically significant differences were also tested among annual mean densities and annual mean biomass (Kruskal–Wallis test, SPSS, Inc., 1988). To examine the stability and persistence of the fish assemblages taken by throw traps and pull traps, annual relative abundance data were calculated. For multiple comparisons of species composition across years, Kendall's coefficient of concordance (W) was used, the significance of which was tested using X^2 values (Siegel, 1956).

RESULTS

Hydrological Data

Water depths in the Everglades fluctuated seasonally and annually, as indicated by central Shark River Slough gage P-33 (Figure 19.3). The closing of the S-12 water control gates north of Everglades National Park in 1962 resulted in 4 successive dry years immediately preceding the original pull-trap study. During the 6 years of that study, water fell below ground surface on two occasions (1966–67 and 1970–71). The 3-year period, from 1967 to 1970, was one without a drydown. Dry conditions returned in the winter of 1970–71 (Figure 19.3). Kushlan (1976) stated that the flooding period was 27 months because water fell below ground during 1967–68. The discrepancy between his estimate of flooding time and the authors' resulted from his use of the incorrect ground-surface elevation for gage P-33. Kushlan

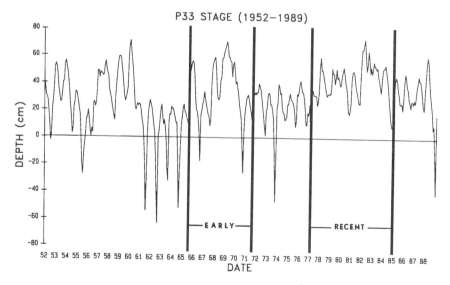

Figure 19.3 Period-of-record water depths at central Shark River Slough recording gage P-33, showing time periods of the two long-term fish studies.

(1976) showed the ground surface at approximately 1.8 m above mean sea level. Because the actual ground surface elevation is 1.56 m, Shark River Slough did not dry in 1967–68; therefore, the flooding period estimate should have been longer (Figure 19.3).

The second study took place from the late 1970s to mid 1980s (Figure 19.3), during a period of prolonged high water brought on by a combination of high rainfall and large water releases to Everglades National Park. Although gage P-33 registered only a moderate decline in depth in 1981, the middle slough throw-trap site actually dried for several days in May. Most Shark River Slough marshes dried completely in May 1985, ending the prolonged high-water period.

Throw-Trap Data

The monthly throw-trap samples produced a pattern of seasonal and interannual fluctuations in density during the 8 years of high water (Figure 19.4). However, there was an increasing trend in fish densities at the upper and middle slough sites over that period, from annual means (±1 SE) of 15.5 ± 1.6 and $17.1 \pm 1.2 \cdot m^2$ in 1977–78 to 30.2 ± 2.8 and $34.5 \pm 2.5 \cdot m^2$ in 1983–84, respectively. The correlation analysis of fish density versus time indicated strong positive relationships among increasing densities and years ($r = 0.90$, $P < 0.01$; $r = 0.892$, $P < 0.03$) for the upper and middle slough sites, respectively. At both sites, annual mean densities were significantly greater later in the study (Kruskal–Wallis: upper slough $X^2 = 25.8127$, $P = 0.0001$; middle slough $X^2 = 33.391$, $P < 0.00001$). Although the sites were separated by 10 km, monthly densities at the sites tracked closely across time, indicating similar responses by the fishes to environmental conditions. The exception occurred in May 1981 (Figure 19.4), when high densities were found at the

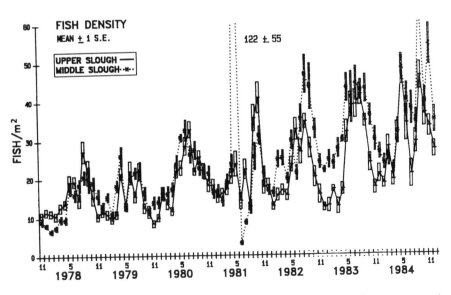

Figure 19.4 Monthly means and standard errors for fish densities from upper and middle Shark River Slough throw-trap sites.

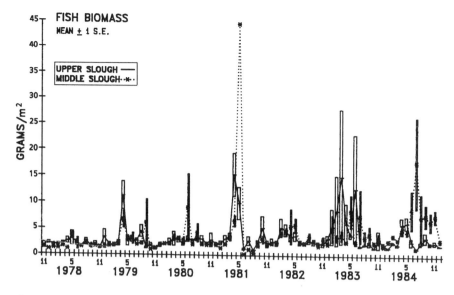

Figure 19.5 Monthly means and standard errors for fish wet biomass from upper and middle Shark River Slough throw-trap sites.

middle slough site but not at the upper slough site. This peak resulted from the concentration of fishes in shallow puddles at the middle slough site, which dried earlier and more completely than the upper slough site.

Annual means for wet biomass at both sites ranged from 2.3 to 4.5 g·m (Figure 19.5). Annual mean biomass increased significantly during the study at the middle slough site (Kruskal–Wallis, $P = 0.05$), but not at the upper slough site ($P > 0.05$). The highest biomass occurred during the concentration event at the middle slough site in May 1981 (Figure 19.5). The variability in biomass estimates increased toward the end of the time series. The fish-size data (Figure 19.6) revealed that much of the variability in biomass was due to the presence of some heavier fishes in the samples. However, average fish size remained small, varying from 0.1 to 0.4 g wet mass per fish. Within years, fish size was usually greatest in late winter and spring (Figure 19.6), before the intense recruitment of juveniles began. Species richness at the upper slough site began with 11 species in 1977–78, reaching a high of 15 species in 1982–83. Species richness at the middle slough site was more constant at 10 to 12 species from 1977–83, reaching a maximum of 13 species in 1984.

Pull-Trap Data

The results of the statistical re-analysis of annual densities, biomass, and fish size data from the earlier study generally matched Kushlan's (1976) results. The highest annual densities occurred in the years with the driest conditions, except in 1970–71 (Figure 19.7a). Influenced by the 3-year absence of a drydown, fish community parameters appeared to change. Annual mean densities declined during that period (Figure 19.7a) and did not rebound immediately when water depth fluctuation returned in 1970–71. Annual mean density did not reach high

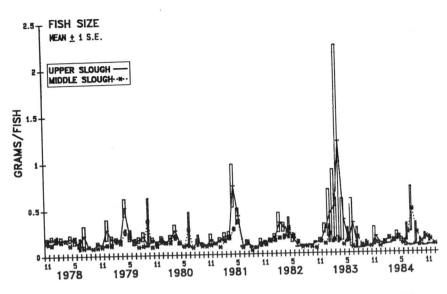

Figure 19.6   Monthly means and standard errors for fish size from upper and lower Shark River Slough throw-trap sites.

levels until 1971–72, which Kushlan (1976) interpreted as confirmation that population levels of small fishes were reduced during high water and required a long time to rebuild. Coincident with the density decline, annual means for wet biomass and fish size increased during high water and then declined with the resumption of the spring drydown (Figure 19.7b and c).

Data from the first pull-trap study (Kushlan, 1976) were not directly comparable to the authors' throw-trap data because of the difference in sampling periods and possibly because the original pull-trap data were collected at night (Figure 19.2). However, comparisons of assemblage responses to hydrological conditions were possible. The authors examined differences in those responses and in species composition by comparing throw-trap data with simultaneously collected pull-trap data from 1977 to 1985. Because density estimates from traps 3 and 4 did not differ significantly from estimates from traps 1 through 10 in the original database (Kolmorgorov–Smirnov test, $P > 0.1$; SPSS, Inc., 1988), it was assumed here that data from traps 3 and 4 from 1977 to 1985 represented what the entire pull-trap array would have produced.

Pull-trap data from 1977 to 1985 (Figure 19.8) showed few discernible trends in density. Monthly pull-trap density estimates were usually much lower than those from nearby throw-trap samples (Figure 19.8) and the patterns of fluctuation did not track closely. The lowest density estimates from the two pull traps occurred in 1984. Unlike the results of the 1965–72 study, in which densities on pull traps did not rebound with the first drydown following the high-water period (Kushlan, 1976), the highest monthly densities in the 1977–85 pull-trap data were recorded during the 1985 drought. The increase indicated that fish population levels were high in the marshes surrounding the traps during the high-water period, leading to fishes concentrating on the traps as the marshes dried. As in the 1965–72 study, the largest fishes in the 1977–85 pull-trap samples were taken near

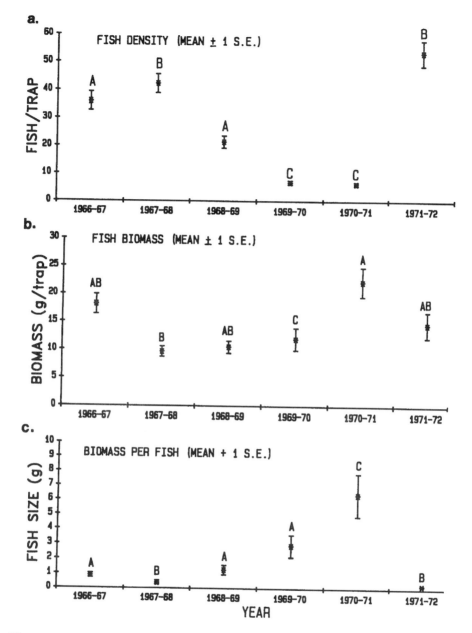

Figure 19.7   (a) Annual means and standard errors for fish densities from the first pull-trap study; (b) annual means and standard errors for wet biomass from the first pull-trap study; and (c) annual means and standard errors for fish size from the first pull-trap study. Similar letters above error bars indicate no significant differences (*P* > 0.05) between means.

the end of the high-water period (Figure 19.9). Species richness increased from 10 species in 1977–78 to 17 species in 1984–85, when the drying of the surrounding marshes concentrated all species into the depressions created by the pull traps (Figure 19.10).

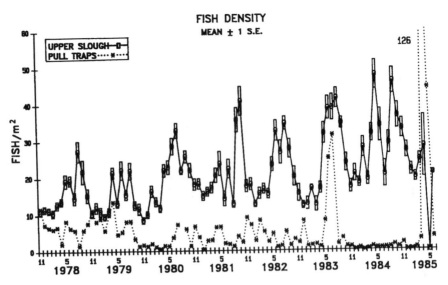

Figure 19.8   Comparison of monthly mean fish densities from concurrently sampled pull traps and throw traps in upper Shark River Slough.

The density increases in the throw-trap samples resulted from increases in the small omnivore/carnivore assemblage of livebearers and killifishes which composed 95% or more of the monthly samples. Comparisons of annual species composition from the two throw-trap sites and the pull traps revealed significant differences in the assemblages sampled (Figure 19.10). Assemblages at both throw-trap sites were similar during every year (Figure 19.10), both in composition and

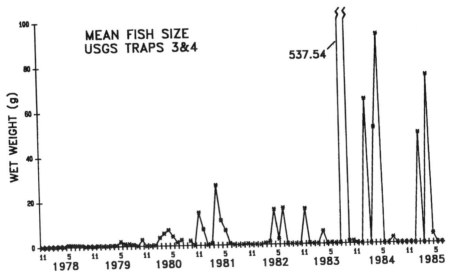

Figure 19.9   Mean monthly fish size from pull traps 3 and 4 in upper Shark River Slough.

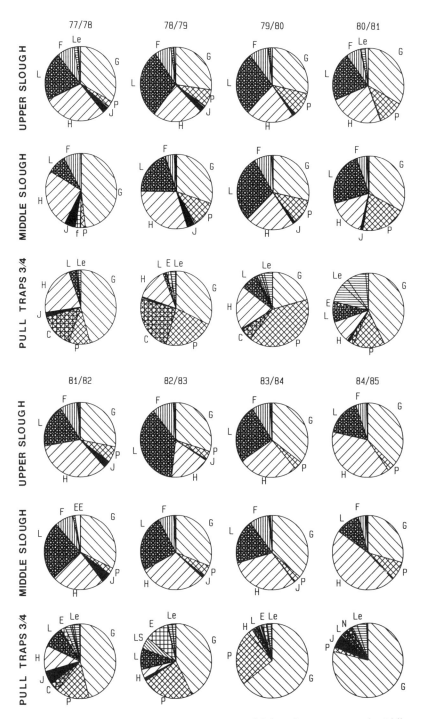

Figure 19.10 Annual relative abundances of fishes from upper and middle slough throw-trap sites compared with those from pull traps 3 and 4 (C = *Cyprinodon variegatus*, E = *Erimyzon sucetta*, EE = *Elassoma evergladei*, f = *Fundulus confluentus*, F = *Fundulus chrysotus*, G = *Gambusia holbrooki*, H = *Heterandria formosa*, J = *Jordanella floridae*, L = *Lucania goodei*, Le = *Lepomis* spp.; LS = *Labidesthes sicculus*, N = *Notropis petersoni*, P = *Poecilia latipinna*).

relative abundances. The composition of those assemblages was relatively stable across time (Kendall's W = 0.91 and 0.90, P < 0.001, respectively, for the upper and middle slough sites), although there was some shifting of ranks by *Lucania* and *Heterandria* (Figure 19.10). There was little correspondence of assemblage composition and abundance data among the throw traps and pull traps. Pull-trap data were more variable among years and the assemblage less persistent (Kendall's W = 0.59, P < 0.001). Note that *Cyprinodon variegatus* was a major assemblage element on the pull traps in some years, yet it was not collected by throw trapping in the adjacent marshes.

DISCUSSION

The major differences between the two long-term studies relate to the responses of the small-fish assemblage to long periods of marsh flooding. The earlier study (Kushlan, 1976) showed that the densities of small fishes, which numerically dominated the fish community during fluctuating water conditions, declined sharply if the marshes did not dry. Kushlan (1976) also concluded that periods of water level stability as short as 2 to 3 years could result in a shift in assemblage dominance from small to large species. Kushlan (1976, 1980, 1987) implied that the mechanism behind the decline in the small fishes during high water was predation by the increasingly numerous large fishes, which had increased survival rates in the absence of drought.

The throw-trap study coincided with an 8-year period of prolonged flooding. If the conclusions of the earlier study were correct, a longer flooding period should have produced even more dramatic shifts toward larger species because of the increased time span of flooding (8 versus 3 years). The two pull traps sampled during the second study did produce results similar to those of the earlier study. For example, the densities of sunfishes (*Lepomis* spp.) and other large fishes increased as the high-water period progressed (Figures 19.9 and 19.10). However, the throw-trap data showed a very different pattern of abundance and composition. Small-fish densities in marshes increased significantly, and the species composition, biomass, and size data provided no evidence for a shift in community dominance toward larger species. Because the small-fish assemblage remained stable over time, there was also no evidence for a functional shift in the marsh food web. The differences between the studies may be explained by the biases inherent in the pull-trap method and in the analytical methods used in the early study (Kushlan, 1976).

Pull traps are stationary devices that alter their sampling sites from the outset. To deploy the traps, all rooted and floating vegetation must be removed before the trap can be placed on the bottom of the marsh (Higer and Kolipinski, 1967). Repeated use of the traps lowers the soil elevation of the trap sites (the traps now rest about 25 cm below the surrounding marsh surface), and years of foot traffic to reach the traps has produced deep cuts in the marsh bottom (Figure 19.11). The submerged netting is rapidly colonized by periphyton on traps set in the open marsh. Rather than making the trap appear more natural (Kolipinski and Higer, 1969), algae on the traps begins to selectively attract species that inhabit open

Figure 19.11 Photograph of pull trap 3 in upper Shark River Slough during the 1985 drydown, illustrating the depression formed at the trap site in the marsh.

habitats with lightly colored substrates; hence the collection of mainly *Cyprinodon variegatus* on pull traps. The zone of vegetation-free deeper water also attracts a different species assemblage, such as sunfishes and silversides (*Labidesthes sicculus*), but is unattractive to the small species that require cover. Those small, cryptic species remain in the densely vegetated marshes and are not sampled effectively by the pull traps at high water. The traps also sample less effectively when the water is deep because periphyton growths occlude the mesh. The traps are then difficult to lift rapidly, allowing fishes to escape. Because of these factors, the assemblage data produced by pull traps during high water are not characteristic of the surrounding marshes.

Pull-trap sites function similarly to small alligator holes (Loftus and Kushlan, 1987) (Figure 19.11). During dry periods (Figure 19.12), the fishes are forced into the deeper trap sites by receding water levels on the marsh surface. The highest density and biomass values on the traps usually correspond to the lowest water levels in the surrounding marshes. Because the early study (Kushlan, 1976) used comparisons of annual mean densities calculated from monthly data, the seasonal biases of the pull trap are very significant. Those years with the lowest water depths had the highest annual mean densities (e.g., 1967–68 and 1971–72), because many fishes moved into the trap sites. Conversely, in high-water years, when small fishes remained in the surrounding marshes, annual mean densities were very low (1969–70). This bias may have led to the erroneous conclusion that low-water years were necessary to produce the highest fish population levels (Kushlan, 1976, 1987). In reality, the pull-trap data bear little relation to actual abundance levels in the surrounding marshes, as shown by the throw-trap data. That fact that the highest densities in the 1977–85 pull-trap database occurred as the marshes dried at the end

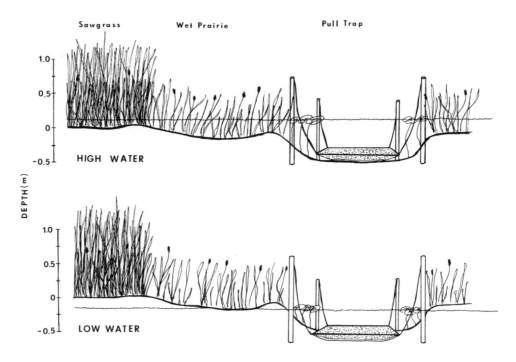

Figure 19.12 Stylized depiction of relative water depths in marsh habitats and pull-trap sites at high and low water.

of the long high-water period demonstrated that fish densities were very high in the surrounding marshes (Figure 19.8).

A second source of bias in the analyses and conclusions of the early study resulted from the treatment of the sample data. Sample sizes for each year varied widely and became progressively smaller with time (Figure 19.13). Because fish densities are highly seasonal, a missing sample for a high-density month would have a large damping effect on the estimated annual mean for that year. Therefore, the annual means for the final two years (1971–72) may have been skewed by missing samples (Figure 19.13).

Although throw-trap data are biased against larger individuals and sparsely distributed species (Kushlan, 1981; Jacobsen and Kushlan, 1987), those biases should not affect the conclusions of the present study concerning the small-fish assemblage. Based on throw-trap sampling, the authors agree with Kushlan (1976) and Loftus and Kushlan (1987) that large-fish species are favored by prolonged high-water periods which foster higher reproductive success and survival. High numbers of large fishes were collected in multiple samples in Shark River Slough alligator ponds until the 1985 drydown (W. F. Loftus et al., unpublished data). Thus, although large fishes were present in high numbers in Shark River Slough during the present study, small-fish densities in the marshes increased. This finding is contrary to the earlier concept that fish predation controls the development of the Everglades small-fish assemblage during long flooding periods (Kushlan, 1976). The authors' data do not suggest that small-fish numbers in the marshes are regulated by larger fishes at high water. The large fishes that increased most under

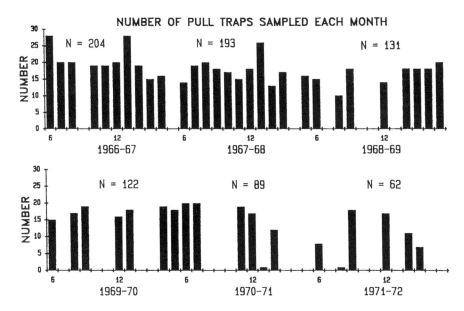

Figure 19.13   Histogram of pull-trap subsamples taken each month during the original pull trap study (Kushlan, 1976), showing months during which no samples were taken.

high-water conditions, sunfishes and yellow bullheads, are mainly invertebrate feeders and are normally rather poor piscivores (Gunderson and Loftus, 1993; Carlander, 1969, 1977). Thus, increases in larger fishes do not necessarily translate into increased predation on smaller fishes in the marshes.

Large, predatory fishes feed most heavily on small fishes during the dry season, when the small fishes are forced into deep-water habitats (Loftus and Kushlan, 1987). The authors believe that the densely vegetated marshes inhibit movement and feeding by large fishes, thereby acting as predator refuges for small killifishes, livebearers, and juvenile sunfishes, similar to the function of lake littoral zones (Werner et al., 1977). Data from two sets of marsh rotenone samples and from the back-calculation of large-fish densities from alligator pond samples (W. F. Loftus et al., unpublished data) produced estimates ranging from 0 to 5 fish, greater than 40 mm S.L., per 100 m^2 of *Eleocharis* marsh. Those data imply that larger fishes are relatively uncommon in the marshes. Sampling in and around deeper, open-water habitats, such as ponds, airboat trails, and canals, has shown that these are the important habitats for larger Everglades fishes (Dineen, 1968; Loftus and Kushlan, 1987; Wiechman, 1987; Loftus, 1988).

The throw-trap data revealed that a relatively stable small-fish assemblage inhabited Shark River Slough *Eleocharis* marshes during the 1977–85 wet period (Figure 19.10). Because the apparent change in assemblage dominance reported by Kushlan (1976) was an artifact of the pull-trap method, if such shifts in composition do occur, a different set of conditions or a longer time span must be required. Based on studies of fishes in the northern Everglades, it appears that the Everglades fish community can respond to different environments by producing different species assemblages. The Water Conservation Areas north of Everglades National Park contain thousands of hectares of impounded marsh, some areas of which had been flooded as much as 2 m deep for more than a decade. In addition to hydrological differences between the Water Conservation Areas and southern Everglades marshes, the vegetation community patterns differed in species composition and stem density (Wood and Tanner, 1990; W. F. Loftus, unpublished data). Water Conservation Area marshes are often classified as white water lily sloughs (Loveless, 1959), with less emergent vegetation and more open-water areas than in southern Everglades marshes (Wood and Tanner, 1990). Limited fish data from the Water Conservation Areas (Dineen, 1968; Morello and Cook, 1984; Morello et al., 1985; Wiechman, 1987; W. F. Loftus, unpublished data) indicate that species composition in some areas differs from Shark River Slough by the presence of larger, heavier species that form most of the biomass. Small-fish density data collected by throw traps in Water Conservation Area 3A from 1979 to 1981 were lower than in Shark River Slough (W. F. Loftus, unpublished data), usually fewer than 12 fish per square meter. The authors hypothesize that it is the ponded character of the Water Conservation Area marshes, in conjunction with the extended hydroperiod and deeper waters, that explains the higher population levels of larger fishes.

The results of the two long-term studies suggest very different options for management of Everglades marshes. The throw-trap study demonstrated that high densities of small species were produced during a long period without a drydown. The large-fish assemblage in alligator ponds also increased during the same period (W. F. Loftus et al., unpublished data). In contrast, the early pull-trap study (Kushlan, 1976) incorrectly indicated that small-fish populations were highest during dry years and declined when marshes remained flooded for 2 or 3 years. The conclusions of that study were used to suggest an idealized hydrograph by which the Everglades might be managed (Kushlan, 1987). The hydrograph emphasized yearly drydowns to prevent shifts in fish community composition and to provide high fish densities for feeding wading birds.

To assess the potential impacts of repeated annual drydowns on fishes, it would be necessary to examine data taken under such conditions. Unfortunately, neither of the long-term databases encompassed a period with more than two successive annual drydowns. However, throw-trap samples taken in the marshes of Northeast Shark River Slough provided insight into the impacts of successive drydowns on fishes. In 1962, the construction of the S-12 water control structures along U.S. Highway 41 diverted water flows from Northeast Shark River Slough, reducing the average annual hydroperiod by several months (Loftus et al., 1990). Before that time, the marshes of Northeast Shark River Slough would have been characterized by long hydroperiods, as shown by the model of unmanaged hydrology (Fennema et al., 1994), and would have remained flooded in most years because the deepest channel of the southern Everglades runs through Northeast Shark River Slough (Everglades National Park, topographic data). Historical soil surveys found extensive areas of peat soils in Northeast Shark River Slough (Gallatin et al., 1958), confirming a history of the long hydroperiods required to produce and maintain peats (Gleason et al., 1984). The hydrograph from gage NE-1 in central Northeast Shark River Slough (Figure 19.1) illustrates the high frequency of drydowns from 1977 to 1985 (Figure 19.14). Fish densities were usually much lower than those from the upper Shark River Slough samples. Even after Northeast Shark River

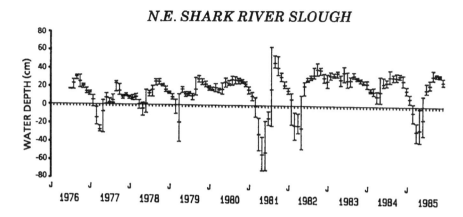

Figure 19.14   Hydrograph from gage NE-1 in Northeast Shark River Slough.

Slough entered a period of prolonged flooding from 1982 to 1985 (Figure 19.14), fish densities remained near the same levels as before (Figure 19.15). Loftus et al. (1990) hypothesized that the years of successive drydowns had reduced the carrying capacity of Northeast Shark River Slough marshes because of effects on the detrital food web. They described a shift in Northeast Shark River Slough soils from the historical peats reported by Gallatin et al. (1958) to organically poor marl soils (calcitic muds) characteristic of shorter hydroperiod marshes (Gleason et al., 1984). Increased spatial sampling across Northeast Shark River Slough (Loftus et al., 1990) showed that low fish densities were typical of that region of the Everglades.

The hydrological changes of recent decades also have affected the top predators of Northeast Shark River Slough. Piscivorous wading birds feed at lower densities

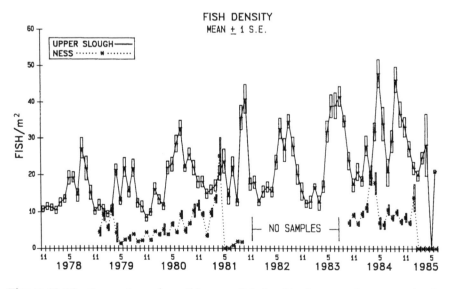

Figure 19.15   Comparison of monthly mean fish densities between throw-trap sites in upper Shark River Slough and Northeast Shark River Slough.

there than would be predicted by the area of marsh available (Fleming et al., in review), possibly because of insufficient prey densities. The number of alligator holes per square kilometer, and presumably the density of alligators, is but a fraction of that found in Shark River Slough (Loftus et al., 1990). The low density of alligator ponds in Northeast Shark River Slough affects fishes by reducing the availability of dry season refugia and the critical habitat for the large-fish assemblage, limiting the number of fishes surviving to recolonize during the next wet season.

The present-day Everglades suffers from the mismanagement of marsh hydrology and the loss or degradation of large areas of marsh, especially along the eastern edges of the system (Gunderson and Loftus, 1993; Loftus et al., 1992). The areal reduction of peripheral marsh habitats due to drainage and development may be critical to the capacity of the system to support large numbers of predators that feed on marsh fishes. The biomass of fishes per unit area is low, even in long-hydroperiod marshes, but the total fish biomass produced across the vast area of the original Everglades was probably many times higher than today. It should be possible to restore the range of Everglades wetlands that have been degraded by water management, from formerly important long-hydroperiod marshes such as Northeast Shark River Slough to drained segments of peripheral marsh. However, restoration activities must be based on the most accurate scientific information available. If the southern Everglades is managed for annual drydowns, the result may be a system similar to present-day Northeast Shark River Slough. The most accurate modeling of the hydrology of the unmanaged Everglades (Fennema et al., 1994) demonstrated the long-hydroperiod nature of central Everglades marshes. Those marshes would have dried completely only during infrequent, severe droughts. Water levels in the original Everglades certainly fluctuated seasonally, helping to concentrate fishes especially along the higher edges of the system. However, it is most unlikely that the original system experienced frequent and complete annual drydowns throughout its history or that such conditions would have sustained a persistent and diverse fish community. The results of the throw-trap sampling agree well with the output from the unmanaged hydrology model in suggesting that both large- and small-fish assemblages respond positively to periods without severe drydowns. In short, fishes do best where there is water!

ACKNOWLEDGMENTS

The authors would like to recognize the efforts of the many sturdy volunteers and technicians, too numerous to mention by name, who assisted in the field collections, sample processing, and data entry for this study. We are grateful to J. A. Kushlan for introducing the senior author to the Everglades and to the throw-trap method. D. Hall of Florida International University provided invaluable help and many long hours in the data summary process using the SIR system. Additional computer work was done by D. Buker and D. Henry. The manuscript was reviewed and improved by comments from R. G. Gilmore, C. C. McIvor, C. F. Jordan, J. C. Ogden, and W. B. Robertson, Jr.

LITERATURE CITED

Carlander, K. D. 1969. *Handbook of Freshwater Fishery Biology,* Vol. 1, The Iowa State University Press, Ames.

Carlander, K. D. 1977. *Handbook of Freshwater Fishery Biology,* Vol. 2, The Iowa State University Press, Ames.

Carlson, J. E. and M. J. Duever. 1978. Seasonal fish population fluctuations in a south Florida swamp. in *Cypress Wetlands for Water Management, Recycling and Conservation,* H. T. Odum and K. C. Ewel (Eds.), 4th annual report to National Science Foundation and Rockefeller Foundation, Center for Wetlands, Gainesville, Fla., pp. 682–705.

Chick, J. H., F. Jordan, J. P. Smith, and C. C. McIvor. 1992. A comparison of four enclosure traps and methods used to sample fishes in aquatic macrophytes. *J. Freshwater Ecol.,* 7:353–361.

Dineen, J. W. 1968. Determination of the Effects of Fluctuating Water Levels on the Fish Population of Conservation Area III, Final report by Florida Game and Fresh Water Fish Commission, Dingell-Johnson Project F-16-R.

Dineen, J. W. 1984. The fishes of the Everglades. in *Environments of South Florida: Present and Past II, Memoir No. II,* P. J. Gleason (Ed.), Miami Geological Society, Coral Gables, Fla., pp. 258–268.

Duever, M. J., J. E. Carlson, J. F. Meeder, L. C. Duever, L. H. Gunderson, L. A. Riopelle, T. R. Alexander, R. L. Myers, and D. P. Spangler. 1986. The Big Cypress Preserve, Res. Rep. #8, National Audubon Society, New York.

Fennema, R. J., C. J. Neidrauer, R. A. Johnson, T. K. MacVicar, and W. A. Perkins. 1994. A computer model to simulate natural Everglades hydrology. in *Everglades: The Ecosystem and Its Restoration,* S. M. Davis and J. C. Ogden (Eds.), St. Lucie Press, Delray Beach, Fla., chap. 10.

Fleming, D. M., J. Schortemeyer, W. Hoffman, and D. L. DeAngelis. In review. Colonial wading bird distribution and abundance in the pre- and post-drainage landscapes of the Everglades. *Oecologia* (submitted).

Frederick, P. C. and M. W. Collopy. 1988. Reproductive Ecology of Wading Birds in Relation to Water Conditions in the Florida Everglades, Tech. Rep. No. 30, Florida Cooperative Fish and Wildlife Research Unit, School for Forest Resource Conservation, University of Florida, Gainesville.

Gallatin, M. H., J. K. Ballard, C. B. Evans, H. S. Galberry, J. J. Hinton, D. P. Powell, E. Truett, W. L. Watts, and G. C. Willson, Jr. 1958. Soil Survey (Detailed Reconnaissance) of Dade County, Florida, Soil Conservation Service Report, Series 1947, No. 4, U.S. Department of Agriculture, Washington, D.C.

Gleason, P. J., A. D. Cohen, P. Stone, W. G. Smith, H. K. Brooks, R. Goodrick, and W. Spackman, Jr. 1984. The environmental significance of holocene sediments from the Everglades and saline tidal plain. in *Environments of South Florida: Present and Past II. Memoir No. II,* 2nd ed., P. J. Gleason (Ed.), Miami Geological Society, Coral Gables, Fla., pp. 297–351.

Gunderson, L. H. 1989. Historical hydropatterns in wetland communities of Everglades National Park. in Freshwater Wetlands and Wildlife, R. R. Sharitz and J. W. Gibbons (Eds.), Savannah River Ecology Laboratory, U.S. Department of Energy, Aiken, S.C., pp. 1099–1111.

Gunderson, L. H. and W. F. Loftus. 1993. The Everglades. in *Biodiversity of the Southeastern United States: Terrestrial Communities,* W. H. Martin, S. G. Boyce, and A. C. Echternacht (Eds.), John Wiley & Sons, New York, chap. 6..

Haake, P. W. and J. M. Dean. 1983. Age and Growth of Four Everglades Fishes Using Otolith Techniques, Tech. Rep. SFRC-83/03, Everglades National Park, Homestead, Fla.

Higer, A. L. and M. C. Kolipinski. 1967. Pull-up trap: A quantitative device for sampling shallow-water animals. *Ecology,* 48:1008–1009.

Hoffman, W., G. T. Bancroft, and R. J. Sawicki. 1994. Foraging habitat of wading birds in the Water Conservation Areas of the Everglades. in *Everglades: The Ecosystem and Its Restoration,* S. M. Davis and J. C. Ogden (Eds.), St. Lucie Press, Delray Beach, Fla., chap. 24.

Hunt, B. P. 1952. Food relationships between Florida spotted gar and other organisms in the Tamiami Canal, Dade County, Florida. *Trans. Am. Fish. Soc.,* 82:13–33.

Jacobsen, T. and J. A. Kushlan. 1987. Sources of sampling bias in enclosure fish trapping: Effects on estimates of density and diversity. *Fish. Res.,* 5:401–412.

Johnson, R. A. and S. C. VonHatten. In preparation. A Review of Water-Management Impacts in the Everglades: 1940 through 1990, Tech. Report, Southeast Region, National Park Service.

Kolipinski, M. C. and A. L. Higer. 1969. Some Aspects of the Effects of the Quantity and Quality of Water on Biological Communities in Everglades National Park, Open File Report 69-007, U.S. Geological Survey, Tallahassee, Fla.

Kushlan, J. A. 1974. Quantitative sampling of fish populations in shallow, freshwater environments. *Trans. Am. Fish. Soc.,* 103:348–352.

Kushlan, J. A. 1976. Environmental stability and fish community diversity. *Ecology,* 57:821–825.

Kushlan, J. A. 1980. Population fluctuations of Everglades fishes. *Copeia,* 1980:870–874.

Kushlan, J. A. 1981. Sampling characteristics of enclosure fish traps. *Trans. Am. Fish. Soc.,* 110:557–562.

Kushlan, J. A. 1987. External threats and internal management: The hydrologic regulation of the Everglades, Florida, *U.S.A. Environ. Manage.,* 1:109–119.

Kushlan, J. A., J. C. Ogden, and A. L. Higer. 1975. Relation of Water Level and Fish Availability to Wood Stork Reproduction in the Southern Everglades, Florida, Open File Report 75-434, U.S. Geological Survey, Tallahassee, Fla.

Kushlan, J. A., S. A. Voorhees, W. F. Loftus, and P. C. Frohring. 1986. Length, mass, and calorific relationships of Everglades animals. *Fla. Sci.,* 49:65–79.

Loftus, W. F. 1988. Fishes of the Everglades: Responses to a seasonally variable environment. in Wildlife in the Everglades and Latin American Wetlands, G. H. Dalrymple, W. F. Loftus, and F. S. Bernardino, Jr. (Eds.), Abstr. Proc. 1988 Everglades Symp., Miami, pp. 3–4.

Loftus, W. F. and J. A. Kushlan. 1987. Freshwater fishes of southern Florida. *Bull. Fla. State Mus. Biol. Sci.,* 31:147–344.

Loftus, W. F., J. D. Chapman, and R. Conrow. 1990. Hydroperiod effects on Everglades marsh food webs, with relation to marsh restoration efforts. in Fisheries and Coastal Wetlands Research, Vol. 6, G. Larson and M. Soukup (Eds.), Proc. 1986 Conf. on Science in National Parks, Fort Collins, Colo., pp. 1–22.

Loftus, W. F., R. A. Johnson, and G. Anderson. 1992. Ecological impacts of the reduction of groundwater levels in the rocklands. in *Proc. 1st Int. Conf. of Ground Water Ecology,* J. A. Stafford and J. J. Simon (Eds.), American Water Resources Association, Bethesda, Md., pp. 199–208.

Loveless, C. M. 1959. A study of the vegetation of the Florida Everglades. *Ecology,* 40:1–9.

Mazzotti, F. J. and L. A. Brandt. 1994. Ecology of the American alligator in a seasonally fluctuating environment. in *Everglades: The Ecosystem and Its Restoration,* S. M. Davis and J. C. Ogden (Eds.), St. Lucie Press, Delray Beach, Fla., chap. 20.

Morello, F. A. and B. A. Cook. 1984. Fish Management Annual Progress Report 1983–84. Everglades Region, Florida Game and Freshwater Fish Commission, West Palm Beach, pp. 2-4 to 2-8.

Morello, F. A., B. Cook, and S. Marshall. 1985. Fish Management Annual Progress Report 1984–85. Everglades Region, Florida Game and Freshwater Fish Commission, West Palm Beach, pp. 1-4 to 1-8.

Ogden, J. C. 1994. A comparison of wading bird nesting colony dynamics (1931–1946 and 1974–1989) as an indication of ecosystem conditions in the southern Everglades. in *Everglades: The Ecosystem and Its Restoration,* S. M. Davis and J. C. Ogden (Eds.), St. Lucie Press, Delray Beach, Fla., chap. 22.

Ogden, J. C., J. A. Kushlan, and J. T. Tilmant. 1976. Prey selectivity by the wood stork. *The Condor,* 78:324–330.

Reark, J. B. 1961. Ecological Investigations in the Everglades, 2nd annual report to Everglades National Park, University of Miami, Miami.

Rose, P. W., M. D. Flora, and P. C. Rosendahl. 1981. Hydrologic Impacts of L-31(W) on Taylor Slough, Everglades National Park, South Florida Research Center Report T-612, Homestead, Fla.

Siegel, S. 1956. *Nonparametric Statistics for the Behavioral Sciences,* McGraw-Hill, New York.

SPSS, Inc. 1988. *SPSS/PC+ V2.0 Base Manual for the IBM PC/XT/AT and PS/2,* SPSS, Inc., Chicago.

U.S. Army Corps of Engineers. 1982. Preliminary Draft Feasibility Study and Draft E.I.S. on Central and Southern Florida Project, Shark River Slough, U.S. Army Corps of Engineers, Jacksonville District, Fla.

Werner, E. E., D. J. Hall, D. R. Laughlin, D. J. Wagner, L. A. Wilsmann, and F. C. Funk. 1977. Habitat partitioning in a freshwater fish community. *J. Fish. Res. Board Can.,* 34:360–370.

Wiechman, J. D. 1987. Abundance, Condition, and Size Structure of Fishes in Relation to Water Level and Water Quality in a Florida Everglades Impoundment, M.S. thesis, University of Florida, Gainesville.

Wood, J. M. and G. W. Tanner. 1990. Graminoid community composition and structure within four Everglades management areas. *Wetlands,* 10:127–149.

Zar, J. H. 1984. *Biostatistical Analysis,* Prentice-Hall, Englewood Cliffs, N.J.

20

ECOLOGY of the AMERICAN ALLIGATOR in a SEASONALLY FLUCTUATING ENVIRONMENT

Frank J. Mazzotti

Laura A. Brandt

ABSTRACT

The American alligator is the only large, abundant, widespread nonmarine carnivore left in the southeastern United States. It is a keystone species within the Everglades and other marsh systems, acting as predator and prey and structuring plant communities. Alligators are important ecologically and are dependent on spatial and temporal patterns of water fluctuations. Patterns of courtship and mating, nesting, and habitat use are all dependent on marsh water levels. Deep water is required for alligator courtship and mating, and water levels help to determine availability of food and therefore patterns of growth and survival. In addition, alligators are able to adjust the height of the nest egg cavity based on spring water levels, which historically indicated whether water levels later in the nesting season (July and August) would be high or low. Because of this, they are an ideal study organism for examining adaptations and responses to seasonally fluctuating hydrological conditions in the Everglades.

How have water management practices and other anthropogenic changes to the Everglades region affected alligators, and can these patterns of changes be used to examine the broader question of what is wrong with the Everglades ecosystem? Historically alligators were abundant in peripheral marshes of the Everglades. Now they are most abundant in central sloughs. However, until recently recommendations regarding managing hydrological conditions for alligators focused on maintaining alligators in central slough habitats. Based on a review of the ecology of alligators in the Everglades ecosystem, it is suggested that hydrological management for alligators should be based on restoring alligators to habitats where they were formerly abundant, rather than maintaining them in habitats where they were not.

It is not possible to have done research on alligators for many years without having gained a reputation for eccentricity as a consequence of the choice of experimental animal. One accepts this and learns to live with it.

Coulson and Hernandez (1983)

INTRODUCTION

No other single organism symbolizes the Florida Everglades as does the American alligator (*Alligator mississippiensis*). Inspiring us with a mixture of respect and awe and conjuring up images of deep, dark, and mysterious marshes and swamps, alligators play a critical role in the Everglades landscape.

Alligators are the top predator in their ecosystem, influencing many populations of their prey items. This is particularly significant because, as a result of the dramatic change in their body size (an extraordinary three orders of magnitude difference from a 50-g [25-cm] hatchling to a 75,000-g [250-cm] adult), alligators eat a particularly wide variety of sizes and taxa of prey. In addition, alligator activities can structure Everglades plant and animal communities. They shape plant communities by excavating ponds and creating trails, resulting in deeper open-water areas, and by constructing nest mounds that provide relatively elevated areas, which may be colonized by plant species not tolerant of seasonal flooding (Craighead, 1968, 1971). The relatively higher ground and deeper water habitats provided by alligators in an otherwise shallow marsh landscape are critical to many wildlife populations dependent on them as nesting, resting, or foraging sites (Craighead, 1968; Deitz and Jackson, 1979; Hall and Meier, 1993; Kushlan, 1974; Kushlan and Kushlan, 1980).

Alligators also demonstrate a clear dependence on hydrological conditions. Courtship and copulation of alligators require open water, deep enough so that the male can mount the female (McIlhenny, 1935; Garrick and Lang, 1975, 1977; Vliet, 1987). Nesting success depends on the survival of eggs during an extended incubation period where they can be exposed to flooding or desiccating conditions (Joanen, 1969; Kushlan and Jacobsen, 1990; Fleming, 1991). Growth and survival of alligators are also dependent on hydrological conditions (Fogarty, 1974; Hines et al., 1968; Wilkinson, 1983).

This combination of ecological importance and hydrological linkages makes alligators an ideal study organism for determining adaptations and responses to a seasonally fluctuating wetland. The potential economic and recreational value of alligators in parts of the Everglades system (David, 1990; Hines, 1990) and the potential usefulness of alligators as indicators of environmental contamination enhance their value as study organisms. For all of these reasons, alligators have been studied in the Everglades ecosystem, as well as throughout their range.

Alligators are not the only crocodilian in southern Florida. The estuarine dwelling American crocodile (*Crocodylus acutus*) reaches the extreme northern end of its range in southern Florida, where it co-occurs with alligators in mangrove-lined ponds and creeks (Ogden, 1976; Mazzotti, 1983). Growth, survival, and reproductive success of crocodiles in southern Florida also have been linked to seasonal fluctuations in hydrological conditions (Mazzotti et al., 1988; Mazzotti and Dunson, 1984; Moler, 1991). The nonnative spectacled caiman (*Caiman crocodilus*) is also established in southern Florida, but with few exceptions is confined to man-made, aquatic habitats (C. M. Sekerak and F. J. Mazzotti, unpublished data).

The purpose of this chapter is to review our knowledge of the ecology of Everglades alligators. Its objectives are to: (1) summarize relevant aspects of the natural history of the American alligator, (2) compare life history traits of Everglades alligators in a seasonally fluctuating marsh landscape with those of alligators in other habitats in Florida and in other geographic areas, (3) compare alligators and crocodiles in the Everglades, and (4) discuss current conservation issues for alligators.

NATURAL HISTORY of ALLIGATORS

The American alligator is found in the southeastern United States along the coastal plain from North Carolina to southern Florida and west through eastern Texas to the Rio Grande River (Figure 20.1). The alligator's distribution extends northward up the Mississippi drainage system to southern Arkansas and southeast-ern Oklahoma. The American alligator and the Chinese alligator (*Alligator sinensis*) are the only species of living crocodilians to occur in warm temperate regions. All other crocodilians occur in subtropical or tropical areas (Ross and Magnusson, 1989).

Within their range, alligators occur in all available fresh or brackish water aquatic habitats: from marshes, swamps, sloughs, and ponds, to rivers, lakes, tidal areas, and on some occasions marine areas (National Fish and Wildlife Laboratory, 1980; Ross and Magnusson, 1989). In recent years, modified and artificial aquatic habitats (canals, impoundments, and borrow pits) have become occupied by alligators (Jacobsen, 1983; Joanen and McNease, 1972; Wilkinson, 1983).

Alligators have a limited ability to tolerate exposure to salt water (Mazzotti and Dunson, 1984, 1989). Sightings of alligators in saline water tend to be of subadult and adult animals in marine areas adjacent to or near freshwater sources (Birkhead and Bennet, 1981; Jacobsen, 1983; Tamarack, 1989). Small alligators are not very tolerant of exposure to salt water. Both field observations (Wilkinson, 1983) and laboratory experiments (Mazzotti and Dunson, 1984) suggest that saline water greater than 25% sea water (9 ppt) adversely affects growth and survival of small alligators.

Both male and female alligators may dig dens or holes (McIlhenny, 1935; Kushlan, 1974). Dens are usually holes in the bottom of ponds or openings in an earthen bank or berm (Craighead, 1968). However, some dens are labyrinths with

Figure 20.1 Range of the American alligator in the United States. (From National Fish and Wildlife Laboratory, 1980.)

tunnels many meters long (Brisbin et al., 1992). Female alligators build nests by scraping up and mounding surrounding vegetation. Nests can be found in shallow marshes, along the shores of lakes and rivers, and on the edges of levees and tree islands (Deitz and Hines, 1980; Joanen, 1969; Ogden, 1976). Alligator nests usually do not occur where water salinity exceeds 10–12 ppt sea water (McNease and Joanen, 1978; Wilkinson, 1983). In constructing nests, alligators are obliged to locate them so that the eggs will be above the seasonal high water level, while remaining close enough to the water's edge to prevent desiccation.

During periods of normal or high water levels, alligators of different sizes and sexes use wetland habitats differently. In general, males prefer open, deeper water areas; females, along with hatchlings and yearlings, are found in small ponds in a marsh landscape; and juveniles and subadults are transient, essentially occurring spatially and temporally where adults are not found (Goodwin and Marion, 1979; Joanen and McNease, 1970, 1972; McNease and Joanen, 1974). This spatial and temporal separation of alligators may serve to reduce agonistic encounters, cannibalism, and competition for food or space (Hunt, 1990; Magnusson, 1986). However, during periods of low water during drought, alligators of all sizes become concentrated in remaining water bodies. The extent that fights and cannibalism between alligators increase during the periods of concentration has not been quantified, but Craighead (1968) reported that larger alligators would eat smaller ones after the food supply was consumed.

Alligators are the only large, abundant and widespread, nonmarine carnivore remaining in the southeastern United States. Alligators eat whatever is available: they eat everything that moves and some things that do not; the larger an alligator is, the larger the prey it will eat. Like other crocodilians, smaller alligators eat primarily invertebrates, as well as small fish and frogs (Hays, 1992). As alligators grow, their diet shifts to primarily vertebrates and includes turtles, snakes, fish, birds, small and medium-size mammals, shellfish, and occasionally other alligators (Delany and Abercrombie, 1986; Fogarty and Albury, 1967; Rootes and Chabreck, 1993; Valentine et al., 1972; Wolfe et al., 1987). Vegetation, rocks, and man-made items, such as cans and shotgun shells, are also ingested, presumably incidentally to consuming normal prey items.

Crocodilians are poikilothermic, that is, they rely on external sources of heat to raise their body temperatures. Thermoregulation of alligators is accomplished by physiological and behavioral means (Lang, 1979; Smith, 1979; Spotila, 1974). Thermal selection (heat-seeking or heat-avoiding behavior) is the primary means of regulating body temperatures in all crocodilians (Mazzotti, 1989b). Particularly striking is the difference in the thermoregulatory behavior of alligators compared with species occurring in tropical areas.

Alligator thermal selection is characterized by heat-seeking behavior (Lang, 1987a, 1987b). They move onto land in the morning to bask. They maintain a more or less constant body temperature during the day by moving into and out of water. They move into relatively warmer water at night, when their body temperature drops slowly until basking is initiated in the morning. Alligators extend their basking hours on cold sunny days and remain in the water on cold cloudy days (Lang, 1987b; Mazzotti, 1989b). Caimans and crocodiles continue to bask even when air temperatures are cold and radiative warming is reduced (Brandt and Mazzotti, 1990; Neill, 1971). As a result, alligators maintain a higher and less variable body temperature than do their more tropical cousins.

Another interesting thermal behavior of alligators is a thermophilic response following feeding. Although observed in all species of crocodilians tested, alligators exhibit heat-seeking behavior following feeding more frequently than do tropical species (Lang, 1979, 1987b). This suite of thermophilic behaviors of alligators can be viewed as an adaptation to a more variable seasonal thermal environment than that found in subtropical and tropical areas.

In addition to seeking heat, alligators also avoid adverse effects of cold temperatures by undergoing periods of dormancy (alligators do not hibernate) in cold weather, frequently in dens or holes (McIlhenny, 1935; Tamarack, 1989). In the warmer parts of their range, especially in southern Florida, alligators remain active during the winter months (F. J. Mazzotti, personal observation).

The complex and well-developed social behavior of alligators (all crocodilians) is an important feature of their life history (Lang, 1989a). Social interactions are mediated by a communication system involving visual, tactile, and acoustic (vocal and nonvocal) signals. Interactions presumably begin when hatchling crocodilians are still in the egg and continue throughout life. Alligator social behaviors include courtship, parental care, dominance, and territorial interactions.

Nest opening and egg manipulation by adult crocodilians are apparently stimulated by vocalizations of hatchlings in the nest, some still in eggs (Herzog, 1975; Kushlan and Simon, 1981). Young alligators vocalize in a variety of contexts. Food, disturbance, danger, and contact with other juveniles all elicit vocalizations, which can also occur spontaneously with no readily apparent cause (Campbell, 1973; Staton, 1978). Juveniles respond to the calls of other juveniles by vocalizing, producing a chorus of vocalizations. Adults respond to vocalizations by approaching the calling juvenile, sometimes picking up the calling animal and carrying it to safety (Kushlan, 1973). If an intruder or predator is causing the vocalizations, the adult will approach the interloper sometimes to threaten or attack it (Lang, 1989a; F. J. Mazzotti, personal observation).

Vocalizations and visual signals among adult alligators establish social interactions and are most pronounced during reproductive activities (Garrick and Lang, 1977; Vliet, 1987). Dominance interactions and territoriality are maintained by social signals. As a result, the frequency of fighting is reduced; fighting rarely occurs, except under crowded conditions (although alligators are sometimes observed with missing limbs and portions of tails). Gaining dominance and establishing territories precede courtship and copulation.

Alligator courtship is a varied sequence of behaviors that can last for minutes or hours (Garrick and Lang, 1977; Vliet, 1987). There are three phases of courtship: attraction (bellowing and head-slapping), pair formation (touching and vocalizations), and copulation, which may occur repeatedly. Courtship behavior may serve to synchronize ovulation and spermatogenesis and peaks 6–8 weeks before nesting. Courtship generally takes place in relatively deep water between receptive females and dominant males that have established territories.

After copulation, nesting takes place. The female alligator can clear out a circle 5–6 m in diameter in building a mound 2 m wide by 1 m tall. The vegetation is cleared and the mound constructed by a process of clipping, scraping, and hauling. The nesting female deposits approximately 20–60 eggs in a depression in the top of the mound. Females frequently stay near the nest, especially late in the incubation period. Under some circumstances, nest guarding may occur, which may serve to deter predators. The incubation period is about 65 days. As a consequence of the seasonal fluctuations of temperatures in temperate climates, alligators have a much shorter breeding season than any other crocodilian (Magnusson et al., 1989). The chronology of reproductive activities is summarized in Figure 20.2.

The sex of alligators is determined by the temperature at which an egg is incubated between the 7th and 21st day. Females come from eggs incubated at or below 30–31°C, males come from eggs incubated at or above 32°C, and mixed sex ratios result from intermediate temperatures (Lang, 1989b). As a result of temperature-dependent sex determination, the location and moisture condition of nests, and even individual eggs within nests, affect the sex ratios of hatchlings. In

Figure 20.2 Nesting chronology of alligators and crocodiles in the Everglades.

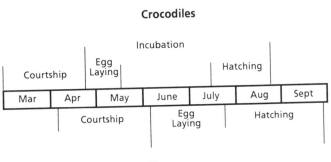

Louisiana, moist marsh nests are cooler and produce primarily females, while dry levee nests are hotter and produce mostly males (Ferguson and Joanen, 1982). Temperature of incubation also has been linked to subsequent individual growth rates of hatchling crocodiles (Lang, 1987a).

In addition to flooding, predation is the major cause of failure of alligator nests (Deitz and Hines, 1980; Joanen, 1969; Metzen, 1977). Again, water levels and locations of alligator nests apparently affect the risk of exposure to failure, with marsh locations and high water levels reducing the probability of predation and levee or shoreline locations and low water levels increasing predation (Fleming et al., 1976). Raccoons are the most common nest predators. Other nest predators include otters, wild hogs, skunks, and bears (Joanen, 1969; McIlhenny, 1935; Metzen, 1977).

Adult alligators essentially have no predators except for man. Potential predators of small alligators include raccoons, otters, wading birds, large fish, bullfrogs, or anything that can subdue them (Fogarty, 1974; McIlhenny, 1935). Intermediate-size alligators have only other alligators, and on rare occasions panthers, to avoid.

Their large size, fearsome aspect, complex behavior, and amphibious nature make alligators ideally adapted for the aquatic habitats that they inhabit and at one time dominated. However, with the arrival of European colonists and the subsequent increase in hunting and drainage of wetlands, alligator numbers began to dwindle. The extent to which the reduced sightings reflected lower populations or increased wariness is unknown. No one knows how many alligators once occurred in the United States, but early authors were clearly impressed by their abundance. Audubon (in Chabreck, 1968) described the abundance of alligators in Louisiana and throughout the southeast as follows:

> ...they are found wherever there is sufficient quantity of water to hide them or furnish them with food and they continue thus in great numbers, as high as the mouth of the Arkansas River, extending east to North Carolina, and as far west as I have penetrated.

Other authors, most notably Bartram (Van Doren, 1955), liked to describe how alligators were so numerous that it was possible to cross bodies of water by walking on their heads (Holt and Sutton, 1926; Romans, 1962; Simpson, 1920). Chabreck (1968) described how old hunters spoke of seeing eyes reflected from the glare of a lamp as being as numerous as stars in the night sky.

Reliable estimates of the current population of alligators are not available. However, in general, the number of alligators in the United States has risen steadily since the Lacey Act was amended, providing conservation agencies with a law to control the interstate movement of alligator hides. Alligator populations have now leveled off or are still increasing in many parts of their range (Joanen and McNease, 1990; Hines, 1979; Palmisano et al., 1974). In recognition of the range-wide recovery of alligators, the U.S. Fish and Wildlife Service has reclassified the federal status of the American alligator to threatened due to similarity of appearance throughout its range (*Federal Register*, 1987). Continued conversion of wetlands to alternative land uses, altered hydrology, and environmental contamination in existing wetlands are currently the greatest threats to specific alligator populations.

EVERGLADES ALLIGATORS

Fluctuating hydrological conditions caused by seasonality of rainfall and runoff are a major factor structuring tropical and subtropical wetlands (Kushlan et al., 1985; Lowe-McConnell, 1975; Tamayo, 1961). A primary focus of Everglades wildlife studies has been evaluating the ecological effects of water level fluctuations on key elements of the ecosystem, especially fishes, wading birds, alligators, and endangered species such as snail kites and crocodiles (Bennetts et al., 1988; Frederick and Collopy, 1988; Jacobsen and Kushlan, 1986; Loftus and Kushlan, 1987; Mazzotti and Brandt, 1989). Their crucial ecological role and sensitivity to hydrological conditions (discussed earlier) make alligators excellent organisms for examining adaptations and responses to the seasonally fluctuating Everglades landscape.

Adaptations of the American alligator to seasonal fluctuations in temperature have already been discussed. Some adaptations (such as heat seeking and a relatively short breeding period) are shared by alligators throughout their range. However, some activities (such as courtship, nesting, and winter dormancy) vary regionally, coinciding with changes in thermal regime. For example, the onset of courtship and nesting is linked with environmental temperatures and occurs later in the season in cooler years (Joanen and McNease, 1979; Kushlan and Jacobsen, 1990; Wilkinson, 1983). Interestingly, the occurrence of warmer temperatures earlier in the spring does not result in an earlier initiation of courtship and nesting in southern Florida. It has been suggested that in subtropical Florida, where temperature conditions are not as restrictive to egg development as in more northern areas, the incubation period is further defined by the hydrological regime (Jacobsen and Kushlan, 1986).

How have Everglades alligators responded to the seasonally fluctuating hydrological conditions found in Everglades wetlands? Are these responses typical of alligators elsewhere, or are they unique to Everglades alligators? Three aspects of alligator biology seem to be both dependent on hydrological conditions and fairly well studied throughout their range: nesting success, growth and survival, and spatial distribution and abundance.

Nesting

Nesting success is perhaps the most striking aspect of alligator biology dependent on water levels. As discussed previously, alligator eggs are exposed for more than 2 months to changing water levels, which may flood or desiccate eggs. Many alligator biologists have linked nest site selection with finding a location that would avoid exposure of the clutch to moisture extremes (Giles and Childs, 1949; Kushlan and Jacobsen, 1990; Ogden, 1976; Rice, 1992).

Most alligator nests in the Everglades are marsh nests located in water less than 25 cm deep (Ogden, 1976). The ability of Everglades alligators to adjust nest height, within engineering limits, to marsh water level was first noticed by Ogden (1976). This accommodation to fluctuating water levels was well documented by Kushlan and Jacobsen (1990), who found that the elevation at which eggs were placed was positively correlated with water level at the time of nesting. Further, they demonstrated that prior to the institution of water management, water level rise was predictably correlated with water level at the time of nest construction. This loss of predictability of hydrological fluctuations in the southern Everglades is coincident with a dramatic fivefold increase in predicted nest losses due to flooding (Kushlan and Jacobsen, 1990). The historical predictability of late-season water levels, especially those causing nest flooding, has been lost because flooding events are now related to water level fluctuations caused by regulatory releases (Kushlan and Jacobsen, 1990; Fleming, 1991).

In the absence of management-caused flooding, egg mortality and nest failure of Everglades alligators are low compared to other alligator populations (Table 20.1). The 27.9% flood-induced egg mortality reported by Kushlan and Jacobsen (1990) and the 25–44% egg mortality reported by Fleming (1991) are exceeded only by egg losses to flooding reported from South Carolina coastal marshes (Murphy and Wilkinson, 1982) (Table 20.1).

Hence, according to Kushlan and Jacobsen (1990), water management policies in southern Florida have changed the nature of egg mortality from low and predictable to high and variable. Life history strategies in addition to nest site selection (e.g., clutch size and parental care) that evolved under the historic egg mortality regime may not be suitable for the new set of hydrological fluctuations in the Everglades (Congdon et al., 1982; Kushlan and Jacobsen, 1990).

The relationship of nest/clutch height and water levels has not been determined for other alligator populations. However, in other locations (e.g., South Carolina) alligators usually do not nest in the marsh, but on the shoreline some distance from water. The distance of alligator nests to water does not vary from year to year, and this may serve to reduce exposure to fluctuating water levels (P. M. Wilkinson, unpublished data).

To summarize, the spatial and temporal orientation of alligator nesting in the Everglades serves to minimize the effects of naturally occurring factors that adversely affect nesting success. The onset of courtship and the initiation of nesting are related to air temperature and water levels (Kushlan and Jacobsen, 1990; Fleming, 1991). Hatching then occurs before naturally occurring peaks in water levels. The location of nests (in shallow water) and variations in the height of nests also serve to compensate for natural fluctuations in water levels. Management-induced changes in water level fluctuations may have seriously compromised the ability of alligators to accommodate seasonal fluctuations in hydrological conditions.

The American crocodile in southern Florida shows many similarities to Everglades alligators in nesting strategy (Mazzotti, 1989a). Crocodile nest failure is caused by predation (13% of 104 nests) and flooding or desiccation (13% of 104 nests). However, Everglades crocodile egg mortality due to flooding has been linked to local rainfall events rather than upstream water management (Mazzotti and Brandt, 1989). As for alligators, the nesting chronology of crocodiles also results in the incubation period being bracketed by seasonally low and high temperatures, desiccating conditions at the end of the dry season (beginning of nesting season), and flooding conditions during peak water levels (Figure 20.2) (Mazzotti, 1989a). Mazzotti (1989a) also concluded that the habitat, structure, and substrate of nests served to buffer the effects of moisture extremes and possibly also to influence incubation temperatures. Interestingly, although crocodiles have

Table 20.1 Comparative Life History Traits of Different Alligator Populations

Location	Age at maturity	Juvenile growth rates (cm/year)	Clutch size and range	Nest predation	Nest flooding	Hatching rate
Southern Everglades	18 s**	34.1, n = 33 l 10.1–15.3 s** 11.9 j	33.1; 4–50; n = 61 j 29.7 ± 7.54; n = 197 t	6% j 6.6% t	27.9% t	59.5% t
Lake Okeechobee			43.6; n = 63 h 42.4; n = 229 u 39.8 m 37.2 m 33.3 m			
North/central Florida	12 female i	21.1; n = 270 o 15.9; n = 80 o 11.9; n = 108 o	37.5; n = 67 g 34.6; n = 21 h 30.3; 1–39; n = 14 k	31%; n = 13 k 62.8%; n = 47 g 50.9%; n = 53 g	8%; n = 13 k	45%; n = 13 k 50.6%; n = 12 g
Louisiana	8 female n* 6 male n*	31.0; n = 38 m 22.0; n = 304 n*	38.9; 2–58 v 32.9; n = 100 x	16.5%; n = 266 v	2%; n = 266 v 4.6% x	58.2%; n = 154 v 91% x
Georgia		4.8 d 9.6–25.2 d	38.7; 27–51; n = 35 d 39.4; 13–55; n = 68 d 30; 12–44; n = 55 c	51.4%; n = 35 d 24.2%; n = 68 d 90% c	0%; n = 35 d 4.4%; n = 68 d 4% c	45.6%; n = 16 d 41.9%; n = 35 d 70% c
South Carolina	10–14 male r 10 female f	16.1 q 11.4 r 23.5 a 15.9 p 21.0 p	44.2; 15–60; n = 203 e 45.8; 35–53; n = 34 r 48.8; 44–52; n = 8 f	13.7%; n = 117 e 11%; n = 14 r 25%; n = 8 f	11%; n = 117 e 54%; n = 14 r 0%; n = 8 f	70.1%; n = 117 e 26.9% l 10% r 82% w 24–68.5% f
North Carolina	14–16 male b 18–19 female b	12.4 b	35.3; 27–40; n = 3 b	12% y		70% y

Note: Nest predation and flooding rates are based on percentage of nests. Hatching rate is based on a percentage of all eggs laid. n = sample size. * predicted from Faben's growth curve. ** predicted from power curves.

Sources: a = Brandt (1991), b = Fuller (1981), c = Metzen (1977), d = Ruckel (1988), e = Wilkinson (1983), f = Brandt (1989), g = Deitz and Hines (1980), h = Woodward (personal communication in Abercrombie (1989), i = Abercrombie (1989), j = Fogarty (1974), k = Goodwin and Marion (1978), l = Hines et al. (1968), m = F. Percival (unpublished data), n = Chabreck and Joanen (1979), o = Deitz (1979), p = Murphy (1977), q = Bara (1972), r = Murphy and Wilkinson (1982), s = Jacobsen and Kushlan (1989), t = Kushlan and Jacobsen (1990), u = Woodward et al. (1993), v = Joanen (1969), w = Bara (1975), x = Carbonneau and Chabreck (1990), y = Klause (1984).

the ability to adjust the height of a clutch (by constructing mound or hole nests), there is no relation between clutches that were elevated in mound nests and their risk of flooding. In fact, marl creek bank nests that were most susceptible to flooding events were usually hole nests, and the biggest mound nests occurred on already elevated beach sites (Mazzotti et al., 1988; Mazzotti, 1989a).

Growth and Survival

Hydrological conditions can influence growth and survival of alligators by affecting the amount and availability of food; exposing alligators to desiccating conditions, resulting in aestivation or sometimes death; or by concentrating alligators of different size classes in remaining water bodies, thus increasing competition for food and risk of cannibalism (Craighead, 1968; Hines et al., 1968; Fogarty, 1974). Unfortunately, intensive field study is required to measure growth and survival of alligators, and the relationship of these life history traits to hydrological conditions is not as direct as for reproductive success.

Jacobsen and Kushlan (1989) were surprised that their hypothesis (first proposed by Fuller, 1981) that southern Everglades alligators should grow relatively faster than alligators in more northern regions as a result of a longer growing season was not supported by their results. In fact, southern Everglades alligators grew relatively slowly, took longer to reach sexual maturity than alligators in most other parts of their range, and were most similar to the population studied by Fuller (1981) in North Carolina (Table 20.1). Jacobsen and Kushlan (1989) hypothesized that high metabolic costs associated with a warmer subtropical environment and a lower food base in Everglades wetlands result in reduced growth rates. In Louisiana, where alligators grow faster and reach sexual maturity at a much younger age, larger alligators eat primarily waterfowl and mammals (McNease and Joanen, 1977, 1981). However, small alligators (upon which most growth rate comparisons are made) eat a similar diet of primarily invertebrates and small aquatic vertebrates in all locations. It is hypothesized here that the slow growth rates of southern Everglades alligators are related not only to the poorer quality of the diet of larger Everglades alligators (primarily fish and snails) (Ogden, 1976; F. J. Mazzotti, personal observation), but also to the decrease in aquatic animal density and biomass caused by prolonged droughts (Loftus et al., 1986). However, it should not be overlooked that the slower growth rates of Everglades alligators could be the result of geographic variation in growth rates, rather than a proximate response to environmental conditions (Ferguson and Brockman, 1980).

By comparison, American crocodiles in southern Florida have relatively fast growth rates when compared to other crocodilians found elsewhere in the world (Mazzotti, 1983). However, Everglades crocodiles grow more slowly than crocodiles found on north Key Largo and at the Florida Power and Light Company, Turkey Point power plant site (Mazzotti, 1983; F. J. Mazzotti and L. A. Brandt, unpublished data; Moler, 1991). It could be that the more estuarine dwelling crocodile has a richer food resource base and/or is more tolerant of warmer temperatures because of its essentially tropical distribution.

The relationship between ambient temperature and food consumption and growth was examined in small alligators and crocodiles by feeding experimental animals *ad libitum* at 20, 25, and 30°C; the results of those experiments are

Table 20.2 Median Percent Initial Mass Change (± SD) of Small Alligators and Crocodiles Fed Mice *Ad Libitum* for 10 Days at Different Temperatures

Species	Temperature (°C)		
	20	25	30
C. acutus	−0.13 ± 0.401	0.97 ± 0.830	1.28 ± 0.677
210–969 g	(8)	(14)	(12)
A. mississippiensis	−0.03 ± 0.182	1.65 ± 0.573	1.25 ± 0.525
370–1100 g	(10)	(10)	(10)

Note: A Dunn's Multiple Comparison test showed that alligator growth rates were different for all temperatures ($p = 0.05$), while for crocodiles growth rates increased significantly from 20 to 25°C , but not from 25 to 30°C ($p = 0.05$). Between species only growth rates at 25°C were different (Mann–Whitney U Test, $p = 0.05$). Sample size in parentheses.

reported here (Table 20.2). Crocodile growth rates and food consumption appeared to increase with temperature, whereas alligator growth and food consumption peaked at the intermediate temperature. Although the observed trend appeared to support the idea that tropical crocodiles are more tolerant of warmer temperatures than alligators, the comparisons between species were statistically significant only at 25°C (Table 20.2).

Slow growth and the subsequent delay in reaching sexual maturity increases the probability of mortality in Everglades alligators. Increased mortality and the longer time required to reach sexual maturity could be compensated by an increase in clutch size (Jacobsen and Kushlan, 1989). However, as for growth rates, the clutch size reported for Everglades alligators is at the lower end of the range reported for alligators (Table 20.1). The same factors that affect alligator growth rates in southern Florida (reduced food supplies and increased metabolic costs) probably influence clutch size. Whatever is affecting clutch size in Everglades marshes is not affecting clutch size in Lake Okeechobee (Table 20.1). Perhaps this is because Lake Okeechobee prey populations are not exposed to the same prolonged drought conditions that marsh species are.

Here again, crocodiles seem to have avoided whatever factors have adversely affected Everglades alligators. Crocodile clutches average 39 eggs (Mazzotti, 1989a). Along with faster growth rates, female crocodiles in Florida have a larger size at sexual maturity and a larger maximum size, both of which are related to clutch size (Greer, 1975; Hall, 1991). Kushlan and Jacobsen (1990) suggested that the smaller clutch size and reduced nesting frequency of alligators in central Everglades marshes compared to crocodiles in Florida Bay was the result of a life history adaptation to a variable environment with high juvenile mortality and low adult mortality. A more conservative explanation may be that the larger clutch sizes, greater nesting frequency, and faster growth rates of crocodiles everywhere in southern Florida are a reflection of a richer prey base in estuaries compared to the central marsh habitat of alligators.

Unlike the marsh-dwelling alligator, crocodiles in southern Florida largely inhabit permanent bodies of water such as ponds and creeks. As a result, crocodiles may not be affected by decreased water levels associated with prolonged droughts in the same manner as alligators. Extended droughts in Florida Bay result in elevated salinities in the estuary, and this may result in adverse effects on food resources, or even on small crocodiles. Moler (1991) found that both hatchling crocodile growth and survival are negatively affected by periods of low rainfall.

Distribution and Abundance

Accurately estimating the current abundance of alligators in the Everglades is difficult; estimating the historical abundance impossible. However, fluctuations in population levels and some of the factors responsible have been recorded (Beard, 1938; Craighead, 1968; Hines, 1979; Jacobsen and Kushlan, 1984). Alligator population levels were low in the late 1930s. Overharvesting was the primary cause of historic population declines, but even in the 1930s Beard (1938) observed that drainage activities had already forced alligators out of peripheral prairie areas. Protection of alligators in the 1940s and the establishment of Everglades National Park in 1947 stopped the sharp decline in alligator populations. Alligators were considered abundant during the 1950s (Craighead, 1968). During the 1960s hurricanes, natural droughts, water management, and poaching combined to cause a second great decline in the Everglades alligator population. Craighead (1968) estimated a greater than 90% reduction in alligator numbers during this period. With the listing of alligators as an endangered species in the early 1960s, the return of surface water flows in the late 1960s, and the enforcement of the Lacey Act in the early 1970s, alligator numbers once again began to increase (Jacobsen and Kushlan, 1984; Hines, 1979).

Kushlan (1990) stated that there was no evidence for a decline in alligator population levels from historic numbers. Although true in the remaining central marshes, there is substantial evidence that alligator populations have declined in peripheral wetland areas. According to Craighead (1968), alligators were historically most abundant in wetland habitats fringing the deeper slough areas, where the limestone bedrock was near the surface, and in freshwater mangrove areas. Fewer alligators were found in the central slough and sawgrass areas characterized by deeper peat deposits. In general, alligators avoided saline areas except during periods of freshwater flows (Craighead, 1968). Today, alligators are most abundant in the central sloughs and canals of the current Everglades landscape (Kushlan, 1990) and are absent or rare in the peripheral wetlands, which have been lost to development or altered hydrologically. Salinization, due to reduced freshwater flows into the estuaries, has resulted in succession of the former freshwater mangrove zone to a saltwater system, changing the pattern of occupancy by alligators by limiting alligator occurrence to periods of freshwater discharge (Mazzotti, 1983). Hence, it can be concluded that the spatial pattern of habitat use by alligators has changed as a direct result of water management practices in southern Florida. It can also be concluded that region-wide there are fewer alligators in the Florida Everglades today than historically due to loss and alteration of wetland habitats.

Kushlan and Mazzotti (1989) came to essentially the same conclusion about the population of American crocodiles in Florida. Where crocodile habitat remains, there is no evidence that the crocodile population density is lower than historic levels. However, extensive areas of historic crocodile habitat have been lost to development on Florida's Gold Coast. Today, reports of sightings of crocodiles in areas north of Florida Power and Light Company's Turkey Point power plant in southern Biscayne Bay are rare, but slowly increasing (F. J. Mazzotti, personal observation).

THROUGH an ALLIGATOR'S EYES

Alligators are not doing well in the Everglades. Lower growth rates, higher age-specific mortality, delayed sexual maturity, small clutch size, and reduced nesting frequency characterize the Everglades alligator population. In addition, alligators are now most abundant where they were once least abundant (central marshes and sloughs) and least abundant where they were once most abundant (peripheral marshes and the freshwater mangrove zone). Do these patterns shed any light on what is "wrong" in the Everglades ecosystem or are they an artifact of geography? If we look through the eyes of an alligator, can we learn how to "fix" the Everglades?

There are two related aspects to the decline of the Everglades system. The basic cause of its deterioration is the drainage of wetlands south of Lake Okeechobee for agriculture (now known as the Everglades Agricultural Area) and along the eastern fringe for residential development (Gunderson and Loftus, 1993; Kushlan, 1990). More than half of the original Everglades system has been irreversibly drained, with up to 80% of peripheral marshes lost in certain areas (Birnhak and Crowder, 1974; Kushlan, 1990). The second factor is the alteration of natural hydrological fluctuations in the remaining marshes.

The present status of alligators in the Everglades reflects the effects of these anthropogenic perturbations. Ogden (1976) noted that alligators were most abundant in the central marshes of Shark River Slough, an area he considered as comparatively less suitable for alligators than higher peripheral marshes or freshwater mangrove areas. Ogden attributed this change in distribution to two factors: (1) fringing marshes and mangrove areas were more accessible to hunting and susceptible to alteration from drainage and (2) Shark River Slough was probably made more suitable for alligators as a result of water management.

In contrast, Kushlan (1990) considered central marshes as the primary habitat of alligators in the southern Everglades. Part of the biological criteria he developed for water management in the southern Everglades was based on maintaining alligator nesting habitat in Shark River Slough (Kushlan, 1987). Based on the discussion in this chapter and recent analyses of the hydrological relationships of wading birds and marsh food webs (Loftus et al., 1986; Walters et al., 1992), it is suggested here that maintaining wet season water levels in Shark River Slough to promote alligator nesting may not be in the best interest of the Everglades ecosystem or possibly even alligators. Alternatively, the recommendation can be made that water management policy for alligators be based on returning them to remnant undeveloped habitats where they were formerly abundant, instead of maintaining them in habitats where they were not.

This recommendation is consistent with the evolution of an experimental approach for restoring characteristic ecological components and processes in the Everglades region (Walters et al., 1992). An important aspect of this recent view of Everglades restoration is that, as an alternative to focusing restoration efforts on interior freshwater marshes (termed the "old approach"), focus instead is on a broader landscape approach of restoring the connectivity of freshwater marshes with downstream estuaries (termed the "new approach"), through re-establishment of historical flow patterns through the central marshes. Protecting and restoring hydroperiods in remnant peripheral marshes should be an important part of any Everglades restoration plan.

Essentially the fault of the old approach is that it tried to maintain the structure and function of the Everglades within what was naturally only a relatively small component of the original system. From a wildlife management perspective, there are two fundamental problems with this. First, for many wildlife populations (especially wide-ranging ones such as wading birds) it is likely that some of the habitat requirements necessary to maintain a healthy population will not be provided in a limited area with reduced habitat diversity. For example, wading birds may require a range of foraging sites from fresh water to salt water to provide a buffer to hydrological fluctuations (Walters et al., 1992). Second, the more limited an area is in terms of size and habitat diversity, the more difficult it is to resolve conflicting demands on the same resource. For example, providing relatively low wet season water levels for alligator nesting, followed by a progressive drydown for wading bird foraging, as recommended by Kushlan (1987), may actually reduce the density, diversity, and biomass of the aquatic prey base (Loftus et al., 1986). This in turn may be related to the slow growth and low reproductive output of Everglades alligators.

The new approach should seek to reintegrate the existing compartmentalized Everglades into a new system that, to the extent possible (and there are ecological, engineering, and economic limits), mimics the range of habitat conditions found in the original Everglades landscape, albeit in a smaller area. In addition, goals should be established to direct ecosystem restoration plans and to provide a basis for evaluating their success.

At the same time that these recommendations are made, the following caveat must also be issued. Even with all of the studies performed on the Everglades system to date, it must be recognized that management decisions are essentially being made based on speculations on incomplete data sets (as is the case here with the evaluation of Everglades alligators in this chapter). Yet, clearly we cannot wait until all of the data have been collected, for instead of gaining the information needed to heal the Everglades, we will have documented its death. Rather, management decisions should be treated as hypotheses of ecosystem response, and monitoring programs should be designed as experiments to test them (Mazzotti, 1991; Mazzotti et al., 1992). This approach would allow management decisions to be revised as necessary in order to meet restoration goals (Holling, 1978).

Just as declines of wading bird and alligator populations foretold the decline of the Everglades system, the decline of the Everglades is now warning of the decline of the southern Florida region. Now that we have seen the Everglades through the eyes of an alligator, will we learn its lessons? We have to.

ACKNOWLEDGMENTS

The authors thank J. Ogden and S. Davis for the opportunity to write this chapter. We would like to recognize the many agencies that have supported crocodilian research in southern Florida: U.S. National Park Service, U.S. Geological Society, U.S. Fish and Wildlife Service, U.S. Army Corps of Engineers, Florida Game and Fresh Water Fish Commission, University of Florida, Pennsylvania State University, South Florida Water Management District, and Florida Power & Light. We are grateful to P. Wilkinson, F. Percival, A. Woodward, D. David, L. Hord, and other members of the Florida Alligator Research Team for providing gator data. P. Sprott, C. Morgenstern, and J. Morgenstern commented on an early draft of this manuscript. This paper benefited greatly from reviews by P. Hall, F. Percival, M. Spalding, and an anonymous reviewer.

LITERATURE CITED

Abercrombie, C. L., III. 1989. Population dynamics of the American alligator. in *Crocodiles: Their Ecology, Management and Conservation,* Crocodile Specialist Group Special Publication, Gland, Switzerland, pp. 1–16.

Bara, M. O. 1972. Alligator Research Project, Annual Progress Report, South Carolina Wildlife and Marine Research Department, Columbia.

Bara, M. O. 1975. American Alligator Investigations, Final study report for the period Aug. 1970–Dec. 1975, South Carolina Wildlife and Marine Research Department, Columbia, 40 pp.

Beard, D. B. 1938. Everglades National Park Project, Wildlife Reconnaissance, unpublished report, National Park Service, Everglades National Park, Homestead, Fla.

Bennetts, R. E., M. W. Collopy, and S. R. Beissinger. 1988. Nesting Ecology of Snail Kites in Water Conservation Area 3A, Tech. Rep. No. 31, Florida Cooperative Fish and Wildlife Research Unit, Gainesville, 174 pp.

Birkhead, W. S. and C. R. Bennett. 1981. Observations of a small population of estuarine-inhabiting alligators near Southport, North Carolina. *Brimleyana,* 6:111–117.

Birnhak, B. I. and J. P. Crowder. 1974. An Evaluation of the Extent of Vegetative Habitat Alteration in South Florida, South Florida Study PB-231-621, U.S. Bureau of Sport Fisheries and Wildlife, Washington, D.C., pp. 1943–1970.

Brandt, L. A. 1989. The Status and Ecology of the American Alligator (*Alligator mississippiensis*) in Par Pond, Savannah River Site, M.S. thesis, Florida International University, Miami, 89 pp.

Brandt, L. A. 1991. Growth of juvenile alligators in Par Pond, Savannah River Site, South Carolina. *Copeia,* (4):1123–1129.

Brandt, L. A. and F. J. Mazzotti. 1990. The behavior of juvenile *Alligator mississippiensis* and *Caiman crocodilus* exposed to low temperature. *Copeia,* 1990:867–871.

Brisbin, I. L., Jr., J. M. Benner, L. A. Brandt, R. A. Kennamer, and T. M. Murphy. 1992. Long-term population studies of American alligators inhabiting a reservoir: Initial responses to water level drawdown. in Proc. 11th Working Meeting of the IUCN/SSC Crocodile Specialist Group.

Campbell, H. W. 1973. Observations on the acoustic behavior of crocodilians. *Zoologica,* 58:1–11.

Carbonneau, D. A. and R. H. Chabreck. 1990. Population size, composition and recruitment of American alligators in freshwater marsh. in Proc. Working Meeting of the Crocodile Specialist Group of the IUCN/SSC, Crocodile Specialist Group, pp. 32–40.

Chabreck, R. H. 1968. The American alligator—past, present and future. *Proc. Annu. Conf. Southeast. Assoc. Game Fish Comm.,* 21:554–558.

Chabreck, R. H. and T. Joanen. 1979. Growth rates of American alligators in Louisiana. *Herpetologica,* 35(1):51–57.

Congdon, J. D., A. E. Dunham, and D. W. Tinkle. 1982. Energy budgets and life histories of reptiles. in *Biology of the Reptilia,* Vol. 13, C. Gans (Ed.), Academic Press, New York, pp. 233–271.

Coulson, R. A. and T. Hernandez. 1983. Alligator metabolism studies on chemical reactions *in vivo. Comp. Biochem. Physiol.,* 74:1–182.

Craighead, F. C. 1968. The role of the alligator in shaping plant communities and maintaining wildlife in the southern Everglades. *Fla. Nat.,* 41:2–7, 69–74, 94.

Craighead, F. C. 1971. *The Trees of South Florida,* Vol. 1, University of Miami Press, Coral Gables, Fla., 212 pp.

David, D. 1990. Florida's alligator management program. in Proc. 9th Working Meeting of the IUCN Crocodile Specialist Group, Gland, Switzerland, pp. 196–205.

Deitz, D. C. 1979. Behavioral Ecology of Young American Alligators, Ph.D. dissertation, University of Florida, Gainesville, 151 pp.

Deitz, D. C. and T. C. Hines. 1980. Alligator nesting in north-central Florida. *Copeia,* (2):249–258.

Deitz, D. C. and D. R. Jackson. 1979. Use of American alligator nests by nesting turtles. *J. Herpetol.,* 13:510–512.

Delany, M. F. and C. L. Abercrombie. 1986. American alligator food habits in northcentral Florida. *J. Wildl. Manage.,* 50:348–353.

Federal Register. 1987. Vol. 52, No. 107.

Ferguson, G. W. and T. Brockman. 1980. Geographic differences of growth rates of *Scleroporus* lizards (Sauria: Iguanidae). *Copeia,* 1980:259–264.

Ferguson, M. W. J. and T. Joanen. 1982. Temperature of egg incubation determines sex in *Alligator mississippiensis. Nature,* 296:850–853.

Fleming, D. M. 1991. Wildlife Ecology Studies, Annu. Rep. South Florida Research Center, Everglades National Park, Homestead, Fla., V-10-1-52.

Fleming, D. M., A. W. Palmisano, and T. Joanen. 1976. Food habits of coastal marsh raccoons with observations of alligator nest predation. *Proc. Annu. Conf. Southeast. Assoc. Game Fish Comm.,* 30:348–357.

Fogarty, M. J. 1974. The ecology of the Everglades alligator. in *Environments of South Florida: Present and Past, Memoir No. 2,* P. J. Gleason (Ed.), Miami Geological Society, Coral Gables, Fla..

Fogarty, M. J. and J. D. Albury. 1967. Later summer foods of young alligators in Florida. *Proc. Annu. Conf. Southeast. Assoc. Game Fish Comm.,* 21:220–222.

Frederick, P. C. and M. W. Collopy. 1988. Reproductive Ecology of Wading Birds in Relation to Water Conditions in the Florida Everglades, Tech. Rep. No. 30, Florida Cooperative Fish and Wildlife Unit, Gainesville.

Fuller, M. K. 1981. Characteristics of an American Alligator (*Alligator mississippiensis*) Population in the Vicinity of Lake Ellis Simon, North Carolina, M.S. thesis, North Carolina State University, Raleigh, 136 pp.

Garrick, L. D. and J. W. Lang. 1975. Alligator courtship. *Am. Zool.,* 15:813.

Garrick, L. D. and J. W. Lang. 1977. Social signals and behaviors of adult alligators and crocodiles. *Am. Zool.,* 17:225–239.

Giles, L. W. and V. L. Childs. 1949. Alligator management on the Sabine National Wildlife Refuge. *J. Wildl. Manage.,* 13:16–28.

Goodwin, T. M. and W. R. Marion. 1978. Aspects of the nesting ecology of American alligators (*Alligator mississippiensis*) in north-central Florida. *Herpetologica,* 34:43–47.

Goodwin, T. M. and W. R. Marion. 1979. Seasonal activity ranges and habitat preferences of adult alligators in a north-central Florida lake. *J. Herpetol.,* 13:157–164.

Greer, A. E. 1975. Clutch size in crocodilians. *J. Herpetol.,* 9:319–322.

Gunderson, L. H. and W. F. Loftus. 1993. The Everglades. in *Biodiversity of the Southeastern United States: Lowland,* John Wiley & Sons, New York, pp. 199–255.

Hall, P. M. 1991. Estimation of nesting female crocodilian size from clutch characteristics: Correlates of reproductive mode, and harvest implications. *J. Herpetol.,* 25:133–141.

Hall, P. M. and A. J. Meier. 1993. Reproduction and behavior of western mud snakes (*Farancia abacura reinwardii*) in American alligator nests. *Copeia,* 1993:219–222.

Hays, L. 1992. Some Aspects of the Ecology and Population Dynamics of American Alligators in Texas, Ph.D. dissertation, Texas A & M University, College Station.

Herzog, H. A., Jr. 1975. An observation of nest opening by an American alligator, *Alligator mississippiensis. Herpetologica,* 31:446–447.

Hines, T. C. 1979. The past and present status of the alligator (*Alligator mississippiensis*) in Florida. *Proc. Annu. Conf. Southeast. Assoc. Fish Wildl. Agencies,* 33:224–232.

Hines, T. 1990. An updated report on alligator management and value added conservation in Florida. in Proc 10th Working Meeting of the IUCN Crocodile Specialist Group, Gland, Switzerland, pp. 186–199.

Hines, T. C., M. J. Fogarty, and L. C. Chappell. 1968. Alligator research in Florida: A progress report. *Proc. Annu. Conf. Southeast. Assoc. Game Fish Comm.,* 22:166–180.

Holling, C. S. 1978. *Adaptive Environmental Assessment and Management,* John Wiley & Sons, Chichester, England, 377 pp.

Holt, E. G. and G. Sutton. 1926. Notes on birds observed in southern Florida. *Ann. Carnegie Mus.,* p. 16.

Hunt, R. H. 1990. Aggressive behavior of adult alligators, *Alligator mississippiensis,* towards subadults in Okefenokee Swamp. in Proc. 9th Working Meeting of the IUCN Crocodile Specialists Group, Gland, Switzerland, pp. 360–372.

Jacobsen, T. 1983. Crocodilians and islands: Status of the American alligator and the American crocodile in the lower Florida Keys. *Fla. Field Nat.,* 11:1–24.

Jacobsen, T. and J. A. Kushlan. 1984. Population Status of the American alligator (*Alligator mississippiensis*) in Everglades National Park, South Florida Research Center Report, Homestead, Fla.

Jacobsen, T. and J. A. Kushlan. 1986. Alligator nest flooding in the southern Everglades: A methodology for management. in Proc. 7th Meeting of the IUCN Crocodile Specialists Group, Morges, Switzerland, pp. 153–166.

Jacobsen, T. and J. A. Kushlan. 1989. Growth dynamics in the American alligator (*Alligator mississippiensis*). *J. Zool. Soc. London,* 219:309–328.

Joanen, T. 1969. Nesting ecology of alligators in Louisiana. *Proc. Annu. Conf. Southeast. Assoc. Game Fish Comm.,* 23:141–151.

Joanen, T. and L. McNease. 1970. A telemetric study of nesting female alligators on Rockefeller Refuge, Louisiana. *Proc. Annu. Conf. Southeast. Assoc. Game Fish Comm.,* 24:175–193.

Joanen, T. and L. McNease. 1972. A telemetric study of adult male alligators on Rockefeller Refuge, Louisiana. *Proc. Annu. Conf. Southeast. Assoc. Game Fish Comm.,* 26:252–275.

Joanen, T. and L. McNease. 1979. Time of egg deposition for the American alligator. *Proc. Annu. Conf. Southeast. Assoc. Game Fish Comm.,* 31:33–35.

Joanen, T. and L. McNease. 1981. Nesting chronology of the American alligator and factors affecting nesting in Louisiana. in Proc. 1st Annu. Alligator Prod. Conf., University of Florida, Gainesville.

Joanen, T. and L. McNease. 1990. Classification and population status of the American alligator. in Proc. 9th Working Meeting of the IUCN Crocodile Specialists Group, Gland, Switzerland, pp. 11–20.

Klause, S. E. 1984. Reproductive Characteristics of the American Alligator (*Alligator mississippiensis*) in North Carolina, M.S. thesis, North Carolina State University, Raleigh.

Kushlan, J. A. 1973. Observations on maternal behavior in the American alligator, *Alligator mississippiensis. Herpetologica,* 29:256–257.

Kushlan, J. A. 1974. Observations on the role of the American alligator (*Alligator mississippiensis*) in the southern Florida wetlands. *Copeia,* 1974:993–996.

Kushlan, J. A. 1987. External threats and internal management: The hydrologic regulation of the Everglades, Florida, USA. *Environ. Manage.,* 11(1):109–119.

Kushlan, J. A. 1990. Wetlands and wildlife, the Everglades perspective in Freshwater Wetlands and Wildlife, R. R. Sharitz and J. W. Gibbons (Eds.), CONF-8603101, DOE Symp. Ser. No. 61, Office of Scientific and Technical Information, U.S. Department of Energy, Oak Ridge Tenn.

Kushlan, J. A. and T. Jacobsen. 1990. Environmental variability and reproductive success of Everglades alligators. *J. Herpetol.,* 24(2):176–184.

Kushlan, J. A. and M. S. Kushlan. 1980. Everglades alligator nests: Nesting sites for marsh reptiles. *Copeia,* 1980:930–932.

Kushlan, J. A. and F. J. Mazzotti. 1989. Population biology of the American crocodile. *J. Herpetol.,* 23:7–21.

Kushlan, J. A. and J. C. Simon. 1981. Egg manipulation by the American alligator. *J. Herpetol.,* 15:451–454.

Kushlan, J. K., G. Morales, and P. C. Frohring. 1985. Foraging niche relations of wading birds in tropical savannas. in *Neotropical Ornithology,* P. A. Buckley, M. S. Foster, E. S. Morton, R. S. Ridgely, and F. G. Buckley (Eds.), Ornithological Monogr. No. 36, Am. Ornithol. Union, Washington, D.C., pp. 663–682.

Lang, J. W. 1979. Thermophilic response of the American alligator and the American crocodile to feeding. *Copeia,* 1979:48–59.

Lang, J. W. 1987a. Crocodilian thermal selection. in *Wildlife Management: Crocodiles and Alligators,* G. Webb, S. Manolis, and P. Whithead (Eds.), Surrey Beatty & Sons Pty. Ltd., Chipping Norton, NSW, Australia, pp. 301–317.

Lang, J. W. 1987b. Crocodilian behavior: Implications for management. in *Wildlife Management: Crocodiles and Alligators,* G. Webb, S. Manolis, and P. Whithead (Eds.), Surrey Beatty & Sons Pty. Ltd., Chipping Norton, NSW, Australia, pp. 273–294.

Lang, J. W. 1989a. Social behavior. in *Crocodiles and Alligators,* C. A. Ross and S. Garnett (Eds.), Weldon Owen Pty. Ltd., Australia, pp. 102–117.

Lang, J. W. 1989b. Sex determination. in *Crocodiles and Alligators,* C. A. Ross and S. Garnett (Eds.), Weldon Owen Pty. Ltd., Australia, pp. 120.

Loftus, W. F. and J. A. Kushlan. 1987. Freshwater fishes of southern Florida. *Bull. Fla. State Mus.,* 31(4):147–344.

Loftus, W. F., J. D. Chapman, and R. Conrow. 1986. Hydroperiod effects on Everglades marsh food webs, with relation to marsh restoration efforts. in Proc. 1986 Conf. on Science in the National Parks, G. Larson and M. Soukup (Eds.), Fort Collins, Colo., pp. 1–22.

Lowe-McConnell, R. H. 1975. *Fish Communities in Tropical Freshwaters,* Longman, New York.

Magnusson, W. E. 1986. The peculiarities of crocodilian population dynamics and their possible importance for management strategies. in Proc of the 7th Working Meeting of the IUCN Crocodile Specialists Group, Gland, Switzerland, pp. 434–442.

Magnusson, W. E., K. A. Vliet, A. C. Pooley, and R. Whitaker. 1989. Reproduction. in *Crocodiles and Alligators,* C. A. Ross and S. Garnett (Eds.), Weldon Owen Pty. Ltd, Australia, pp. 118–135.

Mazzotti, F. J. 1983. The Ecology of *Crocodylus acutus* in Florida, Ph.D. dissertation, The Pennsylvania State University, University Park, 161 pp.

Mazzotti, F. J. 1989a. Factors affecting the nesting success of the American crocodile, *Crocodylus acutus,* in Florida Bay. *Bull. Mar. Sci.,* 44:220–228.

Mazzotti, F. J. 1989b. Structure and function. in *Crocodiles and Alligators,* C. A. Ross and S. Garnett (Eds.), Weldon Owen Pty. Ltd., Australia, pp. 42–57.

Mazzotti, F. J. 1991. Ecological agriculture: A scientific framework for integrating the conservation of biological diversity with agricultural development. in Proc. Environmentally Sound Agriculture Conf., University of Florida, Gainesville, pp. 95–192.

Mazzotti, F. J. and L. B. Brandt. 1989. Evaluating the Effects of Ground Water Levels on the Reproductive Success of the American Crocodile in Everglades National Park, Final Report, Florida Cooperative Fish and Wildlife Unit, Gainesville.

Mazzotti, F. J. and W. A. Dunson. 1984. Adaptations of *Crocodylus acutus* and *Alligator* for life in saline water. *Comp. Biochem. Physiol..* 79A:641–646.

Mazzotti, F. J. and W. A. Dunson. 1989. Osmoregulation in crocodilians. *Am. Zool.,* 29:903–920.

Mazzotti, F. J., A. Dunbar-Cooper, and J. A. Kushlan. 1988. Desiccation and cryptic nest flooding as probable causes of embryonic mortality in the American crocodile, *Crocodylus acutus,* in Everglades National Park, Florida, *Fla. Sci.,* 52:65–72.

Mazzotti, F. J., L. A. Brandt, L. G. Pearlstine, W. M. Kitchens, T. A. Obreza, F. C. Depkin, N. E. Morris, and C. E. Arnold. 1992. An Evaluation of the Regional Effects of New Citrus Development on the Ecological Integrity of Wildlife Resources in Southwest Florida, Final Report, South Florida Water Management District, West Palm Beach, 187 pp.

McIlhenny, E. A. 1935. *The Alligator's Life History,* The Christopher Publishing House, Boston.

McNease, L. and T. Joanen. 1974. A study of immature alligators on Rockefeller Refuge, Louisiana. *Proc. Annu. Conf. Southeast. Assoc. Game Fish Comm.,* 28:482–500.

McNease, L. and T. Joanen. 1977. Alligator diets in relation to marsh salinity. *Proc. Annu. Conf. Southeast. Assoc. Fish Wildl. Agencies,* 31:36–40.

McNease, L. and T. Joanen. 1978. Distribution and relative abundance of the alligator in Louisiana coastal marshes. *Proc. Annu. Conf. Southeast. Assoc. Fish Wildl. Agencies,* 32:182–186.

McNease, L. and T. Joanen. 1981. Nutrition of alligators. in Proc. 1st Annu. Alligator Production Conf., P. Cardeilhac, T. Lane, and R. Larsen (Eds.), University of Florida, Gainesville.

Metzen, W. D. 1977. Nesting ecology of alligators on the Okefenokee National Wildlife Refuge. *Proc. Annu. Conf. Southeast. Assoc. Fish Wildlife Agencies,* 31:29–32.

Moler, P. E. 1991. American Crocodile Population Dynamics, Final Report, Florida Game and Fresh Water Fish Commission, Tallahassee, 23 pp.

Murphy, T. M. 1977. Distribution, Movement and Population Dynamics of the American Alligator in a Thermally Altered Reservoir, M.S. thesis, University of Georgia, Athens, 58 pp.

Murphy, T. M. and P. M. Wilkinson. 1982. American Alligator Investigations: Management Recommendations and Current Research—1982, Unpublished report, South Carolina Wildlife and Marine Resources Department, Columbia, 93 pp.

National Fish and Wildlife Laboratory. 1980. Selected Vertebrate Endangered Species of the Seacoast of the United States. American Alligator, FWS/OBS-80/01.27, Biological Services Program, U.S. Fish and Wildlife Service, Washington, D.C., 9 pp.

Neill, W. T. 1971. *The Last of the Ruling Reptiles: Alligators, Crocodiles, and Their Kin,* Columbia University Press, New York.

Ogden, J. C. 1976. Crocodilian ecology in southern Florida. in Research in the Parks: Transactions of the National Park Centennial Symp., 1971, National Park Service Symp. Ser. No. 1, U.S. Department of the Interior, Washington, D.C.

Ogden, J. C. 1978. Status and nesting biology of the American crocodile, *Crocodylus acutus* (Reptilia, Crocodylidae), in Florida. *J. Herpetol.,* 12:183–196.

Palmisano, A. W., T. Joanen, and L. McNease. 1973. An analysis of Louisiana's 1972 experimental alligator harvest program. *Proc. Annu. Conf. Southeast. Assoc. Game Fish Comm.,* 27:184–208.

Rice, K. G. 1992. Alligator Nest Production Estimation in Florida, M.S. thesis, University of Florida, Gainesville, 57 pp.

Romans, B. 1962. A Concise Natural History of East and West Florida, University of Florida Presses, Gainesville.

Rootes, W. L. and R. H. Chabreck. 1993. Cannibalism in the American alligator. *Herpetologica,* 49(1):99–107.

Ross, C. A. and W. E. Magnusson. 1989. Living crocodilians. in *Crocodiles and Alligators,* C. A. Ross and S. Garnett (Eds.), Weldon Owen Pty. Ltd., Australia, pp 58–75.

Ruckel, S. W. 1988. Productivity of Alligators in Georgia, Unpublished project report, Georgia Endangered Wildlife and Research Surveys, 47 pp.

Simpson, C. T. 1920. *In Lower Florida Wilds,* Putnam, New York.

Smith, E. N. 1979. Behavioral and physiological thermoregulation of crocodilians. *Am. Zool.,* 19:239–247.

Spotila, J. R. 1974. Behavioral thermoregulation of the American alligator. in Thermal Ecology, J. W. Gibbons and R. R. Sharitz (Eds.), Symp. Ser. Conf-730505, Technical Information Series, U.S. Atomic Energy Commission, Springfield, Va.

Staton, M. A. 1978. "Distress calls" of crocodilians—whom do they benefit? *Am. Nat.,* 112:327–332.

Tamarack, J. L. 1989. Georgia's coastal island alligators: Variations in habitat and prey availability. in Proc. 8th Working Meeting of the IUCN Crocodile Specialists Group, Gland, Switzerland, pp. 105–118.

Tamayo, F. 1961. Los Llanos de Venezuela, Instit. Pedagogico, Direccion de Cultura, 189 pp.

Valentine, J. M., Jr., J. R. Walther, K. M. McCartney, and L. M. Ivy. 1972. Alligator diets on the Sabine National Wildlife Refuge, Louisiana. *J. Wildl. Manage.,* 36:809–815.

Van Doren, M. 1955. *Travels of William Bartram,* Dover Publications, New York, 414 pp.

Vliet, K. A. 1987. A Quantitative Analysis of the Courtship Behavior of the American Alligator (*Alligator mississippiensis*), Ph.D. dissertation, University of Florida, Gainesville, 198 pp.

Walters, C., L. Gunderson, and C. S. Holling. 1992. Experimental policies for water management in the Everglades. *Ecol. Appl.,* 2(2):189–202.

Wilkinson, P. M. 1983. Nesting Ecology of the American Alligator in Coastal South Carolina, Study Completion Report Aug. 1978–Sept. 1983, South Carolina Wildlife and Marine Research Department, Charleston, 113 pp.

Wolfe, J. L., D. K. Bradshaw, and R. H. Chabreck. 1987. Alligator feeding habits: New data and a review. *Northeast Gulf Sci.,* 9:1–8.

Woodward, A. R., H. F. Percival, M. L. Jennings, and C. T. Moore. 1993. Low clutch viability of American alligators on Lake Apopka. *Fla. Sci.,* 56(1):52–63.

21

The SNAIL KITE in the FLORIDA EVERGLADES: A FOOD SPECIALIST in a CHANGING ENVIRONMENT

Robert E. Bennetts

Michael W. Collopy

James A. Rodgers, Jr.

ABSTRACT

The snail kite (*Rostrhamus sociabilis*) is a highly specialized raptor whose diet in the Everglades consists almost exclusively of one species of aquatic snail. Yet snail kites have persisted, although tenaciously at times, in an environment that has experienced a myriad of anthropogenic changes in addition to natural fluctuation. Spatial and temporal changes in snail kite populations have accompanied changes in the Everglades ecosystem. Snail kites exhibited a period of substantial decline during the early to mid-1900s, which coincided with large-scale drainage projects in the Everglades. The impounding of the Water Conservation Areas eventually restored longer hydroperiods in some parts of the Everglades, but the hydroperiod itself provides only a partial explanation for the variation in kite numbers and distribution. The endangered status of the snail kite and the immensity of the Everglades ecosystem largely preclude experimental investigations; however, correlative evidence suggests that changes in vegetation, nutrient loadings, and the availability of apple snails also contribute to the spatial and temporal patterns of snail kite populations.

INTRODUCTION

The Everglades is an ecosystem that experiences considerable between- and within-year variation in its major life support component, water. In addition to natural variation, human-induced changes have influenced virtually the entire Everglades ecosystem. Included in these human-induced changes are drainage of large portions of the Everglades, impoundment and compartmentalization of the remaining portions of the system, manipulation of the water flows, and increased nutrient loadings of water flowing through the system. In such a dynamic system, one might expect to find wildlife that is highly adapted to changing conditions. Species that successfully cope with changing environments often are generalists that are able to use a variety of habitats or food sources (e.g., coyotes [*Canis latrans*] and great-horned owls [*Bubo virginianus*]), while it is often the highly specialized species (e.g., spotted owls [*Strix occidentalis*]) that are the first to disappear as habitats are altered. The snail kite (*Rostrhamus sociabilis*), with its diet of almost exclusively one species of aquatic snail, fits the description of a food specialist. In this chapter, the biology of the snail kite is reviewed, and how this specialized species exists within the fluctuating and changing environment of the Florida Everglades is examined.

The SNAIL KITE in the FLORIDA EVERGLADES

Life History

Feeding

The snail kite feeds almost exclusively on the freshwater apple snail (*Pomacea paludosa*) in the Everglades (Howell, 1932; Stieglitz, 1965; Snyder and Synder, 1969; Sykes, 1987a; Beissinger, 1988). Two introduced species of aquatic snail also occur within the range of snail kites in Florida and are potential food sources for kites. Takekawa and Beissinger (1983) presented indirect evidence of kites foraging on *Pomacea bridgesi* in flooded agricultural fields, but there is no evidence of kites foraging on *Marisa cornuarietis* in Florida even though they do so in Colombia (Snyder and Kale, 1983). On rare occasions, snail kites in Florida may feed on prey other than snails, particularly small turtles (Sykes and Kale, 1974; Woodin and Woodin, 1981; Bennetts et al., 1988; Beissinger, 1988, 1990a). Beissinger (1990a) summarized the use of alternative prey to *P. paludosa*. In addition to the alternative prey reported by Beissinger, kites in Florida also have been seen capturing and feeding on crayfish (*Procambarus* spp.) and a 23-cm speckled perch (*Pomoxis nigromaculatus*) (R. E. Bennetts, personal observation). Beissinger (1990a) suggested, and the authors' observations concur, that the use of alternative prey in Florida occurs when apple snails are scarce or unavailable (e.g., during drought) (see also Sykes and Kale, 1974).

The two primary hunting methods used by snail kites in Florida are aerial course hunting and perch or still hunting. These methods have been described in detail by Holgerson (1967), Snyder and Snyder (1969), and Sykes (1987a). Course hunting occurs when kites fly from 1 to 3 m above the marsh in search of snails. Once a snail is detected, the kite drops down to capture it, using its talons. Course hunting is probably the most frequently used hunting method in Florida (Snyder and Synder, 1969). Still hunting occurs when kites search for snails from a perch and descend upon them once detected. Snyder and Snyder (1969) suggested that still hunting by kites in Florida is limited by the availability of suitable perches; however, in habitats where numerous perches are available (e.g., dwarf cypress [*Taxodium* spp.]), it is commonly used (Bennetts et al., 1988). Sykes (1987a) reported no difference in success rates between the two hunting methods.

Snail kites capture snails that are primarily within the top 10 cm of water (Sykes, 1987a). Apple snails may spend much of their time on the marsh bottom (Perry, 1971), where they are not likely to be vulnerable to kite predation. They come to the water surface primarily to respirate (McClary, 1964), lay eggs (Perry, 1973; Hanning, 1978), or to feed (Talbot, 1970 in Perry, 1973). Because most egg laying occurs at night (Perry, 1973; Hanning, 1978) and feeding may be more common on the marsh bottom (Perry, 1971), apple snails probably are most vulnerable to snail kites during respiration.

Apple snails respirate using both gills and lungs (McClary, 1964). Lung respiration (also called surface inspiration) is carried out just below the water surface by extending a siphon to the surface to draw air into the lung (McClary, 1964). It is at this time that snails are vulnerable to kite predation. The frequency with which surface inspiration occurs is inversely related to the amount of dissolved oxygen in the water (McClary, 1964). Colder temperatures also reduce oxygen consumption by snails (Freiburg and Hazelwood, 1977). Cary (1985) found significant relationships between temperature and both the frequency of snail kite foraging bouts and foraging success. He also reported that no captures were observed during the passage of cold fronts when water temperatures dropped below 10°C. Consequently, factors that influence the amount of oxygen in the water (e.g., water temperature and peat depth) or the activity of snails (e.g., temperature) also affect the availability of apple snails to kites.

Nesting

Snail kites have a relatively high reproductive potential among raptors. Kites are capable of breeding as yearlings, and those that do breed at this age can be quite successful (Snyder et al., 1989a). Some snail kites probably do not attempt to breed as yearlings, but most probably become breeders by their second year (Snyder et al., 1989a). In addition to this early age of first reproduction, snail kites are known to have successfully raised two broods per year (Snyder et al., 1989a). Mate desertion (see later) also may increase the reproductive potential of kites by allowing one member of the pair to initiate a second breeding attempt prior to the termination of parental care for the first brood (Beissinger and Snyder, 1987).

Some nesting activity of snail kites has been reported in all months of the year (Sykes, 1987b; Beissinger, 1988; Snyder et al., 1989a), and the length of the breeding season may be related to water conditions (Snyder et al., 1989a). However, an 18-year study by Snyder et al. (1989a) showed that 98% of all nesting initiation occurred between December and July, with 89% of the attempts occurring from January to June.

Snail kites in the Florida Everglades may nest solitarily, but more often they breed in loose colonies (Sykes, 1987c; Beissinger, 1988; Snyder et al., 1989a). These colonies have been reported to contain up to 20 pairs (Beissinger, 1988) and may be in association with herons or anhingas (*Anhinga anhinga*) (Sykes, 1987c;

Snyder et al., 1989a). Snyder et al. (1989a) found no differences in either clutch size or nesting success of colonial and noncolonial nests.

Snail kites have an unusual mating system. During times of food abundance, one of the parents may desert its mate late in the nesting period and attempt to renest with another mate (Beissinger and Snyder, 1987). The remaining parent is usually successful at rearing the young to fledging (Beissinger and Snyder, 1987). Biparental care is usually continued throughout the breeding season, when environmental conditions are less favorable (e.g., during or immediately following droughts) (Beissinger, 1986). When mate desertion occurs, either sex may be the deserter, but the tendency is for females to desert more frequently than males (Beissinger, 1990b).

Roosting

The gregarious nature of snail kites is exemplified not only by colonial nesting, but also by communal roosting (Sykes, 1985). Congregation at communal roosts is more common during nonbreeding periods (Sykes, 1982; Beissinger, 1988); however breeding often occurs within these same sites (Sykes, 1985). Roosts are typically used by 2 to 200 individuals (Beissinger, 1988), but use may vary greatly between years (Sykes, 1985). Takekawa and Beissinger (1989) reported one roost containing 372 kites at an inundated impoundment in Palm Beach County. This high-use period was during 1985, when most areas of the Everglades were dry.

Habitat Requirements

Foraging

The foraging methods of snail kites in Florida necessitate that the snails be located at or near the water surface (Sykes, 1987a; Beissinger, 1988). Thus, the presence of emergent vegetation for the snails to climb is an important component of snail kite foraging habitat. However, relatively open wet prairie or slough communities (e.g., water lily [*Nymphaea* spp.] and eleocharis [*Eleocharis* spp.] flats [Loveless, 1959]) are also important because kites are unable to forage effectively in dense emergent vegetation (Beissinger, 1983b; Bourne, 1985; Sykes, 1987a). These open habitats may also be important foraging areas for apple snails (Hanning, 1978). Consequently, suitable snail kite foraging habitat is characterized by an interspersion of emergent vegetation with open-water communities (Stieglitz and Thompson, 1967; Sykes, 1987c; Bennetts et al., 1988). The proportion of open water to emergent vegetation in suitable kite habitat generally averages about 30–40% (Stieglitz, 1965; Stieglitz and Thompson, 1967; Sykes, 1987c; Bennetts et al., 1988).

Nesting

Snail kites typically nest over water (Sykes, 1987b; Beissinger, 1988) and rarely build nests over less than 20 cm of water (Sykes, 1987c; Bennetts et al., 1988). The complete drying of an area may cause apple snails to aestivate or may kill them outright (Hanning, 1978) and ultimately may reduce apple snail populations

(Kushlan, 1975). In addition, the drying of an area may increase access to nests by terrestrial predators (Beissinger, 1984; Sykes, 1987c). Sykes (1987c) and Bennetts et al. (1988) reported that water depths at nest sites in the Everglades averaged between 40 to 60 cm and typically range from 10 to 115 cm; however, nests at sites such as gator holes may be over water considerably greater than 115 cm (N. Snyder, personal communication). These depths are similar to the optimum range in depths of 45–90 cm suggested by Schortemeyer (1980). The peak months of snail kite nesting activity (January through May) occur when the Everglades typically undergoes a drying period. Because water levels often drop more than 30 cm during this period, nests initially built in less than 20–30 cm of water have a high probability of drying out before the nesting cycle is completed (Bennetts et al., 1988).

Snail kites in the Everglades rarely build nests in areas with greater than 1 m of water (Bennetts et al., 1988); however, there are relatively few sites with this water depth in the Everglades. Hanning (1978) suggested that water depths of ≤1 m have suitable light penetration and buffering of air temperatures to maintain apple snail populations, although it is not clear if depths >1 m are less suitable for *Pomacea paludosa*. Pennak (1953) suggested that water depths greater than 3 m may lack sufficient oxygen to sustain populations of many snail species. In addition to the potential effect of deep water on snail populations, these areas may lack sufficient emergent vegetation for snails to climb near the surface of the water (i.e., where they are vulnerable to kites) or may lack woody vegetation for nest sites (see later).

In the Everglades, snail kites nest almost exclusively in woody vegetation (Bennetts et al., 1988; Snyder et al., 1989a). Southern willow (*Salix caroliniana*) is probably the most frequently used nesting substrate in the Everglades. Bennetts et al. (1988) found that 168 (45%) of 375 kite nests in Water Conservation Area 3A were in willow. However, willow was used less frequently than expected based on availability (Bennetts et al., 1988). Plant species used as nesting substrates in higher proportion than their relative abundance were those found over water (as opposed to strictly on hammocks) and those that offered good structural support in the form of relatively strong lateral branches (Bennetts et al., 1988). These included cypress, pond apple (*Annona glabra*), punk tree (*Melaleuca quinquenerva*), and cocoplum (*Chrysobalanus icaca*). Similarly, J. A. Rodgers, Jr. (unpublished data) found punk tree used most frequently as a nest substrate in Water Conservation Area 2B.

In contrast to the Everglades, kites in the lake regions of central Florida frequently use cattail (*Typha* spp.) (Snyder et al., 1989a) or bulrush (*Scirpus californicus*) (J. A. Rodgers, Jr., unpublished data) and often suffer high nest loss from structural collapse (Snyder et al., 1989a). In the past, this has resulted in a management strategy of placing nests in artificial nest baskets in this region (Chandler and Anderson, 1974; Sykes and Chandler, 1974). The high use of cattail and bulrush in this region probably is due to a lack of woody vegetation. The woody vegetation that is available tends to be close to uplands, where the risk of predation is high (Beissinger, 1986; Sykes, 1987c). Consequently, snail kites tend to use woody vegetation in these areas only during years of high water, when these sites are inundated (Beissinger, 1986).

Influences of Reproductive Success

During the past two decades there has been considerable study of snail kite reproductive ecology (e.g., Sykes, 1979, 1987b, 1987c; Beissinger, 1986; Beissinger and Snyder, 1987; Bennetts et al., 1988; Snyder et al., 1989a). This extensive effort has produced much agreement among researchers about the influences of reproductive success, although considerable disagreement remains. Some areas of disagreement undoubtedly are due to the spatial and temporal variation in kite reproduction or to the differential effort among researchers. Other points hinge more on differences in interpretation of existing data. Here an overview is presented of the research for the past two decades on the influences of reproductive success. An attempt has been made to review the evidence and to point out wherever possible the areas of agreement and disagreement.

The specific causes of nesting failure have not been studied directly. Although speculation abounds (e.g., Beissinger, 1986; Sykes, 1987b; Bennetts et al., 1988; Snyder et al., 1989a), few data are available to support these authors' viewpoints. Of the numerous causes of nest failure reported, predation, structural collapse of nests, and abandonment are the most frequently suggested.

Predation is undoubtedly a widespread cause of nest failure (Sykes, 1987b; Beissinger, 1986; Bennetts et al., 1988). However, with few exceptions (e.g., Toner, 1984; Bennetts and Caton, 1988), the evidence for predation is usually an empty nest (Bennetts et al., 1988; Snyder et al., 1989a). Predators of snail kite nests in the Everglades include snakes (e.g., rat snakes [*Elaphe obsoleta*] and cottonmouths [*Agkistrodon picivorus*]), birds (e.g., boat-tailed grackles [*Quiscalus major*] and crows [*Corvus* spp.]), and mammals (e.g., raccoons [*Procyon lotor*]) (Beissinger, 1986; Sykes, 1987b).

Structural collapse of nests is also considered a major cause of nest failure (Beissinger, 1986; Snyder et al., 1989a). The majority of these failures, however, occur when nests are placed in cattails (Beissinger, 1986; Snyder et al., 1989a). In contrast to lake habitats, snail kites in the Everglades rarely nest in cattails (Bennetts et al., 1988; Snyder et al., 1989a). Consequently, structural collapse is less frequent in the Everglades than in lake habitats, although it still occurs when nests are placed in weak substrates or during periods of high winds.

There is considerable disagreement among researchers regarding nest abandonment, mostly because of what is defined as a breeding attempt. Snyder et al. (1989a) considered a breeding attempt to begin with nest building. In contrast, for the purposes of estimating nesting success, the authors agree with the recommendation of Steenhof (1987) that a breeding attempt begins with the laying of the first egg. This definition is preferred for several reasons. First, nest building is initiated by the male as part of courtship (Beissinger, 1988; Bennetts et al., 1988). Thus, inclusion of "prelaying" failures may include nests in which a pair bond has not even been established between a male and female. Second, researchers find nests in various stages of nest building, but rarely at the placement of the first stick. If failure is high during the initial stages, and many nest starts are missed, then estimates of nesting success would be biased high (Mayfield, 1961; Miller and Johnson, 1978; Johnson, 1979; Hensler and Nichols, 1981; Hensler, 1985; Steenhof, 1987). Similarly, if researchers are not using the same operational definition of what constitutes a nest or if the search effort differs among studies, then the corresponding results would not be comparable. The estimator proposed by Mayfield (1961) reduces these problems but cannot be used when prelaying nests are included due to the variability in duration of nest building (Snyder et al., 1989a). Third, courtship activity during the early part of breeding season is often terminated with the passage of cold fronts and resumed (often at a new location) when temperatures return to prefront conditions (Beissinger, 1988; Bennetts et al., 1988). Thus, if two cold fronts passed before eggs were actually layed, the pair would be considered to have made three separate breeding attempts (with two failures) even if the pair successfully raised a brood. For demographic purposes, we view these early postponements as false starts of one breeding attempt, rather than multiple breeding attempts with each false start being considered a breeding failure.

It should not be inferred that prelaying abandonment does not occur. In contrast, the authors agree with Snyder et al. (1989a) that it may occur often and may also be a biologically important indicator of environmental conditions. During 1987, 28.8% ($n = 66$) of the nest starts observed in Water Conservation Area 2B were abandoned before eggs were layed (J. A. Rodgers, unpublished data). Similarly, Bennetts et al. (1988) found that a minimum (nests were not counted until a recognizable structure was completed) of 11 and 4% of the nests found during 1986 and 1987, respectively, were abandoned before eggs were laid. Although these abandonments occur and are important, as noted by Snyder et al. (1989a), when used to measure nesting success, prelaying abandonments may be inconsistent and misleading. A suggested solution to both viewpoints is to estimate prelaying abandonment when the objectives of a study warrant this information, but that it be treated separately in the demographic analysis so as to maintain comparability and repeatability among studies. The number of young produced per female is what is ultimately important from nesting success data for demographic assessments. This can be estimated with or without inclusion of prelaying abandonments, and for snail kites the latter probably provides a more reliable estimate. Regardless of what approach is used, these differences should be noted when comparing nesting success among studies.

Given the above discussion, Snyder et al. (1989a) suggested that nest abandonment was one of the most frequent causes of nesting failure; however, they included abandonment of nests before eggs were laid in their estimate. Beissinger and Snyder (1987) reported only four cases of nests abandoned with eggs and only one with young over a 5-year period from 1979 through 1983. Similarly, Sykes (1987b) found only four cases of abandonment of eggs or young from 1968 through 1978, and Bennetts et al. (1988) found only 2 of 375 nests abandoned during 1986 and 1987, when water conditions were favorable. During drought years, however, abandonment of eggs or young may be a considerably more widespread cause of nesting failure if food resources are scarce (J. A. Rodgers, Jr., unpublished data; B. Toland, personal communication).

In contrast to the specific cause of a nest failure, nest site characteristics and environmental conditions are relatively easy to measure. Furthermore, the likelihood of a specific causal agent affecting a nest is probably influenced by the conditions surrounding the nest. It is therefore considerably more feasible and perhaps more useful to examine how reproductive success is affected by the environment of the nest.

Effects of Nest Site Characteristics

The type of vegetation in which a snail kite builds a nest greatly influences its susceptibility to collapse. Snyder et al. (1989a) compared nesting success in woody and herbaceous vegetation and found that success was considerably greater when nests were in woody vegetation. Because kites rarely nest in herbaceous vegetation in the Everglades, a more meaningful comparison for the Everglades would be among the various woody substrates used. Snyder et al. (1989a) did not find significant differences between nests built in pond apple and willow, but were unable to compare among additional woody substrates. Bennetts et al. (1988) found no significant differences in success among nests in willow, pond apple, cypress, and melaleuca.

Bennetts et al. (1988) found inconsistent evidence for the effect of nest height on success. During 1986 there were no significant differences among nests of three height classes (0–1, 1–2, >2 m), but the trend was for success to increase with increasing nest height. In 1987, nests in the 1- to 2-m height class were significantly less successful than those that were higher or lower. An additional multivariate analysis (logistic regression) of these data did not indicate that height contributed significantly to the logistic model. The effect of height on success shown by the univariate analysis may therefore have been an artifact better explained by a correlated variable.

Sykes (1987c) suggested that the distance to upland habitats strongly influenced the likelihood of nests being destroyed by predators. While this hypothesis is intuitively appealing, evidence for such a relationship is weak. First, Sykes used a sample of 5 failed nests from a total of 11 nests observed in Loxahatchee National Wildlife Refuge (Sykes, 1979) during 1970. Bennetts et al. (1988) compared nesting success among three categories of distance to upland habitats (<100, 100–500, and >500 m) during 1986 ($n = 148$) and 1987 ($n = 228$) in Water Conservation Area 3A. During 1986, they found that nests in the intermediate distance category had significantly lower success than nests in the farthest distance to upland category. Nests closest to upland had the highest success, although differences were not significant. A multivariate analysis (logistic regression) of these data did not indicate that distance to upland contributed significantly to a model of successful versus unsuccessful nests. During 1987, no significant differences were detected using either univariate or multivariate analyses. These results lead to the suggestion that the importance of distance to upland habitats may have been overstated or at least in need of additional investigation.

Effects of Environmental Conditions

Considerable emphasis has been placed on water levels during studies of snail kite reproduction (e.g., Sykes, 1983b, 1987c; Beissinger, 1986; Bennetts et al., 1988; Snyder et al., 1989a). However, the conclusions from these studies are not in complete agreement. Sykes (1983b) and Beissinger (1986) concluded that snail kite reproductive success is greatly influenced by water levels. In contrast, Sykes (1987c) and Bennetts et al. (1988) found little evidence for an effect of water level on nesting success. Snyder et al. (1989a) found significant differences in nesting success between low and high water years, but attributed this partly to differential use of substrates between these years.

There is general agreement among researchers that during drought years snail kite reproductive effort and success are reduced. Snyder et al. (1989a) found that success was also reduced in years following a drought, probably as the result of a lag time for snail populations to recover. The two years in which Bennetts et al. (1988) conducted their study were relatively high water years. Consequently, they concluded that the effect of water level on nest success is likely a threshold response. Drought years (including at least a 1-year lag time) may greatly reduce snail kite reproduction, but, provided foraging areas remain inundated, there may be relatively little effect from water level on nesting success in the Everglades. This is not to imply, however, that this is true for the lake habitats, where water levels may influence nest site selection and predator access.

Bennetts et al. (1988) examined the influence of weather conditions on nest success by comparison among categorical increments of rainfall, wind, and temperature. Rainfall and wind did not significantly affect nest success. Nest success did decrease with increasing increments of temperature. However, this effect may be an indirect result of a correlation among temperature and other seasonal effects (e.g., food resources and snake activity).

Coping with Drought

Periodic drought probably has the most direct influence on snail kite populations in Florida (Sykes, 1983b; Beissinger and Takekawa, 1983; Beissinger, 1986; Takekawa and Beissinger, 1989). Desiccation of an area may cause apple snails to aestivate (Kushlan, 1975) or cause direct snail mortality (Hanning, 1978). Influences of drought on snail kites include decreased nesting effort or success (Sykes, 1979; Beissinger, 1986; Snyder et al., 1989a), dispersal (Beissinger and Takekawa, 1983; Takekawa and Beissinger, 1989), and mortality (Sykes, 1979; Beissinger and Takekawa, 1983).

The hydrology of the Everglades prior to the 1900s is poorly understood. Although considerable change has occurred since the early 1900s, periodic drought undoubtedly has long been a part of the Everglades system. Snail kites existing in such a system would have been faced with periods of food depletion whenever this occurred. The high degree of mobility exhibited by snail kites (Sykes, 1983b; Beissinger and Takekawa, 1983; Takekawa and Beissinger, 1989) certainly would have been an advantage or even a necessity during these periods. Sykes (1983b) referred to snail kites as "nomadic," and Beissinger and Takekawa (1983) clearly showed the ability of kites to seek out areas of refuge during periods of drought. In recent years, however, refugia used by snail kites are rapidly disappearing (Beissinger and Takekawa, 1983; Takekawa and Beissinger, 1989). Takekawa and Beissinger (1989) even suggest that the availability of refugia may be the primary limiting factor of future kite populations.

Flexibility in reproductive effort is another means by which snail kites cope with drought conditions. During periods of drought, snail kites rarely breed or may abandon a breeding effort (Beissinger, 1988; J. A. Rodgers, Jr., unpublished data). During times of favorable conditions, however, the reproductive effort of kites greatly increases.

To date, there has been little research conducted on the mortality of snail kites, even though it is probably the most important demographic parameter for this species. Snyder et al. (1989b) reported that 13 of 15 (87%) radio-marked nestlings survived more than a year beyond fledgling. Estimates of adult mortality associated with droughts have been suggested to be in excess of 50% (Beissinger, 1984, 1986, 1988; Takekawa and Beissinger, 1989). These estimates were derived primarily from the results of population surveys conducted by Beissinger (1984), Sykes (1983a), and Rodgers et al. (1988). Using these survey results to estimate mortality implies that decreases in annual counts are due to mortality. As Rodgers et al. (1988) pointed out, it is not known whether decreases in kite numbers observed are due to mortality, dispersal, decreased productivity, or a combination of these factors. For example, Snyder et al. (1989b) reported that at least seven of eight radio-marked adult snail kites survived from 1981 to 1982 (the missing bird was of unknown fate and may have survived). Thus, these radiotelemetry data suggested a minimum adult survival of 88% during a drought, when estimates derived from annual count differences were reportedly 38% (i.e., 62% population decline) (Beissinger, 1984, 1988). Whereas some mortality probably occurs during droughts, accurate information on its extent is not available.

SPATIAL and TEMPORAL CHANGES in SNAIL KITE POPULATIONS

Methods

Trends in the counts of snail kites from 1967 until present were examined using the annual winter counts reported by Sykes (1979), Rodgers et al. (1988, unpublished data) and Bennetts et al. (unpublished data). These data were log transformed to meet the assumptions of normality and to stabilize the variance. The null hypothesis that the slopes of kite populations since 1967 have not significantly differed from zero was tested using a t statistic (SAS, 1988).

The continuous recording stations from each region were used to determine water stages. From these data, the minimum annual water stage and the number of consecutive months of inundation for each area were determined. A drying event was considered to have occurred (i.e., an interruption of continuous inundation) when the nearest recording station indicated that the water stage was below the elevation midway through the snail kite distribution of that area. Using this procedure, a drying event in a given year would have preceded the annual count and would have been reflected in that year's minimum water stage. However, because apple snail populations may require time to recover following a drying event, the number of months of continuous inundation was used as an alternative independent variable. Spearman's rank correlation statistic was used to assess relationships between the number of kites recorded during the annual count from each region and the minimum water stage of that region during that year and the number of months of continuous inundation.

During 1986 and 1987, Water Conservation Area 3A was systematically searched for nesting snail kites (Bennetts et al., 1988), with occasional visits to other areas. During 1987, the remaining regions of the Everglades were also searched (J. A. Rodgers, unpublished data). In addition to these searches, reports of snail kite activity were received from numerous managers and researchers working throughout the Everglades. During this time, 426 nests of breeding snail kites (i.e., nests containing at least one egg) were located and monitored. Numerous additional nest starts (i.e., nests were being constructed but no eggs were laid) were located, and reports of approximately 20 additional nests (mostly in known nesting areas) were received. The location of each breeding attempt was determined using a LORAN C navigational unit. Nest locations in Water Conservation Area 3A were entered into a geographic information system and overlayed with a vegetation coverage derived from a classified and geo-referenced 1987 satellite image (SPOT Image Corporation). Nest locations were verified by comparing the position of each nest to its position within or juxtaposition to known tree islands.

The satellite imagery was classified using an unsupervised maximum-likelihood classification (ERDAS, 1987) with 100 vegetation classes. At least ten representations of all classes totaling over 1% of the image were visited on the ground to confirm their identity. Classes totaling less than 1% of the image were lumped with other classes based on a combination of nearest-neighbor analysis and the Euclidian distance of the reflectance values in each of three spectral bands. The classification included considerable detail on the vegetation in Water Conservation Area 3A; however, in this chapter, the 100 vegetation classes have been condensed into 3 broad classes of interest based on the habitat requirements of snail kites, in order to simplify illustration of the relationship between suitable habitat and the distribution of nesting kites. These classes are open water, sawgrass, and tree islands. The open-water classes consist of areas with sufficient open water to enable snail kites to forage. This class includes open sloughs and relatively open wet prairies of spikerush (Loveless, 1959). The sawgrass class includes primarily sawgrass, but also stands of cattail and some denser stands of spikerush. The tree island class includes all woody vegetation types, including willow stands, bayheads, and hardwood hammocks.

Results

Temporal Changes in Snail Kite Numbers

There are few records of snail kite numbers in Florida prior to the 1900s (reviewed by Sykes, 1984). Scott (1881) and Wayne (1895) both described them as abundant in some areas. Howell (1932) probably gives the best indication of snail kite numbers in Florida during the early 1900s by describing "scattered flocks of a hundred or more birds frequently being found." Although concentrations of this magnitude recently have been reported at some of the major roost sites (Sykes, 1985; Takekawa and Beissinger, 1989), the limited accessibility of the Everglades at the time of Howell's report suggests that kites probably numbered in the high hundreds if not thousands (Stieglitz, 1965).

Beginning in the 1930s and continuing through the 1940s and 1950s, virtually all reports indicated that snail kite numbers in Florida had severely declined (Howell, 1932; Bent, 1937; Sprunt, 1945, 1954; Wachenfeld, 1956). Sprunt (1945) estimated that there were between 50 and 100 kites left in Florida and later (Sprunt, 1954) between 50 and 75. Wachenfeld (1956) suggested that the population at that time was 20 birds or less.

During the 1960s, actual counts of snail kites in Florida began. Stieglitz (1965) reported the results of an aerial and ground survey conducted in 1965 of all known major snail kite habitats, including Water Conservation Areas 1, 2, and 3, Lake Okeechobee, marshes of the St. Johns and Indian rivers, Lake Hicpochee, and the canals between Lake Okeechobee and Water Conservation Area 1. A total of 10 birds was reported from that survey, leading them to conclude that the total population in Florida at the time was probably less than 20 individuals. While these early population estimates undoubtedly were influenced by the inaccessibility of the Everglades and by survey methods that were neither standardized nor refined, they consistently indicate that snail kite numbers during the mid-1900s were very low.

In 1967, an annual survey of major snail kite habitats was initiated by Sykes (1979, 1982) and later continued by Rodgers et al. (1988, unpublished data) and Bennetts et al. (unpublished data). This survey consists of a series of systematic transect counts of snail kites at each area by airboat and by counts conducted at communal roost sites (Sykes, 1979, 1982; Rodgers et al., 1988). Total counts (all Florida) from 1967 through 1992 have ranged from 58 to 733 kites. Although the annual counts have shown considerable variability (Rodgers et al., 1988), the overall counts (all Florida) have shown a significantly increasing trend during that time ($t = 7.12$, $P < 0.001$) (Figure 21.1). The Everglades counts (including Water Conservation Areas 1, 2A, 2B, 3A, and 3B, Northeast Shark River Slough, and Everglades National Park) also have shown an increasing trend during this period ($t = 1.83$, $P = 0.08$), although not significant at $\alpha = 0.05$. Recent counts from the Everglades also have declined, probably in response to several drought years, although in 1992 they showed signs of an increase. Several of the declines since 1967 have coincided with major drying events (e.g., 1981 and 1985); however, it is not known to what extent differences among annual counts reflect actual changes in numbers of kites (Rodgers et al., 1988).

Changes in Spatial Distribution

Reports of snail kites in Florida from the 1800s through the 1930s indicate that they were considerably more widespread than today (Sykes, 1984). Reports ranged from the Wakulla River in the Florida panhandle (Howell, 1932) to Cuthbert Lake in the southern Everglades (Dutcher, 1904), with scattered sightings occurring throughout much of peninsular Florida (Howell, 1932; Bent, 1937; Sykes, 1984). Most of these early reports were one-time sightings of single birds, pairs, or small groups. There are a few areas, however, in which the descriptions indicate that kites were common or abundant. These areas include the Wacissa River (Wayne, 1895), Lake Panaskoffkee (Scott, 1881), the south shore of Lake Okeechobee (Howell, 1932), Lake Hicpochee (Howell, 1932), the headwaters of the St. Johns River (Nicholson, 1926; Howell, 1932; Sykes, 1984), the Loxahatchee Marsh (northeast of what is now Arthur R. Marshall Loxahatchee National Wildlife Refuge and Water Conservation Area 1) (Howell, 1932), and marshes near the head of the Miami River (Sykes, 1984). In addition to these areas, Howell (1932) reported that snail kites "bred and wintered abundantly in many parts of the Everglades," but did not provide detailed locations. Bent (1937) reported that snail kites were

Figure 21.1 The number of snail kites recorded during the annual winter surveys from 1967 to 1992 for all Florida (top) and the Everglades (bottom). Counts from 1967 to 1980 were conducted by the U.S. Fish and Wildlife Service (Sykes, 1983a, 1983b), counts from 1981 to 1990 were conducted by the Florida Game and Fresh Water Fish Commission (Rodgers et al., 1988, unpublished data), and counts from 1991 to 1992 were conducted by R. E. Bennetts et al. (unpublished data).

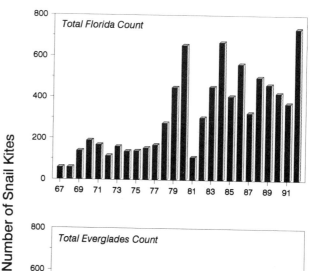

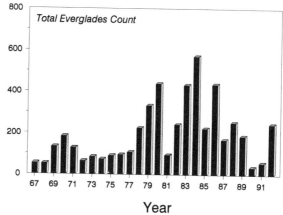

"breeding commonly all through the southern Everglades, west of Palm Beach and back of Miami and Homestead."

By the 1930s the reported distribution of snail kites in Florida was changing. Howell (1932) reported declines or complete absence of snail kites at Lakes Hicpochee and Panasoffkee, the Loxahatchee Marsh, and the southern shore of Lake Okeechobee. He also reported that whereas kites were once abundant in many parts of the Everglades, they were now restricted to a few localities (no locations specified). The major areas used by kites through the 1940s–50s included the marshes on the west shore of Lake Okeechobee (Stieglitz, 1965) and the marshes of the St. Johns River (Sprunt, 1954).

By the early 1960s, the reported distribution of kites in Florida was again changing. Stiglitz (1965) noted that only two reports of kites had occurred north of Lake Okeechobee during the previous 20 years (i.e., 1945–65) and that these were suspected of being wandering individuals. Kite use at this time also was reported to be decreasing at Lake Okeechobee but increasing in the newly formed Water Conservation Areas (particularly the southern portion of Water Conservation Areas 1 and 2A) (Stieglitz, 1965; Stieglitz and Thompson, 1967).

Since 1967, when systematic surveys were begun by Sykes (1979, 1982), the spatial distribution of snail kites in the Everglades has continued to shift. Snail kites

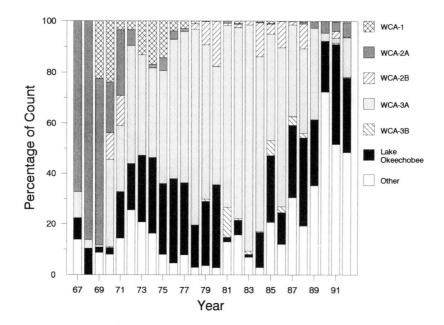

Figure 21.2   Percentage of annual snail kite surveys from 1967 to 1992 that were in each of the major use areas of Florida. Other includes Lake Kissimmee, Lake Tohopekaliga, and the St. Johns marsh.

have increased in some areas but have been virtually eliminated from others (Rodgers et al., 1988). During the late 1960s, the greatest proportion of kites in the Everglades was found in Water Conservation Area 2A (Figure 21.2). The relative abundance of kites has since decreased in Water Conservation Area 2A (although in 1992 kites were known to have nested there for the first time in over two decades), with the majority of kites more recently being found in Water Conservation Area 3A.

The shifts in distribution reflected in the annual surveys can also be seen in changes in the distribution of nesting kites (Figure 21.3). Sykes (1987c) surveyed snail kite nesting areas each year from 1968 through 1978. During the late 1960s, he reported most snail kite nests in the Everglades in Water Conservation Area 2A. During the early to mid-1970s, most were reported in Water Conservation Area 1; however, by the mid to late 1970s most kites were reported in Water Conservation Area 3A. During 1986 and 1987, most of the kite nesting activity was also found in Water Conservation Area 3A; however, Water Conservation Area 2B also received relatively high use. A few nests were found in Water Conservation Areas 3A (north of Alligator Alley) and 3B and Everglades National Park, but no indication of nesting activity was found in Water Conservation Areas 1 or 2A. More recently, most kite nesting activity has occurred in the lake regions of central Florida (J. A. Rodgers, Jr., unpublished data) and in the St. Johns marsh (B. Toland, unpublished data). This recent shift is probably in response to drought conditions that occurred in south Florida from 1989 to 1991.

In addition to long-term shifts in distribution, snail kites also show annual and

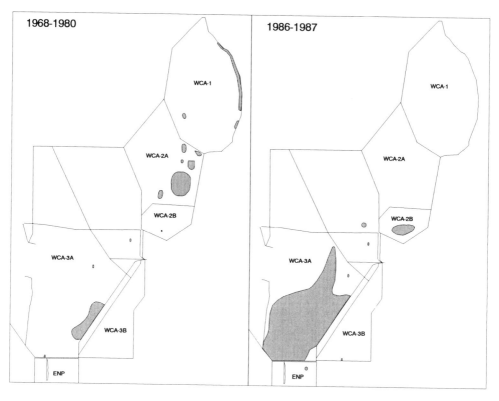

Figure 21.3 The distribution of nesting snail kites reported in the Everglades from 1968 to 1980 (left; after Sykes 1984) and from 1986 to 1987 (right; this study).

seasonal shifts between and within compartments of the Everglades. For example, the distribution of kites may shift considerably between regions among years (Rodgers et al., 1988), between regions within a given nesting season (Beissinger, 1986), and within regions (e.g., Water Conservation Area 3A) within a given season (Bennetts et al., 1986).

Relationships between Spatial and Temporal Patterns of Snail Kite Populations and Environmental Conditions

The long-term trends in snail kite numbers generally coincide with the drainage history of south Florida. In the early 1900s, prior to the major drainage efforts in south Florida (Worth, 1983; DeGrove, 1984), descriptions of snail kite populations suggested that they were quite numerous. As drainage in south Florida progressed, accounts of kite populations indicated that they were declining and moving to areas with longer hydroperiods. Shortly after impounding the Water Conservation Areas, snail kite counts increased substantially, probably as a result of the longer hydroperiods in these impoundments. The majority of snail kites in Florida over the past two decades have been found within the Water Conservation Areas.

In addition to long-term population trends, the general distribution of snail kites in the Everglades also corresponds to areas of longer hydroperiod. For example,

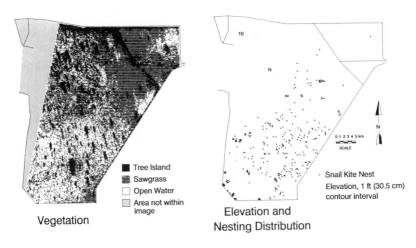

Vegetation · Tree Island · Sawgrass · Open Water · Area not within image

Elevation and Nesting Distribution · Snail Kite Nest · Elevation, 1 ft (30.5 cm) contour interval

Figure 21.4   Classified satellite image (SPOT Image Corp.) of Water Conservation Area 3A showing major zones of tree islands, sawgrass, and open water (left). Also shown is the location of 375 snail kite nests during 1986 and 1987 and approximate 1-ft elevation contours (right). Note the correspondence between nesting distribution and areas with a high proportion of open water.

the distribution of nesting snail kites within regions of the Everglades tends to occur within elevational gradients having relatively longer hydroperiods (Stieglitz and Thompson, 1967; Bennetts et al., 1988) (Figure 21.4). These areas tend to have higher apple snail populations (Kushlan, 1975) and vegetation with a relatively high proportion of open water (Loveless, 1959).

Although long-term trends of snail kite populations and their general distribution within the Everglades coincide with hydrologic conditions, fluctuating water levels can only partially account for annual variation of snail kite numbers over the past two decades (Figure 21.5). There is a weak correlation between the number of kites counted during the annual surveys and the minimum water levels in Water Conservation Area 3A (Spearman $r = 0.29$, $0.05 < p < 0.10$) and less of a correlation for Water Conservation Area 1 (Spearman $r = 0.16$, $0.10 < p < 0.25$) or 2A (Spearman $r = -0.03$, $p > 0.25$). There also is a weak correlation between the number of kites and the months of continuous inundation for Water Conservation Area 1 (Spearman $r = 0.32$, $0.05 < p < 0.10$) and 2A (Spearman $r = 0.27$, $0.05 < p < 0.10$), but not for Water Conservation Area 3A (Spearman $r = 0.12$, $p > 0.25$).

The declines in Water Conservation Areas 1 and 2A also tend to coincide with recent increased nutrient loadings in these areas. Much of the recent water flows into the Everglades system originated as agricultural runoff (Davis, 1994). Water Conservation Areas 1 and 2A were the first in line to receive this high-nutrient water and consequently have had the highest concentrations of total nitrogen and phosphorus (Millar, 1981). The northeast region of Water Conservation Area 3A (southwest of Andytown) also has had relatively high concentrations of phosphorus and nitrogen (Millar, 1981) and has received relatively little kite use even though it has hydrologic conditions similar to other areas used more frequently by kites.

A lack of research precludes any strong conclusions regarding the influence of

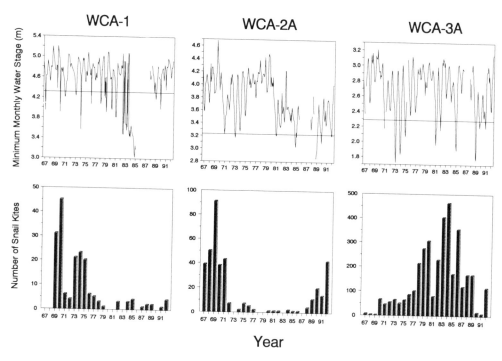

Figure 21.5   Minimum monthly water stages (top) and number of snail kites reported during the annual surveys (bottom) for Water Conservation Areas 1 (left), 2A (middle), and 3A (right). The horizontal lines represent ground elevation at each recording station (stations 1-19, 2-17, and 3-28 for Water Conservation Areas 1, 2A, and 3A, respectively).

increased nutrient loadings on snail kite declines or distribution; however, it is becoming clear that increased phosphorus is altering the habitat in ways that may be detrimental to snail kites. Substantial areas of the Everglades that have received phosphorus increases are being converted from sawgrass to cattail (Davis, 1994). This change may result in a depletion of dissolved oxygen and substantial reductions in the numbers of gastropods (e.g., apple snails) (Davis, 1994).

An alternative hypothesis to explain the declines in kite numbers in Water Conservation Area 2A is that the vegetation has been unsuitably altered from prolonged inundation. A relatively long hydroperiod is required to sustain sufficient populations of apple snails to support kite populations. A relatively long hydroperiod also may be required to maintain the open slough or wet prairie communities that kites need for foraging (Loveless, 1959; McPherson, 1973). These conditions have even led some researchers to conclude that snail kites require areas with continuous inundation (Howell, 1932; Bent, 1937; Stieglitz, 1965; Stieglitz and Thompson, 1967; Beissinger, 1983a, 1988). As hydroperiod is increased, contiguous sawgrass stands tend to convert to open sloughs (Figure 21.6) (Tanner et al., 1986; Wood and Tanner, 1990). This change probably is beneficial to snail kites through increased foraging habitat' and higher snail populations. Continuous flooding, however, eventually results in the loss of woody vegetation (U.S. Department of the Interior, 1972; McPherson, 1973; Worth, 1983; Alexander and Crook, 1984). Willow is more tolerant of prolonged flooding than most woody species in the

Figure 21.6 Areas of relatively short (top), intermediate (middle), and long (bottom) hydroperiod. The short-hydroperiod site is in Everglades National Park west of the Shark Valley Road and has dried out on average every 1.5 years. Note the contiguous stands of sawgrass and numerous tree islands. The intermediate site lies directly across Tamiami Trail from the short-hydroperiod site in Water Conservation Area 3A and has dried out on average approximately every 3 years. Note that the saw-grass stands are interspersed with open-water sloughs and that the tree islands are still numerous but have a higher proportion of willow (the lighter vegetation). The long-hydroperiod site lies adjacent to the L-67 levee near the Dade/Broward county line in Water Conservation Area 3A and has dried out on average about every 5 years. Note the lack of tree islands, except the one large willow stand in the foreground. The intermediate site provides the best combination of both foraging habitat (open sloughs) and nesting substrates (woody vegetation).

Everglades (Loveless, 1959); however, continuous flooding of even water-tolerant species results in detrimental accumulation of toxic compounds in the root zone (Harms et al., 1974; Patrick, 1974). Substantial loss of tree islands from prolonged flooding has occurred in Water Conservation Area 2A (Worth, 1983) and the deeper parts of Water Conservation Area 3A (McPherson, 1973). In the lake habitats where suitable woody vegetation is lacking, nesting failure may exceed 95% due to collapse when nests are erected on cattail or bulrush. Nests in these lake habitats have had to be placed in artificial structures to reduce this high failure rate (Sykes and Chandler, 1974). Consequently, permanent inundation is probably also detrimental to maintaining suitable habitat for snail kites.

Seasonal shifts in kite distribution may follow localized drying events; however, shifts also occur which do not seem to reflect seasonal hydrologic patterns. During both 1986 and 1987 snail kites were found throughout southern Water Conservation Area 3A from January through April. By late June and July (immediately following the primary breeding season), snail kites in Water Conservation Area 3A were found primarily in the cypress region along the L-28 canal in the southwestern portion of the conservation area (R. E. Bennetts, personal observation). Also during this time period, several kites could be found in the vicinity of Lostmans Slough in Big Cypress National Preserve (M. Spier, personal communication; R. E. Bennetts, personal observation), where they had not been seen during earlier months. Much of the cypress region along the L-28 canal had been either completely dry or had little water just prior (<1 month) to increased kite use. Similarly, the Lostmans Slough area probably had been completely dry. Areas along the L-67 canal in southeastern Water Conservation Area 3A remained inundated throughout this period, but received little use.

Discussion

The inferences postulated in this chapter regarding population trends must be considered in light of the bias previously discussed for the annual snail kite surveys. Surveys conducted during nondrought years are likely a reasonable index of overall snail kite numbers but may reflect considerable variability. During drought years, however, kites disperse into habitats which are not normally surveyed, and the resulting counts may be a poor reflection of actual snail kite numbers. This bias should not greatly affect the conclusions here because they are based primarily on long-term trends rather than specific year-to-year variation. Nevertheless, the reader should be aware of this potential bias when considering these (and other) conclusions.

Given the preceding discussion, snail kite numbers in the Everglades have not remained constant across spatial and temporal landscapes. Some areas of the Everglades have had dramatic increases (e.g., Water Conservation Area 3A) in snail kite use over the past two decades, while others have had dramatic decreases (e.g., Water Conservation Areas 1 and 2A). Additionally, within the existing impoundments in the Everglades there have been shifts in the spatial distribution of kites. Although inferences about the cause and effect of these changes are clouded with uncertainty, understanding of the changes in snail kite numbers and distribution in combination with environmental changes enables hypotheses to be formulated about the mechanisms for these changes.

The spatial and temporal patterns of snail kite populations in Florida suggest that different mechanisms influence populations at different scales. For example, long-term population trends and distribution probably reflect hydrologic and nutrient regimes. Shorter term fluctuations may be influenced by such factors as current hydrologic conditions, localized food depletion, and factors that influence dissolved oxygen levels or snail activity (e.g., water temperature, peat depth, etc.).

Long-term habitat suitability in the Everglades probably requires a compromise between a sufficiently long hydroperiod to sustain apple snail populations and open slough communities, but not so long as to eliminate the woody vegetation. This combination is probably best achieved in areas that dry out every 3–5 years (Bennetts et al., 1988). This interval also assumes that most drying events are of short duration and low intensity (i.e., subsurface moisture remains high). How the duration and intensity of drying events affect the survival of apple snails is not yet clear. Despite these effects, maintaining areas with suitably long hydroperiods will not provide a complete solution to maintaining viable populations of snail kites. Any management scenario aimed at long-term survival of both snail kite populations and the Everglades ecosystem also must include refugia to which kites (and other species) can retreat during drying events. In the absence of refugia, the periodic drying events necessary to maintain woody vegetation become detrimental to snail kite populations (Beissinger and Takekawa, 1983; Takekawa and Beissinger, 1989). Historically, refugia for kites probably consisted of lakes, the deeper center of the Shark River Slough, and areas within the Everglades system where pooling of water occurred (Fennema et al., 1994). Current management planning should consider the temporal distribution of hydrologic events and attempt to maintain some inundated habitats during times of large-scale drought.

Several factors could account for the seasonal shifts in kite distribution. For example, some shifts probably reflect localized food availability. The high nesting densities associated with colonies may result in localized depletion of snails. During 1986 and 1987 snail kites in high-density nesting areas tended to forage at greater distances from the colony as the season progressed (Bennetts et al., 1988). The shells of turtles that had been consumed by kites were found at the base of nests later during the nesting season. Turtles tend to be included in the diet of snail kites during times of food shortage (Beissinger, 1990a). The movements of snail kites out of the nesting areas at the end of the 1986 and 1987 breeding seasons were likely a result of decreased food availability.

The availability of snails to kites also is influenced by conditions of the area which influence snail activity and surfacing rates (e.g., water temperature and dissolved oxygen). As dissolved oxygen in the water decreases, the frequency of surface inspiration increases, and thus snails are more vulnerable to predation by kites. As temperature increases, snail activity also increases. Consequently, kites may take advantage of localized conditions which increase the vulnerability of snails to kites.

Under recent water management regimes, a compromise situation exists between water quantity and quality. Sufficient water to maintain a relatively long hydroperiod has recently required flows to be supplemented by agricultural runoff. The high-nutrient loadings from this process are likely detrimental to apple snails through reductions in dissolved oxygen, to snail kites through lower populations of apple snails, and to the Everglades ecosystem via conversion of native sawgrass marshes to cattail.

Resiliency and the Future of Snail Kite Populations in the Everglades

Snail kite populations are quite resilient to the natural fluctuation of the Everglades ecosystem. During times of food shortage they typically reduce their breeding effort and are capable of widespread dispersal in search of food. During times of food abundance, however, they extend their breeding season, raise multiple broods, and use a mating system (i.e., ambisexual mate desertion) that further increases their potential to raise multiple broods.

Although snail kites are highly adapted to a naturally fluctuating environment, the resiliency of kite populations to human-induced changes is less certain. The persistence of snail kite populations in the Everglades hinges on having refugia available to sustain a viable breeding population during times of drought, having a sufficiently long period between droughts to enable populations to recover, and having a high-quality environment that enables high reproductive potential during times of food abundance.

The anthropogenic changes of this century have affected each of these conditions. The potential drought refugia have been rapidly disappearing (Beissinger and Takekawa, 1983; U.S. Fish and Wildlife Service, 1986; Takekawa and Beissinger, 1989). The interval between drying events may be shortening (Beissinger, 1986), and increasing urban and agricultural demands for water in south Florida are likely to further shorten the interdrought interval. Habitat quality has been altered by increased nutrient loadings, invasion of exotic species (e.g., *Melaleuca*), and changes in flow regimes. Consequently, the availability of suitable habitat has shifted spatially as the various compartments remaining in the Everglades are manipulated. The net result is that kites may depend more frequently on refugia that are more widely dispersed and must recover in habitats that are spatially less predictable and of variable quality.

Undoubtedly, compromise solutions will need to be identified in order to accommodate increasing demands for water, habitat for snail kites, and flow systems that will maintain the unique Everglades environment. Almost any proposed solution to the problems of the Everglades and the kite will meet with opposition from individuals or groups with differing objectives or viewpoints. Current restoration planning in the southern Everglades is no exception. Arguments can easily be made for restoring longer hydroperiods in the historic Shark River Slough. It is likely that the deeper areas of the slough and other pools within the Everglades basin were once used extensively by kites. It can also be argued, however, that the impoundments of the Water Conservation Areas now serve this role and that substantial reductions in hydroperiod in these impoundments may, at least in the short term, have a negative impact on kites. It is not even clear that substantial reductions in hydroperiod would occur in the specific areas that are used most heavily by kites. What is certain is that whatever plans are adopted, they will not be unopposed.

It is the hope and intent here that the readers of this volume might provide additional insight into the causes of population changes that have been overlooked in this chapter, to validate (or refute) those that have been postulated, and to help find solutions to the complex problems facing the snail kite and the Everglades ecosystem. Increased cooperation and integration of knowledge has been a goal of this volume from the outset. Sound biological decision making about the snail kite (or any other species) cannot be achieved by working in a vacuum. The Everglades is a relatively simple ecosystem compared to the more diverse tropical forests. Yet even in this simple system, precious little is known about the integrated workings of its various components. We can only hope that decisions based on this knowledge are sufficient to keep the Everglades (including kites) intact while we strive toward a better understanding of this ecosystem.

SUMMARY

The snail kite is a highly specialized raptor whose diet consists almost exclusively of one species of aquatic snail. Yet snail kites have persisted, although tenaciously at times, in an environment that has experienced a myriad of human-induced changes in addition to natural fluctuation. Spatial and temporal changes in snail kite populations have accompanied changes in the Everglades ecosystem. However, the mechanisms that influence long-term spatial and temporal patterns may differ from those that influence short-term variability.

Snail kites exhibited a period of substantial decline and shifts in distribution during the early to mid-1900s, which coincided with large-scale drainage projects in the Everglades. The impounding of the Water Conservation Areas eventually restored longer hydroperiods in some parts of the Everglades, but hydroperiod itself provides only a partial explanation for the variation in kite populations. The endangered status of the snail kite and the immensity of the Everglades ecosystem largely precludes experimental investigations; however, correlative evidence suggests that in addition to hydrologic regimes, changes in vegetation and nutrient loadings also probably contribute to the long-term spatial and temporal patterns of snail kite populations.

Short-term variation in snail kite numbers and distribution is also influenced by hydrologic conditions, particularly drought. Annual and seasonal variation may also be influenced by other factors such as food availability. The abundance of apple snails may generally follow hydrologic patterns or be subject to localized depletion by kites or limpkins. The vulnerability of snails to kites, however, may also be closely associated with how frequently snails breathe at the water surface. This frequency of surface inspiration may in turn be influenced by factors that influence the amount of dissolved oxygen in the water (e.g., water temperature and marsh substrate).

Snail kites are highly adapted to a fluctuating environment. During times of food shortage, they disperse and may reduce their breeding effort. During times of food abundance, they may extend their breeding season, raise multiple broods, and exploit a mating system that further enhances their reproductive output. The strategy that snail kites use to survive in this fluctuating environment requires (1) refugia to survive periodic drying events, (2) enough time between drying events for the population to recover, and (3) habitat of sufficient quality to enable high reproductive output when conditions are favorable. Each of these factors has been undergoing considerable anthropogenic change. Consequently, the future of the snail kite in the Florida Everglades may largely depend on the decisions of this decade.

ACKNOWLEDGMENTS

The authors are grateful to the many people who helped in a multitude of ways. Elaine Caton, Hugh Dinkler, and Nancy Dwyer provided many long hours of assistance in the field during 1986 and 1987. Peter Frederick, Reed Bowman, and Susan Fitzgerald simultaneously worked on a wading bird project and were a constant source of field assistance, logistic support, and companionship. Paul Sykes generously allowed use of his annual survey data. Funding for REB and MWC was provided through a grant to MWC from the U.S. Army Corps of Engineers and South Florida Water Management District. Funding for JAR was provided through the Florida Game and Fresh Water Fish Commission via the U.S. Fish and Wildlife Service E-1 funding and the Florida Nongame Program. Steve Beissinger helped secure funding from the South Florida Water Management District and the project benefitted from his involvement and discussions. We are especially grateful to Jon Moulding and Lewis Hornung of the U.S. Army Corps of Engineers for their constant support. The logistic support of the Florida Cooperative Fish and Wildlife Research Unit is appreciated, particularly Wiley Kitchens, John Richardson, and Leonard Pearlstine for sharing their of knowledge of GIS and satellite image processing. We appreciate the helpful comments from Noreen Walsh, Noel Snyder, John Ogden, and an anonymous reviewer on drafts of this manuscript. To the many other persons, too numerous to name individually, who assisted with our research we are grateful.

LITERATURE CITED

Alexander, T. R. and A. G. Crook. 1984. Recent vegetational changes in South Florida. in *Environments of South Florida: Present and Past II,* P. J. Gleason (Ed.), Miami Geological Society, Coral Gables, Fla., pp. 199–210.

Beissinger, S. R. 1983a. Nest Failure and Demography of the Snail Kite: Effects of Everglades Water Management, Annual Report to the U.S. Fish and Wildlife Service, 22 pp.

Beissinger, S. R. 1983b. Hunting behavior, prey selection, and energetics of snail kites in Guyana: Consumer choice by a specialist. *Auk,* 100:84–92.

Beissinger, S. R. 1984. Mate Desertion and Reproductive Effort in the Snail Kite, Ph.D. dissertation, University of Michigan, Ann Arbor, 181 pp.

Beissinger, S. R. 1986. Demography, environmental uncertainty, and the evolution of mate desertion in the snail kite. *Ecology,* 67:1445–1459.

Beissinger, S. R. 1988. The snail kite. in *Handbook of North American Birds,* R. S. Palmer (Ed.), Yale University Press, New Haven, Conn., pp. 148–165.

Beissinger, S. R. 1990a. Alternative foods of a diet specialist, the snail kite. *Auk,* 107:327–333.

Beissinger, S. R. 1990b. Experimental brood manipulations and monoparental threshold in snail kites. *Am. Nat.,* 136:20–38.

Beissinger, S. R. and N. F. R. Snyder. 1987. Mate desertion in the snail kite. *Anim. Behav.,* 35: 477–487.

Beissinger, S. R. and J. E. Takekawa. 1983. Habitat use and dispersal by snail kites in Florida during drought conditions. *Fla. Field Nat.,* 11:89–106.

Bennetts, R. E. and E. L. Caton. 1988. An observed incident of ratsnake predation of snail kite (*Rostrhamus sociabilis*) chicks in Florida. *Fla. Field Nat.,* 16:14–16.

Bennetts, R. E., M. W. Collopy, and S. R. Beissinger. 1986. Effects of the Modified Water Delivery Program on Nest Site Selection and Nesting Success of Snail Kites in Water Conservation Area 3A, Annual Report to the U.S. Army Corps of Engineers and the South Florida Water Management District, 76 pp.

Bennetts, R. E., M. W. Collopy, and S. R. Beissinger. 1988. Nesting Ecology of Snail Kites in Water Conservation Area 3A, Tech. Rep. No. 31, Florida Cooperative Fish and Wildlife Research Unit, Department of Wildlife and Range Sciences, University of Florida, Gainesille, 174 pp.

Bent, A. C. 1937. Life Histories of North American Birds of Prey. Part 1, U.S. Nat. Mus. Bull., 167 pp.

Bourne, G. R. 1985. The role of profitability in snail kite foraging. *J. Anim. Ecol.,* 54:697–709.

Bovbjerg, R. V. 1975. Dispersal and dispersion of pond snails in an experimental environment varying to three factors, singly and in combination. *Physiol. Zool.,* 48:203–215.

Cary, D. M. 1985. Climatological Factors Affecting the Foraging Behavior and Ecology of Snail Kites (*Rostrhamus sociabilis plumbeus* Ridgeway) in Florida, M.S. thesis, University of Miami, Coral Gables, Fla., 58 pp.

Chandler, R. and J. M. Anderson. 1974. Notes on Everglade kite reproduction. *Am. Birds,* 28:856–858.

Davis, S. M. 1984. Cattail Leaf Production, Mortality, and Nutrient Flux in Water Conservation Area 2A, Tech. Publ. 84-8, South Florida Water Management District, West Palm Beach, 40 pp.

Davis, S. M. 1994. Phosphorus inputs and vegetation sensitivity in the Everglades. in *Everglades: The Ecosystem and Its Restoration,* S. M. Davis and J. C. Ogden (Eds.), St. Lucie Press, Delray Beach, Fla., chap. 15.

DeGrove, J. M. 1984. History of water management in South Florida. in *Environments of South Florida: Present and Past II,* P. J. Gleason (Ed.), Miami Geological Society, Coral Gables, Fla., pp. 22–27.

Dutcher, W. 1904. Report of the A.O.U. Committee on the Protection of North American Birds. *Auk,* 21(Suppl.):97–208.

ERDAS, Inc. 1987. *ERDAS User's Guide,* Version 7.2, ERDAS, Inc., Atlanta.

Fennema, R. J., C. J. Neidrauer, R. A. Johnson, T. K. MacVicar, and W. A. Perkins. 1994. A computer model to simulate natural Everglades hydrology. in *Everglades: The Ecosystem and Its Restoration,* S. M. Davis and J. C. Ogden (Eds.), St. Lucie Press, Delray Beach, Fla., chap. 10.

Freiburg, M. W. and D. H. Hazelwood. 1977. Oxygen consumption of two amphibious snails: *Pomacea paludosa* and *Marisa cornuarietis* (Prosobranchia: Ampullariidae). *Malacologia,* 16:541–548.

Hanning, G. W. 1978. Aspects of reproduction in *Pomacea paludosa* (Mesogastropoda: Pilidae), M.S. thesis. Florida State University, Tallahassee, 149 pp.

Harms, W. R., D. D. Hooks, C. L. Brown, P. P. Kormani, R. P. Schultz, E. H. White, and H. T. Schender. 1974. Examination of the Trees in Rodman Reservoir during October–November 1972, Final Environmental Impact Statement for the Oklawaha River, Ocala National Forest, EIS-FL-73-0075-F-1, Forest Service, U.S. Department of Agriculture, Silver Springs, Fla.

Hensler, G. L. 1985. Estimation and comparison of functions of daily survival probabilities using the Mayfield method. in *Statistics in Ornithology,* B. J. T. Morgan and P. M. North (Eds.), Springer-Verlag, New York, pp. 289–301.

Hensler, G. L. and J. D. Nichols. 1981. The Mayfield method of estimating nesting success: A model, estimators and simulation results. *Wilson Bull.,* 43:42–53.

Holgerson, N. E. 1967. Life History, Ecology, and Management of the Florida Everglades Kite (*Rostrhamus sociabilis plumbeus*), Bureau of Sport Fisheries and Wildlife, U.S. Department of the Interior, Washington, D.C.

Howell, A. H. 1932. *Florida Bird Life,* Coward-McMann, New York, 579 pp.

Jackson, J. A. 1976. Relative climbing tendencies of gray (*Elaphe obsoleta spiloides*) and black rat snakes (*E. o. obsolete*). *Herpetologica,* 32:359–361.

Johnson, D. H. 1979. Estimating nest success: The Mayfield method and an alternative. *Auk,* 96:651–661.

Kushlan, J. A. 1975. Population changes of the apple snail (*Pomacea paludosa*) in the southern Everglades. *Nautilus,* 89:21–23.

Loveless, C. M. 1959. A study of the vegetation in the Florida Everglades. *Ecology,* 40:1–9.

Mayfield, H. 1961. Nesting success calculated from exposure. *Wilson Bull.,* 73:255–261.

Mayfield, H. 1975. Suggestions for calculating nest success. *Wilson Bull.,* 87:456–466.

McClary, A. 1964. Surface inspiration and ciliary feeding in *Pomacea paludosa* (Prosobranchia: Mesogastropoda: Ampullariidae). *Malacologia,* 2:87–101.

McPherson, B. F. 1973. Vegetation in Relation to Water Depth in Conservation Area 3, Florida, Open File Report, U.S. Geological Survey, Tallahassee, 62 pp.

Millar, P. S. 1981. Water Quality Analysis in the Water Conservation Areas 1978 and 1979, Interim Progress Report, Tech. Memo., South Florida Water Management District, West Palm Beach, 63 pp.

Miller, H. W. and D. H. Johnson. 1978. Interpreting the results of nesting studies. *J. Wildl. Manage.,* 42:471–476.

Nicholson, D. J. 1926. Nesting habits of the Everglades kite in Florida. *Auk,* 43:62–67.

Patrick, W. M., Jr. 1974. Evaluation of Soil Properties as Affected by Flooding in the Oklawaha River, Final Environmental Impact Statement for the Oklawaha River, Ocala National Forest, EIS-FL-73-0075-F-1, Forest Service, U.S. Department of Agriculture, Silver Springs, Fla.

Pennak, R. W. 1953. *Fresh Water Invertebrates of the United States,* Ronald Press, New York, 769 pp.

Perry, M. C. 1971. Habitat Requirements of the Apple Snail (*Pomacea paludosa*) at Lake Woodruff National Wildlife Refuge, Progress Rep. No. 1, Lake Woodruff National Wildlife Refuge, DeLeon Springs, Fla., 12 pp.

Perry, M. C. 1973. Ecological studies of the apple snail at Lake Woodruff National Wildlife Refuge. *Fla. Sci.,* 36:22–30.

Rodgers, J. A., Jr., S. T. Schwikert, and A. S. Wenner. 1988. Status of the snail kite in Florida: 1981– 1985. *Am. Birds,* 42:30–35.

SAS Institute. 1988. *SAS/STAT User's Guide,* Release 6.03 ed., SAS Institute, Cary, N.C.

Schortemeyer, J. L. 1980. An Evaluation of Water Management Practices for Optimum Wildlife Benefits in Conservation Area 3A, Unpublished report, Florida Game and Fresh Water Fish Commission, Fort Lauderdale, 74 pp.

Scott, W. E. D. 1881. On birds observed in Sumpter, Levy, and Hillsborough counties, *Fla. Bull. Nuttall Ornithol. Club,* 6:14–21.

Snyder, N. F. R. and H. W. Kale II. 1983. Mollusk predation by snail kites in Colombia. *Auk,* 100: 93–97.

Snyder, N. F. R. and H. A. Snyder. 1969. A comparative study of mollusk predation by limpkins, Everglade kites, and boat-tailed grackles. *Living Bird,* 8:177–223.

Snyder, N. F. R., S. R. Beissinger, and R. Chandler. 1989a. Reproduction and demography of the Florida Everglades (snail) kite. *Condor,* 91:300–316.

Snyder, N. F. R., S. R. Beissinger, and M. R. Fuller. 1989b. Solar radio-transmitters on snail kites in Florida. *J. Field Ornithol.,* 60:171–177.

Sprunt, A., Jr. 1945. The phantom of the marshes. *Audubon Mag.,* 47:15–22.

Sprunt, A., Jr. 1954. *Florida Bird Life,* Coward-McMann, New York, 527 pp.

Steenhof, K. 1987. Assessing raptor reproductive success and productivity. in *Raptor Management Techniques Manual,* B. A. Giron Pendleton, B. A. Millsap, K. W. Cline, and D. M. Bird (Eds.), National Wildlife Federation, Washington, D.C., pp. 157–170.

Stieglitz, W. O. 1965. The Everglade Kite (*Rostrhamus sociabilis plumbeus*), Division of Refuges, Bureau of Sport Fisheries and Wildlife, Boynton Beach, Fla., 32 pp.

Stieglitz, W. O. and R. L. Thompson. 1967. Status and Life History of the Everglade Kite in the United States, Special Sci. Rep., Wildl. No. 109, Bureau of Sport Fisheries and Wildlife, U.S. Department of the Interior, Washington, D.C., 21 pp.

Sykes, P. W., Jr. 1979. Status of the Everglade kite in Florida, 1968–1978. *Wilson Bull.,* 91: 495–511.

Sykes, P. W., Jr. 1982. Everglade kite. in *CRC Handbook of Census Methods for Terrestrial Vertebrates,* D. E. Davis (Ed.), CRC Press, Boca Raton, Fla., pp. 43–44.

Sykes, P. W., Jr. 1983a. Snail kite use of the freshwater marshes of South Florida. *Fla. Field Nat.,* 11:73–88.

Sykes, P. W., Jr. 1983b. Recent population trends of the snail kite in Florida and its relationship to water levels. *J. Field Ornithol.,* 54:237–246.

Sykes, P. W., Jr. 1984. The range of the snail kite and its history in Florida. *Bull. Fla. State Mus.,* 29:211–264.

Sykes, P. W., Jr. 1985. Evening roosts of the snail kite in Florida. *Wilson Bull.,* 97:57–70.

Sykes, P. W., Jr. 1987a. The feeding habits of the snail kite in Florida, USA. *Colon. Waterbirds,* 10:84–92.

Sykes, P. W., Jr. 1987b. Some aspects of the breeding biology of the snail kite in Florida. *J. Field Ornithol.,* 58:171–189.

Sykes, P. W., Jr. 1987c. Snail kite nesting ecology in Florida. *Fla. Field Nat.,* 15:57–70.

Sykes, P. W., Jr. and R. Chandler. 1974. Use of artificial nest structures by Everglade kites. *Wilson Bull.,* 86:282–284.

Sykes, P. W., Jr. and H. W. Kale II. 1974. Everglades kites feed on nonsnail prey. *Auk,* 91: 818–820.

Takekawa, J. E. and S. R. Beissinger. 1983. First evidence of snail kites feeding on the introduced snail *Pomacea bridgesi,* in Florida. *Fla. Field Nat.,* 11:107–108.

Takekawa, J. E. and S. R. Beissinger. 1989. Cyclic drought, dispersal, and conservation of the snail kite in Florida: Lessons in critical habitat. *Conserv. Biol.,* 3:302–311.

Tanner, G. W., J. M. Wood, and R. Hassoun. 1986. Comparative Graminoid Community Composition and Structure within the Northern Portion of Everglades National Park, Northeast Shark River Slough, Water Conservation Area 3A, and Water Conservation Area 3B, Draft Final Report to the U.S. Army Corps of Engineers, Florida Cooperative Fish and Wildlife Research Unit, Work Order No. 41, 67 pp.

Toner, M. 1984. The kite hangs by a thread. *Natl. Wildl.,* 4:38–42.

U.S. Department of the Interior. 1972. A Preliminary Investigation of the Effects of Water Levels on Vegetative Communities of Loxahatchee National Wildlife Refuge, Florida, Bureau of Sport Fisheries and Wildlife, U.S. Department of the Interior, Boynton Beach, Fla., 20 pp.

U.S. Fish and Wildlife Service. 1986. Everglade Snail Kite (*Rostrhamus sociabilis plumbeus* Ridgeway) Revised Recovery Plan, U.S. Fish and Wildlife Service, Atlanta, 60 pp.

Wachenfeld, A. W. 1956. Present status of the Everglade kite. *Linnaean Newsl.,* 10:1.

Wayne, A. T. 1895. Notes on the birds of the Wacissa and Aucilla River regions of Florida. *Auk,* 12:362–367.

Wood, J. M. and G. W. Tanner. Graminoid community composition and structure within four Everglades management areas. *Wetlands,* 10:127–149.

Woodin, M. C. and C. D. Woodin. 1981. Everglade kite predation on a soft-shelled turtle. *Fla. Field Nat.,* 9:64.

Worth, D. 1983. Preliminary Responses to Marsh Dewatering and Reduction in Water Regulation Schedule in Water Conservation Area-2A, Tech. Publ. 83-6, South Florida Water Management District, West Palm Beach, 63 pp.

22

A COMPARISON of
WADING BIRD NESTING
COLONY DYNAMICS
(1931–1946 and 1974–1989)
as an INDICATION of
ECOSYSTEM CONDITIONS in
the SOUTHERN EVERGLADES

John C. Ogden

ABSTRACT

Patterns of nesting for five species of colonial wading birds (Ciconiiformes) in the central and southern Everglades in Florida were compared between two separate periods: an early drainage period (1931–46) and a late drainage period (1974–89). Parameters examined during the two periods were (1) numbers of birds nesting in each colony, (2) locations of colonies, (3) timing of nesting, and (4) colony success. The five species analyzed were great egret (*Casmerodius albus*), tricolored heron (*Egretta tricolor*), snowy egret (*Egretta thula*), white ibis (*Eudocimus albus*), and wood stork (*Mycteria americana*). These analyses were conducted to show changes in patterns of nesting between periods and to examine how these changes may have been caused by broader scale changes in hydrological patterns. A more complete, recent colony database (1953–89) for wood storks was also examined to supplement these analyses.

The total number of nesting wading birds declined from a peak of 180,000–245,000 birds (1933–34) in the early period to a peak of 50,000 birds (1976) in the recent period. For all species except the wood stork, the locations of the largest colonies changed between periods from a headwaters subregion located at the lower end of the Shark River Slough to a central Everglades subregion located north of Everglades National Park. Timing of nesting remained largely unchanged between periods, except for the wood stork, which shifted the average time of colony initiation from early December to late January. The best colony success data were for storks, which showed a change from 7 successful nesting years out of 9 years from 1953 to 1961 to 6 successful years out of 28 from 1962 to 1989.

Reductions in the number of nesting birds and changes in the location of major colonies appear to correlate with the reduction in the total area of wetland foraging habitat, an increased frequency of extensive dry outs in the lower Shark River Slough marshes, and the relocation of the longer hydroperiod marshes into the Water Conservation Area impoundments. Changes in timing of nesting by wood storks and the reduction in stork reproductive effort and colony success rate appear to correlate with a loss in food resources in the early dry season foraging habitats, located in the higher elevation freshwater marshes that flank the major sloughs and in the extensive mainland estuaries. Restoration of more natural patterns of colonial wading bird nesting will require substantial increases in volumes of water flowing into the southern Everglades, re-establishment of longer hydroperiods in the higher elevation marshes, increased flows into the mainland estuaries, and re-establishment of nearly permanent flooding in the deeper central sloughs.

INTRODUCTION

The relationships among the nesting patterns of different species of wading birds (Ciconiiformes) and environmental variables are understood well enough that these species can be used as indicators of change in hydrological and ecological conditions in wetland ecosystems (e.g., Bildstein et al., 1990; Custer and Osborn, 1977; Frederick and Collopy, 1988; Hafner and Britton, 1983; Kushlan, 1976a, 1989a). An indicator species is one which in its biology and status is representative of an array of other species with similar ecological requirements in the same ecosystem and which provides a measure through time of the persistence of one or more significant components or functions of an ecosystem or community.

Wading birds are appropriate indicators in the Everglades region because of (1) the large numbers that once occurred in southern Florida (Beard, 1938), (2) the fact that many species are so conspicuous and relatively easily counted (Kushlan, 1979a; Ogden et al., 1987), and (3) because it has been demonstrated that they respond quickly to both local and regional hydrological changes in the Everglades basin (Hoffman et al., 1989; Kushlan, 1976a, 1979b; Kushlan et al., 1975).

Wading birds are also the group of vertebrates for which the best data are available on population characteristics from a period prior to the creation of Everglades National Park (Robertson and Kushlan, 1974). These data provide an opportunity to understand, in a quantitative sense, the characteristics of an important component of the southern Everglades prior to the time of large-scale drainage and water management projects (pre-1950s). By using information on wading bird biology and ecology in the Everglades (derived primarily from studies and censuses conducted since the early 1970s and reviewed later) to explain the nesting colony patterns reported from the earlier period, important insights can be gained into the way this ecosystem has changed between the two periods.

The objectives in this chapter are to (1) characterize wading bird nesting patterns in the more natural Everglades ecosystem during the early period, (2) show how wading bird nesting patterns have changed between the early period and a recent period since the completion of the major surface water control system, and (3) use the changes that have occurred in colony patterns to suggest how the ecosystem has also changed since the 1930s in terms of hydrological patterns and ecological functions.

METHODS

The analyses presented in this chapter are based on patterns of wading bird nesting activity in the central and southern Everglades basin, south of the present Alligator Alley (I-75 in Broward County). Data were used from all nesting colonies that were either large in size (<100 nesting pairs) or traditional in location (used in more than one decade) and were located in the freshwater herbaceous marshes and in the downstream mangrove swamps in the southern Everglades. These two categories were selected because (1) the smaller colonies generally were not reported during the early period (and thus comparisons between the two periods are most accurate if small colony data are excluded from the latter period) and (2) the small scattered colonies appear to be more erratic in performance and to contain only a small percentage of the total nesting population (Bancroft et al., 1988b; Frederick and Collopy, 1988).

Colonies located on coastal islands adjacent to the southern Everglades (south of Everglades City) were included when the species composition was typical of Everglades colonies and when it had been documented or was reasonably expected that the birds in these colonies were feeding primarily on the mainland. Thus, the mixed-species colonies on islands in extreme northwestern Florida Bay were included, while the colonies in eastern Florida Bay (including Tern, Porjoe and Nest keys) were not. The latter not only have been composed primarily of nesting roseate spoonbills and reddish egrets, but were not censused except in a few years during the early period (Allen, 1942; Powell et al., 1989). Nesting colonies located in the Big Cypress region (primarily in Collier, Hendry, and Lee counties) were excluded from this analysis.

The colony data used here are from two periods: a "natural" period, 1931–46 (referred to as the early period), and a recent, postdrainage period, 1974–89 (recent period). Although the early period certainly occurred well after the initiation of drainage programs in the Everglades basin (Hanna and Hanna, 1948), it preceded the time of major drainage projects in the southern Everglades (the South Florida Water Management Project was initiated in 1947). The recent period occurred after the construction of the major water control system in the southern Everglades region and reflects the impacts of the Minimum Water Delivery (1971–83) and Experimental Water Delivery (1985–present) schedules for Everglades National Park (South Florida Research Center, 1990). The first year of the recent period was identified as 1974 for two reasons: it was the first year of an expanded effort by the Everglades National Park staff to census all southern Everglades colonies (Kushlan and White, 1977), and it was sufficiently into the Minimum Water Delivery schedule (initiated in 1971) to show the impact of this schedule on wading bird nesting efforts. Colony data from these two periods, rather than data from all years since the 1930s, were compared because the colony record from the 1950s and 1960s is much less complete.

Data were used for the five species of wading birds that characteristically made up 90–100% of the birds in typical Everglades colonies: (1) great egret (*Casmerodius albus*), (2) snowy egret (*Egretta thula*), (3) tricolored heron (*Egretta tricolor*), (4) white ibis (*Eudocimus albus*), and (5) wood stork (*Mycteria americana*). Each colony was characterized in each year by four parameters: (1) number or estimate of nesting birds for each of the five species, (2) location of each colony, (3) timing of nesting activity, and (4) nesting success. For consistency, numbers of nesting adult birds are used throughout this chapter, rather than numbers of pairs or nests. Although cattle egrets (*Bubulcus ibis*) are now one of the most common nesting wading birds in some Everglades colonies, this species is excluded from consideration in this chapter for three reasons: (1) this self-introduced species did not occur in south Florida during the early period, (2) it is not a component of the Everglades system in its feeding ecology (Palmer, 1962), and (3) no data suggest that the introduction of the cattle egret into the guild of Everglades wading birds has contributed to changes in nesting patterns within the region.

Hydrological data were supplied by the South Florida Research Center (Everglades National Park). Because no surface water gauges were present in the southern Everglades during the early period and because the nearest rainfall gauge was located at the Homestead Experimental Station, rainfall data from this station were used to help characterize the hydrological conditions during wet and dry seasons during both periods. Using these rainfall values, a cumulative deviation was calculated from normal rainfall for all months in each wet season (May–October) and each dry season (November–March). The rainfall data from this 11-month period were considered to represent a "bird year" (May–March), from the initiation of the previous rainy season to the month (March) by which most colonies had formed. The assumption was that hydrological conditions during this bird year were major factors influencing the timing and magnitude of the annual wading bird reproductive effort (Bancroft et al., 1988b, 1990; Kushlan, 1986; Ogden et al., 1980).

Because rainfall at a single somewhat peripheral station does not adequately characterize hydrological conditions throughout the southern Everglades basin, surface water depth data were used from the P-33, P-36, G-596, and G-789 stations south of the Tamiami Trail and the 3A-4 station (located in Water Conservation Area 3A) as an additional way of understanding general hydrological conditions during the recent period (Figure 22.1). Narrative statements of surface water conditions contained in the Audubon warden reports were used to supplement the hydrological record for the early period. Rainfall data and these reports of surface water conditions were used to suggest an overall hydrological rating for each year: *dry* for bird years that were drier than normal, *wet* for the years that were wetter than normal, and *normal* for those years that fell in-between.

The early period database was produced by the National Audubon Society wardens and sanctuary leaders who protected and documented nesting colonies

Figure 22.1 The central and southern Everglades region of wading bird nesting colonies and the location of five water depth gauge stations (P-33, P-36, G-596, G-798, and A-3-4).

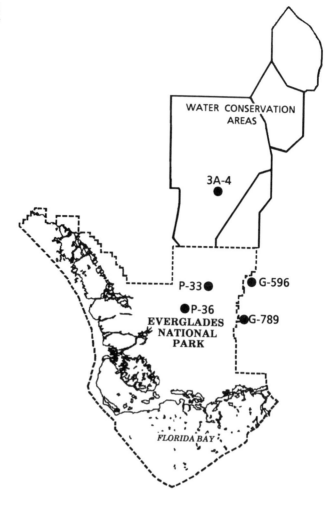

WATER CONSERVATION AREAS

3A-4

P-33   G-596

P-36

EVERGLADES NATIONAL PARK   G-789

FLORIDA BAY

in the southern Everglades region between the turn of the century and the mid-1940s (Graham, 1990). This database, which consists primarily of original daily and weekly warden reports and periodic field inspection reports, is best preserved for the period 1930–46.

The usefulness of the colony data produced by the Audubon wardens has been questioned by Frohring et al. (1988), who suggested that because of a lack of formal training or because of possible ulterior motives, some wardens either seriously erred or exaggerated in their estimates of the number of wading birds in colonies. However, when colony estimates were compared among years and different wardens and at a regional level, they were generally found to be consistent. The approach used to address this issue was to tabulate and compare all estimates of the number of nesting birds for each colony, looking for inconsistencies, especially between years or between visits within a year. No obvious problems were found, with the following exceptions: the Lane River rookery report for 1933 (Holt, 1933) and Robert Porter Allen's reports on his visit to Rookery Branch in April 1934 (Allen, 1934; Allen, original field notebook [National Audubon

Society, Tavernier, Florida]) contain estimates of numbers of birds that far exceed those produced for the same colonies by other Audubon staff in the same or other years during the early period. These particular estimates, therefore, were excluded from the analyses. Somewhat cryptic warden reports of 50,000 birds nesting at East River in June 1937 and 100,000 ibis in East River in June 1941 were also closely reviewed and ultimately accepted; both records were of late (May–June) colony formations in years when water levels were unusually high 2 months earlier, when colonies normally formed. Otherwise, there appeared to be no justification for doubting the estimates contained in the remaining reports. By this process it was also possible to determine which observers included nestlings in their estimates of colony totals (a surprisingly common practice) and which reported pairs versus total adult birds.

In the few cases where the early Audubon Society data were published (Allen, 1934, 1937; Holt, 1933), the original source material was used wherever possible for information on colony characteristics. For example, the unpublished warden reports from the famous Rookery Branch colony in 1934 suggest a more consistent and believable estimate of 225,000 white ibis than the published report of 1 million birds (Allen, 1934).

Like the Audubon data, the colony information for the recent period (1974–1989) largely comes from unpublished material in the Everglades National Park and National Audubon Society (Tavernier, Florida) files. Those that were published appeared in Frederick and Collopy (1988, 1989) and Kushlan and White (1977). The unpublished data were collected by National Park Service and National Audubon Society biologists during aerial and ground surveys and field studies during this period. In common with the early period, the methods used to estimate or count the number of birds in a colony were rarely reported.

The colony data for both the early and recent periods appear to be accurate for locations and timing of colony activity. The counts or estimates of numbers of birds in colonies are less accurate and are subject to the usual problems encountered in dealing with large numbers or in viewing cryptically colored birds from an airplane (Kushlan, 1979a). Thus, aerial estimates of numbers of white-plumaged birds are usually much more accurate than estimates of dark-plumaged species. Almost all of the Audubon warden data were gathered on the ground, whereas many of the recent surveys were conducted from the air. Because many of the colonies during the recent period were also visited on the ground (Bancroft and Jewell, 1987; Bancroft et al., 1988a; Frederick and Collopy, 1988, 1989; Frohring, 1988; Ogden et al., 1987) and most of the birds censused from the air were white rather than cryptically colored, the between-period changes in estimated numbers of breeding birds reported here should accurately reflect regional patterns and trends.

Two useful methods of expressing general levels of colony nesting success are to estimate the number of nests that successfully fledged one or more young compared to the total number of active nests or to estimate the total number of young birds fledged from a colony in relation to the number of pairs nesting. The Audubon wardens during the 1930s and 1940s rarely did either. In fact, quantitative estimates of nesting success in either period are relatively rare. Thus, the analysis of success patterns was supplemented by adding a more dependable wood stork data set for the years 1953–89. A wood stork colony was considered to have been successful if that colony fledged a number of young birds that equaled or exceeded the maximum estimate during the same year for adult pairs in that colony. The more extensive stork data set was also used to examine relationships between colony patterns and hydrological conditions in more detail than was possible with the general wading bird data.

When more precise information on colony success was lacking for other species of wading birds and for all species during the early period, a colony was characterized as successful if it was reported to have produced large numbers of young with no indications of stress or partial failures affecting the site. A colony was characterized as a partial failure if it produced numbers of young, but was known to have had a substantial segment of the nesting pairs fail. A colony was characterized as failed if the site was completely or largely abandoned prior to the completion of nesting. With the exception of the supplemental stork data, these estimates of colony success were included only because they appear to be consistent with larger patterns of change between periods.

REVIEW

The major papers since 1970 (prior to this volume) that dealt with wading bird populations in the Everglades are briefly summarized in this section. The summary comments cover only those aspects of each paper that pertain to the relationship of Everglades regional colony patterns to long-term changes in the spatial and temporal scales of the system.

Prior to Ogden (1978), the long-held view that the number of wading birds nesting in the southern Everglades must have numbered one million or more birds as recently as the 1930s was widely accepted (Robertson and Kushlan, 1974; Kushlan and White, 1977). Estimates of many hundreds of thousands of birds in south Florida "rookeries" were published repeatedly beginning in the 1930s (Allen, 1934, 1937, 1958; Holt, 1933), resulting in widespread acceptance of these numbers. Robertson and Kushlan (1974) summarized the history and status of southern Everglades wading bird populations and included an overview of the environmental factors that caused the post-1930s declines. This was the first report of the adverse impacts of water management practices on Everglades wading bird populations. Kushlan and White (1977) presented the results of the first intensive aerial survey of colonies to be conducted in the Everglades–Big Cypress region (1974–75). This paper was the first to report a shift by nesting wading birds out of Everglades National Park; only 20% of the total south Florida nesting population nested in the park that year.

Ogden (1978), followed by Kushlan and Frohring (1986), Frohring et al. (1988), and Frederick and Collopy (1988), suggested that earlier published estimates on the size of the 1930s population of southern Everglades wading birds were too high. Ogden (1978) and Frohring et al. (1988) specifically questioned the estimate of birds in Lane River rookery in 1933 and at Rookery Branch in 1934. Kushlan and Frohring (1986) reported that no data existed to show that the historical population of nesting wood storks in southern Florida was ever larger than it was in 1967.

Kushlan and Frohring (1986) also concluded that the "maintenance of seasonally excessive water levels in Everglades National Park account for the stork's repeated nesting failures..." and that a recent shift by nesting storks from traditional colony sites to new colonies in the central and northern Everglades "...may well be a response to unnatural but relatively beneficial patterns of water level fluctuations in the northern marshes caused by current water management policy." In this and another report (Kushlan, 1987), it was concluded that increased rates of drying result in improved colony success rates for storks in Everglades National Park.

The major theme presented by Frohring et al. (1988) was that while historical data on wading birds are extremely valuable in understanding population trends, the nature of these data requires that they be interpreted with caution. They argued that the estimates of numbers of wading birds in the 1930s colonies were routinely exaggerated by the Audubon wardens, in large part in support of a "...drive to establish a national park in the Everglades... ."

Kushlan (1976a, 1977, 1979b) reported on seasonal and annual responses by foraging and nesting white ibis to water levels in the Everglades. These were the first reports to show the importance of the central Everglades (Water Conservation Areas) as nesting habitat for white ibis and are important in that they show seasonal and interannual changes in foraging and colony sites for ibis to be related to hydrological patterns.

Frederick and Collopy (1988, 1989) reviewed current opinion regarding the reasons for the regional population declines and included a discussion of the relative importance of reproductive failures and shifts in nesting locations to areas outside of the Everglades. Their papers, plus those by Bancroft and Jewell (1987) and Bancroft et al. (1988a), include some of the few published data on colony success rates for great egrets, small herons, and ibis from the Everglades region. In addition, Frederick and Collopy (1988, 1989) described factors regulating nesting success at colonies in Water Conservation Area 3 in 1986 and 1987.

Bancroft (1989) reviewed wading bird population trends in the Everglades between the 1930s and the present. This report also included a summary of the results of recent systematic aerial surveys of feeding wading birds conducted by the National Park Service and the National Audubon Society.

Robertson (1965) was the first to show that the historical pattern was for relatively larger numbers of wood storks to nest and to be successful in the Everglades during the wetter years. Kushlan et al. (1975) showed that a breakdown in this relationship occurred beginning in 1962, and they demonstrated a relationship between the timing of wood stork nesting and the rate of drying during the dry season. Ogden et al. (1978) reported on regional foraging patterns by wood storks related to colony activity and fish availability in the southern Everglades during a single year (1974).

The status of the wood stork nesting population in southern Florida was summarized by Ogden and Nesbitt (1979) and Ogden et al. (1987). The latter suggested that storks that once nested in the Everglades have shifted their colony locations to the region between central Florida and coastal South Carolina.

Ogden et al. (1988) reviewed the history of wading birds in the Everglades since the 1930s and summarized the factors responsible for the population declines. They speculated that the decline in numbers of nesting birds may have been due in part to a reduction in the range of options where wading birds can forage under a variety of hydrological conditions. The importance of a range of foraging habitats for wading birds, both within and between seasons and interannually, was previously described by Bancroft et al. (1988a), Browder (1976), Kushlan (1979b, 1986), and Kushlan et al. (1975).

Finally, two recent papers reported on wading bird colony patterns and population trends in regions immediately adjacent to the Everglades. Bancroft et al. (1988b) described colony patterns in the Big Cypress region of Collier County from 1982 to 1985. They showed annual colony turnover rates of 30–40% and indicated that colony formation occurred later in a season that followed 18 months of below normal rainfall. Powell et al. (1989) reviewed the population histories for three species of wading birds in Florida Bay that are "specialists" of that estuarine ecosystem (i.e., great white heron [*Ardea herodias occidentalis*], reddish egret [*Egretta rufescens*], and roseate spoonbill [*Ajaia ajaja*]).

RESULTS

Numbers of Breeding Birds

The estimated number of wading birds for five species nesting in the central-southern Everglades declined by an estimated 75–80% between the year when the highest number nested during the early period (1934) and the similar high year in the recent period (1976) (Table 22.1). For the early period, the annual nesting population ranged between no birds in 1939 and approximately 180,000–245,000 birds in 1933 and 1934. The annual mean for the early period, depending on whether the low or high values were used from the years where only an estimated range was possible, was approximately 69,000–88,000 nesting birds.

For the recent period, the annual nesting population ranged between estimates of 5000 birds in 1985 and 50,000 in 1976 (Table 22.1). For the entire 16-year period, the annual mean was approximately 18,000 nesting birds. The suggestion of a continuing decline during the recent period is shown by the mean of 26,000 birds nesting in the first half of this period compared with a mean of 10,000 nesting in the second half.

The estimated maximum number of nesting birds for each of the five species reported for any year during the two periods is shown in Table 22.2. For the early period, the estimate for snowy egrets and tricolored herons is combined, because Audubon wardens often reported them that way. The species estimates from the recent period have been divided into two equal subperiods of 8 years each, to show the recent pattern of population trends.

The most abundant nesting species was the white ibis (Table 22.2). Ibis made up approximately 84% of the total number of nesting birds of these five species in the early period, compared to 52% of the total nesting birds during the recent period.

From the early to the recent period, all species showed substantial declines except for the great egret (Table 22.2). The number of nesting great egrets appears to have remained stable between the early period and the first subperiod of the recent period and then declined between the two subperiods. The white ibis showed the greatest numerical decline from the estimated maxima of 225,000–

Table 22.1 Estimated Total Number of Five Species of Wading Birds (in Thousands) Nesting in the Central-Southern Everglades, Florida for 1931–46 and 1974–89[a]

Year	Birds	Year	Birds
1931	100–110	1974	27.0
1932	40–50	1975	49.0
1933	180–235	1976	50.0
1934	195–245	1977	41.0
1935	<20	1978	9.0
1936	60	1979	18.0
1937	15–55	1980	6.5
1938	<10	1981	10.0
1939	0	1982	17.0
1940	150	1983	5.5
1941	35–105	1984	7.0
1942	?	1985	5.0
1943	?	1986	10.5
1944	?	1987	12.0
1945	?	1988	15.0
1946	>25+	1989	10.0
Mean	69.1–88.7[b]		18.2
			(1974–81 = 26.1)
			(1982–89 = 10.3)

[a] Data are from Kushlan and White (1977), Frederick and Collopy (1988), and unpublished files of Everglades National Park and the National Audubon Society.

[b] Excludes years of insufficient data (1942–46). The annual mean for the early period, depending on whether the low or high values were used from the years where only an estimated range was possible, was approximately 69,000–88,000 nesting birds.

29,000 birds. On a percentage basis, the total maximum nesting population for all species combined declined approximately 80% between the early period and the first subperiod of the recent period and declined 93% to the second subperiod. The decline between the two subperiods of the recent period was approximately 66%.

Table 22.2 Estimated Maximum Number of Nesting Wading Birds for Five Species in the Central-Southern Everglades for 1931–46 and 1974–89

Species	1931–46	1974–81	1982–89
Great egret	5,000–8,000	6,500	4,200
Snowy egret/tricolored heron	20,000–30,000	9,000/7,000	3,000/2,000
White ibis	175,000–225,000	29,000	12,500
Wood stork	5,000–8,000	2,650	750
Total	205,000–271,000	54,150	22,450
Percent decline	—	73.6–80.1	89.1–91.8

Colony Locations

All major nesting colonies that were known to be active during either of the two periods were located in one of five ecologically and geographically distinct subregions within the central and southern Everglades and the downstream mangrove estuary (Figure 22.2). These five subregions were (1) the lower Ten Thousand Islands (a string of natural islands close offshore along the northwestern coast of Everglades National Park between the mouth of Lostmans River and Everglades City); (2) northwestern Florida Bay (a cluster of islands in Florida Bay located south of Flamingo); (3) the southern mainland estuary (located from the Cape Sable peninsula east across the southern mainland mangrove estuary to the lower Taylor Slough basin); (4) the headwaters of the major rivers that feed into Whitewater Bay and the lower Gulf coast, along the mangrove-freshwater marsh ecotone at the lower end of the Shark River Slough (approximately East River to Lostmans River); and (5) the interior freshwater Everglades north of the lower Shark River Slough, overlapping the current Dade–Broward county line.

In the years during the early period when relatively large numbers of birds nested (Table 22.1), the vast majority of the nesting occurred in the headwaters subregion (Table 22.3). The remaining subregions contained much smaller num-

Figure 22.2 The central and southern Everglades region of wading bird nesting colonies and the locations of five wading bird colony subregions: (1) lower Ten Thousand Islands, (2) north-western Florida Bay, (3) southern mainland, (4) headwaters, (5) interior Everglades.

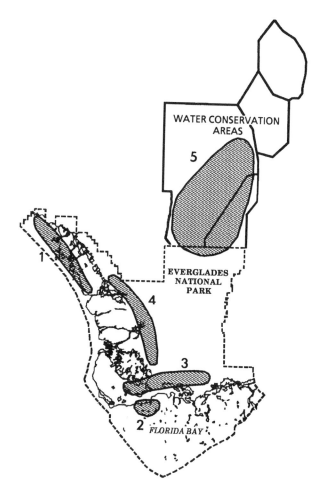

Table 22.3 Five Nesting Area Subregions in the Central and Southern Everglades (Major Colonies and Principal Species within Each) for the Early (1931–46) and Recent (1974–89) Periods

Subregion	1931–46	Percentage of total nesting population and principal species[a]	1974–89	Percentage of total nesting population and principal species[a]
No. 1 Lower Ten Thousand Islands	Duck Rock Gopher Key Pelican Key Plover Key Pumpkin Key	(5%) SE TH WI	Chokoloskee only 1974–75	TH[b]
No. 2 NW Florida Bay	Dildo Key	(1%) GE SE TH	Frank Key Sandy Key	(10%) GE TH SE WI
No. 3 Southern mainland	Alligator Lake Mud Lake Mud Hole Cuthbert Lake	(3%) WS GE	Cuthbert Lake Madeira Rookery	(5%) WS GE
No. 4 Headwaters	East River Lane River Rookery Branch Broad River	(90%) WI SE TH WS GE	East River Lane River Rodgers River Bay	(15%) SE TH GE WI
No. 5 Interior Glades	Blue Shanty	(1%) SE TH	Andytown L-67 Mud Canal Big Melaleuca	(70%) WI GE SE TH

[a] Principal species are listed in approximate order of abundance of nesting birds in the subregion, based on an "average" year. (GE = great egret; SE = snowy egret; TH = tricolored heron; WI = white ibis; WS = wood stork.)

[b] No percentage was calculated because colony was used only one year.

bers of nesting birds, although sharp increases in nesting effort occurred in some coastal colonies in years when relatively few birds nested in the headwaters (e.g., 1936 in the Ten Thousand Islands). The strong year-to-year differences in numbers of nesting birds during the early period were largely due to wide annual variation in reproductive effort by white ibis in the headwaters colonies (Table 22.4).

Table 22.4 Estimated Number of Nesting Birds in Each Everglades Subregion for Each Year during 1931–46[a]

	Year	Subregion[b] 1	2	3	4	5	Year	Subregion[b] 1	2	3	4	5
GE	1931	c	1000		N		1939					
SH		c	4000	N	N							
WI		c			100K							
WS					N							
GE	1932		N[d]		N		1940		1500		N	3500
SH		7000	N[d]	N	N			1000			N	
WI					25–30K						90–100K	
WS				2000						N	N	
GE	1933		N	500	1000		1941					
SH		800	400	5500	20–25K			1000			N	N
WI					150–200K			4000			30–100K	
WS				3000	3000							
GE	1934		1000	100	1500		1942			N	200	
SH		500	500	1500	10K			N			N	
WI				400	175–225K			N			10K	
WS				3000	5000					2500	4000	
GE	1935				N		1943		N	N	N	
SH		N						N[f]	N			
WI								N[f]				
WS				N	4000							
GE	1936		5000	N	3000		1944		500	N	N[g]	
SH		25K	6000		1000				2000	N	N	
WI					15K				N		N	
WS				N	4000					N[h]	3000–4000	
GE	1937				N[e]		1945		N	N	N	
SH		500			N[e]			100			N	
WI					5–40K						N	
WS				3000	N					N	N	
GE	1938				500		1946				N[i]	
SH								N			N	
WI											25K	
WS				200+						N	N	

[a] N = nested, no numbers; GE = great egret; SH = small herons (snowy egret and tricolored heron); WI = white ibis; WS = wood stork.

[b] See Table 22.3 for subregion locations.

[c] Several thousand egrets, herons, and ibis nesting.

[d] 1500 combined.

[e] 5000–10,000 combined.

[f] 5000–10,000 combined.

[g] East River "3 times as many as in 1943."

[h] "Several thousand."

[i] "Largest nesting in years."

Major differences in the distribution of nesting wading birds occurred from the early to the recent period (Table 22.3). These changes included the abandonment of colonies in the Ten Thousand Islands subregion, a substantial decline in the number of birds nesting in the mainland estuary and headwaters subregions, and a large increase in the number nesting in the interior Everglades subregion. The collapse of the headwaters colonies and the parallel growth of the interior Everglades colonies (primarily located north of the park) are the most significant distributional changes between the two periods. This northward shift in nesting distribution has involved a relatively high percentage of the regional nesting population for all species except the wood stork.

Differences in relative importance were found between the different subregions for the different species (Tables 22.4 and 22.5). During the early period, great egrets nested primarily in the northwestern Florida Bay, mainland estuary, and headwaters subregions, whereas wood storks nested in the mainland estuaries subregion and in the southern colonies in the headwaters subregion. The small herons/egrets and ibis nested in large numbers in all subregions except the mainland estuaries. Species distributional patterns during the recent period remained approximately the same, except that ibis continued to be common nesters in the headwaters subregion only at the northernmost colony site (Rodgers River Bay) and wood storks had largely abandoned the headwaters by the end of the recent period.

Timing of Nesting

Information on the average timing of colony formation and completion of nesting for each of the five species and for each of the periods is shown in Table 22.6. For all species in the early period, the average sequence of initiations of nesting was wood storks first (early December), followed by great egrets (mid-January), then the small herons (late February), and finally white ibis (early March). In years during the early period when relatively large numbers of birds nested, the peak numbers nested from March through May. In some years when relatively smaller numbers of birds nested, and the large headwaters colonies did not form, a relative increase occurred in the number of birds that nested later (April–August) and in coastal colonies (e.g., 1932, 1936, 1937). During the early period, the timing of coastal nesting (especially for all species in the Ten Thousand Islands subregion and for white ibis in the interior) averaged later than inland nesting for herons and egrets.

In the recent period, the sequence of nesting initiations was great egrets first (mid-January), followed by wood storks (late January), then small herons and egrets (mid-February), and finally white ibis (late March) (Table 22.6). Coastal nesting occurred on the same schedule as inland nesting in the southern Everglades. Colonies in the central Everglades tended to form slightly later than more southern colonies, although white ibis tended to start nesting earlier in the central Everglades than in the southern. Some years also had strong nesting efforts during the summer, for example at Rodgers River Bay in 1975, 1977, and 1983 and at Frank Key in some years.

Only the wood stork showed a substantial change in timing from the early to recent periods (Table 22.6). On average, wood storks initiated and completed

Table 22.5 Estimated Number of Nesting Birds in Each Everglades Subregion for Each Year during 1974–89[a]

		Subregion[b]							Subregion[b]			
	Year	1	2	3	4	5	Year	1	2	3	4	5
GE	1974		500	350	1200	N	1982			100	950	800
SH		800	2000		2500	N				100	1500	
WI			500		800	17K			300			12.5K
WS				500	1500					400	360	
GE	1975		700	600	1500	4000	1983		1000	500	1400	350
SH		100	300	300	6500	7300			500		100	
WI			1700		2000	22K			400			
WS					2200							
GE	1976		800	300	250	1200	1984	N	900		3000	800
SH			1450	2600		5000					500	
WI			4800		2000	29K						
WS					2500					700	600	
GE	1977		900	400	300	3200	1985	N	100	N		600
SH			500		2600	11.4K		N		N		
WI			1300		2300	16K						3000
WS				120	1200					250		
GE	1978		350	400	800	3000	1986	N	200	700	300	
SH			450	250	1900			N	100	4000	120	
WI			500		300	600				400	3200	
WS										750		
GE	1979		450	900	1350	2100	1987	N		300	550	750
SH			600	250	8000	500			2000		2000	1200
WI			400		200				1200		60	4100
WS				850	1500					300		
GE	1980		300	300	700	1000	1988	N		300	800	3200
SH			400		3200			N		50	3000	2500
WI			500									3000
WS				100						440		
GE	1981		350	350	600	900	1989	N		400	700	2100
SH			30		500	1000			2200	2100	600	1200
WI			500		430	4000						3000
WS				900	80					500		

[a] N = nested, no number; GE = great egret; SH = small herons (snowy egret and tricolored heron); WI = white ibis; WS = wood stork.
[b] See Table 22.3 for subregion locations.

nesting from early December to late April during the early period and from late January to early June during the recent period. Differences between the first and second halves of the recent period suggest that the timing of stork nesting continued to shift to later in the dry season during that period.

Table 22.6 Average Timing of Colony Initiation and Completion by Species, Subregion, and Period for the Florida Everglades

| Species | Subregions | 1931–46[a] | | 1974–89[a] | |
		Initiation	Completion	Initiation	Completion
Great egret	1	7.0 (2; 7–7)	No data		
	2	4.0 (7; 3–7)	7.4 (8; 5–10)	4.0 (8; 2–6)	7.7 (8; 6–9)
	3/4	3.5 (13; 1–6)	7.1 (10; 6–9)	3.6 (15; 2–4)	8.1 (15; 6–10)
	5			4.5 (13; 4–5)	8.2 (13; 7–9)
Snowy egret and	1	7.16 (12; 5–11)	10.57 (7; 10–11)		
tricolored heron	2	5.0 (4; 3–7)	8.4 (5; 5–11)	4.5 (1; 2–6)	8.5 (6; 7–10)
	3/4	4.73 (19; 3–6)	8.12 (16; 7–10)	4.5 (15; 3–6)	8.3 (15; 7–9)
	5			5.0 (10; 4–6)	8.1 (10; 7–9)
White ibis	1	7.3 (3; 7–8)	10.5 (2; 10–11)		
	2	7.0 (1)	9.0 (1)	6.7 (6; 5–8)	10.0 (6; 9–11)
	3/4	5.2 (12; 3–8)	8.7 (8; 7–11)	6.6 (11; 5–8)	9.4 (9; 7–11)
	5			5.8 (11; 5–7)	8.3 (9; 7–10)
Wood stork[b]	3/4	2.1 (9; 1–3)	6.8 (13; 6–8)	3.9 (15; 2–5)	8.2 (14; 7–9)
				3.7 (8; 2–5)	8.0 (8; 7–9)
				4.1 (7; 3–5)	8.3 (7; 7–9)

[a] First number represents average month (1 = November, etc.). Numbers in parentheses represent number of colonies in sample and the range of months in that sample.

[b] Wood stork data from the recent period are shown for the total period and divided into early (1974–81) and late (1982–89) subperiods, respectively.

Success Rates

The Audubon reports for the early period gave the impression that many more colonies fledged large numbers of young than those that did not. A common comment from a warden was that a colony "…contained thousands of hatching birds…" and that 4–8 weeks later the same colony still contained "…large numbers of young…" or "…many young are beginning to fly." Not all colonies were successful, and those that had problems apparently were noted with comments. For example, the large colony at Rookery Branch may have been largely unsuccessful in 1934 due to heavy rains in late May. The warden report for May 17, following five uninterrupted days of rain, contained the observation that they "…found young birds dead in piles. There is [sic] all ages. A foot of water over the nests that are on the ground." Using such qualitative statements as a basis for scoring colonies into broad success or failure categories, Table 22.7 was compiled in an attempt to present general patterns of colony success for the early period. These data are perhaps the best available for this period. Success data from colonies during the recent period are shown in Table 22.8. Data in these two tables suggest that a decline in colony success rates occurred from the early to recent periods (80.0% versus 43.7% for fully successful colonies and 86.6% versus 69.7% for fully successful and partial failures combined).

Table 22.7 Major Colonies during the Early Period (1931–46), Years of Activity, and Success Rates[a]

					Colony				
Year	Lane River	East River	Cuthbert	Alligator Lake	Rookery Branch	Dildo Key	Duck Rock	Mud Hole/Lake	Other
1931	A	S		F(WS)	N	S	A		
1932	A		S(WS)		N	S	S		
1933	A	S(WS) M			S	S	A	A	
1934	A	A		S(WS)	F	S	S	S(WS)	
1935	S(WS)	N			N	N	S		
1936		S(WS)				S	S	A	Pelican = S
1937		F	S(WS)		A	S	A	S(WS)	Pumpkin = S, Pelican = A
1938	N	N		S(WS)	N				
1939	N	N			N		N		
1940	N	A	S(WS)		F(WS)	A			Pelican = A
1941	N	S			N		S		Pelican = A, Plover = S
1942	N	S(WS) M	S(WS)		N		A		Pelican = A
1943	N	S(WS)	S(WS)			S	A		Clive = S
1944	N	S(WS)	S(WS)		F	S			Catfish = S
1945	N	S(WS)	S(WS)			S	A		
1946	N	S(WS) S	A		F		A		

Success/failure summary: Wood stork: 19 S (90.4%), 2 F
All other waders: 24 S (80.0%), 4 F, 2 M

[a] A = active; N = not active; S = active/successful; F = active/failed; M = active/partial failure; WS = wood stork; success or failure unknown for other species present.

Wood Storks

The strongest data on timing patterns, reproductive effort (number of nesting birds), and nesting success over a long period in the Everglades region are for wood storks. Storks formed colonies in either November or December in 14 of 16 years between 1953 and 1969, but only as early as December in 2 of 20 years between 1970 and 1989 (Figure 22.3). Storks were successful in 7 of 9 years between 1953 and 1961, a success rate not too different from the 90% rate estimated for the early period for storks (Table 22.7). Stork colonies in Everglades National Park were successful in only 6 of 28 years between 1962 and 1989. Reproductive effort was greatest in early forming colonies, ranging from a mean estimate of 2250 pairs in November colonies to 450 pairs in March colonies (Figure 22.4).

Colony Dynamics Related to Hydrological Conditions

Rainfall data and the qualitative assessments of the wardens for both surface water conditions and general patterns of wading bird responses during the early

Table 22.8 Major Colonies during Recent Period (1974–89), Years of Activity, and Success Rates[a]

				Colony					
Year	Lane River	East River	Cuthbert	Rodgers River	Madeira	Frank Key	Andytown	L-67	Other
1974	S	A	S	S	S	S			
1975	S	A	S	S	F	S	S	S	
1976	F	F	S	F	F	S	M	F	
1977	F	M	M	M	A	S	M	S	
1978	M	F	S	S	N	S	M	S	
1979	F	F	M	A	F	A	S	S	
1980	A	F	M	A	A	A	M	S	
1981	M	F	S	A	F	A	M	M	Big Mike = S
1982	F	F	F	A		A	A		
1983	A	A	S	M	A	S	S		
1984	F	F	F	M	M	M	M		
1985	A	S	S			A	A	A	
1986	M	S	M	S	A	A	M	S	Coopertown = F Big Melaleuca = S
1987	F	F	M	F		S	S	S	Coopertown = S Big Melaleuca = S
1988	N	F	M	S	A	A	S	S	Big Melaleuca = F
1989	N	F	M	F	F	A	S	M	

Success/failure summary: 42 S (43.7%); 25 M (26.0%); S + M = 69.7%)

[a] A = active; N = not active; S = active/successful; F = active/failed; M = active/partial failure; WS = wood stork; success or failure unknown for other species present.

period are shown in Table 22.9. The annual hydrological patterns that are suggested by these data and reports, when compared with the early period nesting record (Tables 22.1 and 22.4), show that more birds initiated nesting in years that were wetter than normal than in years that were drier than normal, although this was not the case in years when the marshes were unusually deeply flooded during the primary nesting season months. The largest colonies during the early period formed in the relatively normal years, characterized by moderate spring season marsh conditions (1933, 1934, and 1940). In all years that were clearly on the dry side, especially when dry conditions occurred during the main nesting season months (March and April), relatively few wading birds attempted nesting (1932, 1935, 1938, 1939).

The nesting efforts during 1942–46 were not as well reported as the earlier years, especially in terms of numbers of birds. Overall, 1942–46 seems to have been a relatively dry period, with reduced numbers of nesting birds compared with the previous years. Dry conditions during the spring of 1946, for example, were suggested by the wardens as an explanation for the collapse of a large colony at Rookery Branch during March.

The location and timing of nesting during the early period seems to have been influenced by surface water conditions. Years when relatively large numbers nested either in the Ten Thousand Islands or the northwestern Florida Bay subregions were usually the extreme years (1932 and 1943 were dry, 1936 and 1941 were very

Figure 22.3   Timing (by month) of colony formation and colony success for wood storks nesting in Everglades National Park, 1953–89.

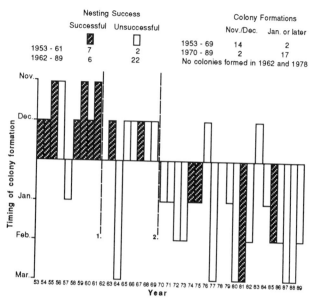

1. Implementation of S-12 water delivery system
2. Implementation of L-67 extension and minimum delivery schedule

wet, 1931 was normal). Years when relatively large numbers of birds nested in the headwaters subregion were either moderately wet (1931, 1941) or normal years (1933, 1934, 1940). In dry years, coastal nesting tended to start late (May or later in 1932, 1935, 1943, and 1945?), while in wet or normal years coastal nesting started in January–April. For small herons, egrets, and ibis in the headwaters colonies, nesting tended to start late in wet years (April or later in 1931, 1937, 1941).

Rainfall and surface water data for the recent period, including a characterization of regional marsh conditions, are presented in Table 22.10. When compared with recent nesting patterns (Tables 22.1 and 22.5), these data show greater numbers of nesting birds in years that were drier than normal, although this

Figure 22.4   Timing of wood stork colony formation related to magnitude of nesting effort and to colony success in Everglades National Park, 1955–89.

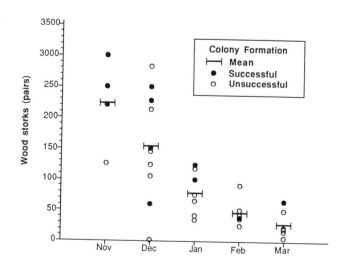

Table 22.9 Early Period Rainfall and Hydrologic Conditions, Including Pertinent Comments from Audubon Warden Reports Related to Surface Water Conditions and Bird Responses, and an Overall Characterization of Each Year in the Southern Everglades

Year	Rainfall[a]		Surface water: warden reports	Bird responses: warden reports	Overall assessment
	Wet season	Dry season			
1931	No data	No data	Feb.: water high; Mar. 29: water 12 in. higher than last year this time	Due to high water, birds late nesting	Wet
1932	−15.57	−2.59	Apr. 2: "terribly dry;" Apr. 27: "fires everywhere near headwaters"	Apr. 12: "looks like a million birds feeding in marsh above Harney;" Apr. 14: "roosting along Rookery Branch for 2 miles;" May 5: Harney Roost deserted	Dry
1933	+9.52	+0.30		April: "estimated 2 million young from 6 rookeries…;" main nesting late March–June	Normal
1934	+7.43	−2.51	Heavy rains began May 17, flooded Rookery Branch	Feb. 9: "Marsh white with birds as far as eye can see at Rookery Branch;" March: "most birds feeding within one mile of Rookery Branch;" nesting mainly late Feb.–May	Normal
1935	+6.63	−7.52	Apr. 26: "glades are dry—been dry for weeks. No birds. Many fires."	Jan. 9: very large roost at Rookery Branch; dropped off after Jan. 26; Apr. 8: "birds leaving headwaters area fast;" nesting May–Sep. on coast	Dry
1936	+8.13	+9.26	March: "High water made normal nesting sites up the rivers unfavorable;" "excessively high water"	"Due to high water, only one rather small ibis rookery near Homestead;" coastal nesting March–June	Wet
1937	+17.70	−0.56	Mar. 16: "deep water" in marshes	Mar. 13 aerial survey: "very few birds in lower glades; most are in open glades to east of Big Cypress;" main nesting April–July	Wet
1938	+6.36	−3.16	Winter: "long spell of fire and drought." "Canals 4 feet below normal in spring;" April: "glades baked hard"	Large roost at Rookery Branch abandoned in early March	Dry

Table 22.9 (continued) Early Period Rainfall and Hydrologic Conditions, Including Pertinent Comments from Audubon Warden Reports Related to Surface Water Conditions and Bird Responses, and an Overall Characterization of Each Year in the Southern Everglades

Year	Rainfall[a] Wet season	Dry season	Surface water: warden reports	Bird responses: warden reports	Overall assessment
1939	−16.82	−4.57	Feb.–April: "fires swept central and eastern sections of glades;" June: "very little water south of Tamiami Trail in glades"	Jan. 18: "due to low water, birds are either on coast or in Big Cypress;" Dec.– Jan.: "only feeding con- centrations are around bays between glades and Gulf: Alligator Bay, Lostman's Bay, Rookery Bay"	Dry
1940	+10.07	−0.14	Apr. 19: "Water conditions and food supply is excellent;" May 7: "very dry around Loop Road"	Main nesting March–May	Normal
1941	+8.89	+8.44	Early Feb.: heavy rains and high water; Mar. 2: "water very high at Rookery Branch"	Main nesting April–July	Wet
1942	−2.42	+3.72		Nesting effort relatively low; large roost at Rookery Branch abandoned in March; main nesting April–June	Normal
1943	−5.92	−1.85		Little nesting except on coast beginning in June	Dry
1944	+2.38	−3.88		"Three times as many birds at East River as in 1943;" nesting Feb.–May	Normal
1945	+0.36	−4.29	Apr. 4: "counted six fires in Shark River area"	Rookeries not large; ibis began in March	Dry
1946	−6.57	+1.66	Mar. 3: "four fires near Rookery Branch"	Rookery Branch colony collapsed in March; East River "largest in years"	Dry

[a] Cumulative deviation from monthly means for all months in each season.

relationship only seems to hold for the 1970s. For 1974–80, the largest reproductive effort occurred in the drier years and the smallest effort in the wetter years. In contrast, between 1981 and 1989, the total reproductive effort was greatly reduced from the previous subperiod and was further reduced in the drier years relative to the normal and wetter years.

Table 22.10 Recent Period Hydrological Conditions for Shark River Slough and Water Conservation Area 3A, with Total Wading Bird Nesting Effort in Each Division

Year	Rainfall[a] Wet season	Rainfall[a] Dry season	P-36 (Shark River River Slough) deviation from normal March high	3A-4 (WCA 3A) deviation from normal March high	Total nesting[b] ENP	Total nesting[b] WCA 3A	Overall characterization
1974	-4.58	-4.58	-0.65	-0.95	10.0	17.0	Dry
1975	-15.16	-3.77	-0.77	-0.89	16.0	33.0	Dry
1976	-3.69	-0.09	-0.56	-0.16	15.0	35.0	Dry
1977	+7.10	-1.37	-0.33	-0.13	10.5	30.5	Normal
1978	+3.29	+5.84	-0.17	+1.02	5.0	4.0	Wet
1979	-2.77	-2.21	-0.20	+0.40	15.0	3.0	Normal
1980	-4.79	+2.56	+0.30	+1.18	5.5	1.0	Wet
1981	+6.15	+4.85	+0.33	+0.07	4.0	6.0	Wet
1982	+14.26	-0.83	+0.07	-0.50	4.0	13.0	Normal
1983	-14.29	+17.09	+1.15	+1.64	5.0	0.5	Wet
1984	-5.80	+1.49	+0.72	+0.43	6.0	0.8	Wet
1985	-6.30	-3.78	-0.03	-0.49	1.5	3.5	Dry
1986	-15.60	+7.74	+0.15	+0.10	7.0	3.5	Normal
1987	-15.82	+2.96	+0.23	+0.10	6.0	6.0	Normal
1988	-11.12	+1.38	+0.32	+0.93	6.0	9.0	Wet
1989	+15.07	-5.13	-0.47	-0.32	3.0	7.0	Dry

	Means[b]			
	1974–80	(No.)	1981–89	(No.)
Everglades National Park				
Dry	14.3	(3)	2.2	(2)
Normal	12.7	(2)	5.6	(3)
Wet	5.2	(2)	5.2	(4)
Water Conservation Area 3A				
Dry	28.3	(3)	5.2	(2)
Normal	16.7	(2)	7.5	(3)
Wet	2.5	(2)	4.1	(4)

[a] Cumulative deviation from monthly means for all months in each season.
[b] Numbers of nesting birds in thousands.
[c] Number of years.

For wood storks, colony timing and reproductive effort changed rather abruptly after 1969 (Figure 22.3). Prior to 1969, most colonies formed during November or December, and the maximum reproductive effort was often in the range of 4000–5000 birds. After 1969, most colonies formed between January and March, and the maximum reproductive effort never exceeded 2500 birds. The magnitude of the reproductive effort before and after 1969 is compared in Table 22.11 with maximum December stages at P-33; before 1969 the average stork reproductive effort was greatest when water levels were relatively high and lowest when levels were low. The reverse was true after 1969, with the larger colonies tending to form in the years with drier December conditions.

To examine the ways that water depth and distribution further influenced

Table 22.11 Maximum December Stages (feet),
Expressed as Deviation from Normal Maximum Stage in December,
at P-33 (Shark River Slough) Related to Number of Wood Storks in Colonies, 1956–89

Year	Deviation from normal	No. birds	Year	Deviation from normal	No. birds
1956	−0.39	2500	1973	+0.05	1800
1957	No data available		1974	−0.39	2000
1958	+0.24	1200	1975	+0.09	2450
1959	−0.19	4400	1976	−0.34	2500
1960	+0.73	5000	1977	−0.10	1300
1961	+0.94	5000	1978	−0.05	0
1962	−0.74	0	1979	+0.30	2350
1963	−0.66	3000	1980	+0.49	100
1964	−0.76	400+	1981	+0.40	1000
1965	−0.68	4250	1982	+0.16	760
1966	−0.58	2900	1983	+1.01	30
1967	−0.12	4550	1984	+0.40	1300
1968	−0.41	5650	1985	+0.23	850
1969	+0.15	2100	1986	+0.04	750
1970	+1.01	800	1987	−0.20	300
1971	−0.18	700	1988	+0.19	440
1972	90.25	1000	1989	−0.05	500

Summary 1956–69   −0.3 or lower (N = 7) = 2670 birds
+0.3 or higher (N = 2) = 5000 birds
Between +0.3 and −0.3 (N = 4) = 3062 birds

Summary 1970–89   −0.3 or lower (N = 2) = 2250 birds
+0.3 or higher (N = 6) = 930 birds
Between +0.3 and −0.3 (N = 12) = 940 birds

colony patterns, maximum December stages at two gauges located in higher elevation marshes that presumably were important early dry season foraging habitats are compared with the timing of wood stork colony formation in Table 22.12. These data show that November colony formations were associated with higher stages than were later formations during 1953–69, when colony schedules continued to be normal (Figure 22.3). A reversal in this relationship occurred after 1969, when early formation was associated with relatively low December stages and progressively later formation was associated with progressively higher stages.

DISCUSSION

This section must be recognized as part speculation and part empirically based discussion. For example, the most likely patterns and relationships that character-ized early period colony dynamics are suggested. To do so, it has been assumed, based on recent studies (Bildstein et al., 1990; Frederick and Collopy, 1989; Hafner and Britton, 1983; Kushlan, 1979b), that the locations, sizes, and timing of colonies

Table 22.12 Timing of Wood Stork Colony Formations in Everglades National Park in Relation to Means of Maximum December Stages at Two High-Elevation, Short-Hydroperiod Marsh Stations East of Shark River Slough, 1953–88

Month of colony formation	1953–69			1970–88		
	G-596[a]	N[b]	G-789[c]	G-596[a]	N[b]	G-789[c]
Nov.	3.37	3	4.78	—	0	—
Dec.	2.67	10	3.64	1.86	2	2.96
Jan.	1.21	1	2.50	2.02	6	3.03
Feb.	—	0	—	2.07	6	3.12
Mar.	1.89	1	3.47	2.33	5	3.34

[a] Located in "rocky glades" in 25°39′37″, 80°30′40″.
[b] Sample size – number of years of colony formation during that month.
[c] Located in headwaters of Taylor Slough in 25°29′28″, 80°33′24″.

during the early period were determined by foraging patterns and prey availability. It must be true that nesting waders fed within foraging range of each colony and that colonies historically, as today, were active at locations where and times when the better food resources were available. Thus, changes in colony patterns through time are assumed to have been largely, if not entirely, due to changes in the timing, location, and strength of the foraging opportunities. It is further assumed that the quality, quantity, and availability of the food base was determined almost entirely by hydrological conditions. This assumption is derived from the above-referenced papers, as well as work by Loftus et al. (1990), Loftus and Eklund (1994), and Loftus in Ogden et al. (1988). In fact, no data on actual early period hydrological and wading bird foraging patterns exist for the Everglades. Therefore, an attempt is made throughout this section to distinguish between those ideas that are speculative and those that are based on empirical data from the two periods.

Nesting Effort

During both periods, the nesting population of wading birds in the southern Everglades varied tremendously in size among the years. In the early period, the years when relatively large numbers of birds nested (≤75,000 in 1931, 1933, 1934, 1940, and possibly 1941) were characterized by either normal or moderately wetter than normal hydrological conditions (Tables 22.1 and 22.9). The years with relatively low reproductive effort (1932, 1935, 1936, 1937, 1938, 1939, and probably all years from 1942 through 1946), with two exceptions, were years that had normal or dry hydrological conditions. The exceptions were two of the wettest years during the early period, including the year of highest dry season rainfall (1936) and the year of highest wet season rainfall (1937). The year 1941 might also belong in this very wet category. Normal hydrological years with large numbers of nesting birds (1933, 1934, and 1940) were on the wet side of normal with a mean rainfall excess of +8.22 in. (209 mm), while normal years with relatively low nesting effort (1942 and 1944) had a mean deficit of −0.10 in. (−2.5 mm) (Table 22.9).

Allen (1958) and Holt (1933) suggested that, even for the 1930s Everglades, the nesting colonies may have been atypically large in 1933 and 1934 due to the concentrating effect that extensive drought and drainage elsewhere would have had on nesting birds. The reports of the Audubon wardens during the early 1930s, however, include references to similarly large colonies during the late 1920s. The rainfall record for the entire Florida peninsula was examined for 1931–1934 to confirm that a serious drought (one extensive enough to have potentially forced a regional population of waders into a few Everglades colonies) had in fact occurred. The record showed that all Florida stations had cumulative rainfall deficiencies of from 3 to 15 in. (76 to 381 mm) from May 1931 to April 1932. The remainder of 1932 and all of 1933 showed either normal or only modest subnormal rainfall patterns. While it may be true that regional environmental conditions may have been at least partially responsible for the large size of the colonies in those two years, the fact remains that regional populations of that magnitude did exist and that the predrainage Everglades was able to support such numbers under those conditions (and possibly in earlier years as well).

Why more birds nested in the normal or wetter years during the early period is a question that cannot be answered with certainty. Presumably, with the natural system lacking levees as boundaries, the wetter years resulted in a larger area of marshland being flooded. It would follow that the greater the area flooded, the greater the area and range of options for potential feeding habitat. In the extremely wet years of the early period, the reduced reproductive effort presumably was because much of the potential foraging habitat within flight range of the traditional colony sites was too deeply flooded for use by feeding birds during the normal February–March colony formation period.

It is possible that during these wetter years, birds not only shifted the timing of nesting, but also shifted colony locations to nontraditional sites and therefore were not located. Ogden et al. (1980) showed that during a wet year great egrets used colony sites that were not used in a drier year, and Kushlan (1976a) and Kushlan and White (1977) reported both inter- and intrabasin shifts in colony locations by white ibis in the Everglades.

What is not easily explained is why the number of nesting birds during the early period was reduced in the drier years. Low water in a healthy ecosystem should concentrate a tremendous food resource earlier in the dry season than would occur during a wetter year. For this reason, relatively early colony formation would be expected by larger numbers of birds than actually occurred in the dry years of the early period. One can only speculate that reduced attractiveness of relatively large areas of higher elevation marshes as feeding habitat during the October–December period during the driest years influenced prenesting season foraging patterns of wading bird such that birds concentrated in other regions. This argument assumes that the range of foraging site options was much greater on a regional basis in 1930s Florida than is the case in modern Florida. Evidence for this regional loss of foraging habitat is shown by Browder et al. (1976), who reported the loss of 35% of the total area of southern Florida wetlands between 1900 and 1973 for five potential foraging habitat categories.

The recent period also showed strong year-to-year variation in the number of wading birds that nested, although the relationship between nesting effort and initial hydrological conditions was the reverse of the early period (Tables 22.1 and 22.10). By far, the largest number of nesting birds during the early years of the recent period (1974–80) occurred during the dry years (mean for three dry years was 42,666 birds). Increased nesting effort by wood storks in the drier years during this subperiod was shown by Kushlan et al. (1975) and Ogden et al. (1978). The four normal and wet hydrological years during this same subperiod averaged less than one-half as many nesting birds (mean of 18,625 birds). The Everglades National Park subregions (combined) and the interior Everglades subregion both showed an increase in numbers of nesting birds in dry years compared to wet years during the early subperiod (Tables 22.5 and 22.10). While the number of nesting birds more than doubled between wet and dry years in the park subregions, the number nesting in the interior Everglades increased by a factor of ten. This between-year pattern suggests that birds increased their reproductive effort at a regional level during the drier years (albeit unevenly among subregions) rather than by shifting between subregions. Because the birds were "missing" from all of the subregions during the wet years, they either did not nest in those years or they nested outside of the central southern Everglades.

Greater nesting effort in the drier years presumably was a response by wading birds to altered hydrological conditions in a compartmentalized wetland system (Bancroft et al., 1990; Gunderson and Loftus, 1993; Hoffman et al., 1989; Johnson et al., 1992; Kushlan, 1989b). With much of the higher elevation edge marshes either partially drained or gone from the system, less remained of the shallow, early dry season feeding habitats, especially in the wetter years when water within the impounded areas was deeper. Thus, nesting would be expected to be initiated earlier in drier years, when relatively low water in the central sloughs would have made prey available to foraging birds earlier in the season. This point of view assumes that overdrained areas outside of the impoundments and disjunct from the central Shark River and Taylor sloughs could not recover a significant prey base during wetter years.

The relationship of increased nesting effort during dry years was not maintained during the second subperiod (1981–89) of the recent period (Table 22.10). For these years, the average total nesting effort increased slightly from that in the previous subperiod during wet years, but declined substantially in normal and dry years. Although the strongest nesting effort tended to occur in hydrologically normal years, the difference in effort among wet, normal, and dry years was much less than during the first subperiod.

Why this relationship between increased nesting effort and lower water levels did not hold during most of the 1980s is another question with no certain answer. The changing patterns of nesting effort during the recent period illustrate that these birds respond to more complex environmental variables than simple water depth relationships. It may be that the change in this relationship between the two subperiods was caused by an increased frequency during the 1980s of extensive dry outs within Everglades National Park and in all impoundments except for the lower portion of Water Conservation Area 3A (Bancroft et al., 1990; Loftus et al., 1990; R. Johnson and W. Loftus, personal communication). These dry outs, characterized by loss of surface water in all or most of the freshwater and estuarine wetlands during the middle and late dry seasons, are thought to have caused substantial degradation of the prey base. The fact that the number of birds that initiated nesting in each year between 1980 and 1989 was reduced from the high numbers of the earlier subperiod and showed less of an interannual relationship with hydrological conditions suggests that the central and southern Everglades basin no longer produced or made available sufficient prey to support large numbers of nesting waders.

Much of the annual variability in numbers of nesting birds during both periods was due to occasional, dramatic seasonal and annual movements by the regional population of white ibis, the species that made up approximately 50–80% of the birds in most large colonies. White ibis apparently nested elsewhere (either elsewhere in the Everglades region or outside the region) in the exceptionally wet or the drier years. The Audubon wardens reported that large numbers of ibis abandoned the traditional headwaters roosts during late April 1932, mid-April 1935, February 1937, and early March 1938 due, in their opinions, to extreme water conditions. Unusually large ibis nesting colonies were reported outside the Everglades at Lake Washington (Brevard County) in 1937 (over 100,000 ibis in a colony that usually contained between 10,000 and 25,000 birds [Allen, 1937]) and at Lake Okeechobee and the upper St. Johns River marshes during the 1938–39 drought (O'Reilly, 1939). During very dry 1974, about 24,000 ibis shifted from Everglades colony sites (used in 1972 and 1973) to nest during late summer and fall in the Okaloacoochee Slough in the Big Cypress region (Kushlan and White 1977; J. Ogden, field notes). The discovery during the early 1970s of large ibis nesting colonies located in herbaceous marsh vegetation in the central Everglades (Kushlan, 1973) and the subsequent regular repetition of this practice by ibis suggest that ibis may have always nested in these "nontraditional" sites in years of extreme hydrological conditions. Thus, it is possible that some ibis colonies during the early period were active in the interior Everglades subregion but were undetected.

The decline in numbers of nesting birds from the early period to the recent presumably was due to the combined effects of reduced nesting effort, an apparent reduction in recruitment, and immigration to other regions for nesting. The relative contribution that each of these factors has made toward the overall population reduction is not known. Reduced nesting effort and success rates have been demonstrated for the wood stork in southern Florida (Ogden et al., 1987) (Figure 22.3) and are suggested for other species as well (Tables 22.8 and 22.10). The fact that wading birds typically forage over large geographical regions throughout a year and opportunistically locate new feeding sites (Kushlan, 1979b) also suggests that many birds could have abandoned the Everglades as colony sites in response to deteriorating feeding conditions. Wood storks established new nesting sites north of the Everglades during the 1970s and 1980s (Ogden et al., 1987), and the same response has been suggested for white ibis (Frederick and Collopy, 1988).

For whatever reasons, the magnitude of the decline between the two periods was largely due to the collapse in the number of nesting white ibis, which may have begun as early as the 1940s. White ibis differ from the other principal species in the Everglades in that they feed primarily on aquatic invertebrates (Kushlan and Kushlan, 1975). The relatively large number of ibis in proportion to the other fish-eating species, approximately 5:1 (Table 22.2), suggests that the natural Everglades system was more of an invertebrate producer than a fish producer. Unfortunately, current incomplete understanding of the dynamics of macroinvertebrates in the freshwater Everglades provides no explanation of how hydrological changes have affected this group of animals (W. Loftus, personal communication).

The water management history of the southern Everglades argues strongly that natural hydrological patterns had already been altered to some extent by the early period (1931–46) and became increasingly so beginning in the late 1940s with the construction of the Central and Southern Florida Flood Control Project. The most reasonable conclusion from this history of both the hydrology of the region and bird populations is that wading birds began to be adversely impacted through reduction in total area of wetlands and alteration in natural hydrological patterns by the 1940s.

Although early estimates of numbers of nesting birds are less reliable than recent estimates, Kushlan and Frohring (1986) were probably incorrect in suggesting that the Everglades nesting population of wood storks did not begin to decline until the late 1960s. The estimated maximum nesting population of storks in the Everglades region was approximately 38% larger during the 1930s (8000 birds) than during the early 1960s (5000 birds in 1960) (Ogden and Nesbitt, 1979). The number of nesting storks used by Kushlan and Frohring for the 1960s is key to their argument that declines did not occur during an earlier period. However, they were incorrect in their assertion that the numbers of storks reported by Ogden and Nesbitt (1979) to be nesting in Everglades National Park in 1959 and 1960 were too high due to a "mix-up" between nest numbers and bird numbers and that the total number of storks nesting in south Florida in 1967 (9425 pairs) represented the highest number that can be "verified" for any year since 1900. The 1959 and 1960 estimates used by Ogden and Nesbitt and in this chapter are based in large part on counts of birds made from high-resolution aerial photographs from both years. The estimate by Kushlan and Frohring for total nesting storks in 1967 was actually too large because it included an incorrect total of 7300 pairs from Corkscrew Swamp; the actual number of pairs in Corkscrew in 1967 was 3680, and the estimated number of young fledged from the colony was 7350 (Paul Henchcliff, personal communication).

Colony Locations

The substantial shift in location of colonies from the early to the recent period was highlighted by the abandonment of most of the once large headwaters sites (subregion 4) and a corresponding increase in number and size of colonies in the interior Everglades subregion (subregion 5). Substantial numbers of all species except the wood stork relocated nesting sites between the two periods. This change in the location of the largest Everglades colonies must have been caused by a shift in the location of the most dependable and/or strongest food base in the remaining Everglades basin during the nesting season (Kushlan 1976a, 1977). The relocated species forage primarily within 20 km of their colony (Bancroft et al., 1988a, 1990), about one-third the usual long-distance foraging range of wood storks in the Everglades region (Ogden et al., 1978). Hydrological analyses presented by the South Florida Research Center (1990) show that water management practices in the southern Everglades since the 1950s have resulted in, among many changes, an increase in the frequency of extensive dry outs in the lower

Shark River Slough and a relocation of the largest long-hydroperiod pool northward approximately 30–40 km from the southern Shark River Slough to the southern end of Water Conservation Area 3A. These hydrological alterations to the system so closely match the major change in colony locations that the relationship may be assumed. Thus, storks from East River or Cuthbert could still reach the newly created long-hydroperiod pool in Water Conservation Area 3A, while smaller wading birds nesting in the headwaters sites could not.

Although the principal stork colonies during the recent period remained in the traditional locations, small percentages of storks nested in the interior Everglades subregion with perhaps increasing frequency since the early 1970s (Table 22.13). These colonies formed primarily during years of relatively extreme hydrological conditions, which suggests that they may represent responses by storks to poor foraging conditions further south. Because these colonies have for the most part been unsuccessful and most of the new sites have been used by storks in only a single year, it appears unlikely that this pattern represents a shift by storks in response to the creation of "beneficial" habitat conditions in the central Everglades, as was suggested by Kushlan and Frohring (1986).

Another change in colony locations between the two periods was the collapse in nesting in the lower Ten Thousand Islands subregion. While not numerically important in terms of nesting birds, this subregion also contained the largest roosts of nonbreeding wading birds known during the early period. Counts ranging as high as 75,000–100,000 birds (mostly white ibis) were repeatedly made at Duck Rock and other island sites during rainy seasons in many years during the early period and as recently as the 1960s (Allen, 1936; Ogden, 1978). Both the nesting and roosting birds fed primarily in mainland estuaries north of Lostmans River. The loss of this colony/roost activity, mainly since the 1960s, suggests that far greater hydrological changes have occurred in the southern Big Cypress and Gulf coast estuaries than has been recognized or suspected.

Table 22.13 Wood Stork Colonies in Subregion 5 and in Loxahatchee National Wildlife Refuge in the Interior Everglades, Florida

Year	Location	No. birds	Success/ failure	Hydrology classification
1971	Central Shark River Slough, Everglades National Park	500[a]	F	Dry
1981	Loxahatchee National Wildlife Reserve	160	F	Wet
1981	Water Conservation Area 2B	96	?	Wet
1981	Water Conservation Area 3A	50	F	Wet
1982	Andytown East, Water Conservation Area 3A	150	F	Normal
1984	L-28 gap, Water Conservation Area 3A	50	F	Wet
1985	L-67, Water Conservation Area 3A	40	?	Dry
1985	Water Conservation Area 3B	50[b]	F	Dry
1989	Water Conservation Area 3A	80	F	Dry
1989	L-67, Water Conservation Area 3A	450	S	Dry

[a] In two willow island colonies.
[b] Two sites.

In contrast, the northwestern Florida Bay subregion appears unchanged numerically or in nesting patterns between the two periods, and as such is the only subregion to remain stable. Birds from these colonies feed primarily in the mainland estuaries from north of Snake Bight to the Cape Sable peninsula (J. Kushlan and G. Powell, personal communication; J. Ogden, field notes). This coastal area, not directly in the path of major freshwater discharges from Shark River Slough, is probably the least changed hydrologically of the five subregions.

Timing

Both periods showed the expected interregional and interannual differences in timing of nesting that almost certainly were due to normal variations in the range of hydrological patterns and fish concentrations that occur between regions and years (Bancroft et al., 1988b, 1990; Loftus et al., 1990; Kushlan and White, 1977). While small herons, egrets, and ibis initiated nesting primarily during February–March and completed nesting by May, there was a strong pattern of summer nesting, especially by ibis, in years when the dry season was unusually wet (1936, 1937, 1941, 1983) or unusually dry (1932, 1974–77). The Audubon wardens suggested that year-round nesting by snowy egrets and tricolored herons at Duck Rock in 1936 was a response to high water in the interior marshes.

During the early period, a wave of coastal nesting by small herons, egrets, and ibis in some years began in the Florida Bay subregion during March–May and in the Ten Thousand Islands subregion in May. The coastal colonies tended to be active much later in the summer and early fall than did the inland colonies, especially in the Ten Thousand Islands subregion. Why the coastal sites were able to support such prolonged periods of nesting, especially through the wet season months, cannot be explained. It is interesting to note that this summer nesting was in the same area where the large summer roosts were located, suggesting that the upper Gulf coast estuaries once made available much better foraging conditions than have occurred during recent years.

The timing patterns of colony activity during the recent period were similar to those of the early period, with the notable exception of the later nesting by wood storks (Table 22.6). White ibis in the headwaters subregion also tended to nest later by about a month during the recent period. The large interior Everglades colonies of the recent period operated on a schedule similar to the headwaters colonies, although they tended to be initiated slightly later, especially by great egrets. The northernmost of the headwaters colonies, at Rodgers River Bay, had substantial summer nesting by small herons, egrets, and ibis in some years and thus operated on a schedule similar to the abandoned Ten Thousand Islands sites. These occasional summer nestings by ibis at Rodgers River, coupled with the absence of ibis nesting in most years at the traditional headwaters sites, were largely responsible for the relatively late mean date for ibis colony initiation during the recent period. Recent studies by Bancroft and Jewell (1987) and Bancroft et al. (1988a) have shown that under certain conditions, birds in the Rodgers River colony do much foraging in the same mainland estuarine zone that was used by the birds that once nested in the Ten Thousand Islands subregion. Thus, the characteristics of this foraging area may explain the pattern of summer nesting at the Rodgers River colony, in contrast to other headwaters colonies.

Wood Storks

The historical pattern for wood storks was for colonies to form between November and January, most often in December. For storks to have formed colonies as early as November or December meant that a strong enough food resource was available within foraging range of the colonies during these months. Stork foraging range during the period of colony formation is estimated to be 20–40 km, much less than the maximum range (Ogden et al., 1978; J. Ogden, field notes).

It is also assumed that the only wetlands available as foraging habitat to wood storks and other wading birds during the early dry season months (November–January) in the more natural Everglades were the higher elevation freshwater wetlands and the more topographically complex mainland estuaries located within flight range of the colonies. Monthly aerial censuses in Everglades National Park during the mid and late 1960s revealed that over 80% of the detected storks were feeding in the mainland estuaries and in areas of early drying, freshwater marshes during November and December (J. Ogden, field notes). The central sloughs presumably remained too deeply flooded to be available to feeding birds in all but the drier years during these months.

The delay since 1969 in the timing of stork colony formation may be related to substantially reduced surface water flows that have occurred in the higher elevation marshes and estuarine wetlands in the southern Everglades region (Everglades National Park, hydrological data). The assumption here is that due to reduced flows and shortened annual hydroperiods, these wetlands, which historically contained the important early dry season foraging habitats, no longer supplied enough food during November through January to support colony formation early in the dry season. Prior to 1970, early (normal) colony formation was apparently more likely to occur in years when December water levels were high enough above sea level for the higher elevation marshes to have surface pools (Table 22.12). As regional water levels became lower during the 1960s and 1970s, the relationship between timing of nesting and water levels in these marshes tended to reverse. Early colonies became associated with relatively low water in December, when there would have been no surface water on the higher elevation marshes, but the deeper sloughs to the west and south would have been shallow enough to supply food for early formation. Relatively high December water levels since 1970 have not stimulated early colonies, most likely because high water only served to disperse a degraded prey base that remained in the early dry season foraging habitats (W. Loftus, personal communication). The recent pattern of colony formation in February and March has apparently occurred as seasonally improved foraging conditions developed in the remaining longer hydroperiod central sloughs and deeper portions of the water impoundments, associated with the more rapid drying rates characteristic in the later months of the dry season (W. Loftus and R. Johnson, personal communication).

The shift in timing by nesting wood storks has been a major factor contributing to a decline in the rate of nesting success and a reduction in nesting effort during the recent period. The wood stork reproductive cycle requires 110–150 days (Kahl, 1964). In order to achieve high rates of nesting success, storks in Everglades colonies must complete the nesting cycle before the initiation of the summer rainy season, which usually begins in late May or June. In addition, newly fledged storks presumably have higher postfledgling rates of survival if they have some period of foraging time in a low-water/high-food concentration environment, rather than in a wetland with rising water levels and dispersed food resources. The ideal nesting schedule, then, would have colony formation occurring between early December and late January, so that young would be fledging no later than mid-May. This schedule would have large nestlings and fledglings utilizing the system at the time when maximum food concentrations presumably occurred along the drying edges of the deeper sloughs of the region.

During the recent period, storks formed colonies in December 1 year, in January 4 years, in February 5 years and in March 5 years (no formation in 1983). Colony formation as late as February or March meant that young were fledged during June–July, when summer rains were dispersing fish concentrations in the sloughs. In years when numbers of young survived to fledge, they entered a relatively hostile environment in terms of food resources. It is assumed that survival of these fledglings has not been at as high a rate as occurred when colonies formed earlier. Thus, 10 out of 15 years had a high probability of colony failure due to the timing of nesting alone. Of the 5 years during the recent period when colonies were successful, 2 were years with January formations (1974 and 1975) and 2 of the 3 with later formations were dry years with delayed rainy season initiations (1985 and 1987).

Late forming colonies tended to contain fewer nesting birds than early colonies (Figure 22.4), and thus these colonies could only produce relatively small numbers of young. Potential breeding birds may be more likely to disperse from Everglades National Park well before the end of the dry season in years when relatively poor feeding conditions occur in the Everglades. For example, approximately 1200 wood storks were located in large soaring flocks over Water Conservation Area 3A during a late January 1969 aerial census, following a period of heavy rain and substantial water level rise in the southern Everglades (J. Ogden, field notes). During 1984, systematic aerial censuses over Everglades National Park showed a sharp decline in the number of storks (from 679 birds on February 29 to 177 on March 27) following a period of heavy rains in late February (M. Fleming, personal communication). Presumably, adult storks that leave the southern Everglades before the end of the dry season disperse northward; some of these same birds may attempt nesting in more northern colonies in the same year (Ogden et al., 1987). Storks typically enter colonies located in central and northern Florida during February and March (J. Rodgers, personal communication).

Although Kushlan (1986, 1987) and Kushlan and Frohring (1986) suggested that drying rates largely or entirely explained wood stork reproductive performance in the Everglades, no consistent relationship was found between stork nesting patterns and drying rates in the present study. Data on drying rates at the P-33 and P-36 stations located in the central-lower Shark River Slough are compared with the timing of stork colony formations in Table 22.14. These data show that between 1953 and 1962, storks formed colonies in almost all years in November or December, regardless of early dry season drying rates. Between 1976 and 1989, storks usually formed colonies in January through March, again without a clear correlation with the drying rate. Only for the intermediate years 1963–75 at P-33

Table 22.14 Early Dry Season (November–January) Drying Rates at Two Shark River Slough Gauges and Months of Wood Stork Colony Formation for Three Separate Periods between 1953 and 1989 for Everglades National Park

	P-33		P-36	
Year	Formation month (no. years)	Mean drying[a] rate	Formation month (no. years)	Mean drying[a] rate
1953–62	Nov. (4)	0.0066		
	Dec. (4)	0.0038	No data	
	Jan. (1)	0.0060		
	No form. (1)	0.0043		
1963–75	Dec. (6)	0.0082	Dec. (1)	0.0134
(P-33)	Jan. (4)	0.0052	Jan. (4)	0.0052
1969–75	Feb. (2)	0.0035	Feb. (2)	0.0026
(P-36)	Mar. (1)	0.0020	—	—
1976–89	Dec. (2)	0.0053	Dec. (2)	0.0047
	Jan. (3)	0.0053	Jan. (3)	0.0060
	Feb. (4)	0.0057	Feb. (4)	0.0051
	Mar. (5)	0.0048	Mar. (5)	0.0041

[a] Drying rate calculated as average daily change in water level (feet per day) between date of highest November stage and date of highest January stage. Period of record for P-33 = 1953–89. Period of record for P-36 = 1969–89. P-36 may be a better indicator of overall Shark River Slough hydrological patterns because of its central slough location, above tidal influence in the lower slough and below the water diversion effects of L-67 extended in the upper slough (P-33).

and 1969–75 at P-36 did there appear to be a correlation between earlier colony formation and more rapid drying (Kushlan et al., 1975; Ogden et al., 1978). Stork nesting effort (numbers of pairs) and drying rates were also examined, but no significant relationship between these parameters was found. The fact that the total number of storks which attempted to nest continued to decline throughout this 33-year period, and especially after 1970, suggests that some factor(s) in addition to drying rates were determining stork nesting effort in the Everglades region.

It is reasonable to expect that more rapid drying in the deeply flooded portions of the system would result in the creation of earlier and greater fish concentrations for feeding birds. Thus, the timing pattern shown for 1963–75 (Table 22.14) is one that intuitively would be expected once the shallower habitats were lost. Why this expected relationship has collapsed since 1976 is not known. The answer that seems most reasonable within the framework of what is known about the southern Everglades is that the prey base in the system has been declining, even in the deeper sloughs. This assumes that generally shortened hydroperiods throughout Everglades National Park have resulted in a chronic pattern of declining production and survival of the primary prey species for storks. The overall pattern of low numbers of nesting birds during the 1980s (Table 22.1) is the strongest circumstantial evidence that a deterioration in the prey base has occurred and that no short-term combination of rainfall and drying patterns and

rates within the current southern Everglades will result in a substantial improvement in reproductive effort.

Water Conditions and Nesting Patterns: An Overview

Changes from the early period to the recent period in the number of wading
birds that attempted to nest (declines by all species), in the locations of major
colonies (egrets, herons, ibis), and in the timing of nesting (storks) appear to have
been direct responses by Everglades wading birds to major alterations in hydrological patterns (Johnson and Fennema, 1989; Leach et al., 1972; South Florida
Research Center, 1990; South Florida Water Management District, 1989). The
substantial reduction in total area of the relatively higher elevation marshes, along
with suspected degradation in estuarine wetlands possibly due to reduced freshwater flow, have almost certainly been the most important factors responsible for
the change in timing of nesting by storks. The abandonment of traditional colony
locations in the southern Everglades by great egrets and the smaller waders appears
to be an adjustment by the birds to the relocation within the basin of the more
permanently flooded surface water pools.

While these changes have operated to reduce overall nesting effort and (for
storks at least) colony success rates, widespread declines in food resources
probably have been the major immediate factor responsible for the between-period
changes in wader colony patterns. The tremendous number of birds in early period
nesting colonies and summer roosts is of itself the most compelling argument that
the central and southern Everglades once produced and made available in the
proper locations and at the proper times a much greater abundance of prey for
waders. Loftus and Eklund (1994) and Loftus et al. (1990) have suggested that
Everglades marshes with annual hydroperiods of less than 9–10 months show
significantly less internal reproduction and long-term survival of fishes than do
marshes with longer hydroperiods. Davis et al. (1994) and Gunderson and Loftus
(1993) have shown that the original Everglades has experienced an approximately
50% reduction in total area, while Fennema et al. (1994) have demonstrated a
reduction in annual hydroperiods in substantial areas of long-hydroperiod wetlands
in the basin. It is almost certain that these longer hydroperiod regions were among
the more important core refugia for the prey of wading birds in the central and
southern Everglades.

A continuing systemwide reduction in prey resources would explain why the
pattern of increased nesting effort in drier years during the 1970s did not hold
through the 1980s. Rapid and extensive drying in a system with a relatively strong
prey base, while reducing the options about where birds might forage, presumably
would also produce high densities and greater availability of prey in the remaining
wetlands than would occur in a wetter or slower drying system. After a number
of years with extensive dry outs, however, the prey base would become regionally
depressed. Years with rapid drying would still reduce the number of foraging
options, but would no longer create high-density foraging sites where surface water
remained. Consistent with this explanation is the fact that (1) it was dry year nesting
effort that declined so sharply between the recent subperiods and (2) the number
of birds nesting during most years in the 1980s remained relatively low regardless
of hydrological conditions. Nesting in drier years rather than in wetter years and

in the interior subregion rather than in the headwaters subregion in fact may only have been transitional patterns in an ecosystem that continues to experience hydrological and ecological change.

The concentration of early period colony sites in the headwaters subregion, coupled with the absence of colonies in the interior Everglades in most years, strongly suggests that the mangrove freshwater marsh ecotone and adjacent estuaries provided better overall nesting and feeding conditions than were found in the interior Everglades. W. Loftus (personal communication) found relatively higher densities and greater species diversity in fishes in ecotone creeks than in interior marshes in Everglades National Park. Kushlan (1979b) and Bancroft et al. (1988a, 1990) reported that nesting ibis as well as small herons and egrets, respectively, shifted foraging from inland to coastal sites following rises in water levels in the interior marshes. The birds studied by Bancroft et al. shifted back to the interior marshes when water levels dropped again. A conclusion from this study was that birds nesting near the ecotone have a greater range of feeding site options under a variety of hydrological conditions than do birds in interior colonies.

Healthy populations of wading birds may be relatively more difficult to recover than many other vertebrates because of their large spatial requirements in the Everglades. Recovery of nesting waterbirds in the southern Everglades will require a return to more permanent flooding in the deeper and southern portions of the Shark River Slough and Taylor Slough, a recovery of predrainage flow volumes into the mainland estuaries, and much more extensive and prolonged flooding in the expansive short-hydroperiod wetlands along the flanks of the basin. Under natural conditions it must have been the deeper marshes of the central sloughs that were the prey refugia for the ecosystem, rather than the much smaller, but numerous, alligator holes and small ponds that occur throughout (Browder, 1976; Kushlan, 1976b). It was probably the prolonged hydrological connection (2–4 months beginning August–October) in most years between the deeper marshes and the higher elevation marshes along the flanks of the sloughs that was largely responsible for the annual reoccupation by fishes and for the November–January timing of dry outs in this later region. At the same time, the longer hydroperiods and infrequent dry outs in the central and southern Shark River Slough maintained a much stronger prey base within foraging range of the historical colony locations than currently occurs. The location of traditional colonies, concentrated along the headwaters of the coastal rivers and within the mangrove estuarine region, suggests that this region also provided a greater range of feeding site options under a variety of climatological and hydrological conditions than did other regions of the Everglades. Recovery of these natural hydrological patterns may be a prerequisite step for the return of nesting wading birds to the traditional colony sites.

ACKNOWLEDGMENTS

Many of the ideas expressed in this chapter are products of the author's 25 years of exposure to the Everglades region. Throughout this period, I have been party to numerous discussions and debates on questions of wading bird colony patterns with (and approximately in this order) W. B. Robertson, Jr., A. Sprunt IV, J. A.

Kushlan, J. T. Tilmant, G. T. Bancroft, P. C. Frederick, W. F. Loftus, D. M. Fleming, and W. Hoffman. Thus, these ideas have evolved from many directions and experiences, and as such, they should be viewed as the contribution of no single person. In addition, I wish to thank W. B. Robertson, Jr., P. C. Frederick, J. A. Rodgers, S. M. Davis, M. A. Soukup, and several anonymous referees for commenting on this manuscript and C. Doffermyre for editorial and computer assistance during the writing phase.

LITERATURE CITED

Allen, R. P. 1934. Inspection of Gulf Coast sanctuaries. *Bird-Lore,* 36:338–342.

Allen, R. P. 1936. The Flock Movements of Herons, Egrets and Ibises, Unpublished manuscript, National Audubon Society, Tavernier, Fla., 16 pp.

Allen, R. P. 1937. News from the sanctuary front. *Bird-Lore,* 39:226–229, 363.

Allen, R. P. 1942. The Roseate Spoonbill, Res. Rep. No. 2, National Audubon Society. New York.

Allen, R. P. 1958. *A Progress Report on the Wading Bird Survey,* National Audubon Society, Tavernier, Fla., 65 pp.

Bancroft, G. T. 1989. Status and conservation of wading birds in the Everglades. *Am. Birds,* 43:1258–1265.

Bancroft, G. T. and S. D. Jewell. 1987. Foraging Habitat of *Egretta* Herons Relative to Stage in the Nesting Cycle and Water Conditions, 2nd Annu. Rep., National Audubon Society, Tavernier, Fla., 174 pp.

Bancroft, G. T., S. D. Jewell, and A. M. Strong. 1988a. Foraging Habitat of *Egretta* Herons Relative to Stage in the Nesting Cycle and Water Conditions, 3rd Annu. Rep., National Audubon Society, Tavernier, Fla., 72 pp.

Bancroft, G. T., J. C. Ogden, and B. W. Patty. 1988b. Wading bird colony formation and turnover relative to rainfall in the Corkscrew Swamp area of Florida during 1982 through 1985. *Wilson Bull.,* 100:50–59.

Bancroft, G. T., S. D. Jewell, and A. M. Strong. 1990. Foraging and Nesting Ecology of Herons in the Lower Everglades Relative to Water Conditions, Final Report, National Audubon Society. Tavernier, Fla., 156 pp.

Beard, D. B. 1938. Everglades National Park Project, U.S. Department of the Interior, Washington, D.C.

Bildstein, K. L., W. Post, J. Johnston, and P. Frederick. 1990. Freshwater wetlands, rainfall and the breeding ecology of white ibises in coastal South Carolina. *Wilson Bull.,* 102:84–98.

Browder, J. A. 1976. Water, Wetlands, and Wood Storks in Southwest Florida, Ph. D. dissertation, University of Florida, Gainesville, 406 pp.

Browder, J. A., C. Littlejohn, and D. Young. 1976. The South Florida Study, Center for Wetlands, University of Florida, Gainesville, and Bureau of Comprehensive Planning, Florida Department of Administration, Tallahassee.

Custer, T. W. and R. G. Osborn. 1977. Wading birds as biological indicators: 1975 colony survey. *U.S. Fish Wildl. Serv. Spec. Sci. Rep. Wildl.,* No. 206.

Davis, S. M., L. H. Gunderson, W. A. Park, J. Richardson, and J. Mattson. 1994. Landscape dimension, composition, and function in a changing Everglades ecosystem. in *Everglades: The Ecosystem and Its Restoration,* S. M. Davis and J. C. Ogden (Eds.), St. Lucie Press, Delray Beach, Fla., chap. 17.

Fennema, R. J., C. J. Neidrauer, R. A. Johnson, T. K. MacVicar, and W. A. Perkins. 1994. A computer model to simulate natural Everglades hydrology. in *Everglades: The Ecosystem and Its Restoration,* S. M. Davis and J. C. Ogden (Eds.), St. Lucie Press, Delray Beach, Fla., chap. 10.

Frederick, P. C. and M. W. Collopy. 1988. Reproductive Ecology of Wading Birds in Relation to Water Conditions in the Florida Everglades, U.S. Fish and Wildlife Service, Cooperative Research Unit, University of Florida, Gainesville, 259 pp.

Frederick, P. C. and M. W. Collopy. 1989. Nesting success of five Ciconiiform species in relation to water conditions in the Florida Everglades. *Auk,* 106:625–634.

Frohring, P. C. 1988. Herons in Everglades National Park, South Florida Research Center, Everglades National Park, Homestead, Fla.

Frohring, P. C., D. P. Voorhees, and J. A. Kushlan. 1988. History of wading bird populations in the Florida Everglades: A lesson in the use of historical information. *Colon. Waterbirds,* 11:328–335.

Graham, F. J. 1990. *The Audubon Ark. A History of the National Audubon Society,* Alfred A. Knopf, New York, 334 pp.

Gunderson, L. H. and W. F. Loftus. 1993. The Everglades. in *Biotic Communities of the Southeastern United States,* W. H. Martin, S. G. Boyce, and A. C. Echternacht (Eds.), John Wiley & Sons, New York.

Hafner, H. and R. Britton. 1983. Changes of foraging sites by little egrets (*Egretta garzetta* L.) in relation to food supply. *Colon. Waterbirds,* 6:24–30.

Hanna, A. J. and K. A. Hanna. 1948. *Lake Okeechobee. Wellspring of the Everglades,* The American Lakes Series, Bobbs-Merrill, Indianapolis, 379 pp.

Hoffman, W., G. T. Bancroft, and R. J. Sawicki. 1989. *Wading Bird Populations and Distribution in the Water Conservation Areas of the Everglades: 1985 through 1988,* National Audubon Society, Tavernier, Fla., 173 pp.

Holt, E. G. 1933. Report of Ernest G. Holt, director of sanctuaries: Florida. *Bird-Lore,* 35:371–372.

Johnson, R. A. and R. J. Fennema. 1989. Conflicts over flood protection and wetland preservation in the Taylor Slough and eastern panhandle of Everglades National Park. in Symp. on Wetlands: Concerns and Successes, American Water Resources Association, Bethesda, Md., pp. 451–462.

Johnson, R., R. Fennema, and T. Bhatt. 1992. Water Management and Ecosystem Restoration in the Everglades, Unpublished manuscript, South Florida Research Center, Everglades National Park, Homestead, Fla.

Kahl, M. P. 1964. Food ecology of the wood stork (*Mycteria americana*) in Florida. *Ecol. Monogr.,* 34:97–117.

Kushlan, J. A. 1973. White ibis nesting in the Florida Everglades. *Wilson Bull.,* 85:230–231.

Kushlan, J. A. 1976a. Site selection for nesting colonies by the American white ibis *Eudocimus albus* in Florida. *Ibis,* 118:590–593.

Kushlan, J. A. 1976b. Wading bird predation in a seasonally fluctuating pond. *Auk,* 93:464–476.

Kushlan, J. A. 1977. Population energetics of the white ibis. *Auk,* 94:114–122.

Kushlan, J. A. 1979a. Effects of helicopter censuses on wading bird colonies. *J. Wildl. Manage.,* 43:756–760.

Kushlan, J. A. 1979b. Feeding ecology and prey selection in the white ibis. *Condor,* 81:376–389.

Kushlan, J. A. 1986. Responses of wading birds to seasonally fluctuating water levels: Strategies and their limits. *Colon. Waterbirds,* 9:155–162.

Kushlan, J. A. 1987. Recovery plan for the U.S. breeding population of the wood stork. *Colon. Waterbirds,* 10:259–262.

Kushlan, J. A. 1989a. Avian use of fluctuating wetlands. in Freshwater Wetlands and Wildlife, R. R. Sharitz and J. W. Gibbons (Eds.), CONF-8603101, Department of Energy Symp. Ser. No. 61, Office of Scientific and Technical Information, U.S. Department of Energy, Oak Ridge, Tenn., pp. 593–604.

Kushlan, J. A. 1989b. Wetlands and wildlife, the Everglades perspective. in Freshwater Wetlands and Wildlife, R. R. Sharitz and J. W. Gibbons (Eds.), CONF-8603101, Department of Energy Symp. Ser. No. 61, Office of Scientific and Technical Information, U.S. Department of Energy, Oak Ridge, Tenn., pp. 773–790.

Kushlan, J. A. and P. C. Frohring. 1986. The history of the southern Florida wood stork population. *Wilson Bull.,* 98:368–386.

Kushlan, J. A. and M. S. Kushlan. 1975. Food of the white ibis in southern Florida. *Fla. Field Nat.*, 3:31–38.

Kushlan, J. A. and D. A. White. 1977. Nesting wading bird populations in southern Florida. *Fla. Sci.*, 40:65–72.

Kushlan, J. A., J. C. Ogden, and A. L. Higer. 1975. Relation of Water Level and Fish Availability to Wood Stork Reproduction in the Southern Everglades, Florida, Rep. 75-434, U.S. Geological Survey, Tallahassee, Fla., 56 pp.

Leach, S. D., H. Klein, and E. R. Hampton. 1972. Hydrologic Effects of Water Control and Management of Southeastern Florida, Rep. of Invest. No. 6, U.S. Geological Survey, Miami.

Loftus, W. F. and A.-M. Eklund. 1994. Long-term dynamics of an Everglades small-fish assemblage. in *Everglades: The Ecosystem and Its Restoration,* S. M. Davis and J. C. Ogden (Eds.), St. Lucie Press, Delray Beach, Fla., chap. 19.

Loftus, W. F., J. D. Chapman, and R. Conrow. 1990. Hydroperiod effects on Everglades marsh food webs, with relation to marsh restoration efforts. in *Proc. Conf. Science in the National Parks,* Vol. 6: Fisheries and Coastal Wetlands Research, U.S. National Park Service and the George Wright Society, Washington, D.C., pp. 1–22.

Ogden, J. C. 1978. Recent population trends of colonial wading birds on the Atlantic and Gulf coastal plains. in *Wading Birds,* A. Sprunt IV, J. C. Ogden, and S. Winkler (Eds.), Res. Rep. 7, National Audubon Society, New York, pp. 135–153.

Ogden, J. C. and S. A. Nesbitt. 1979. Recent wood stork population trends in the United States. *Wilson Bull.*, 91:512–523.

Ogden, J. C., J. A. Kushlan, and J. A. Tilmant. 1978. The Food Habits and Nesting Success of Wood Storks in the Everglades National Park in 1974, Natural Resources Rep. 16, U.S. National Park Service, Washington, D.C., 25 pp.

Ogden, J. C., H. W. Kale II, and S. A. Nesbitt. 1980. The influence of annual variation in rainfall and water levels on nesting by Florida populations of wading birds. *Trans. Linn. Soc. N.Y.*, 9:115–126.

Ogden, J. C., D. A. McCrimmon, Jr., G. T. Bancroft, and B. W. Patty. 1987. Breeding populations of the wood stork *Mycteria americana* in the southeastern United States. *Condor,* 89:752–759.

Ogden, J. C., W. F. Loftus, and W. B. Robertson, Jr. 1988. Wood Storks, Wading Birds, and Freshwater Fishes, South Florida Research Center, Everglades National Park, Homestead, Fla., 21 pp.

O'Reilly, J. 1939. Wildlife protection in south Florida. *Bird-Lore,* 41:128–140.

Palmer, R. S. 1962. *Handbook of North American Birds,* Vol. 1, Yale University Press, New Haven, Conn.

Powell, G. V. N., R. D. Bjork, J. C. Ogden, R. T. Paul, A. H. Powell, and W. B. Robertson, Jr. 1989. Population trends in some Florida Bay wading birds. *Wilson Bull.*, 101:436–457.

Robertson, W. B., Jr. 1965. The 1964–1965 Breeding Season of Wood Ibis in Everglades National Park, Everglades National Park, Homestead, Fla., 8 pp.

Robertson, W. B., Jr. and J. A. Kushlan. 1974. The southern Florida avifauna. *Miami Geol. Soc. Mem.*, 2:414–452.

South Florida Research Center. 1990. An Assessment of Hydrological Improvements and Wildlife Benefits from Proposed Alternatives for the U.S. Corps of Engineers General Design Memorandum for Modified Water Deliveries to Everglades National Park, R. A. Johnson and J. C. Ogden (Eds.), Everglades National Park, Homestead, Fla., 99 pp.

South Florida Water Management District. 1989. Surface Water Improvement and Management Plan for the Everglades, Draft Tech. Rep., South Florida Water Management District, West Palm Beach.

23

NUTRIENT TRANSPORT by WADING BIRDS in the EVERGLADES

Peter C. Frederick

George V. N. Powell

ABSTRACT

The effect of nutrient accumulation resulting from deposition of feces in colonies of colonially breeding and roosting wading birds is estimated in this chapter for breeding and nonbreeding ciconiiform birds in the Everglades ecosystem, by modeling energy consumption and feces deposition rates and by using existing measurements of size, energy, and nutrient content of prey items from the Everglades. Current populations of breeding and nonbreeding birds are estimated to consume 4.9 fewer tonnes of prey (dry mass) per year than the much larger populations of the 1930s and 1940s, equivalent to an estimated 14.6 million fewer prey items per year. This difference translates into 455 fewer tonnes of feces deposited in roosts and colonies per year, roughly equivalent to 59 fewer tonnes nitrogen and 5.6 fewer tonnes phosphorus. Nonbreeding birds are estimated to account for only 1.5% of the difference in nutrient flux attributable to birds between the two periods, indicating that the differences are due to reductions in energy-intensive breeding attempts. Although even the largest historical populations are estimated to have redistributed only a very small fraction of the total annual deposition of phosphorus and other nutrients in the marsh, loading rates at colonies can be extremely high. Loading rates at historical colony sites could have been as high as 120 g phosphorus $m^{-2} \cdot yr^{-1}$ (approximately 3000 times the estimated historic atmospheric deposition rate), while current colonies are estimated to have rates of only 0.9 g phosphorus $m^{-2} \cdot yr^{-1}$ (more than 20 times the historic atmospheric deposition rate). Evidence from the Everglades and other ecosystems suggests that high nutrient concentrations in the vicinity of colonies has a strong effect on the productivity and species composition of aquatic fauna and flora. This

© St. Lucie Press CCC 0-9634030-2-8 1/94/$100/$.50

may have strong feedback effects for survival of young wading birds, which characteristically develop foraging skills at or near colony sites. Recent relocation of large colonies from the estuarine zone to the freshwater Everglades implies that nutrient input to the estuary has decreased significantly. Nutrient-rich colonies probably serve as islands of refugia for nutrient-tolerant species in the oligotrophic Everglades and may serve to significantly affect the variability in biodiversity of the marsh. Sources of error tend to be in the direction of overestimation of nutrients transported, and in this regard, the amount of food required by nestlings is a central and poorly understood variable.

INTRODUCTION

Populations of animals can serve as important vectors of nutrient flow, both by transporting nutrients between ecosystems (Morales and Pacheco, 1986; Bildstein et al., 1992) and by recycling and redistributing nutrients within ecosystems (Meyer et al., 1983; Powell et al., 1991). Both processes are accomplished by the consumption of food in one location and the excretion of nutrients, in the form of feces, in the same or another location.

The redistribution and in many cases concentration of nutrients may have an impact on the nutrient budgets of entire ecosystems, particularly those which are naturally oligotrophic. Meyer et al. (1983) reported that schools of fish which fed away from a coral reef contributed between 30 and 48% of the total ammonia and between 59 and 90% of the total particulate phosphorus to coral heads. Nutrient concentration by animals can also be dramatic in nesting colonies of birds, which often forage at great ranges from their colonies, but deposit a large proportion of their feces at communal breeding sites, often islands. Bildstein et al. (1992) found that white ibises (*Eudocimus albus*) imported 33% as much phosphorus and 9% as much nitrogen into a South Carolina salt marsh as did atmospheric sources. In the oligotrophic Okefenokee Swamp in southern Georgia, Stinner (1983) found that white ibises contributed 10 times as much phosphorus and calcium to the ecosystem as were imported by stream flow and atmospheric deposition.

Such concentration of nutrients also may strongly affect local community structure. Bird colonies have been shown to increase nutrient concentrations and phytoplankton in the Barents Sea (Golovkin, 1967; Golovkin and Garkavaya, 1975). A number of studies have found that locally high concentrations of nutrients due to bird colonies can profoundly alter both the vegetation in the colony site (Dusi et al., 1971; Weseloh and Brown, 1971; Wiese, 1978; Stinner, 1983; Allaway and Ashford, 1984) and the plant and animal communities surrounding the colony. Bosman and Hockey (1986) determined that seabird deposition of feces in South African colonies resulted in the production of thick, intertidal algal mats not present on nearby unenriched intertidal zones. The mats, in turn, created foraging habitat for a number of shorebird species. Onuf et al. (1977) found that mangroves in wading bird (Ciconiiformes) colonies in a Florida estuary were more productive than those on adjacent noncolony islands and that leaves of colony trees were grazed by insects at higher rates than those on noncolony trees. Powell et al. (1991) found that seagrass beds surrounding colony islands in Florida Bay were denser than beds surrounding nearby noncolony islands. In addition, species composition of seagrasses around the colony islands was different from the noncolony islands, probably as a result of the elevated sediment nutrient levels near colonies. Because seagrass density was positively correlated with densities of demersal invertebrates and fishes in Florida Bay (Sogard et al., 1987), these changes in vegetation surrounding marine wading bird colonies have potential for structuring aquatic animal communities.

Wading birds have historically been important consumers of fishes, invertebrates, and anurans in the Everglades ecosystem (Robertson and Kushlan, 1974; Ogden, 1978, 1994; Frederick and Spalding, 1994), and the size alone of even conservatively estimated historic breeding assemblages suggests that these birds played an important role in nutrient relocation and liberation within the ecosystem (Harris, 1988). Because virtually no records exist to indicate the size of Everglades wading bird populations prior to the turn of the century (Kushlan et al., 1984), estimates made during the early 1930s will be used here as indicators of prehistoric populations and ecosystem conditions.

Wading birds have declined precipitously as breeders in the Everglades since the 1930s, although large populations of waders continue to winter in the Everglades (Hoffman et al., 1991). In addition, Ogden (1978) noted that wading birds have shifted the center of their breeding distribution northward within the southeastern United States, as well as toward freshwater areas within the Everglades

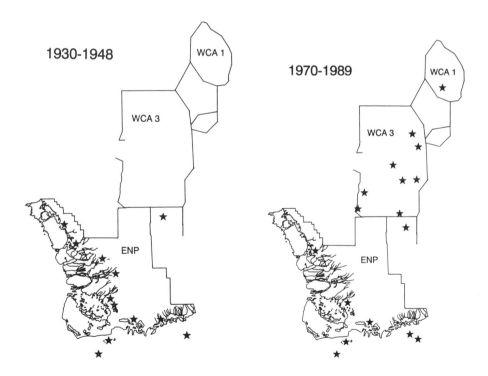

Figure 23.1   Map of the study area, showing boundaries of Water Conservation Areas 1 and 3 and Everglades National Park. Stars represent locations of colonies containing at least 150 nests, reported at any time during 1930–48 and 1970–89. Compared with earlier conditions, colonies are presently shifted away from estuarine sites and centered in the freshwater marshes, particularly the Water Conservation Areas.

(Figure 23.1). The size of breeding assemblages has also decreased. Between 1934 and 1948, at least 10 breeding assemblages were reported to contain in excess of 10,000 nests (Ogden, 1978; Kushlan et al., 1984). During the period from 1978 to the present, no colonies contained in excess of 7000 nests.

These changes in seasonal abundance, breeding status, breeding location, and breeding assemblage size may have had profound impacts on nutrient transport by birds within the ecosystem and, in particular, upon nesting sites as islands of nutrient enrichment within an otherwise oligotrophic wetland. Here, nutrients consumed and deposited at colony and roost sites in the Everglades are estimated for a variety of wading bird species, as well as differences in nutrient flux attributable to wading birds during the past 60 years.

METHODS

Numbers of Wading Birds

Numbers of wading birds breeding in the freshwater and estuarine Everglades (current Water Conservation Areas 1, 2, and 3, Everglades National Park [not including islands in Florida Bay], and northeast Shark River Slough) (Figure 23.1) were estimated from a number of survey reports and historical summaries (Allen, 1958; Kushlan, 1974; Kushlan and White, 1977; Ogden, 1978; Kushlan et al., 1984; Ogden et al., 1987; Frederick and Collopy, 1988; Everglades National Park Research Center records; National Audubon Society Research Department records). Locations of colonies also were compiled from these reports. Recent estimates of numbers of wintering, summering, and nonbreeding wading birds were derived from systematic aerial surveys conducted since 1985 (Hoffman et al., 1991). The numbers of breeding wading birds used for estimation of energetics are shown in Table 23.1.

Nutrient Consumption and Excretion

Nutrient consumption by wading birds was estimated from energetic requirements of adults and young, and excretion was estimated as the nonassimilated portion of energy consumed. Assimilation efficiencies were taken from three studies in which ciconiiform species were hand-raised and fed *ad lib* into adult-

Table 23.1 Numbers of Wading Bird Nesting Attempts per Year
Used in Comparisons of Historical and Current Nutrient Transport (see text for sources)

Decade or year	Wood storks	White ibis	Great egrets	Tricolored herons	Snowy egrets	Little blue herons	Total
1930s	4,000	100,000	4,000	15,000	10,000	2,500	135,500
1940s	4,000	48,000	4,000	15,000	10,000	2,500	83,500
1975	1,330	12,956	3,267	3,500	4,500	2,000	27,553
1986	275	2,503	1,751	1,133	1,319	437	7,418
1987	100	4,130	2,005	537	1,383	723	8,878
1988	220	4,250	2,478	267	200	135	7,550

hood (Kahl, 1964; Kushlan, 1977; Johnston and Bildstein, 1990). In all cases, a 79% assimilation efficiency was used for both adults and young.

Energy Required by Adults

Daily adult energetic needs were calculated for white ibises from Kushlan's (1977) activity budget, for wood storks (*Mycteria americana*) directly from Kahl's (1964) estimates, and for other species from mass-dependent metabolic equations by Wiens and Dyer (1977). Estimates of daily existence energy needs for adults of all species were used directly to estimate nonbreeding energy needs. Total energy consumption by nonbreeding birds (including nonbreeding, wintering, or migrating birds) was estimated as daily existence needs for each species times the number of nonbreeding bird-days for each species. For years since 1985, species-specific abundances were used from the 1985–88 Systematic Reconnaissance Flight surveys for the Water Conservation Areas (Hoffman et al., 1991) and Everglades National Park (M. Fleming, unpublished). For years prior to 1985, there is little useful information concerning numbers of nonbreeding birds. To estimate the month-specific numbers of nonbreeding bird-days during the period prior to 1985, the monthly ratios of nonbreeding to breeding birds from 1985 to 1988 (Hoffman et al., 1991) were applied to the size of the breeding population in each year prior to 1985. Because historic nesting conditions are generally agreed to have been more attractive than during the last decade, and because several of the species expanded their breeding range northward during the period of record, it is reasonable to assume that this method overestimates historic nonbreeding populations relative to current ones.

For wood storks and white ibises, Kahl (1964) and Kushlan (1977), respectively, calculated the increase in energy required for breeding activity by adults. Daily existence needs of other species were then multiplied by these factors to arrive at adult daily breeding energy. Number of days of breeding activity for adults of each species was taken from Bent (1926), Kahl (1964), Wiese (1975), Rudegeair (1975), and Kushlan (1974). Daily adult breeding energy was multiplied by the number of days of nesting to achieve total energy needs of adult breeding birds.

Energy Required by Young

Energy needs of developing young white ibises and wood storks were based on models presented by Kushlan (1977) and Kahl (1964), respectively. Energy needs of young great egrets (*Casmerodius albus*), tricolored herons (*Egretta tricolor*), snowy egrets (*E. thula*), and little blue herons (*E. caerulea*) were calculated by summing weight-specific energy needs over the period of growth, using equations for energy needs of growing birds presented by Kendeigh et al. (1970). Age-specific weights for these species were taken from data presented by Black et al. (1984) and Werschkul (1979).

Excreta and Nutrient Deposition

Energy of excreta was estimated as the nonmetabolizable portion (21%) of energy consumed. It was assumed that young deposited all excreta at colony sites until the development of sustained flight abilities, which was assumed to occur at 80 days for wood storks, 65 days for great egrets, and 45 days for white ibises, snowy egrets, tricolored herons, and little blue herons (Palmer, 1962; Kahl, 1964; Wiese, 1975; Rudegeair, 1975; Werschkul, 1979; Black et al., 1984; Frederick and Collopy, 1988; Bancroft et al., 1991; Frederick et al., 1993). It was also assumed that young of all species had a 20- to 30-day period of postfledgling attachment to the colony, during which the young spent only nights at the colony (13 h), depositing an estimated 54% of their daily (24 h) excreta. A constant rate of excretion was assumed in all calculations of excreta production. Because most nonbreeding, adult wading birds roost communally at night, it was also assumed that 54% of their daily excreta was deposited at roost sites. Using activity budget and breeding phenology data from a number of sources (Bent, 1926; Kahl, 1964; Rudegeair, 1975; Wiese, 1975; Werschkul, 1979; Frederick, 1987; Bildstein et al., 1992), it was estimated that breeding adults deposited 54% of daily excreta in the colony during precourtship and during incubation, 81% during courtship and egg laying, 75% during early nestling stages, and 27% during late nestling and postfledging periods.

Excreta were assumed to contain an energy value of 8372 $J \cdot g^{-1}$ (dry mass), a value intermediate to 5860 $J \cdot g^{-1}$ measured by Stinner (1983) and 8372–10,046 $J \cdot g^{-1}$ measured by Bildstein et al. (1992), to translate excreted energy into dry mass of excreta. Nutrient analyses of white ibis feces by Stinner (1983) and Bildstein et

Table 23.2 Nesting Energy Costs (kJ) and Other Constants Used in Calculation of Nutrient Transport by Wading Birds

	Wood storks	White ibis	Great egrets	Tricolored herons	Snowy egrets and little blue herons
Energy required by adults, per nesting attempt	619,952	101,628	134,140	63,037	63,037
Energy required to grow 1 young	133,533	36,083	66,926	26,937	26,937
Young fledged/nest	2.25	2.0	2.0	2.0	2.75
Total energy per nesting attempt	868,532	173,794	267,992	116,911	116,911
Daily nonbreeding energy per adult	1,884	682	703	322	322
Dry weight (kg) excreta per nesting attempt	23.75	3.22	3.86	1.77	1.77

Other constants			
Composition of feces (%)	Nitrogen	Calcium	Phosphorus
White ibis	5.0	15.0	1.9
Piscivorous waders	13.0	15.0	1.9
Assimilation efficiency	79%		
Energy value of prey (kJ/g dry mass)			
Fish	21.2 J/g		
Crayfish	17.03 kJ/ g		

al. (1992) and of fish-eating waders by Dusi et al. (1971) were used to estimate the nutrients deposited by all wading bird species (Table 23.2).

No attempt was made to estimate the deposition of nutrients in colonies due to mortality of young, because Bildstein et al. (1992) found that carcasses of young contributed less than 1% of total nutrient deposition in a large white ibis colony which exhibited mortality rates similar to Everglades colonies (Frederick and Collopy, 1988).

Energy consumption was translated into weight of prey consumed through caloric measurements of Everglades fishes and crayfishes (Kushlan et al., 1986). Mean sizes of prey consumed by Everglades wading birds were taken from Bancroft et al. (1988), G. T. Bancroft (unpublished data), Ogden et al. (1978), Kushlan and Kushlan (1975), and P. C. Frederick (unpublished) to derive numbers of crayfishes (principal food of white ibises) and fishes (all other species) consumed.

RESULTS

On a per bird basis, nesting wood storks require by far the most energy, consuming at least 4.8 times the amount required by white ibises and over 6 times the amount required by the small *Egretta* herons (Table 23.2).

Estimated consumption of prey necessary to fuel both reproductive efforts and nonbreeding activities is shown in Figure 23.2 for several points in the recent past; Figure 23.3 shows the amounts of nutrients deposited in colony and roost sites by wading birds. Both parameters show dramatic declines since the 1930s, along with breeding population decreases. It is estimated that wading bird populations currently consume 4901 tonnes less prey (89% reduction, dry mass) per year

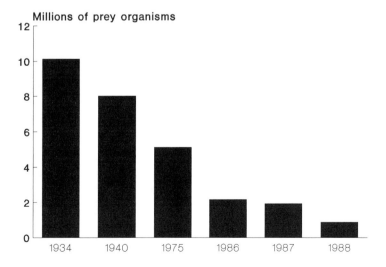

Figure 23.2 Estimated annual consumption of prey by wading birds during breeding and nonbreeding activities in six different years.

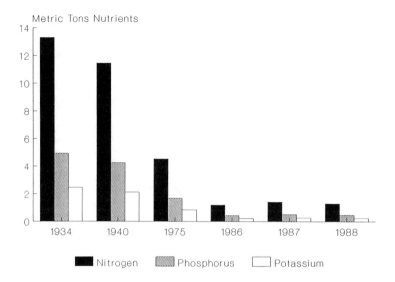

Figure 23.3   Estimated mass of nutrients (in tonnes dry mass) transported to breeding colonies and roosts by wading birds and deposited in the form of feces. Changes in deposition are a function of both changes in numbers of birds using the Everglades and changes in numbers of birds breeding there.

than 1930s populations did, amounting to 14.6 million fewer individual prey items eaten (87% reduction). It is also estimated that current populations of wading birds leave 455 tonnes less feces per year at colony and roost sites, a difference that represents an estimated 59 tonnes of nitrogen and 8.6 tonnes of phosphorus per year.

The majority of this reduction in consumption of prey items and transport of nutrients can be traced to reductions in breeding activity, rather than changes in the size of nonbreeding populations (Figure 23.4). Since the 1930s, reductions in dry mass of excreta transported to roost sites by nonbreeding birds amount to only 1.5% of the total reduction in deposition to colony and roost sites. This stems from the fact that nesting attempts are inherently much more costly in energy than nonbreeding activities. This finding is probably even more robust than stated, because historic nonbreeding populations are likely to have been overestimated relative to current ones. It seems clear that the decline in wading bird breeding has resulted directly in a greater than 90% reduction in predation by the wading bird guild.

The differential reduction in numbers of each species breeding (Table 23.1) has led to changes in the contributions of each to total nutrient transport (Figure 23.5). While white ibises are still the dominant contributor to nutrient flux by wading birds, their annual contribution has decreased by an order of magnitude. Wood storks, originally the second-ranked contributor, have become ranked third, behind great egrets. Although numbers of nesting great egrets have not changed dramatically since the 1940s, their relative importance as nutrient vectors has increased from moving less than 5% of all nutrients moved by waders to more than 40%.

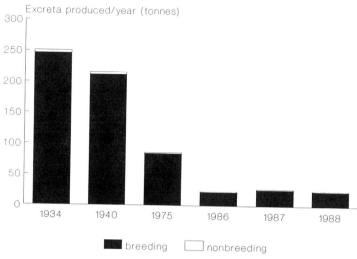

Figure 23.4 Excreta deposited in colonies by breeding birds and growing young (solid) compared with deposition in roosts by nonbreeding birds (open) as a function of time. In all years the vast majority of deposition is contributed through reproductive effort.

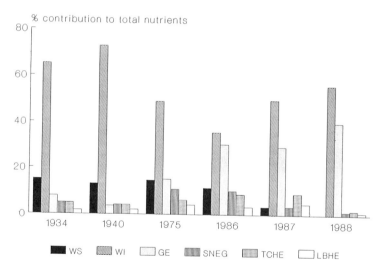

Figure 23.5 Historic and current contribution of individual wading bird species to the deposition of nutrients in the Everglades.

DISCUSSION

The findings presented in this chapter suggest that even large historical populations of wading birds contributed little to the overall flow of nutrients within the Everglades marshes. The estimated 5.07 tonnes of phosphorus annually deposited in colony sites by wading birds during years of peak abundance in the early 1930s is less than the 3% of total historic annual phosphorus inputs to the entire ecosystem from rainfall estimated by Davis (1994). Wading birds, then, can be seen as inconsequential movers of nutrients relative to annual inputs, at least on the scale of the Everglades and associated estuarine areas.

This conclusion is bolstered by the fact that potential sources of error in the estimates in this chapter are likely to lead to an overestimation of nutrient transport by wading birds. In an effort to err consistently, the highest values have been used to estimate energy needs and excretion rates, leading to a liberal estimation of nutrient transport and concentration. This is particularly true of nestling food consumption, which is based largely on studies of captive hand-reared birds, all of which were fed *ad lib*. Because wading birds have been noted for frequent brood reduction, which is largely dependent on available food (Mock et al., 1987), and chronic competition among nestlings for food, it is reasonable to suppose that food satiation of nestlings is rare in the wild. Mock et al. (1987) may have reported the only measurements of food amounts consumed by wild nestlings. Their measurement of 1453 kJ·d^{-1} for broods of great egrets 21 days or less of age is less than 20% of the energy needs calculated through weight-specific growth equations (which are based on *ad lib* feedings). This example suggests that the current estimation of energy consumption by young could lead to overestimation of food consumption and excretion by young of up to 80%. Because individual energy consumption rates of young have a large effect on total estimates of nutrient transport, it is likely that this potential source of error overwhelms other sources. The nesting energy figures in this chapter should therefore be treated as maxima. The huge difference in food consumption between captive fed and wild nestlings also suggests that digestive efficiencies or growth rates are different between the two. If wild birds typically have higher digestive efficiencies than the figures used from captive-reared birds, then production of excreta and probably nutrient content of excreta has been overestimated. However, the digestive efficiency of 79% is at the upper end of measured digestive efficiencies for birds.

Even considering these potential sources of error, it is obvious that large quantities of nutrients can be concentrated in very small spaces through wading bird defecation. Two examples are instructive. In his most conservative estimate, Allen (1958) reported the approximately 16-ha 1934 Shark River colony to contain over 250,000 nests (Ogden, 1978). Using Allen's species breakdowns and excreta deposition factors from the study described in this chapter, it is estimated that this colony site would have received at least 331 g·m^{-2} yr^{-1} nitrogen and 120 g·m^{-2} yr^{-1} phosphorus in 1934. This level of phosphorus deposition is approximately 3000 times the historic annual areal input estimated by Davis (1994). While the Shark River colony was considered at the time to be abnormally large for the Everglades (Ogden, 1978), it may serve as an example of the upper limit of nutrient loading by wading birds. The far smaller, less dense Alley-North (Rescue Strand) colony of 1987 represents a more common and more typical example of a recent nesting aggregation. This 352-ha freshwater colony had approximately 5450 nests in 1987. Using the composition of nests given by Frederick and Collopy (1988), it was calculated that the colony received 20.3 g·m^{-2} yr^{-1} nitrogen and 0.90 g·m^{-2} yr^{-1} phosphorus. This estimated phosphorus input is 20 times the historical annual areal input (Davis, 1994). The difference in phosphorus input between the Alley-North and the Shark River colonies is on the order of 120 times.

The works of Davis (1994) and D. Scheidt (unpublished National Park Service report) suggest that such levels of nutrient loading can have a significant effect on species composition and productivity of freshwater and estuarine marsh macrophyte communities. Browder et al. (1994) have described dramatic changes in freshwater marsh periphyton and invertebrate communities associated with elevated nutrient concentrations. In estuarine mangrove situations similar to many of the large historically important Everglades colony locations, elevated nutrient levels due specifically to bird feces can increase productivity of mangroves (Onuf et al., 1977) and change both species composition and productivity of seagrass beds (Powell et al., 1991; Sogard et al., 1987). Increases in local productivity may have important feedback effects on the survivorship of juvenile wading birds by providing highly productive foraging habitat close to the colony where juveniles develop their feeding skills (Rodgers and Nesbitt, 1979).

Thus, while nutrient redistribution and concentration by wading birds is, and was historically, a relatively small part of the entire Everglades nutrient budget, the local concentrations of nutrients dramatically altered the nutrient character of the colony islands themselves. The extent to which nutrient enrichment at colonies may affect downstream communities undoubtedly varies with size of colony, enrichment concentrations, plant uptake rates, and water flow rates. At a small (400 pair) mixed-species colony in a shallow central Florida marsh impoundment, J. Burney (personal communication) found that increases in phosphorus and ammonia from the colony were not detectable beyond 300 m from the colony edge. In a situation of higher volume and flow of surrounding water, Powell et al. (1991) found that effects of breeding bird colonies on aquatic vegetation could not be detected beyond about 200 m from the colony edge. It seems prudent to propose that the effects of nutrient enrichment, at even very large colony sites, are limited to the immediate vicinity of the colony (certainly less than 1 km downstream) and are best described as "islands" of enrichment.

It is unclear how long such local alteration of plant, and possibly animal, communities in and surrounding colonies might persist following the cessation of excretion inputs. Dramatic changes in species composition of macrophytes are still apparent 6 years after nutrient dosing experiments in central Shark River Slough (D. Sheidt, personal communication). Similarly, the Broad River colony site in coastal Everglades National Park, last used in 1940, is still clearly distinguishable (P. C. Frederick and G. V. N. Powell, personal observation) as an area of luxuriant mangrove growth, despite the passage of over 50 years and at least two major hurricanes. J. C. Ogden (personal communication) reports that the Lane and East River colony sites in coastal Everglades National Park are still distinctly greener than surrounding mangroves even though it has been more than 10 years since the last heavy use of the colony by birds. These observations suggest that locally abundant nutrient input by wading birds is bound quickly into soils and vegetation and is retained and recycled at the site of deposition for long periods.

While this seems plausible for some colony sites, it may not be true of those which have strong surrounding currents, tidal overwash, and quite different surrounding soils and macrophyte communities. Lund (1957) found that nutrient concentrations in sediment samples were not elevated at two estuarine colonies in Everglades National Park (Duck Rock and East River) but were elevated in the immediate vicinity of the Cuthbert Lake colony. Lund attributed the difference to local sediment types.

Because a number of large colony sites are documented to have been used repeatedly over a period of many years (Kushlan et al., 1984), historic patterns of nutrient loading by breeding birds may have resulted in the creation of long-term, i.e., periods of 3–15 years (if continued use by birds is the criterion), islands of refugia for species with high nutrient requirements or tolerances, such as cattail (*Typha* sp.) (Davis, 1994), in an otherwise oligotrophic marsh.

The changes in distribution, colony size, and numbers of breeding wading birds over the last 50 years have undoubtedly had important ramifications for such refugia-dependent species. It is suggested here that the dramatic reductions in numbers of breeding birds have significantly reduced the number of patches where species disadvantaged by low nutrient conditions can survive. The reduction in availability of the special conditions surrounding colonies may have significantly reduced heterogeneity in composition and abundances of both plant and animal communities within the freshwater and estuarine Everglades. The movement of large colonies away from estuarine sites suggests that these effects were most pronounced in the estuarine zone.

ACKNOWLEDGMENTS

The authors wish to thank Keith Bildstein, James Fourqueran, Ron Jones, and John Ogden for comments on earlier drafts of this chapter. Work was supported by grants from Everglades National Park and the John D. and Catherine T. MacArthur Foundation.

LITERATURE CITED

Allaway, W. G. and A. E. Ashford. 1984. Nutrient input by seabirds to the forest on a coral island of the Great Barrier Reef. *Mar. Ecol. Prog. Ser.,* 19:297–298.

Allen, R. P. 1958. A Progress Report on the Wading Bird Surveys, Reports 1 and 2, National Audubon Society Research Department, Tavernier, Florida.

Bancroft, G. T. B., S. D. Jewell, and A. M. Strong. 1988. Foraging habitat of *Egretta* Herons Relative to Stage in the Nest Cycle and Water Conditions, 3rd Annu. Rep. to South Florida Water Management District, West Palm Beach.

Bancroft, G. T., S. Jewell, and A. Strong. 1991. Foraging and Nesting Ecology of Herons in the Lower Everglades in Relation to Water Conditions, Final Report to South Florida Water Management District, West Palm Beach.

Bent, A. C. 1926. Life histories of North American marsh birds. *Bull. U.S. Nat. Mus.,* No. 135.

Bildstein, K. L., E. Blood, and P. C. Frederick. 1992. The relative importance of biotic and abiotic vectors in nutrient processing in a South Carolina, U.S.A., estuarine ecosystem. *Estuaries,* 15:147–157.

Black, B. B., M. W. Collopy, H. F. Percival, A. A. Tiller, and P. G. Bohall. 1984. Effects of Low Level Military Training Flights on Wading Bird Colonies in Florida, Tech. Rep. #7, Florida Cooperative Fish and Wildlife Research Unit, School of Forestry Research and Conservation, University of Florida, Gainesville.

Bosman, A. L. and P. A. R. Hockey. 1986. Seabird as a determinant of rocky intertidal community structure. *Mar. Ecol. Prog. Ser.,* 32:247–257.

Browder, J. A., P. J. Gleason, and D. R. Swift. 1994. Periphyton in the Everglades: Spatial variation, environmental correlates, and ecological implications. in *Everglades: The Ecosystem and Its Restoration,* S. M. Davis and J. C. Ogden (Eds.), St. Lucie Press, Delray Beach, Fla., chap. 16.

Davis, S. M. 1994. Phosphorus inputs and vegetation sensitivity in the Everglades. in *Everglades: The Ecosystem and Its Restoration,* S. M. Davis and J. C. Ogden (Eds.), St. Lucie Press, Delray Beach, Fla., chap. 15.

Dusi, J. L., R. T. Dusi, D. L. Bateman, C. A. McDonald, J. J. Stuart, and J. F. Dismukes. 1971. *Ecologic Impacts of Wading Birds on the Aquatic Environment,* Water Resources Research Institute, Auburn University, Auburn, Ala.

Frederick, P. C. 1987. Responses of male white ibises to their mates' extra-pair copulations. *Behav. Ecol. Sociobiol.,* 21:223–228.

Frederick, P. C. and M. W. Collopy. 1988. Reproductive Ecology of Wading Birds in Relation to Water Conditions in the Florida Everglades, Tech. Rep. No. 30, Florida Cooperative Fish and Wildlife Research Unit, School of Forestry Research and Conservation, University of Florida, Gainesville.

Frederick, P. C. and M. G. Spalding. 1994. Factors affecting reproductive success of wading birds (Ciconiiformes) in the Everglades ecosystem. in *Everglades: The Ecosystem and Its Restoration,* S. M. Davis and J. C. Ogden (Eds.), St. Lucie Press, Delray Beach, Fla., chap. 26.

Frederick, P. C., M. G. Spalding, and G. V. N. Powell IV. 1993. An evaluation of methods for measuring nestling survival in colonially-nesting tricolored herons (*Egretta tricolor*). *J. Wildl. Manage.,* 57:34–41.

Golovkin, A. N. 1967. The effects of colonial seabirds on the development of phytoplankton. *Oceanology,* 7:521–529.

Golovkin, A. N. and G. P. Garkavaya. 1975. Fertilization of waters off the Murmansk coast by bird excreta near various types of colonies. *Sov. J. Mar. Biol.,* 15:345–351.

Harris, L. D. 1988. The nature of cumulative impacts on biotic diversity of wetland vertebrates. *Environ. Manage.,* 12:675–693.

Hoffman, W., G. T. Bancroft, and R. W. Sawicki. 1991. Wading Bird Populations and Distribution in the Water Conservation Areas of the Everglades, 1985–1988, Final Report to South Florida Water Management District, West Palm Beach.

Johnston, J. W. and K. L. Bildstein 1990. Dietary salt as a physiological constraint in white ibis breeding in an estuary. *Physiol. Zool.,* 63:190–207.

Kahl, M. P. 1964. Food ecology of the wood stork *Mycteria americana. Ecol. Monogr.,* 34: 97–117.

Kendeigh, S. C., V. R. Dolnick, and V. M. Gaurilov. 1970. Avian energetics. in *Granivorous Birds in Ecosystems: Their Evolution, Populations, Energetics, Adaptations, Impact and Control,* J. Pinowska and S. C. Kendeigh (Eds.), Cambridge University Press, New York.

Kushlan, J. A. 1974. The Ecology of the White Ibis in Southern Florida: A Regional Study, Ph.D. dissertation, University of Miami, Coral Gables, Fla..

Kushlan, J. A. 1977. Population energetics in the American white ibis. *Auk,* 94:114–122.

Kushlan, J. A. and M. S. Kushlan. 1975. Food of the white ibis in southern Florida. *Fla. Field Nat.,* 3:31–38.

Kushlan, J. A. and D. A. White. 1977. Nesting wading bird populations in southern Florida. *Fla. Sci.* 40:65–72.

Kushlan, J. A., P. C. Frohring, and D. Vorhees. 1984. History and Status of Wading Birds in Everglades National Park, National Park Service Report, South Florida Research Center, Everglades National Park, Homestead, Fla.

Kushlan, J. A., S. A. Vorhees, W. F. Loftus, and P. C. Frohring. 1986. Length, mass, and calorific relationships of Everglades animals. *Fla. Sci.,* 49:65–79.

Lund, E. H. 1957. Phosphate content of sediments near bird rookeries in south Florida. *Econ. Geol.,* 52:582–583.

Meyer, J. L., E. T. Schultz, and G. S. Helfman. 1983. Fish schools: An asset to corals. *Science,* 220:1047–1049.

Mock, D. W., T. C. Lamey, and B. J. Ploger. 1987. Proximate and ultimate roles of food amount in regulating egret sibling aggression. *Ecology,* 68:1760–1772.

Morales, G. and J. Pacheco. 1986. Effects of diking a Venezuelan savanna on avian habitat, species diversity, energy flow, and mineral flow through wading birds. *Colon. Waterbirds,* 9:236–242.

Ogden, J. C. 1978. Recent population trends of colonial wading birds on the Atlantic and Gulf coastal plains. in *Wading Birds,* A. Sprunt IV, J. C. Ogden, and S. Winckler (Eds.), Res. Rep. #7, National Audubon Society, New York.

Ogden, J. C. 1994. A comparison of wading bird nesting colony dynamics (1931–1946 and 1974–1989) as an indication of ecosystem conditions in the southern Everglades. in *Everglades: The Ecosystem and Its Restoration,* S. M. Davis and J. C. Ogden (Eds.), St. Lucie Press, Delray Beach, Fla., chap. 22.

Ogden, J. C., J. A. Kushlan, and J. Tilmant. 1978. The Food Habits and Nesting Success of Wood Storks in Everglades National Park, 1974, Natural Resources Rep. #16, National Park Service, U.S. Department of the Interior, Homestead, Fla.

Ogden, J. C., D. A. McCrimmon, G. T. Bancroft, and B. W. Patty. 1987. Breeding populations of the wood stork in the southeastern United States. *Condor,* 89:752–759.

Onuf, C. P., J. M. Teal, and I. Valiela. 1977. Interaction of nutrients, plant growth and herbivory in a mangrove ecosystem. *Ecology,* 58:514–526.

Palmer, R. S. 1962. *Handbook of North American Birds,* Vol. 1, Yale University Press, New Haven, Conn.

Powell, G. V. N., J. Fourqurean, W. J. Kenworthy, and J. C. Zieman. 1991. Bird colonies cause seagrass enrichment in a subtropical estuary: Observational and experimental evidence. *Mar. Coastal Shelf Sci.,* 32:567–579.

Robertson, W. B. and J. A. Kushlan. 1974. The southern Florida avifauna. *Miami Geol. Soc. Mem.,* 2:414–452.

Rodgers, J. A., Jr. and S. A. Nesbitt. 1979. Feeding energetics of herons and ibises at breeding colonies. *Colon. Waterbirds,* 3:128–132.

Rudegeair, T. J. 1975. The Reproductive Behavior and Ecology of the White Ibis *Eudocimus albus,* Ph.D. dissertation, University of Florida, Gainesville.

Sogard, S. M., G. V. N. Powell, and J. G. Holmquist. 1987. Epibenthic fish communities on Florida Bay banks: Relations with physical parameters and seagrass cover. *Mar. Ecol. Prog. Ser.,* 40:25–39.

Stinner, D. H. 1983. Colonial Wading Birds and Nutrient Cycling in the Okefenokee Swamp, Ph.D. dissertation, University of Georgia, Athens.

Werschkul, D. F. 1979. Nestling mortality and the adaptive significance of early locomotion in the little blue heron. *Auk,* 96:116–130.

Weseloh, D. V. and R. T. Brown 1971. Plant distribution within a heron rookery. *Am. Midl. Nat.,* 86:57–64.

Wiens, J. A. and M. I. Dyer. 1977. Assessing the potential impact of granivorous birds in ecosystems. in *Granivorous Birds in Ecosystems,* J. Pinowski and S. C. Kendeigh (Eds.), Cambridge University Press, Cambridge, U.K.

Wiese, J. H. 1975. The Reproductive Biology of the Great Egret *Casmerodius albus egretta* Gmelin, Master's thesis, Florida State University, Tallahassee.

Wiese, J. H. 1978. Heron nest-site selection and its ecological effects. in *Wading Birds,* A. Sprunt IV, J. C. Ogden, and S. Winckler (Eds.), Res. Rep. #7, National Audubon Society, New York.

24

FORAGING HABITAT of WADING BIRDS in the WATER CONSERVATION AREAS of the EVERGLADES

Wayne Hoffman

G. Thomas Bancroft

Richard J. Sawicki

ABSTRACT

Wading birds form an important component of the Everglades marsh ecosystem and are often used as indicators of the health of the system. Aerial survey data collected in the Systematic Reconnaissance Flight program were used to analyze patterns of foraging habitat use in the Everglades Water Conservation Areas. Habitat use was analyzed on two scales.

First, hierarchical log-linear analyses were used to compare patterns of wading bird distribution and abundance to vegetative communities and concurrent water conditions on the basic 2-km × 2-km scale of the Systematic Reconnaissance Flight data. These analyses were performed for great egrets (*Casmerodius albus*), great blue herons (*Ardea herodias*), wood storks (*Mycteria americana*), and white ibises (*Eudocimus albus*). Concurrent water conditions and previous month water conditions were compared to determine which better explained bird distribution.

Great egrets almost always tended to avoid dense grass habitats. They were seen extensively in transitional water conditions, but usually were dispersed. They foraged less often in wet conditions, but tended to be in larger groups when present. Great blue herons also avoided dense grass habitats. When water levels were low, they also avoided open grass habitats, instead preferring the tree island marsh habitats. Great blue herons avoided dry areas and when feeding in groups tended to use transitional areas. White ibises also tended to avoid dense grass and in some months avoided open grass areas, preferring instead tree island marsh habitats. White ibises tended to forage in large flocks when they used transitional water conditions. In some months wood storks also avoided dense grass habitats.

In the analyses of concurrent versus previous month water conditions, concurrent conditions usually better explained bird distribution when water levels were declining, and previous month conditions often better explained distribution when water levels were rising.

In the second, larger scale analysis the 5 Water Conservation Areas were partitioned into a total of 17 units that exhibited fairly consistent patterns of wading bird use. Four of the units were dense grass habitats with relatively short hydroperiods. They tended to support wading birds only when water was high, and then only in small numbers. Four units were occupied primarily at very low water levels and hosted great numbers of birds during drought years. The nine other units were used extensively under intermediate water conditions and supported the bulk of the birds in most years. Two units, Loxahatchee Tree Island Marsh and L-28 Gap, supported wading birds over an exceptionally wide range of water conditions.

Four major management recommendations are given in this chapter to improve the Water Conservation Areas as wading bird habitat (and hence to improve the health of the marsh overall): (1) promote the interspersion of slough and wet prairie habitats into areas of dense grass, (2) promote the development of tree islands, (3) protect the areas that are currently most productive from alteration, and (4) where possible, manage to lengthen hydroperiods without greatly deepening water levels.

INTRODUCTION

Wading birds (herons, egrets, storks, ibises, and spoonbills) are a conspicuous and valued part of the Everglades. This system formerly hosted the largest concentrations of breeding wading birds in North America (Robertson and Kushlan, 1974). In fact, Everglades National Park was created in large part to protect those great colonies (Beard, 1938; Robertson, 1959).

Wading birds feed primarily on aquatic prey (Bent, 1926; Hancock and Kushlan, 1984). They stand or wade in shallow water and stab, grasp, or filter out their prey (Kushlan, 1978). Thus, they need both adequate prey densities and suitable water depths for foraging. Because such suitable foraging conditions are irregularly distributed and often are short-lived, these birds have large home ranges and change foraging locations frequently (Hafner et al., 1982; Erwin, 1983, 1985; Hafner and Britton, 1983; Dugan et al., 1988). Successful breeding requires the continuous availability of foraging sites within range of the colony sites through the entire nesting cycle.

Wading birds serve as indicators of the health of the Everglades ecosystem (Custer and Osborne, 1977). The long-term decline in breeding populations (described briefly in this chapter and in Ogden, 1994) is one of the several changes in the ecology of the system that have prompted the current concern over its health. Wading birds can serve as indicators of environmental health on a much smaller scale as well. Because they respond behaviorally by moving to local concentrations of food and away from unsuitable areas, their presence and absence can be used to assess very transient conditions (Erwin, 1985; Dugan et al., 1988).

In the 1930s, up to 245,000 pairs of wading birds attempted to nest in the Everglades system during good years (Ogden, 1994). These included 5000–8000 pairs of wood storks (*Mycteria americana*), 25,000–38,000 pairs of herons and egrets (tricolored and little blue herons [*Egretta tricolor* and *E. caerulea*], snowy and great egrets [*E. thula* and *Casmerodius albus*]), and 175,000–225,000 white ibises (*Eudocimus albus*). Total numbers varied substantially between years, and the maximum counts for the various species occurred in different years.

Since the 1930s, the number of pairs in the Everglades has decreased about 50% for great egrets, 70–80% for snowy egrets and tricolored herons, and 90% for white ibises and wood storks (Bancroft, 1989; Ogden, 1994). Although these nesting populations have decreased drastically over the past 50 years, the system remains an important feeding area for many nonbreeding birds. During the dry seasons between 1985 and 1990, combined populations of all wading birds ranged from a low of 60,000 to peaks of 190,000–245,000 birds between January and March.

Foraging habitat use was examined in the Everglades Water Conservation Areas. The Water Conservation Areas are diked and gated and in many areas have peripheral canals, so that water levels and water flow can be managed (Light and Dineen, 1994). During late winter, this portion of the Everglades supports more than half the wading birds in the Everglades/Big Cypress area. In recent years it has also supported more than half of the remaining nesting population of wading birds (Ogden, 1994).

The objective of this study was to explore the environmental variables influencing choice of foraging sites. Because prey abundance and water depth (the two variables most directly associated with foraging site choice) are amenable to study only on small scales, the focus was on variables that are less directly associated but that can be assessed over much larger areas. Specifically, the role of water conditions (extent and coverage of surface water) and vegetation communities in determining foraging habitat choice over the 3600 km² of the Water Conservation Areas was studied. These are readily assessed from the air over large areas. Water conditions are strongly related to water depth, although microtopographical variation prevents detailed mapping of depths from average data. Water conditions change substantially from week to week. Vegetation communities are less labile, normally persisting for months or years at a time (Gunderson, 1994). Vegetation communities are directly related to the long-term hydroperiods of the area. They may affect prey productivity and availability to wading birds.

The Systematic Reconnaissance Flight database for the Water Conservation Areas was used as the source of distributional information. This program is described by Bancroft et al. (1994). Water condition data and vegetation data were also collected from the air during the surveys.

Two approaches were used to analyze foraging habitat choice. The first was a statistical analysis of habitat differences between areas used by birds and areas not used. This analysis used a contingency table approach, the hierarchical log-linear analysis (Sokal and Rohlf, 1981; Norusis, 1988), to identify habitat types and water conditions used more or less frequently than expected by chance. This analysis is not area specific, in the sense that the results do not depend on which patches of a particular habitat type or water condition are used by the birds.

The second approach was area specific. All the monthly bird distribution maps were examined, and a set of areas within the Water Conservation Areas that the birds seemed to treat as units was defined. These units were conceived as potential units of management, areas suitable for individual management prescriptions (obviously hydrological connections will limit the individuality of unit management plans).

Together, these two analyses provide a coherent picture of the environmental conditions preferred by wading birds as foraging habitat and of the spatial distribution of prime foraging areas within the Water Conservation Areas. They also provide insight into the environmental conditions avoided by wading birds and may lead to some recommendations for habitat restoration.

METHODS

The Systematic Reconnaissance Flight surveys are described in detail by Bancroft et al. (1994). For the analyses presented here, surveys from February 1985 through August 1988 for the Everglades Water Conservation Areas were used. The Systematic Reconnaissance Flight grid in the Water Conservation Areas includes 50 transects (numbers 5–54 of the total grid). The surveys were flown in a Cessna 182 fixed-wing aircraft, along east-west transect lines spaced 2 km apart. Altitude was 61 m (200 ft) above ground level (controlled by radar altimeter); 150-m-wide strips were censused on both sides of the plane to provide 15% coverage of the entire area. Position information was provided by an on-board LORAN-C receiver. The wading bird observations were grouped into 2-km × 2-km cells. The Water Conservation Areas portion of the grid included 898 of these cells.

During the surveys, surface water coverage in three categories was recorded. *Wet* conditions were defined as having complete coverage of the ground surface by surface water. *Transitional* conditions had partially exposed ground, but some surface water. Under *dry* conditions, surface water appeared only in ponds, canals, and alligator holes. These water conditions were recorded on a broad scale, such that each 2-km × 2-km cell was assigned a single value in a given month.

During the surveys, the distribution of major vegetation communities was mapped on the same 2-km × 2-km grid. These communities were defined very broadly so that each 2-km × 2-km cell was assigned to a single category and most habitat patches recorded included many cells. Eight general vegetation categories were recorded but for this analysis were combined into five categories (Figure 24.1). *Dense grass* was essentially continuous coverage of sawgrass (*Cladium jamaicense*), cattails (*Typha* sp., mainly *T. domingensis*), sedges, reeds, or grasses. *Open grass* habitats were comprised of stands of sawgrass and other grass-like plants interspersed with at least 25% (by area) open water (or exposed mud). The water typically formed long narrow sloughs and pools, generally oriented north-south. In *tree islands* habitats the marsh was liberally dotted with tree islands, clumps of broad-leaved woody vegetation. The marsh itself generally had sawgrass interspersed with wet prairies or sloughs. *Cypress* was used for areas along the western margin of Water Conservation Area 3A, with cypress trees (*Taxodium*

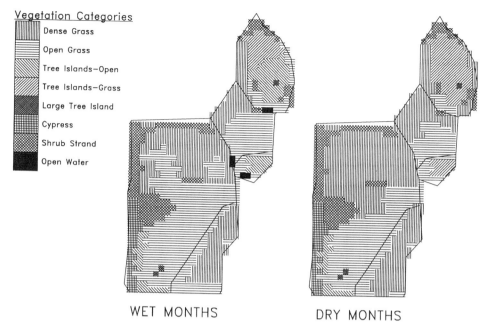

Figure 24.1 Vegetational categories found within the Water Conservation Areas based on aerial surveys in 1987 and 1988.

ascendens) interspersed with sloughs and sawgrass. *Shrub strand* was used for areas of dense grass with large numbers of shrubs (often wax myrtle [*Myrica cerifera*] or willow [*Salix caroliniana*]) interspersed. They differed from tree islands in that the shrubs spread out through the grass rather than clumped.

Two vegetation maps were produced. One map showed conditions during high-water months and the other showed conditions during low-water months. The maps differ in the coverage of open grass versus dense grass. As the system dried, many of the open grass areas were reclassified as dense grass.

Wading bird census data were analyzed in a program written in the dBASE III+ data management programming language. The program sorted the data by species, assigned each observation to its appropriate 2-km × 2-km cell, calculated a density estimate for each cell, and assigned it to a relative density category. The program produced a set of summary statistics, including total population estimates and variance and confidence interval estimates calculated two ways (cell based and transect based). It also produced files containing the density estimates and density categories for each of the 898 2-km × 2-km cells.

For this chapter, relative density categories were calculated by taking the highest cell density value found for the species in the survey and scaling density from it. Category 1 included cells with no birds. Category 2 included cells with density greater than zero but less the one-eighth of the maximum cell density. Category 3 was from one-eighth to three-eighths, and category 4 was from three-eighths to maximum cell density. In the analyses described in this chapter, the relative density categories for the cells were used, along with the water condition and vegetation data, to construct a series of three-way contingency tables.

For this chapter, hierarchical log-linear analysis (Sokal and Rohlf, 1981; Norusis, 1988), a sophisticated approach to contingency table analysis, was used. This is a powerful approach for analyzing categorical data in multi-way contingency tables. Multiple variables and their interactions can be analyzed simultaneously. Hierarchical log-linear analysis can also do a better job than simpler analyses in determining which particular cells in a contingency table contribute most strongly to a significant effect. Thus, for example, it may be possible to determine whether a strong interaction between bird distribution and vegetation categories results from avoidance of dense grass or from a preference for open grass.

The analyses were run on data collected in the 1986–87 and 1987–88 seasons and were conducted using density data for great egrets, great blue herons (*Ardea herodias*), and white ibises. Wood stork data were also used, but only for the months January–April of the two years, because storks were too uncommon to analyze in the other months. Data for the small herons (tricolored and little blue herons, snowy and cattle egrets) were not used because the observations were combined into the categories "small dark heron" and "small white heron" in the data. A concern was that this lumping of species would obscure single-species patterns and possibly create combined-species patterns that were artifacts of the lumping procedure. Great white herons (*Ardea herodias occidentalis*), roseate spoonbills (*Ajaia ajaja*), and glossy ibises (*Plegadis falcinellus*) were considered too uncommon to be analyzed.

The Hi-Loglinear procedure in the SPSS PC+ statistical program (Norusis, 1988) was used for these analyses. For each month, a saturated log-linear model was constructed for a three-way contingency table containing bird density categories, vegetation categories, and water categories (*bird density × vegetation × concurrent water conditions*). By definition, the saturated model included all possible interactions and explained all the variance (heterogeneity) in the table. A series of unsaturated models was then constructed, removing the higher order interaction terms one at a time. The likelihood ratio χ^2 value reported for each model represented the variance (heterogeneity) not explained by that model and hence attributable to the removed higher order terms. Thus the χ^2 value reported for the first unsaturated model (excluding the three-way interaction and including all two-way and one-way terms) was attributable entirely to the three-way interaction. The next model excluded the three-way interaction and one two-way interaction (excluding the vegetation × water interaction). Its χ^2 was attributable to the two excluded interactions, and because the three-way effect was already known, the χ^2 for the vegetation × water interaction could be determined by subtraction. Similarly, as the remaining effects are removed, their contributions can be determined by further subtraction. This approach works for likelihood ratio χ^2 values, but not for the Pearson χ^2 values (also reported by SPSS) because the latter are not additive (Norusis, 1988).

For interactions with significant χ^2 values, the relative contributions of different cells to the interaction can be examined by looking at the parameter values for the contingency table cells. The SPSS printout provides standard errors and z values for these parameters. A positive parameter value with a z value >1.96 indicates that the cell contains a significant associative relationship between the variables. A significant negative parameter value (z value <−1.96) indicates a significant disassociative (avoidance) relationship. Unfortunately, in many cases, a significant interaction may not be attributable to a single cell or set of cells, but may be the result of individually nonsignificant deviation in many cells. In some cases the degrees of freedom (DF) for the χ^2 values are uncertain. SPSS reports "unadjusted" DF and "adjusted" DF, in which one degree of freedom is subtracted for each cell of the multi-way contingency table with an expected value of zero. The unadjusted DF provides an upper bound to the actual DF, while the adjusted DF is generally an underestimate. The unadjusted DF, which is conservative with respect to type I errors (Norusis, 1988), was used in this study. In these applications, the unadjusted DF was probably an overestimate of the three-way interactions, so that type II errors (failing to recognize some significant interactions) may have been made. For the two-way interactions, DF was determined by subtraction, and therefore adjusted and unadjusted values are usually very similar.

Ideally, only foraging birds should be included in these analyses, as birds may choose different habitats for roosting and nesting than for feeding and may fly over still other habitats. All observations were used in this analysis, for two reasons. First, many birds recorded as flying were flushed by the plane and may have been foraging in the immediate area. Second, birds standing on the ground were usually assumed to be foraging, but this often may have been incorrect. In practice, roosting and nesting birds were very uncommon in these surveys. Including all birds may have reduced the power of the analysis but in this circumstance should not introduce biases.

Hierarchical log-linear analysis was also used to test the hypothesis that water conditions 1 month before a survey may be important in determining bird distribution. When water has risen, previous month water conditions may be a better predictor of prey availability because prey are likely to be rare in the most recently flooded areas. For this analysis, data from January–August in 1987 and 1988 were used. The December surveys were not used because previous month (November) water conditions were unknown. For the August surveys, June data were used as the previous survey, because no surveys were flown in July. This analysis was conducted using distribution data for great egrets, great blue herons, and white ibises. The saturated model included the three-way interaction birds × concurrent water condition × previous water conditions. Interaction terms were removed stepwise as in the previous analysis.

In the second approach to habitat preference, maps of the spatial distribution of wading birds in the Water Conservation Areas were examined for the survey years 1985–88. Units of the Water Conservation Areas for which hydroperiod and vegetation combined to produce definable patterns of habitat availability and quality were visually identified and characterized. These units were much larger than the 2-km × 2-km cells but small enough that several fit into each of the larger Water Conservation Areas. The hope is that these units will be useful in considering the relationship of hydroperiod and water management to bird distribution and seasonal habitat use.

The initial units defined were regions that were consistently, over the 4 years, attractive or unattractive to the birds at particular stages of the hydrologic regime. The remaining areas, with less well-defined patterns of bird use, were defined more on the basis of vegetation and hydroperiod to complete the subdivision of the area.

All units defined were contained entirely within single Water Conservation Areas, and disjunct areas with similar attributes were defined as separate units.

Hydroperiod, vegetation type, topography, relation to canals and levees, and relevant water management practices were described for each unit. Also described were the general patterns of bird use for each unit. Wood storks and white and glossy ibises were found more useful in this regard than herons. Great egrets and great blue herons appear to be habitat generalists in the study area and do not respond as strongly to environmental variables. The small *Egretta* herons tend to feed in areas that have feeding concentrations of other birds, and they also feed extensively along canal banks. The tendency of these herons to join concentrations of birds makes their distribution somewhat redundant, and their use of canal banks puts small numbers of them in the units with canals at all stages of the hydrological regime, thus obscuring their responses to marsh conditions.

RESULTS

Analyses of Habitat Preferences

The results of the log-linear analyses of the influences of water conditions and vegetation categories on great egret, great blue heron, white ibis, and wood stork distributions are presented in Tables 24.1 to 24.4, respectively. The tables include likelihood ratio χ^2 values for the third-order interaction and the two second-order interactions involving birds. The other second-order interaction (water conditions × vegetation) is not presented, but was always highly significant. This is to be expected, as the different vegetation types are located in part by their hydroperiod preferences (Gunderson, 1994).

The results for great egrets (Table 24.1) show that the three-way interaction was never significant, implying that great egret choices of foraging locations were not generally the result of complex conditional decisions. The birds × vegetation interactions were significant in all 16 surveys, meaning that the birds were always nonrandomly distributed with respect to the vegetation categories. In January and December 1987 and August 1988, no single cells in the contingency table were identified as making significant contributions to the interaction. In the other 14 months, the analysis showed a significant avoidance of dense grass habitats. The results were more complex for the birds × water conditions comparisons. This interaction was significant in 12 of the 16 months tested. In all 12 months with a significant interaction, at least one cell of the contingency table made a significant contribution, and most months had two or three contributing cells. The strongest trend (all 12 months) was for more cells than expected with transitional water conditions to have great egrets present in small numbers. In other words, this habitat was extensively used, but mostly by dispersed foragers. A related trend for larger concentrations to occur mainly in wet areas was evident in three months. The nonsignificant months were January, April, and December 1987 and August 1988 (Table 24.1). These were the four wettest months in the sample, and evidently the preponderance of wet conditions (and lack of dry conditions) provided the birds too few habitat choices for patterns to be detected.

Table 24.1 Patterns of Interaction between Great Egret Distribution, Vegetation, and Water Conditions in the Water Conservation Areas

	Likelihood ratio χ^2 values		
Month	GE × Wat × Veg (96 DF)	GE × Veg (32 DF)	GE × Wat (12 DF)
December 1986	22.353	52.068*	44.404***
January 1987	3.452	55.311**	0.468
February 1987	6.140	71.102***	27.971**
March 1987	74.814	86.381***	63.170***
April 1987	10.968	59.369**	18.408
May 1987	39.838	86.431***	104.084***
June 1987	47.138	95.705***	92.626***
August 1987	22.014	134.287***	141.028***
December 1987	0.186	81.238*	5.592
January 1988	16.904	75.672***	39.794***
February 1988	13.125	90.997***	26.296**
March 1988	34.943	85.646***	68.319***
April 1988	32.667	76.325***	150.151***
May 1988	66.103	123.921***	156.875***
June 1988	73.331	159.797***	155.057***
August 1988	3.539	111.432***	4.597

Note: $*p < 0.05.$ $** p < 0.01.$ $*** p < 0.001.$

For great blue herons, the three-way interaction also was never significant (Table 24.2). The birds × vegetation categories interaction was significant in 12 of 16 months (December 1986, December 1987, and March and August 1988 were the exceptions). In January 1987, no single cell was found to contribute significantly to the interaction, but in all of the remaining 11 months great blue herons showed a significant tendency to avoid dense grass habitats. They tended to avoid open grass habitats when water levels were low (at those times they were most abundant in the tree island marsh habitats). The birds × water conditions interactions were significant in 11 of the 16 months. These interactions reflected a tendency for great blue herons to avoid dry areas and when feeding in groups (which is uncommon) to use primarily transitional areas. The five months without significant selection of particular water conditions were January, April, and December 1987 and August 1988 (the same months as for great egrets), and December 1986. Again, these were the wettest surveys in the sample.

The white ibis results were rather different (Table 24.3). The three-way interactions again were all nonsignificant. The birds × vegetation interactions were significant in 10 of the 16 months. The six nonsignificant months were primarily high-water months. In the ten months with significant interactions, the birds tended to avoid dense grass areas, and in five they avoided open grass areas (instead they tended to prefer marsh habitats with trees). White ibises also tended to select habitats nonrandomly with respect to water in 9 of the 16 months. Again, significant nonrandom habitat use was detected in drier surveys. White ibises were highly clumped in their use of transitional areas: more cells than expected had no ibises, and more cells than expected had large concentrations.

Table 24.2 Patterns of Interaction between
Great Blue Heron Distribution, Vegetation, and
Water Conditions in the Water Conservation Areas

| Month | Likelihood ratio χ^2 values | | |
	GE × Wat × Veg (96 DF)	GE × Veg (32 DF)	GE × Wat (12 DF)
December 1986	11.537	36.660	13.962
January 1987	1.260	92.178***	2.150
February 1987	5.087	117.707***	28.186**
March 1987	9.737	70.738***	53.754***
April 1987	11.096	53.546***	17.628
May 1987	24.854	67.800***	72.167***
June 1987	26.669	66.583***	47.101***
August 1987	5.808	59.266**	90.132***
December 1987	0.418	43.658	0.956
January 1988	4.731	62.758***	21.810*
February 1988	4.314	51.132*	21.162*
March 1988	14.631	22.776	33.033***
April 1988	22.988	52.052*	73.094***
May 1988	10.089	65.374***	66.473***
June 1988	12.983	85.279***	85.303***
August 1988	0.001	22.103	0.817

Note: * $p < 0.05$. ** $p < 0.01$. *** $p < 0.001$.

Table 24.3 Patterns of Interaction between White Ibis Distribution,
Vegetation, and Water Conditions in the Water Conservation Areas

| Month | Likelihood ratio χ^2 values | | |
	GE × Wat × Veg (96 DF)	GE × Veg (32 DF)	GE × Wat (12 DF)
December 1986	3.283	24.381	22.509*
January 1987	0.946	19.162	3.259
February 1987	6.110	36.877	7.089
March 1987	75.061	173.330***	54.587***
April 1987	28.232	107.250***	5.583
May 1987	16.146	59.761**	62.780***
June 1987	18.593	46.403*	114.797***
August 1987	22.564	98.878***	20.565
December 1987	0.871	42.347	2.342
January 1988	1.555	35.714	50.449***
February 1988	2.878	198.917***	11.364
March 1988	15.075	75.149***	33.162***
April 1988	22.354	92.917***	30.282**
May 1988	21.777	123.250***	69.852***
June 1988	36.899	136.029***	52.201***
August 1988	0.001	23.477	0.213

Note: * $p < 0.05$. ** $p < 0.01$. *** $p < 0.001$.

Table 24.4 Patterns of Interaction between Wood Stork Distribution, Vegetation, and Water Conditions in the Water Conservation Areas

Month	Likelihood ratio χ^2 values		
	GE × Wat × Veg (96 DF)	GE × Veg (32 DF)	GE × Wat (12 DF)
December 1986	—	—	—
January 1987	0.000	32.024	1.385
February 1987	5.450	17.651	3.891
March 1987	12.689	23.720	44.275***
April 1987	3.452	20.501	1.527
May 1987	—	—	—
June 1987	—	—	—
August 1987	—	—	—
December 1987	—	—	
January 1988	3.243	19.317	1.642
February 1988	3.197	49.098*	2.638
March 1988	20.404	50.544*	15.385
April 1988	8.232	35.867	12.407
May 1988	—	—	—
June 1988	—	—	
August 1988	—	—	

Note: * $p < 0.05$. ** $p < 0.01$. *** $p < 0.001$.

The wood stork analyses were run only for January–April of each year, when storks were most common (Table 24.4). Again, the three-way interactions were not significant. Storks were distributed nonrandomly with respect to vegetation conditions only in February and March 1988. In both months they avoided dense grass habitats. Wood storks were found to use the water condition categories nonrandomly only in March 1987 (the month of greatest abundance). Like white ibises, they showed a tendency toward clumped distributions in the transitional areas at that time.

The effects of concurrent and previous month water conditions on great egret distribution are summarized in Table 24.5. The three-way interactions were non-significant in all 14 months. The great egret × previous water conditions interactions were significant for all tested surveys except January and February 1987 and January 1988. This indicates that some aspect(s) of previous month water conditions were significantly correlated with current egret distributions. The parameter estimates indicate that the major effect was avoidance by groups of egrets of areas that had been dry the previous month. The great egret × concurrent water conditions interactions were significant in all months except January 1987 and August 1988. The major effects were avoidance of dry areas and in two months preferences for transitional areas.

Because the great egret × previous water conditions and great egret × concurrent water conditions χ^2 values have the same degrees of freedom, comparison of the χ^2 values gives an indication of the relative importance of effects. Thus, in 1987 the previous water interactions were stronger in January, April, June, and August.

Table 24.5 Association of Great Egret Distribution with
Concurrent Water Conditions and Previous Month
Water Conditions in the Water Conservation Areas

	Likelihood ratio χ^2 values		
Month	GE × Wat × Veg (16 DF)	GE × Veg (8 DF)	GE × Wat (8 DF)
December 1986	—	—	—
January 1987	1.231	12.124	0.306
February 1987	2.779	7.466	26.605***
March 1987	4.927	35.789***	74.271***
April 1987	8.910	43.304***	18.481*
May 1987	5.782	32.393***	103.173***
June 1987	14.554	114.127***	88.103***
August 1987	8.742	164.720***	132.767***
December 1987	—	—	—
January 1988	0.738	2.345	39.905***
February 1988	9.941	35.570***	26.070**
March 1988	5.703	54.476***	69.590***
April 1988	7.103	103.991***	148.505***
May 1988	3.098	197.272***	150.700***
June 1988	13.634	202.862***	146.720***
August 1988	3.865	64.901***	4.610

Note: * $p < 0.05$. ** $p < 0.01$. *** $p < 0.001$.

Because water levels in all four months had risen somewhat from the previous month, the comparison suggests that birds may have been avoiding newly flooded habitat. In 1988, the previous water χ^2 values were larger for February, May, June, and August. These are more difficult to explain, because water levels in 1988 declined rather continuously until June. In August, water levels were high, and most of the area was in wet water condition. The transitional and dry areas in June might well be the shallow areas used by the birds in August.

The relationships for great blue herons are summarized in Table 24.6. Again, none of the three-way interactions was significant. The interactions between heron distribution and previous water conditions were significant in all months except February 1987 and January, February, and August 1988. In most cases the parameter estimates did not indicate that any particular cells of the contingency table contributed strongly to the significance. The χ^2 values for previous water were stronger than those for concurrent water in January and April 1987 and June 1988. Because water levels were higher in all three cases than in the previous month, previous month water conditions may again be a better prediction of bird distribution immediately after a reversal.

The interactions of white ibis distribution with concurrent and previous month water conditions are displayed in Table 24.7. The three-way interactions were significant twice, in August 1987 and June 1988. The importance of this is uncertain, as these are 2 significant results out of 42 tests, and one would expect 1 false-positive result in every 20 tests simply by chance.

Table 24.6 Association of Great Blue Heron Distribution with Concurrent Water Conditions and Previous Month Water Conditions in the Water Conservation Areas

Month	Likelihood ratio χ^2 values		
	GE × Wat × Veg (16 DF)	GE × Veg (8 DF)	GE × Wat (8 DF)
December 1986	—	—	—
January 1987	5.402	24.199**	2.510
February 1987	0.661	3.819	27.819***
March 1987	2.525	18.052*	48.067***
April 1987	2.173	37.893***	17.644*
May 1987	2.374	20.007*	72.144***
June 1987	5.126	44.403***	45.356***
August 1987	6.453	75.194***	88.563***
December 1987	—	—	—
January 1988	2.962	1.352	21.875**
February 1988	2.516	15.050	20.059*
March 1988	2.731	16.175*	32.861***
April 1988	0.603	54.225***	72.748***
May 1988	3.655	53.367***	65.300***
June 1988	13.634	202.862***	146.720***
August 1988	0.001	11.239	0.814

Note: * $p < 0.05$. ** $p < 0.01$. *** $p < 0.001$.

Table 24.7 Association of White Ibis Distribution with Concurrent Water Conditions and Previous Month Water Conditions in the Water Conservation Areas

Month	Likelihood ratio χ^2 values		
	GE × Wat × Veg (16 DF)	GE × Veg (8 DF)	GE × Wat (8 DF)
December 1986	—	—	—
January 1987	3.370	9.248	12.007
February 1987	2.001	0.799	7.130
March 1987	0.416	10.059	47.880***
April 1987	0.629	15.168	5.589
May 1987	6.646	3.950	62.726***
June 1987	22.346	96.570***	113.737***
August 1987	35.882**	26.150***	20.690**
December 1987	—	—	—
January 1988	0.365	0.955	14.698
February 1988	10.386	7.948	11.011
March 1988	3.317	22.333**	32.881***
April 1988	3.683	37.252***	29.955***
May 1988	20.064	47.820***	67.851***
June 1988	26.946*	77.524***	49.644***
August 1988	0.000	4.571	0.263

Note: * $p < 0.05$. ** $p < 0.01$. *** $p < 0.001$.

The white ibis × previous water conditions interactions were significant in June and August 1987 and March, April, May, and June 1988. In most cases the parameter estimates were not significant or were difficult to interpret, but the tendency to avoid dry areas was again apparent. The white ibis × concurrent water interactions were significant in four of the seven months each year. In 1987 they were significant in March, May, June, and August. In 1988 they were significant in March, April, May, and June. In each year these were the driest months. The parameter estimates showed the usual avoidance of dry areas, but also a distinct tendency to prefer transitional areas.

The previous water condition χ^2 values were higher for April and August 1987 and for April, June, and August 1988. Because all of these except April 1988 are months with rising water, previous month transitional areas may correspond to concurrent month shallow wet areas.

Spatial Characterization

The study area was partitioned into 17 regions that seemed to be used as units by wading birds. These units average a bit more than 200 km² in area. Figure 24.2 is a map of the Water Conservation Areas showing the 17 units. The units do not cross any of the boundaries of the five Water Conservation Areas that comprise the study area. Units 1–3 are in Water Conservation Area 1, 4–6 are

1. Loxahatchee High Marsh.
2. Loxahatchee Tree Island Marsh.
3. Loxahatchee Impounded Sloughs.
4. WCA 2A High Marsh.
5. WCA 2A Open Marsh.
6. WCA 2A Impounded Sloughs.
7. WCA 2B.
8. WCA 3A High Marsh.
9. Miami Canal Strip.
10. Alley-North Slough.
11. L-28 Gap.
12. Alligator Alley High Marsh.
13. Cypress Strand.
14. WCA 3A Middle Unit.
15. L-67 Slough.
16. L-67 Pocket.
17. WCA 3B.

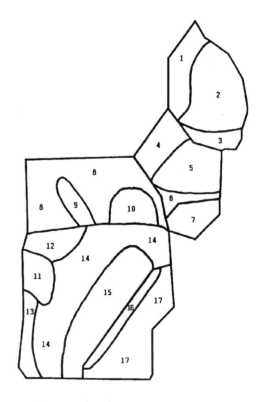

Figure 24.2 Spatial characterization of the Water Conservation Areas.

in Water Conservation Area 2A, 7 includes all of Water Conservation Area 2B, 8–15 are in Water Conservation Area 3A, and 16 and 17 are in Water Conservation Area 3B.

The location, vegetation patterns, and patterns of wading bird use are described for each unit in the following sections. These units should prove to be useful areas to consider in managing the hydrology of the Water Conservation Areas.

Water Conservation Area 1

Loxahatchee High Marsh

This area comprises the northern and western sides of Water Conservation Area 1. It had shorter hydroperiods than the remainder of Water Conservation Area 1. The vegetation included a narrow band of dense willow thicket along the L-7 levee and perimeter canal and a band of dense grass (partly dense sawgrass, partly cattails) about 4 km wide. At the north end, the sawgrass was extensively invaded by wax myrtle. Along the eastern boundary, the dense sawgrass graded into tree island marsh habitats. The dense sawgrass area was broken by a series of open areas that had the appearance of evaporation pans. The Loxahatchee High Marsh did not have extensive use by wading birds. It was used more in periods of high water than when water levels were lower. When it was used, birds were generally found in the "evaporation pans," around isolated alligator holes, roosting in the willow thickets, or on the perimeter canal. The areas of dense grass, and particularly those dominated by cattails, had little use.

Great egrets and white ibises were the most frequent users of this unit. The evaporation pans were used quite a bit by glossy ibises and small herons. Small herons were the most frequent users of the perimeter canal.

Loxahatchee Tree Island Marsh

This unit comprises the bulk of Water Conservation Area 1. Along its eastern margin were areas of open sawgrass and some willow thickets. The bulk of the area, however, was comprised of open sawgrass and wet prairie habitats scattered with great numbers of small tree islands. The unit had a high density of small ponds and alligator holes. In general, it seemed to have more relief than most other units in the study area and remained in "transitional" water conditions for more of the hydrological regime than most units.

Loxahatchee Tree Island Marsh was extremely important as a wading bird habitat. The greater relief seemed to provide foraging opportunities over a wide range of water depths. Early in the dry season, birds used the numerous high spots, often along the margins of the tree islands. As water levels fell, the birds flocked to the centers of the spikerush (*Eleocharis* sp.) sloughs and prairies. When much of the area was dry, the birds concentrated around the numerous alligator holes and ponds. This unit consistently supported high densities of birds and was of prime importance. It was one of the units most important to wood storks and white ibises, but was used extensively by all the birds considered. It also supported the highest density of great blue herons in the Water Conservation Areas.

Loxahatchee Impounded Sloughs

This unit occupies the southern end of Water Conservation Area 1. It contained sloughs dominated by white water lilies (*Nymphaea odorata*) and floating macrophytes, alternating with small patches of sawgrass. In some areas the water from the perimeter canals extended unbroken a kilometer or two into the marsh at high water.

The Loxahatchee Impounded Sloughs were used relatively little by wading birds when water levels were high. As water levels dropped, the birds moved in and fed primarily in the transitional areas. When water was low, this unit was particularly important to glossy ibises and wood storks.

Water Conservation Area 2A

Water Conservation Area 2A High Marsh

This unit occupies the northern end and northwest side of Water Conservation Area 2A. It was composed mainly of dense grass (sawgrass and cattails) with a few large tree islands and had a relatively short hydroperiod. It generally supported fewer birds than the rest of Water Conservation Area 2A, although it hosted large numbers at times in high water (April 1986, December 1987).

Water Conservation Area 2A Open Marsh

This unit comprises the bulk of Water Conservation Area 2A. It was dominated by open sawgrass habitats, with patches and strands of sawgrass alternating with sloughs that dried to expose extensive bare mud. The north end was dominated by cattail and mixed sawgrass and cattail communities (Davis, 1994). This unit had a moderately long hydroperiod.

The Water Conservation Area 2A Open Marsh is an extremely valuable wading bird habitat. It is the single unit most used by wood storks and often hosts great numbers of white ibises. Glossy ibises also use it extensively. It also appears very important to tricolored and little blue herons.

Water Conservation Area 2A Impounded Sloughs

This unit occupies the southern point of Water Conservation Area 2A. It contained extensive areas of white water lily sloughs alternating with patches of sawgrass and some cattails. As water levels became very low, the birds that were foraging in the Water Conservation Area 2A Open Marsh moved into this area. All species used it extensively when water levels were low enough.

Water Conservation Area 2B

Water Conservation Area 2B is treated as a single unit, because it is too small to partition. It had a variety of habitats, including open grass, sloughs, and sawgrass heavily invaded by *Melaleuca quinquenervia*. Hydroperiods in Water Conservation Area 2B appeared to be very irregular and strongly influenced by water management decisions.

Water Conservation Area 2B provided important foraging habitat for wading birds when water levels were suitable. In high-water conditions birds sometimes fed along the northern edge of the area (e.g., September, November, and December 1985, January and December 1988). When water levels were low (e.g., August 1987), the southern end of Water Conservation Area 2B occasionally hosted great concentrations of birds.

Water Conservation Area 3A

Water Conservation Area 3A High Marsh

This unit includes the short-hydroperiod marshes covering much of Water Conservation Area 3A north of Alligator Alley. The habitat was primarily dense sawgrass, in some areas heavily colonized by wax myrtle. This unit supported few tree islands. The presence or absence of wax myrtle in this unit seemed partly determined by hydroperiod and partly by fire history. Alligator holes and natural ponds were rare in this habitat, but a few man-made ponds were present. This unit was being invaded by *Melaleuca* on its east edge and by cattails throughout. In spring 1989, most of this unit was burned in a series of fires. Severe peat fires in the central portion of the unit burned up to 0.5 m into the ground.

The Water Conservation Area 3A High Marsh supported relatively few wading birds, except in the burnouts from the peat fires. In high-water periods great egrets were scattered through the area in low densities, and occasionally (September and December 1985, December 1986, April 1987) white ibises were present in numbers. In low-water periods, great blue herons and great egrets used the man-made ponds. Since the spring of 1989, the burnouts from the peat fires have supported concentrations of wading birds whenever the water table has been high enough to flood them. White ibis have dominated these concentrations, but glossy ibis, great egrets, and small herons have also been present.

Miami Canal Strip

This unit is a narrow strip of marsh along the Miami Canal north of Alligator Alley. It had a much longer hydroperiod than the surrounding Water Conservation Area 3A High Marsh because it was often flooded by water diverted or leaking from the canal. In periods of low water, when most of Water Conservation Area 3A north of Alligator Alley was dry, large volumes of water were transported down the Miami Canal into developed areas of Dade County, and this strip of marsh remained wet. Vegetation in the strip was primarily dense grass, but close along the canal were willow thickets and thickets of *Ludwigia* spp. (willow herb) and some *Eleocharis* marshes.

This unit received much more attention from wading birds than the surrounding high marsh. Dry season maps often showed an irregular line of occupied cells along the Miami Canal in the otherwise empty or sparsely used expanse of Water Conservation Area 3A north (May 1985; February, March, and May 1986; March and June 1987; March 1988). Great egrets and small herons (snowy egrets, tricolored herons, and little blue herons) were the most frequent users of the unit.

Alley-North Slough

This unit is a section of Water Conservation Area 3A north of Alligator Alley and well east of the Miami Canal. It was named for the large Alley-North colony that often formed in a large tree island in the center of the unit. The unit had a series of north-south drainages with strips and patches of *Eleocharis* marsh, separated by ridges covered with dense sawgrass (the difference in elevation between the ridges and sloughs was probably a few centimeters). The unit had a distinctly longer hydroperiod than the surrounding high marsh, but unlike the Miami Canal Strip, it dried out when overall water levels were low.

The Alley-North Slough hosted large numbers of wading birds when overall water levels were moderate, particularly when the Alley-North colony was active. White ibises, great egrets, and small herons were the main users of the unit, although great blue herons and wood storks were also regular. White ibises seemed to feed extensively along the drying edge in the unit, to the north of the areas used at a particular time by the herons. They may have been feeding extensively on crayfish or other mud-inhabiting prey in areas too shallow to retain exploitable fish populations.

The L-28 Gap

This small unit occupies an area along the west side of Water Conservation Area 3A south of Alligator Alley. When the L-28 levee was constructed, the gap was left to preserve a natural drainage gradient (Mullet Slough) from Big Cypress Preserve into Water Conservation Area 3A. In addition, the L-28 interceptor canal conveys some water from the Big Cypress Seminole Indian Reservation and adjacent areas southeast into the unit. The unit itself extends from the gap east for about 8 km and a few kilometers to the southeast. Like the Alley-North Slough unit, the L-28 Gap unit had more relief than surrounding areas. A series of elongate tree islands and patches of dense sawgrass (some invaded by wax myrtle) extended from north-northwest to south-southeast and were separated by sloughs of open grass habitat. This area hads a particularly complex hydrological regimen. The relief extended the period that it spent in the transitional stage, and drainage of water from Big Cypress also kept the unit wetter than surrounding areas.

The L-28 Gap unit hosted large numbers of birds over a wide range of overall water levels. It seemed particularly important in dry periods, when the adjacent Water Conservation Area 3A Middle Unit was largely dry. White ibises, great egrets, and small herons dominated the bird populations, but the unit also provided some of the best habitat for wood storks in Water Conservation Area 3A.

Alligator Alley High Marsh

This unit lies immediately south of Alligator Alley and north of the L-28 Gap unit. It extends from the L-28 levee north of the gap east to the Water Conservation Area 3A Middle Unit. It had a relatively short hydroperiod, apparently shortened artificially by the effects of the L-28 canal and the ditches flanking Alligator Alley. Much of the unit had shrub-strand habitat (dense sawgrass invaded by wax myrtle and other shrubby vegetation), dense sawgrass, and open grass.

The Alligator Alley High Marsh was used fairly extensively when overall water levels were high (August–November 1985, February–April 1986). Great egrets were the most prevalent users of this unit.

Cypress Strand

The Cypress Strand unit is a narrow strip along the L-28 levee south of the L-28 Gap. It was composed of cypress strand habitats, with relatively sparse sawgrass, numerous sloughs, and some cypress domes. It was used by wading birds when overall water depths were moderate. The area immediately west of the unit in southeastern Big Cypress Preserve was used more extensively by wading birds, particularly when water levels in Water Conservation Area 3A were still high. Wood storks used southeastern Big Cypress extensively in mid-winter and a few spilled over into the Cypress Strand unit.

Water Conservation Area 3A Middle Unit

This unit is an area of fairly long hydroperiod which extends through the center of Water Conservation Area 3A from Tamiami Trail to Alligator Alley and east at the north end to U.S. 27. It is bounded on the west by the Alligator Alley High Marsh, L-28 Gap, and Cypress Strand units and on the east by the L-67 Slough unit. The vegetation was fairly heterogeneous, but was dominated throughout by open grass with small shallow sloughs. In the south there were numerous tree islands.

The Water Conservation Area 3A Middle Unit was used extensively from moderately high water levels to fairly low water levels. As the area dried, the birds tended to shift to the southeast, toward the L-67 Slough unit. This unit was used extensively by all species except glossy ibises. Great blue herons and great egrets were dominant early in each drydown, but as water levels dropped, small herons and white ibises moved in.

L-67 Slough

The L-67 Slough unit occupies a broad band along the eastern edge of Water Conservation Area 3A, from a few kilometers south of Alligator Alley to Tamiami Trail. It is bounded on the east by the L-67a levee and on the north and west by the Water Conservation Area 3A Middle Unit. Due to water management practices, this area was more or less permanently impounded and dried only in particularly dry years. This area dried in the spring of 1985 (April–June) but remained flooded from summer 1985 through the end of 1988. The vegetation was a series of sloughs separated by narrow bands and patches of sawgrass, with a scattering of willow heads and other tree islands. In the south the sloughs were filled with water lilies, but in the north they grew extensive mats of periphyton and fewer macrophytes.

The L-67 slough usually had water too deep for extensive wading bird use. The few birds seen here were generally great blue herons and, in the drier months, great egrets. Any ibises or small herons were generally just flying over.

The situation was very different in particularly dry years. The very dry spring of 1985 saw the rest of Water Conservation Area 3A go completely dry and great numbers of birds crowd into the L-67 Slough unit. In April 1985, nearly every cell in the unit was occupied, and large flocks were widespread. By June 1985, birds (and water) persisted only in the northern end of the unit. This was probably the lowest area in the Water Conservation Areas.

Water Conservation Area 3B

The L-67 Pocket

This unit comprises a narrow strip along the western edge of Water Conservation Area 3B. It is enclosed by the L-67a and L-67c levees. The L-67c levee and canal were constructed to trap any water that might leak through the L-67a levee into Water Conservation Area 3B at times when the stage in Water Conservation Area 3A was being maintained higher than the stage in Water Conservation Area 3B. The hydroperiod of the pocket was similar to that of the rest of Water Conservation Area 3B, except that it retained water in extremely dry periods such as May and July 1985 (also spring 1989).

The L-67 Pocket hosted large numbers of wading birds in April and May 1985, when the L-67 Slough also supported many birds. When water levels were higher, patterns of bird use in the pocket were similar to those in the rest of Water Conservation Area 3B.

Water Conservation Area 3B

This unit includes all of Water Conservation Area 3B except the L-67 Pocket. It was composed mainly of dense grass, with some areas of open grass around the periphery. The southern edge has a small canal running east-west parallel to the L-29 levee (along Tamiami Trail) and several north-south canals running down to L-29. The southeastern end tended to dry out first, perhaps because of drainage by these canals. The southern and central parts had some large tree islands. Invasion by *Melaleuca* was underway, especially in the northeastern section.

Water Conservation Area 3B supported great blue herons and great egrets under most water conditions and also supported white ibises and small herons when overall water levels were low. The southeastern corner seemed to be more attractive than the rest of Water Conservation Area 3B, except when water was extremely low. Glossy ibises were almost never recorded in Water Conservation Area 3B.

Patterns of Wading Bird Use

Examples of the bird distribution maps that were used to define these units are presented in Figures 24.3 to 24.6. The distribution of great egrets in December 1985 is shown in Figure 24.3. Water conditions were moderately high. The Loxahatchee High Marsh, Water Conservation Area 2A High Marsh, and Water Conservation Area 3A High Marsh were nearly empty of birds. The birds were concentrated in the Loxahatchee Tree Island Marsh, Water Conservation Area 2A Open Marsh, Alligator Alley High Marsh, Alley-North Slough, and Water Conservation Area 3A Middle Unit. The Loxahatchee Impounded Sloughs and the L-67 Slough, which were wet, were nearly devoid of egrets. The combined density of all species of wading birds

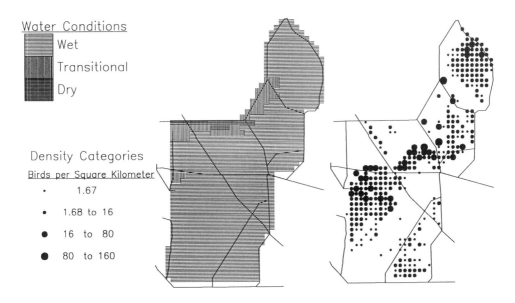

Figure 24.3 Distribution of great egrets in the Water Conservation Areas in December 1985, when water conditions were high.

in June 1987 is displayed in Figure 24.4. Water levels were intermediate overall. The Loxahatchee High Marsh, Water Conservation Area 2A High Marsh, and Water Conservation Area 3A High Marsh units were nearly devoid of wading birds. The Loxahatchee Tree Island Marsh, Water Conservation Area 2A Open Marsh, Alley-North Slough, L-28 Gap, and Water Conservation Area 3A Middle units all hosted concentrations of birds. The Miami Canal Strip unit also had moderate numbers of birds. The L-67 Slough unit had noticeably fewer occupied cells than the surround-

Figure 24.4 Combined distribution of all species in the Water Conservation Areas in June 1987, when water conditions were intermediate between wet and dry.

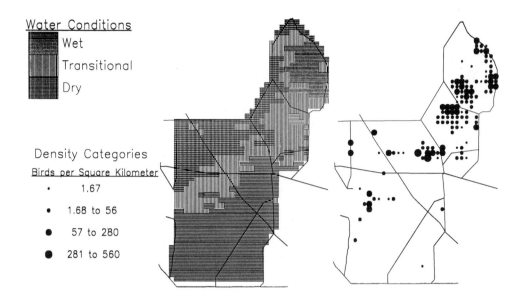

Figure 24.5 Distribution of white ibises in the Water Conservation Areas in May 1986, when water conditions were moderately low.

ing areas. White ibis distribution in May 1986, when water levels were moderately low, is displayed in Figure 24.5. Ibises were concentrated in the Loxahatchee Tree Island Marsh, Water Conservation Area 2A Open Marsh, Alley-North Slough, L-28 Gap, and Water Conservation Area 3A Middle units. The combined distribution of all wading birds in April 1985, a very dry period, is shown in Figure 24.6. The

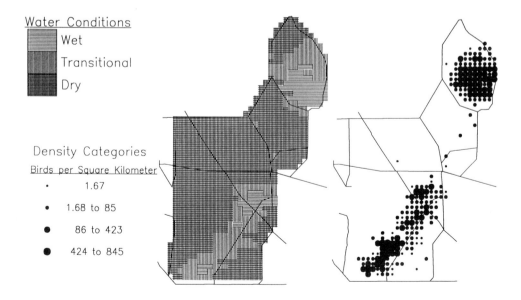

Figure 24.6 Combined distribution of all species in the Water Conservation Areas in April 1985, during a major drought.

Loxahatchee Tree Island Marsh was heavily occupied, but most of the birds in the system were concentrated into the L-67 Slough unit, with some spillover into the L-67 Pocket.

The relationship between water conditions and wading bird use of the units is summarized in Table 24.8. Because water management practices sometimes cause water conditions in different Water Conservation Areas to diverge (water may be impounded in one Water Conservation Area while another is drawn down), these water conditions should be considered separately for each Water Conservation Area. Under *high* water conditions, a majority of the Water Conservation Area is wet. Under *intermediate* conditions, some of the Water Conservation Area is dry, some is wet, and much is transitional. *Low* conditions find much of the Water Conservation Area dry and little, if any, wet. *Drought* conditions occur only in the driest periods of the driest years. The Water Conservation Areas are all nearly dry, with remnant transitional conditions in the impounded areas. These predicted patterns of bird use of the units are approximate. The actual levels of bird use of a particular unit under intermediate water conditions, for example, will be influ-

Table 24.8 Patterns of Wading Bird Use in the 17 Units Comprising the Water Conservation Areas[a]

Water Conservation Area/Unit	Water conditions			
	High	Intermediate	Low	Drought
WCA 1				
1. Loxahatchee High Marsh	*			
2. Loxahatchee Tree Island Marsh	*	****	***	***
3. Loxahatchee Impounded Sloughs	*	**	***	
WCA 2A				
4. WCA 2A High Marsh	*			
5. WCA 2A Open Marsh	*	****	**	
6. WCA 2A Impounded Sloughs		*	**	***
WCA 2B				
7. WCA 2B	*	*	***	
WCA 3A				
8. WCA 3A High Marsh	*	**	**	
9. Miami Canal Strip	*	*	**	
10. Alley-North Slough	*	***	*	
11. L-28 Gap	**	***	***	**
12. Alligator Alley High Marsh	**	*		
13. Cypress Strand	*	**		
14. WCA 3A Middle Unit	*	****	**	
15. L-67 Slough		*	***	*****
WCA 3B				
16. L-67 Pocket	*	**	**	***
17. WCA 3B	*	**	**	

[a] The number of asterisks represents intensity of use by wading birds. A blank represents minimal bird use. One asterisk represents consistent use by small numbers. Two and three asterisks represent moderate to heavy use, and four and five asterisks represent the most important feeding areas found.

enced by time of year, recent hydrological history (and therefore amount of food present) of the unit, and the condition and quality of alternative foraging locations in the Everglades.

DISCUSSION

Habitat Preferences

In most of the log-linear analyses, the wading birds were nonrandomly distributed with respect to water conditions and vegetation communities. This was to be expected and affirms the utility of these environmental attributes in descriptions of foraging habitat for these birds. Because both of these attributes are amenable to modification, management schemes to benefit wading birds can be developed incorporating these results.

When the log-linear analyses began, complex (three-way) interactions among bird distribution, water conditions, and vegetation communities were expected. In other words, bird use of particular community types was expected to vary substantially with water conditions. Such interactions were seen, at a trivial level at least, when bird distribution was examined from month to month, but not in within-months analyses. For example, the higher dense grass areas in the north and west parts of Water Conservation Areas 1, 2A, and 3A are used very little by birds except in months when water is high. The variation in water conditions within vegetation communities within months may be inadequate to demonstrate such an interaction. The vegetation categorization may be too crude to allow recognizing the occurrence of such interactions on a finer scale. Alternatively, the hypothesis may simply be wrong, and water conditions and vegetation communities may influence bird distribution independently.

The two-way interactions of birds × vegetation consistently show that these species avoid dense grass habitats. This avoidance is actually stronger than the analyses show because some of the area classified as dense grass contains artificial ponds (dug to improve habitat conditions for deer), and most of the wading birds recorded in these areas were in fact using the ponds and not the expanses of dense sawgrass. Many, but not all, of these areas of dense sawgrass are in parts of the Water Conservation Areas that have had their hydroperiods shortened by management activities. Alterations in management that would increase hydroperiods in these areas might allow succession back toward the open grass habitats more useful to wading birds (Gunderson, 1994). Any redesign of the structures conveying surface water into the Water Conservation Areas should consider effects on the hydroperiods of these areas.

These analyses did not take into account the role of fire in modifying vegetation communities. In the spring of 1989, much of the surface of the Water Conservation Areas burned. Most of these fires were superficial and did not greatly modify the marshes. In at least four areas, however, long-term modification of the vegetation communities may have occurred. A fire that burned most of Water Conservation Area 2B may well have affected the rate and pattern of succession to *Melaleuca* forest. Two areas in the north end of Water Conservation Area 3A (one east and one west of the Miami Canal) suffered extensive peat combustion.

In the months following these peat fires, wading birds foraged in numbers in the depressions burned into the soil whenever these depressions contained water. Because these fires killed sawgrass in large patches, the pattern of postfire succession may provide different habitats than the former dense grass. Since the fire, the deepest depressions burned into the soil have been flooded several times, when water tables rose to near the former soil level, and use by wading birds was fairly extensive. The fourth area is the northwest side of Water Conservation Area 1. The fire there killed all the trees on thousands of tree islands. These areas may recover as tree islands, or they may be replaced with herbaceous growth. The long-term patterns of wading bird use in all of these areas are likely to be modified, but in unpredictable ways.

Great blue herons and white ibises preferred marsh habitats dominated by tree islands over open grass habitats in many of the analyses. Inspection of the data shows that the preferred tree island marsh habitats were those of Water Conservation Area 1; therefore, it is possible that this is some sort of areal effect. The tree island habitats of Water Conservation Area 1 were extraordinarily attractive to wading birds in general, and the birds commonly roosted in the trees, but it appears that this attractiveness may be only partially a result of the trees. These habitats seem to display more local relief than most of the rest of the Water Conservation Areas. They have more and deeper potholes, ponds, and depressions and more hummocks or high spots. This increased relief results from the formation and stranding of floating peat islands (Gunderson, 1994). These islands provide the better drained high spots needed for the growth of the tree islands. The increased relief allows the prey-concentrating effects of the seasonal drydowns to operate over a longer period each year and also breaks up the drying marsh into many smaller pockets of available food. As a result, food supplies may be more reliable and more consistently available to the birds over a longer part of the year. The effects of dry season rainstorms may tend to be more localized in these higher relief marshes, and food may also become available more rapidly after water level reversals.

Great egrets were widespread but dispersed in transitional areas, and where they occurred in wet areas, they were more often in larger groups (20–100 birds). Presumably they were feeding on different prey, or on prey distributed differently, in these two situations. In contrast, white ibises tended to feed in large flocks (100–800 birds) when in transitional areas and to occupy a small proportion of the transitional cells. Great blue herons displayed a third pattern, tending to feed solitarily in wet areas and in small concentrations (5–25 birds) in transitional areas. Wood storks showed significant patterns in only a few months, tending to cluster in transitional areas.

Herons often establish and defend foraging territories, but they also forage at times in mixed flocks (Kushlan, 1976, 1978; Bayer, 1978, 1982). Individuals searching for a location to forage must often choose between foraging alone and joining flocks. Individuals foraging (successfully) alone may be faced with the options of defending their sites from other herons, relinquishing their sites, or allowing a flock to form. These choices are likely determined on the basis of food availability and abundance, site defensibility, number of other herons present, and recent experience (Bayer [1978] also found that nonterritorial feeding locations were less predictably available in tidal situations). The authors' observations suggest that once the decision is made not to defend a feeding site, additional birds can join without experiencing territorial aggression. It may be that unused areas, areas used solitarily, and areas used by flocks occur in that order along a gradient of increasing prey availability. Sites that are defended by solitary birds, then, would have lower prey availability than sites that are used but not defended. This could occur in two ways. If prey availability is sufficient, territory defense might become unprofitable, simply because prey is locally, although temporarily, unlimited. Alternatively, as the number of competing herons increases, the costs of territory defense may escalate to the point of unprofitability (Brown, 1964).

The differences between species are probably related to their well-documented differences in prey preferences and foraging behavior (Kushlan, 1976, 1978). Because great blue herons feed on larger fish and can wade in deeper water than the other species, the preponderance of great blue heron sightings in wet areas is not surprising. The increased tendency for great blue herons in transitional areas to forage in groups (usually mixed-species groups around deep holes, with great egrets and sometimes other species as well) suggests that only here are larger fish concentrated enough that territoriality breaks down. The tendency of great egrets to spread out and forage solitarily in transitional areas may be a function of their habit of eating smaller prey and the consequent need for higher prey densities. Smaller prey may persist in adequate densities in shallower water than is needed for great blue heron preferred prey. The tendency of great egrets to feed in groups in wet areas is more difficult to understand; this may actually be a statistical consequence of the relative infrequency of solitary great egrets in wet areas.

Because white ibises and wood storks seldom, if ever, defend feeding areas, the dynamics of their choices of foraging locations are very different (Kushlan, 1979, 1986). White ibises are relatively short-legged and thus are unable to feed in most wet areas (Kushlan, 1978; Powell, 1987). In the transitional areas where they do most of their feeding, they commonly occur in flocks of up to several hundred birds and occupy relatively few of the cells. At times, wood storks are seen roosting or standing on the ground far from suitable foraging sites. This behavior limits interpretation of the analyses, but storks, like white ibises, tend to clump in the transitional areas.

Analyses of the relation between bird distribution and concurrent versus previous water conditions show that when water levels are rising, birds tend to feed in deeper water than when levels are dropping. Under these conditions, previous month water conditions better explain bird distribution than do concurrent water conditions. This may result from time lags in the recovery of prey populations. Areas that were wet or transitional in the previous month survey are the refugia that will provide stock to repopulate the newly flooded areas, but evidently that population growth takes long enough that the refugia maintain higher prey populations for at least several weeks.

Spatial Characterization

Spatial characterization of the Water Conservation Area marshes summarizes the findings of this study on habitat preferences of birds. This characterization should be useful in several ways. It may be useful, in combination with water level information, in predicting general patterns of bird distribution at any given time. It may also provide units appropriate in size for consideration in assessing consequences of water management practice for wading birds. A given practice (e.g., lowering the water level in Water Conservation Area 2A) will affect more than one of the units defined and may affect different units differently. By analyzing the effects unit by unit, a clearer overall understanding should be attained.

The 17 units defined all differ somewhat in patterns of wading bird use; together they provide at least some foraging habitat during all phases of most dry seasons. The high marshes (Loxahatchee, Water Conservation Areas 2A and 3A, and Alligator Alley high marshes) provide wading bird habitat primarily when water levels are high, and these units generally support relatively few birds. In future plans to modify water management in the system, it may be profitable to search for ways to rehabilitate some of these areas as wading bird habitats.

As water levels drop, birds become more abundant in the system and move into intermediate habitats: Loxahatchee Tree Island Marsh, Water Conservation Area 2A Open Marsh, Water Conservation Area 2B, Alley-North Slough, L-28 Gap, Cypress Strand, Water Conservation Area 3A Middle, and Water Conservation Area 3B. These units provide the most important habitat, year in and year out, for wading birds. The Loxahatchee Tree Island Marsh and the L-28 Gap unit are exceptional in that they provide foraging habitat over a wider range of water stages than the other units. Any plans to change the configuration of water management in the Water Conservation Areas should be analyzed carefully to ensure that they do not have negative effects on the suitability of these units for wading birds.

The remaining units—Loxahatchee Impounded Slough, Water Conservation Area 2A Impounded Sloughs, Miami Canal Strip, L-67 Slough, and L-67 Pocket—provide foraging habitat primarily at low stages. Most of these provide some habitat in most dry seasons. The L-67 Slough unit provides extensive foraging opportunities only in the driest years, but then hosts great numbers of birds. In fact, these units may support more birds than the entire Water Conservation Area hosts in more normal years, because under drought conditions, these are the best areas of freshwater habitat that remain in south Florida. The normal (seasonal) population of the Water Conservation Areas is joined by birds from elsewhere, presumably Everglades National Park, Big Cypress National Preserve, and the myriad small wetlands throughout south Florida (unpublished Systematic Reconnaissance Flight data confirm very low numbers of these birds in Everglades National Park and Big Cypress Preserve in these drought years). These deeper water habitats are probably important as dry season refugia for the fish and aquatic invertebrates that provide most of the food for the wading birds.

Recommendations

Taken together, the results of these two sets of analyses provide insights into habitat needs of wading birds that may be useful in designing water management and habitat management schemes for the Water Conservation Areas. First, schemes that promote the interspersion of slough and wet prairie habitats into dense sawgrass areas would probably benefit wading birds. Second, schemes that promote the development of tree island dominated marshes, or at least that limit the loss of tree islands, should be beneficial. The benefits are likely to accrue more from the created (or preserved) microtopographic variation than from the trees themselves, although tree islands do provide colony sites. Third, areas that are currently highly productive, especially the Loxahatchee Tree Island Marsh and the L-28 Gap unit, should be protected from alteration. The hydrology of these last areas may deserve special study in order to gain a better understanding of why they remain productive so much longer than the other units. Fourth, management to lengthen hydroperiods without greatly deepening water might greatly improve habitat. Under current conditions, water levels of the long-hydroperiod habitats (except canal banks) are too deep for wading birds most of the time, and they seem to provide good foraging habitat for too short a period to support successful nesting when they are available. Shallow but long-hydroperiod marshes can be established (or restored) by managing water for gradual flow-through over much of the year, in contrast to the current impounded marshes. For example, structures and management systems proposed in the Everglades Agricultural Area to release treated water from the Water Management Areas might be designed to allow gradual releases over much longer periods, in natural "flow-way" areas within the Water Conservation Areas, than current releases from the Everglades Agricultural Area.

ACKNOWLEDGMENTS

The authors thank the South Florida Water Management District for support throughout this study of wading birds in the Everglades. Peter David, Steve Davis, James F. Milleson, Vyke Osmondson, and the late J. Walter Dineen were particularly helpful. Discussions with Reed Bowman, Paul Cavanagh, Martin Fleming, Peter Frederick, Susan Jewell, John Ogden, William B. Robertson, Jr., James Schortemeier and Allan Strong contributed much to the continued success of the Systematic Reconnaissance Flight project. The pilots of Lynch Enterprises and James Wyatt, Inc. successfully and safely flew the surveys. R. Michael Erwin and John C. Ogden provided valuable comments on an earlier version of the manuscript. We are, as always, especially indebted to Nancy Paul, whose editorial and technical skills contributed so much to the quality and appearance of the final manuscript. We thank all these people and institutions for their help.

LITERATURE CITED

Bancroft, G. T. 1989. Status and conservation of wading birds in the Everglades. *Am. Birds,* 43:1258-1265.

Bancroft, G. T., A. M. Strong, R. J. Sawicki, W. Hoffman, and S. D. Jewell. 1994. Relationships among wading bird foraging patterns, colony locations, and hydrology in the Everglades. in *Everglades: The Ecosystem and Its Restoration,* S. M Davis and J. C. Ogden (Eds.), St. Lucie Press, Delray Beach, Fla., chap. 25.

Bayer, R. D. 1978. Aspects of an Oregon estuarine great blue heron population. in *Wading Birds,* A. Sprunt IV, J. C. Ogden, and S. Winckler (Eds.), Res. Rep. 7, National Audubon Society, New York, pp. 213–217.

Bayer, R. D. 1982. How important are bird colonies as information centers? *Auk,* 99:31–40.

Beard, D. B. 1938. Everglades National Park Project, National Park Service, U.S. Department of the Interior.

Bent, A. C. 1926. Life histories of North American marsh birds. *U.S. Nat. Mus. Bull.,* No. 135.

Brown, J. L. 1964. The evolution of diversity in avian territorial systems. *Wilson Bull.,* 76: 160–169.

Custer, T. W. and R. G. Osborn. 1977. Wading birds as biological indicators: 1975 colony survey. *U.S. Fish Wildl. Serv. Spec. Sci. Rep. Wildl.,* 206:1–28.

Custer, T. W. and R. G. Osborn. 1978. Feeding-site description of three heron species near Beaufort, North Carolina. in *Wading Birds,* A. Sprunt IV, J. C. Ogden, and S. Winckler (Eds.), Res. Rep. 7, National Audubon Society, New York, pp. 355–360.

Davis, S. M. 1994. Phosphorus inputs and vegetation sensitivity in the Everglades. in *Everglades: The Ecosystem and Its Restoration,* S. M. Davis and J. C. Ogden (Eds.), St. Lucie Press, Delray Beach, Fla., chap. 15.

Davis, S. M., L. H. Gunderson, W. A. Park, J. Richardson, and J. Mattson. 1994. Landscape dimension, composition, and function in a changing Everglades ecosystem. in *Everglades: The Ecosystem and Its Restoration,* S. M. Davis and J. C. Ogden (Eds.), St. Lucie Press, Delray Beach, Fla., chap. 17.

Dugan, P., H. Hafner, and V. Boy. 1988. Habitat switches and foraging success in the little egret (*Egretta garzetta*). in Acta IXI Congr. Int. Ornithologici, H. Ouellet (Ed.), Ottawa, Canada, pp. 1868–1877.

Erwin, R. M. 1983. Feeding habitats of nesting wading birds: Spatial use and social influences. *Auk,* 100:960–970.

Erwin, R. M. 1985. Foraging decisions, patch use, and seasonality in egrets (Aves: Ciconiiformes). *Ecology,* 66:837–844.

Gunderson, L. H. 1994. Vegetation of the Everglades: Determinants of community composition. in *Everglades: The Ecosystem and Its Restoration,* S. M. Davis and J. C. Ogden (Eds.), St. Lucie Press, Delray Beach, Fla., chap. 13.

Hafner, H. and R. H. Britton. 1983. Changes of foraging sites by nesting little egrets (*Egretta garzetta* L.) in relation to food supply. *Colon. Waterbirds,* 6:24–30.

Hafner, H., V. Boy, and G. Gory. 1982. Feeding methods, flock size and feeding success in the little egret (*Egretta garzetta*) and the squacco heron (*Ardeola ralloides*) in Camargue, Southern France. *Ardea,* 70:45–54.

Hancock, J. and J. Kushlan. 1984. *The Herons Handbook,* Harper and Row, New York.

Kushlan, J. A. 1976. Wading bird predation in a seasonally fluctuating pond. *Auk,* 93:464–476.

Kushlan, J. A. 1978. Feeding ecology of wading birds. in *Wading Birds,* A. Sprunt, IV, J. C. Ogden, and S. Winckler (Eds.), Res. Rep. 7, National Audubon Society, New York, pp. 249–297.

Kushlan, J. A. 1979. Feeding ecology and prey selection in the white ibis. *Condor,* 81:376–389.

Kushlan, J. A. 1986. Responses of wading birds to seasonally fluctuating water levels: Strategies and their limits. *Colon. Waterbirds,* 9:115–162.

Lund, F. and J. W. Dineen. 1994. Water control in the Everglades: A historical perspective. in *Everglades: The Ecosystem and Its Restoration,* S. M. Davis and J. C. Ogden (Eds.), St. Lucie Press, Delray Beach, Fla., chap. 4.

Norusis, M. J. 1988. *SPSS/PC+ Advanced Statistics V2.0,* SPSS, Chicago.

Ogden, J. C. 1994. A comparison of wading bird nesting colony dynamics (1931–1946 and 1974–1989) as an indication of ecosystem conditions in the southern Everglades. in *Everglades: The Ecosystem and Its Restoration,* S. M. Davis and J. C. Ogden (Eds.), St. Lucie Press, Delray Beach, Fla., chap. 22.

Powell, G. V. N. 1987. Habitat use by wading birds in a subtropical estuary: Implications of hydrography. *Auk,* 104:740–749.

Robertson, W. B., Jr. 1959. *Everglades—The Park Story,* University of Miami Press, Coral Gables, Fla.

Robertson, W. B. and J. A. Kushlan. 1974. The south Florida avifauna. in *Environments of South Florida: Present and Past, Memoir No. 2,* P. J. Gleason (Ed.)., Miami Geological Society, Coral Gables, Fla., pp. 414–452.

Sokal, R. R. and F. J. Rohlf. 1981. *Biometry,* W. H. Freeman, San Francisco.

25

RELATIONSHIPS among WADING BIRD FORAGING PATTERNS, COLONY LOCATIONS, and HYDROLOGY in the EVERGLADES

G. Thomas Bancroft

Allan M. Strong

Richard J. Sawicki

Wayne Hoffman

Susan D. Jewell

ABSTRACT

Restoration of wading bird breeding populations in the Everglades requires a better understanding of the relationships among wading bird foraging patterns, colony locations, and hydrology. To address this need, general foraging distribution data from systematic aerial surveys, specific foraging distribution data obtained from following flights, habitat data from U.S. Geological Survey orthophotomaps, hydrological data from gauges and aerial surveys, and colony location, size, and success data from three recent studies were analyzed.

Nesting great egrets (*Casmerodius albus*) and white ibises (*Eudocimus albus*) typically foraged within 9 and 10 km, respectively, of their colonies. Historically, these species bred in large, mixed-species colonies in the mangrove zone of Everglades National Park, whereas currently they breed in much smaller colonies in the Water Conservation Areas. The persisting historic colonies in the mangrove zone are surrounded by a diverse mosaic of habitats and generally have a smaller percentage of freshwater habitats.

In the Water Conservation Areas, great egret and white ibis foraging distributions varied within and among years and were generally correlated with differences in water depth and distribution. Comparison of colony location and size with overall foraging distributions during the months overlapping breeding indicated that colony location for these two species was only a marginal predictor of the location of food resources at the time when they were feeding young. Examination of the formation, growth, and decline of the L-67 colony in the Water Conservation Areas during the drought year 1989 showed that initially the nesting birds were feeding close to the colony. As the area dried out, the overall foraging distribution shifted well south of the colony. Nesting birds gradually had to fly farther to find foraging sites, and the colony experienced high levels of nest abandonment. It can be concluded that wading birds initiate nesting near foraging aggregations, feeding on large and concentrated prey bases near suitable nesting sites, at physiologically appropriate times of year. These simple cues, however, may no longer be adequate indicators of foraging sites that will provide food for the 3–4 months needed to complete nesting.

In addition to influencing the patterns and timing of water flows, it is likely that water management has aggravated the effects of dry season rainfall by increasing the severity and duration of reversals, creating pulsed regulatory releases and reducing water levels so that a given rainfall event has a greater diluting effect. The compartmentalization of the Everglades may have decreased the ability of forage fish to migrate through the system, especially into the deeper sloughs during the dry season, thus decreasing the productivity of these areas for nesting wading birds.

The extent of wetlands should be maximized by restoring degraded marshes wherever possible. The natural connectivity in the system should be increased by reducing compartmentalization and the critical features of natural hydrology should be replicated, especially in the northern ends of the Water Conservation Areas and Shark River Slough. Additionally, the hydrology and productivity of the lower Shark River Slough wetlands and the associated estuaries should be analyzed more thoroughly, and peak flows out of Water Conservation Area 3A and Shark River Slough should be increased to improve habitats in the areas from Nine Mile and Paurotis ponds to Watson, North, and Roberts rivers and into the headwaters of Gum, Dixons, Lostmans, and East sloughs.

INTRODUCTION

Historically, the Everglades supported large numbers of nesting wading birds (Robertson and Kushlan, 1974; Ogden, 1978, 1994; Bancroft, 1989). These birds bred primarily during the winter–spring dry season, when aquatic prey were being concentrated in drying pools. Although the Everglades still provides foraging habitat for large numbers of nonbreeding wading birds, the number of breeding birds has been reduced by approximately 90% (Bancroft, 1989; Ogden, 1994). In addition to this dramatic decrease in nesting numbers, the distribution of nesting birds has shifted. During the 1960s, the majority of the nesting population moved out of Everglades National Park and into the Water Conservation Areas.

Most authors feel that drainage of wetlands and changes in the management of water in the Everglades are the ultimate causes of the declines in breeding populations (Kushlan, 1986a, 1987; Bancroft, 1989; Ogden, 1994). If the quantity, distribution, and timing of water flow have either reduced the biomass of fish or changed how they become available to birds, nesting populations may show differential impacts relative to the location of the colony site and the habitats surrounding these colonies. Knowledge of historic colony distribution, habitat types around colonies, and the spatial scale utilized by white ibises (*Eudocimus albus*) and great egrets (*Casmerodius albus*) at a number of colonies throughout Everglades National Park and the Water Conservation Areas should allow us to evaluate how the current water management strategies may have affected nesting distributions.

The declines in wading bird nesting populations make understanding the factors that lead to colony formation critical. The shift from Everglades National Park to the Water Conservation Areas provides an opportunity to study the interactions between surface water conditions, foraging populations, and colony locations in a more homogeneous environment. Presumably, the pattern of drydown directly affects how prey become available to wading birds in the drying marsh (Loftus and Eklund, 1994); however, the factors leading to colony formation are probably complex and variable among species (Ogden et al., 1980). Wood storks (*Mycteria americana*) primarily use traditional colony sites and are less flexible in their tendency to shift to new sites in response to changing water conditions (Kushlan, 1986a). Only in extremely dry years do some storks shift from traditional sites along the mangrove coast to inland Everglades sites. White ibises, on the other hand, are extremely flexible in their choice of colony sites. Colonies form in different years near large foraging aggregations (Kushlan, 1976). Although these studies provide a framework for understanding colony site selection, the data are qualitative. The exact set of environmental cues used by wading birds to initiate colony formation is unknown. A quantitative analysis of great egret and white ibis colony site selection relative to foraging site selection is provided in this chapter. The objective is to compare colony site fidelity in these two species and determine their relative abilities to gauge conditions of the foraging habitat throughout the nesting season.

Because successful nesting requires 3–4 months of nest attendance (McVaugh, 1972; Werschkul, 1979), the choice of colony site is critical to these central place foragers. Water conditions (i.e., depth, hydropattern, hydroperiod, duration of current coverage) will influence the size and number of aquatic prey in an area (Loftus and Eklund, 1994) and the availability of prey to birds (Custer and Osborn, 1978; Frederick and Collopy, 1989; Hoffman et al., 1994). Several researchers have shown that water conditions and prey populations influence foraging site selection (Erwin, 1983, 1985; Hafner and Briton, 1983; Hafner et al., 1986; van Vessem and Draulans, 1987; Dugan et al., 1988). However, little is known about how these factors influence nesting success. With major changes in water coverage and, therefore, prey availability, wading birds show long-distance shifts in foraging locations (Kushlan, 1979, 1986a; Kushlan et al., 1975, 1985; Hoffman et al., 1994; Frederick and Collopy, 1988; Ogden et al., 1978). Data on colony site selection relative to foraging distributions over a range of hydrological conditions in the Water Conservation Areas are provided here. This information can be used to determine how plastic the response of wading birds is to environmental conditions and whether these responses vary in relation to the available habitat around different colony sites.

Water management changes in the Everglades system have apparently caused a delay of several months in the onset of breeding by wood storks in the Everglades (Kushlan, 1986a; Ogden, 1994). The evidence for a delay in nesting by other wading birds is more controversial (Ogden, 1978, 1994). In the late 1980s, most *Egretta* herons did not lay eggs until the end of March and during April (Frederick and Collopy, 1988; Bancroft et al., 1990). This pattern of timing makes it difficult for most birds to successfully produce independent young by the time summer rains begin in late May or June. It seems unlikely that historically birds would have consistently started this late and been able to maintain populations. Water conditions in the past several decades may have resulted in a delay in the onset of laying for egrets and herons. Therefore, it is necessary to better understand the relationship between water conditions and colony site selection, foraging habitat, and nesting success. This information is critical for the development of effective management strategies because it helps define the conditions that lead to initiation of reproduction and survival of independent young.

METHODS

Distribution and Abundance of Wading Birds

Data on wading bird distribution and abundance and on surface water conditions were taken from the Systematic Reconnaissance Flight Survey program over Water Conservation Areas 2 and 3 (Hoffman et al., 1989, 1994). Surveys were flown during the first week of each month. In mid-April and mid-May 1989, additional surveys were conducted in southern Water Conservation Areas 3A and 3B to supplement data collected at the L-67 colony in Water Conservation Area 3A.

For analyses of bird distribution and water conditions, the survey area (Water Conservation Areas 2A, 2B, 3A, 3B) was divided into seven subregions (Figure 25.1). Within each subregion, the number of cells in each water category (Hoffman et al., 1994) was counted. These counts were used in a contingency table analysis of changes in water conditions between months or years. To compare shifts in bird distribution between years or months, the number of birds counted within the subregions was summed. The observed numbers were compared using two-way contingency tables and the *G*-statistic (Sokal and Rohlf, 1981). In these statistical analyses (contingency tables), actual counts rather than population estimates were used.

For significant *G*-tests, confidence limits around the observed proportion in each cell of the contingency table (Neu et al., 1974) were calculated. The expected proportion to these confidence ranges was then compared. If the expected proportion was above or below the confidence limits of the observed proportion, then significantly fewer or more birds than expected, respectively, used the area for feeding or breeding.

Data for great egrets and white ibises were taken from the February–June

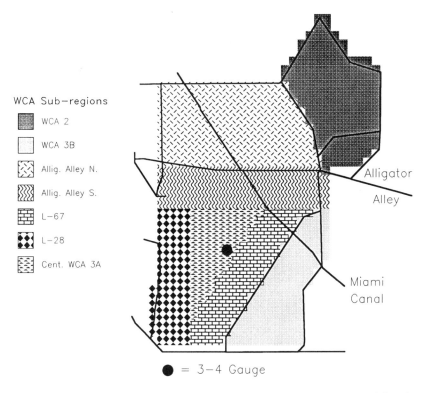

Figure 25.1 Subregions of Water Conservation Areas 2 and 3 used for the analysis of bird and colony distribution and location of the 3–4 gauge.

Systematic Reconnaissance Flight surveys in 1986–89. These two species were chosen because they were easily censused from the air, both on foraging grounds and in nesting colonies. However, their population dynamics and foraging ecology in the Everglades are quite different. Great egrets were relatively permanent residents during the dry season and showed fewer population fluctuations than did white ibises (Hoffman et al., 1989). White ibises were the most abundant feeding and nesting wading birds in the system (Bancroft, 1989), but their numbers varied more than great egrets (Hoffman et al., 1989). Both foraging and nesting white ibises were more nomadic than great egrets. A comparison of these two species would, therefore, show how birds with different life history traits responded to water conditions.

Location of Wading Bird Colonies

Frederick and Collopy (1988) conducted aerial surveys to monitor the location, number of breeding pairs, and species composition of wading bird colonies in Everglades National Park and Water Conservation Areas 3A and 3B during 1986 and 1987. P. Frederick and the National Audubon Society continued the surveys in 1988, and the National Audubon Society and the National Park Service continued them in 1989. The 1988 and 1989 surveys also included Water Conservation Areas 2A and 2B. In 1986 and 1987, surveys were flown at least twice a month, and in

1988 and 1989 they were flown monthly (see Frederick and Collopy [1989] for survey protocol).

Shifts in nesting abundance among the subregions of the Water Conservation Areas were examined. The maximum nest count for each colony was doubled to estimate the number of breeding birds present. These estimates were then summed for all colonies within each section of the Water Conservation Areas in each year. Contingency table analyses were used to identify shifts in breeding concentrations between years.

Foraging Locations of Breeding Birds

Flight-line counts and following flights were used to determine locations of foraging sites for individual breeding birds. Following flights were initiated after it was determined that most birds were feeding nestlings rather than courting, nest building, or incubating. This increased the probability of following birds on foraging trips, rather than birds gathering nest material, and ensured that foraging choices were examined during the most stringent period of nesting. The colony was circled in a Cessna 172 and birds were followed as they left the colony. The birds were followed until they landed and began feeding or had landed in a flock in which all other birds were feeding. Additional following flight data were provided by B. Patty for the Roberts Lake Strand, Andytown, and Rodgers River Bay colonies in 1983 and by Frederick and Collopy (1988) for many colonies in Everglades National Park and the Water Conservation Areas in 1986 and 1987.

Foraging Habitat

Twenty-two U.S. Geological Survey 7.5' orthophotomaps were used to digitize the habitat types around the Lane/East River, Rookery Branch, Frank Key, and Rodgers River Bay colonies into a computer-compatible format. The following eight habitat categories were used to classify foraging sites:

1. *Inland marsh:* Sawgrass (*Cladium jamaicense*) dominated marshes, where hydropattern is influenced by both upstream flow and local rainfall. These marshes are inland from the mangrove zone and interspersed with tree islands, cypress (*Taxodium ascendens*) domes, and deeper water sloughs.

2. *Slough:* Areas of deeper water within the inland marshes. Sloughs often have a higher proportion of *Eleocharis* spp. and *Nymphaea odorata* relative to sawgrass.

3. *Mangrove:* Areas of mangrove (primarily red mangrove [*Rhizophora mangle*], but also black mangrove [*Avicennia germinans*]) forests of varying heights that are essentially a continuous cover of woody vegetation. Some areas of higher elevation harbor less water-tolerant woody species.

4. *Coastal marsh:* Treeless areas of grasses, rushes, and sedges between the inner bays and the coast. They are generally dominated by *Spartina* spp., sawgrass, or cattail (*Typha* spp.) and because of their geographic location are not strongly influenced by upstream freshwater flows. Many small ponds are interspersed throughout the marsh.

5. *Mangrove-coastal marsh:* This is a large, heterogeneous transitional area

between the mangroves and coastal marshes and the inland marshes and sloughs. These areas are dominated by grasses (*Spartina* spp., bulrush [*Scirpus* spp]., *Typha* spp., and sawgrass), red mangroves, and some low elevation trees and shrubs (e.g., willow [*Salix caroliniana*], buttonwood [*Conocarpus erectus*], and wax myrtle [*Myrica cerifera*]). This habitat type also contains many small ponds.

6. *Mangrove-inland marsh interface:* The ecotonal area between the mangrove-coastal marsh and the inland marsh. This is an irregular transitional zone containing numerous mangrove "finger" projections that extend into the marsh. Within 100 m, foraging wading birds often can walk between the two habitat types.

7. *Tidal flat:* Foraging sites along the Gulf of Mexico and in Florida Bay that are available to birds on lunar, wind-driven, or seasonal tidal cycles.

8. *Coastal prairies:* Treeless areas of salt-tolerant, broad-leaved, and succulent plants. Common plants are saltwort (*Batis maritima*), glasswort (*Salicornia* spp.), and sea purslane (*Sesuvium portulacastrum*). These areas border Cape Sable and Florida Bay.

These broad habitat classifications allowed feeding site coordinates (as opposed to observer assessment) to be used to summarize habitat selection. We flew over the area in a Cessna 172 to verify habitat demarcations. Two habitat types could not be defined as discrete areas on the map: mangrove-inland marsh interface and tidal flat habitats. Both of these habitats were ecotonal and difficult to define as mappable units. For these habitat types, observer assessment of habitat type was used.

A 9-km-radius circle around historic colonies in the mangrove zone was used to define foraging area available to breeding wading birds. (This distance contained 75% and approximately 60% of all great egret and white ibis following flights, respectively.) Calculation of area in each habitat type was done using Atlas Draw™ software (Strategic Locations Planning, Inc.). For the mangrove-inland marsh interface, a 0.5-km-"radius" corridor was defined around the digitized line separating inland marsh or slough from mangrove or mangrove-coastal marsh habitats. One-half of the area of the inland-marsh interface was subtracted from the area of mangrove or mangrove-coastal marsh. The remaining half was subtracted from the area of slough and inland marsh habitats in proportion to their linear distance along the ecotone. For tidal flat, the area defined as shoals on the 1986 NOAA Nautical Chart was used. The remainder of the habitat in the 254-km² foraging area was classified as open water. Water depth at the 3–4 gauge (Figure 25.1) was provided by South Florida Water Management District personnel.

RESULTS

Area Used for Foraging by Breeding Birds

When nesting, wading birds are constrained to feed relatively close to the colony site; therefore, successful colonies must be located in areas that provide adequate food for their 10- to 14-week nesting cycle. The distance nesting birds can commute between the colony and the feeding areas varies substantially from species to species. The distribution of potential foraging areas (those sufficiently near the colony sites) between the 1930s and the 1980s was compared in order to determine shifts in habitat use associated with the major declines in breeding wading bird numbers. Recent data on foraging flight lengths were used to estimate the areas close enough to the colonies to be considered potential foraging areas for nesting great egrets and white ibises.

Great Egrets

In 1983 and from 1986 to 1989, great egrets averaged 6.3 km on 422 flights from 12 colonies (Table 25.1). The distribution of flight lengths was highly skewed toward shorter flights (Figure 25.2). Over 45% of the flights were less than 4 km in length, 72% were less than 8 km, and 83% were less than 12 km. The longest flight was at least 40 km and the shortest less than 0.25 km. Considerable variation occurred between colonies in mean flight distance (Table 25.1). For the eight colonies with more than 25 following flights, mean distance varied from 2.4 km at Frog City to 12.0 km at Andytown. The distribution of flight lengths at these eight colonies also varied (Figure 25.3). Three colonies had almost flat distributions with about equal numbers of flights across a range of distances. Four colonies had highly skewed distributions, and one colony had close to a statistically normal distribution (Sokal and Rohlf, 1981).

Circles with a radius of 9.0 km were plotted around colonies of great egrets in the study area to examine the area used by breeding birds (Figure 25.4). This flight distance included 75% of the 422 flights. At 6 of 8 colonies in which more than 25 birds were followed, a 9-km circle included 75–100% of the flights. The area

Table 25.1 Mean Distance (km) of Foraging Flights by Great Egrets in the Everglades

Colony	Year	n	Mean	SD
Alligator Alley North[a]	86, 87	82	7.2	6.1
Andytown[a]	83, 86	32	12.0	9.7
Big Melaleuca[a]	86	2	3.0	3.3
L-67[c]	89	57	3.0	3.3
Frog City[a]	86	38	2.4	1.7
L-28[a]	87	17	4.1	4.2
Roberts Lake[b]	83	64	4.4	2.6
East Everglades[a]	86	1	15.6	—
Stix and Nicks[a]	86	12	3.4	3.2
Rodgers River[b,c]	83, 87–89	57	10.4	5.7
East River[c]	87, 89	26	5.9	5.3
Frank Key[c]	87	34	7.0	5.4
Total		422	6.3	5.2

[a] Frederick and Collopy (1988).
[b] B. Patty, unpublished data.
[c] Bancroft et al. (1990).

Figure 25.2 Foraging flight distances of great egrets nesting in the Everglades.

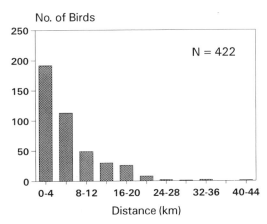

enclosed in these circles indicated that much of Water Conservation Area 3A south of Alligator Alley and Water Conservation Area 3B was used by breeding birds. Relatively little of Water Conservation Area 3A north of Alligator Alley and Water Conservation Areas 2A and 2B was used. Also, very few of the freshwater parts of Everglades National Park were used by breeding great egrets. Most of Shark River Slough was not included within circles or even particularly close to circles. Several small great egret colonies not shown on the map formed in the slough in 1986 and 1987, but they contained relatively few birds (Frederick and Collopy, 1988).

White Ibises

For white ibises, 200 foraging flights from four colonies showed a bimodal pattern of foraging flight distances (Figure 25.5). Over 65% of the flights were less than 10 km in length. The longest flights occurred at colonies that were failing (Frederick and Collopy, 1988, 1989), suggesting that these birds were searching for suitable feeding sites. If these flights are eliminated from the sample, then over 75% of the flights were to feeding areas within 10 km of the colony. In Figure 25.6, circles with radii of 10 km were plotted around colony locations where white ibises nested in 1986–89. Much of the Water Conservation Areas was used in various years, but large expanses of Everglades National Park were not being used. Furthermore, the colonies that formed at Lane River (1986) and Rodgers River Bay (1987) were both small. The Rodgers River Bay colony formed late in the year and did not produce many independent young.

Distribution of Habitats Surrounding Historic Mangrove Zone Colonies

In the 1930s and 1940s, great egrets and white ibises nested primarily in the larger mixed-species colony sites in the mangrove zone, where the Everglades meets tidewater (Ogden, 1994). Neither species was known to nest in large numbers in the northern Everglades. In Figure 25.7, the four major nesting areas in this region are displayed, surrounded by circles with 9-km radii. Two points are evident from comparing Figures 25.4, 25.6, and 25.7. First, the majority of great egret and white ibis nesting activity has shifted out of the freshwater portions of

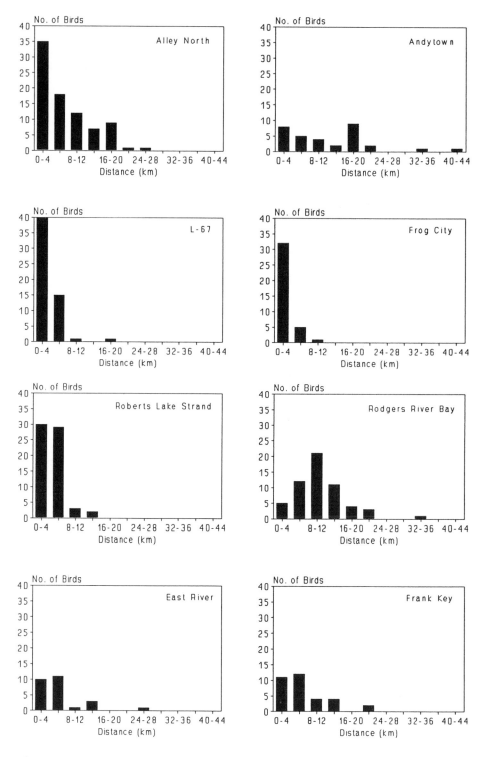

Figure 25.3 Foraging flight distances of great egrets nesting at eight colonies in the Everglades.

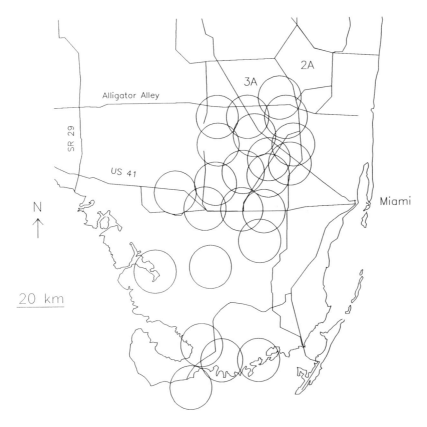

Figure 25.4   Foraging areas within 9 km of 23 great egret colony sites in the Everglades.

Figure 25.5   Foraging flight distances of white ibises nesting in the Everglades.

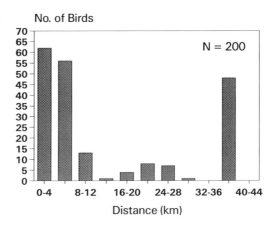

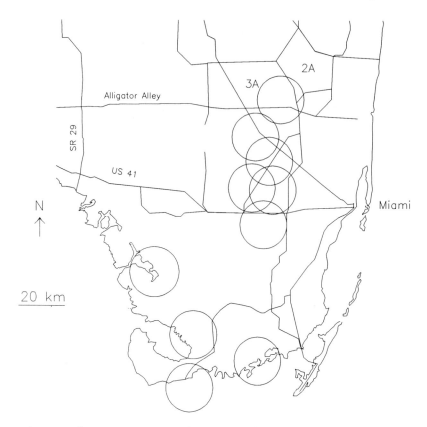

Figure 25.6 Foraging areas within 10 km of ten white ibis colony sites in the Everglades.

Everglades National Park and into the Water Conservation Areas or Florida Bay. Second, the marshes in the lower portion of Shark River Slough, which once hosted the largest nesting concentrations in the system, are no longer being used extensively by nesting birds.

The composition of foraging habitats surrounding the four historic colonies in the mangrove zone of Everglades National Park (Rodgers River Bay, Rookery Branch, Lane/East River [combined here because of their proximity to one another], and Frank Key) was compared. The area in nine habitat types within 9 km (254 km²) of each colony (Table 25.2) was totaled. The habitat composition differed substantially among colonies. Frank Key and Rookery Branch were the most different, with Rookery Branch surrounded by 85% freshwater habitats and Frank Key having only marine and brackish habitats. The Rodgers River Bay and Lane/ East River colonies were intermediate between Frank Key and Rookery Branch. Rodgers River Bay was surrounded by slightly more brackish habitats; Lane/East River was surrounded by slightly more freshwater sites. The brackish sites were different in composition between the two colonies, with Rodgers River Bay containing primarily mangrove-coastal marsh habitat and Lane/East River primarily mangrove.

The composition of habitat types surrounding these colonies seemed to be

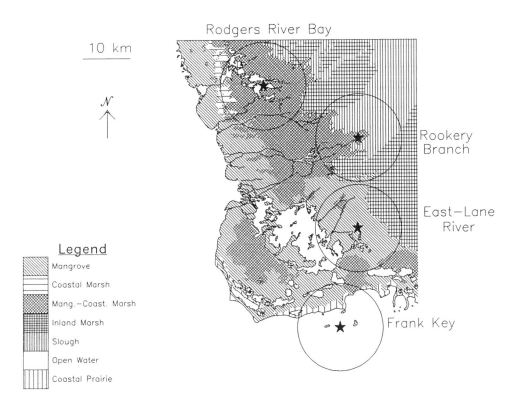

Legend

Mangrove

Coastal Marsh

Mang.–Coast. Marsh

Inland Marsh

Slough

Open Water

Coastal Prairie

Figure 25.7 Foraging habitat within 9 km of four historic nesting areas in the southwest Everglades.

Table 25.2 Percent Area in Various Habitat Types within 9 km (254.5 km^2) of the Rodgers River Bay, Rookery Branch, Lane/East River, and Frank Key Colonies

Habitat type	Rodgers River Bay	Rookery Branch	Lane/East River	Frank Key
Marine				
Tidal flat	—	—	—	47
Total	—	—	—	47
Brackish				
Mangrove	3	—	59	15
Coastal marsh	3	—	—	—
Coastal prairie	—	—	—	1
Mangrove-coastal marsh	71	15	10	6
Total	77	15	69	22
Fresh water				
Mangrove-inland marsh interface	10	23	7	—
Inland marsh	1	27	14	—
Slough	1	35	—	—
Total	12	85	21	—
Open water				
Open water	11	0	10	31
Total	11	0	10	31

correlated with current nesting activity levels. The Frank Key colony, which received fresh water only from rainfall, has had the most nest initiations in Everglades National Park in recent years. The Rodgers River Bay colony has had consistent nesting activity, although only a small proportion of the historic nesting population. This colony was surrounded primarily by brackish habitats with the freshwater habitats well north of the main Shark River Slough drainage. The Lane/East River colonies are still active, but have supported less than 200 nesting pairs from 1987 through 1991. The foraging sites around these colonies were also predominantly brackish. However, the composition was not as diverse as around the Rodgers River Bay colony, being primarily uniform mangrove stands. In addition, freshwater habitats around the Lane/East River colony were probably dependent on flows from Shark River Slough. The Rookery Branch colony has not supported a large nesting population since the 1960s. This colony was surrounded by 85% freshwater habitats, all of which were in Shark River Slough.

The colonies in Everglades National Park that had a smaller percentage of freshwater habitats maintained higher nesting populations. However, the current center of nesting populations shifted to the Water Conservation Areas which are totally surrounded by freshwater marshes. Feeding sites around the current colonies in the Water Conservation Areas are entirely inland marsh and slough habitats. However, these marshes were all impounded in the 1960s and their hydroperiods are substantially different from the historic system.

Water Conditions and Bird Distribution

Water depths and hydropatterns should influence where birds forage and nest. The relationships between water conditions and bird distributions in Water Conservation Areas 2 and 3 will be examined in this section. Water conditions and hydropatterns during March, April, May, and June (1986–89) illustrate the foraging conditions available to birds during the main part of the nesting season. Again, data will be presented for great egrets and white ibises. Similar relationships appeared to occur for snowy egrets (*Egretta thula*) and tricolored and little blue (*E. caerulea*) herons.

Water Coverage and Hydropatterns

In early March, the proportion of the Water Conservation Areas in each water category changed significantly among years (G = 1086, DF (degrees of freedom) = 6, p < 0.001). In March 1986, 1987, and 1988, over 75% of Water Conservation Areas 2 and 3 had continuous water coverage, with 10–20% transitional coverage and about 5% dry (Figure 25.8). In these three years, nearly all of Water Conservation Area 3B and Water Conservation Area 3A south of Alligator Alley were wet (Figure 25.9). In March 1989, Water Conservation Areas 2 and 3 were over half dry, 30% was transitional, and 20% wet.

In early April, the proportion of the Water Conservation Areas in each water category also differed significantly among years (G = 2211, DF = 6, p < 0.001). In 1986 and 1987, water levels rose during March, and continuous water coverage increased to over 90% of the area by early April (Figures 25.8 and 25.10). In 1988,

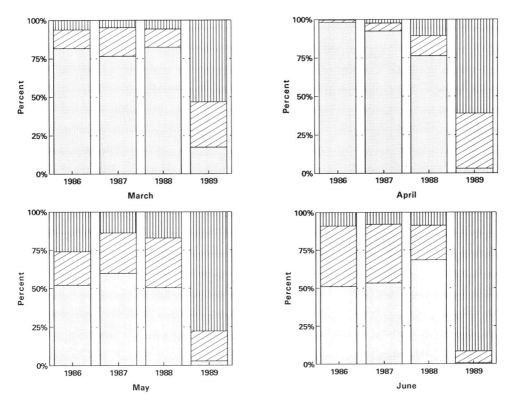

Figure 25.8   The percentage of Water Conservation Areas 2 and 3 covered by continuous water (dots), transitional water (diagonal lines), and dry (vertical lines) during March, April, May, and June 1986, 1987, 1988, and 1989.

the area continued to dry through March, and April began with 75% coverage of continuous water. In April 1989, conditions were much drier. April began with 60% dry, 36% transitional, and 3% wet. The only wet conditions occurred in the bottom of Water Conservation Area 2A and in a few places in the southern end of Water Conservation Areas 3A and 3B.

During all four years, the Water Conservation Areas dried from early April to early May. As in March and April, the proportion of the Water Conservation Areas in each water category changed significantly among years (G = 1064, DF = 6, $p <$ 0.001). In early May 1986, 1987, and 1988, 50–60% of the area was wet, and transitional water coverage in May varied from about 22% in 1986 to 32% in 1988 (Figure 25.8). The southern part of Water Conservation Area 3A and most of Water Conservation Area 3B were wet in these years (Figure 25.11). Conditions were very different in 1989; only 3% of the area was wet, 20% transitional, and over 75% was dry.

June water conditions also differed among years (G = 1969, DF = 6, $p <$ 0.001). In 1986 and 1987, about half of the area was wet, over 40% was transitional, and less than 10% was dry (Figures 25.8 and 25.12). In 1988, about 70% was wet and 20% was transitional, again leaving about 10% dry. In 1989, nearly the entire area was dry.

Figure 25.9 Water conditions in March 1986, 1987, 1988, and 1989 in Water Conservation Areas 2 and 3.

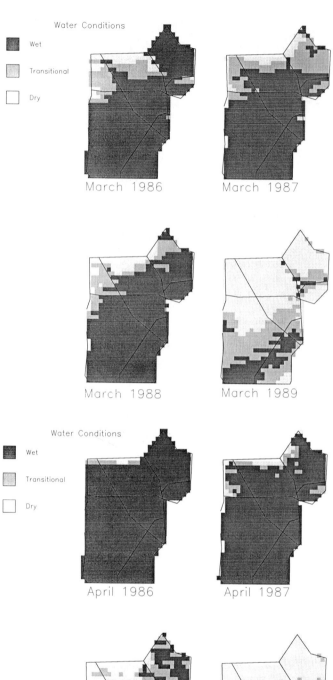

Figure 25.10 Water conditions in April 1986, 1987, 1988, and 1989 in Water Conservation Areas 2 and 3.

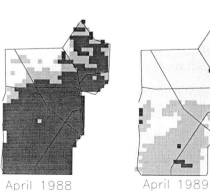
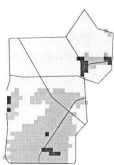

Figure 25.11 Water conditions in May 1986, 1987, 1988, and 1989 in Water Conservation Areas 2 and 3.

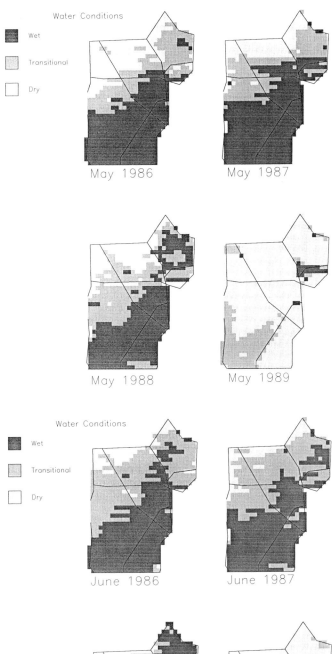

Figure 25.12 Water conditions in June 1986, 1987, 1988, and 1989 in Water Conservation Areas 2 and 3.

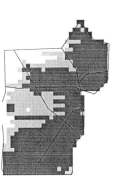

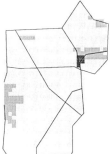

Table 25.3 Estimated Foraging Populations, Peak Breeding Populations, and Breeding Population as a Percent of Total Population for Great Egrets in Water Conservation Areas 2 and 3

	Foraging populations		Peak breeding populations	Percent of total	
	March	April		March	April
1986	15,840	9,820	2,162	14	22
1987	12,633	7,794	3,106	25	40
1988	14,606	10,086	4,548	31	45
1989	28,740	17,941	2,586	9	14

Great Egret Foraging Distribution

Numbers of foraging great egrets varied seasonally and among years within Water Conservation Areas 2 and 3 (Hoffman et al., 1989). Within the Water Conservation Areas, population estimates for great egrets were significantly higher in early March than in early April of each year, indicating that many great egrets left the area during March. In Water Conservation Areas 2 and 3, the estimated March population varied from 12,500–16,000 birds in 1986–88 and was 29,000 in 1989 (Table 25.3). In April, population estimates decreased to between 8000 and 10,000 birds for 1986–88 and 18,000 for 1989.

In no year were foraging great egrets dispersed randomly through the Water Conservation Areas (all $G > 3064$, all $p < 0.01$), indicating that the birds were selecting some subregions and avoiding others. The patterns of selection and avoidance of subregions suggested a close relationship with water levels. In 1986–88, the Alligator Alley North and Alligator Alley South subregions supported a mosaic of water conditions. During these years, Alligator Alley North was selected in all months except one (Table 25.4) (all $p < 0.05$ except April 1988) and the Alligator Alley South subregion was selected in March and April of all years (for all months except April 1988, $p < 0.05$). In contrast, in 1989 the Alligator Alley North and Alligator Alley South subregions were nearly dry. During this year, both of these subregions were avoided ($p < 0.05$) in all but one month.

Selection of foraging habitat by great egrets in the L-67 and L-28 subregions showed the opposite pattern. In the drought year 1989, these two subregions were selected in all months (Table 25.4) (all $p < 0.05$). From 1986 to 1988, they were flooded too deeply for efficient foraging (Figures 25.9 and 25.10); L-67 was avoided in all months and L-28 was avoided in five of nine months (all $p < 0.05$). Selection or avoidance of Water Conservation Area 2 and central Water Conservation Area 3A by great egrets varied, even in years that appeared to have grossly similar water conditions.

Comparison of March–April distributions showed significant shifts for each year (all $G > 26.9$, all $p < 0.005$, all DF = 6). In 1986 and 1987, water levels rose between the March and April surveys and great egrets selected the subregions Alligator Alley North and South. In 1988, however, the area continued to dry and the distribution of great egrets shifted south. In April 1988, only 10% of the great egrets used the Alligator Alley North subregion, compared to about 25% in the previous two years.

Table 25.4 Estimated Population of Great Egrets in Subregions of
Water Conservation Areas 2 and 3 during February, March, and April in 1986–89

	Alligator Alley N	Alligator Alley S	L-67	L-28	Central WCA 3A	WCA 3B	WCA 2	Total
February								
1986	4,460	2,227	307	1,707	1,567	1,440	1,387	13,095
1987	2,920	1,853	180	1,240	960	1,180	4,140	12,473
1988	3,333	2,153	60	680	613	667	2,520	10,026
1989	233	3,900	2,933	2,967	3,287	2,293	3,887	19,500
March								
1986	3,933	3,340	260	1,653	2,233	2,607	1,813	15,839
1987	3,067	2,227	360	887	913	1,767	3,413	12,634
1988	3,913	2,847	413	1,787	2,020	1,747	1,880	14,607
1989	100	747	8,520	5,033	2,927	8,993	2,060	28,380
April								
1986	2,940	2,367	147	893	1,107	1,287	1,080	9,821
1987	2,260	1,547	567	420	486	1,160	1,360	7,800
1988	1,667	1,427	1,153	1,420	1,340	1,713	1,367	10,087
1989	13	413	8,507	4,100	2,647	1,840	333	17,853

Note: Box indicates selection of subregion; shade indicates avoidance; all others indicate no
selection. February estimates were used to produce Table 25.8.

Instead, they fed in the Water Conservation Area 3B and 2 subregions, and
although selection was not apparent, numbers increased in Central Water Conservation Area 3A and L-28.

White Ibis Foraging Distribution

White ibises were substantially more seasonal than great egrets in their use of
the Water Conservation Areas. Foraging populations in the Water Conservation
Areas peaked at more than 70,000 birds between January and March and then
decreased to about one-third the peak number by early April (Hoffman et al., 1989).
Apparently, many northern migrants were present during winter and left the Water
Conservation Areas during March (Hoffman et al., 1989). The April foraging distributions, then, were most representative of the distribution of potential breeders in
the system and June (May in 1989) foraging distributions most representative of
birds feeding young.

April foraging distributions varied significantly among years (Table 25.5) ($G =$
9259, DF = 18, $p < 0.001$). From 1986 to 1988, white ibises avoided the L-67,
L-28, and Central Water Conservation Area 3A subregions, and in 1987 and 1988,
they avoided Water Conservation Area 3B. These subregions had continuous
water coverage, and the water was too deep for effective foraging. They

Table 25.5 Number of White Ibises Counted and the Estimated Population in Subregions of Water Conservation Areas 2 and 3 during April

	Number counted (population estimate)[a]			
	1986	1987	1988	1989
Alligator Alley North	112 (747)	195 (1,330)	157 (1,047)	0 (0)
Alligator Alley South	60 (400)	18 (120)	21 (140)	222 (1,480)
L-67	1 (7)	0 (0)	0 (0)	2,707 (18,047)
L-28	7 (47)	0 (0)	0 (0)	1,453 (9,687)
Central WCA 3A	28 (187)	0 (0)	34 (227)	728 (4,853)
WCA 3B	87 (580)	0 (0)	4 (27)	286 (1,907)
WCA 2	1,398 (9,320)	289 (1,927)	537 (3,580)	81 (540)
Total	1,693 (1,1287)	502 (3,347)	753 (5,020)	5,477 (36,513)

[a] Box indicates selection of subregion; shade indicates avoidance; all others indicate no selection.

selected Water Conservation Area 2 in all three years and selected the Alligator Alley North subregion in 1987 and 1988. In the drought year 1989, Alligator Alley North was virtually dry, and little water remained in Water Conservation Area 2. Instead, they selected L-67, L-28, Central Water Conservation Area 3A, and Alligator Alley South subregions, all of which presented a mosaic of water coverages and depths.

May–June foraging distributions also varied significantly between years ($G = 9799$, df = 18, $p < 0.001$). White ibises selected the Alligator Alley North subregion in 1987 and avoided this subregion in 1988 and 1989 (Table 25.6). They also avoided the Alligator Alley South subregion in 1989. Ibises avoided the L-67 subregion in 1986, 1987, and 1988, but selected it in 1989. They avoided the L-28 subregion in 1986 and 1987, but selected it in 1989.

Table 25.6 Number of White Ibises Counted and the Estimated Population in Subregions of Water Conservation Areas 2 and 3 in May–June

	Number counted (population estimate)[a]			
	June 1986	June 1987	June 1988	May 1989
Alligator Alley North	1,180 (7,867)	1,129 (7,527)	230 (1,533)	25 (167)
Alligator Alley South	777 (5,180)	389 (2,593)	527 (3,513)	122 (813)
L-67	3 (20)	0 (0)	33 (220)	1,588 (10,587)
L-28	738 (4,920)	389 (2,593)	642 (4,280)	1,992 (13,280)
Central WCA 3A	1,674 (11,160)	323 (2,153)	377 (2,513)	258 (1,720)
WCA 3B	76 (5,077)	1 (7)	210 (1,400)	55 (367)
WCA 2	1,099 (7,327)	258 (1,720)	429 (2,860)	57 (380)
Total	5,547 (26,980)	24 (16,593)	2,448 (16,320)	4,097 (27,313)

[a] Box indicates selection of subregion; shade indicates avoidance; all others indicate no selection.

Foraging Distribution and Colony Location

The results in the preceding section indicate that for most subregions of the Water Conservation Areas, foraging distributions of great egrets and white ibises closely followed water conditions. The relationship between foraging distributions and colony locations was investigated next. Many birds choose breeding localities and time the onset of breeding such that young are in the nest where and when food is most abundant (Southern, 1954, 1959; Lack, 1966, 1968; Harrison et al., 1983). The spatial aspects of this hypothesis were tested for wading birds, using data for great egrets and white ibises. For wading birds, the hypothesis can be revised to state that birds should select colony sites located in the areas that will provide the best foraging habitat when nestlings are being fed. Thus, when nestlings are being fed, colony locations should be good predictors of the locations of the best available foraging habitat. Because nonbreeding individuals are not tied to a colony site, it is assumed that their distribution reflects the distribution of favorable foraging sites. Thus, when nonbreeders are more abundant than breeders (as was the case for both species) (Tables 25.3 and 25.9), the general foraging distribution was considered to be dominated by birds that were not tied to a colony site. The test consisted of comparisons of colony locations to foraging distribution before and during the period in which the nesting birds were feeding small young.

Great Egrets

Historically and during this 4-year study, great egrets in the Everglades generally began nesting in February and March and had eggs hatching in late March and early April (Kushlan et al., 1984; Frederick and Collopy, 1988; Bancroft et al., 1990; Ogden, 1994). Therefore, the foraging distribution of great egrets in early April was examined to determine how well it was predicted by the locations of the colonies formed in the preceding February and March. Because the proportion of the total population that was breeding was always less than 50% in the relevant months (it varied from 9% in March 1989 to 45% in April 1988), it was assumed that the foraging distributions were dominated by nonbreeding birds and that their distribution represented the distribution of optimal foraging habitat.

Ten colonies were found in 1986, 21 in 1987, 22 in 1988, and 11 in 1989 (Figure 25.13). The peak breeding populations (calculated by summing maximum nest counts across colonies) varied between years (Table 25.3) from 2162 birds (1081 pairs) in 1986, to 3106 (1553 pairs) in 1987, 4548 (2274 pairs) in 1988, and 2586 (1298 pairs) in 1989.

In 1986 and 1987, when water levels rose between early March and early April, the majority of birds nested in Alligator Alley North (Figure 25.13). In 1988, when conditions continued to dry from early March to early April, more great egrets attempted to nest than in the other three years and the nesting distribution shifted southward. In 1989, the total number nesting decreased to a level similar to 1986 and 1987. Nesting location, however, was significantly different (Table 25.7) ($G = 2789$, $p < 0.001$). No great egrets nested in Alligator Alley North and the number nesting in the L-28 subregion decreased. The greatest numbers (>40%) nested in the L-67 subregion. Water Conservation Area 3B and the Alligator Alley South subregion each had about 25% of the breeding population. Breeding birds in all

Figure 25.13 Colony distribution relative to foraging distribution for great egrets in April 1986, 1987, 1988, and 1989.

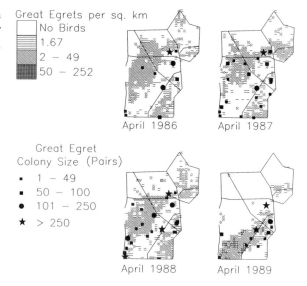

three of these subregions were within easy flying distance of the drying sloughs in the L-67 subregion.

Shifts in the size and location of colonies between years paralleled the shifts in foraging distribution (Table 25.7). In 1986 and 1987, breeding great egrets selected the Alligator Alley North subregion and avoided the L-67, L-28, and Water Conservation Area 3B subregions. In both years, over 50% of the breeders were in Alligator Alley North, whereas less than 15% used the L-67, L-28, and Water Conservation Area 3B subregions. Although more birds nested in Alligator Alley North in 1987 than in 1986, the proportion (52%) of the total nesting population decreased.

In 1988, the correlation between breeding and feeding distributions appeared poorer. Breeding great egrets avoided the Alligator Alley North subregion and selected the subregion along the L-28 canal (Table 25.7). Over 37% of the breeding population nested in the L-28 subregion. This subregion included one colony with

Table 25.7 Number (Percentage) of Great Egrets Nesting in Subregions of the Water Conservation Areas[a]

	1986	1987	1988	1989
Alligator Alley North	1290 (60)	1600 (52)	550 (12)	0 (0)
Alligator Alley South	376 (17)	512 (16)	860 (19)	700 (27)
L-67	220 (10)	470 (15)	866 (19)	1060 (41)
L-28	76 (4)	258 (8)	1690 (37)	170 (7)
WCA 3B	200 (9)	266 (9)	582 (13)	656 (25)
Total	2162	3106	4548	2586

[a] Box indicates selection of subregion; shaded indicates avoidance; all others indicate no selection.

500 great egret pairs and numerous smaller colonies. In 1989, breeding great egrets strongly avoided both Alligator Alley North and L-28 subregions (Table 25.7). Instead, 41% nested along the L-67 canal in Water Conservation Area 3A and 25% nested on the opposite side of the L-67 levee in Water Conservation Area 3B, exploiting the drying sloughs along the L-67 canal in Water Conservation Area 3A. Despite the number foraging in the L-28 subregion during April, few attempted to nest there.

Contingency table analyses for foraging birds and breeding birds are compared relative to subregions in Table 25.8. The third column of the table, which compares breeding sites to the April foraging distribution, is most relevant to the hypothesis that birds should select colony sites located in areas that will provide the best foraging habitat when nestlings are being fed.

Of the 20 comparisons in the third column, 9 showed the expected selection-selection or avoidance-avoidance pattern. Seven involved foraging and/or breeding distributions not differing significantly from random, and four showed the unexpected pattern of selection in one analysis and avoidance in the other. In the two subregions avoided by foraging birds but selected by breeding birds (the Alligator Alley South subregion and Water Conservation Area 3B in 1989), the birds had short foraging flights to the heavily used L-67 subregion. Two subregions (Alligator Alley South in 1987 and L-28 in 1989) were selected as feeding sites but were avoided as nesting areas. These subregions do have suitable colony sites and in fact were heavily used in other years.

The results for February and March foraging distributions (first two columns of Table 25.8) are marginally closer to what was expected for April. For February, 8 of the 20 comparisons did not show significant selection or avoidance in the foraging distributions, indicating a more uniform distribution than in March or April. Of the 20 comparisons, 8 show the selection-selection and avoidance-avoidance patterns expected, but only 3 show the contrary selection-avoidance patterns. For March, 12 of the 20 show the expected selection-selection and avoidance-avoidance patterns and 4 show the contrary selection-avoidance patterns. The comparison is not as clean as would be desirable because birds nesting in one subregion were easily capable of flying to another subregion to forage.

The test results, then, provide only weak support for the hypothesis. It is

Table 25.8 Comparison of Association within Each Subregion for Overall Population and Nesting Population of Great Egrets

| | Foraging great egrets | | | | | | | | |
| | February | | | March | | | April | | |
Nesting great egrets	Sel.	NS	Avoid.	Sel.	NS	Avoid.	Sel.	NS	Avoid.
Selection	3	2	1	4	1	1	3	1	2
No selection	0	2	1	2	0	1	2	1	0
Avoidance	2	4	5	3	0	8	2	3	6

Note: Determinations of selection, no selection, and avoidance are from Table 25.4 for foraging birds and Table 25.7 for breeding birds.

concluded that colony sites picked by great egrets in February often are not good predictors of the best April feeding habitat.

White Ibises

The foraging and breeding dynamics of white ibises differed from those of great egrets in significant ways. In a given year, white ibises typically used only a few nesting sites in the Everglades, and the colonies that did form were usually substantially larger than great egret colonies (Kushlan, 1976, 1986a). The locations of these colonies changed with foraging conditions. The small number of colonies forming each year was inadequate for contingency table analyses of selection and avoidance of subregions comparable to those for great egrets, but the distribution of foraging birds around each colony could be examined.

Breeding populations peaked at around 3000 individuals (1500 pairs) in 1986, 1988, and 1989 and around 8200 individuals (4100 pairs) in 1987 (Table 25.9). The percentage of the total population that was breeding varied significantly between years ($G = 12{,}602$, DF = 3, $p < 0.001$). In 1986, it dropped from 22% in early May to 8% in early June. In 1987, when breeding did not begin until mid to late May, breeders comprised 49% of the 16,600 population estimate for early June. The Systematic Reconnaissance Flight-based total population estimates, however, tend to be low when large numbers of birds are nesting. Because the colony north of Alligator Alley was located between the survey lines, birds present in the colony at the time of the survey were not counted. Adding 4100 birds (50% of the breeders) creates a more accurate estimate of the general population for early June. With this adjustment, the breeding birds make up 40% of the overall population. In 1988, breeding populations were 30% of the total population in early May and 20% in early June. Therefore, in 1986–88, June foraging distributions were compared to colony locations because the June surveys overlapped the peak in nesting (Figure 25.14). In 1989, breeding populations were 8% of the April and 11% of the May populations. By early June 1989, most white ibises had abandoned colonies, and virtually all had left the Water Conservation Areas. Thus, in 1989, May foraging distribution data were used because these surveys immediately followed the peak in breeding numbers (Figure 25.14).

In 1986, two large white ibis colonies formed in Water Conservation Area 3A, one each in the Alligator Alley North and South subregions. Both colonies were

Table 25.9 Estimate of Foraging Populations, Peak Breeding Populations, and Breeding Population as Percent of Total Population for White Ibises in Water Conservation Areas 2 and 3

	Foraging population			Peak breeding population	Time of major nesting	Percent of total	
	April	May	June			May	June
1986	11,287	13,386	36,980	2,906	Late April–mid-May	22	8
1987	3,347	7,433	16,594	8,200	Mid-May–early June	—	49
1988	5,020	11,153	16,320	3,300	Late April–mid-May	30	20
1989	36,553	27,427	3,760	3,070	Early April	11	—

Figure 25.14 Colony distribution relative to foraging distribution for white ibises in June 1986, 1987, and 1988 and May 1989.

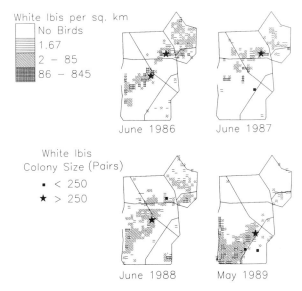

near large foraging aggregations. Almost 8000 white ibises were estimated in Alligator Alley North and over 5000 were feeding in Alligator Alley South (Figure 25.14). Few white ibises were in the L-67 subregion and no colony formed in this subregion.

The distribution of foraging birds was farther north in 1987 than in 1986, with concentrations occurring from an area east of the L-28 gap northeast through Water Conservation Area 2A (Figure 25.14). Over 7500 ibises were feeding in the Alligator Alley North subregion during the first week of June. One large colony formed in Alligator Alley North and a small colony formed in the L-67 subregion. The L-67 colony formed during mid-June. Little rain fell during early June and the area around the colony continued to dry into July, allowing white ibises to forage near the colony in mid-June.

In 1988, a large colony formed in the Alligator Alley South subregion, and a small colony formed in Alligator Alley North. In early June, almost 3500 white ibises were feeding in the Alligator Alley South subregion and 1500 were in Alligator Alley North. Over 2500 were in central Water Conservation Area 3A and within flying distance of the large colony.

During 1989, the number of white ibises feeding in Alligator Alley North was less than 200 and the number in Alligator Alley South was less than 1000. No white ibises attempted to nest in either of these subregions. One large colony formed in the L-67 subregion and two small satellite colonies formed south of the large colony. The L-67 subregion contained more than 10,000 feeding white ibises in early May (Figure 25.14). In the previous three years, white ibises had avoided this subregion.

The relationships between colony formation and the contingency table analyses of foraging distribution for white ibises are summarized in Table 25.10. A total of nine white ibis colony formations occurred in the four years, involving five sites in four subregions. Patterns of foraging distribution were compared to colony

Table 25.10 Comparison of White Ibis Colony Locations
to Patterns of Foraging Habitat Use in April and June (May 1989)

| | Foraging white ibises | | | | | |
| | April | | | May–June[a] | | |
	Selection	NS	Avoidance	Selection	NS	Avoidance
Colony(s) present	3(4)[b]	4	1	4(5)[b]	1	3
No colony	0	3	5	2	0	7

Note: Determinations of selection, no selection, and avoidance of subregions are from Tables 25.5 and 25.6.

[a] Foraging distribution data are for June 1986, 1987, and 1988 and May 1989, to coincide with peak nesting.

[b] Two colonies formed in the L-67 subregion in 1989.

locations prior to the nesting season (April) and at the peak of nesting (June for 1986, 1987, and 1988; May for 1989). Four colonies occurred in subregions that were selected as foraging sites in April, whereas five colonies occurred in subregions selected for foraging in May or June. However, in June three colony locations were in subregions avoided by foraging birds as compared to only one in April.

Four of the nine colony formations occurred in regions that changed status as foraging habitat between April and May–June. These might provide some insight into the prediction that colony sites are chosen to anticipate good foraging conditions when young are being fed. One colony formed in a subregion (Alligator Alley North in 1988) that was selected in April but avoided in June. Two colonies formed in subregions (Alligator Alley North in 1986, Alligator Alley South in 1987) that were neither selected nor avoided in April but were selected in June. One formed in a subregion (Water Conservation Area 3B in 1989) that was neither selected nor avoided in April but was avoided in May. Thus, two sites appeared to suffer deterioration in foraging attractiveness after colony formation and two appeared to experience improvements.

The white ibis analyses, then, also provide only weak support for the hypothesis that birds choose colony locations that will provide good foraging when their young are being fed.

Drought Effects on Nesting and Foraging Phenology

Although the results presented in the previous section provide evidence that the distribution of nesting great egrets and white ibises was not a good predictor of where birds would be foraging when their young needed to be fed, the differentiation between breeding and nonbreeding birds was somewhat equivocal. In the drought year 1989, the formation, growth, and decline of the L-67 colony was studied to document in detail the patterns of foraging site selection of the nesting birds and to compare their foraging distribution to the overall foraging distribution of wading birds in the area. In this period, 145 following flights were conducted, and Systematic Reconnaissance Flight Survey data were supplemented with additional surveys of Water Conservation Areas 3A and 3B south of Alligator Alley on

Figure 25.15 Water conditions in Water Conservation Areas 3A and 3B south of Alligator Alley in 1989. The two circles represent 10- and 20-km flight distances from the L-67 colony (star).

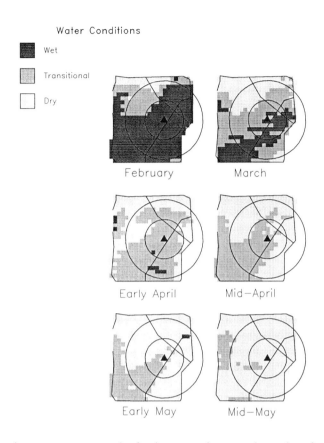

April 19 and May 18. These observations provide further insight into the role of hydrology in colony formation and foraging site selection during nesting. They also provide some insights into mechanisms of colony failure and the decline of nesting populations of wading birds in the Everglades.

The L-67 colony is located in a large tree island in eastern Water Conservation Area 3A near the L-67 canal (Figure 25.15). This colony site is located in a band of sloughs paralleling the L-67 canal that normally hold water impounded by the levee system surrounding Water Conservation Area 3A. These sloughs were flooded continuously from June 1985 until spring 1989. In 1986, 1987, and 1988, this colony supported 122, 430, and 90 nesting pairs of wading birds, respectively (Frederick and Collopy, 1988; G. T. Bancroft and S. D. Jewell, unpublished data). In 1989, drought conditions in south Florida led to a rapid drydown in the Water Conservation Areas. More than 3000 pairs of wading birds attempted to nest at the L-67 colony during the season, with a peak of 2000 pairs present in mid-April (Figure 25.16). The environmental conditions that led to the growth of this colony and its subsequent decline will be discussed in the following sections.

Water Conditions and Nesting Phenology

In 1989, the Water Conservation Areas experienced an exceptionally rapid drydown, which continued until the area had virtually no surface water (Figure

Figure 25.16 The number of active nests of wading birds at the L-67 colony during 1989.

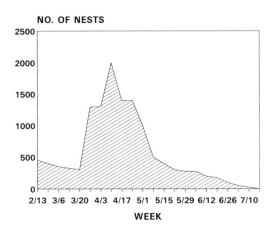

25.15, Table 25.11). The rapidity with which this drydown occurred is illustrated by the hydrograph from the 3–4 gauge in central Water Conservation Area 3A (Figure 25.17). Aerial surveys detected great egrets beginning to nest at the L-67 colony during early February 1989, when 450 active nests were present (Figure 25.16). In late March, other species began to nest in the colony and the total nesting population increased to about 1400 nests. In mid-April, the colony peaked at more than 2000 nests, including 1500 white ibis nests. From mid-April to early May, many nests were abandoned and the colony decreased substantially in size. On May 10 fewer than 500 nests remained active, and the colony gradually decreased in size through June and was empty by mid-July.

Following Flights and Distribution of Wading Birds

Locations of foraging birds followed from the L-67 colony were compared to the general distribution of foraging birds in the southern part of Water Conservation Areas 3A and 3B through the nesting season. Following flight information was collected during the second wave of nesting at the colony. The data were divided into four periods that correspond with the four Systematic Reconnaissance Flight surveys (two full surveys and two mid-month surveys covering Water Conservation Area 3 south of Alligator Alley). Initiation of following flights for various species

Table 25.11 Percentage of Area in Water Conservation Area 3 South of Alligator Alley in Wet, Transitional, and Dry Water Conditions during the 1989 Breeding Season

Period	Wet	Transitional	Dry
February	72	19	9
March	26	44	30
April 6–13	2	57	41
April 14–25	0	43	57
April 27–May 7	1	29	71
May 8–18	0	10	90

Figure 25.17 Water level at the 3–4 gauge in Water Conservation Area 3A during the 1989 nesting season. The 3–4 gauge is located approximately 9.8 km west of the L-67 colony. Zero represents ground level.

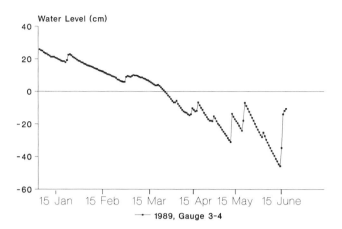

was staggered to coincide with the period when most marked nests on the colony transects contained young.

April 6–13, 1989. In early April, wading birds were concentrated in southern and southwestern Water Conservation Area 3A and western Water Conservation Area 3B (Figure 25.18). The largest aggregations were along the L-67 canal and along the receding water edge in southwestern Water Conservation Area 3A. Three great egrets followed to foraging sites flew only 2.3 ± 1.0 km from the colony (Table 25.12).

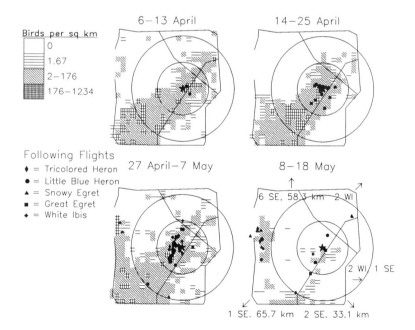

Figure 25.18. Wading bird distribution in Water Conservation Area 3 south of Alligator Alley (hatch patterns) and foraging locations of wading birds nesting at the L-67 colony (symbols). The inner and outer circles represent 10- and 20-km flight distances, respectively.

Table 25.12 Mean Distance Traveled to Feeding Sites
by Wading Birds Nesting at the L-67 Colony in 1989 (n = 145)
(All Means are in km ± SD [Number of Following Flights in Parentheses])

| Species | n | Period[a] | | | | Mean distance by species |
		1	2	3	4	
Great egret	57	2.3 ± 1.0 (3)	1.9 ± 1.8 (38)	5.8 ± 4.6 (13)	6.2 ± 4.6 (3)	3.0 ± 3.3
White ibis	26	—	—	5.1 ± 2.1 (14)	20.4 ± 8.3 (12)	12.1 ± 9.6
Snowy egret	35	—	—	17.2 ± 8.3 (8)	34.3 ± 15.6 (27)	30.4 ± 15.9
Little blue heron	3	—	—	4.1 ± 4.2 (2)	5.2 ± 0.0 (1)	4.5 ± 3.0
Tricolored heron	24	—	—	3.8 ± 1.2 (22)	24.7 ± 0.0 (2)	5.6 ± 6.0
Mean by period		2.3 ± 1.0 (3)	1.9 ± 1.8 (38)	6.4 ± 5.8 (59)	27.7 ± 15.7 (45)	11.7 ± 14.4

[a] Period 1 = April 6–13; period 2 = April 14–25; period 3 = April 27–May 7; period 4 = May 8–18.

April 14–25. During the second half of April, wading bird distribution became further concentrated along the L-67 levee. The largest numbers were well south of areas in which nesting great egrets were feeding (Figure 25.18); 38 great egrets were followed to feeding sites 1.9 ± 1.8 km from the colony (Table 25.12), indicating that good foraging conditions existed very close to the colony.

April 27–May 7. In early May, the concentrations of wading birds began to disperse. Although large numbers of wading birds still existed in Water Conservation Area 3 (42,986 estimated on the early May Systematic Reconnaissance Flight Survey), the majority were over 10 km from the colony. The largest numbers occurred in the southwest corner of Water Conservation Area 3A, the L-28 gap, and a cluster of birds just southwest of the colony. This latter concentration was probably birds from the L-67 colony that were feeding at the northeastern corner of remaining transitional water conditions (Figures 25.15 and 25.18). Average distance (for all species) flown to foraging sites was only 6.4 ± 5.8 km (n = 59). Again, this indicates that foraging conditions close to the colony were still relatively favorable, although most birds feeding in Water Conservation Area 3 were beyond the foraging range of breeders. Significant differences did occur among species (Kruskal–Wallis, χ^2 = 89.3, p < 0.0001). Snowy egrets flew farthest to feeding sites (17.2 ± 8.3 km) (Table 25.12) and tricolored herons fed closest to the colony (3.8 ± 1.2 km).

May 8–18. By mid-May, the largest foraging aggregations were well over 20 km from the colony (Figure 25.18). Birds were clustered at the southern end of Water Conservation Area 3A and the L-28 gap and a few birds were along the L-67 canal. Following flights paralleled this distribution. Average distance to feeding sites was 27.7 ± 15.7 km (n = 45) (Table 25.12). Most flights were either to the L-28 gap or along the L-67 canal. Indicative of the lack of adequate foraging opportunities near the colony, 14 birds were followed outside of the Water Conservation Areas. Six snowy egrets flew to the Everglades Agricultural Area, two snowy egrets flew to near Tamiami Airport, and one flew to the mangrove zone in Everglades National Park. Additionally, four white ibises and one snowy egret flew to areas east of the colony, but were lost in restricted airspace east of the Water Conservation Areas.

Figure 25.19 Nesting chronology of wading birds followed at the L-67 colony in 1989. (Cross hatching = new clutches, single hatching = nests with eggs, stippling = nests with young, line = proportion of active nests failing.)

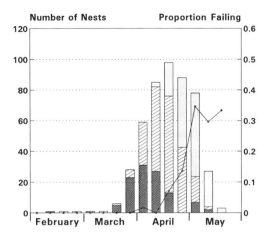

Distance to foraging sites was significantly different among the four periods (Kruskal–Wallis, χ^2 = 91.56, p < 0.0001). Distance to feeding sites was significantly correlated with date (r = 0.78, p = 0.001), indicating much higher costs of travel as the breeding season progressed. The increased distance flown to feeding sites corresponded with nest abandonment at the colony. Peak activity was found during the third week of April and decreased rapidly in May (Figure 25.19). Mean date of clutch completion was March 31 for tricolored herons (n = 35), April 2 for snowy egrets (n = 23), April 9 for little blue herons (n = 4), and April 19 for white ibises (n = 27). These species require 21 additional days for incubation and 40–60 days for brood rearing after clutch completion to successfully raise young to independence (G. T. Bancroft, unpublished data). However, by mid-May, when most young would be less than 30 days old, adults were flying over 25 km to feeding sites. By late April, the proportion of active nests that were failing began to increase and during each of the first three weeks of May, 30% of the active nests failed (Figure 25.19). The large proportion failing caused the size of the colony to decrease rapidly during late April and early May.

DISCUSSION

Over the past half century, total nesting populations of wading birds have decreased by 90% and the majority of the nesting population has shifted from the mangrove-freshwater interface of Everglades National Park to the Water Conservation Areas (Ogden, 1994). The colonies formed in recent years in the Water Conservation Areas do not begin to replace numerically the breeding birds that have disappeared from the southern Everglades (Ogden, 1994), and studies of productivity cast doubt on their ability to be self-sustaining (Frederick and Collopy, 1988; Bancroft et al., 1990). Historically, the Everglades marshes of the Water Conservation Areas did not support large numbers or a high proportion of the breeding population of wading birds (Ogden, 1994). Currently, more white ibises, great egrets, and tricolored herons nest in the Water Conservation Areas than on the mainland part of Everglades National Park (Frederick and Collopy, 1988; Bancroft, 1989; Ogden, 1994). These changes have presumably resulted from the alterations that have occurred in the system. The results of this study extend our understanding of how wading birds have responded to modifications of Everglades hydrology and the difficulties they face in attempting to reproduce.

Wading bird foraging distributions in the Everglades are closely tied to the distribution of water in the system (Kushlan, 1986a; Hoffman et al., 1994). The location of the "drying edge" of transitional wetland, rather than time of year, determines where wading birds will be foraging. Colony formation is more likely to occur in transitional wetland areas that support large numbers of feeding birds and, therefore, is also closely tied to water conditions. Because wading birds span a variety of leg lengths, their tolerance of water depth varies by species (Custer and Osborn, 1978; Erwin, 1983, 1984; Powell, 1987; Hoffman et al., 1994). Thus, not all species react to changes in water level in exactly the same way. However, the results presented in this chapter, which provide information on great egrets, white ibises, and to a lesser extent snowy egrets, tricolored herons, and little blue herons, cover a sufficiently diverse range of species to afford generalizations about their responses to changing water conditions.

Water levels affect foraging habitat for both breeding and nonbreeding birds, but nonbreeding birds are free to shift their feeding locations to increase foraging success. These moves may involve relatively short distances as local water conditions change (moves of 50–100 km within south Florida [G. T. Bancroft, unpublished data]) or abandonment of the area altogether. Breeding birds, however, are committed to foraging within a limited distance from the colony site for the duration of the nesting attempt, up to 4 months (McVaugh, 1972; Werschkul, 1979, unpublished data). Abundant, readily available food sources must be present throughout this period to permit successful breeding.

In the Everglades, suitable foraging conditions throughout the nesting season are no longer being met; the quantity and timing of water flows in the system have become erratic enough to seriously affect the ability of wading birds to raise young (Kushlan, 1987). Recent studies of colony productivity (Frederick and Collopy, 1988, 1989a; Bancroft et al., 1990) (Figures 25.16 and 25.19) indicate that current colony sites in the Everglades often do not provide suitable foraging habitat long enough to complete a nesting cycle, either because they reflood, dispersing the remaining food (e.g., Lane River in 1987, Alligator Alley North for white ibises in 1986 and 1987), or dry out completely (e.g., Alligator Alley North in 1988, L-67 colony in 1989). Temporary reversals of the drying trend also disperse food, at least for wood storks, great egrets, and *Egretta* spp., and often lead to colony abandonment (Frederick and Collopy, 1988, 1989; Frederick and Spalding, 1994). Data from the study described in this chapter support four explanations for the observed declines (Walters et al., 1992): decreased estuarine productivity, altered cues for colony formation, increased susceptibility to dry season rainfall, and decreased food availability during nesting.

Decreased Estuarine Productivity

In the 1930s and 1940s, large colonies containing the majority of the Everglades breeding populations occurred in a strip from the headwaters of Lostmans River south to Florida Bay (Ogden, 1994). The appearance of 9-km circles around the four westernmost historical nesting areas suggests that the entire mangrove-inland marsh interface was used for feeding by breeding birds during the 1930s (Figure 25.7). Although this area is still used as a feeding site by nonbreeding birds, the majority of those nesting in the Everglades have shifted from Everglades National Park to the Water Conservation Areas (Ogden, 1994). There is evidence that decreased freshwater flows to both the inland marshes and estuaries have decreased the productivity of these areas for fish populations (Pool et al., 1975; Schomer and Drew, 1982; Loftus and Eklund, 1994).

Because the productivity of the predevelopment Everglades was never measured, no baseline exists to test the hypothesis that productivity of wading bird feeding areas in the estuarine zone has decreased. Primary and secondary productivity in the freshwater marsh is directly related to hydroperiods, and the productivity in the estuaries and mangrove-inland marsh interface may be related to freshwater flows (Kolipinski and Higer, 1969; Wood and Maynard, 1974; Browder et al., 1980 in Schomer and Drew, 1982; Walters et al., 1992; Loftus and Eklund, 1994). It follows that if hydroperiods and freshwater flow have decreased, then productivity of both freshwater and estuarine habitats is likely to have decreased.

Models of predevelopment hydrology of the Everglades (Walters et al., 1992; Fennema et al., 1994) indicate that peak wet season water levels at the latitude of Tamiami Trail were significantly higher than the current maximum allowed stages. The hydrological consequences of this are interpreted as follows. The greater hydrological head would have increased wet season flow rates in Shark River Slough and increased the wet season depths at the edges of the slough. More water would have spilled southeast from Shark River Slough, near Mahogany Hammock, toward West and Cuthbert lakes. Maximum flow rates and stages would probably also have been higher in the Forty-Mile Bend area, where water moved southwest into East Slough and the other sloughs flowing into the stairstep region. If the shorter hydroperiod marshes fringing Shark River Slough were flooded for longer periods of time, they should have provided much more food early in the dry season along the margins of the slough. Birds preparing to nest at Lane River or East River might have had productive early season foraging to stimulate breeding in the area between Mahogany Hammock and West Lake. For birds setting up at Rookery Branch, both sides of the slough are within easy commuting distance, and the area between Pa-hay-okee Overlook and Mahogany Hammock could be expected to have been more productive.

The hydrology of the freshwater marshes in the stairstep region may have been very different as well, based on simulations from the Adaptive Environmental Assessment Model developed by Walters et al. (1992). Peak water levels should have been higher, and the marshes would have extended farther into the upland areas between the sloughs. Flows probably would have decreased abruptly in the sloughs in autumn, when water levels in the Forty-Mile Bend area dropped below the levels of the divides separating these sloughs from Shark River Slough. Because these drops in flows may not have occurred simultaneously in the different sloughs, good foraging habitat may have appeared sequentially in different sloughs. Under these conditions, the Rodgers River Bay colony should have formed earlier, induced by abundant food in freshwater marshes, probably in areas that now rarely flood.

The analysis of available habitat around the four colonies in the mangrove-freshwater interface supports the contention that the presumed decline in productivity (Schomer and Drew, 1982) has resulted in a decline in breeding numbers in this area. The historic mangrove colonies can be listed in the following order of decreasing percentage of freshwater habitats within 9 km of the colony: Rookery Branch, Lane/East River, Rodgers River Bay, and Frank Key. A decline in the volume of fresh water to the estuaries (and also through the freshwater marshes) would presumably have the greatest effect on those colonies that have the greatest percentage of freshwater habitat within their foraging radius. The Rookery Branch colony was located in the center of the main outflow section of Shark River Slough and has not supported a large colony since the 1960s (Kushlan et al., 1984). The foraging habitats within 9 km of this colony are dominated by freshwater sites (85%) (Table 25.2); thus, degradation of the Shark River Slough estuary would leave birds with long flights to suitable coastal feeding areas. The Lane/East River area receives freshwater flow from Shark River Slough through the North River and Roberts River drainages. This colony has not been successful since 1986 and has not supported large numbers of nesting birds since the mid-1970s (Ogden, 1994). As with the Rodgers River Bay colony, the foraging habitat within 9 km of the Lane/East River colony is primarily brackish. However, the Rodgers River Bay colony receives freshwater flows through the East, Lostmans, and Dixons slough drainages. In addition, the brackish habitats within 9 km of the Rodgers River Bay colony are not influenced by freshwater flows. The diversity of coastal habitats near the Rodgers River Bay colony (Bancroft et al., 1990) probably provides better foraging habitat than the relatively continuous stand of short red mangroves that dominate the brackish habitats surrounding the Lane/East River colony.

The Rodgers River Bay and Frank Key colonies have been the most productive colonies in Everglades National Park in the last 10 years. Both have extensive foraging areas that hydrologically are largely independent of overland freshwater flows. In 1987, birds from Frank Key fed primarily on Cape Sable and the tidal flats in Florida Bay, where water levels in the marshes and mangroves appeared to be controlled by rainfall and tides (Bancroft et al., 1990). Tricolored herons were able to nest successfully (albeit to varying degrees) at the Rodgers River Bay colony during three years (1987–89) with very different water conditions (Bancroft et al., 1990). Thus, these two colonies may be buffered from the effects of decreased flows on the inland marshes and estuaries, especially in the Shark River Slough drainage.

Altered Cues for Colony Formation

Wading birds must rely on a set of environmental cues to determine where and when to initiate nesting. On the basis of natural selection, these cues would be expected to indicate that there would be sufficient food resources for adults and their young for 4 months. Data suggest that whatever cues were used to initiate nesting are no longer reliable indicators of foraging conditions for 120 days. In the Water Conservation Areas, the cue that colonial egrets, herons, and ibises are using to initiate nesting may simply be a large aggregation of birds feeding on a large and concentrated prey base near a suitable nesting site. Season may well play a role; herons are usually unwilling to initiate nesting in late summer and fall, although white ibises will (Kushlan, 1976, 1986a), and the presence of preexisting colonies (e.g., anhingas [*Anhinga anhinga*]) may reduce the colony formation threshold.

The comparisons of colony site selection to foraging habitat selection (Tables 25.8 and 25.10) show that both great egrets and white ibises usually began nesting near concentrations of foraging birds and that by the time they were feeding young, the areas near the colony often had deteriorated as feeding habitat. In 7 out of 20 possible cases, a subregion was selected for nesting by great egrets (Table 25.8). In early March, four of the seven were also selected by foraging birds and only one was avoided, but by April only three were selected and two were avoided. For white ibises, only one of the nine colony formations occurred in a subregion that was avoided in April, prior to egg laying, but by the time of hatching, three of the colonies were in areas that were being avoided.

The formation and fate of the L-67 colony in the drought year 1989 gives further insight into colony site selection. When the colony formed and grew, the surrounding area supported one of the heaviest concentrations of wading birds that the authors have ever seen in the Water Conservation Areas. As the area continued to dry, the concentration of nonbreeding feeding birds shifted south and west toward and into the L-28 subregion, while breeding birds fed much closer to the colony. Because the L-28 subregion had hosted a sizable colony in 1988, a suitable site was available. By the time eggs were hatching at L-67, the L-28 subregion appeared to be a much more convenient nesting area because of its proximity to good foraging habitat. However, no major colony was active in the L-28 subregion during this time. Had birds been better able to predict the appearance of good foraging conditions, they may have nested in the L-28 subregion instead of L-67.

This hypothesis has major implications for the future of wading bird populations breeding in the Everglades. It contends that wading birds decide to nest based on current food availability and are not able to predict future availability accurately. It also implies that restoration goals should be to provide water management practices which ensure that food becomes abundant early enough in the season to induce nesting, in areas where food will remain abundant long enough to complete the nesting cycle. This hypothesis also implies that birds may have more difficulty choosing the optimal time to nest in the freshwater Everglades than in the estuarine zone, where the appearance of abundant food may be a better predictor of consistent availability.

If the hypothesis that the proximate cues for colony formation are abundant prey and a concentration of wading birds is correct, it follows that in the traditional colony sites in the southern Everglades, the appearance of abundant food in winter formerly was an adequate predictor of availability of food for the entire nesting period. It is clear that this is no longer the case. Delays in nesting initiation (Frederick and Collopy, 1988, 1989; Bancroft et al., 1990) would mean that concentrations of food (and birds) adequate to trigger nesting apparently do not appear until much later in the dry season than they did in the past.

The hypothesis may be testable through further analysis of the Systematic Reconnaissance Flight data, looking for the expected foraging concentrations preceding colony formation. Extending the analysis to the data from Everglades

National Park might be particularly useful. However, the 1-month interval between Systematic Reconnaissance Flight surveys may be too great to consistently catch these concentrations.

Wading birds differ in their fidelity to colony sites. Wood storks are extreme traditionalists, using only a few colony sites and returning to them each year (Kushlan, 1986a). Only in extreme years, such as the drought of 1989, do storks shift colony locations in response to water conditions. In that year, some storks nested in the Water Conservation Areas in addition to a traditional site in Everglades National Park. Findings of the authors for white ibises are similar to those of Kushlan (1976, 1986a) for colony site selection in the 1970s. Both studies show that the white ibis population is highly mobile and the location of colonies between years is extremely variable and related to water and feeding conditions.

Great egrets seem to exhibit a pattern intermediate between those described for white ibises and wood storks. Great egrets use many traditional colony sites each year, but the number nesting at each site varies with water conditions. In addition, the number of smaller satellite colonies seems to vary with water conditions. Although intercolony movement by great egrets was not demonstrated directly, the interannual variation in colony occupancy strongly implies such movements. Telemetry studies demonstrated that individual tricolored herons will attempt to nest in different colonies in different years, or even sequentially within years (S. D. Jewell and G. T. Bancroft, in preparation).

Colonies, however, tend to form repeatedly in the same sites (Kushlan et al., 1984; Frederick and Collopy, 1988; Bancroft et al., 1990); therefore, populations may exhibit a collective tradition of fidelity to a group of sites. In the longer term, a collective tradition of fidelity to previously successful sites might eventually concentrate nesting activity in sites where the proximate cues used are good predictors of longer term food availability. Conversely, a history of nesting failure in an area may lead to abandonment of the area and major shifts in breeding range (Ogden et al., 1987). It is likely that great egrets, snowy egrets, and tricolored herons all show this intermediate pattern of colony site fidelity. Analyses of nesting populations and colony sites throughout peninsular Florida (Ogden et al., 1980; Bancroft et al., 1988) support this concept of a fluid population with traditionally used sites.

Increased Susceptibility to Dry Season Rainfall

Currently, dry season rainfall events in the Everglades cause reversals in the drying trend of water levels. These reversals commonly are devastating to wading bird colonies, often causing widespread abandonment (Kushlan, 1987; Frederick and Collopy, 1989; Frederick and Spalding, 1994). Little evidence is seen to indicate that the pattern of dry season rainfall has changed over the last several decades in south Florida (Duever et al., 1994) and it is doubtful that the historic colonies of the Everglades could have suffered the frequency of losses to reversals that is now seen. Therefore, a hypothesis is offered that structural modifications to the Everglades and associated water management practices have aggravated the effects of rain events on wading bird habitat (Kushlan, 1986b, 1987).

Several likely mechanisms exist. First, the use of the Everglades, especially the Water Conservation Areas, as reservoirs for flood waters from surrounding developed areas undoubtedly increases the size and duration of reversals late in the dry season (Kushlan, 1987). This effect may be most severe in Water Conservation Areas 1 and 2A, which receive large volumes of flood water from the Everglades Agricultural Areas.

Second, the diking of much of the Everglades may lengthen the effects of a reversal on foraging habitat. In the historic, undiked system, a dry season reversal would have caused reflooding of recently dried peripheral areas, followed almost immediately by recession, quickly regenerating transitional water conditions. Currently, if a reversal raises water levels to the point that all or part of a Water Conservation Area is flooded from levee to levee, transitional conditions cannot be regenerated in that area until water levels drop back below the level of the base of the levee.

Third, when water is released from Water Conservation Area 3A into Everglades National Park following a reversal, the current release schedules may cause more of a pulse in water levels than would have happened under natural conditions (MacVicar, 1984; Kushlan, 1987; Walters et al., 1992). Fourth, water management clearly has lowered the seasonally adjusted average water levels in much of the Everglades, and the effects of reversals may be more severe when water levels are lower. A given volume of rain water dumped into a small remnant pool will dilute the food resources there more than the same volume of rain water dumped into a larger pool.

Decreased Food Availability during the Nesting Season

A critical unknown parameter in the mechanisms that make food abundant for fish-eating birds is the mobility of small fish under conditions of declining water levels. One possibility is that a significant fraction of the fish might perform extensive movements toward the deeper sloughs when water levels are dropping. Fish do get trapped in peripheral marshes, and some do find their way into refugia (Kolipinski and Higer, 1969; Loftus and Eklund, 1994), but whether a significant number escape is not known.

If significant numbers of these fish manage to escape the drying peripheral marshes into the central sloughs, the effect would be to increase the fish biomass in the central sloughs. If the peripheral marshes are extensive enough, this might increase fish biomass well above the level possible by local productivity. Under this scenario, loss of the peripheral marshes would affect the quality of wading bird habitat in the central sloughs, by both affecting local productivity and reducing this enhancement through migration. Compartmentalization of the marsh could lead to the same result: if areas of peripheral marsh are isolated by levees from the central sloughs, this migratory enhancement of fish populations would be prevented. Without this migratory enhancement, the central sloughs would have to dry down further before the fish were concentrated enough to provide a stimulus to the birds to breed.

If a significant number of fish do not leave the peripheral marshes when water levels are falling, then a significant effect of water management on these fish populations might result from the increased severity of reversals described previously. Frequent major reversals would be expected to draw fish out of refugia and each time many of them would be stranded. The overall effect of the reversals, then, would be to deplete the refugia, slowing down the recovery of populations once the next wet season arrives.

Third, hydrologic changes may have affected the size that fish reach in the Everglades marshes; sailfin molly (*Poecillia latipinna*), marsh killifish (*Fundulus confluentus*), sheepshead minnow (*Cyprinodon variegatus*), mosquitofish (*Gambusia affinis*), and flagfish (*Jordanella floridae*) represent the most important prey for egrets and small herons (Bancroft et al., 1990). These are small fish that rapidly grow to adult size and reach sexual maturity in only a few months (Haake and Dean, 1983; Travis and Trexler, 1987). Presumably, there is a trade-off between growth in these fish and the onset of reproduction. The longer hydroperiod and large flow volumes during the wet season of the historic Everglades (Walters et al., 1992; Fennema et al., 1994) may have resulted in individual fish growing to larger size before they began reproducing. Longer hydroperiods increase the productivity of the periphyton mat and, therefore, would influence the food available to fish both in the local area and downstream (Kolipinski and Higer, 1969; Wood and Maynard, 1974; Browder et al., 1980 in Schomer and Drew, 1982). If fish did grow larger, then the historic Everglades may have had substantially higher biomass per unit area.

Great egrets ate fish that averaged 1 g, and snowy egrets, tricolored herons, and little blue herons ate fish that averaged less than 1 g (Bancroft et al., 1990). At these sizes, the small herons would need to catch more than an estimated 300–400 fish per day to sustain themselves and care for young (unpublished data). A different hydrologic scenario that resulted in a doubling of the average size of fish would substantially increase feeding efficiency and decrease the foraging time required for maintenance and reproduction.

CONCLUSIONS

The structural changes to the Everglades and the water management practices instituted over the past several decades may have had at least four major effects on breeding populations of wading birds (Walters et al., 1992). First, they may have interfered with the processes of colony formation and nest initiation. Second, they may have induced an increase in nesting in the Water Conservation Areas, albeit not with great success. Third, they may have magnified the effects of dry season rainfall on the hydrology of the Everglades, such that a given storm event would now have a larger deleterious effect on nesting birds than it would have had in the past. Fourth, they may have affected the productivity of the marshes, such that less food is now available than in the past. Results of this study suggest that a goal in restoration should be to ensure that food becomes abundant early enough in the season to induce nesting in areas where food will remain abundant long enough to complete the nesting cycle.

RECOMMENDATIONS

1. *Maximize the extent of wetlands in the Everglades system.* The causes of the declines in bird populations are still not absolutely clear, but the loss of wetland area surely contributed. The westward expansion of urban communities into the historic Everglades in Dade, Broward, and Palm Beach counties needs to be halted and, where possible, the degraded marshes east of highway US 27 restored. Any place where a degraded wetland can be reflooded or otherwise restored (e.g., exotic plant removal) should be considered valuable. The East Everglades is the obvious example, but the Storm Water Treatment Areas in the Everglades Agricultural Area also are important as additional wetland acres, as well as for their water quality functions.

2. *Maximize natural connectivity.* Within the Everglades, compartmentalization may be a major problem, particularly if fish migrate, as suggested. Judicious removal of levees within and around the Water Conservation Areas might help greatly. In the restoration of the East Everglades, the L-67 levee south of Tamiami Trail should be removed completely. In the design of structures to move water from Water Conservation Area 3A to 3B, levee removal should be maximized, and the number and size of bridges under highways crossing the system should be increased. The newly rebuilt Alligator Alley (Interstate 75) across Water Conservation Area 3A provides for much more natural flow than most roads in the Everglades. Restoration of more natural flow patterns in the Forty-Mile Bend area of Tamiami Trail is also important, as described below.

3. *Replicate the important features of natural hydrology.* The important features include responsiveness to annual variation in rainfall, buffering of reversals, appropriate seasonality of peak flows and drydowns, and adequate volumes of water. As noted previously, the reversals caused by dry season rainfall appear to be more severe than they were prior to extensive water management. Management to mitigate the effects of reversals should be a priority in the Everglades.

The remaining recommendations relate to the hydrology of specific areas of importance to wading birds within the Everglades.

4. *Water redistribution.* The proposed redistribution of water within the Water Conservation Areas associated with the Everglades Surface Water Improvement and Management (SWIM) Plan (South Florida Water Management District, 1992) calls for restoring the hydroperiods of marshes at the northern end of each Water Conservation Area. This restoration should improve the productivity of these areas, thus providing more fish to wading birds earlier in the nesting season. Conceivably, this management strategy could lead to earlier annual nesting. However, it is critical that this restoration effort not decrease the volume of water available for restoration efforts in Everglades National Park.

5. *Adaptive system management and assessment.* The restoration of Everglades National Park will require restoration of the hydrology and productivity of the freshwater marshes of lower Shark River Slough, the marsh-mangrove interface, and the associated estuaries. The interplay between freshwater flow and the ecology of these areas needs to be re-established and understood through adaptive system management and assessment. The reduced volume of water that flows through the freshwater marshes and into the estuaries may have reduced productivity (Walters et al., 1992). Hydrological conditions that reduce the number of fish or the growth rates of fish would have a major impact on wading bird breeding. The area from Whitewater Bay through Lostmans River deserves particular attention. This area once supported the largest concentration of nesting wading birds in North America (Ogden, 1978, 1994). Numerous birds still forage there, but few nest.

6. *Wetland productivity.* The marshes and mangrove swamps from Nine Mile Pond and Paurotis Pond north and west to the base of Shark River Slough were the primary foraging areas for birds from the Lane/East River colony (Ogden, 1977; Bancroft et al., 1990). The volume of freshwater flow through Shark River Slough will affect these foraging areas. Increased water in Shark River Slough may create a larger "freshwater head" that would increase the flow of water across the marshes and into Watson, North, and Roberts rivers. The increased flow and added depth would probably improve the productivity of these areas. Winter drying would start with water in the marsh at deeper levels, and this might stimulate birds to begin nesting earlier than they currently do by increasing primary and secondary productivity in these wetlands.

7. *Rodgers River Bay Colony.* The area south of the Loop Road in Big Cypress Swamp is an important feeding area for birds nesting at the remaining large colony at Rodgers River Bay. Management efforts should be directed toward ensuring that foraging areas of this colony are maintained and possibly improved. The L-28 levee and US 41 have restricted the flow of water out of southwestern Water Conservation Area 3A and through the Loop Road area. This decreased volume probably affects the hydrology of Gum, Dixons, Lostmans, and East sloughs, all of which are important feeding areas for Rodgers River Bay birds. These sloughs need to receive increased flows during periods of high water in summer. Because the increased volume of water would flood across additional marshes, recession of water in late winter could begin earlier and more rapidly and stimulate birds to begin nesting earlier. Historically, this area may have been a critical feeding habitat for birds just prior to nesting. Increasing its productivity may help to return nesting birds to Everglades National Park.

8. *Water flow management.* The management of the S-12A structure in the L-29 levee and the S-343a, S-343b, and S-344 structures along the L-28 levee plays a major role in determining how much water flows out of Water Conservation Area 3A and into the sloughs south of Tamiami Trail. Because few data are available on the volume of flow out of the latter three structures or how they are regulated, evaluation of the effects of their management on

downstream hydrology is difficult. Revised regulation of these structures may restore this area to more natural conditions, but some structural changes, including removal of at least part of the L-28 levee, will probably be necessary.

ACKNOWLEDGMENTS

This work was supported by grants from the South Florida Water Management District, and the authors are grateful to Peter David, James F. Milleson, Vyke Osmondson, and the late J. Walter Dineen of the district for their help in setting up the project and their generous support during its tenure. We also thank the staff of Everglades National Park for technical support and advice during the duration of the study. Reed Bowman, Allison Brody, Mary Carrington, Paul Cavanagh, and Cynthia Thompson helped with field work. Discussions with Martin Fleming, John Ogden, and William Robertson improved various aspects of this study. Peter Frederick kindly provided access to unpublished data on following flights of great egrets and white ibises. Theo Glenn helped with earlier versions of the figures. We thank the many pilots who flew us safely whenever and wherever we needed to fly. Keith Bildstein, Peter Frederick, and John Ogden improved an earlier version of the manuscript. Finally, we thank Nancy Paul, who typed the various versions of the manuscript and tables. We thank all of the above people and organizations for their help.

LITERATURE CITED

Bancroft, G. T. 1989. Status and conservation of wading birds in the Everglades. *Am. Birds,* 43:1258–1265.

Bancroft, G. T., J. C. Ogden, and B. W. Patty. 1988. Wading bird colony formation and turnover relative to rainfall in the Corkscrew Swamp of Florida during 1982 through 1985. *Wilson Bull.,* 100:50–59.

Bancroft, G. T., S. D. Jewell, and A. M. Strong. 1990. Foraging and Nesting Ecology of Herons in the Lower Everglades Relative to Water Conditions, Final Report, South Florida Water Management District, West Palm Beach, 156 pp.

Custer, T. W. and R. G. Osborn. 1978. Feeding-site description of three heron species near Beaufort, North Carolina. in *Wading Birds,* A. Sprunt IV, J. C. Ogden, and S. Winckler (Eds.), Res. Rep. No. 7, National Audubon Society, New York, pp. 355–360.

Duever, M. J., J. F. Meeder, L. C. Meeder, and L. B. Meeder. 1994. The climate in south Florida and its role in shaping the Everglades ecosystem. in *Everglades: The Ecosystem and Its Restoration,* S. M. Davis and J. C. Ogden (Eds.), St. Lucie Press, Delray Beach, Fla., chap. 9.

Dugan, P., H. Hafner, and B. Boy. 1988. Habitat switches and foraging success in the little egret (*Egretta garzetta*). in *Acta XIX Congr. Int. Ornithologici,* H. Ouellet (Ed.), University of Ottawa Press, Ottawa, Ontario, pp. 1868–1877.

Erwin, R. M. 1983. Feeding habitats of nesting wading birds: Spatial use and social influences. *Auk,* 100:960–970.

Erwin, R. M. 1984. Feeding flights of nesting wading birds at a Virginia colony. *Colon. Waterbirds,* 7:74–79.

Erwin, R. M. 1985. Foraging decisions, patch use, and seasonality in egrets (Aves: Ciconiiformes). *Ecology*, 66:837–844.

Fennema, R. J., C. J. Neidrauer, R. A. Johnson, T. K. MacVicar, and W. A. Perkins. 1994. A computer model to simulate natural Everglades hydrology. in *Everglades: The Ecosystem and Its Restoration*. S. M. Davis and J. C. Ogden (Eds.), St. Lucie Press, Delray Beach, Fla., chap. 10.

Frederick, P. C. and M. W. Collopy. 1988. Reproductive Ecology of Wading Birds in Relation to Water Conditions in the Florida Everglades, Tech. Rep. No. 30, Florida Cooperative Fish and Wildlife Research Unit, School of Forestry Research and Conservation, University of Florida, Gainesville, 259 pp.

Frederick, P. C. and M. W. Collopy. 1989. Nesting success of five Ciconiiform species in relation to water conditions in the Florida Everglades. *Auk*, 106:625–634.

Frederick, P. C. and M. G. Spalding. 1994. Factors affecting reproductive success of wading birds (Ciconiiformes) in the Everglades ecosystem. in *Everglades: The Ecosystem and Its Restoration,* S. M. Davis and J. C. Ogden (Eds.), St. Lucie Press, Delray Beach, Fla., chap. 26.

Haake, P. W. and J. M. Dean. 1983. Age and Growth of Four Everglades Fishes Using Otolith Techniques, Report SFRC-83/03, South Florida Research Center, Homestead, Fla., 68 pp.

Hafner, H. and R. H. Britton. 1983. Changes of foraging sites by nesting little egrets (*Egretta garzetta* L.) in relation to food supply. *Colon. Waterbirds*, 6:24–30.

Hafner, H., P. J. Dugan, and V. Boy. 1986. Use of artificial and natural wetlands as feeding sites by little egrets (*Egretta garzetta* L.) in the Camargue southern France. *Colon. Waterbirds,* 9:149–154.

Harrison, C. S., T. S. Hida, and M. P. Seki. 1983. Hawaiian seabird feeding ecology. *Wildl. Monogr.,* 85:1–71

Hoffman, W., G. T. Bancroft, and R. J. Sawicki. 1989. Wading Bird Populations and Distributions in the Water Conservation Areas of the Everglades: 1985–1988, South Florida Water Management District, West Palm Beach, 173 pp.

Hoffman, W., G. T. Bancroft, and R. J. Sawicki. 1994. Foraging habitat of wading birds in the Water Conservation Areas of the Everglades. in *Everglades: The Ecosystem and Its Restoration,* S. M. Davis and J. C. Ogden (Eds.), St. Lucie Press, Delray Beach, Fla., chap. 24.

Kolipinski, M. C. and A. L. Higer. 1969. Some Aspects of the Effects of the Quantity and Quality of Water on Biological Communities in Everglades National Park, Open File Rep. 69007, U.S. Geological Survey, Tallahassee, Fla.

Kushlan, J. A. 1976. Site selection for nesting colonies by the American white ibis *Eudocimus albus* in Florida. *Ibis*, 118:590–593.

Kushlan, J. A. 1979. Foraging ecology and prey selection in the white ibis. *Condor,* 81:376–389.

Kushlan, J. A. 1986a. Responses of wading birds to seasonally fluctuating water levels: Strategies and their limits. *Colon. Waterbirds,* 9:155–162.

Kushlan, J. A. 1986b. *The Everglades: Management of Cumulative Ecosystem Degradation,* Proc. Managing Cumulative Effects in Florida Wetlands, New College Environmental Studies Prog. Publ. No. 37, Omnipress, Madison, Wisc.

Kushlan, J. A. 1987. External threats and internal management: The hydrologic regulation of the Everglades, Florida, USA. *Environ. Manage.,* 11:109–119.

Kushlan, J. A., J. C. Ogden, and J. L. Tilmant. 1975. Relation of Water Level and Fish Availability to Wood Stork Reproduction in the Southern Everglades, Florida, Open File Report 75-434, U.S. Geological Survey, Tallahassee, Fla.

Kushlan, J. A., P. C. Frohring, and D. Voorhees. 1984. History and Status of Wading Birds in Everglades National Park, National Park Service Report, South Florida Research Center, Everglades National Park, Homestead, Fla.

Kushlan, J. A., G. Morales, and P. C. Frohring. 1985. Foraging niche relations of wading birds in tropical wet savannas. *Ornithol. Monogr.,* 36:663–682.

Lack, D. 1966. *Population Studies of Birds,* Oxford University Press, Oxford, 341 pp.

Lack, D. 1968. *Ecological Adaptations for Breeding in Birds,* Methuen and Co., London.

Loftus, W. F. and A.-M. Eklund. 1994. Long-term dynamics of an Everglades small-fish assemblage. in *Everglades: The Ecosystem and Its Restoration,* S. M. Davis and J. C. Ogden (Eds.), St. Lucie Press, Delray Beach, Fla., chap. 19.

MacVicar, T. K. 1984. The Rainfall Plan: A New Approach to Water Management for the Southern Everglades, South Florida Water Management District, West Palm Beach.

McVaugh, W. 1972. The development of four North American herons. *Living Bird,* 11:155–173.

Neu, C. W., C. R. Byers, and J. M. Peek. 1974. A technique for analysis of utilization-availability data. *J. Wildl. Manage.,* 38:541–545.

Ogden, J. C. 1977. An evaluation of interspecific information exchange by waders on feeding flights from colonies. *Proc. 1977 Conf. Colon. Waterbird Group,* 1:155–162.

Ogden, J. C. 1978. Recent population trends of colonial wading birds on the Atlantic and Gulf coastal plains. in *Wading Birds,* A. Sprunt IV, J. C. Ogden, and S. Winckler (Eds.), Res. Rep. No. 7, National Audubon Society, New York, pp. 137–153.

Ogden, J. C. 1994. A comparison of wading bird nesting colony dynamics (1931–1940 and 1974–1989) as an indication of ecosystem conditions in the southern Everglades. in *Everglades: The Ecosystem and Its Restoration,* S. M. Davis and J. C. Ogden (Eds.), St. Lucie Press, Delray Beach, Fla., chap. 22.

Ogden, J. D., J. A. Kushlan, and J. T. Tilmant. 1978. The Food Habits and Nesting Success of Wood Storks in Everglades National Park in 1974, Natl. Res. Rep. 16, U.S. National Park Service, Washington, D.C.

Ogden, J. C., H. W. Kale II, and S. A. Nesbitt. 1980. The influence of annual variation in rainfall and water levels on nesting by Florida populations of wading birds. *Trans. Linn. Soc. N.Y.,* 9:115–126.

Ogden, J. C., D. A. McCrimmon, Jr., G. T. Bancroft, and B. W. Patty. 1987. Breeding populations of the wood stork in the southeastern United States. *Condor,* 89:752–759.

Pool, D. J., A. E. Lugo, and S. C. Snedaker. 1975. Litter production in mangrove forests of southern Florida and Puerto Rico, in *Proc. Int. Symp. Biology and Management of Mangroves,* G. Walsh, S. Snedaker, and H. Teas (Eds.), Institute of Food and Agriculture Science, University of Florida, Gainesville, pp. 213–237.

Powell, G. V. N. 1987. Habitat use by wading birds in a subtropical estuary: Implications of hydrography. *Auk,* 104:740–749.

Robertson, W. B. and J. A. Kushlan. 1974. The south Florida avifauna. in *Environments of South Florida: Present and Past, Memoir No. 2,* P. J. Gleason (Ed.), Miami Geological Society, Coral Gables, Fla., pp. 414–452.

Schomer, N. S. and R. D. Drew. 1982. An Ecological Characterization of the Lower Everglades, Florida Bay, and the Florida Keys, Off. Biol. Serv. FWS/OBS-82/58, U.S. Fish and Wildlife Service, Washington, D.C.

Sokal, R. R. and F. J. Rohlf. 1981. *Biometry,* 2nd ed., W. H. Freeman, San Francisco, 859 pp.

Southern, H. N. 1954. Tawny owls and their prey. *Ibis,* 96:384–410.

Southern, H. N. 1959. Mortality and population control. *Ibis,* 101:429–436.

South Florida Water Management District. 1992. Surface Water Improvement and Management Plan for the Everglades, South Florida Water Management District, West Palm Beach.

Travis, J. and J. Trexler. 1987. Regional Variation in Habitat Requirements of the Sailfin Molly with Special Reference to the Florida Keys, Tech. Rep. No. 3, Nongame Wildlife Program, Florida Game and Fresh Water Fish Commission, Tallahassee, 47 pp. + iii.

van Vessem, J. and D. Draulans. 1987. Spatial distribution and time budget of radio-tagged grey herons, *Ardea cinerea,* during the breeding season. *J. Zool. London,* 213:507–534.

Walters, C., L. Gunderson, and C. S. Holling. 1992. Experimental policies for water management in the Everglades. *Ecol. Appl.,* 2:189–202.

Werschkul, D. F. 1979. Nestling mortality and the adaptive significance of early locomotion in the little blue heron. *Auk,* 96:116–130.

Wood, E. J. F. and N. G. Maynard. 1974. Ecology of the micro-algae of the Florida Everglades. in *Environments of South Florida: Present and Past, Memoir No. 2,* P. J. Gleason (Ed.), Miami Geological Society, Coral Gables, Fla., pp. 123–145.

26

FACTORS AFFECTING REPRODUCTIVE SUCCESS of WADING BIRDS (CICONIIFORMES) in the EVERGLADES ECOSYSTEM

Peter C. Frederick

Marilyn G. Spalding

ABSTRACT

This chapter analyzes over 40 years of inquiry into the factors which affect reproductive success of wading birds in the Everglades system, with a focus on the past 20 years. It is unlikely that reproductive success has been limited by human disturbance (in the post-plume-hunting era), a depauperate potential nesting population, or lack of suitable colony substrate. Predation of nest contents usually has a minor effect on reproductive success, although predation is clearly a correlate of other stresses, such as drying of the marsh surface, human disturbance, or changes in food availability. The majority of reproductive failures instead appear to be associated with the availability and quality of food and with disease agents. For wood storks and white ibises both the initiation and abandonment of nesting can be associated indirectly with the recession rate of surface water in Everglades marshes. This relationship is almost certainly the result of increased prey availability through entrapment during drying phases. Tricolored herons, snowy egrets, and great egrets all show some susceptibility to abandon nesting during winter storms and reversals in the drying trend, although the mechanism may not entirely be related to dilution of prey. Rising water levels do not apparently contribute to the abandonments in great egrets, while the effects of cool temperatures on prey do.

While surface water recession is undoubtedly a contributing factor to nesting effort and nesting success in the present and historic Everglades, the relationship varies with species and has many weak points as a general rule. The available literature suggests instead that the underlying mechanisms controlling success and abandonment of nests in the Everglades and other locations are through food availability and that the current reliance of reproductive success on rapid drying rates is to a large extent unrepresentative of the historical Everglades. Factors controlling both density and availability of prey are poorly understood and likely to strongly influence the responses of wading birds to hydropattern. Studies of these subjects should be a priority for future research.

Diseases, parasites, and toxins are historically unlikely to have exerted much influence on nesting success, but may occupy a much expanded role as other ecological stresses increase. The effect of eustrongylidosis on nesting success is known to be devastating at some colonies and can only become worse with increasing human populations and associated waste disposal and nutrient-laden runoff problems. Mercury contamination is a fact for wading birds foraging in the central freshwater marshes of the Everglades. The available literature suggests that mercury contamination could have a very large effect on reproduction, through impairment of foraging and reproductive skills, as well as survival.

INTRODUCTION

Perhaps the single most frequently cited indicator of the degradation of the Everglades ecosystem is the decline of the wading bird breeding population. The total numbers of nesting attempts by all species combined have been reduced by a conservative estimate of 90% since the 1940s (Figure 26.1). At the species level, this magnitude of decline applies for wood storks (*Mycteria americana*), white ibises (*Eudocimus albus*), and snowy egrets (*Egretta thula*); nesting records are too incomplete or inaccurate to make the same statement for little

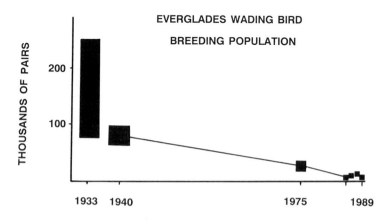

Figure 26.1 History of the wading bird breeding population in the Everglades ecosystem from 1877 to 1987. Because historical population estimates vary widely, the range of sizes is shown by height of bar.

blue herons (*Egretta caerulea*), great blue herons (*Ardea herodias*), and tricolored herons (*Egretta tricolor*). Great egrets (*Casmerodius albus*) appear to be the only species whose nesting numbers have remained relatively stable (Kushlan et al., 1984).

The decline is of interest from a number of perspectives. First, the near complete loss of breeding birds is disturbing on philosophical and aesthetic grounds. Second, the birds play a central role in the food web through predation and nutrient transport (Kushlan, 1976a; Christy et al., 1981; Stinner, 1983; Bildstein et al., 1990; Frederick and Powell, 1994), suggesting that their loss may have already had feedback effects on the ecosystem. Third, because of their position in the food web, the decline of wading birds probably signifies that the health of wetland species upon which the birds depend is in question. Finally, the Everglades is a geographically strategic component of the suite of habitats still utilized by wading birds during the winter and on migration (Hoffman et al., 1991; Bancroft et al., 1992); the management and fate of Everglades habitat could therefore have direct effects on wading bird populations of the entire southeastern United States.

Several lines of evidence suggest that degradation of nesting conditions has played a central role in the decline of Everglades breeding birds. First, the reductions in breeding birds in the Everglades was not matched by similar reductions in nesting in the southeastern United States (Ogden, 1978; Custer and Osborn, 1977), implying that breeding reductions were not unilateral. Second, the declines in Everglades breeding populations have occurred despite the fact that the Everglades region continues to be an important wintering ground for nearly all of the wading bird species concerned. The wintering populations are currently one to two orders of magnitude larger than breeding populations (Figure 26.2) (Bancroft

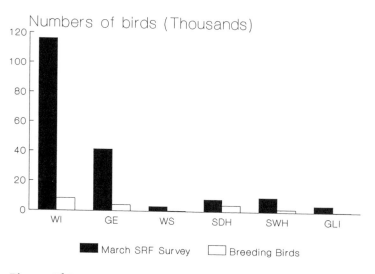

Figure 26.2 Comparison of breeding and wintering wading bird population sizes in 1987. Wintering population estimate is from Systematic Reconnaissance Flight results (Bancroft et al., 1987) in March, which usually represents peak numbers in the winter. Breeding censuses were as reported by Frederick and Collopy (1988).

et al., 1987; Frederick and Collopy, 1988; Bancroft and Jewell, 1987; Hoffman et al., 1990). The lack of spring nesting by this winter audience is therefore dramatic and suggests that breeding conditions are unattractive, at least by comparison with more northerly breeding locations. Third, of the adult birds which do not migrate north and remain in the Everglades during the spring nesting season, a relatively small proportion attempt to breed, strongly implying that breeding conditions in the Everglades are inappropriate in an absolute as well as a relative sense. Finally, there are some indications that breeding success and fecundity are quite severely reduced by comparison with historic conditions and with other southeastern breeding sites (Powell, 1983; Frederick and Collopy, 1988; Powell et al., 1989; J. C. Ogden, personal communication).

The reductions in nesting in southern Florida have been accompanied by increases in breeding by both white ibises and wood storks elsewhere in the southeast (Sprunt, 1922, 1944; Ogden, 1978, 1991; Kushlan and Frohring, 1987; Ogden et al., 1987; Custer and Osborn, 1977). The shifts in breeding occurred at some time during the period 1940–76 for white ibises and 1960–85 for wood storks. These moves by breeders could have been in part a consequence of the declines in southern Florida, but were not likely to have been a cause. First, the reproductive success recorded in most of the northern colonies is not particularly good (Hammatt, 1981; Frederick, 1987; Shields and Parnell, 1986; Coulter, 1987; Post, 1990). Also, if the Everglades birds moved solely because of the discovery of better breeding elsewhere (the "distant magnets" theory) (Walters et al., 1992), one would still expect the few remaining Everglades breeders to show good reproductive success, which they do not.

To summarize, the dramatic reductions in breeding wading birds in the Everglades are a phenomenon of the Everglades, and not the region, and were accompanied in some rough sense by large shifts in the breeding range of several species; responses of wintering and breeding birds imply that local breeding conditions are degraded relative to historic conditions and that this degradation is likely to have been an important cause of breeding declines. Because most species of wading birds are opportunistic nesters, a return to attractive breeding conditions in the Everglades could well result in rapid colonization by a portion of the wintering population. An analysis of current variation in attractiveness and of bottlenecks in breeding success is therefore of considerable use.

In the following sections, several measures of reproductive success will be defined, and the limitations of reproductive parameters as biological indicators will be discussed. Available research on nesting success in the Everglades ecosystem will then be reviewed, and the importance and mechanisms of the biotic and abiotic factors that affect reproduction will be evaluated. Because food availability emerges as a particularly important factor, considerable attention is devoted to the relationship among weather, hydrological conditions, and nesting success for several species.

Wading Bird Reproduction as a Biological Indicator

Ciconiiform birds range from being primary to tertiary predators, and several species have been used as examples of birds that are reproductively limited by food (Lack, 1968). This combination of characteristics, along with their colonial nesting

Table 26.1 Comparison of Nest Success Estimates
Derived from Traditional and Mayfield Methods, Using the Same Data

Year	Species	Traditional[a]	Mayfield[b]	Percentage-point difference
1986	Great egret	0.3846	0.1481	23.7
	White ibis	0.3061	0.2039	10.2
	Tricolored heron	0.7036	0.4613	24.2
	Little blue heron	0.9487	0.7284	22.0
	Snowy egret	0.8925	0.7117	18.1
1987	Great egret	0.3475	0.2064	14.1
	White ibis	0.6035	0.4887	11.5
	Tricolored heron	0.7632	0.6652	9.8
	Little blue heron	0.7941	0.7164	7.77
	Great blue heron	0.2857	0.1766	10.9

Note: All data are from colonies in the freshwater Everglades marshes (see Frederick and
    Collopy, 1988).
[a] Number of nests that succeed per number of nests monitored.
[b] Probability of success, prorated over the entire period; see Mayfield (1975) and Hensler and
    Nichols (1981).

habit, has given this group the potential of being accurate (and easily assessed) biological indicators of aspects of the food web which supports them, such as demographics of prey animals, pollutant load, and degree of parasitism (Custer and Osborn, 1977).

However, the reproductive parameters of birds are often quite removed in space and time from environmental parameters (Temple and Wiens, 1989), and caution must be used in assigning indicator relationships (Morrison, 1986). For example, the use of wading bird reproductive parameters as biological indicators of prey abundance, density, or availability is often implicitly assumed, yet it rests heavily on the assumption that food is the major factor controlling reproductive success. Wading bird reproduction is in fact likely to be strongly affected by intense predation by a variety of mammals, reptiles, and birds (Baker, 1940; Burger and Hahn, 1977; Shields and Parnell, 1986; Rodgers, 1987), as well as human disturbance (Tremblay and Ellison, 1979; Parsons and Burger, 1982) and disease (Wiese, 1977). In addition, the methods used to measure or estimate reproductive success can strongly affect the resulting picture of nesting success (for example, see Table 26.1). It is therefore essential that the factors which actually affect reproductive parameters be identified before reproductive parameters are used as bioindicators of environmental conditions. This must be done on a case by case basis. In addition, different measures of reproductive success may reflect different factors in the environment (fecundity versus nestling health, for example), and some definitions of the most commonly used parameters may be of use to the reader in this regard.

Definitions of Reproductive Parameters

Reproductive effort is defined as the number of nesting attempts (usually those progressing to egg laying) in a given location or year and is the only useful figure

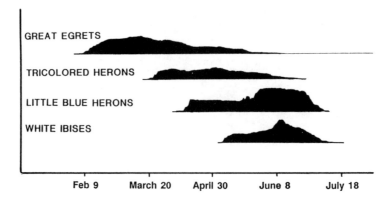

Figure 26.3   Timing of nesting initiations during 1987 for four species of wading birds in the freshwater marshes of the Everglades. Height of any section represents number of active nests.

that can be extracted from the vast majority of historic reports for the Everglades (Kushlan et al., 1984). Reproductive success is defined most accurately as the number of young produced from a nesting attempt which survive to breeding age. Unfortunately, wading bird young do not simply fly away from the nest (fledge) at some well-defined point, but instead pass through several stages of mobility, often spending several weeks of their preindependence period in the tops of trees (McVaugh, 1972; Rudegeair, 1975; Werschkul, 1979; Frederick et al., 1993; DeSanto et al., 1990). Censuses of young during these latter stages become difficult and unreliable (Frederick et al., 1993) and survival is simply measured to a given developmental stage prior to independence. Thus "reproductive success" of wading birds only rarely measures the probability of rearing young to independence, much less breeding age. Nesting success is perhaps more specific, usually defined as the probability of a nest surviving to produce at least one young to a given age of interest (Mayfield, 1961, 1975). Various measures of hatchability, clutch size, and survival of young are also often used to express productivity and fecundity of bird populations. Finally, timing of nesting in the Everglades has often been used for comparison between eras. While this can be a very useful measure when differences are robust (Ogden, 1994), renests within a season are common (Bancroft et al., 1991), and the nesting season within southern Florida is quite extended by temperate standards (Kushlan and White, 1977) (Figure 26.3).

FACTORS AFFECTING NESTING SUCCESS in the EVERGLADES

Human Disturbance

During the plume-hunting period (1860–1910), commercial hunting in wading bird colonies was a major cause of nesting failure and relocation of colonies, both of hunted and nonhunted species (Pierce, 1962). Since the cessation of the plume trade and the creation of Everglades National Park, the vast majority of large colony sites in southern Florida have been effectively protected from hunting and entry by humans. The exception to this rule has been researchers, who have monitored colony activity and breeding success using methods which vary widely in their intrusiveness. Researcher disturbance cannot be invoked to explain the general decrease in nesting success and breeding numbers in the Everglades, because researcher disturbance has only been common during the last two decades of the decline; further, much of the monitoring has been remotely done using aircraft. Overflights by single-engine reconnaissance aircraft (Kushlan, 1979) and military jets (Black et al., 1984) have not been found to result in reduced reproductive success or in behaviors likely to lead to reduced success.

Entry of researchers into colonies, however, can strongly affect nesting success of herons (Tremblay and Ellison, 1979), especially if colonies are entered during the sensitive courtship and egg-laying stages. This is of particular importance because the very act of entering colonies to monitor success may affect the measurement of reproduction itself. In the freshwater Everglades, Frederick and Collopy (1989a) found no differences in five measures of reproductive success between paired colonies of tricolored herons visited frequently (4-day interval) and infrequently (8-day interval). The latter result suggests that there was little fre-quency-of-visit effect. Because success was high in both colonies, it was concluded that carefully conducted nest checks during the post-egg-laying period may not strongly affect nesting success. However, until completely undisturbed nesting success (monitored remotely from a tower or vantage point) can be compared with the effects of a single-disturbance event, it is impossible to determine the effects of visits themselves (at any frequency). Similarly, effects of human disturbance are likely to vary considerably with species and location.

One of the most common effects of researcher disturbance is to increase predation and scavenging of nest contents, usually while researchers are in the colony (Milstein et al., 1970; P. C. Frederick and M. G. Spalding, personal observation). In a North Carolina colony, Shields and Parnell (1986) found that over 40% of all white ibis eggs were taken by fish crows (*Corvus ossifragus*), perhaps largely as a direct result of researcher intrusions. Predation by common crows (*C. brachyrhynchos*) on eggs of herons during researcher visits has been a recurrent problem in colonies in the coastal mangrove region of Everglades National Park, where crows are abundant (Bancroft and Jewell, 1987; Bancroft et al., 1991) and in Florida Bay (G. V. N. Powell, personal communication). The magnitude of the effect of crow predation in the absence of researcher disturbance has never been estimated because it requires (as above) measuring nest success remotely at control sites. A key question in this regard is whether crows can force attending parental waders off their nests in order to steal eggs or young. If not, crows must normally be scavengers in colonies, rather than primary predators. Nest attendance is nearly continuous during egg laying and incubation in several species of wading birds (Jenni, 1969; Rudegeair, 1975; Wiese, 1975; Rodgers, 1978, 1980a; Frederick, 1985), suggesting that the effect of crows might only be important during periods of food stress, abandonment, and disturbance or when crows occur in extremely large numbers (Post, 1990).

Other important disturbance effects include thermal stress on eggs and young and the premature departure of young from nests (Parsons and Burger, 1982). While careful timing of visits can presumably take care of the former problem, the effect of the latter on survival of young has not been systematically investigated for any of the wading bird species which nest in the Everglades. Interestingly, Parsons and Burger found that early habituation of nestlings significantly reduced the early fledging problem, suggesting that frequent visits starting when chicks are small will have less impact than single-time censuses.

Disease and Parasites

Disease, including parasitic disease and those resulting from toxins, is one of the most poorly understood factors which affect reproductive success of ciconiiforms, particularly in the Everglades. This neglect has a variety of sources. Diseases in wading birds are unlikely to be reported, given the vastness and inaccessibility of much of the area and the fact that most diagnoses are difficult without detailed necropsy procedures. Disease in nestlings in particular is likely to be overlooked because wading bird young often die in large numbers, and it is generally assumed that a large percentage of these deaths must be from starvation, intersibling aggression (Lack, 1968; Powell, 1983; Clark and Wilson, 1981; Mock, 1984), and predation.

Despite these popular beliefs, thorough necropsy, parasite screening, histopathologic examination, and metal and toxin analysis are necessary to identify (or rule out) disease as a contributor to death. For example, a major epizootic of eustrongylidosis (described in detail later) in the Everglades during 1986–90 would have gone unnoticed by professional ornithologists if complete necropsy examinations by a pathologist had not been made (Spalding et al., 1993). Disease also may be obscured as a cause of death because the dead or moribund chick has subsequently been evicted from the nest, scavenged, fallen into the water, or even eaten by a nestmate.

In addition to the lack of reporting, the specific effects of diseases are also extremely difficult to isolate. Diseases almost always occur in conjunction with many other disease agents and toxins commonly found in "healthy" birds, along with the nutritional and environmental stresses with which they live. These factors may alter the response and thus the outcome of the "primary disease" as compared with single-factor experimental infection results. Lesions compatible with the initial illness may no longer be present at the time of death because they were obscured by a secondary disease process, thus resulting in an inaccurate assessment of the importance of the primary disease. Aside from mortality, diseases in general might have a number of other effects of note for wading bird species. One is the abandonment of colony sites by adults. There is evidence that severe ectoparasitism causes abandonment by adults in other colonial nesters (King et al., 1977). Abandonments in a small colony in central Florida occurred when nestlings (and possibly adults) were very severely infected with *Eustrongylides* larvae (Spalding et al., 1993; N. Edelson, personal communication). On the other hand, there is evidence that some colonies will reform the following year even though a devastating disease agent is still active (Wiese, 1977; Spalding et al., 1993). Finally, sublethal diseases or contaminants may have subtle effects on wading bird individuals but dramatic effects on foraging and reproductive success or effort; these effects are likely to be quite difficult to demonstrate. For example, the contaminants found in Everglades wading bird tissues (see later) alone could be a plausible explanation for reduction in nesting effort.

Other poorly understood epidemiologic aspects of diseases on populations are the importation of diseases by migratory birds, exposure to contaminants outside of Florida, and human transport and release of exotic and rehabilitated birds along with their diseases.

Emaciation, or the lack of fat and decreased muscle mass, can result from chronic disease or severe parasitism as well as from simple malnutrition. It can also predispose birds to secondary disease. For this reason, the assessment of malnutrition or starvation as a cause of death is difficult. A preliminary survey of 562 nestling carcasses collected from Everglades colonies during 1987–90 (M. G. Spalding, unpublished) revealed that malnutrition (based on the absence of fat and evidence of disease) was likely the primary cause of death in 2% of nestlings; 34% of carcasses fell in the gray zone of both disease and emaciation. If malnutrition is a predisposing factor to disease, as much as 61% of deaths could be ascribed to the malnutrition category; however this is unlikely. Diseases were suspected to be the cause of death in 39% of carcasses in good nutritional condition and in 49% of carcasses in poor nutritional condition (46% of all carcasses, regardless of nutritional status). Thus the assumption that most nestlings found dead without evidence of predation have starved to death is clearly inaccurate. Trauma, including predation, accidents, and intraspecific aggression, accounted for 27% of deaths (this is likely an underestimate because many predators remove the carcass from the site).

A discussion of diseases with known or potential demographic level effects follows. Eustrongylidosis is caused by a large nematode parasite which burrows through the stomach wall and into other viscera. This parasitic disease can have strong debilitating effects and cause death in otherwise healthy birds, particularly nestlings (Spalding and Forrester, 1993). Eustrongylidosis has a serious impact on the reproductive success of wading birds at certain colonies in the Everglades (Spalding et al., 1993a) as well as in other states (Wiese, 1977; Roffe, 1988). More than 50% of nestlings in some Everglades colonies were infected with *Eustrongylides* during 1987–90. The life cycle of the nematode involves an aquatic oligochaete and many species of fish as intermediate hosts. Preliminary evidence suggests that anthropogenic addition of nutrients and physical alteration of foraging sites results in increased numbers of infected fish (Spalding et al., 1993). Eustrongylidosis usually is not obvious from the appearance of a bird, but can be detected by external physical examination (Spalding, 1990). It is unknown what role this disease has played in the decline of the Everglades breeding population, and work on the ecology of this disease is ongoing.

A number of infectious diseases are known to produce very large die-offs of free-ranging birds. Avian cholera, due to infection with the bacteria *Pasteurella multocida,* was reported in the Everglades in 1967–68, when up to 6000 American coots (*Fulica americana*) were estimated to have died. Four other species of waterfowl were involved, although no wading birds were found dead (Klukas and Locke, 1970). Avian cholera has been reported to kill great blue herons in the western United States (Menisk and Botzler, 1989; Rosen and Bischoff, 1949). Generally, avian cholera is a common disease of migratory waterfowl of central and western states and only rarely occurs in the eastern United States. Conditions necessary for mortality from this disease are poorly understood (Brand, 1984).

Enterotoxemia, or mortality due to enteritis and toxin produced by *Clostridium perfringens* type C, was responsible for the death of at least 104 adult wading birds of 7 species near Lake Okeechobee in 1971. This epizootic occurred during the late nesting season, although the effect of the disease on nestlings was not mentioned. The source of the infection was unknown (Jasmin et al., 1972).

Botulism, resulting in mortality due to a toxin produced by *C. botulinum* type C, has occurred in northern Florida (Forrester et al., 1980). Although primarily a disease of waterfowl, one snowy egret was involved in this die-off. It is believed that the conditions produced in a phosphate mining operation in conjunction with high temperatures were responsible for this outbreak, which would otherwise be unexpected in Florida.

Salmonellosis, chlamydiosis, and avian pox occasionally occur in adult and nestling wading birds in the Everglades (Conti et al., 1986; M. G. Spalding, unpublished) and have the potential to cause widespread mortality. No large die-offs due to these diseases have been reported.

Dermestid larvae occasionally cause lesions on nestlings in the Everglades, but the number of affected birds is usually small (Snyder et al., 1984; M. G. Spalding, unpublished). Blood parasites and tumors are relatively insignificant in Florida wading birds (Telford et al., 1992; Spalding and Woodard, 1992).

A number of experimental studies have shown effects of contaminants on the immune system of birds and increased susceptibility to disease (Goldberg et al., 1990; Trust et al., 1990; Rocke and Samuel, 1991); however clear evidence of increased susceptibility in free-flying birds has not been demonstrated.

In the early 1970s a great flurry of work focused on the analysis of pesticide concentrations in birds along the east coast of North America (Ohlendorf et al., 1978, 1981). Occasionally wading birds were included in these analyses. DDE, TDE, DDT, dieldrin, heptachlor epoxide, oxychlordane, *cis*-chlordane, *trans*-nonachlor, *cis*-nonachlor, endrin, toxaphene, mirex, and PCB residues have all been reported in Florida great blue herons (Ohlendorf et al., 1981). In only one case was mortality attributed directly to pesticide intoxication (dieldrin). With the exception of *cis*-nonachlor and heptachlor epoxide, the same array of pesticides was found in black-crowned night heron eggs in Florida (Ohlendorf et al., 1978). Both heavy metals and pesticides in eggs were found to be lower in the southern states when compared to northern states in the early 1970s (Ohlendorf et al., 1978). In black-crowned night heron colonies outside Florida, Custer et al. (1984) found an effect on hatching success in colonies with higher pesticide concentrations, but little effect on overall reproductive success. Embryonic mortality and congenital defects (common symptoms of some contaminants) do not appear to be a severe problem in southern Florida at present. Several authors reported that at least 80% and usually more than 90% of all eggs present in active nests at the expected time of hatching do hatch (Girard and Taylor, 1979; Shields and Parnell, 1986; Rodgers, 1980a, 1980b; Rodgers, 1987; Black et al., 1984; Frederick and Collopy, 1988). Congenital defects were present but uncommon in a preliminary survey of 562 nestlings (M. G. Spalding, unpublished).

Ohlendorf et al. (1981) measured concentrations of mercury in black-crowned night heron eggs at four locations in Florida (none were in the Everglades). Although measurable concentrations were found, they were generally lower than at other sites on the eastern seaboard of the United States. Ogden et al. (1974) found high mercury concentrations in white ibises collected from the Everglades. Very high mercury concentrations have been found in nestlings and adults of most ciconiiform species in southern Florida (Sundlof et al., in review). The effect of mercury contamination on survival and reproduction is understood poorly in ciconiiforms. Experimental dosing with mercury results in decreased reproduction and disturbed locomotion in non-ciconiiform bird species (Borg, 1969; Fimreite, 1971; Fimreite and Karstad, 1971; Spaan et al., 1972; Finley and Stendell, 1978). Decreased hatchability and fledging success was associated with elevated mercury concentrations in eggs in tern colonies (Fimreite, 1974). High mercury concentrations were also found to significantly reduce survival of grey herons (*Ardea cinerea*) during stressful weather conditions (Van der Molen et al., 1982).

In light of work that has been done in other species, it is not unreasonable to assume that high concentrations of mercury found recently in Everglades wading birds could result in the sublethal effects of reduced foraging and courtship ability. Each of these symptoms could result in reduced breeding effort and success and could be a powerful factor in explaining the reduced reproduction observed in the Everglades. The current state of knowledge on the effects of specific concentrations of mercury on wading bird behavior and survival is nonexistent.

Effects of Colony Site Characteristics

Wading bird colony sites fall into two general types within the Everglades: freshwater willow head and estuarine and mangrove islands. At freshwater sites, colonies are most often located in tree islands dominated by willow (*Salix caroliniana*), although cypress (*Taxodium distichum*), melaleuca (*Melaleuca quinuenervia*), cocoplum (*Chrysobalanus icaco*), pond apple (*Anona glabra*), and bays (*Persea* spp.) also are occasionally used (Frederick and Collopy, 1988). Of all the tree island types found in the freshwater Everglades, willow tolerates the longest hydroperiod (Gunderson, 1994), and colonies may be consistently located in willow because they represent relatively deep water within any given locale. Within mangrove-dominated estuarine areas, red mangroves (*Rhizophora mangle*) are primarily used for nesting. Powell et al. (1991) and Frederick and Powell (1994) discuss the magnitude and type of effects that wading bird excreta have on vegetation and environs of the colony.

At the time of formation, most colony sites are inundated, or at least surrounded, by water. Drying out of these sites greatly increases the chances that they will be visited by raccoons (*Procyon lotor*) (see section on nest predation). Mangrove colonies are often permanently inundated by salt or brackish water and thus are not accessible to mammalian predators. During the 1960s and 1970s, the majority of wading birds changed nesting location within the Everglades, moving from large historically used colonies along the mangrove fringe of Everglades National Park to large willow heads located in Water Conservation Area 3 (Frederick and Powell, 1994). Although the changing dynamics of food sources was almost certainly of importance during this relocation, colony sites within Water Conservation Area 3 also may have become more attractive because they were more reliably flooded than previously due to the creation of the water conservation impoundments.

The location of colonies within the Everglades ecosystem appears to have a strong effect on clutch size. Kushlan (1977) found significantly larger white ibis clutch sizes at inland sites than at coastal sites within the Everglades. A comparison of data collected during the late 1980s shows a similar result for tricolored herons (Frederick et al., 1992). This phenomenon has also been reported from other sites for white ibises (Rudegeair, 1975; Frederick, 1987) and *Egretta* herons (Jenni, 1969; Maxwell and Kale, 1977). While the coastal "depression" of clutch size might be a result of differences in food availability (Kushlan, 1977), the phenomenon might also arise as a function of distances traveled to foraging sites (Rudegeair, 1975; Kushlan, 1977) or of physiological stress due to the ingestion of salty prey (Bildstein et al., 1990; Johnston and Bildstein, 1990) (see next section). Kushlan (1977) found that significant geographic differences in white ibis clutch sizes within the Everglades did not result in differences in numbers of young produced during the two breeding seasons in which he studied reproduction; he suggested that clutch size was not very important in determining production of young in white ibises. The possibility exists that at a longer time scale, clutch size differences have an effect on production.

As discussed in the final section, distance from colony to foraging sites can have important energetic consequences for production of young. The location of colonies may be determined by a number of processes which vary between species. White ibises often change colony sites to take advantage of locally abundant food sources and have been considered almost nomadic nesters (Kushlan, 1977; Ogden, 1978; Frederick, 1987). Actual location of white ibis colony sites in any year probably involves an assessment of nearby food sources, photoperiod, and reproductive condition (Kushlan, 1974). At the opposite end of the spectrum are wood storks, which are much more site faithful, a condition probably enabled by their use of low-energy flight techniques, allowing exploitation of a very large foraging range (Kahl, 1964; Kushlan, 1986). Great and snowy egrets appear to be slightly less nomadic than white ibises and often use old colony sites when nesting, although the same sites may not be used every year. Bancroft and Jewell (1987) have shown that within a season, individual tricolored herons may attempt to nest in several different, traditionally occupied colony sites along the Everglades gulf coast. Frederick and Collopy (1988) and Bancroft et al. (1988) found high colony turnover rates within the freshwater Everglades marshes and Big Cypress regions of southern Florida. These results imply that some fidelity may pertain at the regional level for these species, but not for a single colony site.

Past Nesting Success

Does past nesting success (or failure) at a site influence future nesting site selection or nesting success for waders? The answer probably depends on the causes of nest failure. Over 5 years, annual destruction of an average of 61% of white ibis nests through tidal inundation at a colony in South Carolina did not result in decreased use of the colony, even following seasons in which over 90% of nests were destroyed (Frederick, 1987). Similarly, at a Delaware River colony repeated mass nest failures of snowy egrets and great egrets due to parasitic infections did not lead to a reduction in numbers of nesting attempts in ensuing years (Wiese, 1977). Burger (1982) found that nest failure of black skimmers (*Rhynchops niger*) due to nest predation was a reliable predictor of colony site abandonment in subsequent years, while failure due to tidal inundation was not. Post (1990) found no reduction in nesting densities of *Egretta* herons or white ibises following 2 years of poor nesting success at a colony in Charleston, South Carolina, but found eventual desertion of the colony following heavy predation by fish crows and mammals. Wood storks have continued to attempt nesting in Everglades National Park following many consecutive years of complete failure during the 1960s and 1970s (Ogden, 1994).

Under current conditions in the Everglades, the major causes of nest failure appear to be poor feeding conditions and disease. It is unknown whether these conditions affect site fidelity in subsequent years.

Salt Stress

Most adult ciconiiforms are capable of excreting excess salt through an infraorbital salt gland. Johnson and Bildstein (1990) found that young white ibises were unable to gain weight and would ultimately die when fed on diets composed of osmoconforming marine organisms, such as marine crustaceans. These authors were able to demonstrate that this effect was due to salt, rather than nutritional differences between types of prey. Bildstein et al. (1990) found that this salt limitation may severely limit both nesting effort and nesting success of white ibises on the South Carolina coast. Bildstein (1990) presents evidence to support the hypothesis that the salinization of a Trinidad estuary was probably responsible for the complete cessation of breeding by the congeneric scarlet ibis (*Eudocimus ruber*).

The implication for the Everglades system is that marine habitats are probably not good alternative reproductive foraging grounds for ibises, particularly when droughts preclude nesting and foraging inland and brackish habitats become more marine. While ibises often nest successfully on marine islands, their success is dependent upon feeding in freshwater habitats. The progressive salinization of the Everglades coastline has been demonstrated by Walters et al. (1992).

Interestingly, most ardeids probably are not as restricted by salt stress, because these species forage primarily on fishes, which osmoregulate in marine habitats. One exception is the black-crowned night heron (*Nycticorax nycticorax*), whose nestlings are often fed almost exclusively marine crustaceans when in estuarine colonies. Young of this species do not appear to be affected by salt imbalance as a result of eating saline prey (W. Post, personal communication).

Nest Predation

Although nest predation is widely assumed to have been an important selective force in the evolution of colonial nesting (Krebs, 1978; Wittenberger and Hunt, 1985), ciconiiforms are quite vulnerable to many types of predators and display no individual or group nest defense behaviors. In fact, the only antipredator strategy appears to be the habit of nesting in vegetation surrounded by water (Rodgers, 1987).

Raccoons are notorious nest predators, and raids by one or a few animals can cause the disruption and abandonment of entire colonies, often through excess killing (Lopinot, 1951; Rodgers, 1987; Coulter, 1987; Post, 1990). In most colonies, raccoon predation has only been noted when the surface water surrounding the colony has receded (Jenni, 1969; Coulter, 1987; Rodgers, 1987; Frederick and Collopy, 1989c) or the colony island is large enough to permanently support a number of raccoons (Post, 1990).

In the relatively long-hydroperiod marshes of Water Conservation Area 3, raccoons are quite uncommon. Frederick and Collopy (1989c) found that only 2 baited tracking stations were visited by raccoons or other potential mammalian nest predators out of a total of 341 tracking station nights at 27 locations. Almost half of these stations were placed in shallow, drying marshes of less than 5 cm depth, so that even shallowly inundated areas were not visited. The collection of six raccoons for mercury assay in Shark River Slough during 1990 proved extremely difficult (O. L. Bass, personal communication). These findings suggest either that mammalian use of long-hydroperiod marshes is severely limited by even very shallow water or that nest-predatory mammals are quite rare in Everglades freshwater marshes. Further, it is not clear if this sparse distribution is typical of a natural situation or if raccoon populations are currently reduced by mercury contamination or disease.

Frederick and Collopy (1989c) found that colonies in Water Conservation Area 3 experienced very little predation from mammals during 1986 and 1987 (estimated at between 1 and 12% of nests). Because colonies are depredated most often by mammals when dry, it follows that both drought years and drying of the colony during the middle of the nesting season would lead to much more regular nest destruction by raccoons than has been documented in the Water Conservation Areas.

In mangrove regions of the Everglades, raccoons are probably more numerous. Nesting by wading birds is entirely on islands well-separated from the mainland, and W. Robertson (personal communication) has noted that roseate spoonbills (*Ajaia ajaja*) nest only on the few islands in Florida Bay that do not have raccoons on them, suggesting a strong potential effect of predation on nesting success. The island nesting strategy may work. Bancroft et al. (1991) and G. V. N. Powell III and R. D. Bjork (unpublished) have never found evidence of raccoon nest predation at any of the estuarine colonies they studied in the Everglades and Florida Bay.

Jenni (1969) proposed that raccoons might also be deterred from entering Florida wading bird colonies by the alligators (*Alligator mississippiensis*) that usually frequent colonies for scraps and fallen chicks. However, no statistical association was noted between alligator activity at freshwater colonies and lack of visits by raccoons (Frederick and Collopy, 1989c). Many of the freshwater tree island colonies have long perimeters and dense surrounding vegetation. These conditions may make raccoons difficult to detect in freshwater colonies. A case cannot be made at present for alligator deterrence of mammalian predators in Everglades wading bird colonies.

Rat snakes (*Elaphe* spp.) are frequently reported as egg predators in wading bird colonies (Dusi and Dusi, 1968; Taylor and Michael, 1971) and are probably the only group of snakes sufficiently arboreal and aquatic to be a threat in Everglades colonies. The importance of snake predation on nesting success is difficult to measure because of their nocturnal habits, diurnal crypticity, and lack of a calling card at bird nests other than the disappearance of eggs. Using an inert tracking medium applied to the trunks of nest trees to obtain tracks of climbing predators, Frederick and Collopy (1989c) found that snakes might have been responsible for the destruction of as many as 7% of all nests (n = 106) monitored and 23% of the nests which failed. Thus, while snakes contribute an important share of the nest predation which occurs, they probably are not responsible for the failure of a demographically important fraction of wading bird nests.

Common crows, common grackles (*Quiscalus quiscala*), boat-tailed grackles (*Q. major*), red-winged blackbirds (*Agelaius phoniceus*), and both turkey and black vultures (*Cathartes aura* and *Coragyps atratus*) have all been noted as potential predators or scavengers in Everglades wading bird colonies. Because their effects are almost always noted in association with abandonment events, food stress, and human disturbance (Shields and Parnell, 1986; Frederick and Collopy, 1988; W. Robertson, personal communication; Bjork and Powell, 1990; Bancroft et al., 1991), it is likely that these birds become important sources of egg and nestling loss only in the context of some other major stress on wading bird reproduction. This is especially likely because none of these species seems able to dislodge parental wading birds from nests on their own (P. C. Frederick and M. G. Spalding, personal observation). Common crows will challenge several species of wading birds for access to nests once the parent has already been disturbed by researchers (P. C. Frederick and M. G. Spalding, personal observation), and it is suspected that crows and vultures grow bolder after previous successes in colonies where disturbances and stress are regular. In the absence of other stresses on reproduction which affect the tenacity and attendance of adults, these avian predators probably result in little nest predation.

As previously mentioned, common crows are abundant in mangrove colonies in the Everglades (Bjork and Powell, 1990; Bancroft et al., 1991), yet are almost completely absent from interior freshwater colonies (Frederick and Collopy, 1989c). The reasons for this distribution are not obvious. Possibly the high turnover in occupancy of interior colonies (Bancroft et al., 1987; Frederick and Collopy, 1988) makes them too unreliable for crows to establish nesting territories nearby, whereas the coastal colonies are active to some degree in almost every year (Kushlan, 1977; Kushlan and White, 1977). Sources of human refuse may also contribute to stability for crows, and common crows are apparently reliant upon the constant supply of garbage and tourist leavings at Shark Valley and Flamingo (J. C. Ogden, personal communication). The combination of constant food near Flamingo and the presence of stable wading bird colonies in the nearby estuarine zone may have allowed the development of frequent crow use of those colonies.

Black-crowned night herons have been noted taking young ibises and herons from active nests (Frederick, 1985; Frederick and Collopy, 1989c) and can probably be considered predators, rather than opportunistic scavengers, because they will displace adult birds from their nests (P. C. Frederick, personal observation). Night herons have never been found taking chicks greater than 1 week of age, and their depredations probably are limited to this short period. Because black-crowned night herons are relatively infrequent nesters in Everglades colonies, their overall contribution to nestling losses probably is quite small.

Southern bald eagles (*Haliaeetus leucocephalus*) often prey upon waders and occasionally their nestlings (Rudegeair, 1975; Frederick et al., 1993; M. G. Spalding, R. B. Bjork, and G. B. V. Powell III unpublished), although eagles have never been reported at Everglades colonies outside of Florida Bay. Similarly, great horned owls (*Bubo virginianus*) can be very destructive in bird colonies, although the rather characteristic evidence of predation by owls (Pratt, 1972; Nisbet, 1975) has not been reported from Everglades colonies. Red-shouldered hawks (*Buteo lineatus*) are common residents throughout the Everglades and are a potentially important predator of nestling waders. These raptors are known to take nestling birds frequently (Sherrod, 1978). While they have not been reported as predators of wading birds, they would probably not leave telltale evidence at the preyed upon nest.

The emerging picture is that predation on nest contents is usually a minor component of nest failure in Everglades marshes, if colonies remain inundated and if food stress and weather (see later) do not affect parental tenacity and attendance. This picture can be substantially altered if the colony dries out during the nesting period, if food or weather conditions deteriorate, if crows are present during researcher visits, and if mammalian predators increase substantially in abundance. It is suggested here that mammalian nest predation usually leads to episodic, catastrophic losses, rather than the regular, low-frequency predation by arboreal snakes or the opportunistic predation by most birds.

Nest predators could both increase or decrease as an eventual result of the compression and compartmentalization of the Everglades ecosystem. Raccoons and other "meso-mass" mammals are hypothesized to become more abundant in response to a reduction of populations of large predators such as the Florida panther (*Felis concolor coryi*) (Harris and O'Meara, 1989), a shortening of hydroperiods, and an increase in available human garbage. Similarly, the intrusion of roads and accompanying human refuse probably helps crows to seasonally exploit wading bird colonies. Populations of raccoons and nearly all marsh-dwelling nest predators could be considerably reduced by mercury contamination or increased exposure to disease agents with human encroachment (Bigler et al., 1973; Hoff et al., 1974).

Drought

Effects of drought on nesting success have rarely been documented, in large part because droughts have been difficult to define in the Everglades, where partial drying of the marsh is a seasonal event. The most obvious effect of drying of the marsh surface is that freshwater colony sites will lose their protection from mammalian predators, and nest losses will undoubtedly increase. Drying of the colony site may result in immediate abandonment by wood storks, especially if during the early stages of nesting or if accompanied by nest predators (Coulter, 1987; Rodgers, 1987). White ibises do not seem as prone to abandonment during colony drying. During the drought associated with the late onset of summer rains in 1987, over 2000 ibises continued to incubate eggs and raise young at the Alley-North (Rescue Strand) colony, where nearly all surface water had dried and raccoons had begun to depredate nests. Similarly, ibises began nesting in 1974 during a period of rapid water level recession in northeastern Water Conservation Area 3 near Andytown, despite the fact that the site was devoid of surface water. The same phenomenon occurred during the 1990 season at colonies along L-67A and in Water Conservation Area 1 (Arthur R. Marshall Loxahatchee National Wildlife Refuge).

While rapid recession of surface water appears to stimulate nesting and may increase nesting success (see following sections), near complete drying of the Everglades marsh surface can result in large-scale abandonment of nests. During the drought of the 1989 spring, Bancroft et al. (1991) observed that great egrets, tricolored herons, and white ibises at the L-67 colony took relatively short foraging trips to the surrounding marsh during the early part of nesting. As the nearby marshes dried, birds were forced to forage at distant locations in agricultural fields south of Lake Okeechobee and highway edges. Nearly all nests were progressively abandoned during April and May; regurgitant samples collected during this time suggested that tricolored herons were forced to feed on terrestrial insect prey by the almost complete lack of aquatic organisms.

Food Availability and Factors Affecting Foraging

Availability of food may be the single most important factor limiting the distribution and nesting success of wading birds, especially in the relatively oligotrophic Everglades ecosystem. Powell (1983) demonstrated that food-supplemented great white herons (*Ardea herodias*) laid significantly larger clutches and raised more young than did unsupplemented pairs in Florida Bay. In addition, Powell found that clutch sizes of unsupplemented birds were significantly lower than clutch sizes recorded during the early part of this century, suggesting that the availability of food had decreased substantially in Florida Bay. Mock et al. (1987) found that food-supplemented great egrets fledged significantly more young than unsupplemented ones. Of a suite of environmental variables, Frohring (unpublished National Park Service report) found that nestling tricolored heron and snowy egret growth rates were most strongly correlated with the density of fish prey in nearby wetlands. Faster growth of nestlings was also linked to increased survival. These studies suggest a strong association of fecundity and production of young with available food supply.

A number of associations have also been found between nesting success and hydrology; the implication has been that hydrology determines availability of prey by concentrating prey during drying and dispersing them during filling. Kushlan et al. (1975) found that the initiation of wood stork nesting was earlier in years that had a combination of January water levels in Everglades National Park above specific values and rapid rates of surface water recession during the winter and early spring. Wood stork nesting usually failed if it continued into the onset of summer rains, and thus later initiation of nesting in the winter and spring could be associated with greater probability of nesting failure. Kahl (1964) found that the initiation and abandonment of wood stork nesting in Corkscrew Swamp was related to specific water levels in the surrounding marshes, but did not examine correlations with water recession rates. Similarly, Clark (1978) found both nest initiation dates and numbers of young produced by wood storks at a Merritt Island, Florida colony to be associated with water levels and water recession rate in the nearby St. Johns River basin. Finally, wood storks have been found to abandon nesting attempts in Everglades National Park in response to strong rainfall events which abruptly reverse regional surface water recession trends (Bancroft and Jewell, 1987; Frederick and Collopy, 1988; J. C. Ogden, personal communication).

This relationship of stork nesting behavior and success with surface water recession rate has been explained as a dependence of storks on the prey which are concentrated into ponds and gentle depressions by falling water levels; during even short periods of rising water, prey presumably disperse and densities are too low for energetically demanding nesting activities. Although many species feed with storks and consume similar prey species, wood storks are presumed to be particularly vulnerable to prey density effects because of their specialized grope foraging technique (Kahl, 1964).

Nesting initiation and abandonment in several other wader species in the Everglades also may be related to hydrological or weather parameters, although the effect and the mechanism does not seem as obvious as with the storks. During three concurrent studies of wading bird nesting success in southern Florida, relatively synchronous colony failure events were noted in great and snowy egrets, tricolored herons, and roseate spoonbills following strong winter frontal weather patterns during which large amounts of rain often (but not always) fell, marsh surface water levels frequently rose, and temperatures dropped (Figure 26.4) (Bancroft and Jewell, 1987; Frederick and Collopy, 1989b; G. V. N. Powell III and R. B. Bjork, unpublished; Frederick et al., 1992). Similarly, white ibises, tricolored herons, little blue herons, and wood storks abandoned nesting synchronously following the onset of the summer rainy season (Frederick and Collopy, 1989b; Bancroft and Jewell, 1987; P. C. Frederick, unpublished data).

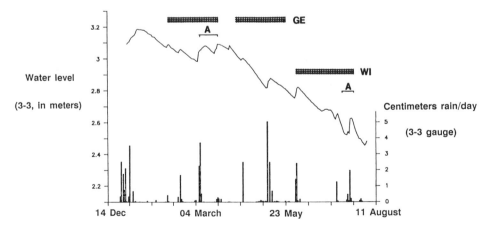

Figure 26.4 Large-scale synchronous abandonments of great egrets and white ibises (A) in relation to surface water dynamics (fluctuating line), rainfall (thin bars), and period of breeding activity (hatched horizontal bars) at Alley-North colony in Water Conservation Area 3 in 1987.

Predation of nests did not explain these nest failures, because predation is unlikely to be so synchronous across the ecosystem. The nest failures appeared to be stimulated directly by the weather events or their hydrological and ecological consequences.

A number of components of these sudden changes could be responsible for the nest failures. Wind damage to nests could almost always be ruled out as a cause of failure, because failed nests were rarely broken up or displaced and eggs were usually intact in nests. Sustained high winds might, however, make it much more difficult for sight-foraging herons and egrets to hunt effectively or for any species to travel to foraging locations. By following nesting wood storks on foraging flights in Everglades National Park before and during an abandonment event in 1973, J. C. Ogden (personal communication) was able to conclude that the birds abandoned because they could not reach their foraging destinations in the high winds.

A drop in temperature is typically associated with winter/spring cold fronts and might influence nesting in two ways. First, temperature drop may be associated with a delay in the initiation of nesting, as has been found for wood storks (Kahl, 1964), great egrets (Simmons, 1959), and white ibises (Rudegeair, 1975), and cooler winters may in general result in fewer nesting attempts (Ogden et al., 1980). Second, low temperatures might induce abandonments by making prey inactive and hence less available. A multivariate analysis of great egret nest success in freshwater colonies of the Everglades showed that probability of failure was strongly associated with severity of cold (Frederick and Loftus, 1993). Surface water level, recession rate, and rainfall were not picked as significant variables in that analysis, suggesting that temperature could be isolated as a factor. These observations also may apply to tricolored herons, which were susceptible to abandonments during both dry and wet cold fronts (Frederick and Collopy, 1989b).

Cold temperatures probably inhibit nesting and result in abandonments by making marsh fishes and invertebrates more difficult to prey upon. Frederick and Loftus (1993) found that visual counts of marsh fishes were significantly lower, and fishes generally were more wary on mornings following a cold snap than on warmer mornings. During simulated cold snaps in the laboratory, seven species of Everglades marsh fishes (*Gambusia affinis, Cyprinodon variegatus, Jordanella floridae, Poecilia latippina, Lepomis gutatus, L. macrochirus,* and *Lucania parva*) showed significantly lower activities at temperatures below 11°C. Five of these species also showed significant tendencies to seek refugia in thick vegetation or mud at low temperatures, and many also showed escape behaviors at greatly increased approach distances compared with warm temperatures. Each of these effects apparently contributes to a scarcity of fish during cold weather, which probably has profound implications for food intake, and hence reproductive decisions, by both sight- and tactile-foraging wading birds.

The disruptive effect of low temperatures on breeding may be modified by a number of factors. For example, no abandonments of estuarine-foraging roseate spoonbills were observed in Florida Bay colonies during the extremely cold, dry weather of December 1989 (R. B. Bjork and G. V. N. Powell III, personal communication). Nesting probably continued because most nests contained newly hatched young, which require little food and which stimulate strong brooding behavior and frequent exchange of parental duties by adults. A number of conditions are likely to modify the thermal responses of marsh fishes as well, especially previous experience of the fish with cold weather events (Frederick and Loftus, 1993).

Neither low temperatures nor sustained high winds are likely to explain abandonments by white ibises, tricolored herons, and wood storks, which typically abandoned during the onset of summer rains. Using a multivariate analysis of environmental conditions at the conclusion of successful and unsuccessful white ibis nests, Frederick and Collopy (1989b) found that nest success was related to early nesting and rapid rates of surface water recession. Interestingly, amount of rainfall was not picked as a significant variable, despite the close timing of abandonments with onset of heavy summer rainfall. These results strongly implied that abandonments by white ibises were directly linked to changes in the rate of water level fluctuation, probably by affecting prey availability. Certainly, it appeared that young ibises were not getting very much to eat during the abandonments (Frederick and Collopy, 1988), and adults flew significantly farther in search of food during the period of abandonments. A parsimonious explanation for all these results is that reversals in surface water recession trends make food difficult to obtain for ibises over large parts of the ecosystem, enough so as to result in nesting failure.

Rate of water level recession also seems to predict the timing and numbers of ibis nesting attempts. Kushlan (1976b) gave several convincing examples of large numbers of white ibises initiating breeding apparently in response to artificial or unseasonal water level recessions. Frederick and Collopy (1989b) analyzed nesting records from 31 foci of nesting and found that those with large numbers of nesting attempts (>2000) had significantly faster water recession rates than did those with small numbers of nesting attempts (<2000), but no difference was found in amount of spring rainfall or initial water level. These results suggest that rapid recession rates both stimulate ibis nesting and are associated with nest success.

The Role of Hydrodynamics in Food Availability

The timing of nest initiations and failures in relation to weather-induced water levels therefore appears to be similar for both wood storks (prior to the decade of the 1980s) and white ibises in the Everglades. Both seem likely to begin nesting during periods of rapid surface water recession and, conversely, may delay initiation in response to slow water recession rates. Both abandon nesting in response to relatively small reversals in receding water trend. For both species, the relationship seems to be the immediate result of food availability. Wood storks feed primarily on small fishes (Ogden et al., 1976), which are known to become concentrated during the dry season in pools and depressions (Kushlan, 1974; W. Loftus, unpublished).

Ibises are primarily demersal, tactile foragers (Kushlan, 1978), which feed largely on crayfish (*Procambarus alleni*) and freshwater shrimp (*Palaemonetes paludosus*) in the Everglades (Kushlan and Kushlan, 1975). Fish become an important food item for ibises only during extremely rapidly receding water trends (Kushlan, 1976b). It is not known whether crayfishes become concentrated during receding water in the same fashion as fishes, because they may burrow in place as surface water recedes. While long-term throw trap data suggest that crayfishes do not become concentrated during receding water trends (W. Loftus, personal communication), this result may be confounded by the inability of throw traps to sample crayfishes living within the substrate. Obviously, ibises would not have this problem.

Alternatively, ibis foraging success might depend on receding water simply to expose new areas of appropriate water depth. In this scenario, the need for new foraging areas could be explained through the local depletion of crayfish, both by ibis foraging and burrowing of crayfish. Conversely, during rising water trends, the only new areas of foraging ground of an appropriate depth would be those that had been previously dry for some time. These recently inundated areas are hypothesized to have depauperate crayfish populations and be poor foraging areas as a result of migration, desiccation, predation, or time-dependent mortality of crayfishes in burrows. Note that this hypothesis explains ibis nest initiations during falling water trends and abandonments during rising trends, but through very different mechanisms than for wood storks.

Rising water trends may also result in increased foraging flight distances for several species of waders in the Everglades (Frederick and Collopy, 1988; Bancroft et al., 1991). However, the energetic costs of this increased flight distance are unlikely to be primarily responsible for nest abandonments. In two studies outside the Everglades, ibises have been found to breed successfully while regularly flying distances similar to those flown during periods of abandonments in Everglades colonies (Bateman, 1970; Bildstein et al., 1990). This suggests that long foraging flights are not sufficient to result in the abandonment of nesting. This hypothesis is further borne out by a consideration of flight energetics. Using the energetics of ibis flight derived by Pennycuick and DeSanto (1989) and the energetics of ibis nesting developed by Kushlan (1977), it is possible to calculate the cost of flight as a proportion of the daily energy requirements of a nesting unit of two adults and their young (Figure 26.5). This analysis shows that even at a distance of 40 km from the colony, the costs of round-trip flight account for less than 10% of the daily energy budget for a breeding pair, an amount unlikely to result directly in abandonment. However, the number of additional prey items required to raise young is large (Figures 26.5 and 26.6), and it is possible that time spent in long foraging trips constrains foraging time excessively. As a result, long foraging flights may only be economic when prey at foraging sites are large and quickly obtainable.

So far, only wood storks and white ibises have been considered here. The extent to which these mechanisms of abandonment and dependencies on hydro-dynamics apply to nesting by ardeids (herons and egrets) is not known. Crayfish may make up a substantial proportion of the diets of great egrets, tricolored herons, and little blue herons, especially during the early part of the nesting season (Telfair, 1981; Frederick and Collopy, 1988). This suggests that any crayfish dynamics which affect ibises would also affect ardeids.

Certainly, ardeids are attracted to many of the pools of concentrated fish that attract wood storks, and several ardeid species probably are able to increase their foraging efficiency if fish are concentrated. The density of fishes near colonies seems related to both growth rate and survival of nestling tricolored herons and snowy egrets (P. C. Frohring, unpublished National Park Service report).

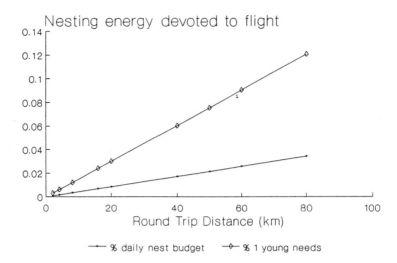

Figure 26.5 Daily energy required by a pair of breeding ibises to fly various distances as a proportion of the total daily nest energy requirements (diamonds) and daily energy requirements of one young (dots). Note that even at a round-trip distance of 80 km, the energy required for flight takes up less than 10% of the energy of either a whole nest or one young.

Little blue herons, however, forage less on fish and more on invertebrates and anurans than any of the other wading bird species (Frederick and Collopy, 1988; Telfair, 1981). This dietary difference may explain why little blue herons were found in several instances breeding successfully in colonies where great egrets, white ibises, and tricolored herons were abandoning (Frederick and Collopy, 1988; Bancroft and Jewell, 1987).

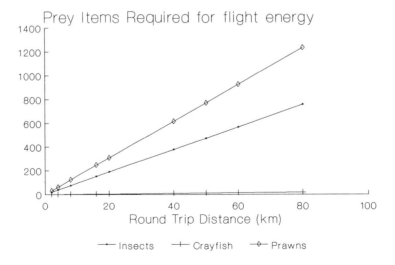

Figure 26.6 Daily energy required by a pair of breeding ibises as a function of distance, expressed as the numbers of prey items required to make up the energy deficit.

A Critique of the Water Recession Model

The preceding information suggests that the rate of water recession in the Everglades is in large part responsible for nest initiations and nest success by white ibises and wood storks and that several of the ardeid species may also benefit from associated increases in availability of prey. However, the notion that wading birds generally benefit from rapid surface water recession is fraught with pitfalls and should be examined in detail, especially since the perceived relationship has strongly influenced past water management plans.

First, a distinction needs to be drawn between surface water recession rates (which concentrate prey) and short-term reversals in falling water (which disperse prey and result in abandonments). Reversals may or may not have a large effect on the calculation of longer term surface water recession rates and in some instances can be considered a separate phenomenon. This suggests that we should be asking whether birds really do well during rapid water recession trends or merely do very poorly during reversals; the latter is the only statistically demonstrated phenomenon (Frederick and Collopy, 1989a; Frederick and Loftus, 1993). Second, the interruption of nesting during reversals cannot be traced to a single mechanism; dispersal of prey is only one, and temperature, wind, and the characteristics of the newly flooded areas have also been shown to strongly affect nesting success.

Finally, there are grounds to believe that the strong association between water recession rate and reproductive success is a relatively recent response to declines in prey populations. Ogden (1994) gives evidence that (1) the only period during which timing and success of wood stork nesting was strongly related to winter and spring water recession rates was the 1970s (the relationship breaks down for earlier and later periods), (2) it has taken successively greater rates of water recession in recent years to produce a successful wood stork nesting year, and (3) the increasing dependence on rapid recession rates has resulted in successively later nesting. There is no reason to believe that water level recession rate is a universal nesting cue for wading birds. While most of the wading bird species nest during the spring months in North America, this is not necessarily a period of receding surface water outside of southern Florida. Indeed, white ibis nesting numbers in coastal South Carolina are directly associated with the amount of rainfall received during the spring season, a response opposite to the association with rapid surface water recession in the Everglades (Bildstein et al., 1990). Water recession rate should therefore be viewed as an indicator of available prey that is peculiar to the freshwater Everglades in its present condition, rather than as an independent and reliable determinant of nesting success.

The alternative explanation for the (probably increasing) dependence on water recession rate, as suggested by J. C. Ogden (personal communication), is that the prey base in the Everglades has decreased drastically, both in areal extent and in densities of prey animals, over the last 30 years. As prey animals become scarcer, it would be expected that wading birds of all hunting guilds would become increasingly dependent on mechanical concentration of prey for reproductive foraging. In this scenario, the increasing dependence on water recession rate can be seen more as an indicator of the degradation of the prey base, rather than as a historically necessary component of nesting. Given more robust prey populations, wading birds might be far less constrained by weather, hydrological variables, and season in their breeding attempts.

Factors Affecting Survival of Fledglings

The preceding discussion has focused largely on effects on initiation and success of nesting attempts, which end with fledging by definition. It should be remembered that first-year mortality in juvenile ciconiiforms is likely to be high and that several species may not breed until their second spring (Palmer, 1962; Bent, 1926). J. C. Ogden (personal communication) estimated over 50% mortality among first-year wood storks during years with poor foraging conditions. Kahl (1963) found over 75% first-year mortality in great egrets, and G. V. N. Powell III and R. B. Bjork (unpublished) found that 86% of great blue herons (*Ardea herodias*) radio-tagged in Florida Bay died within their first year of life. An analysis of over 150 band recoveries of white ibises estimated first-year mortality to be 62% (P. C. Frederick, unpublished data).

In general, fledglings are relatively inefficient foragers. Bildstein (1984) found a steady increase in foraging success with age among first-year, second-year, and adult white ibises foraging in the same place and at the same time. Similarly, Recher and Recher (1969) and G. V. N. Powell III (unpublished) found that juvenile little blue herons and great blue herons, respectively, had significantly lower foraging success than adults. It is likely that young ciconiiforms will require excellent foraging conditions to compensate for their developing abilities, perhaps as dense as is required for nesting. Rodgers and Nesbitt (1979) hypothesized that the high prey densities in the immediate vicinity of colonies were important for the development of juvenile heron foraging skills and critical for the survival of young. Certainly fledglings begin foraging on their own directly under nest trees and gradually expand their hunting to nearby marshes surrounding the colony (Rodgers and Nesbitt, 1979; Rudegeair, 1975; DeSanto et al., 1990; P. C. Frederick, personal observation).

Very few studies have focused on post-fledging survival, despite the fact that mortality during this period appears to have a very large effect on recruitment. It seems that an understanding of the factors which affect survival of juveniles is every bit as important as an understanding of factors affecting nesting.

Summary of Factors Affecting Reproductive Success

In summary, there is reasonable evidence to propose that nesting success of wading birds in the Everglades has not been limited severely by human disturbance, a depauperate potential nesting population, or lack of suitable colony sites. In addition, it is probable that in the absence of other stresses, predation of nest contents usually is a relatively minor component of nesting success. The majority of reproductive failures instead appear to be associated with the availability and quality of food and with disease agents. For wood storks and white ibises, both the initiation and abandonment of nesting can be associated indirectly with the recession rate of surface water in Everglades marshes. This relationship is almost certainly the result of increased prey availability as a result of entrapment during drying phases, although in the case of the white ibis, the mechanism by which prey are affected is not clear. Tricolored herons, snowy egrets, and great egrets all show some susceptibility to abandon nesting during winter storms and reversals in drying trends, although the mechanism may not entirely be related to dilution of prey. Rising water levels do not apparently contribute to the abandonments in great egrets, for example. Alternatively, both sustained high winds and particularly cooler temperatures during winter storms make prey quite scarce, and foraging difficult for most wading bird species, and may lead directly to reproductive failure.

While surface water recession is undoubtedly a contributing factor to nesting effort and nesting success in the present and historic Everglades, this varies with species, and there are many weak points in the association between recession rate and reproductive success. The underlying mechanisms controlling success and abandonment of nests instead are through food availability.

The effect of eustrongylidosis on nesting success is known to be devastating at some colonies and can only become worse with increasing human populations and associated waste disposal problems. Contamination with mercury is a fact for wading birds foraging in the central freshwater marshes of the Everglades. The available literature suggests that mercury contamination could have a very large effect on reproduction, through impairment of foraging and reproductive skills, as well as survival.

EFFECTS of CHANGES in SPATIAL and TEMPORAL PATTERNS in the EVERGLADES on RESILIENCE of WADING BIRD REPRODUCTION

Utilization of a progression of foraging habitats by wading birds seems necessary to supply food throughout the breeding season (Kushlan, 1977; Kushlan, 1986; Bancroft et al., 1991), and the asynchronicity of food availability in these habitats is certainly a key factor in supporting reproduction (Holling et al., 1994). Browder (1978) estimates that up to 30% of wading bird foraging habitat in the Everglades has been lost due to drainage and habitat alteration, and these losses have been concentrated in marshes at the two ends of the hydroperiod spectrum. Short-hydroperiod marshes have been severely reduced in the eastern "finger glades" and wet prairies. Vast areas of the longest hydroperiod marshes have been completely drained along the western side of Miami. In addition, the salinization of the mangrove estuary (Walters et al., 1992) has probably reduced the value of that habitat for use by wading birds during the early part of the breeding season. Thus the habitats that are likely to be used very early and late in the breeding season are those that have been affected the most. In addition, these habitats are also the ones that are most likely to be used during generally very wet and dry years. Therefore, the loss of these habitats has probably put a real constraint on the timing and success of reproduction, as well as on the ability of birds to breed during all but the most favorable of climatic conditions. The problem of constricted timing may be joined by a second effect of reduction of spatial scale. The oligotrophic nature of the Everglades means that accumulations of aquatic organisms can only be derived from vast areas (Davis et al., 1994). Although the present system is certainly large enough to accumulate the energy needed for large populations of nesting birds (Frederick and Powell, 1994; Loftus and Eklund, 1994), the number, size, and consistency of food concentrations will increasingly depend upon hydropatterns as the system grows effectively smaller.

Compartmentalization and controlled water flow have probably had large effects on the prey of wading birds. The compartmentalization of long- (Water Conservation Areas) and short- (much of Everglades National Park, Northeast Shark River Slough, Big Cypress National Preserve) hydroperiod marshes has had measurable effects on aquatic fauna populations (Loftus and Eklund, 1994), and the barriers between them must serve to inhibit normal emigration and seasonal recolonization. The barrier effect is likely to be most extreme following periods of extended and severe drought, when the only populations of fishes surviving will be found within these long-hydroperiod impoundments bordered by deep canals. If this results in increased recolonization times, large sections of the marsh may be rendered inactive, even well into periods when aquatic fauna populations should be rapidly reproducing.

Reductions in spatial scale and resilience are also likely to affect wading birds in less direct ways. The severe reductions in Florida panther populations that resulted from reduction in spatial scale may lead to increases in small mammal populations, which prey on wading birds. Similarly, reductions in the aquatic prey base may tend to further focus predatory efforts by marsh mammals on wading bird colonies.

The reduction in numbers and changes in locations of colonies of breeding birds has in turn probably had effects on the heterogeneity of aquatic flora and fauna within the marsh, because wading bird colonies are islands of heavy nutrient enrichment within an otherwise highly oligotrophic marsh (Frederick and Powell, 1994).

Decreased resilience in wading bird reproduction in the Everglades cannot be seen as an effect isolated to southern Florida. The Everglades is an important overwintering site and concentration point for the breeding population of wading birds in the southeastern United States (Bancroft et al., 1992) and poor conditions for breeding could easily extend to the wintering population if unchecked. Further, the Everglades until recently hosted a substantial part of the southeastern United States breeding population of wading birds (Ogden, 1978; Kushlan, 1977; Osborn and Custer, 1977; Nesbitt et al., 1982), and it is not clear that the new nesting locations farther north can take the place of the Everglades. Every wading bird species that was numerically important in the Everglades now has the distinction of having some kind of federal or state protected status (Florida Game and Fresh Water Fish Commission, 1993). The white ibis, for example, remains the most numerous wading bird in Florida, yet has decreased so markedly as to warrant a suggested threatened status (Frederick, in review). Just as a variety of habitats and sites within the Everglades were important to the resilience of wading bird breeding, the Everglades was probably a very important component in the array of sites available for a regional breeding population (Ogden, 1978). Clearly, decreases in local resilience can cascade up to a regional level.

FUTURE DIRECTIONS

Knowledge of the mechanisms by which nesting success is affected by surface water and weather conditions is basic to attempts to manage both the aquatic ecosystem and the migratory populations of birds. Several major holes are apparent in current knowledge. Why are various aspects of ibis reproduction so strongly associated with surface water recession rate if their main prey, crayfishes, do not become concentrated by falling water? What aspect of little blue heron reproductive ecology allows them to nest successfully during summer rains when nearly all other species are incapable? The effect of the current state of reproduction on demographics, or even the degree of site fidelity shown by the current depauperate nesting population, is not known.

The role that parasites and diseases play in reproductive success has largely been overlooked. How do contaminants such as mercury interact in the susceptibility to diseases and, probably even more important, what role do these play in reproductive effort? The role of anthropogenic factors in producing large numbers of fish infected with eustrongylides larvae is also poorly understood.

There appear to be two possible routes to reconstructing the breeding populations of wading birds in the Everglades. The first is to attempt to thoroughly understand all of the processes through a detailed evaluation of current reproduction. The danger of this method is that we are apparently dealing with a vestigial breeding population in a highly degraded ecosystem that is increasingly unlikely to reflect healthy breeding conditions. Such a situation is likely to breed false leads. The second route is to assume that nearly all species are dependent on relatively high densities of prey for breeding, throughout the entire breeding season, and to focus on aquatic prey dynamics in order to restore the avian populations. The latter seems more plausible, because it takes a broader ecosystematic view and is readily supportable both from what is known and what can be inferred about bird needs, as well as what can be gathered from food web studies.

ACKNOWLEDGMENTS

The authors thank G. T. Bancroft, D. McCrimmon, and J. C. Ogden for insightful comments on drafts of this manuscript and R. Bennetts, M. Collopy, D. Forrester, W. Loftus, and especially J. C. Ogden for contributing central ideas discussed in this chapter. Work in the Everglades has been supported by grants from the Nongame Program of the Florida Game and Fresh Water Fish Commission and the U.S. Army Corps of Engineers and by logistical support from the National Audubon Society, Everglades National Park, and the Florida Cooperative Research Unit of the U.S. Fish and Wildlife Service.

LITERATURE CITED

Baker, R. H. 1940. Crow predation on heron nesting colonies. *Wilson Bull.,* 52:124–125.

Bancroft, G. T. and S. D. Jewell. 1987. Foraging Habitat of *Egretta* Herons Relative to Stage in the Nesting Cycle and Water Conditions, 2nd Annual Report to South Florida Water Management District, West Palm Beach, 174 pp.

Bancroft, G. T., W. Hoffman, and R. Sawicki. 1987. Wading Bird Populations and Distributions in the Water Conservation Areas of the Everglades, Report to the South Florida Water Management District, West Palm Beach, 166 pp.

Bancroft, G. T., J. C. Ogden, and B. W. Patty. 1988. Colony formation and turnover relative to rainfall in the Corkscrew Swamp area of Florida during 1982 through 1985. *Wilson Bull.,* 100:50–59.

Bancroft, G. T., S. D. Jewell, and A. M. Strong. 1991. Foraging and Nesting Ecology of Herons in the Lower Everglades Relative to Water Conditions, Final Report to South Florida Water Management District, West Palm Beach, 156 pp.

Bancroft, G. T., W. Hoffman, and R. Sawicki. 1992. The importance of the Water Conservation Areas in the Everglades to the endangered wood stork (*Mycteria americana*). *Conserv. Biol.,* 6:392–398.

Bateman, D. L. 1970. Movement-Behavior in Three Species of Colonial-Nesting Wading Birds: A Telemetric Study, Ph.D. dissertation, Auburn University, Auburn, Ala., 233 pp.

Bent, A. C. 1926. Life histories of North American marsh birds. *U.S. Nat. Mus. Bull.,* No. 135, 490 pp.

Bigler, W. J., R. G. McLean, and H. A. Trevino. 1973. Epizootiologic aspects of raccoon rabies in Florida. *Am. J. Epidemiol.,* 98:326–335.

Bildstein, K. L. 1984. Age-related differences in the flocking and foraging behavior of white ibises in a South Carolina salt marsh. *Colon. Waterbirds,* 6:45–53.

Bildstein, K. L. 1990. Scarlet ibises and salt tolerances of nesting in the Caroni Swamp of Trinidad. *Biol. Conserv.,* 54:61–78.

Bildstein, K. L., W. Post, J. Johnston, and P. Frederick. 1990. Freshwater wetlands, rainfall, and the breeding ecology of white ibises (*Eudocimus albus*) in coastal South Carolina. *Wilson Bull.,* 102:84–98.

Bjork, R. B. and G. V. N. Powell III. 1990. Studies of Wading Birds in Florida Bay: A Biological Assessment of the Ecosystem, Comprehensive report to the Elizabeth Ordway Dunn Foundation, National Audubon Society, New York.

Black, B. B., M. W. Collopy, H. F. Percival, A. A. Tiller, and P. G. Bohall. 1984. Effects of Low Level Military Training Flights on Wading Bird Colonies in Florida, Tech. Rep. #7, Florida Cooperative Fish and Wildlife Research Unit, School of Forest Resources and Conservation, University of Florida, Gainesville.

Borg, K. S. 1969. Alkylmercury poisoning in terrestrial Swedish wildlife. *Viltrevy,* 6:301–379.

Brand, C. J. 1984. Avian cholera in the central and Mississippi flyways during 1979–80. *J Wildl. Manage.,* 48:399–406.

Browder, J. A. 1978. A modeling study of water, wetlands and wood storks. in *Wading Birds,* A. Sprunt IV, J. C. Ogden, and S. Winckler (Eds.), Res. Rep. #7, National Audubon Society, New York, pp. 325–346.

Burger, J. 1982. The role of reproductive success in colony site selection and abandonment in black skimmers (*Rhynchops niger*). *Auk,* 99:109–115.

Burger, J. and C. Hahn. 1977. Crow predation on black-crowned night heron eggs. *Wilson Bull.,* 89:350–351.

Christy, R. L., K. L. Bildstein, and P. DeCoursey. 1981. A preliminary analysis of energy flow in a South Carolina salt marsh: Wading birds. *Colon. Waterbirds,* 4:96–103.

Clark, E. B. 1978. Factors affecting the initiation and success of nesting in an east-central Florida wood stork colony. *Colon. Waterbirds,* 2:178–188.

Clark, A. B. and D. S. Wilson. 1981. Avian breeding adaptations: Hatching asynchrony, brood reduction, and nest failure. *Q. Rev. Biol.,* 66:253–277.

Conti, J. A., D. J. Forrester, and R. T. Paul. 1986. Parasites and diseases of reddish egrets (*Egretta rufescens*) from Texas and Florida. *Trans. Am. Microsc. Soc.,* 105:79–82.

Coulter, M. C. 1987. Wood Storks of the Birdsville Colony and Swamps of the Savannah River Plant, 1986 Annual Report to Savannah River Plant, Savannah River Ecology Lab, U.S. Department of Energy, Aiken, S.C.

Custer, T. W. and R. G. Osborn. 1977. Wading birds as biological indicators: 1975 colony survey. *U.S. Fish Wildl. Serv. Spec. Sci. Rep. Wildl.,* No. 206.

Custer, T. W., G. L. Hensler, and T. E. Kaiser. 1984. Clutch size, reproductive success, and organochlorine contaminants in Atlantic Coast black-crowned night-herons. *Auk,* 100:699–710.

Davis, S. M., L. H. Gunderson, W. A. Park, J. Richardson, and J. Mattson. 1994. Landscape dimension, composition, and function in a changing Everglades ecosystem. in *Everglades: The Ecosystem and Its Restoration,* S. M. Davis and J. C. Ogden (Eds.), St. Lucie Press, Delray Beach, Fla., chap. 17.

DeSanto, T. L., S. G. McDowell, and K. L. Bildstein. 1990. Plumage and behavioral development of nestling white ibises. *Wilson Bull.,* 102:226–238.

Dusi, J. C. and R. T. Dusi. 1968. Ecological factors contributing to nesting failure in a heron colony. *Wilson Bull.,* 89:456–458.

Fimreite, N. 1971. Effects of dietary methylmercury on ring-necked pheasants. *Can. Wildl. Serv. Occas. Pap.,* No. 9.

Fimreite, N. 1974. Mercury contamination of aquatic birds in northwestern Ontario. *J. Wildl. Manage.,* 38:120–131.

Fimreite, N. and L. Karstad. 1971. Effects of dietary methyl mercury on red-tailed hawks. *J. Wildl. Manage.,* 35:293–300.

Finley, M. T. and R. C. Stendell. 1978. Survival and reproductive success of black ducks fed methyl mercury. *Environ. Pollut.,* 16:51–64.

Florida Game and Fresh Water Fish Commission. 1993. Official Lists of Endangered and Potentially Endangered Fauna and Flora in Florida, Tallahassee.

Forrester, D. J., K. C. Wenner, F. H. White, E. C. Greiner, W. R. Marion, J. E. Thul, and G. A. Berkhoff. 1980. An epizootic of avian botulism in a phosphate mine settling pond in northern Florida. *J. Wildl. Dis.,* 16:323–327.

Frederick, P. C. 1985. Extra-Pair Copulations in the Mating System of White Ibis (*Eudocimus albus*), Ph.D. dissertation, University of North Carolina, Chapel Hill.

Frederick, P. C. 1987. Chronic tidally-induced nest failure in a colony of white ibises. *Condor,* 89:413–419.

Frederick, P. C. In review. White ibis (*Eudocimus albus*). Species account. in *Florida Rare and Endangered Biota Series,* H. Kale II (Ed.), Florida Audubon Society, Maitland.

Frederick, P. C. and M. W. Collopy. 1988. Reproductive Ecology of Wading Birds in Relation to Water Conditions in the Florida Everglades, Tech. Rep. No. 30, Florida Cooperative Fish and Wildlife Research Unit, School of Forestry Resources and Conservation, University of Florida, Gainesville.

Frederick, P. C. and M. W. Collopy. 1989a. Researcher disturbance in colonies of wading birds: Effects of frequency of visit and egg-marking on reproductive parameters. *Colon. Waterbirds,* 12:152–157.

Frederick, P. C. and M. W. Collopy. 1989b. Nesting success of five species of wading birds (Ciconiiformes) in relation to water conditions in the Florida Everglades. *Auk,* 106:625–634.

Frederick, P. C. and M. W. Collopy. 1989c. The role of predation in determining reproductive success of colonially-nesting wading birds (Ciconiiformes) in the Florida Everglades. *Condor,* 91:860–867.

Frederick, P. C. and W. Loftus. 1993. Cold-weather behavior of estuarine and freshwater marsh fishes as a determinant of nesting success in wading birds in the Everglades of Florida. in *Estuaries,* 16:216–222.

Frederick, P. C. and G. V. N. Powell. 1994. Nutrient transport by wading birds in the Everglades. in *Everglades: The Ecosystem and Its Restoration,* S. M. Davis and J. C. Ogden (Eds.), St. Lucie Press, Delray Beach, Fla., chap. 23.

Frederick, P. C., R. B. Bjork, G. V. N. Powell IV, and G. T. Bancroft. 1992. Comparative reproductive parameters of herons in three southern Florida habitats. *Colon. Waterbirds,* 15:192–201.

Frederick, P. C., M. G. Spalding, and G. V. N. Powell. 1993. An evaluation of methods for measuring nestling survival in colonially-nesting tricolored herons (*Egretta tricolor*). *J. Wildl. Manage.,* 57:34–41.

Girard G. T. and W. K. Taylor. 1979. Reproductive parameters for nine avian species at Moore Creek, Merritt Island National Wildlife Refuge, Florida. *Fla. Sci.,* 42:94–102.

Goldberg, D. R., T. M. Yuill, and E. C. Burgess. 1990. Mortality from duck plague virus in immunosuppressed adult mallard ducks. *J. Wildl. Dis.,* 26:299–306.

Gunderson, L. H. 1994. Vegetation of the Everglades: Determinants of community composition. in *Everglades: The Ecosystem and Its Restoration,* S. M. Davis and J. C. Ogden (Eds.), St. Lucie Press, Delray Beach, Fla., chap. 13.

Hammatt, R. B. 1981. Reproductive Biology in a Louisiana Heronry, M.Sc. thesis, Louisiana State University, Baton Rouge, 93 pp.

Harris, L. D. and T. E . O'Meara. 1989. Changes in Southeastern Bottomland Forests and Impacts on Vertebrate Fauna. in Freshwater Wetlands and Wildlife, R. R. Sharitz and J. W. Gibbons (Eds.), DE90005384, National Technical Information Service, Springfield, Va., pp. 755–772.

Hensler, G. L. and J. D. Nichols. 1981. The Mayfield method of estimating nesting success: A model, estimators, and simulation results. *Wilson Bull.,* 93:42–53.

Hoff, G. L., W. J. Bigler, S. J. Proctor, and L. P. Stallings. 1974. Epizootic of canine distemper virus infection among urban raccoons and gray foxes. *J. Wildl. Dis.,* 10:423–428.

Hoffman, W., G. T. Bancroft, and R. W. Sawicki. 1991. Wading Bird Populations and Distribution in the Water Conservation Areas of the Everglades: 1985–1988, Report to South Florida Water Management District, West Palm Beach, 173 pp.

Holling, C. S., L. H. Gunderson, and C. J. Walters. 1994. The structure and dynamics of the Everglades system. in *Everglades: The Ecosystem and Its Restoration,* S. M. Davis and J. C. Ogden (Eds.), St. Lucie Press, Delray Beach, Fla., chap. 29.

Jasmin, A. M., D. E. Cooperrider, C. P. Powell, and J. N. Baucom. 1972. Enterotoxemia of wildfowl due to *Cl. perfringens* type C. *J. Wildl. Dis.,* 8:79–84.

Jenni, D. A. 1969. A study of the ecology of four species of herons during the breeding season at Lake Alice, Alachua County, Florida. *Ecol. Mongr.,* 39:245–270.

Johnston, J. and K. L. Bildstein. 1990. Dietary salt as a physiological constraint in white ibises breeding in an estuary. *Phys. Zool.,* 63:190–207.

Kahl, M. P., Jr. 1963. Mortality of common egrets and other herons. *Auk,* 80:295–300.

Kahl, M. P. 1964. Food ecology of the wood stork (*Mycteria americana*). *Ecol. Monogr.,* 34: 97–117.

King, K. A., D. R. Blankinship, and R. T. Paul. 1977. Ticks as a factor in the 1975 nesting failure of Texas brown pelicans. *Wilson Bull.,* 89:157–158.

Klukas, R. W. and L. N. Locke. 1970. An outbreak of fowl cholera in Everglades National Park. *J. Wildl. Dis.,* 6:77–79.

Krebs, J. R. 1978. Colonial nesting in birds, with special reference to the ciconiiformes. in *Wading Birds,* A. Sprunt IV, J. C. Ogden, and S. Winckler (Eds.), Res. Rep. #7, National Audubon Society, New York.

Kushlan, J. A. 1974. The Ecology of the White Ibis in Southern Florida: A Regional Study, Ph.D. dissertation, University of Miami, Coral Gables, Fla.

Kushlan, J. A. 1976a. Wading bird predation in a seasonally fluctuating pond. *Auk,* 93:86–94.

Kushlan, J. A. 1976b. Site selection for nesting colonies by the American white ibis (*Eudocimus albus*) in Florida. *Ibis,* 118:590–593.

Kushlan, J. A. 1977. Population energetics in the American white ibis. *Auk,* 94:114–122.

Kushlan, J. A. 1978. Feeding ecology of wading birds. in *Wading Birds,* Res. Rep. #7, A. Sprunt IV, J. C. Ogden, and S. Winckler (Eds.), National Audubon Society, New York.

Kushlan, J. A. 1979. Effects of helicopter censuses on wading bird colonies. *J. Wildl. Manage.,* 43:756–760.

Kushlan, J. A. 1986. Responses of wading birds to seasonally fluctuating water levels: Strategies and their limits. *Colon. Waterbirds,* 9:155–162.

Kushlan, J. A. and P. C. Frohring. 1986. The history of the southern Florida wood stork population. *Wilson Bull.,* 98:368–386.

Kushlan, J. A. and M. S. Kushlan. 1975. Food of the white ibis in southern Florida. *Fla. Field Nat.,* 3:31–38.

Kushlan, J. A. and D. A. White. 1977. Nesting wading bird populations in southern Florida. *Fla. Sci.,* 40:65–72.

Kushlan, J. A., J. C. Ogden, and J. L. Tilmant. 1975. Relation of Water Level and Fish Availability to Wood Stork Reproduction in the Southern Everglades, Florida, Open File Rep. 75-434, U.S. Geological Survey.

Kushlan, J. A., P. C. Frohring, and D. Vorhees. 1984. History and Status of Wading Birds in Everglades National Park, National Park Service Report, South Florida Research Center, Everglades National Park, Homestead, Fla., 110 pp.

Lack, D. 1968. *Ecological Adaptations for Breeding in Birds,* Methuen, London, 409 pp.

Lopinot, A. C. 1951. Raccoon predation on the great blue heron, *Ardea herodias. Auk,* 68:235.

Loftus, W. F. and A.-M. Eklund. 1994. Long-term dynamics of an Everglades small-fish assemblage. in *Everglades: The Ecosystem and Its Restoration,* S. M. Davis and J. C. Ogden (Eds.), St. Lucie Press, Delray Beach, Fla., chap. 19.

Maxwell, G. R. II and H. W. Kale II. 1977. Breeding biology of five species of herons in coastal Florida. *Auk,* 94:689–700.

Mayfield, H. F. 1961. Nesting success calculated from exposure. *Wilson Bull.,* 73:255–261.

Mayfield, H. F. 1975. Suggestions for calculating nest success. *Wilson Bull.,* 87:456–466.

McVaugh, W. 1972. The development of four North American herons. *Living Bird,* 11:155–173.

Menisk, J. G. and R. B. Botzler. 1989. Epizootiological features of avian cholera on the north coast of California. *J. Wildl. Dis.,* 25:240–245.

Milstein, P. L., I. Prestt, and A. A. Bell. 1970. The breeding cycle of the grey heron. *Ardea,* 58: 171–257.

Mock, D. W. 1984. Siblicidal aggression and resource monopolization in birds. *Science,* 225: 731–733.

Mock, D. W., T. C. Lamey, and B. J. Ploger. 1987. Proximate and ultimate roles of food amount in regulating egret sibling aggression. *Ecology,* 68:1760–1772.

Morrison, M. L. 1986. Bird populations as indicators of environmental change. in *Current Ornithology,* Vol. 3, R. L. Johnston (Ed.), Plenum Press, Chicago.

Nesbitt, S. A., J. C. Ogden, H. W. Kale II, B. W. Patty, and L. A. Rowse. 1982. Florida Atlas of Breeding Sites for Herons and Their Allies: 1976–78, FWS/OBS-81/49, U.S. Fish and Wildlife Service, 449 pp.

Nisbet, I. C. T. 1975. Selective effects of predation in a tern colony. *Condor,* 77:221–226.

Ogden, J. C. 1978. Recent population trends of colonial wading birds on the Atlantic and Gulf coastal plains. in *Wading Birds,* Res. Rep. #7, A. Sprunt IV, J. C. Ogden, and S. Winckler (Eds.), National Audubon Society, New York, pp. 135–153.

Ogden, J. C. 1991. Nesting by wood storks in natural, altered, and artificial wetlands in central and northern Florida. *Colon. Waterbirds,* 14:39–45.

Ogden, J. C. 1994. A comparison of wading bird nesting colony dynamics (1931–1945 and 1974–1989) as an indication of ecosystem conditions in the southern Everglades. in *Everglades: The Ecosystem and Its Restoration,* S. M. Davis and J. C. Ogden (Eds.), St. Lucie Press, Delray Beach, Fla., chap. 22.

Ogden, J. C., W. B. Robertson, G. E. Davis, and T. W. Schmidt. 1974. Pesticides, Polychlorinated Biphenyls and Heavy Metals in Upper Food Chain Levels, Everglades National Park and Vicinity, U.S. Department of the Interior.

Ogden, J. C., J. A. Kushlan, and J. T. Tilmant. 1976. Prey selectivity by the wood stork. *Condor,* 78:324–330.

Ogden, J. C., H. W. Kale II, and S. A. Nesbitt. 1980. The influence of annual rainfall and water levels on nesting by Florida populations of wading birds. *Trans. Linn. Soc. N.Y.,* 9:115–126.

Ogden, J. C., D. A. McCrimmon, Jr., G. T. Bancroft, and B. W. Patty, 1987. Breeding populations of the wood stork in the southeastern United States. *Condor,* 89:752–759.

Ohlendorf, H. M., E. E. Klass, and T. E. Kaiser. 1978. Environmental pollutants and eggshell thinning in the black-crowned night-heron. in *Wading Birds,* Res. Rep. #7, A. Sprunt IV, J. C. Ogden, and S. Winckler (Eds.), National Audubon Society, New York, pp. 63–82.

Ohlendorf, H. M., D. M. Swineford, and L. N. Locke. 1981. Organochlorine residues and mortality of herons. *Pestic. Monit. J.,* 14:125–135.

Osborn, R. G. and T. W. Custer. 1977. Herons and Their Allies: Atlas of Atlantic Coast Colonies, 1975 and 1976, Biological Report FWS/OBS 77/08, U.S. Fish and Wildlife Service.

Palmer, R. S. 1962. *Handbook of North American Birds,* Vol. 1. Yale University Press, New Haven, Conn.

Parsons, K. C. and J. Burger. 1982. Human disturbance and nestling behavior in black-crowned night herons. *Condor,* 84:184–187.

Pennycuick, C. J. and T. De Santo. 1989. Flight speeds and energy requirements for white ibises on foraging flights. *Auk,* 106:141–143.

Pierce, C. W. 1962. The cruise of the bonton. *Tequesta,* 22:3–63.

Post, W. 1990. Nest survival in a large ibis-heron colony during a 3-year decline to extinction. *Colon. Waterbirds,* 13:50–61.

Powell, G. V. N. 1983. Food availability and reproduction by great white herons *Ardea herodias:* A food addition study. *Colon. Waterbirds,* 6:139–147.

Powell, G. V. N., R. D. Bjork, J. C. Ogden, R. T. Paul, A. H. Powell, and R. B. Robertson, Jr. 1989. Population trends in some Florida Bay wading birds. *Wilson Bull.,* 101:436–457.

Powell, G. V. N., J. W. Fourqurean, W. J. Kenworthy, and J. C. Zieman. 1991. Bird colonies cause seagrass enrichment in a subtropical estuary. *Estuarine Coastal Shelf Sci.,* 32:567–579.

Pratt, H. W. 1972. Nesting success of common egrets and great blue herons in the San Francisco Bay region. *Condor,* 74:447–453.

Recher, H. F. and J. A. Recher. 1969. Comparative foraging efficiency of adult and immature little blue herons (*Florida caerulea*). *Anim. Behav.,* 17:320–322.

Rocke, T. E. and M. D. Samuel. 1991. Effects of lead shot ingestion on selected cells of the mallard immune system. *J. Wildl. Dis.,* 27:1–9.

Rodgers, J. A., Jr. 1978. Breeding behavior of the Louisiana heron. *Wilson Bull.,* 90:45–59.

Rodgers, J. A., Jr. 1980a. Little blue heron breeding behavior. *Auk,* 97:371–384.

Rodgers, J. A., Jr. 1980b. Reproductive success of three heron species on the west coast of Florida. *Fla. Field. Nat.,* 8:37–40.

Rodgers, J. A., Jr. 1987. On the antipredator advantages of coloniality: A word of caution. *Wilson Bull.,* 99:269–270.

Rodgers, J. A., Jr. and S. A. Nesbitt. 1979. Feeding energetics of herons and ibises at breeding colonies. *Colon. Waterbirds,* 3:128–132.

Roffe, T. J. 1988. *Eustrongylides* sp. epizootic in young common egrets (*Casmerodius albus*). *Avian Dis.,* 32:143–147.

Rosen, M. N. and A. I. Bischoff. 1949. The 1948–49 outbreak of fowl cholera in birds in the San Francisco Bay area and surrounding counties. *Calif. Fish Game,* 35:185.

Rudegeair, T. J. 1975. The Reproductive Behavior and Ecology of the White Ibis (*Eudocimus albus,* Ph.D. dissertation, University of Florida, Gainesville.

Sherrod, S. K. 1978. Diets of North American Falconiformes. *Raptor Res.,* 12:49–121.

Shields, M. H. and J. Parnell. 1986. Fish crow predation on eggs of the white ibis at Battery Island, North Carolina. *Auk,* 103:531–539.

Simmons, E. M. 1959. Observations on effects of cold weather on nesting common egrets. *Auk,* 76:239–241.

Snyder, N. F., J. C. Ogden, J. D. Bittner, and G. A. Grau. 1984. Larval dermestid beetles feeding on nestling snail kites, wood storks, and great blue herons. *Condor,* 86:170–174.

Spaan, J. W., R. G. Heath, J. F. Kreithzer, and L. N. Locke. 1972. Ethyl mercury *p*-toluene sulfonanilide: Lethal and reproductive effects of pheasants. *Science,* 175:328–331.

Spalding, M. G. 1990. Antemortem diagnosis of eustrongylidosis in wading birds (Ciconiiformes). *Colon. Waterbirds,* 13:75–77.

Spalding, M. G. and D. J. Forrester. 1993. Pathogenesis of *Eustrongylides ignotus* (Nematoda; Dioctophymatoidea) in ciconiiformes. *J. Wildl. Dis.,* 29:250–260.

Spalding, M. G. and J. C. Woodard. 1992. Chondrosarcoma in a wild great white heron from southern Florida. *J. Wildl. Dis.,* 28:151–153.

Spalding, M. G., D. J. Forrester, and G. T. Bancroft. 1993. The epizootiology of *Eustrongylides ignotus* in wading birds (Ciconiiformes) in Florida. *J. Wildl. Dis.,* 29:237–249.

Sprunt, A., Jr. 1922. Discovery of the breeding of the white ibis in South Carolina. *Oologist,* 39:142–144.

Sprunt, A., Jr. 1944. Northward extension of the breeding range of the white ibis. *Auk,* 61: 144–145.

Stinner, D. H. 1983. Colonial Wading Birds and Nutrient Cycling in the Okeefenokee Swamp, Ph.D. dissertation, University of Georgia, Athens.

Sunlof, S. F., M. G. Spalding, J. Wentworth, and C. Steible. In review. Mercury in livers of wading birds (Ciconiiformes) in southern Florida. *Arch. Environ. Contam. Toxicol.*

Taylor, R. J. and E. D. Michael. 1971. Predation on an inland heronry in eastern Texas. *Wilson Bull.,* 83:172–177.

Telfair, R. C. 1981. Cattle egrets, inland heronries and the availability of crayfish. *Southwest. Nat.,* 26:37–41.

Telford, S. R., M. G. Spalding, and D. J. Forrester. 1992. Hemoparasites of wading birds (Ciconiiformes) in Florida. *Can. J. Zool.,* 70:1397–1408.

Temple, S. A. and J. A. Wiens. 1989. Bird populations and environmental changes: Can birds be bio-indicators?. *Am. Birds,* Summer:260–270.

Tremblay, J. and L. N. Ellison. 1979. Effects of human disturbance on breeding of black-crowned night herons. *Auk,* 96:364–369.

Trust K. A., M. W. Miller, J. K. Ringelman, and I. M. Orme. 1990. Effects of ingested lead on antibody production in mallards (*Anas platyrhynchos*). *J. Wildl. Dis.,* 26:316–322.

Van der Molen, E. J., A. A. Blok, and G. J. DeGraaf. 1982. Winter starvation and mercury intoxication in grey herons (*Ardea cinerea*) in the Netherlands. *Ardea,* 70:173–184.

Walters, C., L. Gunderson, and C. S. Holling. 1992. Experimental policies for water management in the Everglades. *Ecol. Appl.,* 2:189–202.

Werschkul, D. F. 1979. Nestling mortality and the adaptive significance of early locomotion in the little blue heron. *Auk,* 96:116–130.

Wiese, J. H. 1975. The Reproductive Biology of the Great Egret *Casmerodius albus egretta* (Gmelin), M.S. thesis, Florida State University, Tallahassee, 82 pp.

Wiese, J. H. 1977. Large scale mortality of nestling ardeids: A summary. *Colon. Waterbirds,* 2: 163–164.

Wittenberger, J. F. and G. L. Hunt, Jr. 1985. The adaptive significance of coloniality in birds. in *Avian Biology,* Vol. VIII, D. S. Farner, J. R. King, and K. C. Parkes (Eds.), Academic Press, New York, pp. 1–78.

27

LANDSCAPE,
WHITE-TAILED DEER, and
the DISTRIBUTION of
FLORIDA PANTHERS in
the EVERGLADES

Tommy R. Smith

Oron L. Bass, Jr.

ABSTRACT

Data gathered on the distribution and abundance of Florida panthers and white-tailed deer in Everglades National Park between December 1986 and October 1989 were coupled with satellite imagery to evaluate the effect of landscape and prey on the distribution of panthers. Combined home ranges of six panthers radio-tagged in the park covered over 2780 km², extending west into Big Cypress National Preserve and east into undeveloped disturbed lands. All panthers utilized upland forest (either continuous pinelands or hammocks in open prairie) more than expected by chance. Wet open landscapes and mangrove forest were used less than expected. Young and nonbreeding panthers utilized open, disturbed landscape outside the park, whereas resident breeding females occupied relatively small insular ranges centered on the Miami Rock Ridge, apparently due to the availability of deer and hunting cover at the edge of the forest. Availability of large prey appears to limit the breeding density of panthers in the eastern Everglades and, hence, determines its carrying capacity for panthers, estimated at 5–10. Despite low numbers, productivity and subsequent dispersal of panthers from Everglades National Park may contribute substantially to the size and genetic diversity of the total wild population in south Florida.

INTRODUCTION

Although Florida panthers (*Felis concolor coryi*) historically occurred through-out the Gulf and Atlantic coastal plains of the southeastern United States (Young and Goldman, 1946), their distribution appeared to have been largely reduced to south Florida by the mid-20th century (Goldman, 1946; Tinsley, 1970). Systematic investigation of observations and field sign subsequently confirmed this assumption and indicated that although reliable sightings of panthers occasionally were reported across much of Florida, established, potentially viable populations of the species appeared to be limited to the Big Cypress Swamp and Everglades in the vicinity and south of Lake Okeechobee (Schemnitz, 1974; Layne and McCauley, 1976; Nowak and McBride, 1976; Belden, 1978; McBride, 1985; but see Belden and Williams, 1976).

Persistence of a panther population in the Big Cypress/Everglades ecosystem appears to be primarily related to the remote, largely protected nature of the area, rather than its quality per se as panther habitat. Early students of the subject maintained that the extirpation of panthers in Florida and elsewhere in their historic range was due more to persecution by man than to loss of habitat (Tinsley, 1970; Layne and McCauley, 1976; Williams, 1978). Recently, cursory assessment of panther sign in south Florida has indicated that human activity (primarily hunting), rather than abundance of prey, may account for local variation in the distribution and abundance of panthers (McBride, 1985). Compared to upland habitats to the north, wetland types of south Florida certainly support low densities of white-tailed deer (*Odocoileus virginianus*) (Harlow and Jones, 1965; Schemnitz, 1974) and wild hogs (*Sus scrofa*), which are the principal large prey of panthers (Belden, 1988; Maehr et al., 1990) and presumably are essential to their fitness (Williams, 1978; Roelke et al., 1986). Based on the acreage of undeveloped land in south Florida (Belden, 1988) and current population estimates, the density of panthers in the Everglades is low compared to other areas in the range of mountain lions (Anderson, 1983).

It is unlikely that any significant portion of the panther's former range will be recovered, at least in the foreseeable future. There is optimism that the species can be re-established in a few "suitable" areas of central and northern Florida (Belden, 1986), if existing efforts to protect and maintain these habitats are increased. Presently, however, the only realistic theater for a wild panther population is the Big Cypress/Everglades ecosystem. Is this finite area adequate to sustain a popu-lation of panthers in the long term? The general notion left by historical accounts that the Everglades is prime panther habitat appears dubious. Quantitative infor-mation on the carrying capacity of the area for panthers and how it varies in space and time is lacking. Assuming that the current wild panther population includes 50 or fewer individuals (Maehr, 1990) and given the theoretical consequences of small population size (Frankel and Soule, 1981), such information is essential in order to evaluate the potential dynamics of the extant wild population and, in turn, develop conservation strategies for the species.

Observations of radio-tagged panthers in Everglades National Park indicated that their distribution across the landscape was not random and that it varied by sex, age, and reproductive status. In this chapter, preliminary radiotelemetry data on panthers are coupled with satellite imagery of Everglades National Park and adjacent lands to determine the effects of landscape on the distribution and abundance of panthers in the area. Data from a companion study of white-tailed deer in Everglades National Park are then superimposed to determine whether availability of prey may explain observed relationships between landscape and the distribution, as well as fitness, of individual panthers. This, in turn, leads to consideration of the carrying capacity of the area for panthers and the potential contribution of Everglades National Park and the East Everglades to long-term conservation of a wild panther population in south Florida.

STUDY AREA

Field studies of Florida panthers and white-tailed deer initially were focused in a 525-km^2 core area in eastern Everglades National Park, with Long Pine Key (the southern terminus of the Atlantic Coastal Ridge) at its center (Figure 27.1). However, the overall study area, which was defined by the ranges of panthers captured in the core area, included most of Everglades National Park, the East Everglades Wildlife Management Area and adjacent lands east of the park, and the southeastern portion of Big Cypress National Preserve along the western boundary of the park.

Everglades National Park encompasses approximately the southernmost 300,000 ha of mainland Florida. Climate of the region is subtropical, with distinct wet and dry seasons. Mean annual rainfall at Long Pine Key is 146 cm, approximately 80%

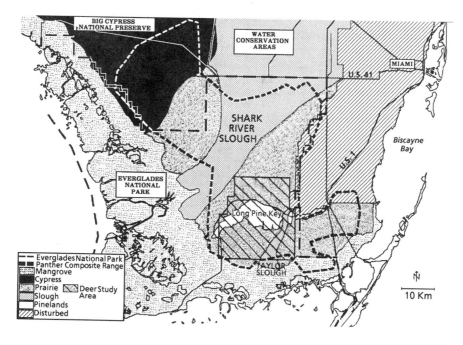

Figure 27.1 Location of Florida panther and white-tailed deer study area in Everglades National Park and adjacent lands in south Florida, December 1986–October 1989.

of which is relatively evenly distributed between May and October. Temperature roughly tracks rainfall, gradually rising and falling between a low monthly mean of 18.5°C in December/January and a high of 27°C in July/August.

Topography of the area is extremely flat. Elevation ranges from approximately 2 m on Long Pine Key to sea level at Florida Bay. Soils are poorly drained peat, marl, or sand underlain by limestone. With the exception of pinelands and hardwood hammocks, which constitute less than 5% of the terrestrial acreage of the park, surface water typically is present across much of the study area between June and November.

Among 16 vegetation types recognized by Davis (1943) in the Everglades, 5 are prominent in the present study area, including:

1. *Miami rockland pine forest* dominated by slash pine (*Pinus elliottii* var. *densa*).

2. *Marsh-prairie,* including the southern Everglades type adjacent to Shark River Slough and the southern coast type between the Miami Rock Ridge and the mangrove zone. These relatively xeric areas are dominated by grasses and sedges (largely *Muhlenbergia filipes* and *Cladium jamaicense,* respectively) interspersed with cypress domes, hardwood hammocks, and bayheads.

3. *Slough,* essentially Shark River Slough and Taylor Slough, which are freshwater marshes, dominated by herbaceous species and subject to relatively long hydroperiods.

4. *Cypress forest,* including bald cypress (*Taxodium distichum*) forests, domes, strands, and prairies that occur on relatively mesic sites.

5. *Mangrove swamp and coastal marsh,* which is brackish wetland associated with estuarine shorelines. This type includes both mangrove forest and herbaceous/shrubby salt marshes, the latter typically occurring as a transition zone between mangrove and freshwater marsh.

Additional descriptions of these and other vegetation associations in Everglades National Park are provided by Davis (1943), Craighead (1971), Gleason (1984), and Myers and Ewel (1990).

METHODS

Field Procedures

Panthers were located and treed using trained hounds; they were immobilized by injection of Ketaset® (ketamine hydrochloride) and Rompun® (xylazine hydrochloride) with a syringe rifle (Telinject USA, Inc., Newhall, Calif.) and were fitted with radio-transmitters affixed to collars (Telonics, Inc., Mesa, Ariz.). Deer were captured using a hand-held netgun (Coda Enterprises, Inc., Mesa, Ariz.) fired from a helicopter (Barrett et al., 1982), restrained without drugs, and equipped with radio-collars. Ages of panthers were estimated based on tooth wear and condition (Shaw, 1979), whereas cementum annuli (Gilbert, 1966) and/or tooth development (Severinghaus, 1949) were used to age deer.

Radio-locations of panthers and deer were made from fixed-wing aircraft (Mech, 1983). With few exceptions, panthers were radio-located daily 1–3 hours after dawn. Deer were located 1–4 times per week, stratifying sampling across daylight hours. Radio-locations were assigned to 1-ha cells to encompass location error, using the Universal Transverse Mercator coordinate system.

In addition to telemetry, information on distribution and relative abundance of deer was obtained by 15 aerial censuses between January 1988 and October 1989. In fixed-wing aircraft at 61 m altitude, transects were flown at 1-km intervals across the 525-km^2 core study area during early morning or late afternoon for 3 consecutive days, and locations of deer observed were assigned to the nearest intersection of Universal Transverse Mercator coordinates.

Classification of Landscape

Digital data on plant communities in south Florida derived from Landsat satellite imagery were obtained from the Florida Department of Transportation. The 22 community (i.e., habitat) types of this classification were grouped into 6 landscape categories based on vegetation structure, hydrology, and land use, closely following the terminology of Davis (1943):

1. *Upland forest pinelands:* Hardwood hammocks and forests and Brazilian pepper (*Schinus terebinthifolius*) forest (i.e., the Hole-in-the-Doughnut)
2. *Lowland forest:* Cypress swamps, hardwood swamps, bay swamps, and shrub swamps
3. *Marsh-prairie:* Southern Everglades and southern coastal types (Davis, 1943)
4. *Slough:* Freshwater marshes and sloughs (i.e., Shark River Slough and Taylor Slough)
5. *Coastal:* Mangrove forests, coastal salt marshes, and mangrove swamps (hereafter referred to as mangrove)
6. *Disturbed:* Urban areas and intensively farmed land or abandoned early successional land

Analytical Procedures

Home ranges (Burt, 1943) of radio-tagged panthers were estimated by the modified minimum area method (Harvey and Barbour, 1965) which, given essentially daily monitoring, appeared to best describe this parameter. Core activity areas of panthers were delineated by the harmonic mean method (Dixon and Chapman, 1980) using the program HOMERANGE (Samuel et al., 1985) and were defined as areas which encompassed 60% of the animals' locations (i.e., the area within the 60% activity area isopleth). Home ranges and distributions of panthers captured as kittens were based on data for the period of independence from their mothers (i.e., subadulthood). Home ranges of deer were estimated by the harmonic mean method using the program TELEM88 (Coleman and Jones, 1988) and were defined as the area within the 80% activity area isopleth.

Distribution of Panthers

A chi-square test for goodness of fit was used to test the null hypothesis that the distribution of panthers in relation to landscape was random (Neu et al., 1974). Simultaneous Bonferroni confidence intervals were constructed for the observed frequency of locations in each landscape category to determine whether usage was greater or less than expected (Byers et al., 1984; White and Garrott, 1990). Hammocks, regardless of size, in open marsh-prairie and slough landscapes were treated in two ways in analyses: (1) as forested landscape and (2) as part of the acreage of open landscapes.

Distribution and Abundance of Deer

The distribution and abundance of deer in the core study area in relation to the forested landscape (i.e., edge) of Long Pine Key were investigated by using two approaches. In the first, the open landscape outside of Long Pine Key was partitioned by drawing contour lines at 1-km intervals from the forest edge. Chi-square analysis was used for an overall comparison of observed and expected numbers of deer observed on aerial censuses in these intervals. Expected numbers were calculated by multiplying the relative area within each interval by the total number of deer observations. Bonferroni confidence intervals (Byers et al., 1984) were constructed to determine whether the observed frequency of observations in each interval was significantly greater or less than expected. In addition, correlation analysis was used to determine if the relative abundance of deer (number observed per unit area) was associated with distance from Long Pine Key.

In a second, less direct approach, correlation analysis was used to determine whether the sizes of home ranges and their proximity to Long Pine Key (i.e., the distance of their geometric centers from the edge of the forest) were associated. This analysis was based on the assumptions that (1) home range size of deer was inversely proportional to habitat quality, (2) abundance of deer was directly proportional to habitat quality, and, therefore, (3) home range size and abundance of deer were inversely correlated (Sanderson, 1966).

RESULTS

Landscape and the Distribution of Panthers

Between December 1986 and October 1989, seven panthers were captured and radio-tagged in Everglades National Park (Table 27.1). Adult females 14 and 15 were captured in December 1986. Two kittens of female 14, a male (16) and a female (21), were captured in January 1987 and March 1987, respectively. Two female kittens (22 and 23) of female 15 also were captured in March 1987. A young female (27), estimated at 2–4 years of age, was captured in April 1988. Following the capture of kittens 22 and 23, kitten 23 did not rejoin her family, and she was intermittently held in captivity until February 1989. Data on this individual were not included in the analysis.

Table 27.1 Home Ranges and Core Activity Areas of Florida Panthers
Captured and Radio-Monitored in Everglades National Park, December 1986–October 1989

ID	Sex	Age	Period of monitoring	Home range[a] (km²)	Core range[b] (km²)
14	F	Adult	Dec. 86–Oct. 89	350	108
15	F	Adult	Dec. 86–Jun. 88[c]	430	150
16	M	Subadult	Apr. 87–Oct. 89	1917	810
21	F	Subadult	Apr. 87–Jul. 88[d]	342	112
22	F	Subadult	Dec. 87–Mar. 89[e]	514	129
27	F	Adult	Apr. 88–Jul. 89[e]	747	352

[a] Estimated by modified minimum area method.
[b] Estimated by harmonic mean method.
[c] Died.
[d] Hit by automobile and taken into captivity.
[e] Radio-transmitter failed.

Dispersal of Panthers in Landscapes

The composite range of the remaining six panthers encompassed 2780 km² approximately bounded by U.S. 41 (Tamiami Trail) on the north, the inland edge of the mangrove zone on the west and south, and the edge of disturbed, intensively farmed lands on the east (Figure 27.1). More than two-thirds (69%) of this area is open landscape, 47% of which is marsh-prairie and 22% slough; 4% of the open landscape acreage actually is forested (hardwood hammocks, bayheads, etc.). Continuous forested landscape, excluding mangrove forest, comprised 22% of the composite panther range (4% pineland and 18% cypress). Disturbed, primarily agricultural lands constituted 3% and coastal mangrove forest 6% of the remaining area.

Home ranges of individual panthers varied considerably in size and landscape composition (Figure 27.2). Breeding adult females 14 and 15 utilized relatively small ranges that lay entirely east of Shark River Slough and were centered around upland forest on Long Pine Key. Only small proportions (3 and 7%, respectively) of their home ranges fell outside the park. In contrast, female 27 (who was physically mature but nulliparous) ranged over a relatively large area that spanned Shark River Slough and extended from Tamiami Trail to Long Pine Key, which it only slightly overlapped; 36% of her home range extended outside the park. Her core activity area consisted primarily of open marsh-prairie landscape along the eastern boundary of the park and into the East Everglades Wildlife Management Area and adjacent private lands. Like their mothers, subadult females 21 and 22 utilized relatively small ranges east of Shark River Slough. Both, however, made exploratory movements east from their natal ranges beyond the park boundary following family break-up. Female 21 continually shifted her activities eastward away from the park before being hit by an automobile (south of Homestead) in July 1988, at approximately 2 years of age. A large proportion (62%) of her home range and her entire core activity area were outside the park boundary. Although female 22 explored an extensive area south and east of Long Pine Key, she returned to her natal range between excursions and eventually settled there after her

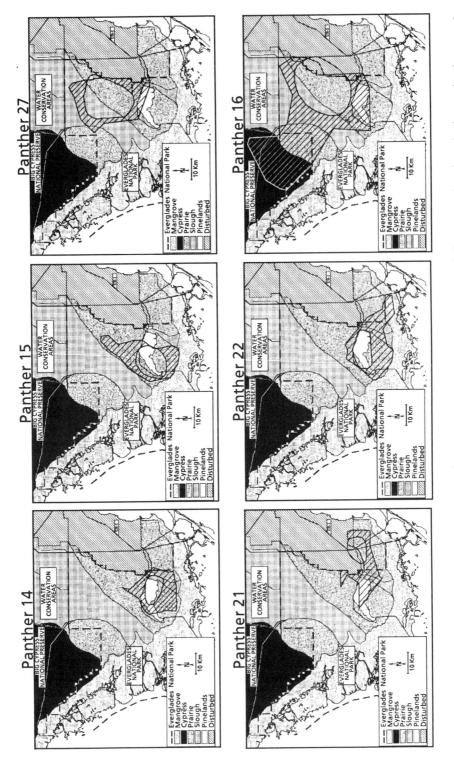

Figure 27.2 Home ranges and core activity areas of radio-tagged Florida panthers in Everglades National Park and adjacent lands in south Florida, December 1986–October 1989.

mother's death in June 1988. Only 10% of her home range fell outside the park, and her core activity area lay completely within the boundary of the park, centered on Long Pine Key. Subadult male 16 ranged widely (1917 km²) during and following dispersal: first into the East Everglades Wildlife Management Area and adjacent disturbed land, where he concentrated his activity for 2 years, and then across Shark River Slough into Big Cypress National Preserve in March 1989 at 3 years of age; his core activity area during this period was centered in marsh-prairie landscape east of Shark River Slough.

Utilization of Landscape

Distinct dissimilarities between the landscape compositions of core activity areas and total home ranges of most panthers (Table 27.2) indicated that landscape categories were not randomly used. This was confirmed by goodness-of-fit tests; without exception, use of landscape categories by panthers in their total home ranges was disproportionate to availability ($p \le 0.001$).

The upland forest landscape was uniformly used more than expected by chance (Table 27.2). In the cases of adult females 14 and 15 and subadult female 22, this preference represented an affinity to Long Pine Key, a continuous pineland bounded on its southern border by an extensive stand of Brazilian pepper and, to a lesser extent, remnant second growth hardwoods on abandoned farmland. These three females used marsh-prairie less than expected. In contrast, preferences for upland forest by adult female 27, subadult female 21, and subadult male 16 (who utilized Long Pine Key only sparingly) were due to their use of hardwood hammocks and bayheads in open marsh-prairie landscape.

Relatively wet open landscape (i.e., slough) within the home ranges of 27, 21, 22, and 16 was used less than expected, and slough/marsh-prairie edge demarcated the north and south boundaries of the range of adult female 15 (Figure 27.2). Adult female 14 used the narrow northern section of Taylor Slough in proportion to availability.

Table 27.2 Composition and Utilization of Landscape Categories in Home Ranges (HR) and Core Activity Areas (CA) of Radio-Tagged Florida Panthers in and Adjacent to Everglades National Park, December 1986–October 1989

| | Percentage composition | | | | | | | | | | | |
| | Upland forest | | Marsh-prairie | | Lowland forest | | Slough | | Coastal | | Disturbed | |
Panther	HR	CA	HR	CA	HR	CA	HR	CA	HR	CA	HR	CA
Adult F 14	27	56 +	63	34 −	0	0	10	0	0	0	0	0
Adult F 15	17	39 +	82	61 −	0	0	0	0	0	0	0	0
Adult F 27	5	0 +	52	67	0	0	41	33 −	0	0	2	0
Subadult F 21	15	0 +	70	60	0	0	5	0 −	0	0	10	40 +
Subadult F 22	22	73 +	61	27 −	0	0	6	0 −	12	0 −	0	0
Subadult M 16	5	0 +	45	63	25	0 −	21	19 −	1	0	3	18 +

Note: + indicates use more than expected by chance ($p \le 0.05$) in home range. − indicates use less than expected by chance in home range.

The mangrove zone also appeared to be avoided by panthers. Subadult female 22, the only panther whose home range included mangrove (12%), used this landscape category less than expected by chance (Table 27.2). The way in which home range boundaries of panthers 14, 15, 16, and 21 skirted the mangrove zone also indicated that this landscape was available to, but unused by, these individuals.

Continuous lowland forest (cypress) occurred only in the range of subadult male 16 (25%), who used this landscape less than expected (Table 27.2). In the ranges of subadult female 21 and adult female 27, there was considerable coverage of lowland forest in wet open landscape. Female 21 utilized willow and buttonwood in disturbed prairie south of Homestead more than expected, whereas female 27 used willow thickets bordering tree islands in Shark River Slough in proportion to availability.

Three panthers ranged into disturbed landscape east of Everglades National Park. Adult female 27 penetrated only the margin of agricultural land along canal L-31W east of the park and the East Everglades Wildlife Management Area; she used this part of her range in proportion to availability. Subadult male 16 ventured further into, and exhibited a preference for, the same area during his first year of independence. Subadult female 21 also used disturbed landscape in her range (east of Everglades National Park and south of Homestead, Figure 27.2) more than expected (Table 27.2).

Distribution and Abundance of Deer in the Core Study Area

Aerial Observations

Between January 1988 and October 1989, 421 deer were observed on the 15 aerial censuses of the 525-km² core study area (Table 27.3). Aerial census was not effective in the 100 km² of continuous forest on Long Pine Key, at the center of the core area.

Numbers observed and expected numbers in ten 1-km-wide intervals radiating from Long Pine Key were significantly different, both overall ($\chi_9^2 = 79.0$, $p < 0.001$)

Table 27.3 Number of Observations (O) and Expected Number of Observations (E) of White-Tailed Deer on Aerial Censuses of Open Landscape in the Vicinity of Long Pine Key, Everglades National Park, January 1988–October 1989

| | | Distance from Long Pine Key (km) | | | | | | | | | |
| | | <1 | 1–2 | 2–3 | 3–4 | 4–5 | 5–6 | 6–7 | 7–8 | 8–9 | 9–10 |
Period	# of censuses	O/E	O/E	O/E	O/E	O/E	O/E	O/E	O/E	O/E	O/E
Jan.–May 88	5	40/23*	24/20	10/18	9/18*	21/17	22/17	9/16	5/13*	10/9	5/4
Jul.–Nov. 88	3	21/12	13/10	3/10*	9/9	8/9	2/9*	16/8	3/7	6/5	1/2
Dec.–Apr. 89	5	33/16*	25/14	7/13	10/12	8/12	5/12*	4/11*	8/9	1/6	6/3
Aug.–Oct. 89	2	23/12*	8/10	8/9	9/9	7/8	8/8	1/8*	4/7	4/4	5/2
Total	15	117/63*	70/53	28/50	37/48	44/46	37/45	30/43	20/36*	21/24	17/12

* indicates significant difference ($p \leq 0.05$) between observed and expected.

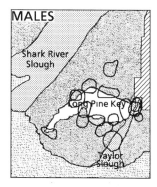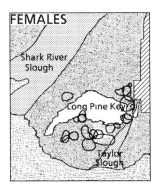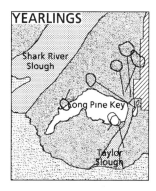

Figure 27.3 Annual home ranges of white-tailed deer radio-tagged in the vicinity of Long Pine Key, Everglades National Park, March 1987–October 1989.

and by season ($\chi_9^2 \geq 23.3$, $p < 0.005$), indicating that the distribution of deer across open landscape in the census area was not random. This was largely due to the effect of the area <1 km from Long Pine Key, in which significantly greater than expected numbers occurred ($p \leq 0.05$) (Table 27.3). Noticeable but not significantly ($p > 0.05$) higher than expected numbers were counted in intervals 1–2 and 9–10 km from the edge of the forest. At other intervals, numbers of deer observed were equal to or less than expected. There was no significant correlation between the relative abundance of deer and distance to Long Pine Key in either dry ($r = +0.32$) or wet ($r = +0.18$) seasons ($p > 0.05$).

Home Range Size

Annual home ranges of 45 radio-tagged deer (16 adult males, 22 adult females, 7 yearling males) resident in the core study area were estimated (Figure 27.3). In all three groups there was a significant ($p < 0.05$) positive correlation between home range size and distance to Long Pine Key (Figure 27.4).

DISCUSSION

The analyses demonstrated that the distribution of Florida panthers across landscapes within and adjacent to Everglades National Park was not random. Variations in size, location, and landscape pattern of panther home ranges were associated with age, sex, and, in particular, reproductive status. After family break-up, young panthers of both sexes dispersed from insular, protected natal ranges centered around continuous forested landscape (Long Pine Key) and ranged widely into disturbed, less protected and largely open landscape near the periphery of the park and beyond. One subadult female was hit by an automobile; another returned to her natal range following her mother's death; a male survived his subadulthood in marginal habitat and subsequently established a large range that encompassed all of Everglades National Park (including his natal range) and a portion of Big Cypress National Preserve (O. L. Bass, unpublished data). A young nonbreeding

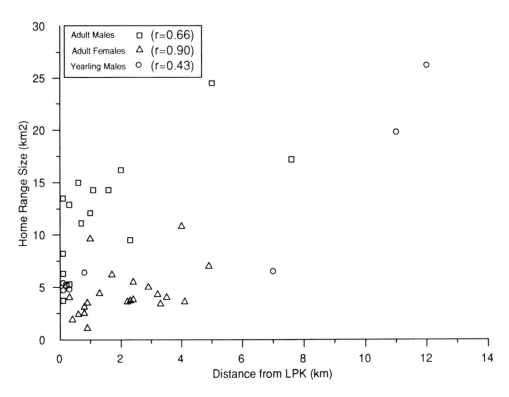

Figure 27.4 Relationship between the size of home ranges of white-tailed deer and the distances of their geometric centers to Long Pine Key in Everglades National Park, 1987–89.

adult female resembled subadults in utilizing a large range in predominantly open landscape, much of which was disturbed and/or lay outside the boundary of the park.

In contrast, the two breeding adult females exhibited strong affinities to insular, forested landscape. Their relatively small core areas encompassed opposite ends and overlapped at the center of Long Pine Key, and they rarely ventured into disturbed landscape or left the park. Furthermore, whereas the two subadult and one nonbreeding females in the sample fed almost exclusively on small prey (e.g., raccoon [*Procyon lotor*], marsh rabbit [*Sylvilagus palustris*], and alligator [*Alligator mississippiensis*]), the breeding females preyed primarily on white-tailed deer (O. L. Bass, unpublished data).

The distribution of mountain lions is limited by the availability of prey, the availability of hunting cover, and/or human interference. In Everglades National Park, acquisition of large prey (i.e., white-tailed deer) appears to explain the dichotomy observed between distributions of nonbreeding and breeding female panthers; the latter focused their activities around Long Pine Key, where the density (i.e., availability) of deer and diversity of vegetative (i.e., hunting) cover were relatively high. The 9 of 13 sites at which female panthers killed radio-tagged deer between November 1987 and December 1989 were within 300 m of the Long Pine Key/prairie edge. This edge effect (Leopold, 1933) appeared to extend 1 or possibly 2 km from Long Pine Key into open landscape, where the density of deer and availability of hunting cover decreased. Although panthers killed radio-tagged deer in open areas away from Long Pine Key (two at edges of hammocks, one in open sawgrass, one in a willow thicket), they apparently were unable to do so with the frequency and regularity required by breeding female mountain lions (Seidensticker et al., 1973).

It appears that the availability of edge between upland forest and open prairie landscapes determines the breeding density of panthers in Everglades National Park and, in turn, largely determines carrying capacity. The present size and composition of the panther subpopulation in the park and adjacent land likely approaches carrying capacity, which is estimated to be five to ten individuals, exclusive of dependent kittens. During this study, two breeding females, three subadult and/or nonbreeding females, and one transient subadult male occupied the area, and, undoubtedly, at least one adult male visited the area in 1986 and bred females 14 and 15. Recently (1990–92), adult male 16 established residency in the area and frequently crossed Shark River Slough, ranging throughout Everglades National Park and southern Big Cypress National Preserve (O. L. Bass, unpublished data).

It also appears that there may be little temporal variation in the carrying capacity of Everglades National Park and adjacent land for Florida panthers. The white-tailed deer population in the area is highly K-selected; density and rates of productivity and mortality are unusually low for the species, and home ranges are unusually large (Smith, 1991), which is indicative of a predictable (i.e., stable) environment with low primary productivity. These parameters varied very little during the period of study, despite dramatic environmental variation between 1987 (normal), 1988 (unusually wet), and 1989 (unusually dry with a wildfire that burned half of the core study area).

Given the ecological characteristics, finite acreage, and existing land use practices in the eastern Everglades, it is unlikely that the carrying capacity of Everglades National Park and vicinity for Florida panthers can be substantially increased by management. Topography, hydrology, soils (or the lack of), and vegetation of the area do not lend themselves to conventional methods of habitat management, particularly given the spatial scale at which panthers interact with habitat. Moreover, management of the natural dynamics in these resources may conflict with policies of the National Park Service. Fire and water, which drive the Everglades system, appear to have little effect on the long-term dynamics of the Everglades National Park deer population (T. Smith, unpublished data), and it is doubtful that any politically feasible changes in fire or hydrologic management would significantly increase carrying capacity of the area for panthers.

Protection of undeveloped lands outside (east) of Everglades National Park, which are suitable panther habitat, may currently be the best strategy for panther management in the eastern Everglades. The north unit of the East Everglades Wildlife Management Area and adjacent private lands have a relatively high availability of upland forest/prairie edge in the vicinity of Grossman Ridge, where numerous hardwood hammocks occur; white-tailed deer are common, and wild hogs are occasionally observed. Authorization has been granted to incorporate this area into Everglades National Park. Forthcoming elimination of controlled hunting, poaching, use of all-terrain vehicles, firing ranges, and free-ranging dogs should positively impact the area, which potentially is suitable range for a breeding female panther. A second area which merits attention is the south unit of the East Everglades Wildlife Management Area and contiguous privately owned marsh-prairie to the east, across U.S. 1 and Card Sound Road south of Homestead to Biscayne Bay (Figure 27.1). Use of this area by two subadult females during this study indicates that it provides suitable panther habitat, at least for nonbreeding females who can survive without regular use of large prey. Together, these two areas east of Everglades National Park may serve as habitat for a reservoir of subadult panthers, particularly females, to replace losses of breeding adults from the interior of the park.

Although the potential number of Florida panthers east of Shark River Slough represents a relatively small proportion (possibly 10–20%) of the total wild population, this subpopulation may play an important role in the long-term conservation of the species. Clearly, Long Pine Key and vicinity in Everglades National Park can support breeding females which consistently produce cubs. Given the finite carrying capacity of the area, some of these young are capable of (and likely to) dispersing across Shark River Slough into Big Cypress National Preserve, as they did during this study. Regardless of the origin of panthers in Everglades National Park (O'Brien et al., 1990), such dispersal undoubtedly would enhance the genetic diversity of the total wild population in south Florida.

LITERATURE CITED

Anderson, A. E. 1983. A Critical Review of Literature on Puma (*Felis concolor*), Spec. Rep. 54, Colorado Division of Wildlife, Denver, 91 pp.

Barrett, M. W., J. W. Nolan, and L. D. Roy. 1982. Evaluation of a hand-held netgun to capture large mammals. *Wildl. Soc. Bull.,* 10(2):108–114.

Belden, R. C. 1978. Florida panther investigation—A progress report. in Proc. Rare and Endangered Wildlife Symp., R. R. Odom and L. Landers (Eds.), Tech. Bull. WL4, Georgia Department of Natural Resources, Athens, pp. 122–133.

Belden, R. C. 1986. Florida panther recovery plan implementation. in *Cats of the World, Biology, Conservation and Management.* Proc. 2nd Int. Cat Symp., S. D. Miller and D. D. Everett (Eds.), National Wildlife Federation, Washington, D.C., pp. 159–172.

Belden, R. C. 1988. The Florida panther. in *Audubon Wildlife Report. 1988/1989,* W. J. Chandler (Ed.), National Audubon Society, New York, pp. 514–532.

Belden, R. C. and L. E. Williams. 1976. Survival status of the Florida panther. in *Proc. Florida Panther Conf.,* P. C. H. Prichard (Ed.), Florida Audubon Society and Florida Game and Fresh Water Fish Commission, Orlando, pp. 78–98.

Burt, W. H. 1943. Territoriality and home range concepts as applied to mammals. *J. Mammal.,* 24:346–352.

Byers, C. R., R. K. Steinhorst, and P. R. Krausman. 1984. Clarification of a technique for analysis of utilization-availability data. *J. Wildl. Manage.,* 48:1050–1053.

Coleman, J. S. and A. B. Jones III. 1988. User's Guide to TELEM88: Computer Analysis System for Radio-Telemetry Data, Res. Ser. No. 1, Department of Fisheries and Wildlife, Virginia Polytechnic Institute and State University, Blacksburg, 49 pp.

Craighead, F. C. 1971. *The Trees of South Florida,* University of Miami Press, Coral Gables, Fla., 212 pp.

Davis, J. H., Jr. 1943. The natural features of southern Florida, especially the vegetation and the Everglades. *Fla. Geol. Surv. Geol. Bull.,* No. 25, 311 pp.

Dixon, K. R. and J. A. Chapman. 1980. Harmonic mean measure of animal activity areas. *Ecology,* 61:1040–1044.

Frankel, O. H. and M. E. Soule. 1981. *Conservation and Evolution,* Cambridge University Press, Cambridge, 327 pp.

Gilbert, F. F. 1966. Aging white-tailed deer by annuli in the cementum of the first incisor. *J. Wildl. Manage.,* 30:200–202.

Gleason, P. J. 1984. *Environments of South Florida: Past and Present II,* Miami Geological Society, Coral Gables, Fla., 551 pp.

Goldman, E. A. 1946. Classification of the races of the puma. in *The Puma, Mysterious American Cat,* S. P. Young and E. A. Goldman (Eds.), American Wildlife Institute, Washington, D.C., 358 pp.

Harlow, R. F. and F. K. Jones, Jr. 1965. The White-Tailed Deer in Florida, Tech. Bull. 9, Florida Game and Fresh Water Fish Commission, Tallahassee, pp. 175–302.

Harvey, M. J. and R. W. Barbour. 1965. Home range of *Microtus ochrogaster* as determined by a modified minimum area method. *J. Mammal.,* 46:398–402.

Layne, J. M. and M. N. McCauley. 1976. Biological overview of the Florida panther. in *Proc. Florida Panther Conf.,* P. C. H. Prichard (Ed.), Florida Audubon Society and Florida Game and Fresh Water Fish Commission, Orlando, pp. 5–45.

Leopold, A. 1933. *Game Management,* Charles Scribner's Sons, New York, 481 pp.

Maehr, D. S. 1990. The Florida panther and private lands. *Conserv. Biol.,* 4:167–170.

Maehr, D. S., R. C. Belden, E. D. Land, and L. Wilkins. 1990. Food habits of panthers in southwest Florida. *J. Wildl. Manage.,* 54(3):420–423.

McBride, R. 1985. Population Status of the Florida Panther in the Everglades National Park and Big Cypress National Preserve, Final Report, Contract #5280-84-04, Everglades National Park, Homestead, Fla., 57 pp.

Mech, L. D. 1983. *Handbook of Animal Radio-Tracking,* University of Minnesota Press, Minneapolis, 107 pp

Myers, R. L. and J. J. Ewel. 1990. *Ecoystems of Florida,* University of Central Florida Press, Orlando, 765 pp.

Neu, C. W., C. R. Byers, and J. M. Peek. 1974. A technique for analysis of utilization-availability data. *J. Wildl. Manage.,* 38:541–545.

Nowak, R. M. and R. McBride. 1976. Status of the Florida panther. in *Proc. Florida Panther Conf.,* P. C. H. Prichard (Ed.), Florida Audubon Society and Florida Game and Fresh Water Fish Commission, Orlando, pp. 116–121.

O'Brien, S. J., M. E. Roelke, N. Yuhki, et al. 1990. Genetic introgression within the Florida panther, *Felis concolor coryi. Natl. Geogr. Res.,* 6(4):485–494.

Roelke, M. E., E. R. Jacobson, G. V. Kolias, and D. S. Forrester. 1986. Medical Management and Biomedical Findings on the Florida Panther (*Felis concolor coryi*), Annual Performance Report 1985/86, Florida Game and Fresh Water Fish Commission, Tallahassee, 65 pp.

Samuel, M. D., D. John Pierce, E. O. Garton, L. J. Nelson, and K. R. Dixon. 1985. User's Manual for Program Home Range, Tech. Rep. 15, Forest, Wildlife and Range Experiment Station, University of Idaho, Moscow, 70 pp.

Sanderson, G. C. 1966. The study of mammal movements—A review. *J. Wildl. Manage.,* 30: 215–235.

Schemnitz, S. D. 1974. Populations of bear, panther, alligator and deer in the Florida Everglades. *Fla. Sci.,* 37(3):157–167.

Seidensticker, J. C., M. G. Hornocker, W. V. Wiles, and J. P. Messick. 1973. Mountain lion social organization in the Idaho Primitive Area. *Wildl. Monogr.,* No. 35, 60 pp.

Severinghaus, C. A. 1949. Tooth development and wear as criteria of age in white-tailed deer. *J. Wildl. Manage.,* 13:195–216.

Shaw, H. G. 1979. A Mountain Lion Field Guide, Spec. Rep. No. 9, Arizona Game and Fish Department, Phoenix, 27 pp.

Smith, T. R. 1991. Recruitment and loss in an unhunted subtropical white-tailed deer population. in *The Biology of Deer*, R. D. Brown (Ed.), Springer-Verlag, New York, p. 76.

Tinsley, J. B. 1970. *The Florida Panther*, Great Outdoors Publ., St. Petersburg, Fla., 60 pp.

White, G. C. and R. A. Garrott. 1990. *Analysis of Wildlife Radio-Tracking Data*, Academic Press, San Diego, 383 pp.

Williams, L. E., Jr. 1978. Florida panther. in *Rare and Endangered Biota of Florida*, Vol. I: Mammals, J. N. Layne (Ed.), University Presses of Florida, Gainesville, pp. 13–15.

Young, S. P. and E. A. Goldman. 1946. *The Puma: Mysterious American Cat*, Dover Publ., New York, 358 pp.

28

The FAUNAL CHAPTERS: CONTEXTS, SYNTHESIS, and DEPARTURES

William B. Robertson, Jr.

Peter C. Frederick

ABSTRACT

Salient themes of the chapters on animal ecology are discussed here, and aspects of the historical/biogeographical setting essential to understanding the Everglades fauna and the prospects for its restoration are described. The Everglades is located at the distal end of a long peninsula, at an important biogeographical boundary, and in a region where sea level rise has been the dominant geological process of the past 15,000 years. Interaction of these forces determined the characteristics of its present biota: relatively species poor, of mixed continental and Antillean origin, and with endemism and small populations strongly concentrated in the reduced area of uplands. In the 1920s through 1960s, the Everglades probably served as the major source of repopulation of wading birds throughout the eastern United States. Now, as chronicled in many chapters of the faunal section, nesting in the area has become a demographic sink for wading birds produced elsewhere, although the Everglades still functions as an important feeding area for wintering and transient populations. Re-establishment of productive wading bird nesting populations is a key measure of the success of Everglades restoration. Hypotheses and interpretations of the restoration needs of Everglades wading birds center on revitalization of short-hydroperiod marshes to cue earlier breeding and repair of the diminished productivity of the estuarine system. However, despite considerable study, wading bird restoration remains hampered by inadequate information, especially concerning the ecology of food organisms. At least several hundred species of nonnative animals are established in developed, coastal areas adjoining the Everglades. More and more of these species appear to be extending into undisturbed habitats where the effects of their presence are at present largely unknown. Information on movementt and home range suggests that the Everglades system, in and of itself, is probably too small to sustain viable populations of many of its more mobile animals. This is particularly true of the reduced and still shrinking upland areas, many species of which may be virtually impossible to restore and maintain.

INTRODUCTION

The original purpose of this chapter was to extract the essence of the preceding series of chapters presenting research on animal ecology. However, it is necessary to consider new information revealed by the onward flow of events and study. Most of the new observations highlighted in this chapter were made by the researchers who presented chapters on the same subjects.

In addition to new data from ongoing studies, abrupt changes of scene in the Everglades itself have also demanded comment. During most of the preparation of this volume, southern Florida was in the second year of a 3-year drought and had not experienced a significant tropical storm in 25 years. The drought was broken in the 1991–92 hydrological year, and now a remarkably wet El Niño winter has followed the epic autumn of hurricane Andrew. Although it is not yet clear that these perturbations will be studied with the sophistication they deserve, their mere occurrence has provided insights into Everglades ecology.

Finally, in preparing this chapter it was decided not to simply rehash the constituent chapters and attempt to connect them. The chapters, after all, engage a wide diversity of subjects and speak capably for themselves. Instead, they are used as springboards or guides to other discussion. Thus, in the material that follows, one section examines the concepts and questions that arise more or less directly from the suite of chapters on wading birds that make up most of the faunal segment. The other four sections are stand-alone essays that do not follow in a natural sequence. Rather, they engage subjects that are discussed little or not at all in this volume, but are central to consideration of the Everglades fauna and critical to any views of restoration. While this may present the reader with a sort of karst topography of subjects, it is more useful to fill gaps than provide continuity of reading. If there is a unifying theme, it is to provide historical and geographical context to the reader's view of the faunal ecology of southern Florida.

HISTORICAL FACTORS and the PRESENT EVERGLADES FAUNA

Aspects of Holocene geographical and ecological change in southern Florida that have largely determined the composition and status of the present Everglades fauna are reviewed here. In particular, the following are considered: (1) Florida's peninsular landform, initially four times wider and much more elevated than it is now (Webb, 1990, Figure 4.8), (2) Florida's geographical location near a significant biogeographical boundary, and (3) Florida's postglacial history of steadily rising sea level, the proximate force that drives the ecogeographical change. These influences have combined to create a south Florida landscape that has been highly unstable at the upper temporal scale of a few thousand years (deAngelis and White, 1994), with a rather dynamic "succession" of communities from upland to freshwater wetland to coastal swamp to estuarine to marine, as sea level rose. These ongoing changes are central to understanding the composition and future of the present regional fauna.

Peninsularity

It has frequently been noted that the species diversity of vertebrate animals often decreases from proximal to distal regions of long peninsulas. Simpson (1964) termed this phenomenon the "peninsular effect" and reasoned that it might occur because extinction rates tended to increase and immigration rates to decrease with increasing distance from the mainland along a peninsula. The enunciation of this concept sparked examination of peninsular faunas, and Florida, being conspicuously a peninsula, has figured extensively in this discussion.

All four classes of vertebrates in Florida exhibit progressively declining species density (i.e., the number of native breeding species in a representative area) from north to south in the peninsula. This result has been reported from analyses of the faunas of fishes (Loftus and Kushlan, 1987), reptiles and amphibians (Duellman and Schwartz, 1958; Kiester, 1971; Busack and Hedges, 1984), birds (Cook, 1969; Tramer, 1974; Robertson and Kushlan, 1974), and mammals (Simpson, 1964; Layne, 1974). The pattern and degree of north-to-south decline in species density varies considerably between groups. For example, the number of species of breeding landbirds declines precipitously from >70 in the extreme northern peninsula to about 25 at the extreme southern end of the peninsula and <20 in the Lower Florida Keys (Robertson and Kushlan, 1974, Figure 1). Conversely, Dalrymple (1988) pointed out that the principal deficiency in the species composition of herpetofauna of Long Pine Key, Everglades National Park was the lack of salamanders, while species density in groups such as snakes and anurans was more similar to that found in comparable areas much farther north in the peninsula.

Although Florida's vertebrate fauna seems to exhibit the hypothesized peninsular effect, the universality (even the reality!) of the phenomenon has come to be widely questioned. Specifically for Florida, several authors (Tramer, 1974; Wamer, 1978; Busack and Hedges, 1984; Means and Simberloff, 1987) have argued that the southward decline of species density principally reflects a similar decline in habitat diversity and productivity from top to bottom of the peninsula. Attention is sometimes drawn to the much reduced area of uplands in southern Florida, known as "the Everglades effect" (Means and Simberloff, 1987), as a factor contributing to reduced species diversity of terrestrial vertebrates.

Two lines of evidence based on change in the breeding ranges of Florida landbird species over the past 60 years or so seem to bear on questions of peninsularity. First, about 25 species have shown significant southward extensions of their breeding ranges in the Florida peninsula within the period of scientific record (Robertson and Kushlan, 1974; Stevenson, 1976; Robertson and Woolfenden, 1992). Most of these breeding range expansions have been closely associated with anthropogenic or extensively altered habitats new to the peninsular landscape. Thus, it appears reasonable to suppose that many of these species had previously reached a habitat-determined southern range limit in the peninsula. Second, about ten bird species (plus several mammals) have apparently disappeared from their southern breeding range extremities in the Florida peninsula within the period of scientific record. Most of these species were close or obligate associates of the diminished and fragmented stands of upland vegetation (especially pine forest) near the southern end of the peninsula. These instances would appear to represent the type of local extinctions without replacement which Simpson (1964) visualized as the mechanism of his "peninsular effect." (See the final section of this chapter and Tables 28.1 and 28.2 therein for additional discussion.)

In summary, the present Everglades system has fewer vertebrate species than occur in areas of similar extent farther north in peninsular Florida. Two principal explanations have been suggested to account for these differences: (1) effects of the narrowed peninsular landform tend toward an increased likelihood of local extinction and a decreased likelihood of colonization and recolonization, as in entirely insular environments, or, specifically for Florida, (2) the southward decrease in habitat diversity and productivity adequately explains the reduced species density of vertebrates in southern Florida. Documented recent changes of breeding range by landbird species in Florida (as discussed later in this chapter) lend some support to each of these explanations. However, comment on this subject must consider the probable state of the peninsula itself and its biota throughout the postglacial cycle of sea level change. The Florida peninsula at earlier stages in this approximately 20,000-year process differed greatly from the peninsula of today (Webb, 1990, Figures 4.8 and 4.9) and peninsular effects on the fauna doubtless also differed.

Location at a Zonal Boundary

As De Pourtales (1877) noted more than a century ago, southern Florida is unusual in that it has predominantly West Indian plant communities (especially uplands and coastal swamps) whose vertebrate inhabitants are mostly North American species. His comment calls attention to a second pervasive influence on the present Everglades fauna, namely, southern Florida's location at the biogeographical frontier between temperate North American and Antillean tropical biotas. Through the various cycles of Pleistocene glacial–interglacial sea levels, the water gap that separated outermost Florida from Cuba and the Bahamas apparently was never much narrower than it is today (Hoffmeister, 1974). Thus, any biotic interchange between the two juxtaposed land masses occurred across a seaway roughly 65–125 km wide at its narrowest. Prevailing winds and ocean currents at present certainly favor transport from the West Indies toward Florida, and these, along with the contribution of migrant birds, no doubt account for the large West Indian component in the flora and the invertebrate fauna of southern Florida (Snyder et al., 1990). Few vertebrates other than birds figured in the exchange, however. For example, the Antillean mammals self-introduced to southern Florida (Layne, 1974) are comprised of only two rare species of bats and the West Indian manatee (*Trichechus manatus*); the Antillean reptiles native to southern Florida (Duellman and Schwartz, 1958) are the reef gecko (*Sphaerodactylus notatus*) and the American crocodile (*Crocodylus acutus*). It appears probable that no North American mammals or reptiles crossed in the opposite direction.

Only in the case of the land avifauna do present distributions indicate sufficient traffic in both directions to give some insight into the history of the exchange process. Robertson and Kushlan (1974) found evidence for an earlier (last glacial maximum?) flow of landbird species from the continent into the Bahamas and Greater Antilles and a smaller (currently ongoing) movement from the Bahamas and Cuba into southern Florida. The former movement resulted in the establishment in the West Indies of 25–35 landbird species originally of North American origin. Examples include the sandhill crane (*Grus canadensis*), ivory-billed woodpecker (*Campephilus principalis*), brown-headed nuthatch (*Sitta pusilla*), blue-gray gnatcatcher (*Polioptila caerulea*), and pine warbler (*Dendroica pinus*). These West Indian populations are now considered at least subspecifically distinct from their closest continental relatives (Bond, 1956), and ecologically, the group is associated predominantly with pine forest and savanna habitats in the West Indies. The apparently more recent influx in the opposite direction has established about a dozen West Indian landbirds on the continent, with most species more or less restricted to southern Florida. Examples include the snail kite (*Rostrhamus sociabilis*), white-crowned pigeon (*Columba leucocephala*), gray kingbird (*Tyrannus dominicensis*), and black-whiskered vireo (*Vireo altiloquus*). Without exception, these species are considered identical to West Indian populations, and ecologically, they tend to be associated with coastal habitats. Recent records of breeding in southern Florida by the West Indian cave swallow (*Hirundo fulva*) (Smith et al., 1988) and probable breeding by the shiny cowbird (*Molothrus bonariensis)* (Smith and Sprunt, 1987) strongly suggest that species flow toward the continent is continuing. The present breeding land avifauna of southern Florida appears to be unsaturated (i.e., it lacks representative species of a number of trophic guilds) and presumably is open to further invasion (Robertson, 1955; Robertson and Kushlan, 1974).

In summary, evidence from the presence and differentiation of North American species of landbirds in the West Indies suggests that this influx occurred during Pleistocene glacial maxima, especially the most recent glacial cycle. During glacial intervals, uplands on either side of the Florida Straits were near their greatest extent, and most of the invading landbirds (at least those that survive at present) were apparently species adapted to upland (in some cases montane) and xeric habitats. In contrast, the lesser but continuing flow of West Indian landbirds to southern Florida appears to be much more recent and comprised of coastal species. Because sea level rise and anthropogenic effects have so greatly reduced the extent of native upland habitats, North American upland species in southern Florida have been lost more rapidly than new West Indian landbirds have been acquired (see the last section in this chapter). All or parts of the Everglades uplands now lack several bird species typical of southeastern U.S. pine forests that are still numerous in the much more extensive pine areas of the northern Bahamas.

Rising Sea Level

In this brief segment, two points will be emphasized: (1) although sea level rise is often a slow, ongoing geological process, its effects on biotic communities are detectable on a temporal scale of decades to a few centuries in low-lying coastal areas such as southern Florida, and (2) overall, rising sea level in the past few thousand years has driven biotic succession in a single general direction—reducing the proportion of area in upland communities and increasing areas of freshwater wetlands, coastal swamps, and estuaries as southern Florida has gradually submerged (see also the final section of this chapter).

Egler (1952) may have been the first to suggest that sea level rise might be of concern to practicing plant ecologists in the region, and more recent studies have amply confirmed this insight. Several studies show that the rate of sea level rise in southern Florida has been most rapid in the intermediate, postglacial past at roughly 7000–2000 years before present (YBP). Study of sediments also indicates this interval as a time of major changes in the regional landscape.

Evidence from peats (Gleason and Stone, 1994) indicates that the Everglades wetland first existed in its present condition and location in about 5000 YBP. At about the same time, buried beds of mangrove peat extending up to several miles offshore mark a major estuarine or marine zone of vegetation along the Gulf coast (Spackman et al., 1966). Conversely, freshwater peats at the bottom of the sedimentary column suggest that an earlier "Everglades" once existed in Florida Bay (Davies and Cohen, 1989). This proto-Everglades evidently met its demise due to rising sea level at between 4500 and 3000 YBP (Wanless and Tagett, 1989).

These results suggest that although relatively youthful in its present setting, the freshwater wetlands and estuaries of the Everglades have a somewhat longer history of more or less orderly retreat in the face of rising sea level. Although there is little description of upland plant material from the sediment studies, upland vegetation of some sort presumably was the predominant plant cover of southern Florida at the lowest sea stages of the present cycle. The present Everglades uplands appear to be fragments of this former landscape, now perched upon the highest remaining elevations.

The retreat of uplands before the sea's advance is currently observable and very much a factor to be considered in the near future of the Everglades. Alexander (1974) found dead trunks of still-rooted slash pine (*Pinus elliottii*) in a red mangrove (*Rhizophora mangle*) swamp on northern Key Largo, and Ross et al. (in press) measured reduction in area of a pine upland on Sugarloaf Key from successive series of aerial photographs. Ross et al. also provide evidence that the loss of upland plants is specifically due to the effects of saltwater intrusion. Given the rapid rate of effective sea level rise in the southern Florida area (Wanless et al., 1994), the continued loss of uplands and freshwater areas is quite likely to have major effects on the remaining fauna, particularly those communities associated with uplands.

WADING BIRDS:
REGIONAL CONTEXT and ECOSYSTEM INTERACTIONS

The study of wading birds has special relevance for the Everglades for a number of reasons. Wading birds were arguably the focus of the first scientific investigations in the Everglades, and the sheer length of the breeding record has provided powerful clues to understanding how ecological relationships have changed in the ecosystem. Wading birds have historically been a numerically dominant group of predators in this wetland and have likely had effects on the ecosystem that may be measured at the community level. Finally, the human empathy evoked by the grace of these animals has been (and remains) a telling force in mobilizing support for preservation and restoration of the Everglades, in ways that other wetland animals do not.

The Everglades: A Critical Link in a Far-Flung Wetland Chain

Wading birds are highly mobile and even nomadic animals and form perhaps the most obvious linkages between the Everglades and other ecosystems in the southeastern United States and the Caribbean. It is essential to realize that the birds using the Everglades are not a population specific to that ecosystem, but are wide-ranging consumers, with variable breeding site fidelity. The large ambit and plastic breeding site fidelity in many wading bird species is a strong indication that these species have evolved to take advantage of highly variable wetland conditions, over ranges much larger than the Everglades. Certainly the relatively young age of the Everglades (see earlier) largely excludes the possibility that wading birds would show adaptations that are specific to a southern Florida existence.

The present-day choices for breeding by the wading bird species seen in the Everglades include the coastal plain of much of the southeastern and mid-Atlantic United States, as well as large parts of the Mississippi valley (Byrd, 1978). However, this range is in large part a recent phenomenon. Wood storks (*Mycteria americana*) and white ibises (*Eudocimus albus*) were virtually unknown as breeders in states north of Florida prior to the 1920s (Wayne, 1922; Sprunt, 1922; Ogden et al., 1987; Rodgers et al., 1987), and glossy ibises (*Plegadis falcinellus*) were extremely rare as breeders in the United States before the turn of the century (Bent, 1926; Ogden, 1981). Since the 1970s, a large proportion of the southeastern wood stork population has begun breeding in South Carolina and Georgia (Rodgers et al., 1987; Ogden et al., 1987), and glossy ibises have by now expanded their breeding range into virtually every Atlantic coastal state, including Maine (Hancock et al., 1992). Most of the breeding activity by North American white ibises was probably centered in the Everglades during the 1930s and 1940s (Ogden, 1978; Kushlan et al., 1984), but shifted to central and northern Florida and the Carolinas during the late 1970s and 1980s (Custer and Osborn, 1975; Ogden, 1978; Kushlan et al., 1984; Post, 1990). At present, the majority of breeding white ibises in the United States have relocated yet again to southern Louisiana (B. Fleury and T. Sherry, personal communication). Extensive northward range expansions of great (*Casmerodius albus*) and snowy egrets (*Egretta thula*) have also been documented since the plume-hunting era at the turn of the century (Byrd, 1978; Ogden, 1978). Although it is possible that the recent range shifts toward the north were set in motion by environmental degradation in the Everglades and other Florida wetlands, uncertainties in the timing and magnitude of the range shifts make this hypothesis eternally vulnerable.

Regardless of the causes of these changing breeding ranges, it is quite possible that for several species, the present breeding situation may actually offer more flexibility than the historic one. The current range of breeding locations available for wood storks and white ibises is now geographically much more dispersed than in the early part of this century and must surely encompass a greater range of possible breeding conditions than formerly. In addition, this greater dispersion of the population is predicted to buffer these populations from catastrophic events such as severe weather and outbreaks of disease.

Among the potential breeding sites in the southeast, the Everglades remains unique in that it is annually guaranteed a large wintering group of wading birds because of its geographic location and climate. This situation gives the Everglades both an inherent advantage in attracting breeding wading birds if conditions are appropriate and a special potential for affecting the condition and health of a wintering population from the entire southeast (Bancroft et al., 1992). These properties are largely a result of physical location—there are few other large wetlands in the United States that stay warm all winter, and none other than the Everglades that are en route to the wintering locations in the Greater Antilles. Thus, the presence of an annual wintering population in the Everglades will probably be relatively robust to all but the most drastic of anthropogenic changes.

Given the large numbers of white ibises and wood storks that were documented breeding in the Everglades during the early and middle part of this century, it is not unreasonable to suspect that the Everglades was a major source of recruitment for these species in the southeast. Similarly, the Everglades had to have been a critical source for the regeneration of populations of great and snowy egrets following the decimations of the plume-hunting era and for roseate spoonbills following the extremely low population levels recorded in the 1930s (Allen, 1942; Powell et al., 1989; Kushlan et al., 1984). Thus, there is a strong possibility that the Everglades has served as a springboard for both population and range expansions.

The current situation seems to be the opposite, with breeding in the Everglades so frequently unproductive or catastrophic for many of the species that the ecosystem may instead be a demographic sink for birds produced from other areas. The exceptions to this rule are the great egret and the roseate spoonbill (*Ajaja ajaja*), both of whose populations seem to be stable or increasing in the Everglades. A related question is whether the larger southeastern populations of other species are also declining. There is good evidence that the wood stork population in the southeast has declined since the 1930s and 1940s (Ogden and Nesbitt, 1979; Ogden et al., 1987), and both white ibises and snowy egrets declined by as much as 50% as breeding birds in the state of Florida between the late 1970s and the late 1980s (Runde, 1991). For all but the wood stork, however, regionwide censuses have been so infrequent and geographically spotty that true population sizes are impossible to determine.

Wading birds also connect the Everglades with Caribbean wetlands. Band recoveries suggest that large numbers of great and snowy egrets, glossy and white ibises, and little blue herons (*Egretta caerulea*) breeding in the United States depend heavily on the Greater Antilles (particularly Cuba) for wintering habitat (Byrd, 1978; Ryder, 1978; P. C. Frederick, unpublished data).

This information suggests that the Everglades has had, and to some extent retains, critical functions for wading bird populations ranging over much of the southeastern United States and the Caribbean. While the Everglades is often accorded recognition for the numbers of wading birds it attracts, it is rarely appreciated that changes in the Everglades could have direct effects on wetland bird populations from as far away as New Jersey, Louisiana, and Cuba.

Population Declines and Functional Relationships: Have Wading Birds Been Studied to Death?

The long history of research and monitoring of wading birds in the Everglades is so often mentioned that one might easily suppose a near perfect understanding of breeding dynamics. It should be emphasized that wading bird ecology in the Everglades is understood largely from a relatively few reliable records of breeding numbers and foraging distributions (Kushlan et al., 1984; Hoffman et al., 1990; D. M. Fleming, unpublished; Ogden, 1994) and that little is in fact known about the ecological relationships and specific mechanisms which explain the resultant patterns.

The evidence concerning this latter statement is embarrassingly abundant. Despite the intensive and extensive evaluations of present and past breeding conditions presented in this book, we still have no firm idea of which conditions and processes once allowed an order of magnitude more birds to breed in the ecosystem than do at present.

One hypothesis suggests that the historical Everglades landscape had sufficiently large amounts of short-hydroperiod marshes in almost any year that birds could find food in all but the most extreme surface water conditions (Browder, 1976; Bancroft et al., 1994; Ogden, 1994; Fleming et al., in press; D. M. Fleming, personal communication). These marshes are hypothesized to have offered considerably better early winter feeding conditions than are now available and thus were able to cue earlier breeding by the birds. This is an attractive explanation in part because there are estimated to have been large amounts of short-hydroperiod marsh lost to anthropogenic changes (Browder, 1976; Gunderson and Loftus, 1993; D. M. Fleming, personal communication; Fleming et al., in press) and because it is known that wading birds (at least in the presently degraded system) frequently use higher elevation marshes prior to breeding (Kushlan, 1976; Hoffman et al., 1990).

One difficulty with this idea is that there are still enormous tracts of shallowly flooded, short-hydroperiod freshwater marshes available in most years in northern Water Conservation Areas 1 and 3, the Rotenberger tract, the wetlands east of Krome Avenue and west of developed Miami, the Pennsuco wetlands, most of Big Cypress National Preserve, the Holey Land, Northeast Shark River Slough, and much of Shark River Slough and the Stairstep area. It is not clear that the short-hydroperiod wetlands which have disappeared had different attributes or dynamics than the present ones do, nor why the latter do not seem to perform an early cueing function for the birds. Loftus et al. (1992) suggest that the lowering of the water table and the consequent effects of repeated drying in the rocky glades to the east of Shark River and Taylor sloughs during this century have reduced the capacity of these wetlands to function as refugia for aquatic animals. This lack of hydrological connectedness between refugia and higher elevation marshes is probably mimicked in many areas by the barriers presented by the presence of levees and canals and may well have impacts on survival of, and speed of recolonization by, aquatic animals. These "unconnected" higher elevation marshes could be sufficiently impacted to be of little value for wading birds.

However, even if "unconnected" short-hydroperiod wetlands no longer function as they once did, the remaining area of well-connected short-hydroperiod wetlands appears to be quite large (Big Cypress National Preserve, Northeast Shark River Slough, Stairstep area, northern Water Conservation Areas 3, 2, and 1), and it remains unclear why these rather large chunks of land cannot fulfill the former role of the short-hydroperiod wetlands now lost. It seems likely that short-hydroperiod wetlands must have performed a central cueing function for multispecies wading bird breeding in many years in the predrainage Everglades, especially those years with rapid drying (as currently occurs). However, unless short-hydroperiod wetlands have all been similarly degraded in some fashion since historical times, it is unclear why early breeding should not continue to be cued by the remaining acreage. The drastic reduction in numbers of wading birds and the absence of early breeding therefore seem difficult to pin on the loss of short-hydroperiod wetlands alone.

A second group of hypotheses suggests that it was some set of attributes of the now degraded coastal zone which allowed large numbers of birds to breed (Walters et al., 1992; Bjork and Powell, 1993). This "coastal degradation" hypothesis derives support from several points. First, all of the large historical colonies were in the estuarine zone, and accounts exist of prebreeding foraging aggregations in estuarine and coastal areas (Kushlan et al., 1984; Ogden, 1994; W. B. Robertson, Jr., unpublished aerial survey results). Second, it is becoming obvious that the present estuarine zone has been and remains severely stressed by reduced freshwater flows (McIvor et al., 1994); the corollary is that historical freshwater flows might well have generated vastly greater secondary productivity than is now evident. While this part of the hypothesis is to some degree lacking empirical evidence in the Everglades, it derives much support from studies of estuaries elsewhere (Walters et al., 1992; Bildstein, 1990). Third, the several computer-generated views of historical hydrology agree that flows to the estuary were vastly greater than at present and that most of the freshwater marshes were also much wetter (Walters et al., 1992; Fennema et al., 1994). The length of hydroperiod and depths of water backcast to be in much of the natural Everglades system even suggest that all but the highest elevation marshes of the freshwater area were in most years too deep for wading bird foraging. Nesting may have been concentrated on the coast in part because it had both high productivity and consistently shallow foraging opportunities early in the breeding season.

The estuarine zone may have offered other advantages for feeding which have since degraded. Bjork and Powell (1993) have recently demonstrated that annual changes in sea level are critical for allowing rapid surface water recession in the estuarine zone of northern Florida Bay in the early winter months, a phenomenon which attracts large aggregations of wading birds. Bjork and Powell note that the rise in effective ocean level since the 1930s has been large (Wanless et al., 1994), perhaps large enough to have lessened the extent or changed the timing of this early breeding season drying dynamic. This hypothesis could be tested as future modeling efforts develop the ability to backcast water behavior in the coastal zone.

Other changes may also have been at work. Bancroft et. al (1994) suggest that the average size of fishes available to wading birds may have decreased as a result of shortened hydroperiod in much of the coastal zone. This hypothesis fits with the fact that the present estuarine zone can still cue early and successful breeding by roseate spoonbills in most years, yet fails to do so for wood storks feeding in the same locations. It may be that the average size of fish available is currently too small for wood storks to capture efficiently (Ogden et al., 1976). Ogden (1994, personal communication) has repeatedly asked which aspect of the coastal zone allowed early summer colonies and large summer roosts of ibises to occur well into the 1960s. According to current wisdom, this would have been a season of high and rising water, when prey in marshes should be extremely disaggregated and foraging conditions most difficult. Clearly, key information about prey movements, prey population behavior, and foraging dynamics is missing.

Although coastal and inland processes have been treated separately for clarity of presentation, they should by no means be seen as mutually exclusive. Indeed, in most coastal/freshwater wetland mosaics that have been studied, the alternating availability of food by year and season at inland and coastal sites seems to be critical to maintaining breeding populations of wading birds (Kushlan, 1977; Bildstein et al., 1990; Bildstein, 1990), and there is no apparent reason why this was not true in the predrainage Everglades.

A final set of hypotheses suggests that one or more effects of droughts set the stage for large amounts of food to become available to wading birds in ensuing seasons, in either fresh or estuarine situations. This hypothesized wading bird response occurs some period of time *after* the drought and is distinct from wading bird response to rapid drying and prey concentration at the *beginning* of a period of drying (Kushlan et al., 1975; Frederick and Collopy, 1989).

At least half of the years in the decade 1930–40 were extremely dry in the lower Shark River Slough drainage, and many of these drought years also produced extensive fires in the estuarine/freshwater ecotone (National Audubon Society warden reports; Ogden, 1978, 1994). These descriptions do not seem to fit the profile of the extremely wet, long-hydroperiod Everglades of the natural system computer scenarios, and Robert Porter Allen's contention (in Ogden, 1978; W. B. Robertson, personal communication) that the conditions during these years may have been anomalous is reiterated here. These dry years (1930, 1932, 1935, 1938) were interspersed with the wetter years (1931, 1933, 1934, 1940) in which very large numbers of wading birds bred (Ogden, 1994), giving the impression that the strong nesting effort was in fact associated with some aspect of the preceding droughts.

The surprisingly large nesting response by wading birds noted in the freshwater Everglades in 1992 may be a modern-day example of this process. The 1992 spring nesting season followed the first full wet cycle after the severe drought of 1989–91 and showed by far the largest nesting response recorded in 17 years (over 27,000 nests) (Frederick, 1993). This remarkable nesting season included early and successful breeding by white ibises, rare nestings by wood storks in Water Conservation Area 3, and the first freshwater nestings recorded for roseate spoonbills in Florida since 1910 (R. Bjork, personal communication). Drying patterns and weather during 1992 were favorable for nesting, but surface water recession rates were by no means exceptional in the context of the preceding 10 years and therefore do not seem to offer an explanation for the strong response. The most obvious difference about the season was that white ibises were feeding predominantly on small fishes in the Water Conservation Areas. This species is normally very poor at capturing fishes and may only do so during periods of very high fish abundance (Kushlan, 1974).

One explanation for this abundance of fish is that the severe drought killed off most of the large predatory fishes and other aquatic predators, allowing the proliferation of the "forage" fishes that have much shorter reproductive cycles (Kushlan, 1976). Another hypothesized mechanism is that the extended drying of the marsh surface enhanced primary productivity of the marsh through the release of nutrients by oxidation and fires, leading to increased productivity of wading bird prey. The regularity of drought and large fires in the lower Shark River Slough drainage during the 1930s, and the ensuing large nesting response, would certainly fit with this mechanism. These fires and drying may also explain poorer responses by wading bird in future years by longer term damage to the marsh surface (W. Loftus, personal communication). Perhaps the most extreme form of this hypothesis was voiced by Robert Porter Allen, who suggested more than once (W. B. Robertson, personal communication; Kushlan et al., 1984) that the large headwaters aggregations of wading birds in the 1930s were a direct result of environmental changes in the Everglades through "excessive" drying and fires.

Fires may also benefit birds by altering plant communities and even by creating soil depressions (Gunderson and Snyder, 1994; Davis et al., 1994). In the northern Everglades in 1992, wading birds fed extensively in burned-out depressions in the peat, following the fires of the preceding drought period (Frederick, 1993; Hoffman et al., 1994). These depressions are much shallower and more extensive than alligator ponds and may possess unique qualities for concentrating and holding prey animals. The relationship between drought, fires, secondary productivity, and wading bird foraging seems to demand future research.

It should be noted that fire and drought can only be viewed as a nutrient-liberating disturbance (and a possible primer for wading bird production) when it occurs infrequently in the context of longer hydroperiod cycles. Several pieces of evidence show that longer flooding periods lead to higher biomass, densities, and species richness of marsh fishes (Loftus and Eklund, 1994).

Several of these potential explanations also begin to beg larger questions about how food becomes available. Prey becomes available to wading birds through a mixture of two processes: prey abundance (density or standing stock of animals) and prey availability (how easily animals are caught). Were the large historical aggregations fueled by vastly higher standing stocks of prey animals, allowing successful foraging under a range of conditions, or were hydrodynamics (depths, drying patterns) simply more conducive to the capture of animals? Although neither process can be completely discounted, rapid drying rates do not seem to have been important in cueing breeding during the early and middle part of this century in southern Florida (Ogden, 1994). Similarly, the summer roosts and colonies in the coastal zone (mentioned earlier) seem contrary to the importance of hydrodynamics in making food available.

It is obvious that these questions are embarrassingly fundamental to an understanding of wading bird breeding ecology and that wading birds are far from being "studied to death" in the Everglades. It is also painfully clear that many of these questions can be traced to a poor understanding of the ecology of aquatic prey animals in the estuarine and freshwater parts of the ecosystem. There is currently no systematic monitoring of aquatic animal populations outside of a few sites in Everglades National Park, and very little research is devoted to understanding the ecology of small fishes and macroinvertebrates. As a dramatic example, next to nothing is known about the only species of crayfish (*Procambarus alleni*) in the Everglades, which is a central food item for most wading birds and many fishes, anurans, and reptiles. Until work on prey animals can be expanded, our power to predict wading bird responses in the ecosystem must remain in a rudimentary state.

It should also be emphasized that the much publicized declines of breeding wading birds in the Everglades have probably been paralleled by declines of many other kinds of more poorly monitored aquatic fauna. Anecdotal accounts of collectors and hunters and personal recollections suggest that many species of snakes, turtles, and anurans are far less abundant now. Similarly, next to nothing is known about trends in insect populations, a remarkable lack in one of the only subtropical protected areas in the United States. The lack of data on these animals reveals (as previously) the strong bird- and mammal-centric emphases in faunal monitoring and research, which is an unhealthy situation.

EXOTIC ANIMALS and the EVERGLADES ECOSYSTEM

Since about 1970, the remnant fauna of native vertebrates in southern Florida's overdeveloped coastal strips has been extensively replaced by assemblages of exotic species. Large-scale importing of animals (especially fishes and birds, but also reptiles and amphibians), a thriving cottage industry in wild animal culture, and many private and public collections and exhibits provide opportunities for escape or deliberate introduction. Mild climate and an endless suburban sprawl vegetated with plantings from throughout the tropics tend to maximize a free animal's chances of survival. Although biologists were quick to anticipate the developing problem, their concerns and pleas for regulation (Courtenay and Robins, 1973; Owre, 1973) have been thoroughly overrun by events. Thus, "The fresh waters of southern Florida may host the greatest diversity of non-native fish species of any comparably sized region on earth" (Loftus, 1987), and "Unfortunately, few places on earth rival suburban southern Florida in number and variety of free-flying, non-native birds" (Robertson and Woolfenden, 1992).

A large proportion of the free-living exotic animals encountered in southern Florida are of more or less tropical origin. For example, about half of the reported exotic avifauna is made up of species from the New World tropics, with the remainder divided about equally among the Palearctic, Afro-tropical, Oriental, and Australian zoogeographic realms (Robertson and Woolfenden, 1992). The ongoing faunal introductions into southern Florida constitute an immense and unsupervised experiment in community ecology, but one whose opportunities for study have been little exploited. With few exceptions (e.g., Belshe, 1961; King, 1966; Carleton and Owre, 1975; Wenner and Hirth, 1984), the history and ecological relations of exotic animal populations in southern Florida have not been studied in detail. Interactions with native fauna tend to be poorly known and speculative. Observations, especially of fishes and birds (summarized by Loftus and Kushlan [1987] and Robertson and Woolfenden [1992]), suggest that the species composition of the exotic community and the status of individual populations are highly unstable. Several instances are known in Florida of nonnative species which flourished and expanded their ranges and then, more or less unaccountably, suffered abrupt declines or disappeared. Examples include the Jack Dempsey (*Cichlasoma octofaciatum*) among fishes (Loftus and Kushlan, 1987) and the budgerigar (*Melopsittacus undulatus*), canary-winged parakeet (*Brotogeris versicolurus*), and spot-breasted oriole (*Icterus pectoralis*) among birds (Robertson and Woolfenden, 1992). In addition to coping with a new environment, every exotic doubtless confronts the possibility that the next introduction may be a species that either preys on it or competes more successfully for the same resources. The following accounts summarize the diversity and status of the nonnative element in the Everglades region for each class of vertebrates. A number of nonnative invertebrates are also reported to be established in southern Florida, but (to the authors' knowledge) no comprehensive list of species exists.

Fishes

Most of the exotic fishes known from southern Florida have been found first in canals near the Atlantic coast. The obvious reason is that most of the facilities from which such fishes might escape or be released are in the coastal area. In addition, exotics may be able to establish more readily in the disturbed and often polluted canal habitats. Today, nonnatives dominate the ichthyofauna at many such sites. Once established in a canal, it is possible for fishes to reach almost any part of southern Florida without leaving the canal system. Much of the concern about the presence of nonnative fishes has centered on whether they would tend to stay confined to canals or would spread into the Everglades marshes. It now seems evident that many of the exotic fishes are indeed invading the marshes, perhaps at an accelerating rate since the dual disturbances of severe drought followed by a strong hurricane.

In their review of the fishes of southern mainland Florida, Loftus and Kushlan (1987) reported that 10–12 species of exotic fishes were thought to be established in southern Florida and noted that three of these species had occurred in natural wetlands, albeit in relatively small numbers. They commented hopefully, "…most exotic species seem unable to successfully colonize the marsh system of the southern Everglades." By the early 1990s (Courtenay et al., 1991; W. F. Loftus, personal communication), 15–17 species of exotic fishes were considered established in southern Florida and 7 or more species were in the Everglades. Two of the nonnatives most widely distributed in the southern Everglades, the predacious pike killifish (*Belonesox belizanus*) and the omnivorous Mayan cichlid (*Cichlasoma urophthalmus*), may be somewhat preadapted, because their natural ranges include similar marsh systems on the Yucatan peninsula (Loftus, 1987; Loftus et al., 1992).

Reptiles and Amphibians

As with the fishes, the exotic herpetofauna of southern Florida has increased in recent decades, but many fewer species have shown the ability to penetrate unaltered natural habitats. Duellman and Schwartz (1958) reported 12 exotic species (9 lizards and 3 frogs), of which only the greenhouse frog (*Eleutherodactylus ricordii*), the Cuban treefrog (*Osteopilus septentrionalis*), and the brown anole (*Anolis sagrei*) were commonly found in natural areas. Thirty-five years later (Wilson and Porras, 1983; W. Meshaka, personal communication) the list of more or less established exotics has more than doubled to include 20 lizards, a snake (*Ramphotyphlops*), a turtle (*Psuedemys scripta*), a crocodilian (*Caiman crocydylus*), and four frogs. However, only the same three species previously known (Dalrymple, 1988), plus possibly a Cuban gecko (*Sphaerodactylus elegans*) (W. Meshaka, personal communication), occur to any extent in native habitats.

Birds

Southern Florida's exotic avifauna is remarkably diverse, but (as of 1993) still closely limited to the heavily developed coastal strips. In Florida at large, about 150 species of nonnative birds have occurred in the wild and about 60 species have reportedly nested in the wild. In the urbanized coastal uplands that embrace the Everglades, the respective totals are about 130 and 45 species (Robertson and Woolfenden, 1992). In many coastal neighborhoods, especially of southeastern Florida, exotic landbirds now are probably more speciose and abundant than the remaining native avifauna. However, the nonnatives are still almost entirely absent from extensive tracts of wild lands in the Everglades–Big Cypress system. Thus, no more than about 20 species of nonnative birds are known to have occurred in Everglades National Park (Robertson et al., 1984; Robertson and Woolfenden, 1992), and only two species, the European starling (*Sturnus vulgaris*) and house sparrow (*Passer domesticus*), both limited to developed enclaves, are known to nest within the park. Preliminary observations suggest that hurricane Andrew did not have strong adverse effects on the populations of exotic birds in its path. In fact, individuals that probably escaped from damaged aviaries made at least a temporary addition to the free-flying avifauna. For example, a Homestead neighborhood in the central storm path still had most of its prehurricane exotics plus records of four or five additional species within 6 months after Andrew (J. C. Ogden, P. W. and S. A. Smith, personal communication).

Mammals

The small size of southern Florida's fauna of nonnative wild mammals may mainly reflect a lower level of input from the import trade and fewer amateur enthusiasts than do other vertebrates. Layne (1974) listed only ten species of introduced mammals that may have had persistent populations, and fully half of these were of somewhat uncertain occurrence in southern Florida or were very localized. However, three species of nonnative mammals, all present in southern Florida for at least many decades, are widespread in natural habitats. Nine-banded armadillos (*Dasypus novemcinctus*), which probably spread from introductions in central Florida in the 1920s (Neill, 1952), are common in sandy uplands of the Big Cypress but scarce or absent in wetter parts of the system. The black rat (*Rattus rattus*), frequent around rubbish dumps and buildings, is also extensively naturalized in coastal plant communities, including mangrove swamps. Feral populations of domestic hogs (*Sus scrofa*), commonly augmented by stock released for hunting, are generally distributed in the Big Cypress and elsewhere in the system usually occur around the fringes of uplands.

Concluding Comments: Effects and Prospects

Although the data are patchy and mostly qualitative, some information exists on the impact of nonnative vertebrates in the habitats of southern Florida. As Loftus and Kushlan (1987) pointed out for fishes, the degree of dominance of exotic species at a given site tends to be inversely proportional to the distance of the site from the Atlantic coast. In the developed coastal strips, the fauna of freshwater fishes, lizards, and landbirds is extensively dominated by nonnative forms. Kushlan (1986a) reported on the decline of native fishes and the greatly increased preponderance of exotic fishes in one Dade County canal between 1964 and 1982. In both developed and more natural habitats, predator–prey interactions that involve native and nonnative species are commonly observed. Thus, Takekawa and Beissinger (1983) reported that snail kites (*Rostrhamus sociabilis*) feeding over flooded farm fields apparently preyed on *Pomacea bridgesi,* an aquatic snail introduced from South America. They suggested that this snail might become a significant alternative food source for the kites, because of its ability to thrive in areas with reduced water quality. Geanangel (1986) noted that ospreys (*Pandion haliaetus*) concentrated at phosphate pits in central Florida to feed on exotic *Tilapia* sp., and a number of observers (Loftus, 1979) have reported aggregations of wading birds actively feeding on the exotic walking catfish (*Clarias batrachus*). The frequent occurrence of pike killifish, Mayan cichlids, and spotted tilapia (*Tilapia mariae*) in regurgitations of nestling great egrets and tricolored herons suggests that these exotic fishes make up a substantial part of the food biomass of young wading birds in some Everglades heronries (P. C. Frederick, personal observation) On the other side of the coin, nine-banded armadillos are known to prey extensively on terrestrial invertebrates and small vertebrates; feral hogs at times root into alligator (*Alligator mississippienis*) and turtle nests to feed on the eggs (D. M. Fleming, personal communication), and the introduced Cuban treefrog regularly preys on native species of *Hyla* (W. Meshaka, personal communication).

The possible competitive interactions between exotic and native vertebrates have been the topic of much speculation, but have been studied relatively little. *Anolis sagrei* is thought to compete with the native green anole (*A. carolinensis*) and to have excluded it in some areas. Competitive exclusion probably also figured in the increasing dominance of exotic vertebrates in developed areas, but direct disturbance of native habitats must have played the major role. Some widespread species groups of exotics, such as fishes of the family Cichlidae, often have biological characteristics (e.g., advanced parental care, herbivory) that may enable them to outcompete native analogues, such as largemouth bass (*Micropterus salmoides floridanus*) and sunfish (Centrarchidae) at least locally (W. F. Loftus, personal communication). However, compared to the native fishes, the exotics may produce more available food per unit area of marsh in prey packages more effectively handled by wading birds.

Any present attempt to assess the overall threat posed by nonnative animals to the integrity of the Everglades ecosystem seems futile. Information available on most individual cases is scanty and largely anecdotal, and the subject suffers from a dearth of useful hypotheses. In addition, thought may tend to become paralyzed by the obvious, perhaps insurmountable, difficulty of effective countermeasures. For example, how would one go about removing a well-established, widely dispersed, exotic fish from among all the other fishes in the Everglades? In a hazy, undetailed, nonrigorous sense, one generality seems to emerge from the sparse database: the Everglades, where it is functionally more or less intact, shows considerable resistance to invasion by nonnative species, but with continuing challenges, this faculty tends to break down. However, it is not even clear that the martial images are fully appropriate. The Everglades region has been considered species poor for various groups of animals. Perhaps some exotic species do not so much "invade" as sift into the cracks. In the end, two points are rather obvious: (1) nonnative animals are present by human agency, and at best they are intrusive and unnatural, to be deplored and, if possible, removed, and (2) as long as the "challenge" species continue to be drawn erratically from a grab bag of all the world's fauna, major displacement of native species in natural areas, if not already underway, will surely occur.

ANIMAL MOBILITY and the SCALE of the EVERGLADES LANDSCAPE

Despite being considerably reduced from its original extent and variously degraded, an extensive area of wild land still exists in the Everglades–Big Cypress region. The most hopeful aspect for Everglades restoration is simply the presence of a large, contiguous block of country that has not been ruinously diverted to other uses. As of 1993, the relatively intact (or at least restorable) portion of the Everglades–Big Cypress system measured about 140 × 190 km, with an area of roughly 20,000 km². Thus, the region is, or still has the potential to become, one of the larger natural preserves on earth. Yet, despite its size, one inevitably wonders to what extent the Everglades region is large enough to encourage the activities of its principal animal populations.

For animals with very small range size, the Everglades seems big enough to provide habitat for the foreseeable future. Tree snails of the genus *Liguus,* for example, occur in isolated hammocks and occasionally in undisturbed pine forests. These animals probably rarely move (except by human purpose) between hammocks or traverse large distances across more open habitat (Pilsbry, 1946). The limits to their existence seem to be catastrophic events that occur within the boundaries of the Everglades, such as fires, hurricanes, and rising sea level. Similarly, it is likely that robust populations of many insects and other invertebrates can be maintained in some degree of perpetuity within the scale of the Everglades.

The Cape Sable seaside sparrow (*Ammospiza maritima mirabilis*) is an example of an endemic vertebrate whose lifetime needs may well be served by the short-hydroperiod marshes and prairies that fringe uplands of the southern Ever-glades and Big Cypress. Adults (at least adult males) are so sedentary that they rarely move more than a few hundred meters unless forced to do so by flooding or marsh fires (Kushlan et al., 1982; Werner, 1975; Werner and Woolfenden, 1983). Even dispersing juveniles may move no more than several kilometers to form new territories (Werner, 1975). Because the historical range of this species is contained almost entirely within Everglades National Park, the ecosystem is by definition large enough to contain a viable population. The major threat for the Cape Sable seaside sparrow is probably habitat loss due to rising sea level. It remains to be seen whether new habitat will be formed at successively higher elevations and whether the birds will move into it.

It is unclear how large an area is needed by most of the species of freshwater fishes, because very little information on their movements and ecological needs is available. Although it is obvious that many of the smaller fish species are able to colonize previously dry areas from local alligator holes and depressions with considerable speed, it is unclear what the upper limits to these recolonization movements might be and whether this occupation of new territory is accomplished by long-distance movement of individuals or by the leap-frogging of successive generations from an original group of survivors. It is therefore difficult to guess how large an Everglades is needed to maintain healthy populations of most freshwater fishes and, specifically, whether dikes and canals constitute important barriers to movement. Many of the more wide-ranging vertebrates in the Everglades may be genetically vulnerable because of reduced chances of interchange with other populations. Even those with relatively sedentary habits and small territory sizes may be at risk. The breeding land avifauna of Long Pine Key in Everglades National Park has lost about a quarter of its former species during this century (see next section). Although individual breeding territory sizes of these species are generally small, the roughly 30 km^2 of pine forest on Long Pine Key may simply be below the minimum area that many landbird populations require for a high probability of survival. For example, no historical record of any of the characteristic breeding birds of southern Florida mainland pine forests exists for the approxi-mately 9-km^2 pine area of the Lower Florida Keys (Ross et al., in press; Howell, 1932; Robertson and Woolfenden, 1992). Conversely, the approximately 400-km^2 pine forest of Big Cypress National Preserve (J. R. Snyder, personal communication) appears to lack only one species (American kestrel [*Falco sparverius*]) of the expected breeding avifauna (Patterson and Robertson, 1981; D. Jansen, personal communication; K. D. Meyer, personal communication).

The Everglades in its present extent seems far too small to support extremely wide-ranging mammals such as the Florida panther (*Felis concolor coryi*). Six radio-collared individuals from eastern Everglades National Park showed home ranges of about 350–750 km^2 for five females and a home range of about 2000 km^2 for a semi-nomadic subadult male, and subadult males at times have undertaken cross-country forays that spanned straight-line distances of up to approximately 100 km (Smith and Bass, 1994; D. S. Maehr, personal communication). These movements frequently bring panthers into contact with roads and other boundaries of the Everglades, resulting in a high incidence of roadkills and other accidents (Smith and Bass, 1994) (see next section). The greater Everglades ecosystem in its present state appears to be simply too small to maintain a genetically diverse and sufficiently populous deme of Florida panthers.

The solution to these problems of isolation seems to be the ability to move to other locations for foraging and breeding. Thus, manatees (Anon., 1989), wading birds (Kushlan, 1986b; R. B. Bjork and G. V. N. Powell, unpublished data; see also previously in this chapter), and snail kites (Bennetts et al., 1994) all possess the ability to move from the Everglades into more suitable locations as habitat conditions warrant. All three show seasonal patterns of movement, but it is known that movements at least for the birds may also be triggered by foraging or breeding conditions (Bancroft et al., 1994; Beissinger and Takekawa, 1983; R. B. Bjork and G. V. N. Powell, unpublished data). These habits not only provide flexibility in adjusting to conditions within the Everglades, but they also must provide greatly increased opportunities for genetic mixing with a larger population. These movements allow snail kites and wading birds to function within a wetland mosaic from several times to an order of magnitude larger than the Everglades, respectively, and manatees to move through an aquatic mosaic one order of magnitude larger than that of the Everglades.

It is not surprising that examination of these species suggests that the present Everglades ecosystem is too small to meet the lifetime requirements of viable populations of a number of wide-ranging species. It is equally true, however, that the Everglades remains a key core area for many of the species for which it is strictly too small as an island reserve (panthers, wading birds, and migratory songbirds). It can only be concluded that the persistence of many of the more widely ranging animals of the Everglades depends at least as much upon linkages with areas outside the Everglades as it does upon changes and management within the Everglades. The restricted nature of present and potential future corridors for overland movement, however, seems to favor only animals that can leave the system entirely in response to seasonal or annual conditions.

RESTORING the EVERGLADES: WETLANDS and UPLANDS

The highly variable wetland-upland complex of the Everglades forms an integrated, heterogenous landscape. However, it is often forgotten that today's Everglades is a composite of two terrains that differ greatly in age and history. It was noted earlier that the Everglades wetland is relatively youthful in its present setting. Its oldest peats date back to about 5000 YBP (Gleason and Stone, 1994). In contrast, the upland component of the landscape consists of eroded remnants of a much older upland which reached its greatest extent at the last glacial maximum about 20,000 YBP (Webb, 1990). The total upland area and vegetation of southern Florida has been shrinking as a result of rising sea level since about 14,000 YBP (Robbin, 1984), and recent data (Alexander, 1953, 1974; Ross et al., in press) indicate that the process is continuing.

Everglades restoration is the dominant theme of conservation in southern Florida, and, understandably, thinking about restoration has focused on hydrological remedies for the diminished productivity and size of regional wetlands. Thus, most of the chapters in the faunal section deal with aspects of the ecology of wetland animals. Their authors are almost unanimous in asserting that population recovery for various species requires a return to freshwater hydropatterns more like those attributed to the original Everglades in the natural system hydrology model (Fennema et al., 1994). More specifically, they prescribe (1) increased wet season flows to flood and revive productivity of the short-hydroperiod marshes of the southern Everglades, along with (2) sufficient dry season flow to hold surface water in deeper sloughs in most years (Fleming et al., in press; Loftus and Eklund, 1994; Loftus et al., 1992; Ogden, 1994). Only one faunal chapter (Smith and Bass, 1994) addresses animal populations of the uplands and upland fringes in any detail. Some more general aspects of the ecology of upland animal species in the Everglades system will be discussed briefly here, and two case histories will be presented. It will be suggested that the regional wetland and upland habitats and faunas differ markedly in their history, their present condition, and especially their potential for restoration.

Setting the Scene

For purposes of this discussion, wetlands (mainly sawgrass [*Cladium jamaicense*], spikerush [*Eleocharis* spp.], and water lily [*Nymphaea odorata*] marshes and cypress [*Taxodium distichum*] strands) are defined as areas with fairly deep peat and/or marl substrates flooded for 6 months or longer in recent average years. Uplands (mainly pine forest and hardwood hammocks) are areas of limestone rockland (Snyder et al., 1990) seldom (or only peripherally) flooded in average years. Between the wetlands and uplands is a series of transitional, predominantly herbaceous communities (wet prairie, mixed prairie, scrub cypress, *Muhlenbergia* prairie, rocky glades, etc.) with hydroperiods in recent average years ranging from about 6 months to less than 1 month. Hydrological restoration presumably would tend to shift the average wetland/transitional interface (i.e., the line of 6-month flooding) a certain distance upslope and tend to further reduce the extent of uplands.

Some of the salient differences between the Everglades wetlands and uplands are outlined in Table 28.1. As suggested therein, the historical differences between the two landscape elements are reflected by striking differences in the characteristics of the upland and wetland biota. Thus, endemism of both plants and vertebrate animals is strongly concentrated in the limited, older upland area and is little evident in the expansive, much younger wetlands. Moreover, vertebrate species represented by relatively small populations appear to be much more numerous in the upland habitats. Finally, the uplands, but not the wetlands, are known to have lost vertebrate species within the period of scientific record (Table 28.1).

Wetland Faunal Restoration Prospects

Biologically, the restoration of Everglades wetlands appears eminently feasible (see Table 28.1). Although considerably smaller than it used to be, the natural area is still very large and is potentially one of the largest wetland preserves on earth. Parts of the area are degraded in various ways, but most is relatively intact or at least restorable. Also, the area exists as a contiguous piece virtually without internal

Table 28.1 Some Comparisons of Wetland and Upland Components
of the Everglades–Big Cypress Ecosystem

Parameter	Wetlands	Uplands[a]
Historical		
Time of most recent greatest extent	Mid-19th century prior to drainage	ca. 20,000 YPB
Estimated percent natural area lost since 1900	ca. 40–50%	ca. 80%
Physical		
Estimated present size of restorable natural area	ca. 20,000 km^2	ca. 1,000–1,500 km^2
Internal habitat continuity	One continuous area	Very fragmented; internal gaps up to ca. 50 km
External habitat continuity	Narrowly disjunct; gaps of 25–40 km	Widely disjunct; gaps of 100–200 km
Biological		
Botanical endemism[b]	None known [?]	ca. 40 taxa[e]
Vertebrate endemism[c]	2–3 subspecies	20–25 subspecies and species[e]
Known local loss of breeding vertebrates[d]	None known	ca. 10 species
Relict and sedentary populations	Few	ca. 75, most of the endemics plus a number of others

[a] Pine forests, hammocks, and higher herbaceous rocklands; wetlands = the rest of the landscape.
[b] Avery and Loope (1980), Snyder et al. (1990).
[c] Duellman and Schwartz (1958), Layne (1974), Stevenson (1976), Snyder et al. (1990).
[d] Robertson (1955), Layne (1974), Robertson and Kushlan (1974).
[e] Includes Florida Keys.

barriers other than levees and other water control structures. While these significant anthropogenic barriers may restrict movements of fishes and other truly aquatic animals, the blocks of land within each basin seem large relative to the probable movements of fishes.

Most importantly, the Everglades wetland still has its full historical complement of vertebrate animals. Although the two federally listed endangered species of these wetlands, the wood stork (ca. 6000 adults) (J. C. Ogden, personal communication) and snail kite (733 individuals in 1992 census) (Bennetts, 1993), are now rare as breeders in Everglades National Park, their populations are still relatively large and have shown signs of being able to save themselves by their mobility (see preceding section). This same mobility demands, however, that effective conservation of these species must take place at scales much larger than the Everglades. It seems that there is no obvious biological reason why restoration of Everglades wetlands should fail. Given time and the experimental flexibility afforded by the system of water control structures, it does not appear that wetland restoration should even be particularly difficult. The only real problem is political—the need to resolve competing uses of resources and competing visions of the future.

Upland Restoration Prospects

Quite unlike the wetlands, the prospects for restoring or even maintaining many of the species of the Everglades uplands seem dim and stem largely from relatively recent reductions in the amount and continuity of upland habitats in southern Florida. The recent history of the Florida panther in the southeastern Everglades illustrates the abrupt changes of fortune that can befall small populations in fairly constrained habitat. Barely two years after concluding that the nine individuals in the southeastern Everglades constituted a stable element in the Florida panther population (Smith and Bass, 1994), the Florida panther effectively became extinct (no known reproductive females) in southern Florida east of Shark River Slough (Bass, 1991). Population biologists had been very concerned about the threat that close inbreeding and genetic depression posed to this small, semi-isolated population, but the actual extinction event seems to have been a series of isolated accidents.

Undoubtedly, this disappearance of the Florida panther from a part of its remnant range is only temporary. The species may reintroduce itself by crossing Shark River Slough from the southern Big Cypress (see preceding section), or, if natural dispersal fails, reintroduction of animals from the captive population or from other wild populations would seem likely. However, the point of interest in this event for the panther and other high-risk upland vertebrates is that local extinctions seem likely to be fairly frequent in populations at the ends of peninsulas of fragmented habitat, such as the Everglades uplands. Recruitment along coastal corridors from the more extensive uplands farther north in Florida once must have tended to balance deficits in these terminal populations. No shred of a functional Atlantic upland corridor exists today. Hurricane Andrew eliminated even the illusion of widely spaced stepping stones of native upland habitats along a swath 25 mi wide in southeastern Florida. A rather tenuous and interrupted corridor of uplands still present on the western edge of the Everglades is becoming increasingly fragmented.

Breeding birds of the rockland pine forest of Long Pine Key in Everglades National Park provide perhaps the clearest case of a progressive loss of species diversity in Everglades uplands. Long Pine Key is the extreme terminus of the limestone ridge along the southeast coast of Florida. It is an archipelago of pine-forested, rocky islands surrounded by Everglades marsh. Avifaunal data are available from the 1920s (Holt and Sutton, 1926; Howell, 1921, 1932), from the early 1950s (Robertson, 1955), and more or less continuously in more recent decades (Robertson and Kushlan, 1974). These data indicate that in the 1920s Long Pine Key and nearby pine areas to the east supported practically all the breeding landbirds expected in pine forests of the coastal plain of the southeastern United States, but that by the early 1990s at least ten species appeared to be gone from the area. This amounts to a 26% loss from the known breeding avifauna of 38 species. The landbirds thought to have become extinct on Long Pine Key within the period of scientific record and their probable dates of disappearance are listed in Table 28.2.

Robertson and Kushlan (1974) found that habitat disturbance on Long Pine Key seemed totally inadequate as an explanation for most species losses there. Of the species concerned, only the red-cockaded woodpecker is likely to have been displaced by lumbering in parts of Long Pine Key in the 1930s and 1940s. Near total

Table 28.2 Bird Species Formerly Breeding in Pine Forests
of Long Pine Key, Everglades National Park, and Now Apparently Extirpated

Species	Last reported	Comment
American kestrel (*Falco sparverius*)	ca. 1940	Also absent from Big Cypress pinelands
Wild turkey (*Meleagis gallopavo*)	Mid-1970s	History in area not clear; reintroduced once
Ground dove (*Columbina passerina*)	1980s?	Always scarce in area; a few still present in local farmlands
Hairy woodpecker (*Picoides villosus*)	1991[a]	Becoming rare in Big Cypress
Red-cockaded woodpecker (*Picoides borealis*)	Early 1940s	Persisted near Homestead to ca. 1964; still about 40 clans in Big Cypress
Eastern kingbird (*Tyrannus tyrannus*)	ca. 1990?	Nested commonly in 1950; still present in area?
Brown-headed nuthatch (*Sitta pusilla*)	Uncertain	Present in 1920s; gone by early 1950s
Eastern bluebird (*Sialia sialis*)	Mid-1960s	Fairly common and nesting in early 1950s
Loggerhead shrike (*Lanius ludovicianus*)	Uncertain	Nested in 1950s; a few still present in Homestead area
Summer tanager (*Piranga rubra*)	1920s?	Nested in Miami area up to ca. 1970s; becoming rare in Big Cypress

[a] E. Lewis, personal communication.

removal of pine forest (for agriculture) on private lands east and northeast of Long Pine Key may have contributed to species loss by eliminating the possibility of local recruitment.

Upland Restoration Prospects: Summary

Loss of species from the southern end of the peninsular uplands may well have started soon after the upland area attained its fullest extent at the last glacial peak. The process may have been accentuated around 5000 years ago, when the origin of the Everglades separated the southern uplands into eastern and western branches. The progressive loss of species diversity in Everglades uplands seems rooted inexorably in a natural phenomenon: postglacial sea level rise. A local anthropogenic element became significant only within the past four or five decades, as human occupation of southern Florida effectively eliminated recruitment along the northward-connecting corridors of upland. It is not certain what the appropriate conservation response to this kind of environmental degradation may be.

One approach might be to undertake re-establishing the corridors, as Harris (1984) and others have suggested and as Noss (1987) outlined for peninsular

Florida. If this objective were pursued urgently, it might still be possible to establish useful upland wildlife corridors west of the Everglades. It is doubtful that upland corridors are feasible in southeastern Florida, where many areas of the former coastal uplands are densely settled well into the edge of the Everglades.

Another alternative, perhaps the only other action alternative, is a program of intensive, long-term, many-species management. Such an effort would need to include frequent monitoring and the augmentation or re-establishment of failing populations. With some 25 species of vertebrates alone to be considered just as the start, this would be an immense undertaking, but, if adequately supported, could succeed. However, one inevitably wonders how prominent a place on the conservation agenda should be accorded to the re-establishment of outpost populations of species that may be fairly common elsewhere. Philosophically, the answer may be elusive, but practically, such sustained and intensive management is likely to occur only for a few high-profile species.

It seems that it should be readily possible to restore a fully functional Everglades wetland system that will retain viability well into the next century and perhaps beyond. As for the Everglades uplands, a continuing loss of species diversity of vertebrates is anticipated until the uplands at length become principally preserves for rare plants, invertebrates, and those vertebrates that either use very small ranges or are mobile enough to access alternative breeding opportunities outside the Everglades ecosystem.

LITERATURE CITED

Alexander, T. R. 1953. Plant succession on Key Largo, Florida, involving *Pinus caribaea* and *Quercus virginiana*. *Q. J. Fla. Acad. Sci.*, 16:133–138.

Alexander, T. R. 1974. Evidence of recent sea level rise derived from ecological studies on Key Largo, Florida. in *Environments of South Florida: Present and Past*, P. J. Gleason (Ed.), Miami Geological Society, Coral Gables, Fla., pp. 219–222.

Allen, R. P. 1942. The Roseate Spoonbill, Res. Rep. #2, National Audubon Society, New York.

Anon. 1989. Florida Manatee Recovery Plan, Florida Manatee Recovery Team, U.S. Fish and Wildlife Service, Atlanta.

Avery, G. N. and L. L. Loope. 1980. Endemic Taxa in the Flora of South Florida, Rep. No. T-588, South Florida Research Center, Everglades National Park, Homestead, Fla.

Bancroft, G. T., W. Hoffman, and R. Sawicki. 1992. The importance of the Water Conservation Areas of the Everglades to the endangered wood stork (*Mycteria americana*). *Conserv. Biol.*, 6:392–398.

Bancroft, G. T., A. M. Strong, R. J. Sawicki, W. Hoffman, and S. D. Jewell. 1994. Relationships among wading bird foraging patterns, colony locations, and hydrology in the Everglades. in *Everglades: The Ecosystem and Its Restoration*, S. M. Davis and J. C. Ogden (Eds.), St. Lucie Press, Delray Beach, Fla., chap. 25.

Bass, O. L., Jr. 1991. Ecology and Population Dynamics of the Florida Panther in Everglades National Park, 1991 Annu. Rep., South Florida Research Center, Everglades National Park, Homestead, Fla., pp. V-4-i–V-4-13.

Beissinger, S. R. and J. E. Takekawa. 1983. Habitat use and dispersal of snail kites in Florida during drought conditions. *Fla. Field Nat.*, 11:89–106.

Belshe, J. F. 1961. Observations on an Introduced Tropical Fish, *Belonesox belizanus*, in Southern Florida, M.S. thesis, University of Miami, Coral Gables, Fla.

Bennetts, R. E. 1993. The snail kite, a wanderer and its habitat. *Fla. Nat.,* 66(1):12–15.

Bennetts, R. E., M. W. Collopy, and J. A. Rodgers, Jr. 1994. The snail kite in Florida Everglades: A food specialist in a changing environment. in *Everglades: The Ecosystem and Its Restoration,* S. M. Davis and J. C. Ogden (Eds.), St. Lucie Press, Delray Beach, Fla., chap. 21.

Bent, A. C. 1926. Life histories of North American marsh birds. *U.S. Nat. Mus. Bull.,* No. 135.

Bildstein, K. L. 1990. Status, conservation and management of the scarlet ibis (*Eudocimus ruber*) in the Caroni Swamp, West Indies. *Biol. Conserv.,* 54:61–78.

Bildstein, K. L., W. Post, J. Johnston, and P. C. Frederick. 1990. Freshwater wetlands, rainfall, and the breeding ecology of white ibises in South Carolina. *Wilson Bull.,* 102:84–98.

Bjork, R. B. and G. V. N. Powell. 1993. Relationships between Hydrologic Conditions and Quality and Quantity of Foraging Habitat for Roseate Spoonbills and Other Wading Birds in the C-111 Basin, Report to Everglades National Park, Homestead, Fla.

Bond, J. 1956. *Check-List of Birds of the West Indies,* Academy of Natural Sciences, Philadelphia.

Browder, J. A. 1976. Water, Wetlands and Wood Storks in Southwest Florida, Ph.D. dissertation, University of Florida, Gainesville.

Busack, S. D. and S. B. Hedges. 1984. Is the peninsular effect a red herring? *Am. Nat.,* 123:266–275.

Byrd, M. A. 1978. Dispersal and movements of six North American Ciconiiforms. in *Wading Birds,* A. Sprunt IV, J. C. Ogden, and S. Winkler (Eds.), Res. Rep. #7, National Audubon Society, New York, pp. 161–185.

Carleton, A. R. and O. T. Owre. 1975. The red-whiskered bulbul in Florida. *Auk,* 92:40–57

Cook, R. E. 1969. Variation in species density in North American birds. *Syst. Zool.,* 18:63–84.

Courtenay, W. R., Jr. and C. R. Robins. 1973. Exotic aquatic organisms in Florida with emphasis on fishes: A review and recommendations. *Trans. Am. Fish. Soc. Bull.,* No. 1021.

Courtenay, W. R., Jr., D. P. Jennings, and J. D. Williams. 1991. Exotic fishes. in *Common and Scientific Names of Fishes from the United States and Canada,* 5th ed., C. R. Robins et al. (Eds.), Spec. Publ. No. 20, American Fisheries Society, Bethesda, Md., pp. 97–107.

Custer, T. W. and R. G. Osborn. 1975. Wading birds as biological indicators: 1975 colony survey. *U.S. Fish Wildl. Serv. Spec. Sci. Rep.,* No. 206.

Dalrymple, G. H. 1988. The herpetofauna of Long Pine Key, Everglades National Park, with relation to vegetation and hydrology. in *The Management of Amphibians, Reptiles, and Small Mammals in North America,* R. C. Azaro, K. E. Stevenson, and D. R. Patton (Eds.), U.S. Forest Service Symp., Gen. Tech. Rep. RM-166, U.S. Department of Agriculture, Flagstaff, Ariz., pp. 72–86.

Davies, T. D. and A. D. Cohen. 1989. Composition and significance of the peat deposits of Florida Bay. *Bull. Mar. Sci.,* 44:387–398.

Davis, S. M., L. H. Gunderson, W. A. Park, J. Richardson, and J. Mattson. 1994. Landscape dimension, composition, and function in a changing Everglades ecosystem. in *Everglades: The Ecosystem and Its Restoration,* S. M. Davis and J. C. Ogden (Eds.), St. Lucie Press, Delray Beach, Fla., chap. 17.

DeAngelis, D. L. and P. S. White. 1994. Ecosystems as products of spatially and temporally varying driving forces, ecological processes, and landscapes: A theoretical perspective. in *Everglades: The Ecosystem and Its Restoration,* S. M. Davis and J. C. Ogden (Eds.), St. Lucie Press, Delray Beach, Fla., chap. 2.

De Pourtales, L. F. 1877. Hints on the origin of the flora and fauna of the Florida Keys. *Am. Nat.,* 11:137–144.

Duellman, W. E. and A. Schwartz. 1958. Amphibians and reptiles of southern Florida. *Bull. Fla. State Mus. Biol. Sci.,* 3:181–324.

Egler, F. E. 1952. Southeast saline Everglades vegetation, Florida, and its management. *Vegetatio,* 3:213–265.

Fennema, R. J., C. J. Neidrauer, R. A. Johnson, T. K. MacVicar, and W. A. Perkins. 1994. A computer model to simulate natural Everglades hydrology. in *Everglades: The Ecosystem and Its Restoration,* S. M. Davis and J. C. Ogden (Eds.), St. Lucie Press, Delray Beach, Fla., chap. 10.

Fleming, D. M., W. F. Wolff, and D. L. DeAngelis. In press. The importance of landscape heterogeneity to wood storks in the Florida Everglades. *Environ. Manage.*

Frederick, P.C. 1993. Wading Bird Nesting Success Studies in 1992 and 1993, Report to the South Florida Water Management District, West Palm Beach.

Frederick, P. C. and M. W. Collopy. 1989. Nesting success of five ciconiiform species in relation to water conditions in the Florida Everglades. *Auk,* 106:625–634.

Geanangel, C. 1986. Apparent recent population shift of wintering ospreys from south Florida to central Florida. *Lake Region Nat.,* 1985–1986:22–23.

Gleason, P. J. and P. A. Stone. 1994. Age, origin, and landscape evolution of the Everglades peatland. in *Everglades: The Ecosystem and Its Restoration,* S. M. Davis and J. C. Ogden (Eds.), St. Lucie Press, Delray Beach, Fla., chap. 7.

Gunderson, L. H. and W. F. Loftus. In press. The Everglades. in *Biodiversity of the Southeastern United States: Lowland Terrestrial Communities,* W. H. Martin, S. G. Boyce, and A. C. Echternacht (Eds.), John Wiley & Sons, New York, pp. 199–255.

Gunderson, L. H. and J. R. Snyder. 1994. Fire patterns in the southern Everglades. in *Everglades: The Ecosystem and Its Restoration,* S. M. Davis and J. C. Ogden (Eds.), St. Lucie Press, Delray Beach, Fla., chap. 11.

Hancock, J. A., J. A. Kushlan, and M. P. Kahl. 1992. *Storks, Ibises and Spoonbills of the World,* Academic Press, San Diego.

Harris, L. D. 1984. *The Fragmented Forest: Island Biogeography Theory and the Preservation of Biotic Diversity,* University of Chicago Press, Chicago.

Hoffman, W., G. T. Bancroft, and R. J. Sawicki. 1990. Wading Bird Populations and Distributions in the Water Conservation Areas of the Everglades: 1985 to 1988, Report to the South Florida Water Management District, West Palm Beach.

Hoffman, W., G. T. Bancroft, and R. J. Sawicki. 1994. Foraging habitat of wading birds in the Water Conservation Areas of the Everglades. in *Everglades: The Ecosystem and Its Restoration,* S. M. Davis and J. C. Ogden (Eds.), St. Lucie Press, Delray Beach, Fla., chap. 24.

Hoffmeister, J. E. 1974. *Land from the Sea—The Geologic History of South Florida,* University of Miami Press, Coral Gables, Fla.

Holt, E. G. and G. M. Sutton. 1926. Notes on birds observed in southern Florida. *Ann. Carnegie Mus.,* 16:409–439.

Howell, A. H. 1921. A list of the birds of Royal Palm Hammock, Florida. *Auk,* 38:250–283.

Howell, A. H. 1932. *Florida Bird Life,* Coward-McCann, New York.

Kiester, A. R. 1971. Species diversity of North American amphibians and reptiles. *Syst. Zool.,* 20:127–137.

King, F. W. 1966. Competition between Two South Florida Lizards of the Genus *Anolis,* Ph.D. thesis, University of Miami, Coral Gables, Fla.

Kushlan, J. A. 1974. The Ecology of the White Ibis in Southern Florida, A Regional Study, Ph.D. dissertation, University of Florida, Miami.

Kushlan, J. A. 1976. Environmental stability and fish community diversity. *Ecology,* 57:821–825.

Kushlan, J. A. 1977. Population energetics of the American white ibis. *Auk.*

Kushlan, J. A. 1986a. Exotic fishes in the Everglades, a reconsideration of proven impact. *Environ. Conserv.,* 13:67–69.

Kushlan, J. A. 1986b. Responses of wading birds to seasonally fluctuating water levels: Strategies and their limits. *Colon. Waterbirds,* 9:155–162.

Kushlan, J. A., J. C. Ogden, and A. L. Higer. 1975. Relation of Water Level and Fish Availability to Wood Stork Reproduction in the Southern Everglades, Open-file Rep., U.S. Geological Survey, Tallahassee.

Kushlan, J. A., O. L. Bass, Jr., L. L. Loope, W. B. Robertson, Jr., P. C. Rosendahl, and D. L. Taylor. 1982. Cape Sable Sparrow Management Plan, Rep. No. M-660, South Florida Research Center, Everglades National Park, Homestead, Fla

Kushlan, J. A., P. C. Frohring, and D. Vorhees. 1984. History and Status of Wading Birds in Everglades National Park, Everglades National Park, National Park Service, Homestead, Fla.

Layne, J. N. 1974. The land mammals of south Florida. in *Environments of South Florida,* P. J. Gleason (Ed.), Miami Geological Society, Coral Gables, Fla., pp. 386–413.

Loftus, W. F. 1979. Synchronous aerial respiration by the walking catfish in Florida. *Copeia,* 1:156–158.

Loftus, W. F. 1987. Possible establishment of the Mayan cichlid, *Cichlasoma uriophthalmus* (Gunther) (Pisces: Cichlidae), in Everglades National Park, Florida. *Fla. Sci.,* 50:1–6.

Loftus, W. F. and A.-M. Eklund. 1994. Long-term dynamics of an Everglades small-fish assemblage. in *Everglades: The Ecosystem and Its Restoration,* S. M. Davis and J. C. Ogden (Eds.), St. Lucie Press, Delray Beach, Fla., chap. 19.

Loftus, W. F. and J. A. Kushlan. 1987. Freshwater fishes of southern Florida. *Bull. Fla. State Mus. Biol. Sci.,* 31:147–344.

Loftus, W. F., R. A. Johnson, and G. H. Anderson. 1992. Ecological impacts of the reduction of groundwater levels in short-hydroperiod marshes of the Everglades. in 1st Int. Conf. on Groundwater Ecology, U.S. Environmental Protection Agency and American Water Resources Association, Washington, D.C., pp. 199–208.

McIvor, C. C., J. A. Ley, and R. D. Bjork. 1994. A review of changes in freshwater inflow from the Everglades to Florida Bay including effects on biota and biotic processes. in *Everglades: The Ecosystem and Its Restoration,* S. M. Davis and J. C. Ogden (Eds.), St. Lucie Press, Delray Beach, Fla., chap. 6.

Means, D. B. and D. Simberloff. 1987. The peninsula effect: Habitat correlated species decline in Florida's herpetofauna. *J. Biogeogr.,* 14:551–568.

Neill, W. T. 1952. The spread of the armadillo in Florida. *Ecology,* 33:282–284.

Noss, R. F. 1987. Protecting natural areas in fragmented landscapes. *Nat. Areas J.,* 7:2–13.

Ogden, J. C. 1978. Recent population trends of colonial wading birds on the Atlantic and Gulf coastal plains. in *Wading Birds,* A. Sprunt IV, J. C. Ogden, and S. Winckler (Eds.), National Audubon Society, New York.

Ogden, J. C. 1981. Nesting distribution and migration of glossy ibis *Plegadis falcinellus* in Florida, U.S.A. *Fla. Field Nat.,* 9:1–6.

Ogden, J. C. 1991. Nesting by wood storks in natural, altered and artificial wetlands in central and northern Florida. *Colon. Waterbirds,* 14:39–45.

Ogden, J. C. 1994. A comparison of wading bird nesting colony dynamics (1931–1946 and 1974–1989) as an indication of ecosystem conditions in the southern Everglades. in *Everglades: The Ecosystem and Its Restoration,* S. M. Davis and J. C. Ogden (Eds.), St. Lucie Press, Delray Beach, Fla., chap. 22.

Ogden, J. C. and S. A. Nesbitt. 1979. Recent wood stork population trends in the United States. *Wilson Bull.,* 91:512–523.

Ogden, J. C., J. A. Kushlan, and J. T. Tilmant. 1976. Prey selectivity by the wood stork. *Condor,* 78:324–330.

Ogden, J. C., D. A. McCrimmon, Jr., G. T. Bancroft, and B. W. Patty. 1987. Breeding populations of the wood stork in the southeastern United States. *Condor,* 89:752–759.

Owre, O. T. 1973. A consideration of the exotic avifauna of southeastern Florida. *Wilson Bull.,* 85:491–500.

Patterson, G. A. and W. B. Robertson, Jr. 1981. Distribution and Habitat of the Red-Cockaded Woodpecker in Big Cypress National Preserve, Rep. No. T-613, South Florida Research Center, Everglades National Park, Homestead, Fla.

Patterson, G. A., W. B. Robertson, Jr., and D. E. Minsky. 1980. Slash pine-cypress mosaic, breeding bird census. *Am. Birds,* 34:61–62.

Pilsbry, H. A. 1946. *Land Mollusca of North America,* Monogr. #3, Vol. II Part I, Academy of Natural Sciences, Philadelphia, pp. 37–102.

Post, W. 1990. Nest survival in a large ibis-heron colony during a three-year decline to extinction. *Colon. Waterbirds,* 13:50–61.

Powell, G. V. N., R. D. Bjork, J. C. Ogden, R. T. Paul, A. H. Powell, and W. B. Robertson, Jr. 1989. Population trends in some Florida Bay wading birds. *Wilson Bull.,* 101:436–457.

Robbin, D. M. 1984. A new Holocene sea level curve for the Upper Florida Keys and the Florida reef tract. in *Environments of South Florida: Present and Past II,* P. J. Gleason (Ed.), Miami Geological Society, Coral Gables, Fla., pp. 437–458.

Robertson, W. B., Jr. 1955. An Analysis of the Breeding-Bird Populations of Tropical Florida in Relation to the Vegetation, Ph.D. thesis, University of Illinois, Urbana.

Robertson, W. B., Jr. and J. A. Kushlan. 1974. The southern Florida avifauna. in *Environments of South Florida: Present and Past,* P. J. Gleason (Ed.), Miami Geological Society, Coral Gables, Fla., pp. 414–452.

Robertson, W. B., Jr. and G. E. Woolfenden. 1992. *Florida Bird Species: An Annotated List,* Spec. Publ. No. 6, Florida Ornithological Society, Gainesville.

Robertson, W. B., Jr., O. L. Bass, Jr., and M. Britten. 1984. *Birds of Everglades National Park (Checklist),* Everglades Natural History Association, Homestead, Fla.

Rodgers, J. A., A. S. Wenner, and S. T. Schwikert. 1987. Population dynamics of wood storks in north and central Florida, U.S.A. *Colon. Waterbirds,* 10:151–156.

Ross, M. S., J. J. O'Brien, and L. S. L. Sternberg. In press. Sea level rise and the reduction in pine forests in the Florida Keys. *Environ. Appl.*

Runde, D. E. 1991. Trends in Wading Bird Nesting Populations in Florida, 1976–1978 and 1986–1989, Final Performance Report, Non-Game Section, Florida Game and Freshwater Fish Commission, Tallahassee.

Ryder, R. A. 1978. Breeding distribution, movements and mortality of snowy egrets in North America. in *Wading Birds,* A. Sprunt IV, J. C. Ogden, and S. A. Winckler (Eds.), Res. Rep. #7, National Audubon Society, New York.

Simpson, G. G. 1964. Species densities of North American recent mammals. *Syst. Zool.,* 13: 57–73.

Smith, T. R. and O. L. Bass, Jr. 1994. Landscape, white-tailed deer, and the distribution of Florida panthers in the Everglades. in *Everglades: The Ecosystem and Its Restoration,* S. M. Davis and J. C. Ogden (Eds.), St. Lucie Press, Delray Beach, Fla., chap. 27.

Smith, P. W. and A. Sprunt IV. 1987. The shiny cowbird reaches the United States. *Am. Birds,* 41:370–371.

Smith, P. W., W. B. Robertson, Jr., and H. M. Stevenson. 1988. West Indian cave swallows nesting in Florida, with comments on the taxonomy of *Hirundo fulva. Fla. Field Nat.,* 16:86–90.

Snyder, J. R., A. Herndon, and W. B. Robertson, Jr. 1990. South Florida rockland. in *Ecosystems of Florida,* R. L. Myers and J. J. Ewel (Eds.), University of Central Florida Press, Orlando, pp. 230–277.

Spackman, W., C. P. Dolsen, and W. Riegel. 1966. Phytogenic organic sediments and sedimentary environments in the Everglades-mangrove complex. Part 1: Evidence of a transgressing sea and its effects on environments of the Shark River area of southwestern Florida. *Palaeontogr. Abt. B,* 117:135–152.

Sprunt, A. 1922. Discovery of the breeding of the white ibis in South Carolina. *Oologist,* 39: 142–144.

Stevenson, H. M. 1976. *Vertebrates of Florida,* University Presses of Florida, Gainesville.

Takekawa, J. E. and S. R. Beissinger. 1983. First evidence of snail kites feeding on the introduced snail *Pomacea bridgesi. Fla. Field Nat.,* 11:107–108.

Tramer, E. J. 1974. Latitudinal gradients in avian diversity. *Condor,* 76:123–130.

Walters, C. J., L. Gunderson, and C. S. Holling. 1992. Experimental policies for water management in the Everglades. *Ecol. Appl.,* 2:189–202.

Wamer, N. O. 1978. Avian Diversity and Habitat in Florida: Analysis of a Peninsular Diversity Gradient, M.S. thesis, Florida State University, Tallahassee.

Wanless, H. R. and M. G. Tagett. 1989. Origin, growth, and evolution of carbonate mudbanks in Florida Bay. *Bull. Mar. Sci.,* 44:454–489.

Wanless, H. R., R. W. Parkinson, and L. P. Tedesco. 1994. Sea level control on stability of Everglades wetlands. in *Everglades: The Ecosystem and Its Restoration,* S. M. Davis and J. C. Ogden (Eds.), St. Lucie Press, Delray Beach, Fla., chap. 8.

Wayne, A. T. 1922. Discovery of the breeding grounds of the white ibis in South Carolina. *Bull. Charleston Mus.,* 17:17–30.

Webb, S. D. 1990. Historical biogeography. in *Ecosystems of Florida,* R. L. Myers and J. J. Ewel (Eds.), University of Central Florida Press, Orlando, pp. 70–100.

Wenner, A. S. and D. H. Hirth. 1984. Status of the feral budgerigar in Florida. *J. Field Ornithol.,* 55:214–219.

Werner, H. W. 1975. The Biology of the Cape Sable Sparrow, Report to the U.S. Fish and Wildlife Service, Everglades National Park, Homestead, Fla.

Werner, H. W. and G. E. Woolfenden. 1983. The Cape Sable sparrow: Its habitats, habits, and history. in *The Seaside Sparrow, Its Biology and Management,* T. L. Quay et al. (Eds.), North Carolina Biological Survey and North Carolina State Museum, Raleigh, pp. 55–75.

Wilson, L. D. and L. Porras. 1983. The Ecological Impact of Man on the South Florida Herpetofauna, Spec. Publ. No. 9, Museum of Natural History, University of Kansas, Lawrence.

V

TOWARD
ECOSYSTEM RESTORATION

29

The STRUCTURE and DYNAMICS
of the EVERGLADES SYSTEM:
GUIDELINES for
ECOSYSTEM RESTORATION

C. S. Holling

Lance H. Gunderson

Carl J. Walters

ABSTRACT

The Everglades system is configured by processes operating across a wide range of scales in space and time. Major land use conversions associated with development have occurred during the past century, transforming a once vast wetland landscape scarcely 5000 years old. The development of the system has occurred in a syndrome of crises and responses that solved the momentary crisis but exaggerated the subsequent crisis. The natural controls of the system have been replaced by human ones, resulting in a deterioration of the valued natural features. The natural system was hierarchically structured with discontinuous attributes. Distinct processes define objects and structures at different hierarchical levels, which provide choices and opportunities for organisms across a wide range of space and time scales. The results of a series of AEAM workshops indicate that key features of this resilient landscape can be restored. Restoration goals can be defined, as can designs to achieve those goals. The guidelines for restoration include (1) tinkering does not work, (2) single quick-fix structural solutions do not work, (3) composite policies do not work, and (4) there are a number of composite policies.

INTRODUCTION

Some 5000 years ago, southern Florida was dominated by an oak savanna landscape. Suddenly, over a few hundred years, the landscape became transformed from the dry conditions characterizing such savannas to a wetter climate with extensive wetlands. This was the continuation of a planetary-wide process of retreat from glaciation. At least 6000 years earlier, human occupation of North America had begun in a massive way as humans invaded in a wave front from Alaska to Patagonia at an average rate of 160 km per decade (Mosimann and Martin, 1975). Eddies of that wave occupied Florida at least 10,000 years ago.

Hence, Floridians witnessed the extraordinary transition from a dry landscape to a wet one. Some relict local plant and animal species survived and combined with invading temperate species from the north and tropical species from the south to establish a potpourri of plant assemblages that characterized the Everglades system up to 100 years ago.

Early in this century, the beginning of another extraordinary transition was witnessed, wrought not by nature this time but by human actions. Perceived from a distance, the transformation has led to extensive urban development of the eastern shores of the peninsula, highly profitable agriculture in the deep peatlands south of Lake Okeechobee, straightening of the meandering Kissimmee River that fed the lake, and partitioning of half of the southern portion into areas for water management, a unique national park, and a national preserve. Coincident with that transformation, however, has been accelerating erosion of the ecosystems that sustain people, economic activities, and the remarkable flora and fauna in the region.

The motivation for this book has been largely determined by the results of that process of transformation. What were the landscape and ecosystem like in the southern third of Florida 100 years ago? What are they like now and what are the forces that made them so? What will they be like in the future?

The book was developed as a process of integrating existing pieces of knowledge. That process was paced and focused over a period of almost 5 years by a series of workshops in which scientists and engineers from different agencies and universities cooperated to integrate their separate knowledge of parts of the system into a simulation model that integrated the whole. The process has been called Adaptive Environmental Assessment and Management, or AEAM (Holling, 1978). Such models are never complete. Some parts become more credible than others, and in the process, gaps in knowledge and method are identified.

A year into the process, a major symposium became a critical venue to formally review knowledge, set priorities, and initiate early drafts of chapters. The result is now a more integrated understanding of the certainties, uncertainties, and unknowns, of which the model represents only one small part. The full body of understanding represents the present scientific representation of the *River of Grass*—that poetic and vivid image of an integrated ecosystem that earlier had mobilized public and political attention in the writings of Marjory Stoneman Douglas and the advocacy of Arthur R. Marshall, Jr.

One focus of this work has been to reconstruct views or images of the ecosystem prior to intensive management, which led to a clearer view of what the system may have looked like in the first few decades of this century. The image came from historical reconstruction of the landscape and physiographic forms of the vegetation and its origins and the status of wading birds in early decades. In addition, the hydrology models for the first time allowed a quantitative reconstruction of the patterns of water movement in space and time by using them to backcast to conditions with no development and no water control facilities. Those reconstructions will constitute a major part of this chapter. Before considering them, the sequence that led to the present state of the system will be encapsulated.

The DEVELOPMENT SYNDROME

The present status of the Everglades is less clear than that of the past because it is such a rapidly moving target as it evolves or devolves under the pressures of development. Those developments have been driven by opportunity for economic development and have been paced by responses to unexpected crises or to anticipated crises that emerged from the unappreciated interaction between development and nature. It is a sequence of crises and responses in which each solution solved the problem of the moment, but at the expense of the integrity of the entire interacting system.

In south Florida, some of the key steps in the sequence were as follows (Blake, 1980; Light et al., in preparation):

- *1926 and 1928:*   Hurricanes during these years led to flooding and major damage. The response was to construct high dikes to isolate Lake Okeechobee and keep water from overflowing into the areas under development in the dense sawgrass peat beds to the south.
- *1947:*   The unexpectedly extreme floods of 1947 came not from overflow from the lake, but from torrential rains coupled with unusually high seasonal water levels. That led to construction of a series of levees and canals to partition the region into an agricultural zone south of Lake Okeechobee (Everglades Agricultural Area), water conservation areas (from north to south: Water Conservation Areas 1, 2A and 2B, 3A and 3B), and a natural area (Everglades National Park). The goal of this massive project was to control the hydrologic variation for flood protection.
- *Early 1960s:*   An unexpected shift in weather led to droughts and fire, the effects of which were exaggerated because the final construction phases of the project kept water away from Everglades National Park. The result was exaggerated fire hazards and intrusion of salt water into the wells serving the growing urban demand. The response was to commit to supplying a defined amount of water for the park and to maintain a freshwater head in the Water Conservation Areas as a freshwater dam to impede saltwater intrusion.
- *Early 1980s:*   Eutrophication of Lake Okeechobee began to be evident as nutrients in runoff from dairy farms appeared in waters supplying the lake from the north and because nutrient-rich water was backpumped from the southern agricultural areas for storage in the lake. This had become possible because of the dikes built after the hurricanes of the 1920s. The solution was to apply the best management practices to control pollution from dairy farms, to stop backpumping, and to move water south, thereby also increasing the quantity of water delivered to Everglades National Park.
- *Late 1980s to early 1990s:*   The nutrient-rich water triggered pollution of

portions of the Everglades system, causing a flip in vegetation from a species-rich sawgrass, wetland, tree island ecosystem to a species-impoverished succession starting with cattails (*Typha domingensis*). This exaggerated the war between "big sugar" and environmentalists and between the federal government (representing Everglades National Park interests) and the state (represented by the Water Management District). Storage of water for urban demand and flood control unexpectedly improved the conditions for one endangered species (the snail kite) and degraded conditions for another (the wood stork). This put the U.S. Fish and Wildlife Service (pro single-species management) at war with the National Park (pro ecosystem management). Because of these battles, the fundamentally important water quantity problem was ignored. Wading bird population decline continued past 90% of the levels recorded during the 1950s. Urban demands for water continue to rise, although restoration efforts for the Kissimmee River and East Everglades have been delayed.

This cycle of crisis and response forms a syndrome with distinct characteristics. The solutions proposed for the crisis of the moment ignored the consequence for the full system, assumed certainty in the future, and succeeded in solving the momentary crisis, but set in motion conditions that exaggerate future crises. The original natural controls, the roles of which were unknown at the time, were replaced by human controls, the functions and controllability of which seemed to be totally apparent. The system became partitioned into apparently manageable pieces of territory. This initiated border wars between different users occupying different territories and the involvement of an increasing number of organizations whose goals began to shift to institutional survival and self-protection.

This is a syndrome in which momentary individual repairs are implemented to address problems that are actually systemic. Political capital is wasted, polarization develops, and public trust in governance wanes. The goal too often seems to be driven by a need to identify enemies and defeat them, rather than to devise systemic solutions that are robust, that create win/win combinations, and that adaptively generate understanding and opportunity (Walters, 1986).

Three positive developments arose against this depressing backdrop. The first was public awareness of the Everglades as a unique treasure and of the sustaining values of the natural system as an integrated regional entity. The second was an accumulation of sufficient scientific understanding of that integrated entity to allow considered action. The third was the emergence of the nucleus for an informal collegium of cooperating scientists and engineers who began to evolve shared values, to develop similar goals for restoration, and to explore ways to achieve those goals jointly. It is significant that the cooperation was informal, drew together people from different agencies and universities, and added a harmonizing force to what at times seemed to be a battlefield with no codes of conduct! This book, the symposium, and the parallel workshops are part of the genesis of that collegium.

It is now clear that critical ecosystem functions can be revived. Both the knowledge and the reserve capacity in the ecosystem are available. What happens depends on people. The Everglades system, albeit reduced to half its size, can be revitalized while maintaining urban demands and transforming agriculture. Some broadbrush policies to achieve this goal were developed by the collegium in an

April 1991 workshop in which the hydrology submodel was used as a policy screening device. The purpose was not to define a single policy, but rather to screen alternatives and to characterize the ingredients for a solution. The full economic, social, and ecological details of a practical policy have yet to be developed, but the general conclusions deserve emphasis in light of the sad history of development and remediation:.

- *Tinkering does not work.* No conceivable pattern of manipulating water delivery schedules can alone return all the critical functions in a sufficiently diverse manner.
- *Single quick-fix structural changes do not work.* Parts of the system may be restored by single structural changes, but often at the price of compromising other restoration goals and flexibility.
- *Composite policies do work.* An integrated set of structural and operational changes can be devised to satisfy the set of restoration goals and provide alternative uses for water.
- *There is more than one composite policy.* The system is not so compromised that only one set of unforgiving solutions is possible. There are several such composites, each with different costs, different degrees of flexibility, and different potentials for adaptive learning.

The goals for the restoration that now clearly seems possible have emerged from understanding the dynamics and structure of the natural ecosystem, which is the next subject of this chapter.

The NATURAL SYSTEM

Overview

The Everglades is one of the flattest and most extensive wetland ecosystems in the world. One image is of a flat limestone plate, overlain with peat and marl, stretching from the southern edge of Lake Okeechobee 200 km south to sea level at Florida Bay. The plate is gently tilted, but decreases in elevation by only 5.3 m over that 200-km stretch (Gunderson and Loftus, 1993). The plate is about 80 km wide and is upturned at its eastern Atlantic side by the rocky Atlantic Ridge and more gently upturned at its western Gulf of Mexico side by a mangrove and marl dam. Those modest barriers are breached in places by transverse depressions through which some water could flow to the northwest and east; most, however, flows south into sloughs feeding Florida Bay.

Water flowed over that plate during floods in a broad expanse of sheet flow at a maximum speed of $36 \text{ m} \cdot \text{h}^{-1}$, not much faster than one one-hundredth the speed of a leisurely walk. If a molecule of water were able to survive the full journey in that torrent, it would take almost 8 months to travel from the lake to Florida Bay, but the chance of that happening was diminishing. The slow speed of movement and the high rate of evapotranspiration meant that the water was continually evaporated, to be replaced by rainfall from convective storms in the summer and frontal systems in the winter.

This does not mean, however, that the southerly parts of the system are not connected to the more northerly. The incline is so gentle and the terrain so flat that in wet seasons the head of water in the north can become great enough to overcome frictional resistance and suddenly push the water more rapidly southward. It is a domino effect, with upstream dominoes initiating movement in a chain southward, without moving far themselves.

The flatness of the terrain means that the slightest (from a human perspective) depression or elevation can have extraordinary biological consequences. The elongated tree islands, for example, are at most only 1.5 m higher than the submerged vegetation in neighboring ponds. Hence a vertical zonation of vegetation communities is produced over a remarkably narrow range of elevations. The "mountains" of the system now are the buildings in the four major urban aggregations on the east coast—100 times the natural range of elevation.

This image of a flat plate is useful but seriously incomplete. Both much larger and much smaller processes are also integral parts of the reality of the Everglades.

At one extreme, planetary processes have led to an abrupt increase in the rate of sea level rise from 4 cm per century for over 3000 years to approximately 40 cm per century since 1930 (Wanless et al., 1994). This rate could increase still further over the next few decades because of the possible effects of global warming from the accumulation of greenhouse gases. Such increases in sea level, coupled with decreased freshwater flows, could account for the erosion in the productivity of the southern estuary and the eastern wetlands (Loftus et al., 1990), thereby reducing feeding options for the large wading birds (Ogden, 1994; Walters et al., 1992). At some point, the rate could become great enough that the dike-building process of siltation and mangrove development could not keep up. Sudden breaching of the mangrove dike during severe storms and hurricanes would then flood large regions of the very low-level land in the southern region.

At the other extreme, microscale processes provide the trophic underpinnings for the ecosystem. Periphyton mats, for example, together with the complex of invertebrates and small fish and the fine-grained vegetation that provided protection from predation, represent the source of aquatic productivity either through a grazing or detrital pathway (Gunderson and Loftus, 1993) that sustains the system.

Between these extremes of scale is a discontinuous gradation of ecosystem entities and processes. A series of aerial photographs and satellite images highlight this discontinuous nature (Plates 10–17).* The first photograph portrays sawgrass (*Cladium jamaicense*) surrounding an alligator hole in the center of the natural Everglades system. Each succeeding picture is approximately one order of magnitude larger, in linear dimension. The sequence covers a range in area of almost 13 orders of magnitude, from approximately 10 m² to the planet as a whole. The first three photographs were taken from a helicopter and have a linear window of 3 m (Plate 11) to 1 km (Plate 13). The remaining images were obtained from archived satellite images and aerial photographs.

The photographic sequence (Plates 10–17) demonstrates three points. First, the ecosystem and landscape are hierarchically structured and have discontinuous architectural attributes. Second, the scale at which the system is observed deter-

---

* Plates 10–17 appear following page 432.

mines the phenomena perceived, the hypotheses devised, and the data collected. Third, the scale at which environmental, economic, or social problems are diagnosed determines the solutions conceived, the degree of conflict engendered, and the cost and robustness of resulting policies. Each of these points is dealt with in the following sections.

The Landscape Hierarchy

Distinct scale breaks in the sequence of images clearly separate different classes of objects and different textures. As one example, what is seen in part of one sequence as a series of pictures of tree islands, sawgrass strands, and wet prairie (Plates 12–14) suddenly flips into representations of aggregates of those plant associations that represent edaphic and use entities-deep peat and marl wetlands, mangrove forests, upland complex, and agriculture (Plate 15). Those breaks in spatial scale identify shifts in hierarchical levels.

Each level is defined by distinct objects which aggregate to form new ones. They not only occur over a distinct range of spatial scales, but are also defined by a distinct range of temporal scales. These temporal scales are determined by the rates of the unique processes that shape each level. These objects and scales are summarized in Table 29.1, and some examples of processes at each level are presented.

Such landscape structures have recently been described for other ecosystems (such as the boreal forest and short-grass prairie) and have been attributed to a small set of keystone variables and keystone processes (Holling, 1992). The same is true for the Everglades. The keystone variables include not only animals such as the alligator (*Alligator mississippiensis*), whose activities maintain holes that provide some refugia for aquatic organisms during the dry season, but also a set of plant and abiotic variables and processes, each operating over different range of scales. Plant processes dominate over the microscale range (Table 29.1), abiotic processes of fire and flood over mesoscales, edaphic processes over macroscales, and geomorphological and evolutionary processes over subcontinental to planetary scales.

There are many species in the landscape other than keystone species. They are affected by the structure and exist because of the opportunities provided. These "entrained" species (Holling, 1992) can therefore be important indicators of the state of the ecosystem. In the Everglades, for example, five species of large wading birds (the great egret [*Casmerodius albus*], snowy egret [*Egretta thula*], tricolored heron [*Egretta tricolor*], white ibis [*Eudocimus albus*], and wood stork [*Mycteria americana*]) make up 90–100% of typical colonies (Ogden, 1994). They have little direct effect on physical architecture, except where import of nutrients can affect local vegetation around rookeries (Frederick and Powell, 1994), but the location, timing, and success of nesting are dramatic indicators of the condition of the landscape. For example, the number of birds nesting during the 1930s and 1940s was ten times the current number nesting. Something has changed the opportunities provided by the landscape.

Just as the landscape is organized into hierarchical classes, so too are the choices that animals make. By comparing these two hierarchies, critical attributes

Table 29.1   Ecosystem Levels, Scales, and Sample Processes

Descriptive level	Window of observation		Scale classes	Processes
	Time (years)	Space		
Leaves, invertebrates, small fish	0.1–2	<10 cm	Micro	Photosynthesis and respiration; decomposition; nutrient uptake
Plants, bushes, periphyton mat	1–10	10 cm–3 m	Micro	Plant growth; seed production; foraging, predation
Alligator hole, pond, tree, battery island	10–100	3–30 m	Micro	Tree growth, competition and mortality; phosphorus immobilization and mobilization; vegetation effects on hydrology; invertebrate and fish population dynamics
Tree island, wet prairie sawgrass strand, slough	30–500	100 m–10 km	Meso	Disturbance dynamics (fire, insect, and disease); seed dispersal; meso-hydrology; fish dispersal
Marl, peat, mangrove, upland regions	500–5000	10–1000 km	Macro	Erosion, deposition, macro-hydrology; silting and dike formation
Land zones	>5000	>1000 km	Planetary	Evolution, geomorphology; planetary dynamics

of the natural system that provide the range of opportunities needed for a viable Everglades system can be identified.

Hierarchy of Choices, Hierarchy of Opportunities

As an example, the hierarchy of choices by wading birds has been reconstructed by Holling (1992). At the largest scale, wading birds potentially have available a number of different wetland regions ranging from the extensive Everglades system itself to the marshes, ponds, and lakes of north-central Florida and even on to those in the Carolinas. Tagging records (Byrd, 1978) indicate that choices of such regions are made over several hundred to 2000 km. Such scales describe the spatial extent for choosing a region. Just as the information in a satellite image of a landscape has not only spatial extent but also resolution, or pixel size, so too does the information utilized at each level of choice have a resolution or grain for that choice (Wiens, 1989, 1990). In this case of locating in a region, the grain is the size of the smallest region of wetlands that can be consistently occupied, probably measured on the order of 100 km.

The choice is strongly influenced by past experience during previous breeding periods. As a consequence of such fidelity to place, changed environmental or

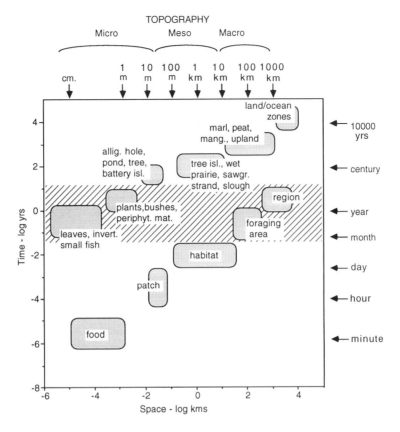

Figure 29.1   Scales in space and time of the landscape hierarchy in the Everglades and of the hierarchy of choices made by wading birds.

landscape attributes can leave a population "trapped" for some years in an environment adequate for survival but inadequate for reproduction. This is what has happened in the Florida Everglades as a consequence of changes in water delivery. Thus, there is not only a spatial scale defining choice, but also a temporal scale for which extent (or time horizon) can be defined as the time taken to extinguish the fidelity for place and for which resolution can be defined as the time since the last experience. In this example, these scales are measured in decades for the time horizon and in years for the time resolution.

Choice of area is only one of a nested set of choices, each of which has its own spatial extent and grain and its own time horizon and resolution. The levels and their scales for this illustrative example are suggested in Figure 29.1, where the landscape hierarchy described in Table 29.1 is included for comparison. In developing this example, feeding has been emphasized as the principal need, but the same hierarchy applies to the choices made for protection, nesting, mate selection, exploration, etc.

At the next finer scale is choice of a foraging area within the spatial extent of the few hundred kilometers of area chosen. The choice is made on the basis of immediate success in acquiring food and establishing undisturbed roosts and presumably on the basis of other attributes of water bodies that have become associated with future resource availability through a rule-of-thumb decision.

A reasonable foraging radius for many of these birds is approximately 20 km (range 3–30 km), so that the spatial grain for choice of territory is about 20 km and the spatial extent is a few hundred kilometers. Some birds have been tracked for up to 60 km of unswerving flight in their search for new foraging possibilities when success within the normal foraging area is poor. Such decisions are made within a year and are likely based on foraging success as integrated over one to a few weeks by some internal measure of nutritional status. Hence, the time horizon can be estimated as a year and the time resolution as a week.

The next finer set of choices is for the habitats within the foraging area where searching for food is concentrated. The extent for choice is now defined by foraging distance, about 20 km. Each day choices are made within this area for wetlands (shallow ponds, slough edges, marshes, estuaries) at depths suitable for foraging and containing prey in adequate concentrations. The grain for choice is set by the minimum size of pond chosen (some observations suggest 50–100 m for great egrets). The choice is again determined by some immediate signal, such as the presence of other birds, but also by location-specific success in the previous days. The time horizon is therefore a few days to perhaps a week and the resolution 1 day.

Food is rarely distributed evenly within a pond, nor is ease of access. The next choice is therefore for the forageable patches whose extent and grain are about 50 and 10 m, respectively. The time for abandonment can be the length of a foraging bout (at the most a few hours) and the time resolution perhaps a few tens of minutes.

The final choices are for the prey themselves, where the extent is set by the reaction distance of the predator (about a meter) and the grain by the smaller prey sizes (about a centimeter). The time horizon is in the tens of minutes and the resolution under a minute.

This information, as summarized in Figure 29.1, highlights three points. First, the range of spatial scales for the full set of choices by wading birds covers the same range of spatial scales for the set of hierarchical levels of the landscape. That is, the status of wading bird species in North America can potentially be affected by microscale transformations in trophic structure measured in centimeters all the way to subcontinental changes that affect availability of wetland regions.

Second, the relationship between the two hierarchies in time is very different from that in space. The overall choice hierarchy operates four orders of magnitude faster than the landscape/ecosystem hierarchy. It is well known that in dynamic hierarchical systems, slow phenomena can be dealt with as constants and fast phenomena as noise. In this case where two separate hierarchies are compared, on the one hand the overall set of choices by the wading birds can develop as if the landscape properties are fixed and, on the other, the overall landscape dynamics can develop independently of the wading bird impacts.

It is not quite that simple, however, which that leads to the third point. There is a range of times from about one month to one decade where the upper portion of the choice hierarchy overlaps the lower portion of the landscape hierarchy (note the horizontal hatched band in Figure 29.1). That is, choice of a region to locate and of a nesting/roosting/foraging area must respond to very local changes in the trophic structure and productivity of the ecosystem measured from fractions of centimeters to a few tens of meters. This is why wading birds are good indicators of a pattern of fine-scale change that occurs over decades.

Reconstructions of wading bird nesting effort during the period 1931–46 show that a yearly average from 80,000–96,000 birds attempted nesting (Ogden, 1994). The numbers were highly variable, however. In 1938 less than 10,000 successfully nested, in 1939 none nested, and in the following year the effort bounced back to 150,000 (Ogden, 1994). This suggests that local trophic productivity could be seriously eroded in the natural system over a few months to 1 or 2 years, presumably as a consequence of variability in rainfall, and yet wading bird nesting could recover.

It was a highly resilient system (*sensu* Holling, 1973), not because of the speed of recovery, but because of the extensive range of variability that could be sustained without the system beginning to slide into a different and less productive stability region. It is likely, in fact, that high climatic variability itself contributed to continual renewal of micro- and mesoscale ecosystem structure through such processes as fire. Ogden (1994) shows, however, that once poor conditions persist over several years to decades, major shifts in the location of roosts and in the choice of regions must occur. This is precisely what the data show during the last few decades of responses to development.

What then should the ecosystem goals be in order to restore a resilient landscape? These ecosystem restoration goals are the focus of the next section.

ECOSYSTEM RESTORATION

A landscape that is both productive and resilient depends not only on the mosaic of opportunities provided at different scales, as emphasized earlier, but also on the interaction between the landscape and the weather. The way that interaction is transformed by development, water control, and management determines the impact on the ecosystem. As a start to exploring the relationship, the range of atmospheric variability is summarized and compared with the landscape hierarchy in Figure 29.2. The figure demonstrates that the range of atmospheric variation operates at a much faster rate than the ecosystem hierarchy. Except where there is an overlap in speeds, the ecosystem can therefore moderate environmental variation by slowing the movement of water over and through the topography and vegetation.

Simulation models provide an effective way to evaluate that interaction between topography, vegetation, and patterns of rainfall and water delivery. Two models have been developed: one to plan water policy and operations by the South Florida Water Management District and one to preliminarily screen policies and integrate the science (the Adaptive Environmental Assessment and Management, or AEAM, model) (Walters et al., 1992). The latter is particularly helpful for the purposes of this section, because alternative policies can be quickly tested, modified, and evaluated by comparing the results of any new policy with the behavior of the natural system and that of the system under present management.

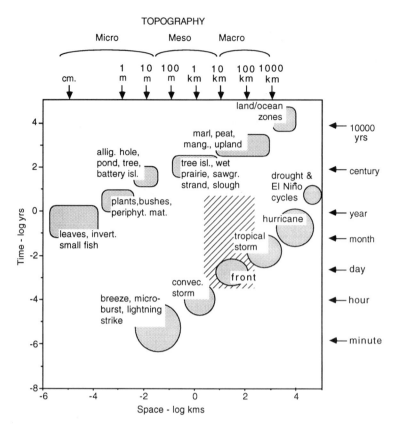

Figure 29.2 Scales in space and time of the Everglades landscape hierarchy and ranges of atmospheric variation.

The spatial and temporal scales of the AEAM model are shown by the hatched area in Figure 29.2. Because it can handle fast processes in a spatially aggregated way, the model has provided a credible dynamic representation of spatial and temporal patterns of both water quality (phosphate pollution) and water quantity in Lake Okeechobee and in a 2-km × 2-km gridded representation of the area south of the lake to Florida Bay.

In a broad sense, the natural system operated in three seasonal modes. In the center of the typical rainy season, from July to November, the wide band of sheet flow over the nutrient-poor, flat landscape functioned as a diffuse but very extensive solar energy collector. The energy collected in primary production was stored in the invertebrates and fish through trophic interactions. Around October or November, the landscape typically started to dry down and shifted into the second mode, which then functioned as a concentrator of the dispersed energy, to deliver it to predators such as wading birds over a 5- or 6-month period. In the third and final mode, the process was reversed as rains in the rainy season rewetted the landscape to re-establish sheet flow.

The drydown phase was critical in maintaining successful wading bird populations. It operated in three successive phases, over three ranges of textural scale.

In the first phase, the contraction of the wetted area provided an initial pulse of concentrated food for wading birds at the water edges along the macrotopographic elevations such as the Atlantic Ridge and upland domes. In the natural system, this transition habitat along the flanks of the Everglades may have been particularly important in providing a staging area for birds (Bancroft et al., 1994; Hoffman et al., 1994), where they were able to accumulate sufficient energy stores prior to nesting.

In the second phase, mesotopographic and vegetation forms, such as tree islands, wet prairies, and sawgrass strands, began to emerge to increase the amount of water edge where there were food concentrations sufficient for feeding. This is the important phase for bird nesting.

In the third phase of drydown, which occurred only in periodic drought years, central sloughs, ponds, and alligator holes appeared as isolated entities and generated a sudden and explosive increase in edges at microtopographic scales under a few tens of meters. As these small water bodies dried, food became more concentrated. Because the bodies dry at different rates and fluctuate asynchronously, they provided a shifting set of feeding concentrations somewhere within the foraging territory throughout the nesting period. One key to a resilient landscape at this scale is such spatial asynchrony of feeding opportunity.

Major alternative feeding options were also available during the drydown in the sustained and rich estuarine production at the freshwater/seawater interface. The model suggests that even during years of reduced rainfall, persistent pools of fresh water were maintained in the dry season along a belt of the mangrove estuary interface on the southern part of the southwest coast. This is consistent with the observation that the major roosts and colonies were located in the estuary prior to major water resource developments. Subsequent construction of major facilities to control water movement eliminated persistent freshwater estuarine pools, and the roosts in those areas have disappeared.

As the rainy season began again, the process was reversed. Refugia for fish had been provided by the ponds, alligator holes, and central sloughs that were deep enough to have retained water, by the persistent pools and sloughs, and by underground chambers in the porous limestone. As the connections became established, fish and invertebrates could disperse and emerge to re-establish full trophic relationships over broader and broader areas (Loftus and Eklund, 1994).

It is likely that each rewetted and initially isolated depression first fostered only primary production as a kind of locally well-fertilized and uncropped organic "reactor vessel." By the time aquatic production was well underway, water connectivity became re-established so that dispersing and emerging invertebrates and fish could reintroduce the energy storage capacity.

The ability of such a nutrient-poor landscape to sustain such an abundance of wading birds depended on the alternation between the three seasonal modes—extensive collection of energy, concentration of energy, and re-establishment of energy production. Modification of any one is likely to have major impacts on wading birds, which are the indicators of ecosystem integrity. Decreasing the spatial extent of the energy collector reduces the gross energy collection. Removal of topographic structure at macro-, meso-, or microscales reduces alternative options for delivering concentrations of energy for feeding throughout the winter

Table 29.2 Restoration Goals for the Everglades Ecosystem
as Outlined during the April 1991 AEA Workshop

1	Increase hydroperiod in eastern marshes, which currently have a 0- to 2-month hydroperiod, to a 3- to 7-month hydroperiod in the average year.
2	Increase water depth in late summer and fall along the freshwater marsh/mangrove interface (from stairstep area of Everglades National Park through Taylor Slough, C-111 basin).
3	Restore persistent depths through Shark River Slough during dry season in an average year.
4	Reduce depth in southern end of Water Conservation Area 3A, maintain hydroperiod, and increase pool size.
5	Remove phosphorus upstream of Water Conservation Areas.
6	Spread regulatory spikes in space and time.
7	Allow water depths and flows to be coupled with natural interannual and seasonal variation.
8	Maintain mosaic of native tree island/sawgrass/slough communities.
9	Maintain hydrologic contiguity between peripheral marsh and deeper water areas.
10	In a general sense, do not attempt to restore the system to what it supposedly "was" where it "was," but attempt to restore critical functions and structures.
11	Restore trophic structure and functions in space and time in the freshwater/saltwater interface and in the areas that were dried out for extensive periods.
12	Make the urban areas and Everglades Agricultural Area more self-sufficient and less dependent on Everglades for supply of water and for flood control.

and spring. Persistent drying of rewetted areas oxidizes earlier organic accumulations so that the organic "reactor vessels" start off in an impoverished state.

Productivity of the system was maintained by the persistence of water during the dry season, by the rewetting sequence, and by the spatial extent of the sheet flow. That productivity now has been reduced by the removal of about 50% of the system for urban and agricultural development, by the reduction in water connectivity, by oxidation of organic accumulations because of reduced hydroperiods, and by phosphate pollution from agricultural and urban areas.

Resilience of the system was maintained by the spatial and temporal heterogeneity during the drydown. That variability has been reduced by removal of transition habitat on the flanks of the system, by death of tree islands in the Water Conservation Areas that contained impounded flood waters for some years, by reduction of microscale depressions in areas receiving reduced amounts of water, and by retreat of the freshwater/seawater interface.

In conclusion, it is now sufficiently clear for policy purposes how a set of processes operating over different ranges of scales in space and time shaped and maintained the natural Everglades system. It has now been so changed by human activity that it can never be returned to its original pattern of functions in the original places and times they occurred. It is clear, however, that the reserve capacity exists to return all the critical functions somewhere within a true Everglades system that is reduced in size. This requires some expansion of its size by acquisition and rehabilitating the eastern flanks of the system and by setting aside large areas in the agricultural and urban fringe areas for phosphate removal and containment.

The knowledge is now available to define the goals for restoration and begin the designs to achieve those goals. This accumulation of understanding has been the result of the growing process of cooperation among individual scientists and engineers from the different agencies and universities. It came to a clear focus in an April 1991 workshop where policies were screened and the ingredients for composite policies identified. The restoration goals that provided the targets for those policies are summarized in Table 29.2.

Now the stage is set for careful design and evaluation of a number of feasible composite policies. Their costs need to be defined. The impacts on industry and citizens need to be determined. Their robustness to changing social, biophysical, and economic conditions nationally, internationally, and globally needs to be assessed. The continuing limits to our knowledge need a continuing emphasis on policies that are adaptive. The only way to achieve this is through cooperation, communication, and integration among the collegium of technical experts, business and environmental interests, policymakers, and, above all, the citizens who benefit from or endure the policies.

LITERATURE CITED

Bancroft, G. T., A. M. Strong, R. J. Sawicki, W. Hoffman, and S. D. Jewell. 1994. Relationships among wading bird foraging patterns, colony locations, and hydrology in the Everglades. in *Everglades: The Ecosystem and Its Restoration,* S. M. Davis and J. C. Ogden (Eds.), St. Lucie Press, Delray Beach, Fla., chap. 25.

Blake, N. 1980. *Land into Water—Water into Land: A History of Water Management in Florida (Everglades),* University Presses of Florida, Gainesville.

Byrd, M. A. 1978. Dispersal and Movements of Six North American Ciconiiforms, Res. Rep. No. 7, National Audubon Society, New York.

Frederick, P. C. and G. V. N. Powell. 1994. Nutrient transport by wading birds in the Everglades. in *Everglades: The Ecosystem and Its Restoration,* S. M. Davis and J. C. Ogden (Eds.), St. Lucie Press, Delray Beach, Fla., chap. 23.

Gunderson, L. H. and W. F. Loftus. 1993. The Everglades. in *Biodiversity of the Southeastern United States,* W. H. Martin, S. C. Boyce, and A. C. Echternacht (Eds.), John Wiley & Sons, New York.

Hoffman, W., G. T. Bancroft, and R. J. Sawicki. 1994. Foraging habitat of wading birds in the Water Conservation Areas of the Everglades. in *Everglades: The Ecosystem and Its Restoration,* S. M. Davis and J. C. Ogden (Eds.), St. Lucie Press, Delray Beach, Fla., chap. 24.

Holling, C. S. 1973. Resilience and stability of ecological systems. *Annu. Rev. Ecol. Syst.,* 4:1–23.

Holling, C. S. 1978. *Adaptive Environmental Assessment and Management,* John Wiley & Sons, London.

Holling, C. S. 1992. Cross-scale morphology, geometry and dynamics of ecosystems. *Ecol. Monogr.,* 62:447–502.

Light, S. S., L. H. Gunderson, and C. S. Holling. In preparation. The Everglades; Evolution of management in a turbulent environment. in *Barriers and Bridges in Co-evolving Ecosystems and Institutions,* L. H. Gunderson, C. S. Holling, and S. S. Light (Eds.).

Loftus, W. F. and A.-M. Eklund. 1994. Long-term dynamics of an Everglades small-fish assemblage. in *Everglades: The Ecosystem and Its Restoration,* S. M. Davis and J. C. Ogden (Eds.), St. Lucie Press, Delray Beach, Fla., chap. 19.

Loftus, W. F., J. Chapman, and R. Conrow. 1990. Hydroperiod effects on Everglades marsh food webs, with relation to marsh restoration efforts. in Proc 4th Triennial Conf. on Science in National Parks and Equivalent Reserves, Fort Collins, Colo.

Mosimann, J. E. and P. S. Martin. 1975. Simulating overkill by Paleoindians. *Am. Sci.,* 63:304–313.

Ogden, J. C. 1994. A comparison of wading bird nesting colony dynamics (1931–1946 and 1974–1989) as an indicator of ecosystem conditions in the southern Everglades. in *Everglades: The Ecosystem and Its Restoration,* S. M. Davis and J. C. Ogden (Eds.), St. Lucie Press, Delray Beach, Fla., chap. 22.

Walters, C. J. 1986. *Adaptive Management of Renewable Resources,* McGraw-Hill, New York.

Walters, C., L. H. Gunderson, and C. S. Holling. 1992. Experimental policies for water management in the Everglades. *Ecol. Appl.,* 2(2):189–202.

Wanless, H. R., R. W. Parkinson, and L. P. Tedesco. 1994. Sea level control on stability of Everglades wetlands. in *Everglades: The Ecosystem and Its Restoration,* S. M. Davis and J. C. Ogden (Eds.), St. Lucie Press, Delray Beach, Fla., chap. 8.

Wiens, J. A. 1989. Spatial scaling in ecology. *Funct. Ecol.,* 3:385–397.

Wiens, J. A. 1990. On the use of "grain size" in ecology. *Funct. Ecol.,* 4:720.

30

A SCREENING of WATER POLICY ALTERNATIVES for ECOLOGICAL RESTORATION in the EVERGLADES

Carl J. Walters

Lance H. Gunderson

ABSTRACT

During a series of workshops focused on ecological restoration of the Everglades, simulation models were used to estimate how the Everglades has changed in terms of water quantity (spatial distribution and seasonality) and to screen strategic alternatives for ecosystem restoration. Conclusions from the workshops suggest that hydrologic restoration should increase the amount of water moving through the system and re-establish aspects of spatial continuity and seasonality of depth patterns. Major new sources of water for quantity restoration include Lake Okeechobee and the eastern coastal ridge. Both of these sources pose water quality problems that will require substantial marsh areas for cleanup prior to movement into the natural areas. Two key quantitative/qualitative management options have emerged from the policy screening process: (1) develop a marsh buffer strip east of the Water Conservation Areas for urban runoff recovery and quality improvement and (2) capture regulatory water releases from Lake Okeechobee that presently become tidewater. Adding water to the Everglades system would have to be accompanied by substantial changes in water management structures and operating schedules. Partitioning water to the natural system need not be at the expense of all other sectors; there are likely win–win situations. Urban water demands over the long term can be met by delivering more water to the south. Alternative water management in the agricultural areas should promote conservation of water and soils.

INTRODUCTION

This volume and the workshops and symposium associated with it have created an opportunity for scientists from many institutions to work together in the development of conceptual and quantitative models for evaluating how to restore the Everglades ecosystem. An ongoing Adaptive Environmental Assessment process of communication and cooperation has been established, using predictive models as communication devices to ensure that common processes, issues, and opportunities are utilized (Holling, 1978). Hydrological models have been particularly useful in helping to define what it would mean to restore the Everglades system in terms of its water characteristics and in identifying strategic options for restoring water quantity, quality, and distribution.

Here, key results and conclusions from the workshops and model-based screening process through 1992 are reviewed. It must be emphasized that there is, as yet, no broad consensus on what restoration means, let alone how to achieve it. This review summarizes discussions and debates that involved more than 50 concerned scientists over five Adaptive Environmental Assessment workshops held during 1989–91; it cannot do proper justice to all of the ideas that emerged from those sessions, nor can every individual whose insight has been crucial in shaping the process be acknowledged. Many of their ideas appear elsewhere in this volume; only a few of the broadest issues that have emerged from the process are summarized in this chapter.

ESTABLISHING a NATURAL BASELINE and TARGETS for RESTORATION

Early in the planning process for this volume it was realized that a key need for Everglades restoration is a clear definition of what the natural system was like in terms of its hydrology. That definition was not as easy to devise as one might expect; there have been substantial debates about such basic points as whether the lower part of the system (Shark River Slough, Taylor Slough, mangrove zone) is now wetter or drier than it was originally and the extent to which water diversions and regulatory (flood control) releases through levees and canals have altered the basic seasonality and pattern of interannual variation created by rainfall and localized overland flow (Leach et al., 1971; Parker, 1984; Kushlan, 1987; Walters et al., 1992).

To help provide better "reconstructions" of the natural system and to help screen future policy options, two large hydrological simulation models have been developed. Very realistic and detailed simulations have been created by the South Florida Water Management District and Everglades National Park (Fennema et al., 1994) to represent daily water dynamics over the system in relation to rainfall, evapotranspiration, overland flow, and groundwater movement. A simplified but computationally much faster version of this model (larger grid cells, cruder rainfall and evapotranspiration calculations, no groundwater dynamics, simplified representation of regulatory policies) was developed at the University of Florida to permit timely screening of broad water management options in workshop settings

(Walters et al., 1992); hereafter, this simplified strategic screening model is referred to as the UFAEA model. Both models have been calibrated (by varying evapotranspiration and overland flow parameters) to give reasonably good fits to historical water depth patterns at key gaging stations in Everglades National Park. Both are "driven" by a historical time series of rainfall (1960–90 period) to provide assessments of interannual as well as seasonal water patterns. The basic outputs of both are water depth maps and flows across index sections of the marsh system. Depth maps are particularly useful summaries of hydrological conditions in terms of the space-time distribution of ecologically important features such as deep water pools and shallow water "transitional" habitats where processes such as wading bird feeding are concentrated.

Scenario development has focused on simple indicators comparing simulated water depth patterns in the natural system to water depth and phosphorus distribution patterns under various management and restoration strategies. For example, Figure 30.1 shows model depth results for the natural system over a typical year (1964 rainfall pattern) and depth difference maps for the same year for two management scenarios. Visual examination of the difference maps quickly shows how "current management" has broken up the water pooling pattern, pushed water westward, and made the lower part of the system drier over much of the year. Under the extreme "add-more" scenario, water from Lake Okeechobee is restored to the Water Conservation Areas via flow ways through the Everglades Agricultural Area, urban runoff water is added to the eastern part of the system via a buffer strip to improve its quality, and Water Conservation Area 3A structures and releases past Tamiami Trail are modified to produce depth patterns with much smaller differences from natural in the lower half of the system. The "add-more" scenario is by no means a current assessment of the best way to restore the system; it is presented in Figure 30.1 only to illustrate how visual comparisons can be helpful in the scenario development process.

The workshops stimulated a debate regarding the use of models for both a reconstruction of what the system may have been and for screening restoration policies. One viewpoint requires documentation and validation of model output in order to evaluate the realism or precision prior to utilization for policy determinations. Another viewpoint argues that no single model can simultaneously meet conditions of generality, realism, precision, and manageability (Levins, 1968; Holling, 1978; Clark et al., 1979; Costanza and Sklar, 1985). Because the rules of water movement are well accepted, the properties of generality and manageability appear to have become implicit in hydrologic modeling, and hence the focus for model development and use is on precision and reality. The UFAEA model emphasizes generality and manageability in order to meet the dual uses of historical representation and preliminary screening of restoration policies, as well as to improve communication among workshop participants. It must be emphasized that both models may be quite inaccurate in reconstructing the natural Everglades hydrology, due to some basic uncertainties about evapotranspiration, historical land surface elevations (and slopes), velocities of overland flow, losses from the system to the east and south (especially groundwater flows), and contributions to the natural system from Lake Okeechobee. The point here is that different models should be used for different purposes and that policies should be robust to the

NATURAL SYSTEM DEPTHS

DIFFERENCE BETWEEN NATURAL AND MANAGED SYSTEMS

DIFFERENCE BETWEEN NATURAL AND ADDMORE SCENARIOS

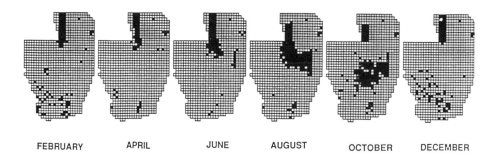

FEBRUARY APRIL JUNE AUGUST OCTOBER DECEMBER

Figure 30.1 Example of UFAEA hydrology model results used to compare and screen water management options. Map cells represent 4-km × 4-km areas used for overland flow calculations, extending from Lake Okeechobee in the north to Florida Bay in the south and from the Big Cypress eastward to Miami. The top row of maps shows simulated water depths for the reconstructed natural system for a fairly typical rainfall year (1964). The lower two rows of maps show depth differences from the natural system simulation (black = wetter, stippled = drier, clear = equal) for two scenarios. The "managed" scenario represents roughly the current water management regime, except that flows to Everglades National Park are reduced somewhat to represent effects of water quality standards that may be used to avoid phosphorus accumulation in Shark River Slough. The "add-more" scenario is an extreme diagnostic simulation, where all Everglades Agricultural Area water was cut off from entering the system (made the area self-sufficient), flow ways were added to convey Lake Okeechobee water to the system, and a marsh strip was added west of Miami to convey urban runoff to the system.

uncertainties inherent in both approaches. Only those aspects of the reconstructions that appear to be robust to variations in uncertain parameter values and can be supported with independent qualitative evidence such as orientation of tree islands (Gleason and Stone, 1994) and patterns of fluorescence banding in Florida Bay corals, which are indicative of higher freshwater inputs (Smith et al., 1989), will be referred to in this chapter.

The natural system apparently had the following key functional features. (1) There was substantial input to the more northerly marsh areas from Lake Okeechobee, but the relative importance of the lake (for overland flows and marsh water depths) diminished toward the south so as to have little direct impact (compared to rainfall and runoff from farther south) at the southern mangrove zone. (2) During most wet seasons, there was essentially continuous water surface and sheet flow over the entire system from Lake Okeechobee southward, with water heads and depths at every point helping to push water southward (and eastward at places) in spite of minimal land slopes. (3) A large and persistent natural pool (with year-round surface water in most years) formed at the upper end of Shark River Slough, eastward from the present site of Water Conservation Area 3A to the eastern coastal ridge. (4) The natural pool acted to stabilize seasonal flows into, and increase hydroperiods in, the lower system (especially Shark River Slough) and perhaps also promoted groundwater flows into and longer hydroperiods in Taylor Slough. (5) Water movement velocities (probably important for tree island formation and maintenance) were higher over extended periods of the annual wet season for areas such as Water Conservation Areas 1, 2A, and 3A, where managed ponding now retards natural overland flow (Fennema et al., 1994; Walters et al., 1992). Taken together with land development patterns, these features imply that the Everglades of today is (1) smaller; (2) shallower in many areas; (3) broken up into a beaded pattern of pools rather than a continuous "river of grass;" (4) subject to more violent seasonal depth fluctuations, with peaks most often occurring earlier than natural; (5) drier, especially late in the dry season, in the southern and eastern "rocky glades" areas; and (6) delivering less fresh water into the estuarine mangrove zone.

General agreement has been reached through screening restoration options to search for policies that will (1) restore flows (of clean water) to the upper part of the system (Water Conservation Areas) lost due to the regulation of Lake Okeechobee; (2) establish a more continuous pattern of overland flow southward through the system, with more natural timing; (3) maintain a major pooling area for water north and east of Everglades National Park; and (4) provide higher total flows to Shark River Slough and Taylor Slough so as to both provide higher freshwater inputs to the mangrove zone and Florida Bay and longer hydroperiods in the transitional (marl, rocky glade) marshes bordering these sloughs. To achieve each of these points, note that a key shared requirement is to increase the *quantity* of water entering the Everglades marsh system. Options that would recover more of the total size of the natural system by converting agricultural areas back into marshes were discussed and analyzed, but these options have mainly involved using such marshes as treatment areas to remove agricultural fertilizers and other chemical contaminants.

It is important to note that in the policy screening process, little time has been devoted to defining biological indicators or objectives for restoration, such as increasing biodiversity or maintaining particular endangered species. The consensus has been that by restoring natural hydrological function to the system, these various objectives will be met through natural ecological processes in response to the hydrological restoration.

While there is consensus that restoration will involve increasing the quantity of water entering the system, the restoration screening process has helped to recognize key uncertainties and realize potentially difficult trade-offs regarding how that water will be distributed. The difficulties arise in part from attempting to meet the dual objectives of trying to restore a natural hydrologic regime to critical locations in the remnant system and the need to replace functions lost as a result of past management practices and land converted to other uses. Ideally, within the remnant system, a natural hydrologic regime should be established in a continuous north-south marsh complex, in the eastern "rocky glades," in a central pool and have substantial outflow into the mangrove zone.

The uncertainties involved are exemplified by how water might be partitioned among Water Conservation Areas 3A and 3B, Northeast Shark River Slough, and Central Shark River Slough. A key feature of the historic system was a persistent pool of water in the area that is now within Water Conservation Area 3B. A pool now occurs to the west due to the impoundment of water in southern Water Conservation Area 3A. The current pool provides some functional attributes such as deep water habitat suitable for apple snails and snail kites (Bennetts et al., 1994). The current pool also has undergone vegetation changes to a more aquatic assemblage. There are key uncertainties as to whether the current pool can replace the lost functions of the historic pool. Another key uncertainty involves rewetting the areas of Water Conservation Area 3B and Northeast Shark River Slough. If water released into these areas is lost to evapotranspiration and groundwater flows to the east, then further estuarine degradation could occur due to a reduction of flows to the south.

Recognition of these trade-offs and uncertainties should not stall restoration efforts. Indeed, carefully constructed polices can be used to experimentally probe and winnow management actions that are robust. For a discussion of this and other experimental management opportunities, see Walters et al. (1992).

QUANTITY SOURCES and the QUALITY PROBLEM

Two major sources of water have been identified for Everglades restoration: Lake Okeechobee and runoff from the eastern developed areas. These sources are interrelated in terms of current consumptive uses and demands (e.g., both the Everglades Agricultural Area and urban areas now use Lake Okeechobee water). Water from all of these sources carries nutrient loads high enough to likely cause vegetation changes (cattail [*Typha domingensis*] development) if discharged on native marshes. Each source could potentially provide an annual average of at least 400,000 acre-ft ($4.93 \cdot 10^8$ m^3) of water to the system, assuming efficient capture and conveyance of regulatory (flood control) releases and some modest improvement in water conservation practices.

Delivery of Lake Okeechobee water to the system on a regular basis would require improved conveyance facilities through the Everglades Agricultural Area and some reduction in consumptive demands during dry years (again mainly by the Everglades Agricultural Area, which demands more than one-half of an acre-foot of Lake Okeechobee water per acre of farmland (Abtew and Khanal, in preparation). An obvious way to meet several needs at once would be to purchase Everglades Agricultural Area lands to form marshland "flow ways." These flow ways would lessen the consumptive demands (by decreasing area in cultivation), act as a water quality treatment system, and reshape pulsed regulatory releases from the lake into a more natural seasonal pattern of flow by the time the water reached the Water Conservation Areas. It should be noted that the concept of flow ways from Lake Okeechobee to the Water Conservation Areas is not new; a mile-wide flow way was suggested by the Corps of Engineers as early as 1955 (U.S. Army Corps of Engineers, 1955).

In the long run, urban water management will be the key to the future of the Everglades. Unless the urban water system is better managed as the coastal region grows, increased water delivery to the Everglades from the north (Lake Okeechobee, Everglades Agricultural Area) may simply end up being drained to the east for wellfield recharge. Fortunately, there may again be a way to solve several problems at once, by establishing a substantial "buffer strip" of marshland just to the east of the L-33/30/31 levee system, from the lower end of Water Conservation Area 2 south to at least the Tamiami Trail. Urban and eastern area agricultural runoff would be pumped or drained into the north end of the strip and allowed to flow slowly south (for water quality treatment). Urban runoff delivered to the buffer strip would (1) help to recharge wellfields (and hence reduce urban demand from western and northern areas); (2) maintain a groundwater head to prevent seepage losses from the Water Conservation Areas, especially Water Conservation Area 3B (see uncertainties mentioned earlier); and (3) if quality at the south end of the strip proves adequate, provide an easily conveyed source of water for restoration of flows to the Taylor Slough area (via C-111 canal system). Notice how the buffer strip concept would "unlock" a number of options for both urban development and Everglades restoration. In addition to major land availability and acquisition costs, a major issue would of course be whether the urban runoff water contains unacceptable levels of contaminants that are not removed by marsh treatment processes. The potential gains, however, are very substantial; a drainage system that could capture and deliver to the buffer strip even 10% of the rainfall from the urban/agricultural lands from Fort Lauderdale to Homestead would deliver nearly 400,000 acre-ft ($4.9 \cdot 10^8$ m^3) of water in average years.

Ambitious programs for water quantity restoration will require correspondingly large areas for water quality restoration. A critical policy issue is where to obtain such large marsh areas for quality treatment. There are two basic options: (1) acquisition of private lands in the Everglades Agricultural Area and just east of the Water Conservation Areas and/or (2) "write off" a large existing area, most likely Water Conservation Area 2A, which is already substantially degraded. Adverse effects on marsh productivity and function due to phosphorus loading (Davis, 1994) along with the broader restoration objective of no further reduction of functional marsh area (Davis and Ogden, 1994) make utilization of the Water Conservation Areas for phosphorus removal a poor restoration option.

Major engineering changes and challenges are involved in these concepts for developing water storage and conveyance facilities to make more clean water available to the Everglades. At a minimum, they involve restructuring conveyance facilities from Lake Okeechobee to the Water Conservation Areas and developing drainage systems to recover more water from eastern areas.

CHANGES in the OPERATING SYSTEM

Water storage and conveyance facilities throughout the Everglades system, from Lake Okeechobee to the Tamiami Trail, have been operated according to storage and release policies (operating rules, schedules) that emphasize water supply and flood protection objectives. There is little doubt that these policies have caused ecological damage by reshaping seasonal and spatial patterns of water pooling and release. Particularly visible have been sharp regulatory releases that cause sudden flooding (water level "reversals") with deleterious effects on feeding and nesting of various vertebrates. The canals and levees have caused breaks in sheet flow and barriers to aquatic animal movement. When the Adaptive Environmental Assessment workshop process began and before model reconstructions of the natural hydrology became available, it was frequently mentioned that change in operating schedules was *the* key need for Everglades restoration. That view is no longer widely held; it appears that restoration should involve *both* increases in total quantity of water and more than "tinkering" with existing operating policies and structures.

The "rainfall formula" (MacVicar, 1985) for water releases to Everglades National Park from Water Conservation Area 3A has established an important precedent or model to show that water management and conservation interests can work together to define operating policies that better meet restoration needs, while not seriously jeopardizing water storage and supply objectives. Especially important for the future will be careful examination of whether regulatory releases from storage pools (from Lake Okeechobee southward) for flood protection, especially releases via canals to tidewater, are really necessary. Eliminating regulatory schedules in favor of rainfall-driven storage and downstream flow policies would certainly be more natural from an ecological point of view. With increased quantities of water moving into the system from Lake Okeechobee, storage requirements at each water supply pool in the system should be reduced; these reduced local storages should produce reduced probabilities of requisite regulatory releases in the first place.

Using the UFAEA model, simulations of various storage release policies have been created for the Water Conservation Area 1, 2A, and 3A storage pools, using a range of rainfall-driven options for release (release 45% of flow that would have occurred naturally, release 75% of that flow, etc.). These screening tests hint that existing storage policies, which can be ecologically deleterious, may do little to improve water supply circumstances for east coast service areas beyond helping to maintain groundwater heads. With such shallow "reservoirs" and high evaporation rates, it does not appear to be possible to store enough water during wet periods to make any significant difference to supplies during dry periods, especially on an interannual basis. Insofar as storage is needed to deal with variation in rainfall and water demands, it appears that Lake Okeechobee has been and will continue to be the key storage unit in the system. Trimble (1986) made similar comments in his documentation for the South Florida Water Management District's South Florida Regional Routing Model for water management planning.

While it appears practical to restore more natural timing of overland flows between the Water Conservation Areas and into Everglades National Park, it would be a more formidable task to re-establish the spatial continuity of water pooling from north to south in the system during wet seasons. As noted earlier, this continuity might be critical to the productivity and seasonal response dynamics of many aquatic organisms that are important food items for wading birds and alligators. The only way to restore natural continuity entirely would be to remove the major structures (canals, levees, highways) that define the Water Conservation Areas and somehow repair elevation changes that may have occurred due to subsidence (oxidation of peat soils) near these structures. Such a destructive policy would probably not work and would reduce future options for using the existing structural system for experimentation and focused restoration activities. It remains to be seen whether there is some way to engineer spatial continuity of water pooling patterns without major destructive changes.

SCENARIO DEVELOPMENT PROCESS: PROGRESS and PROSPECTS

Since early 1990, the scenario development process has moved from relatively narrow concerns about operating policies and seasonal hydroperiod restoration to a much broader strategic analysis of water sources and related changes in land use to improve water quality. From an initial group representing mainly scientific interests and concerns (biology, geology, hydrology), participation in the process has expanded to include key personnel from other government agencies and conservation organizations. The process of trying to define strategies for water quantity restoration has helped to stimulate imaginative thinking about approaches to water storage, conservation, and quality improvement, such as the urban buffer strip.

Perhaps most importantly, there has been enhanced communication among individuals representing water supply and ecological restoration interests. This communication has led to a clearer perception of shared objectives and possible win–win options. There are, and always will be, some real conflicts of interest, both between water supply and conservation interests and within the conservation community itself; the process has helped to clarify and pinpoint these conflicts and to stimulate discussion about imaginative, positive ways to deal with them.

Participants in Adaptive Environmental Assessment workshops have developed a variety of scenarios combining the major water quantity sources (Lake Okeechobee, urban) with changes in operating policies and structures. No obvious best strategy has emerged from comparing these scenarios. Indeed, it appears that general restoration objectives (quantity reaching lower system, spatial distribution, seasonality) can be achieved by a variety of water source/operation changes.

An important point that has emerged from the scenario development process is that even removing all water management structures from the Everglades Agricultural Area southward (except the eastern levees of the Water Conservation Areas) and converting the entire Everglades Agricultural Area back into a marshland with regular Lake Okeechobee inflow would not ensure full restoration of natural Everglades hydrology. This extreme option would in fact produce a considerably wetter system than natural, due to water ponding along the eastern levees and moving southward along a narrower, deeper "channel" (with less evapotranspiration loss). Thus, continued intervention and management of the Everglades water system appears to be an inevitable requirement for restoration.

CONCLUSIONS

While there is as yet no consensus on the specifics of hydrological changes needed to effect Everglades restoration, the following broad principles emerged from the series of workshops and will likely be important in reaching that consensus:

1. Tinkering with existing structures and operating policies (storage, release) will not be sufficient to ensure restoration.
2. Single restoration actions, such as removal of particular structures or changes in operational rules for individual canals, have the potential to do more harm than good unless they are part of a coordinated program that begins with restoration of larger quantities of water to the system.
3. Adding substantial water quantity will have a very high price tag in terms of marsh area required for water quality management, but will create opportunities for improving urban water supplies as well as meeting conservation needs.
4. Lake Okeechobee and marsh buffer strips west of the Miami urban area will likely play key roles in the long-term restoration strategy.

LITERATURE CITED

Abtew, W. and P. Khanal. In preparation. A Preliminary Water Budget for the Everglades Agricultural Area, Tech. Rep. (Mimeo), South Florida Water Management District, West Palm Beach.

Bennetts, R. E., M. W. Collopy, and J. A. Rodgers, Jr. 1994. The snail kite in the Florida Everglades: A food specialist in a changing environment. in Everglades: *The Ecosystem and Its Restoration,* S. M. Davis and J. C. Ogden (Eds.), St. Lucie Press, Delray Beach, Fla., chap. 21.

Clark, W. C., D. D. Jones, and C. S. Holling. 1979. Lessons for ecological policy design: A case study of ecosystems management. *Ecol. Model.,* 7:1–53.

Costanza, R. and F. Sklar. 1985. Articulation, accuracy and effectiveness of mathematical models: A review of freshwater wetland applications. *Ecol. Model.,* 27:45–68.

Davis, S. M. 1994. Phosphorus inputs and vegetation sensitivity in the Everglades. in *Everglades:*

The Ecosystem and Its Restoration, S. M. Davis and J. C. Ogden (Eds.), St. Lucie Press, Delray Beach, Fla., chap. 15.

Davis, S. M. and J. C. Ogden. 1994. Toward ecosystem restoration. in *Everglades: The Ecosystem and Its Restoration,* S. M. Davis and J. C. Ogden (Eds.), St. Lucie Press, Delray Beach, Fla., chap. 31.

Fennema, R. J., C. J. Neidrauer, R. A. Johnson, T. K. MacVicar, and W. A. Perkins. 1994. A computer model to simulate natural Everglades hydrology. in *Everglades: The Ecosystem and Its Restoration,* S. M. Davis and J. C. Ogden (Eds.), St. Lucie Press, Delray Beach, Fla., chap. 10.

Gleason, P. J. and P. A. Stone. 1994. Age, origin, and landscape evolution of the Everglades peatland. in *Everglades: The Ecosystem and Its Restoration,* S. M. Davis and J. C. Ogden (Eds.), St. Lucie Press, Delray Beach, Fla., chap. 7.

Holling, C. S. 1978. *Adaptive Environmental Management and Assessment,* John Wiley & Sons, New York.

Kushlan, J. A. 1987. External threats and internal management: The hydrologic regulation of the Everglades, Florida, USA. *Environ. Manage.,* 11(1):109–119.

Leach, S. D., H. Klein, and E. R. Hampton. 1971. Hydrologic Effects of Water Control and Management of Southeastern Florida, Report of Investigations 60, Florida Bureau of Geology, Tallahassee.

Levins, R. 1968. Ecological engineering: Theory and technology. *Q. Rev. Biol.,* 43:301–305.

MacVicar, T. K. 1985. A Wet Season Field Test of Experimental Water Deliveries to Northeast Shark River Slough, Tech. Publ. 85-3, South Florida Water Management District, West Palm Beach.

Parker, G. G. 1984. Hydrology of the pre-drainage system of the Everglades in southern Florida. in *Environments of South Florida: Present and Past II,* P. J. Gleason (Ed.), Miami Geological Society, Coral Gables, Fla.

Smith, T. J., III, H. Hudson, G. V. N. Powell, M. B. Robblee, and P. J. Isdale. 1989. Freshwater flow from the Everglades to Florida Bay: A historical reconstruction based on fluorescent banding in the coral *Solenastrea bournoni. Bull. Mar. Sci.,* 44:274–282.

South Florida Water Management District. 1989. Surface Water Improvement and Management Plan for the Everglades, South Florida Water Management District, West Palm Beach.

Trimble, P. 1986. South Florida Regional Routing Model, Tech. Publ. 86-3, South Florida Water Management District, West Palm Beach, 146 pp.

Trimble, P. and J. Marban. 1988. Preliminary Evaluation of the Lake Okeechobee Regulation Schedule, Tech. Publ. 88-5, South Florida Water Management District, West Palm Beach, 73 pp.

U.S. Army Corps of Engineers. 1955. General Design Memorandum for Modifications to Lake Okeechobee Outflows, U.S. Army Corps of Engineers, Jacksonville, Fla.

Walters, C. J., L. H. Gunderson, and C. S. Holling. 1992. Experimental policies for water management in the Everglades. *Ecol. Appl.,* 2(2):189–202.

31

TOWARD ECOSYSTEM
RESTORATION

Steven M. Davis

John C. Ogden

INTRODUCTION

In this chapter, the results, conclusions, and hypotheses provided in the preceding chapters are integrated into a collective technical framework and vision for ecological restoration of the Everglades. As such, this chapter serves as a discussion for the volume as a whole. A case is built for a restoration strategy by synthesizing the information from the chapters into sections that examine (1) the physical and ecological characteristics of the Everglades prior to the major management activities of this century, (2) how drainage and management have negatively impacted the ecosystem, (3) the characteristics of the system which make it necessary to design a restoration strategy aimed at a moving target, and (4) the uncertainty inherent in any ecological restoration strategy for the Everglades. Following this characterization of the ecosystem, the specific restoration recommendations contained in the chapters of this volume are summarized. In conclusion, the above considerations are used as a basis to pose a set of hypotheses as guidelines for ecosystem restoration in the Everglades. This chapter has been reviewed by the senior authors of the preceding chapters in order to ensure that the conclusions and recommendations represent a collective point of view.

HOW the EVERGLADES WORKED

Disturbance, Renewal, and Diversity in a Vast Wetland

The Immense Wetland Mosaic

Prior to drainage, Everglades wetlands covered almost 1.2 million ha (Davis et al., Chapter 17), but that represented only a fraction of the spatial extent of south Florida's contiguous wetlands in basins surrounding the Everglades. Light and

Dineen (Chapter 4) calculate the size of the Kissimmee/Okeechobee/Everglades system to be more than 2.8 million ha. Adding the contiguous 640,000 ha of Big Cypress Swamp and the 220,000 ha of Florida Bay, the Everglades was part of a greater wetland system of more than 3.6 million ha, which resulted in most of southern Florida being inundated at least during wet seasons. The preceding chapters suggest that the sheer immensity of the Everglades played three roles in supporting the diversity and abundance of flora and fauna which characterized the predrainage system. As summarized by Robertson and Frederick (Chapter 28), it provided enough space to support genetically viable numbers of individuals and subpopulations for those species with large home ranges (Florida panther [*Felis concolor coryi*]) or with narrow habitat requirements (Cape Sable sparrow [*Ammospiza maritima mirabilis*]). It also enabled the systemwide aquatic production in a nutrient-poor ecosystem necessary to support large populations of wading birds, snail kites (*Rostrhamus sociabilis*) and other consumers dependent upon aquatic food webs, as discussed later in this chapter. Finally, the size of the ecosystem allowed for the perpetuation of habitat diversity through the processes of natural disturbance.

DeAngelis and White (Chapter 2) demonstrate that environmental heterogeneity in the form of patchiness at a variety of scales functions in combination with ecosystem size to enable persistence and resilience of populations of plants and animals. Spatial extent spreads the risks of seasonal environmental changes and disturbances that affect organisms and populations. White (Chapter 18) elaborates on the consequence of the loss of ecosystem spatial extent for vegetation dynamics, noting that constantly changing vegetation patterns reach a dynamic equilibrium in systems with a large spatial extent compared to disturbance patch size. Spatial extent thus acts as a buffer to prevent patchwise population changes from creating species or population extinctions in the area as a whole. White identifies a key issue in Everglades ecosystem restoration as "whether the reduction in the size of the ecosystem will itself affect pattern and process in the ecosystem once natural processes are restored."

Patchworks of vegetation and topographic variations, at scales from plant clusters to physiographic landscapes, appear to be closely integrated with the diversity and persistence of Everglades wildlife populations. Loftus and Eklund (Chapter 19) demonstrate that a mosaic of wetland habitats is necessary to provide protection for smaller fishes from predators during high water and refugia from drought during low water. Bennetts et al. (Chapter 21) portray the tenuous combination of permanently flooded pools and willows which require periodic drying as essential to the persistence of snail kite populations. Bancroft et al. (Chapter 25), Frederick and Spalding (Chapter 26), and Hoffman et al. (Chapter 24) offer evidence to show that wading birds in the system require large areas of wetlands to provide adequate foraging habitat during seasonal and annual variations in rainfall and hydrological conditions. Hoffman et al. illustrate the support function of habitat heterogeneity as it pertains to wading birds in the Everglades. They depict regions with patchworks of relatively high topographic relief and plant community diversity, from ponded depressions to terrestrial hammocks, as extraordinarily attractive to wading birds that roost in the trees and feed in the aquatic depressions. Such areas provide foraging habitat over a wider range of water stages than elsewhere in the Everglades. Smith and Bass (Chapter 27) report that the availability of edge between large tracts of upland forest and surrounding open marsh communities determines the carrying capacity for Florida panthers in the Everglades. The edge habitat is critical to the success of panthers in the acquisition of white-tailed deer (*Odocoileus virginianus*) as prey, which determines the breeding density of panthers in Everglades National Park.

A Patchwork of Change

DeAngelis and White (Chapter 2) emphasize the organizing role of natural disturbances in large ecosystems, in the sense of creating diversity at the level of microscale and mesoscale patchiness. White (Chapter 18) explains how disturbance acts as one of the key processes by which patches are formed in a landscape mosaic. Full vegetation recovery tends to be slow compared to the duration of a disturbance event, producing an asymmetry of relatively fast rates of disturbance and slow rates of recovery. Such time lags become important in creating vegetation diversity.

The dislodging and floating of patches of water lily marsh peat under deeper water conditions is demonstrated by Gleason and Stone (Chapter 7) to contribute to microscale patchiness by forming bayhead tree islands, as well as slough habitat in the depressions that are left. Mazzotti and Brandt (Chapter 20) describe the role of alligator (*Alligator mississippiensis*) activities in structuring the microscale patchwork of Everglades plant and animal communities by excavating ponds and creating trails, resulting in deeper open-water areas, and by constructing nest mounds that provide relatively elevated areas. Gunderson (Chapter 13) notes that this activity creates plant community diversity by sculpting peat into nest mounds, which serve as colonization sites for willow (*Salix virginiana*) and other swamp hardwoods and may later develop into larger tree islands, and by excavating open-water ponds as alligator holes which may later support a slough community. Mazzotti and Brandt review the critical role of this microscale patchwork of higher ground and deeper water habitats in an otherwise shallow marsh landscape as nesting, resting, or foraging sites for many animal species.

Duever et al. (Chapter 9) describe how periodic freezing temperatures can damage or kill plants in complex ways, thus creating mosaics in local and regional vegetation patterns. They show how a single freeze event can produce spatially different impacts on vegetation due to local differences in surface water and plant cover patterns and due to micropatterns in the duration and intensity of a freeze.

Hurricanes represent an example in which a functionally necessary or relatively benign role of a natural disturbance regime in shaping the ecosystem may change to a destructive role due to alteration of global climate and sea level. Duever et al. (Chapter 9) conclude that in the past, hurricanes have primarily impacted growth forms of plant communities and individual organisms rather than the long-term viability of populations or communities. However, the potential synergistic role of hurricanes in combination with sea level rise, as predicted by Wanless et al. (Chapter 8), will very likely escalate the scale and severity of the impact of hurricanes, resulting in the erosion of natural coastal levees in the southern Everglades and major ecosystem shifts from freshwater to marine environments.

The pervasiveness of fire in the Everglades is noted by Gleason and Stone (Chapter 7) in the geologic record and by Gunderson and Snyder (Chapter 11) in the historic record. In their review of fire regimes in the Everglades, Gunderson and Snyder indicate the wide range of spatial and temporal scales over which fire creates patchworks of habitat diversity in the system. Evidence is cited for return intervals of 3 years for fire years when 4000–8000 ha burn, 6–7 years for severe fire years when more than 8000 ha burn, and a longer term cycle in which larger areas burn every 10–15 years. They document a large variability in size of individual fires from square meters to nearly 75,000 ha during a 1-in-100-year drought. Gunderson (Chapter 13) attributes much of the variation in spatial distribution of Everglades plant communities to disturbances (such as severe fires accompanying droughts) that operate at intermediate scales of tens of kilometers and decades.

Gunderson and Snyder (Chapter 11) portray fire as working synergistically with drought to maintain mesoscale patchiness through peat combustion, lowering of soil elevation, and creation of a mosaic of wet prairie and slough in areas of otherwise dense sawgrass (*Cladium jamaicense*). Reports by Hoffman et al. (Chapter 24) of peat combustion in Water Conservation Area 3A during the 1989 drought support this hypothesis. Sawgrass was killed in large patches, and upon reflooding, depressions that were burned into the soil formed open-water habitat which was extensively used by wading birds.

White (Chapter 18) notes that while severe fire can result in conversions among sawgrass, wet prairie, slough, and tree island communities, the natural periodicity of low-intensity fire that is associated with rainfall periodicity serves to maintain communities at near stable composition and structure. An example cited by Gunderson is the adaptation of sawgrass to burn intervals of 5–7 years, which set back detritus accumulation and kill back hardwoods.

In addition to its influence on fire regimes, drought also creates mesoscale patchiness through direct influences on vegetation communities, which may enhance habitat values for wildlife species. Bennetts et al. (Chapter 21) note that despite the benefits of a relatively long hydroperiod to snail kites, continuous inundation is detrimental because of the resultant loss of woody vegetation, particularly willow, which the kite depends upon as nesting habitat. Thus, dispersal ability and refuge availability are critical to the snail kite and other species that depend on both flooding and drying, as discussed in the following sections.

Sanctuary

DeAngelis and White (Chapter 2) demonstrate that dispersal ability plays a key role in plant and animal population resilience in ecosystems by allowing individuals to reach sanctuaries as annual water cycles or disturbances exclude them from their former territories. White (Chapter 18) adds that recolonization after disturbance depends on the recolonists being present somewhere in the landscape; hence, spatial configuration of the disturbed areas relative to refugia and the relationship between disturbance patch size and landscape area become critical.

An important role of refugia in the Everglades during droughts is the presence of flooded habitats for aquatic organisms and feeding grounds for the predators that eat them. The natural system hydrology model (Fennema et al., Chapter 10) identifies two major permanently flooded pools that served this function in the undeveloped Everglades, one in what is now Water Conservation Area 3B extending southward into northeast Shark River Slough and another in lower Shark River Slough. To the north, the topographic depression of the Hillsborough Lake region, now mostly in Water Conservation Area 1, provided similarly persistent flooded conditions.

Examples abound of flora and fauna in the Everglades that depend upon dispersal and refugia as strategies for coping with the wet-dry cycle periodicity and disturbances. Browder et al. (Chapter 16) comment that some of the algal species in periphyton communities may survive dry periods in sinkholes and deeper ponds, while repopulation after drydown by other species may originate from desiccation-resistant heterocysts, akenetes, or resting spores. Loftus and Eklund (Chapter 19) conclude that persistence of populations of marsh fishes requires refugia of permanent water under dry conditions. Connectivity of the various marsh pools and parts appears to have been essential to the successful retreat of fishes and other aquatic fauna into drought refugia and to their recolonization of surrounding marshes upon reflooding. Alligator holes serve as dry season refugia, and the low density of alligator holes in overdrained areas of northeast Shark River Slough (Mazzotti and Brandt, Chapter 20) may limit the number of fishes surviving into the next wet season. Bennetts et al. (Chapter 21) review findings that the high degree of mobility of snail kites and their nomadic ability to seek out flooded areas of refuge during periods of drought represent an important adaptation for survival in the highly variable hydrologic conditions of the Everglades. They stress that the availability of refugia may be the primary limiting factor of future snail kite populations because in the absence of refugia, the periodic drying events necessary to maintain woody vegetation as nesting habitat become detrimental due to the drying of flooded areas for feeding habitat. Hoffman et al. (Chapter 24) describe breeding wading birds as needing continuously available foraging sites with surface water within the range of colony sites through the entire nesting season; however, because suitable foraging conditions change in space and time, the birds have large home ranges and change foraging locations frequently.

At the other end of the hydroperiod scale, Robertson and Frederick (Chapter 28) note that the upland pine and tropical hammock habitat of the rock ridge that once extended from Miami to Long Pine Key represents a critical terrestrial refugium within the Everglades. This habitat harbors most of the endemism and a large part of the species richness in a system that was much more extensively terrestrial in the recent geologic past before the Everglades formed (Gleason and Stone, Chapter 7).

At a larger spatial scale, the Everglades appears to have functioned as a refugium, supporting wading bird populations at a regional level. Robertson and Frederick (Chapter 28) suggest that while the species of wading birds that occur in south Florida have always operated over much larger geographical scales, the natural Everglades system was a major production center and as such "…served as a springboard for both population and range expansions."

The Hydrologic Pulse

DeAngelis and White (Chapter 2) propose that the annual periodicity of wet-dry cycles plays an organizing role in many ecosystems including the Everglades, in that the reproductive success of many plant and animal populations is keyed to such cycles. The risk to reproductive success due to variations in flooding or drying patterns is moderated in large ecosystems with high habitat diversity in the form of microscale and mesoscale topographic irregularities.

The seasonal progression of the wet and dry cycle dominates nearly every aspect of Everglades ecosystem function. Duever et al. (Chapter 9) document that approximately 60% of rainfall is concentrated from June through September mostly due to thunderstorm activity in the Everglades. In response to the seasonal rainfall pattern, natural system water levels rose and fell, flooded areas and water depth distributions expanded and contracted, and flow volumes surged and ebbed. In animated monthly time steps simulated by the adaptive environmental assessment model (Walters and Gunderson, Chapter 30), this rhythm bears a visual similarity to a heartbeat. The annual pulse appeared to be characterized by relatively steady rises and falls in water level during many years. Reversals in falling water depths due to dry season rainfall events or in rising water levels due to low rainfall periods during the wet season tended to be dampened due to the slow, steady conveyance of water through the marsh system. Snapshots of this pulse at the ends of wet and dry seasons, as it might have looked in the system prior to man's modifications this century, are illustrated in the depth distribution maps (Plates 2–6)* from the natural system hydrology model (Fennema et al., Chapter 10). The preceding chapters suggest that the seasonal pulse in fluctuating water levels worked in synchrony with cycles of energy collection and concentration, with reproductive cycles of fauna and flora, and with fire cycles to produce the diversity of habitats and abundance of wildlife that characterized the predrainage Everglades. Vital attributes of the hydrologic pulse that sustained the diverse and prolific predrainage Everglades ecosystem included flow volume and distribution, the spatial extent of flooding, water depth fluctuation, and hydroperiod prolongation.

The River of Grass

The natural system hydrology model indicates that "Overland flow in the natural system appears continuous from the south shore of Lake Okeechobee through what is now the Everglades Agricultural Area and Water Conservation Areas to Shark River Slough" (Fennema et al., Chapter 10). Gleason and Stone (Chapter 7) review evidence for once-greater flows from Lake Okeechobee to the Everglades. A sedimentary layer southward into the marsh from the lake indicates the transport of suspended mineral clays for considerable distances downstream at times in the past. The natural system hydrology model supports this conclusion, simulating large surface water outflows from the lake during years of above normal rainfall. Temporal flow patterns out of the sawgrass plain that is now the Everglades Agricultural Area were similar to those observed from Lake Okeechobee. Other

---

* Plates 2–6 appear following page 432.

water entered the Everglades from the southern portion of the Atlantic Coastal Ridge westward into the headwaters of the historic Shark River Slough and Taylor Slough basins. Furthermore, the model portrays surface water and groundwater flows well into the dry season from the southern portion of the Atlantic Coastal Ridge westward into the Shark River Slough basin, supplying more uniform flows than occur today and producing a smooth transition between the wet and dry season, compared to the much more rapid drops in flow and water level that occur today.

In addition to maintaining hydroperiods and water depths throughout the freshwater Everglades, freshwater flow apparently contributed to the productivity of the Florida Bay estuary with which the Everglades ecosystem is strongly interdependent. McIvor et al. (Chapter 6) delineate the freshwater flow patterns entering Florida Bay through the southern marl marshes, creek systems fed by Taylor Slough, the C-111 basin, and Shark River Slough via Whitewater Bay and around Cape Sable into the western bay. They review simulations indicating that currently hypersaline areas in northeast Florida Bay probably rarely experienced hypersaline conditions in the natural system, in which salinities were 20–30 ppt lower than in the current managed system. McIvor et al. further cite evidence suggesting that freshwater flow and the resulting moderation of salinity levels provided an important support function for estuarine productivity in Florida Bay, including recruitment of pink shrimp (*Penaeus duorarum*), snook (*Centropomus undecimalis*), spotted sea trout (*Cynoscion nebulosus*), and red drum (*Sciaenops ocellatus*); reproductive success of ospreys (*Pandion haliaetus*), great white herons (*Ardea herodias occidentalis*), roseate spoonbills (*Ajaia ajaja*), and many of the wading birds that historically nested in the estuarine ecotonal area; and utilization of the bay by manatees (*Trichechus manatus latirostris*) and American crocodiles (*Crocodylus acutus*). Ogden (Chapter 22) postulates that early wading bird colony sites were concentrated along the mangrove-freshwater marsh ecotone at the lower end of Shark River Slough because the ecotone and adjacent estuaries provided overall better nesting and feeding conditions than were found in the interior Everglades.

The Expansive Solar Collector

Holling (Chapter 29) characterizes the natural Everglades system as operating in contrasting seasonal modes keyed to hydrology. During the wet season the wide band of sheet flow functions as a diffuse but expansive solar panel that stores the energy collected through primary production in invertebrates, fish, and amphibians. During the dry season the system transforms into a second mode when falling water levels concentrate stored energy along the "drying front," as referred to by DeAngelis (Chapter 12), to deliver it to wading birds and other predators. The onset of summer rains shifts the system back from the concentrator to collector mode. Browder et al. (Chapter 16) suggest that the production of higher level consumers is related to this pulsed expansion and contraction of water area.

The natural system hydrology model (Fennema et al., Chapter 10) indicates that almost the entirety of the original 1.2 million ha of the predrainage Everglades was inundated with at least 30 cm of water at the end of the wet season, regardless of whether it was a relatively wet or dry year. The Everglades does not appear to be highly efficient at transforming solar energy inputs into net primary or secondary production, as might be expected in the oligotrophic peatland system portrayed by Davis (Chapter 15). Thus, the size of the spatially expansive energy collector appeared to play a critical role. Browder et al. (Chapter 16) report that low rates of aquatic net primary production are typical of shallow Everglades periphyton areas, despite high rates of gross primary production, because emergent plants account for much of the gross production but contribute little to net aquatic primary production. Loftus and Eklund (Chapter 19) report that the biomass of fishes per unit area is low in the Everglades compared to other wetlands. Mazzotti and Brandt (Chapter 20) review evidence that alligator growth rates are low in the Everglades compared to more northern wetlands and to those of the American crocodile in nearby estuaries, due in part to a less abundant food base in Everglades wetlands.

The Rhythm of Depth

The vertical rise and fall of water affect rates of productivity and the distribution and abundance of animal populations throughout the Everglades. Browder et al. (Chapter 16) propose that water depth determines the presence and species composition of periphyton mats due to the effects of depth on photosynthesis and water chemistry. They note that the periphyton community at the base of the Everglades food web appears to be adapted to persist through wet and dry cycles. White (Chapter 18) describes one organizing role of the annual wet-dry cycle in maintaining nearly stable plant community structure and composition through the relatively low-intensity fires associated with the hydrologic cycle. Loftus and Eklund (Chapter 19) report that the combination of depth and hydroperiod influences the size and species composition of fishes in the Everglades. Mazzotti and Brandt (Chapter 20) review evidence for the accommodation by nesting alligators to fluctuating water levels in the Everglades. They cite reports that prior to the institution of water management, water level rise during the wet season was predictably correlated with water level at the time of nest construction. Bennetts et al. (Chapter 21) report relationships between water depth and nesting and foraging conditions for snail kites.

Frederick and Spalding (Chapter 26) review information suggesting that the rate of water recession in the Everglades is in large part responsible for nest initiations and nest success by white ibis (*Eudocimus albus*) and wood storks (*Mycteria americana*) and that several other wading bird species may also benefit from associated increases in availability of prey. They note that the utilization of a progression of foraging habitats by wading birds seems necessary to supply food throughout the breeding season, and the asynchronicity of food availability in these habitats is certainly a key factor in supporting reproduction. Bancroft et al. (Chapter 25) provide examples of how depth distribution patterns influence the location and size of wading bird nesting colonies. The importance of the drying front is reflected in the distribution of foraging wading birds which Hoffman et al. (Chapter 24) report in transitional, intermediate-depth habitats of the Water Conservation Areas during most years.

The reproductive risk to wading birds and snail kites associated with drying patterns probably was moderated in the Everglades due to the large spatial variability in rainfall and drought, as described by Duever et al. (Chapter 9). They note that this variability produces a mosaic pattern of drydown that prolongs suitable feeding conditions for mobile predators such as wading birds and also results in the occurrence of refuges for aquatic organisms that can colonize the more severely drought-impacted areas after the return of rains. However, such meso- to macroscale variability in rainfall and hydrology would only provide refugia and recolonization centers within the mobility parameters of the organisms affected (i.e., ideal water levels in Water Conservation Area 1 would be of no use to a heron trying to nest at Rodgers River in the southern Everglades and vice versa). Hence, White (Chapter 18) emphasizes the importance of spatial relationships of disturbed (in this case dried) areas to refugia.

The Enduring Water

The preceding chapters suggest that prolonged periods of flooding in the natural Everglades system played a key role in the support of populations of aquatic fauna and the consumers that fed on them. Prolonged flooding appeared to result from a lag response of flows into the next dry season, generated from wet season rainfall inputs. Resistance to flow by the emergent vegetation and the sheer flatness of the landscape, in combination with the large water storage capacity of the natural system, appeared to create the lag response and enable prolonged hydroperiods.

The natural system hydrology model (Fennema et al., Chapter 10) suggests that patterns of water depth in the natural Everglades depended more on overflow from Lake Okeechobee and the flow-through and carry-over storage characteristics of the marsh than on the more immediate effects of wet season rainfall. The model projects substantial flows in the early dry season out of the sawgrass plain that is now the Everglades Agricultural Area into marshes now within the Water Conservation Areas. As a result, a significant portion of the natural Everglades system had between 10- and 12-month hydroperiods. The marshes that now lie in the Water Conservation Areas would have been completely inundated during all but the driest years. The natural system model also indicates that areas now well to the east of the Water Conservation Areas and Everglades National Park maintained 12-month hydroperiods during normal to wet years, but experienced drydowns during lower rainfall years. Shark River Slough would have been inundated continually during the 10-year simulation period.

Browder et al. (Chapter 16) show that hydroperiod is a major environmental factor affecting the species composition of periphyton mats: diatoms, desmids, and filamentous green algae associated with higher levels of ecosystem production are characteristic of long-hydroperiod marshes. Loftus and Eklund (Chapter 19) report that assemblages of both large and small marsh fishes respond positively to periods without severe drydowns. They demonstrate that high densities of small fish species are produced during multiple years of continuous flooding, while much lower densities are sustained in areas with a high frequency of annual drydowns. Larger marsh species such as sunfishes (*Lepomis* spp.) and yellow bullheads (*Ameiurus natalis*) similarly are favored by prolonged high-water periods which foster higher reproductive success and survival. They draw the remarkable conclusion that, at least in the Everglades, fishes do best where there is water! The findings of Loftus and Eklund mark a significant departure from viewpoints held for more than a decade that annual drydowns increase the density and availability of small fishes as food for wading birds in the Everglades. Loftus and Eklund contend that "water levels in the Everglades certainly fluctuated seasonally, helping to concentrate fishes especially along the higher edges of the system. However, it is most unlikely that the original system experienced frequent and complete drydowns throughout its history or that such conditions would have sustained a persistent and diverse fish community." Their conclusion is supported by simulations from the natural system hydrology model (Fennema et al., Chapter 10).

Bennetts et al. (Chapter 21) summarize findings to indicate that during drought years, snail kite reproductive effort and success are reduced. Complete drying during the January–May nesting season may reduce food availability by causing apple snails (*Pomacea paludosa*) to aestivate or by killing them outright and may increase access to nests by terrestrial predators, resulting in widespread nesting failure. They conclude that the persistence of snail kite populations in the Everglades hinges on having sufficiently long periods between droughts to enable populations to recover. Additionally, refugia of flooded habitat must be available during droughts.

Ogden (Chapter 22) proposes that a large prey base of fishes and other aquatic fauna, which was enabled by the prolonged hydroperiods characteristic of the natural system, in combination with its large spatial extent, was a prerequisite for the large populations of wading birds and wood storks that the natural ecosystem once supported. The natural system hydrology model (Fennema et al., Chapter 10) shows that the distribution of the longer hydroperiod pools in the Everglades was, not surprisingly, in the lower elevation sloughs. Ogden suggests that most of the large, traditional wading bird colonies were located close to the lower Shark River Slough/mangrove ecotone because of the more dependable foraging conditions created by the juxtaposition of a long-hydroperiod freshwater pool next to a highly productive estuarine region. Bennetts et al. (Chapter 21) report that snail kites were both widely and commonly distributed through the interior and along the eastern flank of the predrainage Everglades, in areas shown by the natural system hydrology model to have had long hydroperiods. In contrast, Mazzotti and Brandt (Chapter 20) suggest that alligators were relatively less abundant in the central sloughs than they are today, because these areas may have been too deeply flooded for nesting during the predrainage period.

Other landscapes in the predrainage ecosystem apparently dried most years and functioned in a different support capacity. Gleason and Stone (Chapter 7) cite evidence that the southern marl marshes persist under water depths averaging 8–51 cm during hydroperiods of 6–7 months typically between May and November. This agrees with simulations from the natural system hydrology model (Fennema et al., Chapter 10) which suggest that the Taylor Slough watershed experienced a seasonal drydown most years. Browder et al. (Chapter 16) add that a 6- to 7-month hydroperiod and relatively shallow water depths in Taylor Slough provide conditions favorable to the formation of the calcareous periphyton that produce the marl substrate of this landscape. This algal community is adapted for exposure to the rigorous environmental extremes associated with annual drydowns. Such communities form only under shallow water depths conducive to $CaCO_3$ saturation. Mazzotti and Brandt (Chapter 20) review evidence that peripheral shallow water wetland areas, particularly those flanking the southern Everglades, supported the largest populations of alligators in the system prior to drainage. Ogden (Chapter 22) suggests that the availability of foraging habitat for wood storks in early drying, short-hydroperiod marshes in the southern Everglades was one factor enabling storks to form colonies during the early dry season. Robertson and Frederick (Chapter 28) review information which suggests that the higher elevation marshes may have provided prey for foraging wading birds primarily by means of fish dispersal from deeper sloughs during high water years. However, Robertson and Frederick correctly add that the dynamics of fishes and wading birds in these higher elevation marshes remains little studied, and the role of these wetlands is little more than speculative. Gunderson (Chapter 13) and Davis et al. (Chapter 17) indicate that the shorter hydroperiod wetlands support landscape and plant community associations different from those found in the deeper wetlands. These landscapes also support species such as the endangered Cape Sable sparrow which do not occur elsewhere in the ecosystem (Robertson and Frederick, Chapter 28).

WHAT WENT WRONG

Disturbance, Renewal, and Diversity in a Remnant Wetland

The Diminished Wetland Mosaic

The Everglades today represents a remnant ecosystem. Davis et al. (Chapter 17) document a reduction in spatial extent during this century to only half of the original 1.2-million-ha wetland and demonstrate a reduction in the heterogeneity of physiographic landscapes through the loss of three of seven predrainage landscapes and the near elimination of a fourth. Mazzotti and Brandt (Chapter 20) observe that for many wildlife populations, especially wide-ranging ones, it is likely that some of the habitat requirements necessary to maintain a healthy population will not be provided in a limited area with reduced habitat diversity. Moreover, the more limited an area is in terms of size and habitat diversity, the more difficult it is to resolve conflicting demands on the same resource. Bancroft et al. (Chapter 25), Bennetts et al. (Chapter 21), Frederick and Spalding (Chapter 26), Mazzotti and Brandt (Chapter 20), and Smith and Bass (Chapter 27) all suggest that the greatly reduced areal extent of the Everglades has significantly reduced the habitat options and long-term population survival for species of vertebrates with large spatial requirements.

Frederick and Spalding (Chapter 26) conclude that the loss of foraging habitats that are likely to be used by wading birds very early and late in the breeding season and during very wet years (i.e., the shorter hydroperiod marl marshes and the persistent deep water pools of Shark River Slough, respectively) has probably put a real constraint on the timing and success of reproduction. Reduction in these habitats has also hampered the ability of the birds to breed during all but the most favorable of climatic conditions. These changes, plus the previously cited observations of Hoffman et al. (Chapter 24), support Ogden's (Chapter 22) proposal that the decline in numbers of nesting birds may have been due in part to a reduction in the range of options where wading birds can forage under a variety of hydrological conditions. Holling et al. (Chapter 29) summarize that the reduction of topographic heterogeneity at various scales has narrowed options for delivering concentrations of the collected energy for wading bird feeding.

The sole remaining habitat with a forested area large enough to support panther populations in the Everglades is Long Pine Key (Smith and Bass, Chapter 27; Robertson and Frederick, Chapter 28). Loss of landscape diversity due to urban and agricultural development of the string of other forested uplands that once extended diagonally through the Everglades from Miami to Long Pine Key, which comprised an upland landscape at least 1.5 orders of magnitude larger than the remnant of Long Pine Key, appears to have placed a fundamental restriction on the breeding population of panthers in the system. Robertson and Frederick suggest that loss of pineland and adjacent marsh habitats near the boundaries of Everglades National Park has also adversely impacted (or has the potential to impact) populations of less mobile species in the park, such as several pineland birds and the Cape Sable sparrow. These once contiguous habitats presumably served as corridors for gene flow, as well as habitats for terrestrial fauna and flora.

Compartmentalization of much of the remaining Everglades basin should be viewed as one form of spatial reduction, in that at certain spatial scales it has created a series of relatively small, independent wetland systems. The natural system hydrology model (Fennema et al., Chapter 10) shows that compartmentalization has resulted in overdrying in the northern portion and excessive flooding in the southern portion of each of the Water Conservation Areas. Gunderson (Chapter 13) and Gunderson and Snyder (Chapter 11) report that this altered hydrological pattern has changed plant communities and altered the frequency, distribution, and intensity of fire. Loftus and Eklund (Chapter 19) and Robertson and Frederick (Chapter 28) suggest that compartmentalization has resulted in disruptions in dispersal and other movement patterns by fishes and aquatic invertebrates. Bancroft et al. (Chapter 25) and Hoffman et al. (Chapter 24) propose that the wet-dry patterns created by compartmentalization have reduced the dependability of foraging habitat surrounding wading bird nesting colonies.

Management toward Monotony

Accompanying the reduction in ecosystem size has been a decline in environmental heterogeneity in the Everglades as a result of water and fire management. Within the mosaic of sawgrass, wet prairies, sloughs, and tree islands, Davis et al. (Chapter 17) show a reduction in plant community heterogeneity with the conversion of more open-water wet prairie and slough habitats to sawgrass, entailing a loss of sites of aquatic productivity and wading bird feeding habitat. This trend may stem from water management activities that lowered water levels beginning early this century. Although additional aquatic habitat has been created due to impoundment of water in the lower ends of Water Conservation Areas 1, 2A, and 3A (Fennema et al., Chapter 10), impounded areas also lack the habitat diversity of the natural mosaic because of drowned sawgrass and tree islands (Gunderson, Chapter 13).

Gunderson and Snyder (Chapter 11) indicate that the role of Everglades fire regimes may have changed from increasing environmental diversity in the natural system to reducing diversity in the current managed system, due to altered seasonal patterns accompanied by overdrainage of wetlands. An unknown factor, however, is the prevalence of anthropogenic fires set by native Americans relative to lightning-ignited fires in past centuries. The cause and annual periodicity of fire appear to have changed from naturally occurring, lightning-ignited fires during unusually dry early summer months of the wet season to fires lit by man during the November–May dry season. Since the late 1940s, naturally occurring lightning-ignited fires have made up only 40% of the fires covering 25% of the area burned in Everglades National Park. Increased occurrence of burnouts of tree islands, due to hotter and more frequent dry season fires, has reduced habitat heterogeneity (Gunderson and Snyder, Chapter 11). Fire patterns have also changed due to prescribed burning, a practice that has been shown by Gunderson and Snyder to dampen the annual and interannual variability in the number and severity of fires. White (Chapter 18) notes that man's tendency to replace natural variations and extremes in disturbances such as fire with a more monotonous variation can lead to loss of biological diversity, in that species tend to be uniquely adapted to natural variations in environmental conditions, and any regularization of physical driving forces will favor some species over others.

Decimated Sanctuary

The natural system hydrology model (Fennema et al., Chapter 10) indicates that both deep and shallow water refugia in the currently managed Everglades basin are much less extensive in area and are relocated and subdivided, compared to patterns prior to drainage. Today, only areas at the southern ends of Water Conservation Areas 1, 2A, and 3A and in the southern C-111 basin experience flooding at the end of a dry season, and diking of the system impedes dispersal of aquatic organisms to and from these pools as the surrounding marsh dries and refloods. Hoffman et al. (Chapter 24) identify the deeper water, impounded pool areas of the Water Conservation Areas as refugia during drought for wading bird foraging habitat. Loftus and Eklund (Chapter 19) and Robertson and Frederick (Chapter 28) propose that shortened hydroperiods in the deeper peatlands and lowered groundwater levels in the higher rocky glades in the southern Everglades have probably greatly reduced the dry season survival capacity of fishes and invertebrates in these two landscapes. Bennetts et al. (Chapter 21) surmise that the naturally fluctuating, long-hydroperiod wetlands once characteristic of the Everglades were of more value to snail kites as refugia than the current deep water impoundments in the southern ends of the Water Conservation Areas.

Robertson and Frederick (Chapter 28) suggest that species of vertebrates with broad spatial and habitat requirements in south Florida have become increasingly stressed by the fragmentation and loss of habitats and that the persistence of such species as manatee, panthers, wading birds, and snail kites will depend on "linkages with areas outside the Everglades...." Expansion of breeding ranges and shifts in population centers away from south Florida by such species as white ibis and wood stork may be responses to deteriorating habitat conditions in the Everglades. The overall impression is that the current Everglades no longer maintains the strong role it once played as a regional population center for many larger vertebrates. The Everglades may now be a sink rather than a source for wading birds (Robertson and Frederick, Chapter 28).

The Hydrologic Pulse

The Stagnated River

The natural system hydrology model (Fennema et al., Chapter 10) indicates that water management produced a significant change in the overland flow patterns in the Everglades. The Everglades was generally more of a flowing system with greater spatial extent and longer periods of inundation than exist today under managed conditions. Regional sheet flow patterns have been significantly disrupted and overland flow volumes reduced by the impoundment of Lake Okeechobee, construction of the Water Conservation Area levees and irrigation/drainage canals in the Everglades Agricultural Area, and the loss of dry season lag flows from the dense sawgrass plain that formerly covered the present Everglades Agricultural Area. Flow patterns out of Lake Okeechobee have shifted from primarily wet season flows in response to rainfall to dry season flows in response to urban and agricultural water supply demands.

Impoundment of water in the Water Conservation Areas and diversion of surface water flows to the east, combined with groundwater and levee seepage losses eastward in the modified system, have significantly contributed to reduced flows and the resultant loss of persistent hydroperiods in the southern Everglades (Fennema et al., Chapter 10). The primary direction of surface water and groundwater flows along the eastern flank of the Everglades, including the western side of the Atlantic Coastal Ridge, has changed from westward flows into the Everglades to eastward flows which carry large volumes of water out of the system to the Atlantic Ocean. The natural system hydrology model shows that by reducing the area of the system, large volumes of rainwater must be drained to the sea annually, in the absence of adequate water storage capacity in the remaining wetlands. Eastward outflows from the natural system through the North New River, Miami River, the transverse glades, and springs in Biscayne Bay remain unknown, but apparently were nowhere near the magnitude of the annual discharges today. Light and Dineen (Chapter 4) cite statistics that as much as 3.3 million acre-ft ($4070 \cdot 10^6$ m^3) may have been discharged to the Atlantic Ocean between 1980 and 1989, whereas the combined discharge to Shark River and Taylor sloughs and C-111 during that period totaled only 813,000 acre-ft ($1003 \cdot 10^6$ m^3). The natural system model (Fennema et al., Chapter 10) indicates that diversion of this water eastward represents a reduction of flows to the downstream Everglades on the order of 665 million m^3 (540,000 acre-ft) during a normal water year. This is in addition to an estimated average annual flow reduction of approximately 340 million m^3 (280,000 acre-ft) into Shark River Slough from marshes upstream which now lie in the Water Conservation Areas. These changes have combined to cause flows in the marshes to drop off much more rapidly following the end of the wet season. Water releases from the Everglades Agricultural Area to the Water Conservation Areas and from the Water Conservation Areas to Everglades National Park have also produced a much greater frequency of sharper peaked water deliveries during both the wet and dry seasons.

In addition to reducing hydroperiods and water depths throughout the fresh-water Everglades, diminished freshwater flow apparently has severely impacted the Florida Bay estuary into which the Everglades empties and with which the Everglades ecosystem is strongly interdependent. McIvor et al. (Chapter 6) review evidence that impoundment of water in the Water Conservation Areas, combined with diversion eastward to urban areas and tidewater, has significantly reduced the volume of freshwater flow to Florida Bay via Shark River and Taylor sloughs, particularly during the dry season. McIvor et al. cite evidence from patterns of fluorescent banding in hard coral to indicate that freshwater flow from Shark River Slough into Florida Bay may be reduced by more than half of the flow volumes that occurred early this century. The conclusion that freshwater flows into the coastal estuaries and Florida Bay have been reduced is supported by the natural system hydrology model (Fennema et al., Chapter 10) and is reiterated by Walters and Gunderson (Chapter 30). McIvor et al. also present evidence which strongly suggests that as a result of diminished freshwater inputs, Florida Bay waters are more saline over more extensive areas and for longer periods of time than under natural predrainage conditions. Particularly intriguing is their review of an assessment of salinity–freshwater relationships based on natural system hydrology model simulations, which predicts that Little Madeira Bay in northeast Florida Bay has salinities under managed conditions that are 20–30 ppt higher than those in the natural system.

McIvor et al. (Chapter 6) suggest that decreased freshwater flow and increased salinity have contributed to the deterioration of estuarine productivity in Florida Bay. "Effects of altered flow include reduced recruitment of pink shrimp, snook, and redfish; lowered reproductive success of ospreys, great white herons, and roseate spoonbills; and shifts in distribution of manatees, American crocodiles, and many of the wading birds that historically nested in the estuarine ecotonal area." Ogden (Chapter 22) notes that the collapse of wading bird colonies along the mangrove-freshwater ecotone corresponded to a reduction in freshwater flow to the estuary in addition to the drying of the once-persistent pool in lower Shark River Slough. McIvor et al. also conclude that "…Reduced freshwater inflow is also implicated as one of a complex series of factors involved in recent mass mortality of seagrasses in the bay. Similarly, hypersalinity is likely a factor in dieback of mangroves in some bay localities."

The Dismantled Solar Collector

In an analogy of the Everglades as an expansive wet season solar energy collector, Holling et al. (Chapter 29) conclude that decreasing the spatial extent has further impacted the system by reducing the gross energy collection that transforms into aquatic productivity. Loftus and Eklund (Chapter 19) add that the total fish biomass produced across the vast area of the original Everglades was probably many times higher than today, partly because of the areal reduction of peripheral marsh habitats by drainage and development and partly because of more frequent drydowns in wetlands remaining in the present system. The decrease in fish biomass may have been paralleled by approximately a 90% reduction in prey consumption by wading birds in the Everglades (Frederick and Powell, Chapter 23). Ogden (Chapter 22) notes that the reduction in the number of nesting wading birds, along with changes in the location of major colonies, appears to correlate with the reduction in the total area of wetland foraging habitat. He suggests that widespread decline in food resources probably has been the major factor responsible for reduced numbers of nesting birds and changes in colony patterns. Frederick and Spalding (Chapter 26) likewise conclude that availability of food may be the single most important factor limiting the distribution and nesting success of wading birds, especially in the relatively oligotrophic Everglades ecosystem. Their review suggests the strong association of fecundity and production of young with available food supply. The reduction in aerial extent of the wetland system by half may place a fundamental limitation on its support capacity for populations of wading birds that once utilized the ecosystem in greater numbers.

The Disrupted Rhythm

Water management has disrupted the annual periodicity of rising and falling water in the remaining Everglades. In particular, the effects of dry season rainfall have been aggravated by increases in the depth and duration of reversals in drying patterns, created by pulsed regulatory releases and reduced water levels such that a given rainfall event has a greater diluting effect. Such reversals probably occurred naturally in the system prior to drainage and may have contributed to the infrequent inland nesting pattern noted by Ogden (Chapter 22). However, the magnitude of the reversals appears to have been amplified. Mazzotti and Brandt (Chapter 20) indicate that the loss of predictability of hydrologic fluctuations in the southern Everglades during most years since the flood control project was completed is coincident with increases in alligator nest losses due to flooding. Reversals in drying patterns due to dry season rainfall events are a major factor contributing to wading bird nesting colony failure, as reviewed by Bancroft et al. (Chapter 25) and Frederick and Spalding (Chapter 26).

Bancroft et al. (Chapter 25), Frederick and Spalding (Chapter 26), and Ogden (Chapter 22) suggest that nesting wading birds have become more sensitive to dry season drying patterns and rates than may have been true in the natural system. Ogden argues that there are grounds to believe that the strong association between water recession rate and wading bird reproductive success is a relatively recent response to declines in prey populations. He reviews evidence for an increasing dependence of wading bird reproductive success on water recession rate that corresponds to a drastic reduction in the prey base, both in aerial extent and in density per unit area. Frederick and Spalding summarize the hypothesis that "as prey animals become scarcer, it would be expected that wading birds of all hunting guilds would become increasingly dependent on mechanical concentration of prey for reproductive foraging." Successful nesting by wading birds in other wetland systems in southeastern North America, without the benefit of drying patterns, supports Ogden's hypothesis. In this scenario, the increasing dependence on water recession rate can be seen more as an indicator of the degradation of the prey base, rather than a historically necessary component of nesting.

This concept unravels a key inconsistency in perceived relationships between Everglades hydrology and the mechanisms by which aquatic production is utilized by wading birds. One line of evidence identifies prolonged hydroperiod as a requirement for sustained production of aquatic prey populations. Another indicates that rapid water level recession during the dry season is needed to concentrate prey populations to densities that enable wading birds to capture them efficiently. It is difficult to have both, in that a rapid rate of water level recession during the dry season in most years would reduce the populations of prey that it is supposed to concentrate. The proposed pattern of increasing wading bird dependency on drying patterns as ecosystem productivity declines underscores the thesis of Holling et al. (Chapter 29) that modification of either one of the modes of energy collection and concentration, through reduction in spatial extent of the energy collector or changes in drying patterns, is likely to have major impacts on ecosystem productivity and food webs.

The Fleeting Water

The natural system hydrology model (Fennema et al., Chapter 10) portrays reduced hydroperiods in most of the remnant Everglades wetlands compared to predrainage conditions. "These results suggest that the reduction in surface water inflows has had a significant impact on the Shark River Slough watershed, leading to lower water depths and substantially shorter periods of continuous inundations in the central slough." Furthermore, areas well to the east of the Water Conservation Areas and Everglades National Park that have experienced a complete loss of surface water due to water management maintained 12-month hydroperiods during normal to wet years. In comparison, only limited areas of the remnant system experience such hydroperiods today. Such conditions now exist only in the impoundments in the southern end of the Water Conservation Areas, particularly 3A, while the remainder of the Water Conservation Areas are drier.

Davis et al. (Chapter 17) propose that agricultural development of the sawgrass plain, formerly the major landscape of the northern Everglades, has accelerated the rate of southward water flow, resulting in sporadic conveyance and increased frequency and spatial extent of drying of the marshes within the Water Conservation Areas. Davis (Chapter 15) suggests that agricultural development of the sawgrass plain landscape further contributes to diminished hydroperiods in the Water Conservation Areas through the eutrophication caused by farm runoff. He cites recent evidence that eutrophication and the related spread of cattail (*Typha domingensis*) are resulting in increased water loss via transpiration from the remaining marsh.

Reduced hydroperiods appear to adversely affect aquatic production at all levels of the food web. Browder et al. (Chapter 16) indicate that periphyton community changes caused by shortened hydroperiod may reduce habitat quality and carrying capacity. They cite evidence that green algal components, particularly desmids, require year-round flooding, and shortening the hydroperiod appears to result in a decrease in diatoms, desmids, and other green algae associated with periphyton. They propose that such a community shift decreases the food value of periphyton, which in turn results in lower secondary productivity and overall ecosystem productivity. Loftus and Eklund (Chapter 19) conclude that fish densities are much lower in overdrained formerly deep slough habitats in northeast Shark River Slough compared to similar habitats with more natural hydroperiods. They report that the larger fishes associated with longer hydroperiod, deeper water marshes become relatively scarce where depths and hydroperiods are reduced. Mazzotti and Brandt (Chapter 20) suggest that decreased density of aquatic prey populations, due to more frequent and prolonged droughts, has contributed to the slow growth rates of alligators in the Everglades relative to more northern populations and to crocodile growth rates in Everglades estuaries.

Ogden (Chapter 22) attributes the abandonment of traditional wading bird colony locations in the southern Everglades to an increase in the frequency of drying in lower Shark River Slough and to the relocation of the largest long-hydroperiod pool northward to the southern end of Water Conservation Area 3A: "...These hydrological alterations to the system so closely match the major change in colony locations that the relationship may be assumed." Bancroft et al. (Chapter 25) report that the newer colony locations in the Water Conservation Areas appear to be areas where nesting success rates are lower than occurred in the traditional locations. They demonstrate that current wading bird colony sites in the Water Conservation Areas often do not provide suitable foraging habitat for long enough into the dry season to complete a nesting cycle. They chronicle the L-67 colony during the 1989 drought to demonstrate how nesting colonies fail as the surrounding marsh goes dry. Hoffman et al. (Chapter 24) report that areas of dense sawgrass in extensive parts of the Water Conservation Areas that have had their hydroperiods shortened by management activities are used little by wading birds as foraging habitat, except during infrequent months when water is high.

Agricultural and urban development of portions of the marl marsh landscape in the southern Everglades and most of the peripheral wet prairie landscape in the northern Everglades (Davis et al., Chapter 17) has extensively reduced the spatial coverage of the higher elevation wetlands that had shorter hydroperiods than the lower peatland landscapes. The natural system hydrology model (Fennema et al., Chapter 10) further shows that select hydrological stations along the eastern edge of the Everglades basin, in remnants of the wet prairie and marl marsh landscapes delineated by Davis et al., have experienced substantial reductions in annual hydroperiods and depths. Mazzotti and Brandt (Chapter 20) report that alligators have declined in the higher elevation marshes of the peripheral wetland areas, and "...alligators are now most abundant where they were least abundant (central marshes and sloughs) and least abundant where they were most abundant (peripheral marshes and the freshwater mangrove zone)." Ogden (Chapter 22) notes that the reduction in spatial extent and hydroperiod of the marl marshland flanking the eastern side of the southern Everglades corresponds with the trend of delayed initiation of wood stork nesting during past decades. Ogden concludes that some combination of changes, including the substantial reduction in total area of the higher elevation marshes along with the suspected degradation of estuarine wetlands, has almost certainly been the most important factor responsible for the change in timing of nesting by storks. Although Robertson and Frederick (Chapter 28) cite the findings of Loftus and Eklund (Chapter 19) that lowered water tables in the higher glades have reduced the survival rates of fishes and aquatic invertebrates, they also qualify that the role of these shorter hydroperiod wetlands in the dynamics of wading birds in the Everglades and the effects of their substantial reduction in area require more study. Certain portions of the higher elevation marshes are essential habitat for the endangered Florida panther (Smith and Bass, Chapter 27) and Cape Sable sparrow (Robertson and Frederick, Chapter 28). Deterioration in habitat suitability that has occurred due to reduced hydroperiods could reduce the long-term survival potential for these populations.

MOVING TARGET

Changes in climate and sea level, organic soil loss, and evolving land use patterns in adjacent agricultural and urban areas are examples in the Everglades of the characterization of ecosystems by DeAngelis and White (Chapter 2) as constantly changing rather than in a steady state. This notion implies that ecosystem restoration must be aimed at "moving targets," as coined by Holling et al. (Chapter 29), rather than "putting it back the way it was" at some point in time.

The continuing evolution of environmental conditions in an Everglades ecosystem that is very young in geologic age is highlighted by Gleason and Stone (Chapter 7). They provide evidence to suggest a strong interdependence among sea level fluctuation, water table, and freshwater wetland environments in the southern Everglades. They demonstrate the occurrence of distinct environmental shifts within the late Holocene period long after the establishment of the Everglades peatland as we know it, including a drier or more strongly seasonal hydrologic regime 2000–3000 years ago and other hydrologic fluctuations before and after, continuing locally up to the present time. These fluctuations resulted in marsh habitats alternating between marl marshes and peatlands. Gleason and Stone caution that the continuation or acceleration in the current rate of sea level rise could result in a significant reduction in the area of freshwater Everglades in the future.

Sea level rise must be a key consideration in any ecological restoration plan for the Everglades during the 21st century. Wanless et al. (Chapter 8) explain the role of the natural coastal dam of mangrove peat and storm-levee marl in separating the landward freshwater environments of the Everglades from the sea. An increase in the rate of sea level rise to 38 cm/100 years since about 1930 has set the stage for dramatic storm-driven modifications in these coastal deposits. Wanless et al. conclude that if the historic rate of the past 60 years continues, the protective coastal barriers between marine waters and the freshwater marshes will be eroded, narrowed, and dissected during hurricane pulses, low-lying freshwater wetlands will become saline, and increased saltwater intrusion will diminish freshwater resources. The time frame for these changes will be during the 21st century, when Everglades restoration efforts envisioned today will have been in progress.

Snyder and Davidson (Chapter 5) demonstrate an example of how the moving target concept applies not only to natural process, but also to the role of man in modifying the ecosystem. They propose that conventional agriculture in the Everglades Agricultural Area will change at an accelerating rate within decades due to the subsidence of peat soils caused by drainage, unless there is a shift to crops that are tolerant of flooding or high soil water tables. These changes could impact a major investment for creating wetland systems for the treatment of polluted runoff of water from the agricultural lands.

UNCERTAINTY

DeAngelis and White (Chapter 2) suggest that it is overly simplistic to predict population changes of species based solely on the fundamental biological characteristics of the species and the abiotic characteristics of the environment. This is due to a multitude of complicating factors that are difficult to anticipate, including the importance of population histories, biological feedback on abiotic driving forces, species interactions, and species (including exotics) invasions. White (Chapter 18) reiterates that we rarely have full understanding of the nature of driving forces in an ecosystem, the interaction among driving forces, or the contribution of biological components to the overall dynamics. He also cautions that the effect of driving forces and the factors that complicate our interpretation of them change depending on their position in space and time in an ecosystem; thus, accurate models must deal with the explicit location and timing of ecological relationships within the system. Maltby and Dugan (Chapter 3) further emphasize our limited knowledge of how wetland ecosystems function, particularly in the tropics and subtropics. However, they stress that waiting for the perfect science base for restoration will result in continued degradation and progressively more difficult rehabilitation.

There is also risk associated with attempting to understand natural functions of an ecosystem that has already experienced substantial change. Davis et al. (Chapter 17) note the uncertainty of response to restoration measures by an Everglades system that has suffered such drastic loss of landscapes and spatial extent. Frederick and Spalding (Chapter 26) caution that attempting to thoroughly evaluate the processes affecting current wading bird reproduction as a basis for restoring reproductive success may be misleading because "...we are apparently dealing with a vestigial breeding population in a highly degraded ecosystem that is increasingly unlikely to reflect healthy breeding conditions. Such a situation is likely to breed false leads." The widely held conclusion that rapid rates of water level recession are inherently necessary for wading bird nesting success, as challenged by Ogden (Chapter 22), may serve as an example.

This book carries a strong underlying message to the effect that although our collective knowledge of the Everglades ecosystem may be much stronger than many of us realized before these chapters were prepared, this baseline of information still contains many important gaps. Thus, at the beginning of the endeavor to restore the Everglades, we cannot define specific ecological restoration goals and the precise restoration process, or predict species or community responses, with an assurance regarding the outcome that can be viewed with great comfort. Although little elaboration of this point is necessary, a few good examples are worthwhile. We know virtually nothing about the relative contributions of periphyton, macrophytes, and detritus to Everglades food webs. Similarly, we lack even rudimentary information (other than the pioneering work of Loftus and Eklund [Chapter 19]) on the ecology and population dynamics of some of the most abundant and presumably important organisms of intermediate trophic levels, such as crayfish (*Procambarus alleni*), freshwater shrimp (*Palaemonetes paludosus*), apple snail, amphibians, and small fishes. Because basic questions regarding the driving forces behind primary and secondary production in freshwater and estuarine marshes remain unanswered, our ability to predict with a high level of accuracy the reproductive patterns by even such intensively studied top-of-the-food-chain guilds as the colonial wading birds continues to elude both our modeling and intuitive approaches. This uncertainty of information becomes another force in support of a broad restoration premise: restoration of hydrology and natural environmental fluctuations is an appropriate target in the stepwise and still imprecise process of attempting to produce ecological restoration.

Mazzotti and Brandt (Chapter 20) summarize: "Even with all the studies performed on the Everglades system to date we must recognize that our management decisions are essentially being made based upon speculations or incomplete data sets...Yet clearly we cannot wait until all the data have been collected, for instead of gaining information needed to heal the Everglades we will have documented its death. Rather, we should treat our management decisions as hypotheses of ecosystem response and design our monitoring programs as experiments aimed at testing them...This approach would allow us to revise management decisions as necessary to meet restoration goals (Holling, 1978)." Hence, we propose guidelines for ecological restoration of the Everglades as "action hypotheses," emphasizing that the time to act is now, but also acknowledging that we have and always will have incomplete knowledge of the ecosystem and its restoration.

RECOMMENDATIONS for RESTORATION
from INDIVIDUAL CHAPTERS

Several of the chapters in this book contain specific recommendations for restoration of more natural conditions or for improved management practices in the Everglades. Restoration recommendations also are extrapolated here from the discussion sections of several chapters. These recommendations are based on each author's understanding of environmental conditions in a more natural Everglades, how these conditions have changed as a result of human intervention in the system, and the environmental requirements of a species, group of species, or community that is dependent on the Everglades. The recommendations fall into two categories: those that propose the re-establishment of more natural patterns of hydrology and natural disturbances as a basis for ecosystem restoration and those that suggest more beneficial management practices focused on the needs of species or communities. These two categories are not necessarily in conflict.

These recommendations are compiled in Table 31.1. The specific recommendations provide the framework for the more general recommendations for systemwide restoration proposed in the following section. Although these recommendations

Table 31.1 Specific Recommendations for Restoration and Improved Management of the Everglades, Derived from Chapters in this Book

Author(s)	Chapter no.	Recommendations
A. Spatial Scale		
Bancroft et al.	25	Maximize extent of total restorable wetlands
Bancroft et al.	25	Restore northern marshes in Water Conservation Areas
Bancroft et al.	25	Maximize natural marsh connectivity
Davis et al.	17	Prevent further reductions in spatial extent
Smith and Bass	27	For Florida panther, reduce potential disturbances by humans in the Northeast Shark River Slough and East Everglades Wildlife Management Area
Wanless et al.	8	Develop restoration plans and programs designed to protect both the availability and quality of freshwater, surface water, and groundwater resources
Mazzotti and Brandt	20	Restore connectivity of the freshwater marshes with downstream estuaries
B. System Heterogeneity		
Hoffman et al.	24	Manage to increase the areal extent of mosaics of slough, wet prairie, and sawgrass habitats
Hoffman et al.	24	Manage to promote/enhance tree-island-dominated marshes; eliminate tree island "drowning" in southern portions of Water Conservation Areas
Hoffman et al.	24	Protect areas in Water Conservation Areas already highly productive for foraging wading birds, including the Loxahatchee Tree Island and L-28 Gap regions
Bennetts et al.	21	For snail kites, recognize the importance of maintaining drought-year refugia within the Everglades basin
C. Natural Disturbances		
Davis et al.	17	Allow natural fire to operate as a means for reversing wet prairie/slough losses
Davis et al.	17	Maintain and enhance habitat diversity by allowing disturbances (hydrology and fire) to occur naturally
D. Dispersal and Refugia		
Robertson and Frederick	28	Re-establish upland wildlife corridors along eastern side of the Everglades
Robertson and Frederick	28	Restore and improve management of remnant uplands and their peripheral short-hydroperiod marshes to stem loss of species diversity
E. Hydroperiod/Surface Water Distribution Patterns		
Ogden	22	Establish longer, more permanent flooding in deeper and southern portions of Shark River and Taylor sloughs
Ogden	22	Provide more expansive and prolonged flooding in shorter hydroperiod wetlands that flank the major sloughs

Table 31.1 (continued) Specific Recommendations for Restoration and
Improved Management of the Everglades, Derived from Chapters in this Book

Author(s)	Chapter no.	Recommendations
Bancroft et al.	25	Replicate natural hydrology, including appropriate seasonality of peak flows and drydowns
Bancroft et al.	25	Re-establish longer hydroperiods in formerly important wading bird feeding habitat between Nine-Mile Pond, Paurotis Pond, and base of Shark River Slough in Everglades National Park
Hoffman et al.	24	Manage to lengthen hydroperiods without greatly deepening water
Frederick and Spalding	26	Manage to re-establish high densities of populations of fishes and aquatic invertebrates
Mazzotti and Brandt	20	Protect and restore hydroperiods in remnant peripheral marshes
Loftus and Eklund	19	For natural populations of both large and small fish species, manage to re-establish extensive areas of deeper sloughs without severe drydowns
Davis et al.	17	Re-establish lag flows from the Everglades Agricultural Area; re-establish extended hydroperiods and flows

F. Depth and Drying Patterns

Bancroft et al.	25	Replicate natural hydrology, including rainfall responses and buffered dry season reversals
McIvor et al.	6	Develop a water management system that delivers sufficient fresh water to the marshes to mimic historic hydropatterns and that avoids sudden pulses of fresh water into the coastal bays

G. Short-Hydroperiod Marshes (see Section E)

H. Flow Volumes

Ogden	22	Recover predrainage flow volumes into mainland estuaries
Bancroft et al.	25	Restore proper volumes of freshwater flow along interface with estuaries
Bancroft et al.	25	Restore flow volume, distribution, and timing into area south of Loop Road, including Gum, Dixons, Lostmans, and East sloughs
McIvor et al.	6	Develop a water management system that delivers sufficient fresh water to the marshes to mimic historic hydropatterns and that avoids unnaturally long periods of below normal flows to Florida Bay

I. Agricultural Practices

Snyder and Davidson	5	For the Everglades Agricultural Area, shift to crops that can grow under conditions of flooding or high soil water levels to reduce organic soil loss due to oxidation

are grouped in the table by subheadings that correspond to subheadings used in the text of this chapter, many of the recommendations appropriately apply to more than a single category of issues.

RESTORATION ACTION HYPOTHESES

Six hypotheses are proposed here, which represent the major conclusions from the collective chapters in this volume. These hypotheses present an ecological framework for designing a full restoration program for the Florida Everglades, when combined with considerations of water quality, mercury, and the control of exotics.

The importance of spatial extent as a crucial aspect of ecosystems indicates that **the reduction in ecosystem size and compartmentalization of the remaining system are trends that must be reversed in any Everglades restoration initiative.** One immediate application is the acquisition of all contiguous remnant Everglades wetlands that are not currently in public ownership and prevention of further degradation or development of these areas in order to preserve options for restoring spatial extent and ecological functions. Key undeveloped areas in question outside the boundaries of the Water Conservation Areas and Everglades National Park include the Holeyland, Rottenburger, Strazula, Compartment D (of Arthur R. Marshall Loxahatchee National Wildlife Refuge), Brown's Farm, Broward Buffer Strip, Pennsucco, East Everglades, and C-111 tracts (Plate 1).* Snyder and Davidson (Chapter 5) further propose that changes in land use in the Everglades Agricultural Area open a potential opportunity to restore some of the spatial extent of the wetland system that was previously lost. They cite R. V. Allison's vision that the creation of marshes in this region "could serve as a gradual bridge between present-day agriculture and the return of the Everglades agricultural area to a wildlife area."

The role of disturbances in maintaining ecosystem heterogeneity suggests a restoration guideline to **allow environmental fluctuations and extremes in hydrology and fire to proceed as they would have in the natural Everglades system.** DeAngelis (Chapter 12) contends that in ecosystems of sufficient spatial extent, the diversity at the mesoscale to macroscale level ensures sufficient availability of essential habitats over the entire region to offset local disruptive effects and that disturbances contribute to that diversity. This concept may apply to fire and hydrology in the Everglades. Allowing natural patterns of infrequent, severe fires to lower soil elevation, expand wet prairie and slough coverage, and offset sawgrass encroachment merits consideration as an experimental guideline to restore plant community heterogeneity in a landscape where wet prairies and sloughs appear to be disappearing. This guideline appears feasible in a wetland system of 600,000 ha with the frequency distribution of fire sizes documented by Gunderson and Snyder (Chapter 11), if it is done in combination with hydrology restoration. Similarly, abolishing the control of naturally ignited fires, which has traditionally been practiced by prescribed burning and fire abatement, would

---

* Plate 1 appears following page 432.

appear to increase diversity in disturbance size and severity, as well as the resulting patchwork of topography and vegetation.

The importance of hydrology to almost every aspect of Everglades ecosystem function, both in the annual pulse of wet and dry cycles and in stochastic deviations as they relate to disturbance, suggests a restoration guideline to **develop water delivery plans based upon antecedent rainfall for all major areas of remnant Everglades marshland, including the Water Conservation Areas.** The rainfall-based plan for water delivery to Everglades National Park, developed by T. K. MacVicar and described by Light and Dineen (Chapter 4), is a first attempt to restore natural hydrology patterns to the Everglades. Application of this approach to other undeveloped areas of marsh, including the Water Conservation Areas, may represent the best strategy for restoring spatial and temporal patterns of hydrology to the remaining ecosystem. The conclusions in this volume indicate that in the formulation of rainfall-based water delivery plans for Everglades ecosystem restoration, it is essential to build in three elements that are lacking in the current plan for the park, as proposed below.

The central role of freshwater flow in re-establishing hydroperiods, configuring vegetation mosaics, and sustaining productivity in Florida Bay suggests a restoration guideline to **incorporate components into rainfall-based water delivery plans that will restore flow volumes and distributions in time and space, as simulated by the natural system hydrology model** (Fennema et al., Chapter 10), in the context of the other hydrologic restoration caveats noted below. Issues of water allocation to the natural system, as affected by surrounding land uses, are central in any Everglades restoration strategy, although they involve the management of areas (i.e., Lake Okeechobee, the Everglades Agricultural Area, and the lower east coast urban and agricultural areas) that are outside of the remaining marsh system. Likely sources of fresh water appear to include the capture of regulatory releases from Lake Okeechobee and the Water Conservation Areas to tidewater and runoff from the Atlantic Coastal Ridge, which now drains eastward to the Atlantic Ocean. Of concern is the possibility of adverse ecological impacts due to the water quality of the captured runoff, which may require extensive flow ways or other treatment systems (Walters and Gunderson, Chapter 30). Another concern is that re-establishing predrainage flow volumes within the spatially reduced, remnant system might have deleterious impacts from too much water. However, the landscape maps of Davis et al. (Chapter 17) indicate that the major flow corridors of the peatland landscape mosaic and the Taylor Slough and Shark River Slough basins are for the most part intact, albeit overdrained.

Spatial and temporal patterns of water depth are requisite to the collection and concentration of energy in food webs and in the reproductive cycles and survival patterns of many species of flora and fauna in the Everglades. This suggests a restoration guideline to **build components into rainfall-based water delivery plans that will restore depth patterns in time and space, as simulated by the natural system hydrology model** (Fennema et al., Chapter 10). Algorithms would restore the annual rhythm of rising and falling water by attenuating sporadic depth fluctuations and drying pattern reversals and by moderating rates of flooding and drying, as simulated for the natural system. They would also provide for restoration of the spatial distribution of predrainage water depths, as they would have been driven by climate and flow, throughout the remaining wetland system. Concern has been expressed that restoration of depth patterns as they occurred in the natural system might be detrimental because most of the remnant system is comprised of deeper water landscapes and thus support functions of higher elevation marshes would not be regained. However, the proportion of remnant shallow water marsh landscapes to deeper water landscapes appears identical now to what it was in the predrainage Everglades. In both situations, shallower marshes (southern marl marshes and peripheral wet prairies combined) comprise 31% of the Everglades wetland according to landscape analyses of Davis et al. (Chapter 17).

Prolonged hydroperiods are essential to support aquatic production and the consumers dependent on aquatic food webs and to maintain peatland landscapes. This suggests a restoration guideline to **integrate elements into rainfall-based water delivery plans that will mimic extended periods of flooding as they would have occurred in the remnant Everglades marshes under predrainage conditions.** In the development of water delivery models for the Water Conservation Areas, extension of hydroperiods into the dry season should be based on the lag response to wet season rainfall and Lake Okeechobee overflows in the natural system as if the sawgrass plain (now the Everglades Agricultural Area) continued to function in the capacity of slowly discharging water to the south. Similarly, the lag response of water flow across the marshes that now lie within the Water Conservation Areas, as if conveyance systems and water regulation schedules were absent, provides a basis for refinement of rainfall-based water delivery in order to prolong hydroperiods within Everglades National Park.

One element of hydroperiod prolongation within the park is the restoration of hydroperiods on the order of 6–8 months during relatively normal water years in the marl marsh and rocky glades landscape flanking the eastern and western sides of Shark River Slough. An important aspect in the restoration of shorter hydroperiod marshes is re-establishing the interactive roles of short-hydroperiod and long-hydroperiod landscapes in the production and concentration of aquatic organisms and their utilization by wading birds. This involves strategically locating persistent pools as drought refugia for aquatic fauna that seasonally inhabit shorter hydroperiod marshes and re-establishing surface water connectivity by removing barriers to the dispersal of aquatic organisms between shorter and longer hydroperiod landscapes.

The RESTORATION PROCESS

The implementation of a long-term restoration program for the Florida Everglades must be tempered by at least two cautionary considerations which deal with the task of identifying restoration goals and the restoration process. The first consideration is that implementation of restoration projects that attempt to integrate the objectives defined by the restoration hypotheses will require a regional ecosystem-level planning process. Such a process assumes a fairly high level of agreement on a vision of what the Everglades and south Florida should be well into the next century. This approach will require that each action be designed and implemented within the framework of a larger regional plan, one that considers the relationships among separate actions and recognizes the interrelatedness of the entire system.

The information in this book, both what is said about the Everglades and what is currently unknown about the system, argues strongly that restoration must be viewed as an open-ended process rather than as having a discrete goal. There are two good reasons for this, as discussed earlier: (1) because of gradually changing conditions at regional levels, restoration must deal with moving targets and an evolving ecosystem that is not likely to resemble either the present or the past Everglades and (2) there is, and always will be, a significant degree of uncertainty in predictions of ecological responses to management or restoration strategies in an ecosystem reduced to half its original size.

The open-ended restoration process that has been termed "adaptive" (Holling et al., Chapter 29; Walters and Gunderson, Chapter 30) suggests regional visions and strategies to be incrementally and experimentally implemented. As previously stated, each project must be designed within the framework of a full regional plan, with consideration for the relationships among separate actions and subunits of the remaining ecosystem. Because of the changing conditions and uncertainties, ecosystem stability can only be viewed as a short-term objective. Long-term restoration must be an ongoing process whereby restoration implementation becomes a continuing series of management decisions. Each decision should be based upon a growing pool of research information, updated measurements of ecosystem responses, and evaluations of degrees of progress in reaching a set of goals or targets that have been identified as indicative of ecosystem vitality. This process and the restoration hypotheses presented earlier suggest that modification of the current water management system for ecosystem restoration should emphasize simplicity and austerity in construction and removal of structures, where possible, if these actions are consistent with a program of operational flexibility dictated by ecosystem responses.

FINAL WORDS

We recently reread portions of two old Everglades reports. One, prepared by Charles F. Hopkins, describes the 1883 Times Democrat Expedition of the Everglades between Lake Okeechobee and the Harney River. The other was prepared by John T. Stewart for the Department of Agriculture's 1907 survey run from west to east across the Everglades, from Brown's Trading Post to Pompano. With remarkable clarity, these two reports describe not only the former vastness and mystery of the natural Everglades at the turn of the century, but also provide wonderfully qualitative descriptions of the region's landscape and flow patterns and surface water and soil depths. At the time of the earlier survey, Lake Okeechobee was more than 22 ft (6.7 m) above sea level, and numerous open "bayous" carried water through the custard apple forest along the south shore of the lake into a broad belt of white water lily marsh. That lily marsh was the transition between the shoreline forest and a 30-mile- (48-km-) wide sawgrass plain that was the major feature of the northern Everglades. Puzzling deep water "gullies" that "had no beginnings and no ends" were the only breaks in this sea of sawgrass.

Only when the surveyors were south of the sawgrass plain did they encounter the first tree islands; on one island near the east side the only rookery was seen, which contained "quite a number of Long Whites" (great egrets). Few alligators were seen, perhaps because Brown's Store claimed to have handled 50,000 gator skins in the 6 years prior to 1907.

Many people interviewed by Stewart were opposed to extensive drainage of the Everglades. They thought that the southern glades was too rocky to ever be farmed and that the peat soils in the central sloughs would require irrigation if drained; more than one person thought that the waters of the glades moderated the effects of winter freezes. The Everglades described by Hopkins and Stewart, and the attitudes of those early Dade County settlers, are part of what William B. Robertson has recently called "a lost world"—one that we will never know as well as we would like.

Going back in time to that lost world is not an achievable goal of Everglades restoration, although many of its attributes and functions are achievable. Perhaps a more realistic vision is going forward, into the 21st century, toward the creation of the next generation of the Everglades. The new Everglades will be different from that of the past, but it must (if restoration is successful) be one that rekindles to the extent possible the wildness and richness of the former system. Stewart and Hopkins witnessed the natural Everglades at the turn of the century, and each of us has a vision of what the Everglades may have been and what it should become as we enter the next century. Let us hope that this volume contributes to the realization of these visions.

INDEX

A

Abiotic driving forces, 307–319
Abiotic environment, 9, 10–12
  classification of driving forces, 12
  disturbances, 15–18
  gradually changing environmental conditions,
    14–15
  interactions between driving forces, 19–20
  natural periodicities, 18–19
  responses of species changes to, 20–24
  temporal and spatial scales of driving forces,
    14
Acrostichum danaeifolium, see Leather fern
Adaptive Environmental Assessment and
    Management model, 751–752
Adaptive Environmental Assessment Model, 128
Adventitious roots, 369
African wetlands, 31–32, 35–36, 37, 39, 43
Agelaius phoniceus, see Red-winged blackbird
Agenda 21, 34
Agkistrodon picivorus, see Cottonmouth
Agricultural runoff, 77
  cattail and, 368
  periphyton and, 393
  snail kite and, 526
Agriculture, 63, 85–115, see also specific topics
  corporate farming, 92–94
  drainage and, 85, 87–89
  economic impact of, 106–107
  effect on international wetlands, 31, 32, 36
  environmental problems, 111–112
  Everglades Agricultural Area, 85, 98–100
  Everglades Experiment Station, 94–95
  future, 107–113
  Hoover Dike, 95
  irrigation, Water Conservation Areas and, 51
  livestock, 94, 97, 100–101
  past, 85–97
  present, 98–107
  rice, 85, 104–106
  settlements and, 86
    on Lake Okeechobee, 92
    in sawgrass, 91–92
  sod, 85, 104, 105
  soils and, 89, 107–109
  subsidence, 107–109
  sugarcane, 85, 93, 94, 96–97, 101–103
  sugar industry, 109–111
  vegetables, 85, 97, 103–104

Ajaia ajaja, see Roseate spoonbill
Algae, 169, 207, 334, 358, 372, 380, 398, 399,
  400
  repopulation after drydown, 388
  water management and, 455
Algal bloom, 111, 140
Algal mat, 380, 383, 389, 396, 408, see also
    Calcareous periphyton; Microalgae;
    Periphyton
  repopulation after drydown, 388
Alley-North, 580, 675
Alley-North Slough, 602, 604, 605, 606, 607, 611
Alligator, 672, 704, 724, 747, see also American
    alligator
  changes in productivity, 436
  ecological restoration and, 771, 776, 779, 784,
    786, see also Restoration
  nests, 488
  ponds, fish communities and 476, 477, 480
  vegetation patterns and, 336
Alligator, America, see American alligator
Alligator Alley, 434, 653
  foraging habitat of wading birds and, 601, 602,
    603
  sea level and, 221
  wading bird foraging patterns, 623
Alligator Alley High Marsh, 602, 604, 607
Alligator Alley North, 632, 633, 634, 636, 638,
  639
Alligator Alley South, 632, 633, 634, 636, 638,
  639
Alligator flag, 330
Alligator hole, 317, 726, 746, 771, 773
  fish communities and, 463
  foraging habitat of wading birds and, 599, 601
  in peatlands, 150
  vegetation pattern and, 330
Alligator Lake, 544, 549
Alligator mississippiensis, see American alligator
Alligator sinensis, 487
Allison, R. V., 95, 112
Alvaradoa, 328
Alvaradoa amorphoides, see Alvaradoa
Amazon, 33
Ambrosia spp., see Ragweed
Ameiurus natalis, see Yellow bullhead
American alligator, 5, 485–505, 672, 704, 724,
  747
  alternation of natural hydrology, effect on,
    489–499

M

Y

Z